The Structures of Practical Knowledge

Matteo Valleriani
Editor

The Structures of Practical Knowledge

Springer

Editor
Matteo Valleriani
Max Planck Institute for the History of Science
Berlin, Germany

ISBN 978-3-319-45670-6 ISBN 978-3-319-45671-3 (eBook)
DOI 10.1007/978-3-319-45671-3

Library of Congress Control Number: 2016960669

Printed on acid-free paper

This Springer imprint is published by Springer Nature
The registered company is Springer International Publishing AG
The registered company address is: Gewerbestrasse 11, 6330 Cham, Switzerland

Foreword

Quel ramo del lago di Como…

Alessandro Manzoni's famous poem—*I promessi sposi* (*The Betrothed*)—begins with these words. This poem, which is *the* milestone of modern Italian literature, led many generations of Italians to reflect upon their habits, their fears, their social identity, and, in one word, their culture. It established this culture and even shaped the terminology used to discuss it.

This volume obviously presents the results of a much more modest research endeavor, but it nevertheless has at least one thing in common with Manzoni's literary project: it also originated at Lake Como. It was at the splendid setting of the Villa Vigoni that I first discussed the basic idea for this project with Eileen Reeves and John Heilbron during a walk in the woods. The occasion was a workshop organized in 2012 by Andrea Albrecht, Giovanna Cordibella, and Volker Remmert titled "The Inkwell and the Telescope: Galileo Galilei at the Crossroads of Scientific, Literary, and Visual Cultures in the 17th Century." At the conference, I presented the subject of my own contribution to this book—my study of the tradition of treatises on *The Sphere*—for the first time and I discussed with them the general frame in which I was collocating my research: the structures of practical knowledge.

My intention was to discuss an epistemology of practical knowledge using a bottom-up approach: from case studies to reflections of epistemic character. I therefore needed help in collecting a series of heterogeneous case studies, all of which should reflect on the role of practical knowledge in shaping abstract knowledge systems. I needed case studies and I also needed people who were willing to discuss this epistemic issue with me. Eileen and John helped me to shape my idea more systematically so that I was better equipped to approach other colleagues for help.

Before this event, however, and in spite of the fact that the idea had not fully materialized, it was already anchored in a more general frame. During several discussions with Jürgen Renn, director of Department 1 of the Max Planck Institute for the History of Science, we envisioned what Jürgen at that time called "the Matrix."

The Matrix is a stable structure of epistemic economy. It considers useful aspects of knowledge and spillover effects that enable knowledge produced in a specific context to be used in a different one. The Matrix has both a cognitive and a social and material dimension, and links early modern science and its organization, to the phenomenon of the Industrial Revolution, considered as a process made sustainable by the positive feedback between material and epistemic economy. The link between the early modern period and the Industrial Revolution is represented by the continuity of knowledge-intensive projects, which required explicit knowledge management, and thus also ruling institutions and scientific academies. Examples of such knowledge-intensive enterprises are building projects, military campaigns; journeys of exploration, mining and shipbuilding enterprises, technological gardening, manufacturing, glass, paper, and book production, activities in the frame of metallurgy and brewing, and construction of machines and water supply systems.

In other words, we were looking for those constant elements of knowledge economy that were developed during the early modern period and that can be considered as elements fundamental to the explanation of the emergence of the historical phenomenon we call the Industrial Revolution. We both had the centralized management of prominent early modern building sites in mind, such as the cathedral of Florence, where the Opera della Lana succeeded in organizing not only the labor but also the knowledge needed to realize that venture. The knowledge was produced, preserved, transformed, and readapted; it was used to conceive and distribute tasks while determining who were the right people, or groups of people, to perform specific tasks. Institutions such as the Opera della Lana were administrations able to handle knowledge in a way that could induce innovations, as in the case of the famous cupola of Filippo Brunelleschi. Jürgen and I had been working on the challenging introduction to a three-volume work on the history of architecture—*Wissensgeschichte der Architektur* (Edition Open Access, Berlin, 2014)—and while investigating the role that institutions played in this history, we recognized that innovations, such as those introduced by Brunelleschi, were only possible on the basis of a specific administration of vast knowledge resources. Returning to the Matrix, we asked if this could be a model to explain the installation of the first industrial plants. Had the industrialization process benefitted from meanwhile established systems of knowledge management, which went back to the early modern period?

Fascinated by this idea, I set out to better understand how knowledge was produced in a form that enabled such institutions to manage it. After all, the Opera della Lana could have limited its work to the selection of the right people, without, for instance, initiating a public context according to whose rules architects had to codify their ideas in the form of models to be used for public discussion. From my previous research on Galileo's work and its relation to the knowledge of the professionals working at the Venetian Arsenal, I already knew that during the second half of the sixteenth century, the production coordinators in the Arsenal had developed a way of codifying the fundamental elements to describe the construction process of all the models of ships of the North Adriatic Sea by means of drawings and written texts. The codification process of practical knowledge that they undertook had resulted from an intrinsic necessity to manage the production of ships.

Thus, it seemed natural to ask why the early modern period testifies to a continuously increasing production of externalizations of practical knowledge, whether these are artifacts such as models, drawings, or texts. This was certainly not done in order to improve the social status of the authors of such texts. This aspect certainly existed, and, if the book happened to help them obtain a better position, it was certainly a welcome effect. But looking at the entire process, which continued over centuries, this cannot have been the primary reason. The explanation must lie in the interplay between those institutions, their evolution and rules, and the people who carried and made use of the practical knowledge—in other words, the management and therefore the knowledge economy. From this perspective, this volume is intended as a first contribution to the project we call the Matrix. Whether the Matrix will ever become a book, I do not know, but it is certainly our muse in looking for continuities in history—my favorite *passatempo*—rather than breaks and disruptions.

From a different perspective, however, an even more general question had to be faced to achieve this goal. How can an externalization of practical knowledge be considered a constituent of knowledge whose primary goal is no longer the production of an object but rather the understanding of natural phenomena? How did this happen? On an individual level, all possible motivations can be found in history to explain why people, whether professionals, managers, or philosophers, began compiling texts on practical knowledge. Micro case studies are vital in helping us to understand not only why these texts were compiled but also how they were written, which form of codification of knowledge they represent, and which means and methods were used to conceive and design them. But what is the exact role of models, artifacts, and objects in the process of the formation of abstract knowledge systems? What relations exist between texts, images, and objects in the process of the codification of practical knowledge? And, my main question, what exactly happened to this knowledge once it was first codified?

Once I was finally able to clarify these questions for myself, I thought of using the overly broad term "structure(s)" to frame all my ideas. If artifacts, images, and texts are managed, it means they are moved from one context to another. They are not restructured in themselves but are connected to other knowledge sources; they enter a different knowledge system and as a consequence may receive different meanings and values. At this level, the term "structures" is, therefore, also used instead of "knowledge system," and for this reason, the reading of this volume can sometimes be somewhat demanding. But the term "structure" has one great advantage in that it also naturally relates to the structure of practical knowledge itself, for instance, the workflow, the precise series of actions, used to produce something, and as well to the social structure of those who produced and of those who managed knowledge. "Structures," therefore, ensure that we do not forget that practical knowledge, its social dimensions, and the conceptual system in which it is embedded are one and the same phenomenon, a phenomenon that should be analyzed in all its complexity.

The Structures of Practical Knowledge is an epistemological narrative. It investigates the nature of practical knowledge: why, how, when, and by whom it was codified and how its codified aspects were structured into knowledge systems. In the first chapter, I discuss at length the fundamental aspects of the epistemology of practical knowledge that emerged from this entire work.

Relevant results from all the contributions are integrated into this general view, and this chapter, therefore, functions also by way of introduction. The contributions touch upon all of the aspects considered relevant for an epistemology of practical knowledge to differing degrees. They are therefore ordered based on a certain predominance of one or the other aspect, so as to begin with the practical knowledge and those forms of its codification that directly emerged from the praxis but then moved away toward those conceptual structures less directly linked to the praxis itself. It continues with the social structures of professionals and those who codified practical knowledge (including the actors' perspective) and ends with contributions that deal more directly with the formation of different conceptual structures.

The contributions could also have been grouped according to the activity or scientific disciplines in which the case studies were developed, such as astronomy, mining, or architecture, but this was not within the scope of the volume. Instead, it reflects the output of a working group that was set up in 2013 to discuss aspects of the epistemology of practical knowledge. In total, the group met twice: in 2013 and at the end of 2014. Pre-circulating drafts of papers were discussed during the lectures of the first meeting, and a first complete version of the papers was discussed during the second meeting. A second version of the essays was sent to review in July 2015. The final version was delivered in Spring 2016. While I take full responsibility for the choice of what is featured in this volume, I would like to emphasize that any interest generated by it will be thanks to the work shared equally by all participants and contributors.

Berlin Wilmersdorf, Germany Matteo Valleriani
May 28, 2016

Acknowledgments

First of all, I would like to thank the contributors to this volume for their willingness to discuss my ideas. What I have learned from this experience will certainly provide a solid background for further research initiatives. Special thanks go to Jürgen Renn for his intellectual proximity and orientation and, a not irrelevant aspect, for his generosity in giving financial support to the working group and its output. I would then like to thank John Heilbron for his continuous support during the long process of producing this book and also for his precious advice, comments, and critique. Not without some remorse, I would like to thank Jim Bennett who was initially part of the group but unfortunately had to leave as further participation would have implied a new research endeavor that was incompatible with his schedule. During the first meeting, our discussion benefited from the ideas of a series of discussants: Johanna Biank, Anna Jerratsch, Gunthild Storeck, Maria Avxentevskaya, Alan Chalmers, and Antony Malet. Concerning the preparation of the first introductory chapter, I would like to thank Matthias Schemmel and Massimiliano Badino for their extremely valuable critical reading. Throughout the project, the organization of the meetings, the preparation of the readers, and the work on the publication were carried out by Anna Siebold, who not only supported my research but also gave me many new intellectual impulses during the last three years and without whom it would have certainly been impossible for me to achieve this and many other results. The organization of the meetings received the decisive support of Petra Schröter and the office of Department 1. As a guide through the mysterious paths that lead to the creation of a material book, I would like to thank the editorial manager of Department 1, my office mate and friend, Lindy Divarci. For the index of the volume, I am very grateful to Olga Potschernina. The almost insurmountable work of retrieving all the rights of reproduction for the many images contained in this volume was accomplished by Urte Brauckmann of the Library of the Max Planck Institute for the History of Science. The Library covered the expenses for the acquisition of electronic files of images and permissions for their use. In this respect, I would like to also thank both the former and current directors of the institute's Library, Urs Schoepflin and Esther Chen, respectively.

Contents

1 The Epistemology of Practical Knowledge

Matteo Valleriani

Abstract

The relation between practical and theoretical knowledge, which is usually perceived as one of the motors of scientific development in the early modern period, is redefined here as the relation between different structures of knowledge, where the qualitative difference between the different structures is specified according to the degree of abstraction and the range of connections between the different fields of knowledge. The investigation begins by identifying practical knowledge and the continuous process of its reorganization into new structures. In this way, the research aims to understand how the transfer of practical activities transitioned to a circulation of practical literature and, finally, how codified practical knowledge became part of the theoretical and conceptual structures that were being established during the early modern period. As an introduction to the entire volume, a heuristic diversification of knowledge production mechanisms is defined on three levels: (1) the knowledge structure of practical activities; (2) the social structuring of practical knowledge; and (3) the conceptual structures of knowledge. The subsequent chapters are discussed and introduced according to these definitions.

Practical knowledge is the knowledge needed to obtain a certain product—for instance, an artistic or mechanical artifact, or specific outputs, such as healing practices or mathematical results—that follows a defined workflow. The workflow can be a construction procedure, a recipe, or even an algorithm, which are, from a formal point of view, all equivalent to one another.

Most historians of science agree on the fundamental significance of practical knowledge in the early modern period, and according to the general historical understanding, on the fact that practical knowledge represents the background against which the astonishing intensification of scientific

M. Valleriani (✉)
Max Planck Institute for the History of Science, Berlin, Germany
e-mail: valleriani@mpiwg-berlin.mpg.de

M. Valleriani (ed.), *The Structures of Practical Knowledge*, DOI 10.1007/978-3-319-45671-3_1

research, which characterizes this period, took place. If, according to Richard Sennett, the concept of 'skill' is enlarged to such an extent that it embraces all human activities, also the theoretical ones, then, all skills, also the most abstract ones, begin as bodily practice (Sennett 2008, 10). Nevertheless, the historical framework is usually defined in terms of a dichotomy between practical knowledge on the one side and theoretical knowledge on the other, and within the scholarship, such a dichotomy between practice and theory is heuristically employed as a method of investigation.

Behind such a heuristic approach lies a distinction between practical and theoretical knowledge that does not directly consider the knowledge itself but rather its external representations—mostly its written codification. For this reason, the history of science and, specifically, the history of science concerned with antiquity up until the early modern period, implicitly divides treatises and historical sources between those that belong to the realm of practical knowledge and those which instead express theoretical considerations, such as, for instance, commentaries to Aristotelian works.

In recent years, the increasing focus on the material aspects of science and of scientific endeavors, has helped to shift attention away from this apparent dichotomy towards other, lesser studied characteristics and processes inherent to the experience of knowing and making. The attention has also been shifted to the workflows, the organizations, and the institutional frameworks that allowed for the achievement of a final material product (Chap. 10). This means that practical activities have been recognized as also being characterized by a structure of knowledge that connects materiality and action on which such activities were based. This tendency obviously results in an association between highly abstract conceptualizations based on practical knowledge and the practical activities that made use of such organized knowledge. However, the fact that practical activities show a sort of embedded knowledge structure does not yet explain how this knowledge became relevant within the process of the emergence of new science.

To fill this gap and, therefore, to understand the epistemic value of practical knowledge, it is necessary to investigate how and through which steps the structure of knowledge used in the frame of practical activities was reflected, conceptualized and externalized, for instance, by means of written treatises—and how this allowed for the creation of more abstract structures of knowledge that linked heterogeneous intellectual and practical fields to each other, so as to build new knowledge systems. Although written codification of at least some aspects of practical activities have existed since ancient times, during the early modern period the codification of practical activities in the form of written texts, drawings, and models exponentially increased (Hall 1979; Dubourg Glatigny and Vérin 2008).

From a purely epistemic perspective, the relationship between practical and theoretical knowledge, which is usually identified as one of the motors for scientific development in the early modern period, is here redefined as the relationship between different structures of knowledge, where the qualitative difference between the different structures is given by the degree of abstraction and by the range of connections between different fields of knowledge. Practical knowledge, like that employed by a Venetian shipwright in producing a galley in the sixteenth century, stands at the core of the theoretical developments that led to the emergence of Galileo Galilei's (1564–1642) theory regarding the resistance of materials. Yet, only those specific aspects of that practical knowledge which were codified, and therefore reflected and conceptualized, became relevant for such a scientific development (Valleriani 2010). But this codification was not undertaken to support Galileo's investigations. Instead, it happened for reasons related to the practical activity itself. This, in turn, implies that the economy, the forms, and the actors of the processes of codification of practical knowledge are what need to be investigated first. This will explain how practical activities transitioned into the circulation of practical literature happened and, finally, how codified practical knowledge became part of the theoretical and conceptual structures that were being established during this epoch.

To achieve this goal, the present research insists on a heuristic diversification of knowledge production mechanisms on three levels. These are: (1) the knowledge structure of practical activities; (2)

the social structuring of practical knowledge; and (3) the conceptual structures of knowledge. The first, concerns the practical knowledge itself structured following the dictates of the workflow and the series of actions needed to achieve the desired final product. At the other end of the scale, there is the layer of knowledge systems, which in their historical dynamics, tended to expand extensively and reflectively, drawing an analogy with cognitive structures of abstraction and representation (Damerow 1996, 29–70). In between, there is the network of people working within a framework of practical activities, whose social organization continuously changed, and that were shaped by institutional bodies and within new, increasingly influential social environments. These analytical layers do not mirror one another in a consistent manner. Rather, they trade with one another and develop dynamics that lead to an association of different aspects of knowledge, that, finally, can also explain how, in a later epoch, scientific knowledge had re-impacted practical activities.

1 Overview

In the early modern period, practical knowledge was continuously re-organized, technological innovations were introduced, the size and the scale of technological enterprises, such as building sites, substantially increased. Correspondingly, new forms of territorial organizations, and, therefore, also new forms of economy, led to the emergence and the development of new habits and workflows for more efficient production. In spite of practical knowledge's elusiveness for historians of science and technology (Chap. 14), it is, nevertheless, possible to investigate and determine some characteristics of established series of actions that constitute workflows. For example, by examining the organization and disposition of a workshop, we might come to understand the workflow involved in the facture of the artifacts it produced. The organization and structure of practical knowledge is closely related to the social context in which such knowledge is used. Practical knowledge exists in concomitance with practical activities, their social regulations, and their organizations. It is, therefore, also re-structured on a social and economic level. The social context of practical knowledge denotes a great variety of the different aspects of a given process, such as the division of labor within a building site, or the economic reasons that stimulated increased production in a specific sector and region. It is at this level of social contextual analysis that our understanding is enriched with the reasons, methods, and means by which practical knowledge was codified. On top of this heuristic subdivision there is the structure represented by the knowledge system which reflects codified practical knowledge. This is the level of the purely conceptualized knowledge where it is also possible to detect what is usually called scientific knowledge.

Codified knowledge can be structured in different ways. The basic structure might be a genuine description of a workflow or just a drawing, for instance, of a building. From this point onward, this knowledge can be integrated and linked to aspects of different areas of knowledge. An obvious example of this historical process is related to mechanics in the early modern period. Descriptions of machines and their building procedures became the basis of the so-called science of machines, and the original machines were analyzed in terms of five simple machine types. This schema, in turn, allowed for the connection to the static formulation of the law of the lever, which finally meant the beginning of theoretical mechanics. At this stage, mechanics itself became the subject of discussion, concerning the role of practical-mathematical disciplines within the more general framework of natural philosophy. Finally, this process contributed to the emergence of the fundamental metaphysical question concerning the gnoseological value of a mathematical approach to understanding nature.

Through this process, different areas of knowledge are integrated and brought together. From the perspective of practical knowledge, the process can be described as an evolution of knowledge from a lesser to a more abstract structure, which is simultaneously able to expand as a growing knowledge system. In this way, more fundamental, but also more abstract principles were formulated, which helped to explain an increasing number of phenomena both connected and not to the practical activities. Not only mechanics, but also other areas of knowledge went through a similar process. From the encounter between medical practice and mechanics, for example, the iatro-mechanical school of physiology emerged, exemplified by Sanctorius Sanctorius's (1561–1636) publication of his *De statica medicina* in 1614; theories regarding the constitution of matter developed in the fields of mechanics, alchemy and pneumatics, and later chemistry.

According to this methodological approach, scientific developments based on practical knowledge can, therefore, be explained as a series of changes within the structure of the knowledge systems. Such changes maintain a connection with the structure of practical knowledge but the connection itself is determined by the social and institutional context, because it is this context that determines which aspects of practical knowledge were codified and thus were made more accessible. In sum, the methodology suggested here foresees, first, explaining the mechanisms that led to the codification processes of practical knowledge and, second, the continuous re-organization of knowledge systems from one structure to the next. A first structure of theoretical knowledge is produced as soon as the codification process starts.

This approach seems to be appropriate for investigating the early modern period, as this epoch is characterized by what could be called a "peak of density of activity." This term defines a constant increase in the intensity of practical activities due to the material wealth that spread thanks to the accumulation of capital as a result of trade activity. As a consequence, an increased codification of knowledge took place. Almanacs, treatises, mathematical instruments, written recipes, and woodblock prints, are only a few of many possible examples which testify to such a density of activity. Generally speaking, practical activities stood at the heart of positive economic progression in this period. They were often activated by architectural enterprises, or by activities related to metallurgy such as mining activities, or in the diffusion and use of heavy artillery. Depending on the practical activity, impressive peaks of "density of production" took place in various fields throughout Europe in the sixteenth, seventeenth and eighteenth centuries. This means that production increased quantitatively.

Since P. W. Anderson's seminal article from 1972, *More is Different* (Anderson 1972), it has also become clear that systems—be they systems analyzed in chemistry, sociology or history—work according to different rules whose validity depends on the size of the system and the greater network of interrelated systems. This idea was previously expressed by Friedrich Hegel (1770–1831) in his *Logic*, according to which, quantitative changes provoke qualitative changes already on a formal level. Against this background, it becomes evident that the well-documented process of the codification of practical knowledge in the early modern period was a consequence of the positive economic development, which was driven in large part by technological innovation and the implementation of these new technologies. Because of the quantitative increase of practical activities, knowledge underwent a qualitative metamorphosis via codification: knowledge in the form of the goods of production changed into knowledge as goods of exchange, and knowledge as a means of building conceptual connections. The economic value, then, was no longer exclusively associated with the final product, but increasingly with the knowledge necessary to produce that final product. At this level, an increasing need to exchange knowledge might be traced to phenomena of knowledge transfer, circulation, and accumulation. Such processes are mirrored, for instance, in the unification of the different aspects of codified knowledge in treatises such as Johannes de Sacrobosco's *De sphaera*, which was reiteratively published and enriched over centuries (Chap. 16).

The process of codification, therefore, principally began due to economic reasons. In a phase of positive economic development dictated by practical activities, the necessity to transfer and preserve knowledge eventually emerged into a new modality which required a faster circulation of knowledge and its more accessible display. Not only did the final products need to be evaluated, used and appreciated, but so did the knowledge necessary to obtain them. This knowledge was given form through treatises, written recipes, drawings, illustrations, tables, letters, commentaries, and models, which displayed not only the final products they sought to elucidate, but also the basic knowledge of the artisan who sought to obtain it. All these forms of codification became relevant because they lowered the costs of access to acquiring specific competences (Mokyr 2005, 1–27) and re-organizing practical activities when this was necessary.

The increasing density of the process of knowledge codification required in turn an increasing number of institutions to rule and manage the knowledge itself. Institutions, such as government bodies, schools, academies and even guilds, also required that knowledge to be codified, in order for it to be properly administered, preserved, archived and accumulated. During such processes of codification, the institutions further developed the framework in which additional knowledge could be produced, and innovations could be made. The institutional task of controlling work output and maintaining a balance between financial investments and workflow was particularly important, as this required that knowledge be transferred, and thus appropriately translated and codified, from the professional practitioners to the ruling body (Renn and Valleriani 2014).

When the size, occurrence and ambitions of practical enterprises increased, there were multiple occasions for bringing together professionals of varying expertise. The great architectural work sites, the large-scale glass manufactures of Venice, the Renaissance tradition of gardening, and the rise of printing houses, are just a few of a great number of possible examples of new, large-scale enterprises of the early modern period that brought together different professionals, with different experience and geographic backgrounds, and provided them with the opportunity to exchange ideas, and perhaps to merge their expertise.

Again, the different sectors of practical skills and expertise were linked by means of institutional rules, which provided not only a system of management, but also centers for contact, that is, for accumulation, transmission and preservation, and, therefore, transformation. Following Wolfgang Lefèvre, these institutions created proto-industrial systems of labor division, and at the same time, provided new head bodies that lay between the professional bodies of the producers and the managing institutions themselves (Chap. 9).

These new head-body representatives, such as great architects and engineers, shared responsibility with, for instance, the organizers of practical activities. These individuals were responsible for disseminating knowledge processes that took place among the professional practitioners, and they worked in physical proximity to the place where practical knowledge was accumulated. These new institutional chairs could come from professional backgrounds but did not have to. They were alike because they did not produce, but rather organized production, evaluated projects and final products, and facilitated production processes. These "outstanding experts" as Ursula Klein defines them (Chap. 11), stood between the ruling institutions and the working practitioners, and were often required to translate and communicate practical activities for individuals outside of the practical world. They acted as points of accumulation, integration, unification and management, and, therefore, also represent the centers in which the same processes of codification often started.

Once practical knowledge is codified, it is no longer just practical knowledge. As mentioned above, it already constitutes a conceptual knowledge system, a conceptual structure. As soon as this is the case, the knowledge changes its structure toward higher levels of abstraction. There are multiple reasons for this change.

First of all, every process of codification requires the means to codify. To take a banal example, if codification takes the form of a simple drawing of an object, a drawing technique is required to accomplish the act of codification, which itself is not required for the production of the object drawn. Drawings and recipe-like texts seem to have been the fundamental, almost primordial means of codification of practical knowledge. However, in early modern history, the codification of practical knowledge almost always required the existence of an elaborated knowledge system. The very fact that single aspects of practical activities were selected for codification confirms that a knowledge system was then in place. Everything that could be vaguely associated to what historians call the Aristotelian worldview, in the revised versions between the late Middle Ages and the early modern period, was eventually used to frame the processes of the codification of practical knowledge. As many of the essays of this volume show, codification processes always made use of aspects of shared knowledge systems, such as geocentric cosmology and its implicit astrology; medicine and agriculture; the theory of elements concerning matter, materials, physiology and biology; physics and mechanics; and finally the Aristotelian taxonomic organization of knowledge.

Fields of knowledge were connected in this period which had originally been codified in completely different frameworks. The definition of the trajectory of projectiles is a good example for this. Originally a theoretical problem, it became an issue of urgent importance with the spread of heavy, mobile artillery in the sixteenth century (Chap. 5). The discussion of this issue, however, could not rely solely on the codified knowledge of the artillerists. If Nicolò Tartaglia's (ca. 1500–1557) *Nova scientia* is analyzed from this perspective, it is clear that in creating his theory of ballistics, Tartaglia also needed, in addition to the experience of the artillerists, cosmological notions, high-level geometry, and Aristotelian dynamics, revisited as a cinematic theory according to the late Medieval interpretations. The analogies he used to explain his reasoning, moreover, were taken from a variety of areas of knowledge, such as astronomy, hydrostatics and everyday experience. The economic motive of Tartaglia's book is also notable: he requested a *privilege* (patent) for the design of an artillerist's instrument to measure angles and distances by sight (Valleriani 2013).

Thus, once again, in connection with large-scale institutions ruling territorial entities, it might have been financial reasons that were the motor for collecting, linking, integrating and unifying different aspects of knowledge or even different knowledge systems on a more abstract level. Moreover, scholarly research initiatives, which emerged on the basis of the already established knowledge systems and which were not necessarily dictated by economy, further contributed to the creation of new structures of knowledge. These epistemic tendencies of transformation, integration and unification of knowledge finally resulted in sharing processes between different knowledge systems.

It was not only external reasons, however, that ruled this process. Once in motion, questions immediately arose and with them, an intrinsic dynamic of knowledge began as well. In other words, knowledge necessarily requires further knowledge as it continuously reformulates the initial problems (Renn and Hyman 2012). The processes of appropriation and transformation of ancient knowledge in Western culture, therefore, should be interpreted from this point of view. Ancient knowledge did not possess any peculiar ontological status in the early modern period, and was integrated into the growing knowledge systems in the form of codified knowledge as much as necessary, and as selectively and eclectically as was required. Because ancient knowledge in certain fields was highly theoretical, these works were highly valued. In the case of mathematics, for example, the introduction of mathematical frameworks to the mechanical arts prompted extensive scholarly studies on ancient works in this field (Valleriani 2014).

In the pages that follow, the structures of practical knowledge, the process of codification, the relation between the processes of codification and the increasing intensity of practical activities, and the conceptual structures of knowledge, are analyzed in greater detail. The topics of this discussion relate to the essays contained in this volume.

2 The Structures of Practical Knowledge

Debates on the relation between practical and theoretical knowledge are not an invention of modern historians of science. These issues were also discussed in the early modern period, and these reflections invite historiographical investigation of the role of the actors in the codification and classification of knowledge. The field of astronomy is particularly well suited for this type of investigation. In spite of its glorious theoretical history—astronomy has always been considered the queen among the theoretical sciences and has a rich tradition extending back to antiquity—many early modern scholars debated the practical usefulness of such a discipline, ultimately identifying practical knowledge as the epistemic motor of scientific development. As Pietro D. Omodeo describes in his essay (Chap. 12), scholars such as Girolamo Cardano (1501–1576) recognized close connections between practical activities and the science of astronomy, and described how the first generated innovations in the second. Petrus Ramus (1515–1572), moreover, suggested that only pure mathematical astronomy should be retained, that which he considered to be the practical knowledge of the discipline. According to his suggestion, all physical interpretations of astronomy should be dismissed and basically forgotten, allowing for the development of a discipline that embraced only mathematical algorithms and specific calculation techniques. For both Cardano and Ramus, the roots of astronomical knowledge were certainly found in practical knowledge, but it was logic which enabled this material to be systematized so as to achieve the compilation of the computational workflows necessary for the discipline.

The history of Cardano's and Ramus's idea to consider computational techniques—practical mathematics—as the practical knowledge of astronomy is investigated by Richard L. Kremer (Chap. 13). The early modern period experienced the emergence of new genres, as for example the Almanacs and the *Practica*, both of which were founded on computational techniques and whose production was in large part motivated by the printers' economic profit. As Kremer states: "*practica* deal with the quotidian, the specific, the operative, with medicines and diseases, with measuring and surveying, with manual manipulations at the hearth, with computational problems of quantity, measure, and value." To these activities, should be added the traditional fields related to calendric calculations of festivities and to meteorology, as well. As Kremer discusses, almanacs displayed "bloodletting calendars," informing users when such practices could be executed, and when not. Almanacs and *practica* therefore represent the useful mathematical output of the medieval knowledge system resulting from the profound and inherent amalgamation of cosmology, astronomy astrology, and medicine.

Investigating the relation between sixteenth-century mining activity and the emergence of theories of metal generation and of the earth's origins, Pamela H. Smith unequivocally shows that everything that is gained in exploring past practical knowledge is related to the concrete materiality of what, in fact, practical knowledge produced (Chap. 14). The way in which extracted materials were worked, or alternatively unworked—in order to show the processes of nature (or God), as in the case of the *Handsteine*—provide a glimpse of the practical knowledge of early modern miners and artists that produced such artifacts. In addition to the objects, the locations and regions where practical activities were executed offer insight into the actual structure of practical knowledge. The organization and processes of production, instruments, and the physical spaces where the actors worked might allow for the reconstruction of the production workflow.

Finally, as Elaine Leong demonstrates in her analysis of knowledge codification in relation to ale brewing in seventeenth-century England, part of the practical knowledge necessary to brew is embedded in the construction characteristics of the "brew house" itself (Chap. 3). Moreover, because brewing was an elaborate process that required a well-established and organized means of production, and consequently also a certain degree of labor division and specialization, the management of the work itself entailed aspects of the practical knowledge, too.

3 Forms of Codification in Their Social Context

The codification process is both dependent on the material workflow to achieve a certain product and on the social organization of the practical labor. Therefore, this process also has two different implications. On one side, it means a direct translation into signs (externalizations) and, on the other it implies the existence of social agreements based on normative rules, for instance, concerning teaching the skills in workshops and schools.

The early modern process of codification of practical knowledge is indeed intimately interconnected with the process of the formation of professions on a social level. As George W. McClure showed, the process of the professionalization of secular activities began in the fourteenth century on the basis of the professionalization that had already taken place in the realm of vocation and theology. Fifteenth-century works such as Archbishop Antonino's (1389–1459) 1455 *Summa theologica* and Francesco Patrizi's (1413–1494) 1460 *De institutione reipublicae*, created the political, normative and social background against which each sort of profession, including the manual ones, could be recognized for its utility. In turn, during the sixteenth century, other works such as *Dello specchio di scientia universale* (1564) by Leonardo Fioravanti (1517–1588) and *La piazza universale di tutte le professioni del mondo* (1585) by Tommaso Garzoni (1549–1589) show the increased need for a taxonomy of all professions with particular regard to the artisanal ones. These books testify that the circulation of artifacts was being replaced by the need to circulate knowledge on how to produce them. They are the consequence of the new phase characterized by a greater circulation of the professionals themselves and the fact that they were beginning to increasingly dedicate their time to the production of practical literature. Tommaso Garzoni, for instance, is quite explicit on the function of his book. Dedicated to Alfonso II d'Este (1533–1597), the book was intended as a useful instrument to discern between good and bad practices. While describing the professions, therefore, Garzoni did not refrain from codifying many aspects of the practical knowledge related to them (McClure 2004, 12–119).

The analysis of the practical literature of the period shows that practical knowledge was codified in a variety of forms that ranged from artifacts, such as models, to highly elaborate textual productions, which in turn could be taxonomically organized according to the genre to which such texts belonged or aimed to belong. Among the means of codification used during the early modern period, two are of particular relevance in identifying cases where codified practical knowledge was particularly close to the practical activities themselves, that is, to the actors personally involved in such activities: recipe-like texts and drawings.

Drawing as a means of codification relates directly to the work of Elizabeth M. Merrill (Chap. 2), in which she analyzes the so-called pocket-size notebooks of late-fifteenth-century Italian architects—figures who were also practitioners of the mechanical arts in general. Also, due to the diffusion of paper, pocket-size notebooks developed as instruments of mobile codification, used for personal reference and for transmitting knowledge on an individual basis. In spite of their diversities, the five notebooks Merrill considers show the emergence of a new structure of architectural knowledge, in which drawings were paramount, but text also worked to codify and conceptually systemize practical material. This first structure achieved by codification is epistemically divided into three groups of notes that can be found in all the sources. These are determined as operational knowledge; shared knowledge, mostly for educational purposes; and experimental knowledge, which will be discussed below.

It is in particular the operational knowledge that helps us understand how the codification processes worked. Operational knowledge is specific, local and project-based, such as the construction of a large-scale building or the realization of a machine. The codification of operational knowledge, therefore, was primarily driven by managerial needs, and frequently took heterogeneous aspects of the project into consideration. This corresponds to the formation of a structure of knowledge that considers many different aspects of real practical work. Accordingly, operational knowledge was codified by making use of different means. Images were indeed annotated and the notebooks also contained recipe-like textual formulations. The image, however, maintained a dominant role, and early processes of practical knowledge codification were, therefore, closely related to the ability to visually reproduce the practical work itself.

In codifying operational knowledge, the early modern architect also fulfilled his educational mission, which was part of the management of workshops. Primary source analysis shows that a high degree of codified knowledge produced for educational purposes was circulated and, ultimately, came to constitute a common shared knowledge, providing standards, that is, the foundation for the emergence of the discipline. The codification accomplished in this respect corresponded to the reproduction of widely circulated model drawings, whether they be of machines or buildings. But even the individual process of copying a standard drawing produced new knowledge, as no two drawings are exactly identical to one another. Instead, what one finds are copy-drawings of equivalent constructions, which display personal variations. Copying and general note-taking for training purposes took place at an individual level, or at the level of a ruling institution, such as a workshop in charge of an educational mission. The function of this codification process was to produce mnemonic tools.

Already with this first step of codification through copying, it is possible to observe the inherent tendency of selective codification processes with regard to all subjects which might possibly be codified. For instance, drawings of widely-circulating and therefore well-known machines, tended to be less accurate, the more the machines were known. By contrast, newer innovative machines were drawn in greater detail and accuracy. Quite clearly, the historical actors perceived that the codification of all aspects of practical knowledge was an impossible task. The codification of practical knowledge is selective, with many aspects of the real practical knowledge simply being ignored or placed on a level of secondary importance during the process of codification. Thus, the codification of practical knowledge at this stage retained operative relevance only at an individual level, through the recording of personal notes, which in turn was combined with the practical experience of the individual who actually wrote the notes.

The relevance of drawing skills, and the natural connection between draftsmanship, technical drawing, and geometry, is further confirmed by Jochen Büttner in his study of the emergence of pre-classical ballistics and early modern artillerists' processes of codifying practical knowledge (Chap. 5). The practical knowledge of gunnery, for example, was codified in reference to one specific aspect that later became pivotal for the emergence of a theory of ballistics: the visual representation of the trajectory of a cannon ball. As the knowledge and the experience required to practice gunnery was obviously not solely limited to knowledge of the trajectory, it is possible to recognize that through codification, an imbalance is born between practical knowledge and its representation. Treatises on ballistics, in fact, deliberately ignored or even violated well-established practical knowledge of the artillery. Due to this, within the field of ballistics, no practical application of theory is known until much later in history.

Just as individual drawing skills and the shared tradition of the *disegno* influenced the way in which an object was represented, and, thus, also the way it related to selected aspects of practical knowledge, the relation between practical knowledge and the systems in which practical knowledge is codified fundamentally defines which particular aspects of knowledge are codified. The trajectory representing the motion of a cannon ball was fundamental in early modern theoretical ballistics because a discussion on the motion of projectiles already existed in concomitance with Aristotelian dynamics. In Büttner's terms, "the flight of a shot is perceived as an epistemic object beforehand."

Codifications assumed a specific form according to the required function of the codified knowledge itself. While scholarly discussions tended to codify practical knowledge through texts, in the case where such knowledge was needed to exert control, a less mobile form of codification might have been used. As Dagmar Schäfer shows in her discussion, in eighteenth-century China practical knowledge was preferably codified in models and embedded in artifacts (Chap. 10). Collected, kept and protected in the empire's capital, the models allowed a class of technocrats to exert full control over practical knowledge, including its aesthetic and moral values. Close to the ruling authority, the knowledge was codified only in the form of models, sketches and, obviously, household records. The model was a means of codification that allowed for the internal control of practical knowledge, and fulfilled the bureaucratic needs of the state machine.

To satisfy the operative needs of the state machine—whether in building or repairing a palace, or creating a model of the then fashionable jade statues—ad hoc forms of knowledge codification were accomplished by means of paper labels. These strips were glued onto the models and bore additional pieces of information, such as the number of rooms in an architectural model, the function of these spaces, their measurements and even how much was to be expended on their construction. This was interchangeable knowledge that did not leave a trace once the work had been accomplished.

More than transforming practical knowledge into scholarly knowledge, this form of codification transformed the emperor Qianlong (1711–1799) into an erudite artisan, who eventually even wrote a treatise on the jade goblet. This form of codification was no longer in use when it was decided that, instead of keeping a centralized collection of models, knowledge should be controlled within an organization of apprenticeships. Naturally, this new system of codification, led to different forms of externalization.

The forms of codification are characterized not only by the envisioned function of the knowledge obtained, but also by the limits of the available means (mostly material means) for such a process. To respond to the demand for more colored prints, developments were made in the early modern period in the diffusion of the chiaroscuro print. As Eileen Reeves shows, the underlying technique of the chiaroscuro woodblocks actually impeded the realization of portraits, a characteristic that clearly shaped the content of these prints (Chap. 7). Yet, this limitation was ultimately overcome, because it "was eventually codified as an aesthetic tendency." Because of the "fixity" of the prints, moreover, and probably due to the fact that prints were long lasting and widely circulated, early modern scholars and artisans came to codify more aspects of practical knowledge than they would otherwise have done, had they been limited to manuscripts and hand-rendered texts. The printed book was in essence conceived in order for knowledge to be able to enter places disconnected from those where the practical knowledge was executed.

On the basis of this idea, Sven Dupré analyzes the work of the erudite glass-maker Johannes Kunckel (1630–1703) at the service of the court of Frederick William, Elector of Brandenburg (1620–1688), considering practical knowledge as codified in the form of recipes (Chap. 6). Recipes, as any other form of codification of knowledge, make the latter mobile, ready for circulation, and, therefore, also ready to be adopted, and transferrable to other production centers. Due to their high mobility, recipes and recipe collections were codifications that took place in a multilingual frame, offering links to further the practical knowledge of a given subject across social, cultural and language borders. New aspects of practical knowledge were therefore continuously codified in varying forms and according to different literary genres. The testing, adaptation, and new formulation of the recipes was codified, as were the errors—the explicit mention of mistakes—that accompanied many of the written prescriptions. Similar to the codification by means of drawing, this form of codification shows a certain closeness to the actual practical activities being codified.

Codification of practical knowledge might also have taken place as a reaction to already circulating knowledge, as Viktoria Tkaczyk illustrates in her study on the relation between theatrical engineering and natural philosophy (Chap. 4). Theater was not an invention of the early modern period. Although many technical aspects of early modern theater were indeed re-invented, for instance the flight-effects, and some of these effects were entirely original, theater itself grew from a long tradition. Closely related to religious events, theater referred back to the Middle Ages, and in spite of the knowledge gaps, even to antiquity.

Because of this long tradition, when the early modern theater began to develop and innovate, a history of thought that reflected such activity already existed, although it was markedly different in certain respects. In particular, within medieval thought the theatrical arts were considered no more than a specific *ars imitandi*—an activity that was associated with mechanics, and therefore it was epistemologically situated at a lower intellectual level. The best example of the theoretical gap regarding theatrical art and its practice is given by Leon Battista Alberti (1404–1472). In his fifteenth-century *De re aedificatoria*, Alberti encourages theatrical activity, but refers only to the classical Roman tradition, despite the fact that he was living in a time when the theatrical performances of courtly and public character had already gained great relevance.

The gap between theory and practice of the theatrical arts was addressed in the foundation of new schools in Florence, wherein the mechanical knowledge developed in the frame of the theatrical activity was fostered. The school of Giulio Parigi (1571–1635), author of the famous intermezzi of the pastoral *Il giudizio di Paride*, was one such organization. The creation of schools and informal academies initiated a process that ultimately led to textual codification, as in this case with Nicola Sabbattini (1574–1654), who prepared a treatise that also reflected Parigi's technical realizations, entitled *Pratica di fabricar scene a machine ne' teatri* (1637–1638). Sabbattini, who had been a pupil of Guidobaldo del Monte (1545–1607), and, therefore, had a mathematically-based education in mechanics, succeeded in translating the practical knowledge of the theater, which was until then most often transmitted orally, and linking it to mechanics and geometry.

In addition to managerial and educational purposes, and besides the need to adapt existing knowledge or to appropriate ancient knowledge, the codification of practical knowledge took place in the presence of technical innovations. In the field of architecture, a peculiar form of innovation was represented by the increasing size of architectural endeavors. According to Wolfgang Lefèvre, such large-scale architectural projects could be led by means of very heterogeneous skills and therefore required a division of labor that mirrored the basic structure of the practical knowledge (Chap. 9). On a social level, such structure was pivoted around the central figure of the architect, whose tasks were ever expanding. In order to organize and codify his knowledge, and to systemize relations between building-site processes and directives of the ruling body, the architect often, although not always, made use of erudite knowledge—whether of scientific or artistic nature. Under the pressures of such a social system, innovative technical solutions led to the emergence of innovative theoretical reflections, such as a new branch of geometry directly related to the building-site experience.

Beginning from the specific social context of individual architects, this approach may be expanded to cast light on the social conditions that allowed, and eventually demanded, the codification of practical knowledge. In her essay, Pamela O. Long calls attention to the "trading zones" (Chap. 8). Taking a biographical approach to five fundamentally significant architect-engineers of sixteenth-century Rome, Long shows how trading zones functioned as incubators for new professional figures whose activities merged the practical and the theoretical in erudite codified knowledge. Notably, within these case studies, theoretical knowledge is univocally identified in textual knowledge originating from antiquity. Scholars active within trading zones were characterized by their mobility, as well as their educational and disciplinary flexibility, which allowed them to merge different knowledge backgrounds. Because of the attention traditionally given to the processes by which ancient knowledge was appropriated, Long primarily considers the merging process which took place within trading zones, as involving fields of knowledge expressed in vernacular on the one hand, and Latin on the other.

Although sixteenth-century historical sources do not furnish an exact definition of the trading zones—aside from their identification in great production centers, such as building sites—investigations focused on later periods are able to more precisely identify the areas of activities in which the codification of knowledge, and, therefore, also the creation of new knowledge, took place. In her contribution, Ursula Klein clearly associates such areas with "large technological enterprises," "luxury markets in urban centers," and the "medical-pharmaceutical systems" (Chap. 11).

Within her analysis of Prussian mining officials, Klein identifies the figure of "technical-scientific expert" as being responsible for the process of codification, while at the same time, unifying scientific knowledge with knowledge directly acquired in the field. In their own intellectual and career profile technical-scientific experts indeed unified the practical mining experience with theoretical approaches. Reminiscent of the sixteenth-century architect, the technical-scientific expert also fulfilled a normative function on behalf of the ruling institution. Educational, economic and even disciplinary aspects constituted the social context of the technical-scientific experts.

These cases indicate that the codification of practical knowledge very rarely took place on an individual basis. In most cases, in fact, such processes were a part of, or even the consequence of, a specific social milieu that might be more precisely identified with an institution. Whether they were educational institutions, for instance, schools or workshops, or courtly centers, ruling bodies of great production centers, or big state machines, the codification of practical knowledge fulfilled a function that went beyond intellectual curiosity, the willingness to participate in the scholarly discourse, or even the desire for augmented social status. The codification of practical knowledge was mostly an aspect and a consequence of economic activity that made use of that practical knowledge: it had an operative function.

4 The Economy of Codification

The codification of practical knowledge in the early modern period was primarily a result of a shift in economic value from the product itself to the knowledge needed to produce it. This shift may in turn be explained by the phase of economic expansion based on technological innovation that characterized early modern Europe. Not all regions in Europe were uniformly influenced by this economic growth, and over time technological leadership—and thus also scientific leadership—geographically moved from south to north, and with the journeys of explorations, eventually involved the western coastal areas (Epstein 1998; 2000).

According to Andrew Pettegree (Pettegree 2002, 1–47), in the early modern period, borders were mostly represented by natural borders. Infrastructural improvements therefore allowed for a relatively free circulation of people and goods even during conflicts. In this frame, the undisputable privileged positions were taken by large urban centers. They were in contact with each other by means of a great variety of highly-circulating social figures who, therefore, created a solid and dynamic capillary network. The internal mobility of large parts of populations was, indeed, very high, which obviously also had a positive economic effect. The strong process of urbanization was accompanied by a general demographic increase while the large-scale economy became increasingly organized in terms of its financial aspects thanks to the rise of a solid bank system over large-scale territories.

The wealth of this period is still recognizable today in the sixteenth-century Italian palaces of Rome, the seventeenth-century Dutch city centers, and those English country houses furnished with well-stocked "brew-houses" of ale and beer, which were maintained due to the questionable potability of drinking water. Such wealth was accompanied by the development of diverse, yet highly-connected, constellations of political, economic and geographic entities, such as the Dukedom of Florence, or even the smaller Este Family principalities in Ferrara and Modena. The economic policy of these geographically smaller political entities was oriented toward the best possible autonomy concerning the fundamental means of life and, at the same time, toward the achievement of European leadership in at least one segment of technological production. Policies for economic autonomy contributed to the better exploitation of agricultural resources, which implied the control of the waters and the emergence of modern botany. The focus on economic leadership, in turn, contributed to the development of more efficient, large-scale activities of soil exploitation, as well as to a remarkable concurrence in technological innovations.

Obviously such a segmented political situation and aggressive economic policy could only be achieved in a cultural context where populations identified themselves with the city in which they were living and its local rulers. This was mostly achieved by means of self-fashioning representations of the ruling boards, in particular of the courts. This, ultimately, explains the growing relevance of theatrical representations in principal European cities in the early modern period, and, therefore, also the mechanics that allowed those increasingly ambitious performances described by Viktoria Tkaczyk (Chap. 4).

The strong relationship between practical knowledge and economy was also recognized by the same scholars of the sixteenth century. Girolamo Cardano perceived and positively evaluated the ongoing process of mathematization, which involved activities of a practical nature and, therefore, the knowledge underlying them as well. But he also recognized that this created a profound dependence on specialists of practical mathematics. The cultural role of the Abacus schools might have been underestimated by historians of science, as demonstrated in the following dramatic denunciation of Cardano, taken from Pietro D. Omodeo's essay (Chap. 12):

> We have come to the point that orators, poets, jurisprudents, physicians, farmers, architects, army leaders and actually all those who are somehow reputed for their works or for some other reason cannot accomplish anything without that [science of numbers or arithmetic]. In fact, many become wealthy only thanks to that activity [of counting numbers], among them bankers, money exchangers and traders. None grows rich more rapidly and more easily than they! Physicians and orators work their whole life to spare a quarter sesterce (and they do not often succeed), whereas those people can amass hundreds of sesterces in very few years. (Girolamo Cardano, *Artis arithmeticae tractatus de integris*) (Cardano 1966, 117)

Focusing on the example of sixteenth-century Rome, Pamela O. Long shows that it was not only the architectural reshaping of the city that led to codification and therefore to a higher circulation of knowledge (Chap. 8). This was also due to the involvement of antiquarian investigations and a "flourishing print industry." Print technology certainly was the most reliable and effective ally of codification processes and the diffusion of practical knowledge. This also implies that printed media and written texts—in the form of books and prints in general—became part of the material production in order to support the market of immaterial knowledge. The more the knowledge became a demanded product, the more printed media, in all its possible forms, was demanded as well.

The intensification of printing gave rise to a process of knowledge codification concerned with this same activity, as examined by Eileen Reeves (Chap. 7). Indeed, the intensity and density of production played a central role in driving codification, as is nicely exemplified by the fact that the chiaroscuro woodcut was developed as a technical solution to the increasing market demand for colored prints. As this demand increased, the chiaroscuro woodcut became a valid alternative to manual coloring and served as "the printmaker's most significant source of wealth." Woodblock techniques, as Bruce T. Moran explains, were among the most efficient means to visually replicate natural forms, such as those of plants (Chap. 15). The business of woodblock-based printing experienced a great increase in intensity during the sixteenth century, with an augmented desire "for individualized experience through private reading and evaluation." Early modern printers developed a tradition and an economic model that allowed them to investigate which knowledge should be codified and in which form. Richard L. Kremer shows how the genres of both the *practica* and the almanacs were born with the advent of printing technology, basically within the printers' workshops (Chap. 13).

One of the leading sectors of economy was certainly linked to architectural enterprises. The early modern expansion of civil and military building sites, as well as production centers—as, for example, the naval shipyards of Venice—was unprecedented. Such places brought together new managerial skills and all forms of mechanical arts. Technological innovation and strong economic and political interests were often framed in the areas related to military activities. The diffusion of a new generation of heavy, mobile artillery, literally required each city and territorial entity to develop new forms of defense. This construction gave rise to a new discipline of military architecture, profoundly related to geometry, and to the emergence of a new figure, the military officer, trained in the mathematical arts. The so-called early modern revolution of the art of war, and the great economy it generated, showed a profound need to keep, organize, protect and develop practical knowledge, and therefore also, to codify it (Valleriani 2010, 3–113).

Increased social mobility found its most relevant historical expression in the age of exploration. Different and innovative ship models, new navigation techniques, a new nautical astronomy with new computational techniques, and the re-development of mathematical instruments were only a few of the great many new practical activities generated during this age. The evident fundamental economic relevance of such knowledge, in spite of the need to protect it, created the conditions for the emergence of one of the widest and most long-lasting scientific discourses over the Continent, mirrored in the curriculum of studies at the European faculties of liberal arts and at the academies shaped over the model of the Florentine *Accademia del disegno*.

The early modern period is strongly characterized by innovations of technological character but maybe the most influential of the innovations of this period concerns the rise of a general need for and attention to, education and training, which in turn, created the conditions and the demand for the increased codification of knowledge. This cultural change was obviously both driven and supported by the diffusion of the printed book. New genres emerged such as pamphlets, catechisms, almanacs and books of prodigies and sensations. Great audiences emerged for books in history and poetry and for romances. And finally, special interest emerged for subjects like cartography, law and, of course, manuals of technical and scientific character (Pettegree 2002, 321; Badino and Navarro 2013).

The process of mathematization implicit in the radical changes experienced in the field of the art of war, created a new educated audience that was increasingly technically oriented—the military officers. At the same time, the rise of the internationally well-connected social group of humanists created an increasing demand for prints of ancient works. Both groups were mostly rotating around the early modern courts, which also became established as educational centers and created a network parallel, to the one of the universities and the academies. These institutions appeared in addition to those already established in the late Middle Ages, such as guilds, that also had a strong educational vocation. Moreover, new large-scale forms of knowledge economy were built in concomitance with the age of discoveries. Not only the discoveries themselves were the result of such an organization but of course the economic exploitation and the political ruling of the new territories required new organizations for education in all fields of knowledge and concerning all kinds of activities. A clear example of such a new organization is the network of Jesuit schools, colleges and chairs intimately connected to the missionary activity of the order.

5 The Structures of Knowledge

In her *Building the House of Knowledge*, (Findlen 2001), Paula Findlen suggests investigating early modern structures of knowledge by exploring historical material structures such as buildings, real or imaginary, like Francis Bacon's (1561–1626) *Solomon's House* or the Renaissance gardens and theaters. Alternatively, books, such as corpora of early modern commentaries on a specific work, could represent the material structure of Findlen's idea.

One of the most astonishing examples that shows how new conceptual structures of knowledge are continuously re-created and expanded on the basis of codified practical knowledge, is represented by the tradition of commentaries on the treatise *De sphaera* (Chap. 16). Compiled by Johannes de Sacrobosco in the thirteenth century, this tract became a center of knowledge accumulation, around which an entire information system was externalized over the course of nearly four centuries.

Beginning with its original section, which provides a qualitative introduction to geocentric cosmology, the knowledge contained in *De sphaera* underwent an incremental process of transformation. More and more subjects were added to the treatise, consistently building on the description of the cosmos. When the era of exploration began, an extraordinary range of subjects related to travel and navigation made their way into the treatise in the form of codified written texts. This included the computational knowledge necessary to calculate the positions of the planets; descriptions of the use and the construction of mathematical and astronomic instruments for the calculation of time and position, including dials and clocks; introductions to metric geography; and proofs and explanations of advanced geometry.

The treatise itself, which was published several hundred times, ultimately, bore the name of its original author, Sacrobosco—*The sphere of Johannes of Sacrobosco*—which in effect became a stable label for the continued accumulation and mutation of knowledge. In addition to the subjects pertaining to navigation, the tract included material on astrology (both in terms of general celestial influence and as a discipline to make predictions concerning human life); formulas and charts belonging to the original conceptual structure of medieval cosmology, such as calendric calculation techniques; and in certain instances, medical texts. In the final phase of its production, the treatise of Sacrobosco contained not only the theoretical knowledge related to travel, but also the practical knowledge accumulated by those who actually traveled. There were long chapters on practical optics (techniques to measure by sight) that were of great benefit to the military officer, while a merchant could use the treatise to learn accounting and bookkeeping techniques.

The early modern treatises on *De sphaera* are, therefore, the material representation of a knowledge system constituted primarily of codified practical knowledge, as understood within the ancient geocentric worldview. Notably, however, the pilots, farmers, astronomers, physicians and artillerists who constituted the primary population of those who owned this treatise, probably did not make extensive use of it in their everyday professional activities. But the treatise did offer these individuals, and also the society at large, a structure of knowledge which their knowledge and activity was positioned in, and to which they could possibly even contribute. Once practical knowledge was codified, it could easily find its place in a comprehensive knowledge system thanks to the shared and cooperative work of scholars and printers, who continuously re-produced this treatise.

The structure of *De sphaera* represents a particularly large example of how knowledge was produced through the combination of existing knowledge, previously codified in the form of written texts. Knowledge structures, however, could be created by combining knowledge in various forms. The most striking example of this is offered by Bruce T. Moran, who traces the life and evolving function of a huge collection of sixteenth-century woodblocks to print botanical images (Chap. 15). These blocks were originally produced to illustrate the 1578-publication on Paracelsian alchemy of the alchemist and physician Leonhard Thurneisser (1531–1596). Thurneisser, however, only printed 36 of the almost 2000 wood-block images. The knowledge embedded in the surviving and unused woodblocks, visually codified from the careful observation of plants, was then epistemically re-contextualized and re-structured. The blocks were re-used in a series of different publication projects, each with a unique scholarly purpose. The Thurneisser woodblocks were first re-used in a 1654-publication by Thomas Panckow (1622–1665), whose goal was to produce a sort of catalogue or reference work for apothecaries, surgeons, gardeners, *Hausväter*, and students of medicine. As Moran puts it, "traveling woodblocks … play a physically fixed but culturally mutable role among the processes and practices that help revise and reshape the perception of the natural world."

Just like the core cosmological text section of *De sphaera*, the woodblocks remained unchanged in their continued conceptual re-contextualization. And in both examples, this re-contextualization bore new structures of knowledge. These structures were created by the work of scholars, but were also fundamentally based on the social organizations constituted by artisans, draftsmen, cutters, and printers.

The knowledge and the skill of such professionals, once codified, could also simultaneously furnish relevant contributions to other areas of scientific inquiry.

For example, as Eileen Reeves shows (Chap. 7), discussions concerning the best materials for making chiaroscuro woodblocks became an integral part of the emerging scientific discussion on the strength of building materials, which featured in the 1554 work *First Four Books on Architecture* of Pietro Cataneo (1510–1569). The method for realizing chiaroscuro prints was first codified by Giorgio Vasari (1511–1564) in his 1550 edition of the *Lives of Artists*. Here, as also in later sources, particular attention is given to the best possible materials needed to produce chiaroscuro woodblocks.

The production of Almanacs and *practica* provide another example for the fundamental role played by printing technology, and by printers themselves, in the creation of new large knowledge structures. Within such texts, a new structure was materialized based on classical, often very elaborate, computational astronomic knowledge, in which astrological knowledge was combined with practice-oriented medicine, agriculture and meteorology. While "almanacs had roots in the computational procedures of the fourteenth-century Alfonsine astronomy," the creation of the entire knowledge structure was in part mediated by a bookish tradition that included genres such as the *iudicum* and the *taccuinum sanitatis*, which, in turn, drew from the tradition of the *Astrologia Ypocratis*.

By analyzing the single knowledge aspects brought together in almanacs and *practica*, Richard L. Kremer shows that the formulas and equations of almanacs required learned skill, and that, likewise, the computation of syzygies were based on advanced mathematics (Chap. 13). However, the more or less implicit rules that allowed computational work to be associated with medical precepts were (and could only be) derived from the experiences of the almanac makers, or from those who furnished the makers with information. It was once again the printer, therefore, who created a new knowledge structure by bringing together very different aspects of knowledge.

As previously discussed, the codification of practical knowledge never takes place as a process that begins from a sort of *tabula rasa*. It always relates the codified knowledge with knowledge of a more general character: characteristically, the processes of practical knowledge codification extend back to the study of the universe; the cosmological, astronomic and astrological study of the geocentric system; the theory of the constitution of matter concerned with the four elements; medicinal science; the four humors of Galen; and virtually everything that has to do with the human body. Clearly, the ancient knowledge system often played a relevant role. This intrinsic characteristic naturally led not only to the formation of a different structure, but also to a structure of knowledge based on a wider range of topics. As Jochen Büttner comments, "representations of practical knowledge typically exceed its scope." Finally, the formation of new structures of knowledge introduced the possibility of associating specific theoretical reflections with different problems. In this way, pre-classical theoretical ballistics came to be determinant in the study of phenomena such as the falling of heavy objects and the shape of water jets.

While the Aristotelian background of codified practical knowledge is particularly evident in those tracts produced in reference to metallurgy and mining activity, primarily against the background of physicians' knowledge, as Pamela H. Smith shows (Chap. 14), the profound influence of ancient knowledge is also evident in the work of the five architect-scholars discussed in Pamela O. Long's investigation (Chap. 8). In the frame of a continuous research for technological innovations, new knowledge was produced that eclectically used the knowledge of antiquity. The same ancient knowledge was thus appropriated—tested, selected and transformed—through the workbench of practical knowledge. But the ancient knowledge system that allowed for codification processes in these cases was not only represented within the context of Aristotelian natural philosophy. It was also activated in humanists' glorification of ancient Rome, and in the knowledge codified in the works of Pliny, Frontinus, and Vitruvius, as well as, the learned tradition that these works gave rise to.

Narrowing the social scale which furnished the formation of new knowledge structures, we might also recognize such structures on a very local level, as in personalized collections of recipes. As discussed by Elaine Leong, recipe-like codification might start as a "straightforward reflection of practice on the ground," although it is also immediately inserted into a wider system of knowledge (Chap. 3). In debating the relevance of the so-called "water boiling step" in brewing ale, the protagonists of Leong's essay came to consider ancient medicinal knowledge, represented by the works of Hippocrates, as well as that given in contemporary manuals on brewing. The issue was later discussed in regimen guides, and was even considered in the frame of the experimental research processed at the Royal Society.

Even single individuals can create significant new structures of knowledge while codifying practical knowledge, as in the case of Leonardo da Vinci (1452–1519). Viktoria Tkaczyk shows how Leonardo, while working on the technical aspects of the flying machines for the purposes of theatrical engineering, also developed his famous studies concerning the phenomenon of flight (Chap. 4). Calling our attention to the epistemic connection between these two different activities, Tkaczyk's essay illustrates how the early modern codification of practical knowledge introduced contemporary reflection on knowledge itself. No longer considered a product of the imitation of nature, practical knowledge came to be understood as a dynamic means of perfectly reproducing nature. This presupposes the careful study and observation of nature, that is, scientific investigation.

Leonardo is one of the first scholars in whose works this shift is recognizable, and it is exactly this new conception of practical knowledge that allowed him to redirect contemporary attention toward natural philosophical issues within the context of practically-based experience. In fact, it was Leonardo and his contemporaries who began the practice of the "technical experimentation on paper." This structure of knowledge, identified by Elizabeth M. Merrill as experimental knowledge, deliberately disregards aspects of practical knowledge in order to achieve other ends (Chap. 2). A remarkable example of this is represented by Francesco di Giorgio Martini's (1439–1502) infinite, illustrated variations of certain machine types.

At the courts, where Johann Kunckel was active, recipe knowledge met and erudite knowledge (Chap. 6). Kunckel's text analyzed by Sven Dupré contains not only a German translation of Antonio Neri's (1576–1614) recipe collection, *Arte vetraria.* It also contains Christopher Merrett's notes on the same book and a short history of glass. It also makes use of scholarly knowledge to organize subjects directly codified from practical knowledge, for instance, by means of the textual structure typically exhibited in the medical *observationes*. Thus, Kunckel's book might be regarded as a sort of commentary on the classic tradition, in which the results of the author's own testing are incorporated. The book, moreover, codifies errors. In historic terms, the codification of errors might be viewed as an indication of the development of practical knowledge structures from their initial form, into more elaborate knowledge systems.

6 The Narrative

All the essays collected in this volume face a great variety of the subjects discussed in this chapter. Nevertheless, they have been grouped according to the epistemology of practical knowledge that they singularly develop the most.

The first two essays, by Elizabeth M. Merrill and Elaine Leong (Chaps. 2 and 3) perfectly show the mechanisms and the reasons for the very first act of codification from within the activity and the workflow of actions that constituted the actual structure of practical knowledge. How such structures of knowledge immediately begin acquiring a conceptual independence from the practical activities themselves is exemplarily shown in the essays written by Viktoria Tkaczyk and Jochen Büttner (Chaps.

4 and 5). How the codification process can also be caused by limitations, be they related to the materiality of the product or to the production workflow, is discussed by Sven Dupré and Eileen Reeves (Chaps. 6 and 7). The dimension of the social structure that norms the practical knowledge and whose rules determine which aspects of practical knowledge are codified and, therefore, build a new structure of knowledge, is discussed by Pamela Long, Wolfgang Lefèvre, Dagmar Schäfer and Ursula Klein (Chaps. 8, 9, 10, and 11). Schäfer's essay shows that similar mechanisms are also at work outside the European continent, specifically in China. By focusing on a later period, Klein's essay allows the trajectory of development of the social structures built during the sixteenth century and up until the eighteenth century to be sketched. Out of the frame of the social structure, the voice is then given to the actual early modern actors while discussing the relation between practical and theoretical knowledge and while applying these categories. This aspect is discussed mainly in the essays by Pietro D. Omodeo and Richard L. Kremer (Chaps. 12 and 13). On the basis of the fundamental overview of the structures of knowledge based on practical knowledge, offered in the essay of Pamela H. Smith (Chap. 14), the last two essays—Bruce T. Moran and Matteo Valleriani (Chaps. 15 and 16)—explore the mechanisms that led to the continuous conceptual re-structuring of knowledge, also by means of a heuristic that makes use of models and techniques developed in the frame of social network analysis.

The epistemological narrative described by this book has the aim to search for continuities in scientific developments and does not require any "crisis of the sciences" for it to function. It only appeals to the fact that, thanks to the enlarging of the spatial horizon, trade and manufacture increased to such an extent that material wealth spread over the continent. The incremental transformation of the early modern period is what brought technology into the realm of science, thus, contributing to the process of mathematization.

References

Anderson, P.W. 1972. More is different. *Science New Series* 177: 393–397.

Badino, Massimiliano, and Jaume Navarro. 2013. Pedagogy and research. Notes for a historical epistemology. In *Research and pedagogy. A history of quantum physics through its textbooks*, ed. M. Badino and J. Navarro, 7–25. Berlin: Edition Open Access.

Cardano, Girolamo. 1966 [1663]. *Opera omnia*. Lugduni: Sumptibus Ioannis Antonii Huguetan & Marci Antonii Ravaud. Reprint, Stuttgart-Bad Cannstatt: Fromann-Holzboog.

Damerow, Peter. 1996. *Abstraction and representation: Essays on the cultural evolution of thinking*. Dordrecht: Springer.

Dubourg Glatigny, Pascal, and Hélène Vérin. 2008. *Réduire en art. La technologie de la Renaissance aux Lumières*. Charenton-le-Pont Cedex: Éditions de la Maison des sciences de l'homme.

Epstein, Larry. 1998. Craft guilds, apprenticeship and technological change in pre-modern Europe. *The Journal of Economic History* 53: 684–713.

Epstein, Larry. 2000. *Freedom and growth, markets and states in Europe, 1300–1750*. London: Routledge.

Findlen, Paula. 2001. Building the house of knowledge: The structures of thought in late Renaissance Europe. In *The structure of knowledge: Classifications of science and learning since the Renaissance*, ed. Tore Frängsmyr, 5–52. Berkeley: University of California Office for History of Science and Technology.

Hall, Bert S. 1979. Der Meister sol auch kennen schreiben und lesen: Writings about technology ca. 1400- ca. 1600 A.D. and their cultural implications. In *Early technologies*, ed. Denise Schmandt-Besserat, 47–58. Malibu: Undena Publications.

McClure, George W. 2004. *The culture of profession in late Renaissance Italy*. Toronto: University of Toronto Press.

Mokyr, Joel. 2005. *The gifts of Athena. Historical origins of the knowledge economy*. Princeton: Princeton University Press.

Pettegree, Andrew. 2002. *Europe in the sixteenth century*. Malden: Blackwell.

Renn, Jürgen, and Malcolm Hyman. 2012. The globalization of knowledge in history: An introduction. In *The globalization of history*, ed. Jürgen Renn, 27–51. Berlin: Edition Open Access.

Renn, Jürgen, and Matteo Valleriani. 2014. Elemente einer Wissensgeschichte der Architektur. In *Wissensgeschichte der Architektur*, vol. 1, ed. J. Renn, W. Osthues, and H. Schlimme, 7–53. Berlin: Edition Open Access.
Sennett, Richard. 2008. *The craftsman*. New Haven: Yale University Press.
Valleriani, Matteo. 2010. *Galileo engineer*, Boston studies in the philosophy of science. Dordrecht: Springer.
Valleriani, Matteo. 2013. *Metallurgy, ballistics and epistemic instruments. The «Nova scientia» of Nicolò Tartaglia. A new edition*. Berlin: Edition Open Access.
Valleriani, Matteo. 2014. Appropriation and tranformation of ancient science. *Nuncius* 29.

2 Pocket-Size Architectural Notebooks and the Codification of Practical Knowledge

Elizabeth M. Merrill

Abstract

Based on the study of five small-scale Italian architectural manuscripts, dated between 1470 and 1520, this article uses the practitioners' drawings to examine early modern artistic education, and the means by which architectural knowledge was systematized, developed, shared, and retrieved. To this end, the paper consider the books' contents in terms of differing divisions of architectural knowledge and modes of visualization. The function of the manuscripts' drawings—which include generic models, building diagrams, design "recipes," and exercises in architectural *invenzione*—varied significantly depending on context, and although no single image served as a "working" plan, all were in some way immediately useful to the practitioner. The drawings may thus be understood as records of architectural practice, and as such, relate to the codification of design-based knowledge and the evolving language of early modern architectural representation.

In book six of his *Libro Architettonico*, Antonio Averlino "Filarete" (1400–1469) advises the student architect to carry with him a notebook, in which to record impressions, measurements and ideas. The drawings and annotations collected in the book, Filarete notes, will help the architect clarify and communicate his building projects, and in the case that details are forgotten, he "can turn to the book and find them again" (Filarete 1972, 148). The ubiquity and utility of such notebooks among fifteenth- and sixteenth-century artists and architects is amply evidenced. Correspondence records and last testaments from the period are filled with citations to *taccuini*, *libretti* and *vacchette* ("little books"). Similarly, contemporary chroniclers and theorists, figures like Marcantonio Michiel (1484–1552) and

E.M. Merrill (✉)
Max Planck Institute for the History of Science, Berlin, Germany
e-mail: emerrill@mpiwg-berlin.mpg.de

M. Valleriani (ed.), *The Structures of Practical Knowledge*, DOI 10.1007/978-3-319-45671-3_2

Giorgio Vasari (1511–1574), report directly on practitioners' reliance on such books.[1] A well-known illustration by Maso Finiguerra (1426–1464) also captures the importance of drawing books in architectural training. The image presents a young artist diligently drawing in a small sketchbook, and bears the inscription: "I would like to be a good draftsman and I would like to become a good architect."[2] Within the study of Italian Renaissance architecture, significant attention has been given to the practitioner's drawing books, in particular to workshop model books and volumes of antiquarian studies.[3] Yet, one category of early modern architectural drawing books has been substantially overlooked: the personal, "pocket-sized" reference book of the architect-technician. Scholarly neglect of these books is variously explained. For one, due to the small size of the volumes—conceived to be readily transferred between the workshop and building site—as well as to the high degree of wear they endured, relatively few of the pocket-sized manuscripts remain intact (Elen 1995, 55–6). Critical examination of the portable books has also been a casualty of the compositional and formal qualities of the volumes themselves. As manuals compiled for the purposes of personal reference, the drawings contained in the handbooks often lack the refinement and finish of those the architect prepared for patrons or meticulously copied into a shared, workshop model book. Thus, within the scholarship, the pocket-sized volumes have been considered "disorderly and chaotic," prosaic, haphazard collections of loosely related remarks (Zwijnenberg 1999, 1).

But this study aims to show otherwise. Examining five exemplary, small-scale manuscripts of the fifty-year period between 1470 and 1520, this article argues that the often overlooked early modern practitioner's manuals are in fact among the richest and most revealing documents of early modern architecture. The five manuscripts are those of Buonaccorso Ghiberti (1451–1516), grandson of the famed Florentine goldsmith and architect Lorenzo Ghiberti (1378–1455) (*Zibaldone* c. 1480–1490); the Sienese military architect Francesco di Giorgio 1439–1501 (*Codicetto Vaticano*, c. 1465–1475); Giuliano da Sangallo (1443–1516), favored architect of Lorenzo de' Medici (1449–1492) (*Taccuino Senese*, c. 1490–1513) Leonardo da Vinci 1452–1519 (Codex Forster III, c. 1493–1497); and the mechanic and topographer Benvenuto della Volpaia (1486–1532) (*Libro di Macchine Diverse*, c. 1520–1524).[4] These volumes, although each entirely unique in regards to form, media and subject matter, are alike in their essential function as personalized collections of architectural knowledge. The books codified the practice of those individuals who used them, and in the sense that they were carried with the architects, shared with collaborators or passed between generations, they were transmitters of knowledge. More so than building contracts, wooden models, and even the final constructions, the personal notebooks offer insight into how the early modern architect structured his thoughts, developed

[1]On early modern citations to artists' drawing books, see (Rushton 1976, 16–7). On M. Michiel's references to drawing books, as well as on the collecting of drawing books in the sixteenth, seventeenth and eighteenth centuries, see (Elen 1995, 128–33).

[2]Maso Finiguerra's "Apprentice Drawing" is conserved in the Gabinetto Disegni e Stampe degli Uffizi, U 115 F. See (Bambach 2003, 6–7).

[3]Notably, there has been considerable scholarship dedicated to early modern scholarly note-taking and academic notebooks. Principal recent studies include: (Blair 2010), (Campi et al. 2008), and (Yale 2011). The distinction between the artist notebook, model book, drawing book and album is addressed by (Elen 1995, 26–32). The definitive study on Renaissance drawing books of antiquities remains that of (Nesselrath 1986). On workshop model books and artist's copybooks, see (Ames-Lewis 1981, 63–72).

[4]Buonaccorso Ghiberti, *Zibaldone*, codex Banco-Rari 228, Biblioteca Nazionale Centrale, Florence; Francesco di Giorgio, *Codicetto Vaticano*, codex Urb. Lat. 1757, Biblioteca Apostolica Vaticana; Giuliano da Sangallo, *Taccuino Senese,* S.IV.8, Biblioteca Comunale di Siena; Leonardo da Vinci, Codex Forster III, Victoria and Albert Museum, London; Benvenuto della Volpaia, *Libro di Macchine Diverse,* codex Italiani IV 41, 5363, Biblioteca Marciana, Venice.

building concepts and executed them.[5] Their value as such has enormous importance in regards to our complicated understanding of the Renaissance architect, whose role was notoriously diffuse, adhering to neither a standard course of training or a set of technical requirements, and varying substantially on a case-by-case basis.[6]

The architecture presented in the books of Buonaccorso Ghiberti, Francesco di Giorgio, Giuliano da Sangallo, Leonardo and Benvenuto della Volpaia primarily concerns building mechanics—machines for pulling, lifting and transporting great weights, hydraulic works and mills, military structures, and urban defense systems. These are not comprehensive books on early modern architecture, and in no way provide a definitive image of the contemporary architect. Yet, examined together, they outline underlying principles that structured the discipline of architecture, as well as the means by which practical knowledge was systemized, developed, shared, and retrieved. Read carefully, the books' contents also suggest different divisions of architectural knowledge, how through varying modes of visualization, this material was translated into practice—such as in the copy of generic, building models, the diagrammatic representation of design "recipes," or the realization of new architectural *invenzione*. None of the drawings presented in the books served as "working" documents as we understand them today, but all were in some way immediately useful to the practitioner. That the architects took the time to meticulously record and compile the figures, often including annotations rendered in miniscule script, is highly significant, and confirms that these were more than just idle sketches or cursory notes. Moreover, the fact that some of these figures were copied in more than one of the books indicates that the knowledge being transmitted belonged to a greater, collaborative field of contemporary architectural practice.

In the pages that follow, the notebooks of Ghiberti, Francesco di Giorgio, Giuliano da Sangallo, Leonardo da Vinci and Benvenuto della Volpaia will be examined in relation to the structuring of practical architectural knowledge at the turn of the sixteenth century. The intent is not to enumerate the contents of the five books, or to systematically outline the projects to which individual designs relate. Rather, this discussion explores the means by which the books were assembled, the types of knowledge they present, and how this knowledge was codified and used by the architect in practice. To this end, the knowledge illustrated in the books is divided into three principle categories: (1) shared and cumulative knowledge, which was developed within a given workshop or family, and disseminated through copy drawing; (2) operational knowledge, typically allied to specific project or problem, and pertaining to an understanding of when and where to access certain facts and employ certain procedures; and (3) experimental knowledge, that which was born in the creative mind of the artist. Notably, each of these divisions of knowledge was complemented by a different aspect of architectural *disegno*—a term used to refer to both the practice of drawing and the resulting product. The *disegno* which accompanied shared knowledge was largely rote; the *disegno* of operational knowledge was didactic and preparatory; and the *disegno* of experimental knowledge was creative. Still, it must be emphasized that the three categories of knowledge explored here are neither definitive nor absolute. Although they follow a certain hierarchical progression—shared knowledge being the most mundane, and experimental embodying the highest level of cognition—as evident in the notebooks, the three categories were integral to practicing architects at every stage in their careers. The categories "shared," "operational," and "experimental" refer not to the level of knowledge, but to the degree to which it was known and circulated, and the means by which it was generated within contemporary practice. In this way, they are useful lenses with which to consider the function of the notebooks, and are also instructive in delineating the underlying structures of early modern architecture.

[5]On the different media of early modern architectural representation and documentation, see (Goldthwaite 1980, 376–9).

[6]The nuanced character of the early modern architect is discussed in (Chaps. 8 and 9).

1 Sizing Up the Notebooks

The notebooks of Buonaccorso Ghiberti, Francesco di Giorgio, Giuliano da Sangallo, Leonardo and Benvenuto della Volpaia may all be classified as pocketbooks in the sense that they are relatively compact reference volumes, and were assembled by their owners as *aide-memiore*.[7] The small size of the manuals was directly related to their function as personal notebooks, easily added-to and amended, and readily transferrable. Yet in physical form and content, the five books are actually quite different. To begin, of the books examined here, only those of Francesco and Leonardo are truly pocket-sized. Measuring 59×81 and 62×94 mm, respectively, both of the books are small enough to be held in the palm of one's hand. The volumes of Buonaccorso, Giuliano and Benvenuto are slightly larger, measuring between 120 and 145 mm in width, and 180–210 mm in length, an appropriate size to be carried in a saddle bag. It is worth considering also the relationship between the size of the volumes and how the architects went about recording their drawings and annotations. Within Francesco di Giorgio's *Codicetto*, for example, the script is often so small that it is difficult to read without a magnifying device. In this case, the utility of such a compact a reference book presumably compensated for the compositional challenges imposed by its small size. In addition to the physical dimensions of the books, other features of their material composition also bear on the intended use of the volumes. The following section will provide a codicological overview of the five books, focusing on the media of the drawing support (paper or parchment), the binding and arrangement of quires, and the architect's preference of pen, pencil or charcoal.

The architect's decision to compose his notebook of either parchment or paper folios is perhaps the most telling material element of a volume's intended use. Parchment, made of the skins of sheep, goats or calves, is an extraordinarily durable material. In the early modern period, when parchment was cured and cut by hand, and supply was often limited due to poor weather conditions and livestock shortages, the acquisition of even a dozen parchment folios was quite expensive (Elen 1995, 32). Among frugal artists, therefore, parchment was seldom employed. In the thirteenth and fourteenth century, artists were known to render their ideas on boards or wax-tablets, or, if working with parchment, customarily scraped clean and re-used the folios until they virtually disintegrated. The artist's turn to paper occurred in the mid-fifteenth century, as the medium became increasingly affordable, and for the first time, he had the means to draw more, and to save his drawings (Ames-Lewis 1981, 19; Hiscock 2000, 173). Paper, moreover, was more suited to sketching than parchment, and opened up the possibility for artists to record rapid impressions (Ackerman 2002, 294). Still, for the purpose of model books and reference manuals, parchment was infinitely superior to paper. Francesco di Giorgio's and Giuliano da Sangallo's choice to construct their notebooks using parchment folios, as opposed to those made of paper, was most likely made in light of heavy use the books were to endure.

Francesco's *Codicetto*, at once both model book and notebook, originally contained approximately 235 folios, held together within two wooden cover plates, and secured with metal clasps.[8] Because the book was disassembled in the 1980s for the purpose of making a facsimile volume and has since been made inaccessible to scholars, it is difficult to know how the volume was initially composed. Still, based on sequential ordering and grouping of folios, such as the commentary on cannons (fols. 42v–44r),

[7]Following Ann Moss's analysis of early modern commonplace-books, Alexander Marr characterizes the copybooks assembled by sixteenth- and seventeenth-century artists as "aide memoire"—customized collections of visual and textual information systematically copied and assembled in personal notebooks. See (Marr 2013).

[8]Nearly fifty of the folios are today missing. Beyond a few references to its contents and remarkable pocket-size, very little has been said about the *Codicetto*, why Francesco assembled it, or how exactly it was used. L.M. Tocci published a facsimile version of the *Codicetto*, accompanied by a short, somewhat speculative commentary. For codicological information on the book, see (Martini 1989, 22). The *Codicetto* has also been summarily reviewed by (Galluzzi 1991). See also (Prager and Scaglia 1972, 191–5) and (Scaglia 1992, 51).

or the exploration of siphon-pumps (fols. 142r–153r), it appears that at least a good portion of the *Codicetto* was pre-bound. Although the little book is undated, scholars generally agree that Francesco assembled it in the early 1460s, when he was first exposed to architecture in Siena, and continued to make additions through the mid-1470s (Mussini 2003, 25). Working with pen and ink, Francesco filled the *Codicetto* with a series of copy-drawings of design prototypes—over one-thousand drawings in total—illustrations of pumps, dams, bridges, aqueducts, mills, balances, construction and transportation mechanisms, as well as military machines and fortification plans. Most of this material was derived from the notebooks of Jacopo Mariano Taccola (1382–1453), the so-called "Archimedes of Siena," although as we will see, Francesco substantially improved upon many of his predecessor's ideas. The form and media of the *Codicetto* were compatible with its function as a personal, pocket-sized encyclopedia, which Francesco might easily carry between the local university, where Taccola's books were held, the city's workshops, and job-sites (Taccola 1984, 28–31). We might imagine that Francesco—still in his early stages of training, inundated with a large quantity of new information, and unable to differentiate the key points from the more minor details—copied everything into the notebook. Along with drawings, he translated passages from Taccola's Latin commentary, which he recorded in an almost undecipherable minuscule script. As one turns page after page, it seems as if the composition of the *Codicetto* was a personal challenge for Francesco, as he endeavored to crowd into each page as much information as possible.

In terms of its composition and intended use, Giuliano da Sangallo's *Taccuino Senese* shares much in common with Francesco's little book. Not only is Giuliano's volume also composed of parchment folios—fifty-one in total—but like the *Codicetto*, it also appears to have been pre-bound. The presence of double-drawings, such as the plan and cut-way elevation view of the Colosseum on folios 6v and 7r, and the study-drawings of the Ionic columnar order on folios 34v and 35r, indicate that these figures were executed in previously bound quires (Hülsen 1910, LIII–LVI).[9] Also like the *Codicetto*, the contents of the *Taccuino* appear to have been recorded over a prolonged period, between c. 1490 and 1513, and were compiled for the architect's reference when traveling or on-site. But the *Taccuino* drawings are without question more refined than those of Francesco di Giorgio. The numerous parallels between the *Taccuino* and Sangallo's generously scaled codex Barberini (ca. 1465–1510)—a collection of models and polished study drawings assembled on luxurious parchment folios—suggests that the former book was to serve as a condensed version of the more elaborate workshop volume. Like the codex Barberini, the *Taccuino* served as reference manual and display book, containing project drawings of the architect's previously executed constructions, illustrations of machinery, significant structures and ornaments, all neatly rendered with dimensions and other relevant details. The images related to Sangallo's antiquarian studies were not rendered *in situ*, but likely executed after-the-fact, with compass and rule and using the notes and sketches previously collected on-site. Still, a good portion of the material included in the *Taccuino* may be traced to existing model books and drawings—figures Giuliano copied from others. Taken as a whole, therefore, the neat little book may be understood as an accumulation of Giuliano's lifelong study, offering insight into the diverse practices of the multiply talented architect (Elen 1995, 413; Brown and Kleiner 1983, 323).

The remaining three books of our set—those of Leonardo da Vinci, Buonaccorso Ghiberti, and Benvenuto della Volpaia—are composed of paper folios. For Ghiberti and Volpaia, the choice of paper was likely economical. Likewise for Leonardo—who was renowned for the breadth of his interests, fecundity of imagination, and proclivity to compile notebooks—paper was clearly the most desirable support media. Paper folios, moreover, were perfectly suited for red pencil or chalk, Leonardo's favored drawing media, so that at a moment's notice he could jot down an impression, image or idea.

[9] Like Francesco di Giorgio's *Codicetto*, the *Taccuino Senese* is also incomplete, with approximately twenty-one folios missing. According to Hülsen's reconstruction, the *Taccuino* was originally formed of seven quires.

Leonardo initiated the Codex Forster III in the early-1490s as a type of daybook, in which he recorded designs, recipes, and technical problems. In size, the Codex Forster III is comparable to Francesco's *Codicetto*, and just like Francesco, Leonardo used the pocket-size book both as mnemonic aide and didactic tool—a means to remember certain facts or details, and also to explain these to others. Yet, unlike the *Codicetto*, which is unique within Francesco's oeuvre, the Codex Forster III is just one of nearly a dozen pocket-sized notebooks composed by Leonardo (Zwijnenberg 1999, 7–8).[10] Within these manuscripts, he recorded his thoughts, observations, studies and dreams—both exceptional and mundane—and in doing so, created an extraordinarily rich document of his intellectual activities. The Codex Forster III contains an assortment of material, from studies of portals, to notations on firearm projectiles, calculations for weighting pulleys, and sketches of horse legs, hats, and costumes. The codex does not solely concern architecture, but it might be considered an architect's notebook in the sense that it was composed by Leonardo, a practicing architect, and is exemplary of his proclivity to skip between disciplines. The cursory little notebook is an excerpt of his "*scrittura infinita*" (infinite writing), in which the boundaries between forms of expression, languages, intellectual disciplines and experiences are eroded (Vecce 2003, 75).

The notebook of Volpaia, composed of ninety-six folios and measuring 142×210 mm, was executed between 1520 and 1524. The book's contents are largely derived from the notes and drawings of Benvenuto's father, Lorenzo della Volpaia (1446–1512), a goldsmith, architect, mathematician, machine-designer and clock-maker who completed projects for the *signori* of Florence around 1500 and was a close collaborator of Leonardo da Vinci.[11] Benvenuto followed after his father, taking over his workshop, and led a fruitful career as instrument and machine designer, topographer and fortifications consultant. He appears to have composed the book in order to ensure he always had his late-father's knowledge on-hand. The volume opens with an unusual memorandum—a reminder to himself to devise a certain home security system when staying in Rome—and then continues, page after page, with recipes (for gun-powder, soap, painting varnishes), instructions for smelting, astrological charts, mathematical equations and formula, designs for wheels and mills, hydraulic instruments, and notes on geography and astrology.[12]

The final codex left to review is that of Buonaccorso Ghiberti, the so-called "*Zibaldone*" (or literally "jumble"), which dates to between 1480 and 1490. Like the Volpaia notebook, Ghiberti's was built upon material assembled within the family workshop, namely that of his grandfather, Lorenzo Ghiberti. The *Zibaldone* functioned as a personal model book and notebook, and is remarkably heterogeneous in its material and structural composition. The paper folios display two distinct watermarks, and are bound in quires of varying size—of three, four, five and eight bi-folios (six, eight, ten and sixteen leaves) (Elen 1995, 47, 411). From this complicated, and mixed arrangement of quires, scholars have been able to ascertain that the book was assembled piecemeal by the architect as he worked. Portions of the manuscript were pre-bound, while others were constructed from individual quires with drawn or written notes (Elen 1995, 47–8). As the book expanded, its function shifted as well. Whereas Ghiberti might have initiated the *Zibaldone* as a portable reference manual—the folio

[10]Over the course of his lifetime, Leonardo filled approximately 15,000 folios in total with text and drawings. Other notebooks of Leonardo comparable in pocket-size dimensions to the Codex Forster III include: Codex Forster I and Codex Forster II (Victoria & Albert, London); and the Leonardo da Vinci Paris Manuscripts E-M (Bibliothèque de l'Institut de France, Paris).

[11]On Lorenzo and Benvenuto della Volpaia, see (Pedretti 1957, 23–6).

[12]On folio 1r, Benvenuto della Volpaia recorded: "To remind me when I go to stay in Rome to make a man of wood who stands inside the doorway, and when one opens the door, that man of wood confronts him either by lowering a stick on his head or catching and tying him with a cable" ("Per ricordarmi quando io vengo a stare a Roma di fare un uomo di legno che stia dentro a una porta e quando uno apre la porta quell'uomo di legno segli faccia innazi con un bastone per dargli in sulla testa hovero con una fune per pigliarlo e legarlo").

dimensions measure a relatively modest 145 × 197 mm—expanding to include nearly 240 folios in its final form, it would have been too large for the architect to comfortably carry in hand. Still, the extensive wear evident within the book—most of the pages have finger stains on the lower-right corners—indicates that Buonaccorso, and likely dozens of other assistants and architects, made frequent use of the practical manual. Most of the material here, although quite varied in terms of subject matter, relates in some way to projects undertaken by the Ghiberti workshop. The volume opens with twenty-three folios, containing the partial transcription of Lorenzo Ghiberti's translation of Vitruvius' *De Architettura*. Then follows a series of images of reconstructed antique monuments, views of tombs and sculptural projects conceived by Florentine artists, machine designs, and annotated drawings of bronze-works—the field in which Buonaccorso appears to have been most active.[13]

The *Zibaldone* is the book of an architect whose commissions and interests continued to evolve, and like the volumes of Leonardo, Francesco, Giuliano and Benvenuto, the diversity of its contents speaks to the architect's technical dexterity and wide ranging interests. Although architecture would become increasingly differentiated from civic, military and hydraulic engineering in the sixteenth and seventeenth centuries, at the turn of the Cinquecento the occupations of the architect, fortification designer, instrumentalist, and machine engineer still existed side by side as aspects of the same mathematically-based design practice, and all of the architects examined here were adept in all or most of these fields.

2 Shared Knowledge

On the most basic level, early modern artist notebooks were educational tools. Just as schoolboys of fifteenth- and sixteenth-century Italy were known to assemble commonplace books, filling them with exceptional passages to be memorized and then paraphrased in their writings, artists of this period compiled notebooks with the rudiments and "grammar" of their discipline (Ames-Lewis 1981, 16; Zwijnenberg 1999, 14–5). This material was then reiterated through extensive practice in *disegno*, by which the apprentice was tasked with the reproduction of his master's works, and came to serve as the foundation for his subsequent career.

The manual process of copy-drawing, considered essential in the memorization and mastery of exemplary models, extended beyond the age of apprenticeship (Ames-Lewis 1981, 15). In the fifteenth and sixteenth centuries, artists and architects at all stages in their careers were known to assemble personal copy-books, the contents of which are as varied as the authors' interests. These books often grew out of those the artist initiated in his early years of apprenticeship—replete with workshop formulas, recipes, and the master's signature compositions. To this foundation, the artist added material drawn from his own practice, as well as selected material found in the notebooks of his contemporaries and collaborators—his access to which may have been brief or limited (Ames-Lewis 1981, 63–4). By means of such accumulation and assembly, the knowledge contained in the personal copybook—primarily conveyed through drawing—invariably extended beyond a single workshop. Although at their origin the designs were born out of either experience or experimentation, over time and through copy and dissemination, they became general and mundane. Within the practitioner's copybook, they belonged to a far more vast, widely circulated and collective body of practical knowledge, material which, in effect, provided a common foundation for the diffuse discipline.

[13] Gustina Scaglia has completed extensive research on the *Zibaldone*. See (Scaglia 1976) and (Scaglia 1979).

With the exception of Leonardo's Codex Forster III, all of the volumes examined here contain, in varying quantities, traces of such shared or collective architectural knowledge.[14] Within the drawing collections of Buonaccorso Ghiberti, Francesco di Giorgio, Giuliano da Sangallo and Benvenuto della Volpaia one finds not only reproductions of the same designs, but also varied representations of the same design concept. In fact, the accumulation and transcription of shared knowledge was rarely literal or direct, and no two drawings recorded by our architects are exactly alike. When it came to copying, more often than not, the authors modified the material they studied, amending and restructuring it according to their own needs and interests. The following discussion will trace examples of copy-drawing in the notebooks of Buonaccorso, Francesco, Giuliano and Benvenuto, considering how such images' functioned, the value of the shared knowledge they contained, and how this knowledge was applied to practice.

Folio 104r of Buonaccorso Ghiberti's *Zibaldone* features an illustration of Filippo Brunelleschi's revolving lantern crane, a construction device engineered by the celebrated architect while at work on the Duomo of Florence (Fig. 1). While the concept of a building crane was in itself not unique, Brunelleschi's device improved upon the basic model in its ability to rotate, thus transporting materials to exactly where they were needed, on all sides of the dome. The essential component of Brunelleschi's crane was the grooved, horizontal beam, along which an engaged block could slide. Within this block, a screw mechanism could be used to lift and lower weights by means of worm-gear hangers—which Ghiberti illustrated in detail alongside the crane (Prager and Scaglia 1972, 74). The novel crane is included in an almost identical illustration in the *Taccuino Senese*, and also appears in Francesco di Giorgio's notebook, although in a slightly different version (Figs. 2 and 3). Leonardo also made a copy of the crane as well, although not in the Codex Forster III.[15]

Clearly the drawings of Buonaccorso Ghiberti, Francesco and Giuliano were based on second-hand knowledge, reproduced from a previously recorded, "original" design. Although Ghiberti and Francesco had indirect ties with Brunelleschi—Buonaccorso's grandfather, Lorenzo, was co-architect of the Duomo, while Francesco's intellectual mentor, Taccola, was a close friend and confidant of Brunelleschi—in terms of the formal qualities of the drawings alone, which of the three is most accurate is not readily apparent. And while Ghiberti's representation has been traditionally granted the most authority—significantly, the *Zibaldone* features several additional mechanisms devised by Brunelleschi, including his "secret hoist" and specific notes on the Florentine Duomo—what is noteworthy here is the degree to which designs of the crane were disseminated (Prager and Scaglia 1972, 67–70). Due to Brunelleschi's celebrity, and the awe inspired by his great dome, his designs were extensively studied by contemporary practitioners, and then circulated and reproduced by subsequent generations of architects. Considering only the *Zibaldone*, the *Codicetto*, and the *Taccuino Senese*, one can see how quickly an authorial invention like the rotating crane came to be incorporated into a more diffuse body of shared knowledge.

In addition to Brunelleschi's crane, the books of Buonaccorso, Giuliano, Francesco and Benvenuto also contain analogous representations of roller-beam hoists, obelisk haulers, column-lifting devices, water-wheels, siphon-pumps, and load positioners. Yet, the significant degree of imprecision evident in many of these drawings introduces new questions in regards to the function of the images, and the quality of knowledge they contained. Such imprecision is highlighted in the comparison of two machines illustrated in Benvenuto's *Libro*—one of a clock designed by his father, shown in section with detailed annotations regarding the diameters, placement and division of the wheels (fol. 31v) and

[14]This is not to say that Leonardo never executed copy-drawings. In fact, his Codex Madrid II (Ms. 8937, Biblioteca Nacional, Madrid), Leonardo reproduces numerous designs of Francesco di Giorgio.

[15]Brunelleschi's lantern crane appears on folio 808v of the codex Atlanticus (dated c. 1478–1480). For discussion on the lantern crane, and other devices of Brunelleschi copied by Leonardo, see (Reti 1965).

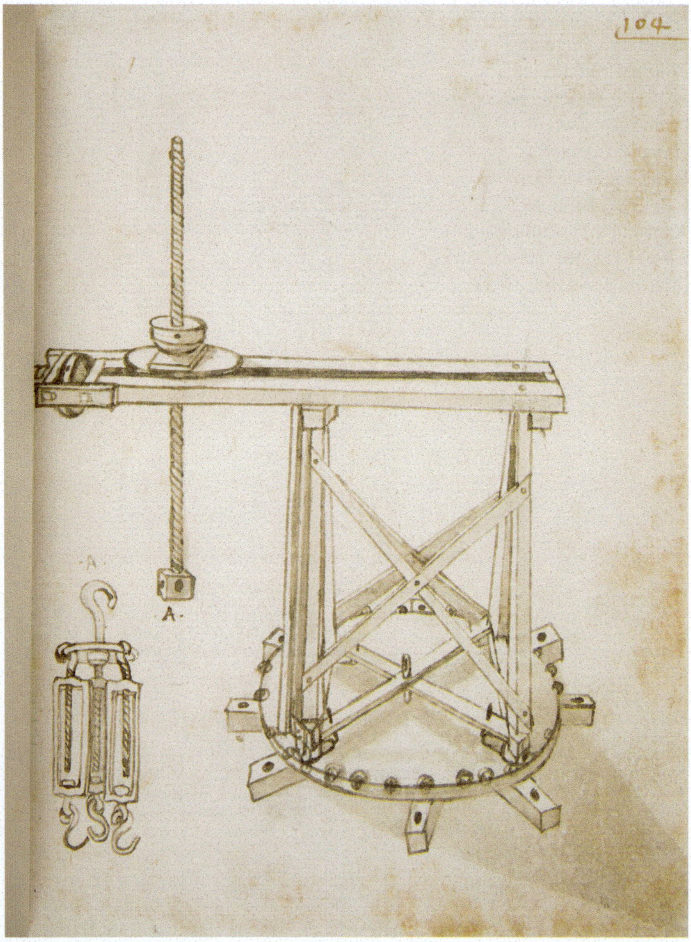

Fig. 1 Filippo Brunelleschi's revolving lantern crane. From (Ghiberti c. 1480–1490, fol. 104r). Courtesy of *Ministero dei beni e delle attività culturali e del turismo – Biblioteca Nazionale Centrale. Firenze*. Prohibition of further reproduction or duplication by any means

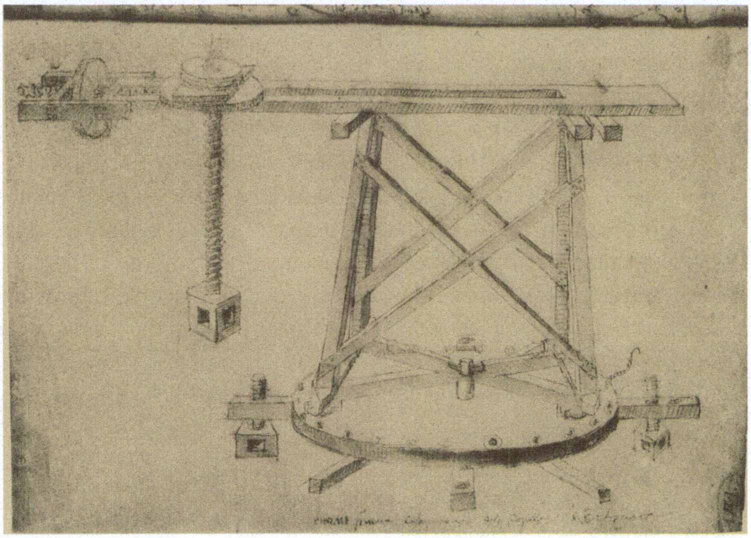

Fig. 2 Filippo Brunelleschi's revolving lantern crane. From Giuliano da Sangallo, *Taccuino Senese*, Codex S.IV.8, Biblioteca Comunale, Siena, c. 1490–1513, fol. 12r (Falb 1902, unnumbered plate)

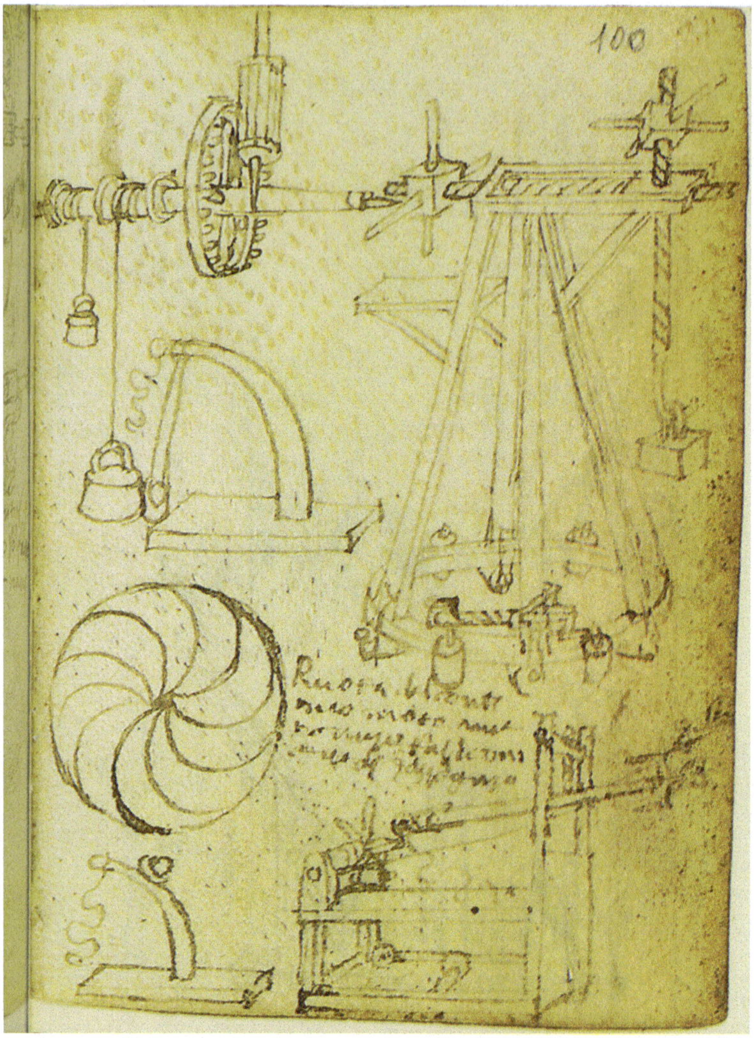

Fig. 3 Construction machinery, including Filippo Brunelleschi's revolving lantern crane. From Francesco di Giorgio, *Codicetto Vaticano*, Codex Urb. Lat. 1757, Biblioteca Apostolica Vaticana, c. 1465–1475, fol. 100r (Martini 1989, fol. 100r)

one general copy-drawing of a column-lifting device, in which the parts of the worm-drive are merely suggested with spiraling lines (fol. 52v) (Figs. 4 and 5). Likewise, the machines represented in similar drawings in the *Zibaldone*, the *Codicetto*, and the *Taccuino Senese* lack specific details, and the devices themselves appear inoperable. If these devices were so important—as their frequent reproduction suggests—why were they so often rendered in such a cursory manner?

An explanation to this query may be derived from the ubiquity of the designs, and the highly collaborative nature of early modern architecture. The authority of the "must-have" architectural models—such as those of column-lifting devices, obelisk haulers, water-wheels and siphon-pumps, which appear in abundance in copy-books—was directly tied to the degree to which they were reproduced.[16] Somewhat counterintuitively, it seems that as the designs became increasingly circulated and known, the degree of precision by which they were rendered became increasingly less important. But this

[16] In addition to the examples of these drawings in the notebooks of Buonaccorso, Francesco, Giuliano and Benvenuto, analogous images also appear in numerous other drawing and model books of the fifteenth and sixteeth century. Dozens of these are discussed by (Scaglia 1992).

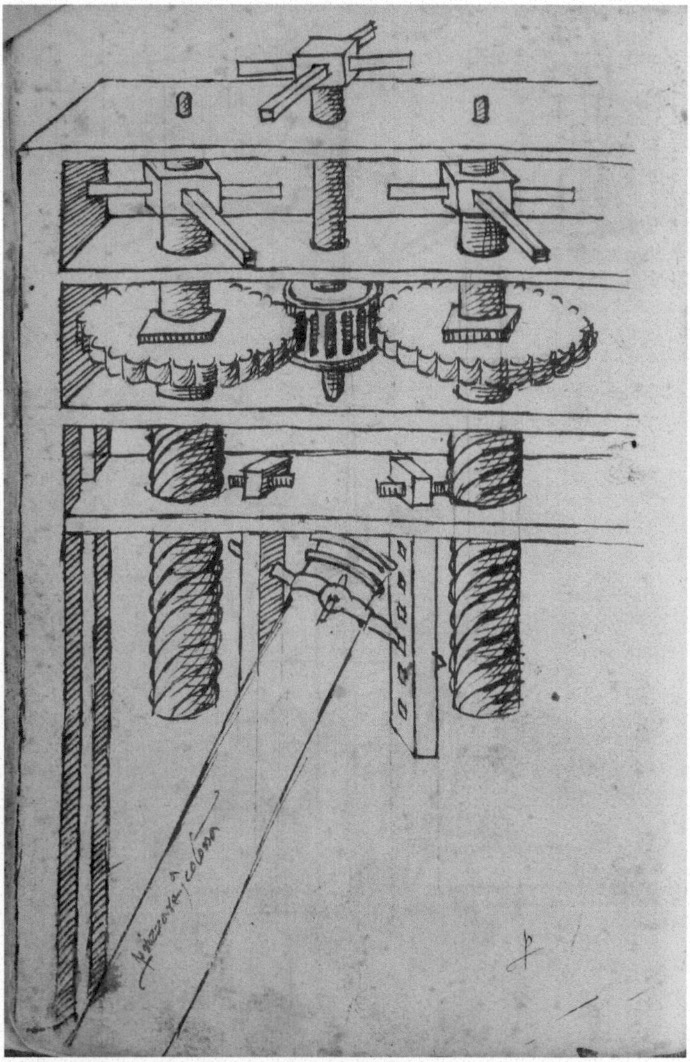

Fig. 4 Column-lifting device. From (della Volpaia c. 1520–1524, fol. 52v). Courtesy of Ministero dei Beni e delle Attività Culturali e del Turismo – Biblioteca Nazionale Marciana. Prohibition of further reproduction or duplication by any means

does not mean that the knowledge contained in the designs was lost or in any way compromised. As elucidated in the studies of David McGee and Wolfgang Lefèvre, machine drawings like those rendered by Francesco on *Codicetto* folio 100r, or by Benvenuto on folio 52v of his *Libro*, were never intended to function as blueprints or comprehensive instructional guides (Lefèvre 2004, 223–4; McGee 2004, 58–63) (Figs. 3 and 4). These were shorthand models, which practitioners could identify at a glance, and which, if the need arose, they could reference in the realization of one of the devices. Even the "original" machine renderings, therefore, would not have been terribly detailed.

The function of architectural drawings as reference tools is elucidated by Francesco di Giorgio in his *Trattato di Architettura*, following his chapter on building machinery. Summarizing his intent for the treatise, which includes general descriptions of design processes and simple, undetailed images, he explains: "This work is ultimately not a set of instructions … but rather it is for those who think and have some intelligence in *disegno* … and because the complete architect must invent in many unforeseen circumstances … and because I cannot clarify everything, these parts are left to the

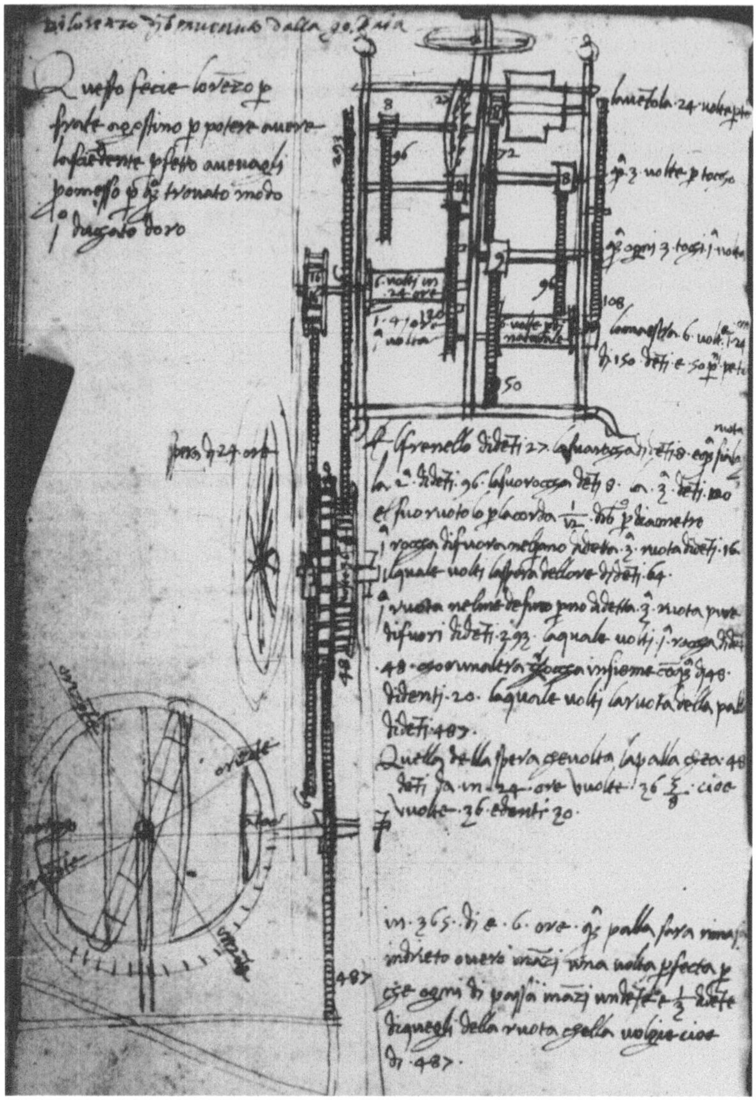

Fig. 5 Section of weighted clock designed by Lorenzo della Volpaia. From (della Volpaia c. 1520–1524, fol. 31v). Courtesy of Ministero dei Beni e delle Attività Culturali e del Turismo – Biblioteca Nazionale Marciana. Prohibition of further reproduction or duplication by any means

discretion of the architect" (Martini 1967, 505–6).[17] That is to say, working with a team of trained craftsmen, the architect would ultimately have to tailor his designs according to the limitations and needs of each specific project. For the purpose of communicating a machine or building type to workmen, moreover, a drawing such as that of *Codicetto* 119r was perfectly adequate. From this, they could comprehend the general composition of the device and arrangement of its parts. And even if the architect did develop more detailed orthogonal and section drawings, it is not clear that these would have been legible to the workmen.

[17] "Uno documento ultimatamente non è da pretermettare al quale dieno avere avvertenzia quelli che di questa mia operetta desiderano consequire alcuno frutto, e questo è che questi tali s'ingegnino avere qualche intelligenzia del disegno, perché senza quello non si può bene intendare le composizioni e parti dell'architettura perché le superficie esteriori comprano le interior e d'ogni parte longo saria dare esempli, e perché il complete architetto richiede la invenzione per molti casi occurrenti indescritti che senza disegno é impossibile consequire, e perché non possendo ogni minima parte dichiarare, quelle che restano sono nella discrezione dell'architetto."

The lack of detail evident in the copy-drawings may thus be traced to the high degree of technical know-how possessed by early modern architects, as well as to the collaborative nature of their practice. The Renaissance architect never worked alone, and a great portion of his work involved the development of designs to be executed by others. The shared knowledge he recorded in his pocketbook, therefore, should be understood as having been the complement to that of the many technicians and building specialists who were employed within his workshop. In the notebooks Buonaccorso Ghiberti and Benvenuto della Volpaia, the knowledge recorded was that of the family workshop—material which had accumulated over generations, enriched or refined through contributions of numerous practitioners, from a variety of different projects. The knowledge of Francesco's *Codicetto*, in turn, came from the Communal building workshop of Siena, where specialists from all fields of design and construction came together, and where copies of Taccola's *De Ingeniis* (ca. 1433) and *De Machinis* (ca. 1449) were widely studied and reproduced by young practitioners (Taccola 1984, 28–31). The contents of Taccola's books, which relate primarily to defensive architecture and hydraulic systems, constituted the general canon of knowledge required of any Sienese architect, and as evidenced in the *Codicetto*, Francesco's appropriation of this material formed the knowledge basis of his own practice.[18]

The shared knowledge recorded in the notebooks may also be regarded as autobiographic. This is particularly evident Giuliano da Sangallo's *Taccuino* (Nesselrath 1986, 127). Like Buonaccorso's *Zibaldone*, Francesco's *Codicetto* and Benvenuto's *Libro*, the *Taccuino* appears also to have been born out of a workshop model book, and reveals strong ties to its author's practice. Yet, unlike these volumes, the contents of which may be substantially linked to a single workshop, the material assembled within the *Taccuino* appears to derive from a broader range of sources. Trained within the thriving workshop of the Florentine woodworker and military architect Francesco di Giovanni "il Francione" (1428–1495)—who was active in Rome and Florence, and completed major commissions for the city of Florence and the Medici—Giuliano was exposed at a young age to diverse body of workshop knowledge. Correspondingly, the material of his *Taccuino* was also assembled from diverse origins. While some of this is original—collected by Giuliano during his investigations in Rome—much was taken from secondary sources. The plans of the Santo Spirito and San Lorenzo in Florence, for example, were likely copied from model books held within the workshop, as were the drawings of Brunelleschian machinery.[19] The copy-drawings after Francesco di Giorgio on folios 49v indicate that Giuliano must have had also access to one of the Sienese architect's notebooks, or a copy of the notebooks. The origins of the architect's studies of the columnar orders are less obvious, but these too derive from sources of shared knowledge. On folio 31v, Giuliano paraphrases Alberti's commentary on the proportions of the orders, and on folios 34r/v and 35r illustrates variations of Ionic bases, capitals and entablatures according to Alberti's instruction (Günther 1988, 130–1). The profile illustration of an Ionic entablature on the right-hand side of folio 34v is also a copy. The image here is akin to that included in the Sélestat manuscript of Vitruvius' *De Architettura*, which, significantly, also appears in Buonaccorso Ghiberti's *Zibaldone*.[20] Even the design of the roasting machine given on folio 50r appears to have been shared knowledge—as it was copied in contemporary drawing-books, including

[18] As overseer of the Siena's aqueducts (1469–1472), Francesco likely made good use of the illustrations and notes he had recorded on pumps, siphons, and door locks, modifying and elaborating upon the models, just as he explained in the *Trattato*, according to the "many unforeseen circumstances" of the work. Likewise, traveling between the dispersed fortifications of the duchy of Urbino, the fortification plans, schemes of earthwork defenses and recipes for gunpowder recorded in the *Codicetto* would have been invaluable references.

[19] Within the *Taccuino*, the plans of Santo Spirito and San Lorenzo appear on folios 5r and 21v, respectively.

[20] These folios are reproduced in (Scaglia 1979, 9, 11, 18).

that of Leonardo (Borsi 1985, 332; Hülsen 1910, LIX).[21] But in the copying process, the shared knowledge recorded by Giuliano in the *Taccuino* became personalized. In its visual reformulation, knowledge which was formerly anonymous and unbound was given authorship, while at the same time being incorporated into a body of specialized knowledge of the individual practitioner.

The repetitive nature of copy-drawing meant that the shared knowledge contained in the images very rarely retained an authorial connection.[22] The image of Brunelleschi's revolving lantern crane was an exception, due to the monumental celebrity of its inventor, and even in this case, it is not certain that everyone who copied the crane knew of its exalted origin. In most instances, as individual copy-books gave way to new copybooks, the authors' names were lost, but the designs they contained were incorporated into an expanding corpus of general knowledge. Thus it might happen that an architect working in the sixteenth century would reproduce an image according to a signature variation by Giuliano da Sangallo or Francesco di Giorgio, without having any conception of these architects.

3 Operational Knowledge

Whereas the shared knowledge employed by early modern architects was essentially receptive—copied, referenced and exchanged through notes and drawings within the workshop—operational knowledge was born out of active practice—whether from the architect's work on the building site, his direct examination of a form or structure, or his longtime study of certain natural processes. Such knowledge, an understanding of when and where to access certain measures, and when and where to employ certain procedures was often situational, in its associated with a given place and circumstance. At the same time, the operational knowledge of the early modern architect might also be characterized as strategic, in that it often corresponded to a specific skill set or process of design.[23] Leonardo's efforts to re-direct the course of the Arno River, for example, meant that he acquired a significant amount of site-specific information regarding the waterway—its tight bends, dramatic variations in level, and numerous sub-basins (Heydenreich 1980, 36). But at the same time, he would also learn about trench digging, the design of sluice gates and locks, and the natural movement of water—knowledge which he in fact recorded in the Codex Forster III, and which might be applied to other circumstances.

According to the contemporary theorists, knowledge acquired through practice and experience was invariably superior to that found in a book. Architects who wished to excel were thus encouraged to learn through direct observation, and to make notes and drawings in the field to record their findings. As explained by Filarete, in reference to the preparation of a preliminary site survey, the material gathered by the architect in situ could later be re-worked, synthesized in the creation of knowledge-based designs (Spencer 1965, 158).[24] Francesco di Giorgio also emphasized the importance of the architect's direct, first-hand knowledge of the building environment, reminding his reader that this fundamental context not only determined the strength of the building, but also the lives of its inhabitants

[21] In the codex Atlanticus (fol. 5v) Leonardo includes a drawing of the roasting device with the captain: "This is the true way to cook roasts, as according to whether the fire is mild or strong, the roast is turned slowly or quickly" ("Questo è il vero modo di cuocere gli arrosti, imperocché secondo che'l fuoco è temprato o forte, l'arrosto si volge adagio o presto").

[22] The unauthorized reproduction of printed images in the early modern period is an interesting corollary to this. On the efforts that were made to discourage or inhibit the re-use of woodblocks and engravings, see (Chaps. 7 and 15).

[23] Here, I am following the essential classification system provided by (de Jong and Ferguson-Hessler 1996, 106–7).

[24] *Libro Architectonico*, Book XII. Addressing the Prince, Filarete comments on the importance of making a preliminary site survey: "Before we go on, I would like to draw this countryside in order to have it better in my mind…. I will draw it roughly on a tablet so that I will not forget it." The Prince, after seeing the drawing replies, "That is good. When we send it off, make it more finished and on a sheet of good paper."

(Martini 1967, 303). "The architect," he warned his reader, "must be alerted to these conditions and the causes [of the site], just as he is to the time, place, and possibilities of the locations, so that he may mitigate the despised elements, which produce disease in men and lead to destruction" (Martini 1967, 310). Leonardo also wrote extensively on the importance of the artist's first-hand experience and observation of nature, noting in one of his notebooks that "sciences which begin and end in the mind cannot be considered to contain truth, because such discourses lack experience, without which nothing reveals itself with certainty" (Zammattio et al. 1980, 7). Echoing this same sentiment in the pocket-sized Codex Forster III, he reminded himself that "knowledge is the daughter of experience."[25]

Within the five practitioner's notebooks of our survey, operational knowledge is codified in two principal modes: (1) in recipes and formulae, which provide step-by-step explanations of specific procedures; and (2) in annotated figures, in which text and image complement one another to elucidate original designs and practices. The first mode of knowledge transmission may be related to the practical, instructive texts of the early modern period, such as the letters on ale written by Colonel Edward Harley and the recipe book of Johannes Kunckel, discussed by Elaine Leong and Sven Dupré, respectively (Chaps. 3 and 6). Like these texts, the recipes of the architects' notebooks either provide instructions, for the preparation of materials or certain working techniques, or direct the user to secondary sources which might provide this information. Although the recipes were inextricably tied to practice—experience which is discernable in the degree of detail they take, and in the authors' knowledge of specific sources of secondary expertise—in their condensed, textual form, they retain an anonymity. By contrast, the second form of operational knowledge evident in the notebooks—the annotated figure—is more frequently tied to the practice of an individual architect, and the information it conveys is more complex than that transferred in text or image alone. Here, the two media work hand-in-hand, the drawings making immediately visible the architect's ideas and discoveries, and the text supplying the details which inevitably could not be contained in a single image (Vecce 2003, 69–70). Still, as discussed by Pamela Smith in her contribution to this volume (Chap. 14), the documents which come down to us record only the traces of the architect's empirical knowledge. The notebook drawings and annotations are more narrow and specific in what they tell us, relating to the on-site investigations, mechanical training and workaday processes that were of greatest concern to the practitioner at a given time and place.[26]

The recipes included in the architects' notebooks—which varying considerably in both form and content—are in essence shorthand for certain practices, procedures and formula (Vecce 2003, 64). Whereas some recipes appear cryptic and incomplete—omitting instructions or material that was considered "general knowledge"—others provide more extensive, detailed instructions (Clarke 2001, 7). Recipes of the former type are found throughout the *Zibaldone*. The book opens with one such recipe, a rather common formula for imitating gold by applying a yellow glaze to silver. It reads: "one-half *denaro* of salt. One-fourth *denaro* of rubber. One-fourth *denaro* of both sulfur and saffron. Put to boil."[27] The Codex Forster III includes similarly epigrammatic recipes—formula for, among other things, producing oil from mustard seed, using spit as ink, making vermilion, and drawing shadows. On casting bronze in plaster Leonardo writes: "For every two bowls of plaster take one scoop of burnt

[25] Codex Forster III, fol. 14r: "La sapienzia è figliola della sperienzia."

[26] These notes might well not address the issues most relevant to a modern audience, answers for which may need to be found in other documentary sources.

[27] "Per colorire airento dorato. 1/2 denaro di sale comune. 1/4 denaro di groma. 1/4 denaro e zolffo e zafferana. E metti a bolire." For transcription and notes, see (Scaglia 1976, 496). Recipes for imitation gold are also referenced by (Clarke 2001, 3).

pitch [*bo'*], mix together and cast."[28] The brevity of these recipes—which are more like ingredient lists than detailed explanations for making—may be related to the practical function of the books. Both Buonaccorso and Leonardo were accustomed to using and preparing materials, and thus neither needed to be told where to find ingredients, how to mix them or for how long. They only needed to be reminded of the elements required, and the ratios or amounts which were necessary.

In contrast to such cursory, schematic recipes, the notebooks of Francesco di Giorgio and Benvenuto della Volpaia provide longer, more detailed instructions for making and doing. In both cases, the authors' elaborated instructions relate a common theme or set of procedures. For Francesco, the subject is firearms and ammunition—recipes for which fill the final fifteen-folios of the *Codicetto*. The inconsistent quality of the author's handwriting, the presence of varying inks, and inclusion of several blank and half-filled folios, suggests that the content of these pages was compiled over an extended period of time. Most of the recipes are a variation on a given type, and it is likely that Francesco had not tested them all. But in compiling them, he amassed a compendium of practical knowledge, so that, for example, at any given moment he had at his immediate disposal a handful of variations by which he might combine sulfur with carbon and salpeter to make gunpowder.[29]

The recipes compiled by Benvenuto in the *Libro* also form a set, in that all may be tied to projects undertaken in the Volpaia family workshop. The first dozen folios of the notebook contain a series of recipe-alternations, related to both processes—for welding silver and engraving letters, for example—and to the production of materials—such as tin, soap, and vermilion. Some of these are rendered shorthand, similar to those recorded by Leonardo and Buonaccorso, while others are elucidated in full paragraphs. Benvenuto's recipe for salpeter, for example, specifies where the soil should be taken from, how long it is boiled, what the resulting crystals should look like and how they are to be extracted.[30] Benvenuto was, moreover, quite conscientious to record the source of his information. Here as elsewhere, he included the name of the individual who provided the formula. As much of the *Libro* derives from material within the Volpaia family workshop, many of his citations are to its progenitor—"*Lorenzo nostro*," or "*mio padre*." Other recipes are attributed to unknown collaborators of the workshop, such as "*maestro Jacopo*," "*maestro Piero da Chassa*," and "*Andrea di Piero*." The care with which Benvenuto recorded these citations suggests that the value of the information was to a certain degree contingent on its source. The efficacy of a recipe had to be proven in practice, and thus its origin carried significant ramifications. If the named practitioner was held in low esteem or thought to have less experience, the value of the knowledge he provided was somewhat diminished.

Within the *Libro*, an analogous system of valuation is conveyed in annotated illustrations—designs for clocks, pumps, mills and wheel mechanisms. Just as the recipes, almost every device surveyed in the *Libro* is accompanied by a name. In addition to the numerous citations to Lorenzo della Volpaia, one also finds references to the well-known inventors Domenico Boninsegni, Lorenzo Serristori, Lorenzo Bartolini, and Leonardo da Vinci.[31] By citing these exceptional figures, Benvenuto provided the authority for the inventions while also supplying the knowledgeable practitioner with a general context for each project, such as approximate date, place of origin, purpose. But unlike the recipes—well-worked formulae that could be transmitted with simple text alone—the customized mechanical devices Benvenuto sought to record in his *Libro* were far more complicated, and the means by which he codified these were correspondingly more sophisticated. Here, the image was key. Using various

[28] Codex Forster III, fol. 39v. "Per gittare bronzo in gesso. Togli per ogni 2 scodelle di gesso una di corno di bo' bruciata, e mischia insieme e gitta."

[29] Codex Urb. Lat. 1757, fol. 182v.

[30] Codex Italiani IV 41, 5363, fol. 15r.

[31] For overview of the *Libro* and references it contains, see (Pedretti 1957, 26–8).

modes of technical drawing—transparent and cutaway views, abstract orthogonal views and sections—Benvenuto presented each mechanism as an illustrated diagram.

Such annotated illustrations are a staple within early modern architectural manuals and feature in every one of the personal notebooks of this study. These figures were almost always composed in the same manner—image first, text second—with the drawing invariably occupying the center of the page, and the writing confined to the periphery. The essential structure of the annotated illustrations, therefore, gives testimony to the primacy architects bestowed on drawing as a didactic and expository tool. As emphasized by architects like Mariano Taccola, Francesco di Giorgio, and Buonauito Lorini, among others, drawing was the principal means by which an architect communicated his ideas, and translated abstract experiences into a tangible, readily comprehensible format.[32] But unlike work-site templates or polished presentation drawings, the annotated illustrations of the architects' notebooks were intended as personal references, and are thus characteristically rather elementary in terms of technique and execution. The schematic contours, cursory over-drawings and quick notes contained in these illustrations might be read as the traces of the architect's systemization of practical knowledge.

Comparing folios from Benvenuto della Volpaia's *Libro*, with those of Buonaccorso Ghiberti's *Zibaldone* and Giuliano da Sangallo's *Taccuino*, one can discern some of the principle modes by which such images codified knowledge. Folio 31v of the *Libro* shows in section one of Lorenzo della Volpaia's weight-driven clocks, which according to the note in the upper left corner, was made for one "father Agostino" and provided a perfect system of timing (Fig. 5). The illustration is one of a learned practitioner, and reads as an encyclopedic entry on an exemplary clock type. Within the single, self-contained page, simple lines and brief annotations provide Benvenuto with the information necessary to manufacture the clock: the distribution of the fifteen wheels, the circumference and number of dentils of each, and the frequency of their rotations. The compact, unpunctuated explanation, which includes few verbs, and references the different clock wheels by their number of dentils, assumes a thorough knowledge of clock-making. The diagram does not explain how to assemble a clock, nor does it detail the component parts. Rather, it elucidates the mathematics of a specific mechanism for someone who already had extensive training in the field.

The illustrations of the *Zibaldone* are also those of a specialist, and despite their crude, almost amateurish appearance, they relay knowledge achieved from first-hand practice. But the way in which this knowledge is codified, within the images and within the book as a whole, is substantially different than in the *Libro*. Let us take for example folio 82v, a page which features three seemingly unremarkable cannon drawings and several brief annotations (Fig. 6).

In comparison to Bevenuto's intelligent composition of folio 31v, in which the elongated clock section structures the form and content of the page, laying out the clock's technical framework while also leaving ample space for the addition of corresponding annotations, Buonaccorso's page is simplistic. But for the practitioner, the annotated drawing bears significantly more information than an untutored glance would comprehend. Illustrated here is a novel, three-part cannon, the segments of which are connected by means of screw heads, visible on the left-hand side of the two lower pieces. As early modern bronze cannons typically weighed several tons, making their transport a slow and grueling process, the partible gun was a decisive innovation.[33] The increased speed of cannon transport meant

[32] Taccola's affirmation of the efficacy of drawing as a didactic tool is found within the pages of his *De Ingeneis*, which includes numerous references such as "as seen in the design" or "although obvious from the drawing." In his *Trattato di Architettura*, Francesco di Giorgio dismisses theorists who have written on architecture without images, commenting that: "Although to the writers it seems that they have elucidated their designs according to their intentions, to us it seems that through lack of drawings there are few who can understand them." See (Martini 1967, 489). On Lorini's discussion of *disegno* see (Henninger-Voss 2004, 149–54).

[33] With carriage, a sixteenth-century cannon weighed over 8000 pounds or four tons. See (Pepper and Adams 1986, 11).

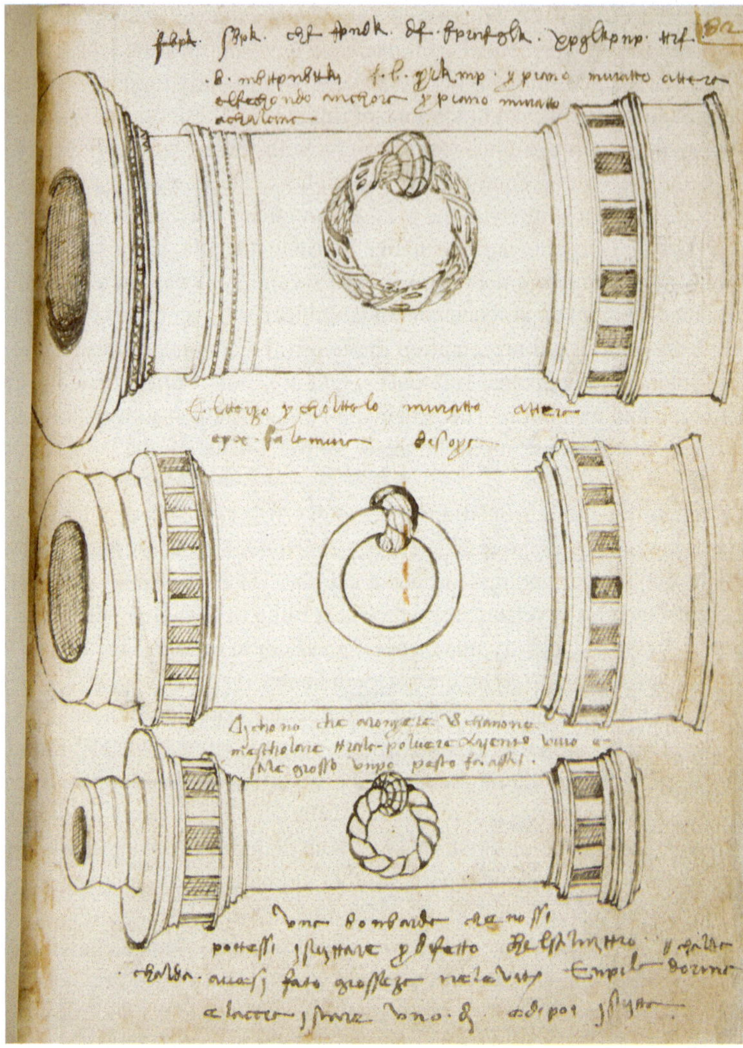

Fig. 6 Annotated illustration of three-part cannon. From (Ghiberti c. 1480–1490, fol. 82r). Courtesy of Ministero dei beni e delle attività culturali e del turismo – Biblioteca Nazionale Centrale. Firenze. Prohibition of further reproduction or duplication by any means

also increased agility of a military unit and the pace of warfare. From the text surrounding the drawing—which relates not cannon design specifically, but to the broader practical framework of their manufacture and function—we learn of Buonaccorso's expertise with such guns. At the top of the page he notes the basic structure and materials of a three-part furnace, the chambers of which were used for melting, working and temper metal, and at the base, he explains how to disengage the nut and screw of two units of the cannon in the case they are stuck. The annotation between the second and third cannon parts—apparently squeezed in as an after-thought—offers an alternative for breaking a cannon into pieces using a paste of "*ariento vivo e sale grosso*" (Scaglia 1976, 503–4). This information is augmented with the information provided in the adjacent folios, which taken together, provide step-by-step instructions for designing, manufacturing and using cannons. As seen here, the codification of practical knowledge is an additive process. Unlike Benvenuto, who acquired his experience and knowledge through his father, and thus with the benefit of time could assemble this information as a series of distinct case studies, Buonaccorso's knowledge as presented in the *Zibaldone* is the product of his ongoing explorations in armaments. The accumulative form of the *Zibaldone*, therefore, might be understood as mirroring Buonaccorso's own incremental acquisition of practical knowledge.

The annotated illustrations of Giuliano da Sangallo's *Taccuino*—careful renderings of entablatures, columns and capitals, building plans, sections and elevations that derived from his investigations of antiquities—provide one final example of how architects employed this expository format to record their empirical findings. In terms of form and subject matter, the *Taccuino* illustrations clearly differ from those previously highlighted in the *Libro* and *Zibaldone*. Whereas both Buonaccorso and Benvenuto paired drawings with significant portions of text, Giuliano's annotations rarely extend more than a few words. The exceptional refinement displayed in the drawings is also noteworthy. Giuliano's are not figures jotted down by the architect offhand, or pieced together from surviving workshop notes, but derive from the author's own existing study- and copy-drawing. As many of these also appear in the codex Barberini, it is safe to believe that Giuliano rendered a good portion of the *Taccuino* images more than once.[34] The subject of Giuliano's investigation—classical antiquities—is the final, definitive feature of his work that distinguishes it from the more technically-oriented explorations of Buonaccorso and Benvenuto. As we have seen, the problems confronted by these architects—such as the functioning of a clock or the design of a three-part cannon—found clearly defined solutions through the practitioners' hands-on experience. Giuliano's studies, by contrast, sought to answer a larger and more abstract question: the principal forms, materials, proportions, and techniques of excellent, *all'antica* architecture. Although Giuliano was certainly not alone in this investigation—the study of the Vitruvian orders and ancient remains was a definitive component of Renaissance architecture, and significantly, the notebooks of Buonaccorso and Leonardo also include material regarding the classical orders—the *Taccuino* is unique among the books of our survey in the extent and degree to which it addresses this material. What is significant here, however, is not so much the content of his studies, but the correspondence they find with the other notebooks of our study in the manner in which they structure and transmit the architect's operational knowledge.

Folio 33v of the *Taccuino* in many ways exemplifies Giuliano da Sangallo's process of antiquarian research (Fig. 7). The page derives from a series of study-drawings executed by the architect in Rome between 1488 and 1492, and illustrates, from top to bottom, a Doric column of the Theater of Marcellus, the plan of a sepulcher near Sant'Agnese, and a capital of the Basilica Aemilia (Hülsen 1910, LIV-LV). The unusual composition of these figures—with no concern for scale, the plan drawing shares contour lines with the two capital studies—as well as the presence of stylus under-drawing, indicates that Giuliano assembled the page some time after he completed his on-site investigations. The fact that he took the time to so painstaking re-render the figures, applying wash shading and noting dimensions, is in itself a clear indication of the studies' importance. Further affirmation of this is found in the secondary annotations. Just as Buonaccorso did with the *Zibaldone*, Giuliano returned to the illustrations of his little notebook on multiple occasions, each time emending the information he had previously recorded.[35] At the top of the page, he added the location of the tomb: "*fuora dela porta a. s. agnesa*." His notations elsewhere are more cursory—he modifies the measurements of the capitals, crossing some out, and includes notes about the supporting columns, and the proportions of the abacus and echinus of the two capitals. In as much as these marks record the architect's expanding understanding of the monuments, they also convey his process of empirical study.

Giuliano da Sangallo's ongoing investigatory process is again evident on *Taccuino* folio 45v. This page presents seven profile drawings of Doric entablatures—four in dark ink with place-name annotations ("*i[n] Roma*" and "*i[n] Monte Chavalo*"), and three at the base, rendered with quick, light pen

[34] The order in which Giuliano da Sangallo recorded drawings in the *Taccuino* and codex Barberini is unclear. Scholars generally conclude that the two books were assembled concurrently.

[35] Clear proof of this is found on folio 7r, which contains a plan drawing of the Colosseum. Based on what is known of Giuliano's travel and antiquarian study, the drawing had been dated to between 1488 and 1492. However, the annotations included on the page record the date of 1513. See (Borsi 1985, 257).

Fig. 7 Mixed folio of antiquarian studies; Doric column of the Theater of Marcellus (top left); plan of a sepulcher near Sant'Agnese (center); Doric capital of the Basilica Aemilia (bottom). From Giuliano da Sangallo, *Taccuino Senese*, Codex S.IV.8, Biblioteca Comunale, Siena, c. 1490–1513, fol. 33v (Falb 1902, unnumbered plate)

strokes, and likely added at a later date. As with folio 33v, these classical models are examined in regards to their measurements and proportions, and notably, the means by which this information is structured bears strong parallels to the examples previously discussed. To begin, we might consider Giuliano's application of place-names. On folio 45v, as elsewhere in the *Taccuino*, the inclusion of place-names served not only to document the figures' locations, but also to recall the architect's experience with the monuments. Like the individual names so carefully recorded by Benevuto della Volpaia, the place-associations added by Giuliano bestowed authority on what would otherwise be a schematic, disconnected rendering. The appellations "*fuora dela porta a. s. agnesa*," "*i[n] Roma*" and "*i[n] Monte Chavalo*" were signifiers of practice that was incommunicable with pen and ink. Folios 33v and 45v also relate to those of the other architects of our study in illustrating the architect's proclivity to compile data. Just as Francesco di Giorgio sought to amass an inventory of gunpowder recipes, formula he could quickly reference when in the field, Giuliano assembled a catalog of variations of the Doric entablature, which at a moment's notice he could show to workmen or a potential patron.

Similar examples of the architects' near-obsessive recordkeeping, and continuous compilation of data, clues and exempla are found throughout all five books of our study. In every one of these cases it is evident that the value of the knowledge recorded was inextricably linked to the personal, referential quality of the notebooks in which it was contained. The seven Doric entablatures Giuliano rendered on folio 45v carried a significance which only he could fully discern. Why he appended the three additional studies at the base of the folio is as unclear to the modern eye as it would have been to one of Giuliano's contemporaries. What is clear, however, is that each of these drawings corresponded with the architect's practice, and that even as fragmentary studies, lacking consistent scale and without comprehensive dimensions, they functioned as "working" drawings.

4 Experimental Knowledge

On folio 4v of the Codex Forster III, beneath a diagram of a weighted A-frame beam, Leonardo poses the question: "If the beams and the [hanging] weight *O* is 100 lb, how much must the length *AB* weigh in order to resist this weight and so not to sag?"[36] The question, as posed here, is rhetorical, and complements the exploratory studies of the adjacent folios concerning framework and basic structural systems. Within many of his notebooks, Leonardo employed such statements as an analytical structuring device, and as presented here, the query reflects a complex reciprocal relationship between empirical observation and thinking on paper (Kemp 2006, 100).[37] While the study of structural forces belongs to the physical world, requiring observation of nature and an understanding of the physical laws of gravity, the calculations it entails are conceptual and might be worked out by the architect off-site. Such paper-based explorations, which took into account just enough empirical knowledge so as to retain relevance for future application, were the definitive traces of experimental knowledge. These previously unknown designs proposed to change the form, composition and function of architecture. The early modern architect's ability to devise and discover was considered all-important, and within the context of his greater training, such artistic exploration was deemed complementary to his imitation of exemplary models, empirical work and examination of nature (Kemp 1977, 350).

With few exceptions, the architect used *disegno* to explore his ideas. These images produced within this process were not model-drawings or project studies, but abstract manifestations of the architects' *invenzione*, freewheeling searching, and generative thought processes. The concept of *invenzione*, as understood and employed by early modern architects, was closely linked to discovery and creation, and ultimately, served as the conduit for the uncovering of truth. Accordingly, contemporary theorists emphasized invention and experimentation as vital to good architecture, routinely echoing Vitruvius' assertion that "the inventive architect always cleverly arrives at an original and correct solution to an old or new problem." Still, architectural theorists were emphatic in pointing out that *invenzione*—which was frequently associated with *ingegno*, skill, and diligence—was not something that could be taught, precisely conveyed or described. Thus also, the knowledge born from *invenzione*, founded upon the architect's discretion and judgment could never be fully transcribed (Kemp 1977, 353). The following discussion then focuses not on architectural knowledge per se, but the processes by which experimental knowledge was generated, the form it took in drawing, and how it was applied to practice. Examples from the notebooks of Leonardo, Francesco and Giuliano demonstrate three different forms of experimental knowledge, which, to varying degrees, incorporate shared knowledge, empirical knowledge, and pure fantasy.

The questions Leonardo addressed in the Codex Forster III, were, with few exceptions, essentially the same as those which concerned Ghiberti, Francesco, Benvenuto and Giuliano. Like these four architects, Leonardo endeavored to find solutions for excavation works, building machinery, hydraulic systems, the proportions of the columnar orders, and military devices. Yet, Leonardo was less reliant than his peers on the general information which was available to him, and almost never reproduced models verbatim. He wanted to find and confirm the answers for himself, and within his notebooks, he presented these ideas in a manner which made most sense for his own understanding. Leonardo's study of the screw on folio 82r, a standard mechanical element that also features in more cursory form in the four other notebooks considered here, is a perfect example of how he approached common architectural problems independently (Fig. 8).

[36] The Italian reads: "Si i travi e 'l peso o sia 100 libbre, quanto peso sarà in ab a fare resistenzia a esso peso che non caggi in basso?"

[37] Other folios of Codex Forster III that include rhetorical questions are 19v, 20r, 50v, 53v 73r.

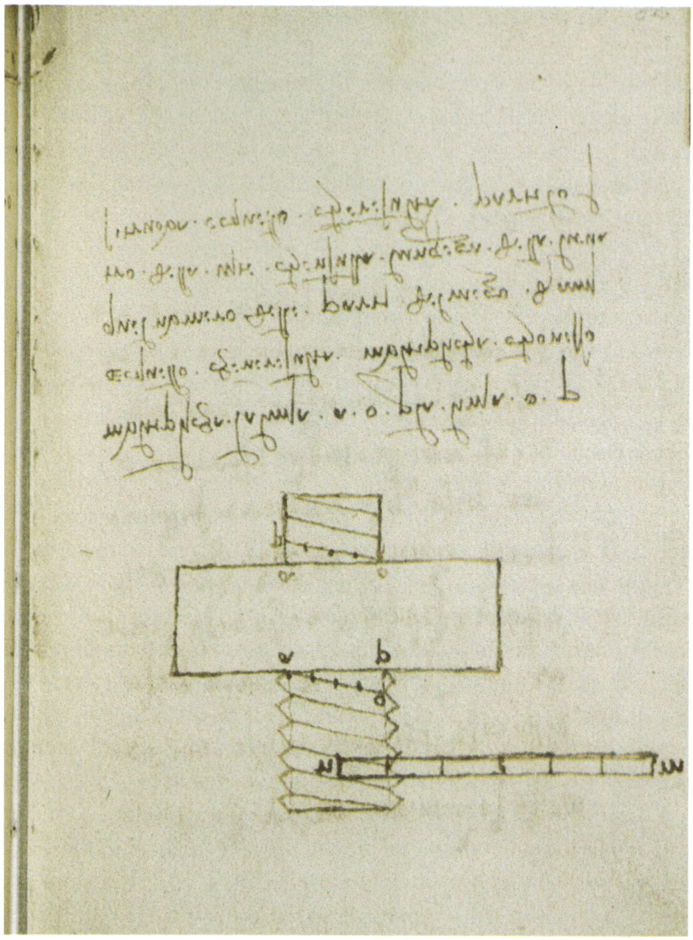

Fig. 8 Study of a screw. From Leonardo da Vinci, *Codex Forster III*, Victoria and Albert Museum, London, c. 1493–1497, fol. 82r (da Vinci 1992, vol. 3, fol. 82r)

Leonardo placed great emphasis on the screw in his mechanical investigations, due to its infinite uses and the various ways it could be modified and adopted. Thus, even though his exposition on the screw on folio 82r is not particularly novel, the illustrated study displays experimental knowledge in its original form, as well as in its correlation to Leonardo's empirical understanding of the element. The purpose of the diagram was to elucidate how to calculate the spacing of a screw's threads, without "armatures or other structures that might hinder … study." And Leonardo achieved this in his simplified, orthogonal projection (Reti 1980, 144–8).[38]

Within the Codex Forster III, we also see traces of Leonardo's unfettered, probing exploration, which resulted in the creation of entirely new devices. Original projects referenced here include those for a sluice gate, mechanical loom, method of relief etching, and excavation machinery, devices Leonardo was working on in the final decade of the fifteenth century, and which he further developed in the codex Atlanticus (Galluzzi 1987, 84, 87; Reti 1999, 378–86).[39] It is possible, therefore, that the studies of the Codex Forster III represent Leonardo's initial thoughts—his on-site inspiration—which

[38]Quote from Codex Madrid, fol. 82r.

[39]Drawings of Leonardo's weaving machine appear on folios 43v, 60v, and 61r. Folios 14v, 15r, 17r and 18r concern excavation processes and machinery. Leonardo's innovative process for relief etching is referenced on folio 59r.

Fig. 9 Study of sluice gate. From Leonardo da Vinci, *Codex Forster III*, Victoria and Albert Museum, London, c. 1493–1497, fol. 57v (da Vinci 1992, vol. 3, fol. 57v)

he subsequently developed in his workshop, or as he often did, when sitting by candle-light in the evening at home (Kemp 2006, 100).

Leonardo's studies for the sluice gate (fols. 57v and 58r) are particularly remarkable in that they represent his excellence in hydraulic engineering—expertise which helped him gain highly prestigious commissions from the city of Florence, Cesare Borgia, and King Louis XII of France (Figs. 9 and 10).

Within the Codex Forster, moreover, the two folios of sluice gate designs may be related to Leonardo's diagrammatic schemes for the diversion of the Arno (fols. 32v and 33r), and designs for a double-pulley excavation machine, which were to be employed in the construction of new canals. As mentioned previously, Leonardo had an abiding interest in extending the course of the Arno River, a project which would open Florence to the sea, bringing the landlocked city extensive economic and agricultural benefits (Heydenreich 1980, 28). The purpose of sluice gate was to help control water levels within a river or canal system by closing the flow of water from one section to another. The essential element of the gate was a trap door, which as rendered by Leonardo on folios 57v and 58r might take different proportions in relation to the gate as a whole. The trap door was operated by a latch mechanism, placed at the top of the gate, which when opened would allow water to flow into the canal, equalizing water pressure on both sides of the gate so that it could be fully opened (Reti 1980, 136).

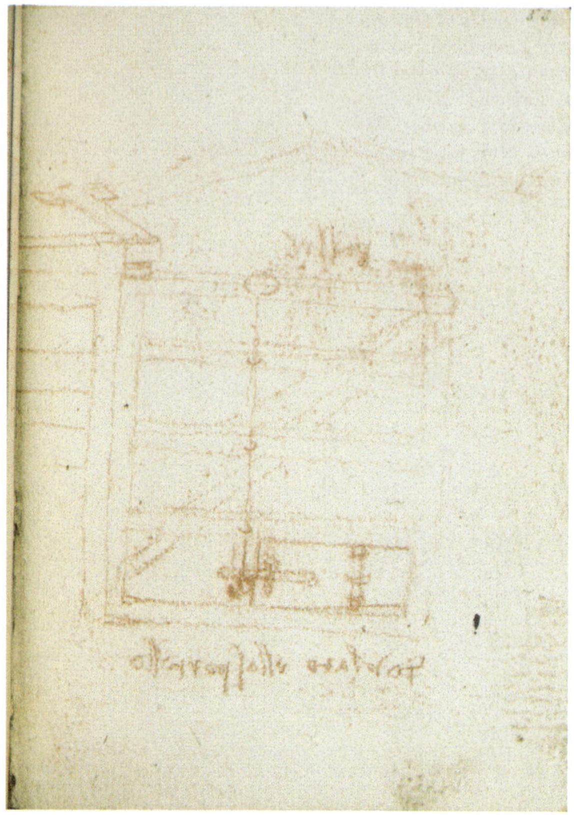

Fig. 10 Study of sluice gate. From Leonardo da Vinci, *Codex Forster III*, Victoria and Albert Museum, London, c. 1493–1497, fol. 58r (da Vinci 1992, vol. 3, fol. 58r)

The drawings themselves, rendered in red chalk—the ideal media for laying down thoughts in a quick, cursory manner—reveal the architect's exploratory process.[40] As characteristic of Leonardo's notebook studies, this exploration concentrated on the intrinsic components of device in question, and how these elements would respond to the forces exerted upon them by nature. On folio 57v, where the opening of the trap door comprises half of the gate, Leonardo considers the force of the water pressing against the gate. The joint of the center gate opening buckles slightly in response to the water's impact, which is illustrated by an arch on the right hand side of the folio. Radiating lines, labeled "*sasso*"—"thrust"—indicate the great force of the surging water.[41] The image of folio 58r, presents the sluice gate in an entirely different form. The gate shown here is akin to the figure given in the codex Atlanticus, but only the door with the opening of the trap door is illustrated. The tentative lines of the figure suggest that Leonardo was still very much working out his concept for the weir. He barely traces out the walls of the canal, which extend above the left-hand side of the gate at a 45° angle, and hesitantly renders the contours of the gate. The only component of the illustration which Leonardo appears certain of is the trap door, which he outlines with a heavier hand, including the note: "On one side is the trap door."[42] The annotation above the latch at the top of the gate, never much more than a scribble, is now indecipherable. Taken together, then, folios 57v and 58r illustrate the

[40]On the ease of drawing with chalk and artists' preference of this medium, see (Brothers 2008, 213–4).

[41]"Sasso" is literally translated as "stone" but as used by Leonardo, refers to the pressure of a heavy force or impulsion.

[42]The Italian reads: "Da un lato è lo sportello."

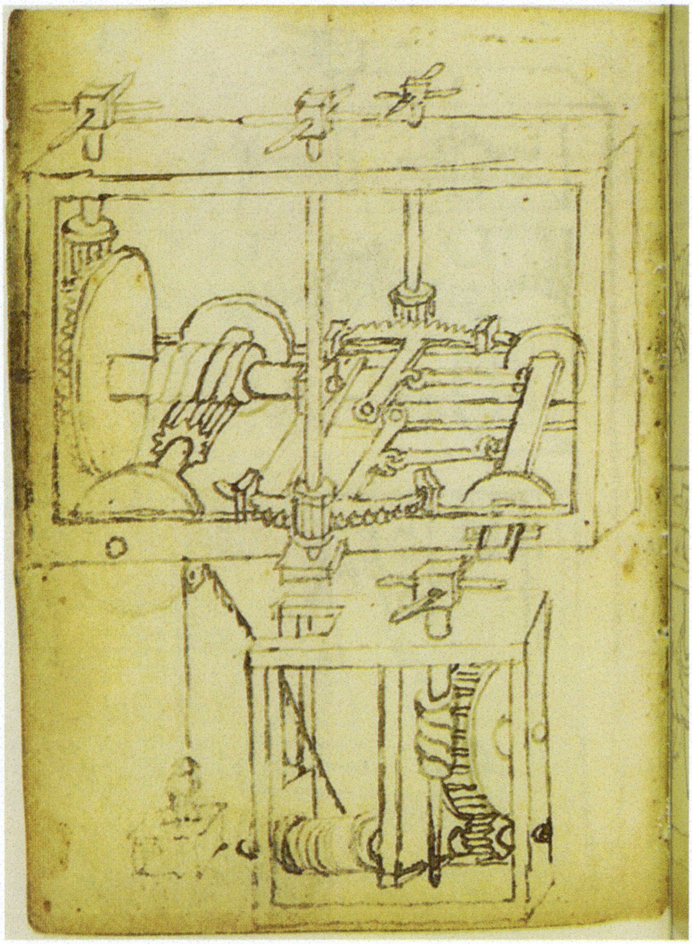

Fig. 11 Design for car (top), and roller-beam hoist. From Francesco di Giorgio, *Codicetto* Vaticano, Codex Urb. Lat. 1757, Biblioteca Apostolica Vaticana, c. 1465–1475, fol. 119v (Martini 1989, fol. 119v)

architect working out the formwork of his sluice gate, using his knowledge of the river as a means to spur invention.[43]

Francesco di Giorgio's *Codicetto* also served its owner as a vehicle for working out mechanical problems and devising new solutions. But as evident in comparing the minute drawings of the *Codicetto* with those of the Codex Forster III, Francesco's process of creation was much different from that of Leonardo. As previously discussed, the *Codicetto* originated from Francesco's study of Taccola's notebooks, and even though much of this material is original, the essential design concepts—aqueducts, mills, canals, roof-beam supports, and even mobile attack dragons—may be traced back to Taccola. But the *Codicetto* is by no means a rote copy-book. The hundreds of drawings that fill the little book attest to Francesco's approach to analytic experimentation, a rational process of design development pursed in the reproduction of the same devices in drawing (Vecce 2003, 71). Thus, beginning with one of Taccola's models—for example, a four-wheel motorized car, which as rendered in *De Ingeneis* includes few mechanical details—Francesco explored the form and function of the device in a series of innovations. This method closely corresponds to his theoretical understanding of architect's inventive capacity and ability to produce an almost infinite number of works, of infinite variety ("*quasi infinite, infinito varia*") (Martini 1967, 505; Kemp 1977, 354) (Figs. 11 and 12). Within the abundance of

[43] The idea that experience could stimulate invention is discussed by (Kemp 1977, 364).

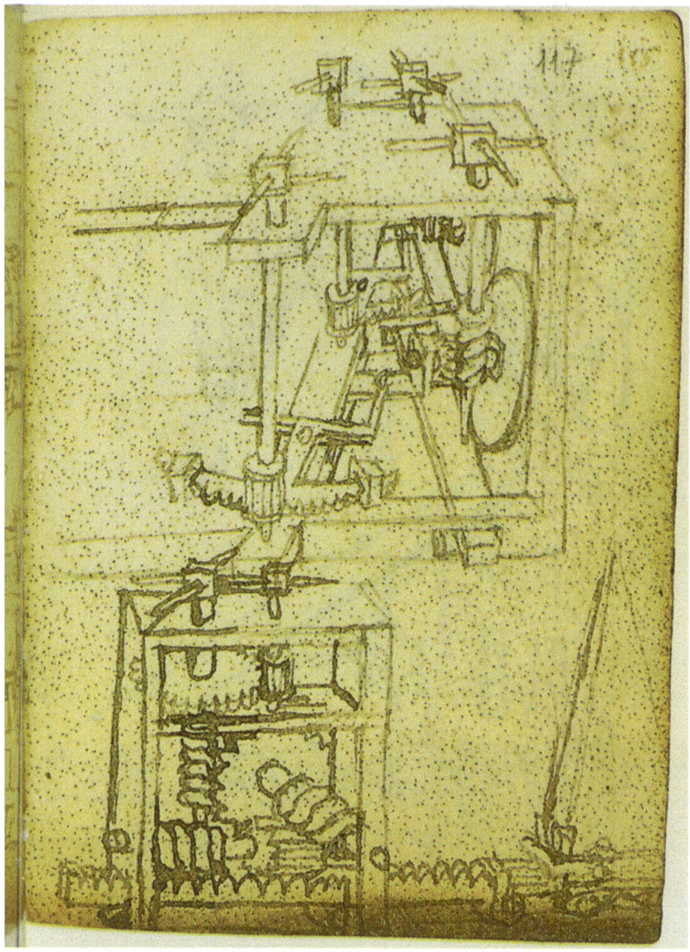

Fig. 12 Detail of foreshortened view of car design. From Francesco di Giorgio, *Codicetto* Vaticano, Codex Urb. Lat. 1757, Biblioteca Apostolica Vaticana, c. 1465–1475, fol. 117r (Martini 1989, fol. 117r)

car-variations recorded on *Codicetto* folios 114r–122r, Francesco tests different configurations of the transmission mechanism—using toothed wheels, worm-gears, pin-and-axle crank system, and click wheels—and also modifies the type of wheels the car might take.

Among the car-variations illustrated on folios 114r–122r, those which display Francesco's experimentation with new formats of technical representation are particularly noteworthy. For Francesco, just as for Leonardo, drawing was not only a mode of thinking, but was also a means of visualizing things previously known only in the abstract. The architect's understanding of the mechanism—not just as a generic concept, but as a complex, multi-part construction—might therefore be correlated to the visual language by which he transmitted it. Illustrating a car akin to that given on 119v in foreshortened perspective (folios 117r/v), Francesco demonstrated the internal structure of the mechanism in a manner which was not possible in the standard, elevation projection. Whereas the elevation view usefully displays the horizontal transmission apparatus, the new view shows the distribution of the axels and connecting rods of the crankshaft.

Francesco di Giorgio's ability to develop a concept through drawing, varying its formal characteristics as well as the manner by which it was transmitted, is also evident in his studies of mills. Within folios 133v to 164v, Francesco renders approximately eighty different mill types—variously formed and powered by running water, pumped water, wind, animals and humans. As with the car designs, these mills are not in themselves exceptional—comparable constructions appear in contemporary architectural copybooks—but are distinguished in the means by which they are represented. Working carefully with pen and ink, Francesco framed the mill-variations in foreshortened "open boxes." The boxes do not represent a literal formwork, but rather provide an abstract visual structure for the machines. By removing the mills from their environmental context, condensing and delineating their parts with box-partitions, the architect could better demonstrate the relationships of the different parts (Kemp 1991, 108). The focus of these drawings is not explicitly mechanical, but concerns the degree to which a given mechanism might be modified and reformed. In contrast to Leonardo, whose inventions express an immediate concern for the physical forces of nature, Francesco took environmental conditions as a given, leaving this aspect of the design process to be resolved by the individual architect, according to parameters of the site.

The distinction between invention born within a specific context, and that which considers empirical evidence, but is developed only to exist on paper, is a subtle one, but is critical to our understanding of the architect's experimental knowledge. In Leonardo's process of invention, the objects of his creation were always subjected, implicitly or explicitly, to the forces of nature. Thus, the designs of even his most daring contrivances intrinsically convey an astute understanding of the natural environment (Marinoni 1987, 130). The strength of Leonardo's invention lay in his extensive empirical knowledge, experience which he continually referenced as a means to spur his creative facility. Francesco's experiments also relied on cognitive thought and empiricism, but as realized in the *Codicetto* were merely conceptual. The primary function of the *Codicetto* drawings was to provide the architect with ideas and inspiration while on site. It is not that Francesco did not have the practical knowledge to realize the machines he rendered—there is ample documentary evidence that he did—only that as represented in the *Codicetto* they were not intended to be realized.

We find a similar correspondences between drawing content, empiricism, and experimental processes in Giuliano da Sangallo's *Taccuino*. Giuliano's inventive facilities are presented in his reconstruction of classical buildings, approximately twenty of which feature in the *Taccuino*. Many of these may be correlated with images of the codex Barberini, and it is by means of comparison—between the representation of a given monument in *Taccuino*, with that provided in the codex Barberini and other contemporary sources—that we come to understand the experimental knowledge contained in Giuliano's drawings. *Disegno* was for Giuliano, just as it was for Leonardo and Francesco, a means by which to represent that which is absent and a catalyst for creative thought. But whereas Leonardo visualized something that might be, and Francesco represented conceptual hypotheses, Giuliano sought in drawing to illustrate structures which no longer existed. His drawings were to discover, or to uncover, a lost truth. Moreover, unlike the designs of Leonardo and Francesco, Giuliano's reconstructions did not require even a pretense of functionality, and thus allowed for a greater degree of artistic license.

Giuliano's restored plan on the Septizonium, a large nymphaeum erected by the Emperor Septimius Severus at the base of the Palatine Hill, illustrates the degree of license with which he approached his study of the antiquity. Although building's remains were entirely disassembled by the end of the sixteenth century, in period leading up to this the Septizonium attracted significant attention from artists and architects, who frequently studied its monumental, three-level columnar façade. From the surviving study-drawings of Maarten van Heemskerck (1498–1574), Giovannantonio Dosio (1533–1609), Vincenzo Scamozzi (1548–1616) and Etienne Du Perac (1520–1607), among others, all showing three bays of the façade elevation, we may reasonably assume that this portion of the structure remained intact through the Cinquecento.[44] There is no indication that enough of the theater survived

[44]Giovannantonio Dosio, 1774Ar (ca. 1557–1570), Gabinetto Disegni e Stampe degli Uffizi; Etienne Du Perac, Du Perac Sketchbook, fols. 22v and 23r (ca.1570), The Morgan Library, New York; Vincenzo Scamozzi, fol. 9r, Ms. It. IV 149 (ca. 1560–1590), Biblioteca Marciana, Venice. For references and images, see (Nesselrath 1984).

to understand its plan—which from recent excavation efforts is known to have been rectilinear, approximately ninety-three meters in length—a supposition confirmed by the Heemskerk sketches that show the theater's ruinous surroundings (Gorrie 2001, 653–7). Yet this is not the impression one gets from the plan Giuliano renders in the *Taccuino*. Here, the Septizonium is shown as a perfect square, encircled by a portico which includes ten columns per side, and centered on an open, porticoed atrium. This is not a precise study-drawing, but an imaginary recreation; a plan loosely based on the architect's empirical findings and significantly enriched with his greater knowledge, or notions, of ancient architecture.

The inventive character of Giuliano's plan is further elucidated when it is compared to a study-sheet of the same structure composed by Francesco di Giorgio. Francesco's studies are more fragmentary and exploratory than Giuliano's, and were likely rendered on-site by the Sienese architect during one of his visits to Rome.[45] Within the page, Francesco notes some of the same interior walls apparent in the *Taccuino* drawing, but also includes a partial plan of the theater with measurements, as well as a smaller, schematic plan and an accompanying elevation sketch. The startlingly distinction between Giuliano's and Francesco's studies—not only in terms of finish, but also in the degree to which Giuliano completed the ruinous structure—prompts reflection about the credibility of the other antiquarian drawings in the *Taccuino*. How do these relate to what Giuliano actually saw? And what function did the reconstructed, partly fantastical vision fill that a more literal depiction could not?

Giuliano's elevation drawing of the Arch of Argentarii, folio 22r of the *Taccuino*, offers insight into these questions, as unlike the Septizonium, the Trajan era arch was largely intact when Giuliano studied it (Fig. 13).

As presented in the *Taccuino*, the Arch of Argentarii, which in the seventh century was partially incorporated into San Giorgio al Velabro in Rome, is freestanding and surmounted by an unusual quadrangular portico. This squat portico is shown open on all four sides, each centered by an arched portal and ornamented with symmetrically placed niches. The image includes two annotations in the form of inscriptions, which provide information of the monument's location—"*di Sangiorgo i[n] Roma*"—and origin—"*Questo è si dicie che è l'archo di Decio in Roma*."[46] If this were Giuliano's only rendering of the Arch of Argentarii, we would assume that it was a relatively accurate representation of the structure, based on his direct study. However, the illustrations he provides of the arch in the codex Barberini confirm that the *Taccuino* representation was not completely factual (Fig. 14), as here, the overall character and form of the arch is entirely changed.

In the significantly more detailed Barberini study, a triangular pediment sits atop the arch, not a portico, and elaborate relief carvings encircle the base. The absence of the ornamental detail in the *Taccuino* image is an implicit indication of Giuliano's limited knowledge of this section of the monument. Yet, as both studies attest, the architect clearly believed that some aspect of the arch was missing. It is likely that both the portal crown of the *Taccuino* version and the pediment of the Barberini study derived from monuments known to Giuliano. However, these pieces did not necessarily originate from the Arch of Argentarii. As applied to this monument, they were inventions, pieced-together by Giuliano in an attempt to "correct" the ruined monument (Brown and Kleiner 1983, 325, 333).

The comparison of Giuliano's differing versions of the Arch of Argentarii in the *Taccuino* and the codex Barberini also illustrates the architect's process of cognitive experimentation. Like Leonardo and Francesco di Giorgio, Giuliano used repetition in drawing as means to examine and develop the objects of his study. As evidenced in the examples previously discussed, all three architects were accustomed to render their subjects multiple times, introducing and testing new ideas with each version.

[45]Francesco di Giorgio, UA 323r/v Gabinetto Disegni e Stampe degli Uffizi, Florence. The drawing is reproduced in (Burns 1993, 337). Giuliano also includes a plan of the Septizonium in the codex Barberini on folio 30r.

[46]"Decio" refers to the Roman Emperor Decius, who ruled from 249–251 AD.

Fig. 13 Arch of Argentarii, Rome. From Giuliano da Sangallo, Taccuino Senese, Codex S.IV.8, Biblioteca Comunale, Siena, c. 1490–1513, fol. 22r (Falb 1902, unnumbered plate)

The development of the sluice gate, just as the invention of new mill types or the discovery of antiquarian models, was a systematic and analytic process. It would thus be incorrect to dismiss Giuliano's renderings of the Septizodium and Arch of Argentarii as mere fantasies, or to say that in reconstructing the ruins he demonstrated a disregard for empirical observation. Giuliano, just like Francesco, Leonardo, Benvenuto della Volpaia, and Buonaccorso Ghiberti, was looking for answers, and beyond this, for faithful models that he could apply in practice. He found these in accumulating and copying secondary material (shared knowledge), making first-hand observations (operational knowledge), and in the case that he was still unsatisfied, which was not infrequently, he found his answers through invention and experimentation.

5 Conclusion

As noted in the introduction, one of the intentions of this study is to shed light on the elusive role of the early modern architect, who lacked both an established system of training and a distinct professional affiliation. To this end, the notebooks of Buonaccorso Ghiberti, Francesco di Giorgio, Giuliano

Fig. 14 Arch of Argentarii, Rome. From Giuliano da Sangallo, Codex Barberini, Codex Vaticano Barberiniano Latino 4424, Biblioteca Apostolica Vaticana, c. 1465–1510, fol. 33r (Hülsen 1910, fol. 33r)

da Sangallo, Leonardo da Vinci and Benvenuto della Volpaia are informative not in the degree to which they differ—as personal handbooks of individual practitioners, variation should be taken as a given—but in their implicit agreement on certain fundamental principles of architecture. Within the notebooks, we find affirmation of the architects' fundamental reliance on *disegno*, a tool which served them in all aspects of his work. The manuals also illustrate the architects' dependence on models and exempla, their propensity to accumulate these, along with other sources of information, and their habit of creating handy, readily accessible records of the material related to their practice. For the architect of the twenty-first century, who has access to countless images, databases, computational aids and reference guides, and is generally subject to the unprecedented information overload of our age, such recording and codification of knowledge is relatively routine. As noted by the distinguished architect and scholar Bill Lacy, "Architects are known for carrying sketchbooks and journals wherever they go to record fleeting ideas and impressions" (Serrazanetti and Schubert 2009, 11). Yet, prior to the fifteenth century, there is little surviving evidence of this practice.[47] In fact, the five notebooks of this study are among the earliest known examples of practitioners systematic writing and recording. The books, therefore, are relevant not only in their form and contents, but also by their essential character as personal reference volumes. They mark a turning point in the history of architecture, and in the character of the practicing architect.

[47] Surviving architectural manuscripts from pre-1450 are primarily either lodge books, kept within the workshop and passed between generations of architects, or theoretical texts, written in Latin and dedicated to distinguished patrons. The Late Hellenistic bronze statue of an artisan, conserved today in the Metropolitan Museum of Art, is perhaps the earliest reference to architects' reliance on pocket-sized notebooks. This figure, who carries a notebook in his belt, has been identified as an architect (possibly Daidalos) due to the prominently placed notebook. See (Picon and Hemingway 2016, 161–62).

Considered within the greater framework of early modern architecture—which on the one hand received unprecedented theoretical attention with the "re-discovery" of Vitruvius' *De Architectura* and spate of architectural tracts that followed, while on the other, was substantially driven by the ever-increasing need for new technologies—the notebooks of Buonaccorso Ghiberti, Francesco di Giorgio, Giuliano da Sangallo, Leonardo da Vinci and Benvenuto della Volpaia may be understood as the architects' attempts to mediate the two, seemingly incongruous aspects of their discipline. Was the architect a theorist, as Leon Battista Alberti argued, who provided the design concept but stayed at arms-length from the worksite? Or was he a practitioner, developing stronger defenses, better mill-systems, and quicker building processes using the empirical knowledge he gained in the field? Beyond this, what constituted the basis of architectural training? Was it enough to be a skilled draftsman with an eye for pleasing forms and proportions? Or did the architect need to know structural systems, material properties and force loads? Despite individual architects' attempts to define their discipline, and to establish it as an autonomous profession, at the turn of the sixteenth century, the state of architecture remained as obscure as ever. Inundated with information, and faced with rising expectations regarding their humanistic knowledge and technical expertise, architects of the period eagerly sought to structure their diffuse practice. The personal note- and drawing books offered a solution. In composing these manuals, the architects became authors, ascending to the level of humanists, while at the same time elevating technical processes of their discipline (Galluzzi 2003, 50).

This is not to say that the architects were entirely secure in their professional role. All five of the books considered here display a certain self-consciousness, one might even say an uncertainty, on the authors' behalf. Benvenuto della Volpaia is quite explicit in this regard, systematically referencing the names of the many individuals with whom he was in contact, and whose collaboration affirmed the validly of his own practice. Likewise, the drawings of Giuliano da Sangallo's *Taccuino* are those of an architect who is questioning both his subject matter and his role in relation to it. Giuliano's inscriptions are replete with miss-spellings, which he oftentimes subsequently corrected in a different ink. The cursory annotations added to many of the more polished antiquarian studies are also suggestive of his irresolution. As conceived, the *Taccuino* was to convey Giuliano's artistic mastery and knowledge of antiquities. Yet as he matured, this understanding of antiquarian architecture was shown to be increasingly precarious and the value of his early, meticulous study drawings became increasingly unclear. The extremely small size of the *Codicetto* and the Codex Forster III may also be indicative of the authors' extreme self-awareness. For the sake of portability alone, the books could have been twice or three times as large. Yet, both Francesco di Giorgio and Leonardo deliberately chose far more compact formats for their books, and we should assume that they did so for a reason. Were Francesco and Leonardo trying to hide their knowledge? Did Francesco seek to disguise the fact that he carried with him a reference manual of design prototypes? Or was Leonardo anxious that someone might steal his ideas, claiming his *ingegno* and all the rewards which it bestowed? As attested to in the writings of Mariano Taccola and Francesco di Giorgio, such appropriation or plagiarism of artistic material was not uncommon in the Renaissance, and evidently, Buonaccorso also felt some threat of this, as several annotations in the *Zibaldone* are recorded in cryptogram.[48] While this is not the place to delve into early modern intellectual property and authorial secrecy, it is worth reiterating that Buonaccorso, like Francesco, Leonardo, Benvenuto, and Giuliano, were doing something substantially new in composing their notebooks. Although still uncertain about the status of the architect, and the supposed scope of his expertise, they could be sure of their own practices and investigations. In codifying this material, and in turn the knowledge it produced, they began to delineate the contours of a new structure of architectural knowledge.

[48] On Francesco di Giorgio's and Taccola's reluctance to share information (Long 2001, 112–23). On the "secret" character of material in the *Zibaldone*, see (Prager and Scaglia 1970, 67–9).

References

Ackerman, J. 1991. Architectural practice in the Italian Renaissance. In *Distance points. Essays in theory and Renaissance art and architecture*, 361–384. Cambridge, MA: The MIT Press.

Ackerman, J. 2002. The conventions and rhetoric of architectural drawing. In *Origins, imitation, conventions*, 293–318. Cambridge, MA: The MIT Press.

Ames-Lewis, F. 1981. *Drawings in early Renaissance Italy*. New Haven: Yale University Press.

Bambach, C. 2003. Introduction to Leonardo and his drawings. In *Leonardo da Vinci: Master draftsman*, ed. C. Bambach. New York: The Metropolitan Museum of Art.

Blair, A. 2010. The rise of note-taking in early modern Europe. *Intellectual History Review* 20: 303–316.

Borgherini, M. 2001. *Disegno e progetto nel caniere medievale. Esempi toscani del XIV secolo*. Venice: Marsilio.

Borsi, S. 1985. *Giuliano da Sangallo: I Disegni di architettura e dell'antico*. Roma: Officina Edizioni.

Brothers, C. 2008. *Michelangelo, drawing and the invention of architecture*. New Haven: Yale University Press.

Brown, B.L., and D.E.E. Kleiner. 1983. Giuliano da Sangallo's drawings after Ciriaco d'Ancona: Transformations of Greek and Roman Antiquities in Athens. *Journal of the Society of Architectural Historians* 42: 321–335.

Burns, H. 1993. I disegni di Francesco di Giorgio agli Uffizi di Firenze. In *Francesco di Giorgio Architetto*, ed. F.P. Fiore and M. Tafuri, 330–357. Milan: Electa.

Campi, E., S. De Angelis, A. Goeing, and A. Grafton (eds.). 2008. *Scholarly knowledge: Textbooks in early modern Europe*. Genève: Librairie Droz S.A.

Clarke, M. 2001. *The art of all colours: Mediaeval recipe books for painters and illuminators*. London: Archetype.

da Sangallo, Giuliano. c. 1465–1510. *Codex Barberini*. Codex Vaticano Barberiniano Latino 4424, Biblioteca Apostolica Vaticana.

da Sangallo, Giuliano. c. 1490–1513. *Taccuino Senese*. S.IV.8, Biblioteca Comunale di Siena.

da Vinci, Leonardo. c. 1493–1497. *Codex Forster III*. London: Victoria and Albert Museum.

da Vinci, Leonardo. 1992. In *I codici Forster del Victoria and Albert Museum di Londra*, ed. Augusto Marinoni. Firenze: Giunti-Barbéra.

Degenhart, B., and A. Schmitt. 1968–1982. *Corpus der italienischen Zeichnungen, 1300–1450: Süd- und Mittelitalien*, vol. 4. Berlin: Gebrüder Mann.

della Volpaia, Benvenuto. 1520–1524. *Libro di Macchine Diverse*. codex Italiani IV 41, 5363. Venice: Biblioteca Marciana.

di Giorgio, Francesco. c. 1465–1475. *Codicetto Vaticano*. codex Urb. Lat. 1757. Vatican: Biblioteca Apostolica Vaticana.

Elen, A.J. 1995. *Italian late-medieval and Renaissance drawing-books*. PhD dissertation. Leiden: Rijksuniversiteit

Ettlinger, L. 1977. The emergence of the Italian architect during the fifteenth century. In *The architect: Chapters in the history of the profession*, ed. S. Kostof, 96–123. New York: Oxford University Press.

Falb, Roberto. 1902. *Il Taccuino Senese di Giuliano da Sangallo*. Florence: Leo S. Olschki.

Ferguson, E.S. 1992. *Engineering and the mind's eye*. Cambridge, MA: The MIT Press.

Filarete, A.A. 1972. In *Trattato di Architetettura*, vol. 2, ed. A.M. Finoli and L. Grassi. Milan: Il Polifilo.

Galluzzi, P. 1987. The career of a technologist. In *Leonardo da Vinci: Engineer and architect*, 41–110. Montreal: The Montreal Museum of Fine Arts.

Galluzzi, P. 1991. Le macchine senesi. Ricerca antiquaria, spirito di innovazione e cultura del territorio. In *Prima di Leonardo. Cultura delle macchine a Siena del Rinascimento*, ed. P. Galluzzi, 15–46. Milan: Electa.

Galluzzi, P. 2003. Art and artifice in the depiction of Renaissance machines. In *The power of images in early modern science*, ed. W. Lefèvre, J. Renn, and U. Schoepflin, 47–68. Basel: Birkhäuser.

Ghiberti, Buonaccorso. c. 1480–1490. *Zibaldone autografo con disegni a penna*. Codex Banco Rari 228. Florence: Biblioteca Nazionale Centrale.

Goldthwaite, R. 1980. *The building of Renaissance Florence*. Baltimore: John Hopkins Press.

Gorrie, C. 2001. The Septizodium of Septimius severus revisited: The monument in its historical and Urban context. *Latomus* 60: 653–670.

Günther, H. 1988. *Das Studium der antiken Architektur in den Zeichnungen der Hochreinassance*. Tübingen: Wasmuth.

Henninger-Voss, M. 2004. Measures of success: Military engineering and architectonic understanding of design. In *Picturing machines 1400–1700*, ed. W. Lefèvre, 143–172. Cambridge, MA: The MIT Press.

Heydenreich, L.H. 1980. The military architect. In *Leonardo the inventor*, 10–71. New York: McGraw-Hill.

Hiscock, N. 2000. Medieval architectural practice. In *The wise master builder: Platonic geometry in plans of Medieval Abbeys and Cathedrals*, 171–204. Aldershot: Ashgate.

Hülsen, C. 1910. *Il Libro di Giuliano da Sangallo: Codice Vaticano Barberiniano Latino 4424*, vol. 2. Lipsia: Ottone Harrassowitz.

Jong, T., and M. Ferguson-Hessler. 1996. Types and qualities of knowledge. *Educational Psychologist* 31: 105–113.

Kemp, M. 1977. From 'Mimesis' to 'Fantasia:' The Quattrocento vocabulary of creation, inspiration and genius in the visual arts. *Viator* 8: 347–398.

Kemp, M. 1991. 'La diminutione di ciascun piano:' la rappresentazione delle forme nello spazio di Francesco di Giorgio. In *Prima di Leonardo. Cultura delle macchine a Siena nel Rinascimento*, ed. P. Galluzzi, 105–112. Milan: Electa.

Kemp, M. 2006. *Leonardo da Vinci: Experience, experiment and design*. Princeton: Princeton University Press.

Lefèvre, W. 2004. The emergence of combined orthographic projections. In *Picturing machines 1400–1700*, ed. W. Lefèvre, 209–244. Cambridge, MA: The MIT Press.

Long, P. 2001. *Openness, secrecy, authorship: Technical arts and the culture of knowledge from antiquity to the Renaissance*. Baltimore: The John Hopkins University Press.

Marinoni, A. 1987. Leonardo's impossible machines. In *Leonardo da Vinci: Engineer and architect*, 111–130. Montreal: The Montreal Museum of Fine Arts.

Marinoni, A. (ed.). 1992. *Il Codice Forster III*. Florence: Giunti Barbèra.

Marr, A. 2013. Copying, commonplaces, and technical knowledge: The architect-engineer as reader. In *The artist as reader. On the education and non-education of early modern artists*, ed. H. Damm, M. Thimann, and C. Zittel, 421–446. Leiden: Brill.

Martini, F.di.G. 1967. In *Trattati di architettura ingegneria e arte militare*, vol. 2, ed. C. Maltese. Milan: Il Polifilo.

Martini, F.di.G. 1989. In *Das Skizzenbuch des Francesco di Giorgio Martini, Vat. Urb. lat. 1757*, ed. L.M. Tocci. Zürich: Belser Verlag.

McGee, D. 2004. The origins of early modern machine design. In *Picturing machines 1400–1700*, ed. W. Lefèvre, 53–84. Cambridge, MA: The MIT Press.

Mussini, M. (ed.). 2003. *Francesco di Giorgio e Vitruvio. Le traduzioni del "De architectura" nei codici Zichy, Spencer 129 e Magliabechiano II.I.141*, vol. 2. Florence: Leo S. Olschki.

Nesselrath, A. 1984. Septizonium. In *Census of antique art and architecture known in the Renaissance*. Berlin-Brandenburgische Akademie der Wissenschften and Humboldt-Universität zu Berlin. www.cenus.bbaw.de. Accessed 16 Oct 2014.

Nesselrath, A. 1986. I libri di disegni di antichità. Tentativo di una tipologia. In *Memoria dell'antico nell'arte italiana*, vol. 3, ed. S. Settis, 87–147. Turin: Giulio Einaudi editore.

Nesselrath, A. 2004. Disegni di Francesco di Giorgio Martini. In *Francesco di Giorgio alla Corte di Federico da Montefeltro. Atti del convegno internazionale di studi. Urbino, 11-13 ottobre 2001*, ed. F.P. Fiore, 337–368. Florence: Leo Olschki Editore.

Pedretti, C. 1957. Il Codice di Benvenuto di Lorenzo Golpaja. In *Studi Vinciani: Documenti, Analisi e Inediti leonardeschi*, 23–41. Geneva: Librarie E. Droz.

Pepper, S. and N. Adams, 1986. *Firearms and Fortifications: military architecture and siege warfare in sixteenth-century Siena*. Chicago: University of Chicago.

Picon, C. and S. Hemingway. 2016. *Pergamon and the Hellenistic Kingdoms of the Ancient World*. New York: The Metropolitan Museum of Art.

Prager, F., and G. Scaglia. 1970. *Brunelleschi: Studies of his technology and inventions*. Cambridge, MA: The MIT Press.

Prager, F., and G. Scaglia. 1972. *Mariano Taccola and his book "De Ingeneis."* Cambridge, MA: The MIT Press.

Reti, L. 1963. Francesco di Giorgio Martini's treatise on engineering and its Plagiarists. *Technology and Culture* 4: 287–298.

Reti, L. 1965. *Tracce dei progetti perduti di Filippo Brunelleschi nel codice Atlantico di Leonardo da Vinci*. Florence: Barbèra.

Reti, L. 1980. The engineer. In *Leonardo the inventor*, 124–185. New York: McGraw-Hill.

Reti, L. 1999. Leonardo da Vinci and the graphic arts: The early invention of relief etching. In *Leonardo's science and technology: Essential readings for the non-scientist*, ed. C. Farago, 378–386. New York: Garland Publishing.

Rushton, J.G. 1976. *Italian Renaissance figurative sketchbooks, 1450–1520*. PhD dissertation. Minneapolis: University of Minnesota.

Scaglia, G. 1966. Drawings of machines for architecture from the early Quattrocento in Italy. *Journal of the Society of Architectural Historians* 25: 90–114.

Scaglia, G. 1976. A miscellany of Bronze works and texts in the "Zibaldone" of Buonaccorso Ghiberti. *Proceedings of the American Philosophical Society* 120: 485–513.

Scaglia, G. 1979. Translation of Vitruvius and copes of Late Antique drawings in Buonaccorso Ghiberti's Zibaldone. *Transactions of the American Philosophical Society* 69: 1–30.

Scaglia, G. 1992. *Francesco di Giorgio. Checklist and history of manuscripts and drawings*. Cranbury: Associated University Press.

Scheller, R.W. 1963. *A survey of medieval model books*. Haarlem: De Erven F. Bohn.
Serrazanetti, F., and M. Schubert (eds.). 2009. *The hand of the architect: Three hundred and seventy-eight signed drawings by some of the greatest contemporary architects*. Milan: Moleskine.
Spencer, J. 1965. *Filarete's treatise on archiecture*, vol. 2. New Haven: Yale University Press.
Taccola, M. 1984. In *Mariano Taccola's De Ingeneis*, vol. 2, ed. G. Scaglia and F. Prager. Weisbaden: Reichert.
Vecce, C. 2003. Word and image in Leonardo's writings. In *Leonardo da Vinci master draftsman*, ed. C. Bambach, 59–78. New Haven: Yale University Press.
Yale, E. 2011. Marginalia, commonplaces, and correspondence: Scribal exchange in early modern science. *Studies in History and Philosophy of Biological and Biomedical Sciences* 42: 193–202.
Zammattio, C., A. Marinoni, and A.M. Brizio. 1980. *Leonardo the scientist*. New York: McGraw-Hill.
Zilsel, E. 2003. *The social origins of modern science*. Dordrecht: Kluwer Academic Publishers.
Zwijnenberg, R. 1999. *The writings and drawings of Leonardo da Vinci: Order and chaos in early modern thought* (Trans. van Eck, C.). Cambridge: Cambridge University Press.

Brewing Ale and Boiling Water in 1651

3

Elaine Leong

Abstract

In the autumn of 1651, Colonel Edward Harley (1689–1741) and his uncle Edward Conway (1594–1655), second Viscount of Conway and Kiluta, spent considerable time discussing Sir John Tracey's (died 1648) recipe to brew ale. High on their list of concerns was a production step which called for water to be boiled on its own for 3 h. This essay takes an in-depth look at Sir John Tracey's ale recipe and the 'boiling water step' in order to explore how and where household recipes might fit into discussions on 'structures of practical knowledge.' Through examining the myriad of reasons why early modern men and women might consider carrying out the 'boiling water step,' I reconstruct a series of contextual frames around the creation of the ale recipe. These frames include the social structure of the early modern English country house and changing contemporary ideas about water quality, diet, health and the human body. I argue that practical knowledge such as recipes are continually framed or 'structured' by shifting social, cultural and intellectual contexts. The longevity of particular recipes lies in their flexibility to connect with changing contextual frameworks.

Early modern Europe was awash with recipes. Whether detailing a miraculous cure for fevers, the best way to make cheesecakes or a secret formula for the philosophers' stone, these short bits of instruction were circulated in letters, manuscript notebooks and in vernacular and learned print. Recipe and recipe collections constitute one of the longest standing epistemic genres and in recent years, medical, culinary, artisanal, alchemical and art technological recipes have received significant attention from

E. Leong (✉)
Max Planck Institute for the History of Science, Berlin, Germany
e-mail: eleong@mpiwg-berlin.mpg.de

M. Valleriani (ed.), *The Structures of Practical Knowledge*, DOI 10.1007/978-3-319-45671-3_3

historians of science and medicine, technical art historians and literary scholars alike.[1] This interest in recipes came as part of the current focus upon practices and materiality in histories of knowledge.[2] Alongside these studies, the early modern household has emerged as a central site for the creation and codification of knowledge.[3]

Within the early modern English context, manuscript recipe books are one of the richest sources for the study of home-based medicine and of knowledge-making processes in the household. Many English men and women diligently gathered and wrote down medical know-how in bound notebooks leading to the creation and survival of hundreds of examples. Often collaboratively compiled by family collectives over generations and filled to the brim with medical, culinary and veterinary recipes and instructions for household tasks, these books offer historians unusually detailed documentation of a series of practices. For the most part, these notebooks of household recipes appear to be records of the everyday. They offer short instructions for all kinds of tasks such as the making of a wound plaster or the brewing of beer, ale and mead, how to clean household linens and how to make pies. Within the books themselves, there is plenty of evidence that many of these recipes were actually used. Compilers leave behind statements like 'this is ye salve I allwaies make,' they add and take away ingredients in the margins, cross out recipes that did not work for them and provide anecdotes for recipes which yielded the required results.[4] With household recipes, then, there is always the temptation to read them as direct representations or records of quotidian activities.

In his introduction to this volume, Matteo Valleriani defines 'practical knowledge' as 'knowledge needed to obtain a certain product—for instance, an artistic or mechanical artifact, or specific outputs, such as healing practices or mathematical results—that follows a defined workflow. The workflow can be a construction procedure, a recipe, or even an algorithm, which are, from a formal point of view, all equivalent to one another' (Chap. 1). Following this definition, recipes are surely one of the key sources for the study of practical knowledge. After all, just what kinds of knowledge did recipes encode? Are recipes a straightforward reflection of practice on the ground? How can we speak of 'structures' of recipe and household knowledge? In this essay, I take an in-depth look at one recipe—Sir John Tracey's (died 1648) recipe to brew ale—in order to explore how and where household recipes might fit into discussions on 'structures of practical knowledge.' In particular, I examine one production step recommended by Tracey—the much debated 'boiling water step.' By exploring the myriad of reasons why early modern men and women might have considered carrying out the 'boiling water step,' I reconstruct a series of contextual frames around the creation of the ale recipe and show

[1] I am grateful to my colleagues in the 'Structures of Practical Knowledge' working group, Max Planck Institute for the History of Science Department II (Daston) and the early modern medicine and science reading group at the Department of History and Philosophy of Science, Cambridge for their insightful suggestions on drafts of this article. Further particular thanks go to John Christie and Alisha Rankin and to the editor of this volume, Matteo Valleriani, for their detailed comments.

The literature on the recipes genre is fast growing, rich and too large a body of works to fully list here. The pioneering studies on these kinds of text are (Ferguson 1898) and (Eamon 1994). For an overview see (Leong and Rankin 2011) and (DiMeo and Pennell 2013). Particularly relevant to the discussion in this essay is (Smith 2012, 4–31) and (Pomata 2013).

[2] See, for example, (Klein and Spary 2010).

[3] For households as spaces of natural inquiry see, (Cooper 2006), (Hunter and Hutton 1997), (Harkness 1997), (Pennell 2004), (Rankin 2007) and (Leong 2008).

[4] A wonderful example for this kind of practice is a recipe for 'Flor Unguentorum' which was labeled 'a most excelent salve.' In the margin are the following notes: 'my mothers receipt is to take of Rosin and perRosin each halfe a pound which makes ye salve much harder' and 'this is ye salve I allwais make puting in half a pound of each sort of Rosin and a pinte less of wine:' Wellcome Western MS 3712, Recipe Book associated with Elizabeth Okeover Adderely fol. 114r. Lady Ann Fanshawe often crossed out recipes which she rejected in the course of her assessments and practice. See, for example, Wellcome Western MS 7113, 6–7.

that the various social, cultural and intellectual contexts or, to echo Pamela Smith's essay in this volume (Chap. 14), 'structures' which the recipe is framed within continually shift. The ale recipe, though, remains largely constant amid this sea of change.

1 Sir John Tracey's Recipe to Make Ale

> For a Hogshead of Ale Eight bushells of Malt, the water must boile by it self 3 houres, then powred scalding Hot upon the Malt, so they must continue together untill the water hath taken the strength of the Malt, when It is fully cold Trenne it up and put yest to it to make it work untill it be a month old it is not good [to] drink.[5]

One September evening in 1651, Colonel Edward Harley (1689–1741) and Sir John Tracey raised tankards of homebrewed ale to toast the health of Harley's uncle Edward Conway (1594–1655). Harley was so impressed with the 'excellent' ale that he requested the recipe from his host and included it in his next missive to his uncle. For the past year, since Conway's retirement to Petworth in 1650, the two had established a weekly correspondence of which a handful of letters, almost all penned by Conway, have survived.[6] These letters covered a range of topics from politics to theology to quotidian household management. Both men shared an active interest in practical know-how and so recipes of all kinds were frequent features in their letters. As this was the heyday of country house brewing, it was not unusual to exchange recipes for ale and beer amongst friends and family (Sambrook 1996; Cliffe 1999).[7] However, Tracey's recipe to brew ale turned out to be a rather controversial one. In fact, this seemingly simple recipe became a recurring topic of conversation in subsequent letters between Harley and Conway.

Conway's initial reaction on receiving the recipe was one of doubt. He queried the recipe on two accounts. Firstly, he felt that eight bushels of malt for one hogshead seemed a bit much. Secondly, and perhaps more worringly for Conway, was the step where water was boiled by itself for 3 h. Conway thought that boiling water for this long would neither improve nor worsen the 'quality' of the water—why then should one waste resources by doing so? Despite his misgivings about Harley's recipe offering, Conway still returned the favor by sending his nephew a new recipe to make mead.[8] From this letter onwards, this simple step—boiling water on its own for 3 h—became the focus of Conway and Harley's discussions.

While awaiting Harley's reply to his letter, Conway continued to ponder upon and investigate Tracey's curious ale recipe. The next step in his inquiry was to seek out an expert. In this case, he consulted the Petworth brewer for further advice on the production method. Not bothering to wait for Harley's reply, Conway sent his nephew another letter on 14 October 1651 which detailed the brewer's views.[9] Like Conway, the brewer felt that the 'the boyling of water three howers cannot adde any thing to the water so that if it be made hot as it ought to be whether it be done in halfe a quarter of an

[5]National Archives, State Papers, Domestic, SP18/16 fol. 56. Letter from Edward Harley to Viscount Conway at Petworth dated September 25 1651. My grateful thanks go to Bernard Capp for bringing the Conway/Harley correspondence to my attention.

[6]The reconstruction of the conversations surrounding the Tracey ale recipe detailed below is based on six letters—five penned by Conway in October and November 1651 and a single letter penned by Harley dated 25 September 1651. See notes below for details and references.

[7]For contemporary advice literature on this issue, see, for example, (Markham 1631).

[8]Letter from Edward Conway to Edward Harley, 7 October 1651, London British Library, Additional MS 70006, fol. 232r.

[9]Letter from Edward Conway to Edward Harley, 14 October 1651. London, British Library Additional MS 70006, fol. 237r.

hower or a quarter, or three howers it is all one.' Additionally, the brewer politely suggested that perhaps Harley's instructions referred to the boiling of the wort—the mix of the water and malt—which was a fairly common step in that period. As the recipe stood, though, the brewer was unwilling to put it to the test. Conway begged his nephew to further enquire into the 'certainty' of the recipe.

After this, the letters between uncle and nephew took a decidedly natural philosophical turn. Conway's letter from 21 October 1651 suggests that Harley had been discussing the effects of boiling on the quality of water with a few physicians in London. The 'boiling water step' was recommended, as they believed that the 'highest' water might be obtained by such an action. And it is with the 'highest' water that Tracey and Harley wanted to brew ale.

At this juncture, then, Conway and Harley's discussion began to touch upon what might constitute the 'best' water and how one might obtain it. Here uncle and nephew held differing views on both points. On the topic of water quality, Harley believed that when physicians argued that the 'best' water was the lightest water, they were referring to the actual weight of water. Within this schema, the 'worse' water weighed more than 'good' water. Conway, on the other hand, felt that this rather referred to the speed at which the water 'goe out of the stomach.'[10] That is, for Conway, the 'best' water is that which is the most easily digested. Evidently, Harley was not convinced by the arguments which Conway put forward on 21 October, for in the next weekly letter (dated 28 October 1651), Conway strongly advised Harley to reconsider his position. Acknowledging that there were others who took Harley's side, Conway argued that physicians 'doe holde that water is not to be iugded by the balance' and countered his nephew's claims with quotes from Johannes Butinus' edition of *Hippocrates Aphorisms* (1625), Louis du Gardin's *Institutiones medicinae* (1634) and Nouncius' *De re libaria.*[11]

Conway's efforts to bring Harley over to his way of thinking succeeded. The final letter in the series, penned by Conway and dated 25 November 1651, reveals that Harley indeed paid heed to his uncle's arguments. In what appears to be a complete turn around, he now thought that 'in the mutch boyling of water the thinnest and the finest parte of the water doth goe away and that you leave only the most grosse, unholsome and earthy parte.' Therefore, he 'will try to make Ale with the water which is but little boyled.'[12] Uncle and nephew, thus, finally saw eye to eye on the issue of water quality and on the futility or even harmfulness of the 'boiling water step.' Yet, despite this agreement, both still considered putting the recipe to trial. Or at least, Conway continued to encourage Harley to put the recipe to trial and let him know whether he was able to make 'good and excellent ale.' Upon Harley's assurance that this was indeed possible, Conway promised that he himself would also 'make trial of it.' After all, wrote Conway, 'if you did what can you doe more, there is not anything to be beleaved in naturall things but experience....' At Petworth, though, Conway was unable to persuade anyone to make it as the brewers said that 'mutch boyling doch neither good nor hurte, but that not to boile the wort will certainly spoile the Ale.'

After this letter, the trail goes cold and the conversations between uncle and nephew move onto other matters. Conway and Harley continued their weekly (on the side of Conway, at least) correspondence until Conway went abroad a few years later. Conway died in Lyon in 1655, leaving behind a set of wonderfully gossipy letters and a vast library which historians and literary scholars alike are just beginning to study (Smith 2012 and 2014). Over the next 50 years, Harley played an active part in

[10]Letter from Edward Conway to Edward Harley, 21 October 1651. London, British Library Additional MS 70006, fol. 239r.

[11]Letter from Edward Conway to Edward Harley, 28 October 1651. London, British Library Additional MS 70006, fol. 241r.

[12]The letter is dated 25 November 1651: London, British Library, Additional MS 70006, fol. 247r. All quotes in this paragraph are taken from this letter.

English politics and died in his country estate of Brampton Bryan in 1700.[13] As for the 'boiling water step,' from what I can tell from household recipe books, this resource-draining process proposed by Tracey and Harley never got fully incorporated into popular domestic know-how. Ale and beer recipes appear with regularity in contemporary manuscript recipe collections but none of these recommend the 'boiling water step' as part of the production process.[14]

Yet, the issue of whether to boil the water used to brew ale and beer did not quietly vanish from contemporary discussions. Tracey, Conway and Harley's concerns over the 'boiling water step' were shared by a number of early modern authors working in different epistemic contexts. The following sections explore household management and country house brewing in mid-seventeenth century England, debates on water quality and drinkability in contemporary regimen texts and a series of brewing tracts touching on the benefits (or not) of the boiling of the liquor (the water used for brewing) published in the late 1690s.

2 Country House Brewing in Mid-Seventeenth Century England

> When our *English House-wife* knowes how to preserve health by wholesome Physicke, to nourish by good meate, and to cloath the body with warme garments, shee must not then by any meanes bee ignorant in the provision of Bread and Drink; shee must knowe both the proportions and compositions of the same.
>
> Chapter 8, Of the Office of the Brew house, and the Bake house, and the necessary things belonging to the same. (Markham 1631, 243)

As Joan Thirsk has recently argued, despite the growth of commercialized 'common brewers,' the brewing of ale and beer at home continued to be common practice in English households up to the mid-eighteenth century (Thirsk 2007, 305–7). In particular, large country estates such as Petworth, where Conway was based at the time, commonly had a separate brew house and a team of workers to supply the entire households' drinking needs. As Gervase Markham (ca. 1568–1637) writes in *The English House-wife*, the stocking of the household cellar fell to the office of the English housewife. Like many conduct book writers of his time, Markham recommended his readers to acquire knowledge and skills in a number of areas such as health and medicine, botany and gardening, cookery and food preservation, distillation and brewing, wool dying and dairy work (Markham 1631, sig. A4r–A6v).

While Markham's *The English House-wife* is prescriptive in nature, there is ample evidence to suggest that early modern housewives took on the many tasks outlined by Markham in his work. A series of letters sent between Lady Johanna St. John and her steward Thomas Hardyman in the 1650s and 1660s provide unusually detailed and rich information on quotidian early modern household management practices. St. John, who lived for the most part in London, repeatedly requested Hardyman, based at Lydiard Park near Swindon, to instruct her household staff in completing a range of tasks. Some of these tasks were geared towards achieving immediate results, such as the baking of a particular pie for her husband's dinner or the production and sending up of a large volume of butter to London

[13] For biographical information on Conway and Harley, see: James Knowles, Conway, Edward, second Viscount Conway and second Viscount Killultagh (bap. 1594, d. 1655). In *Oxford Dictionary of National Biography*, online ed., edited by Lawrence Goldman. Oxford: OUP. http://www.oxforddnb.com/view/article/55441 (Accessed 2 May 2016) and Gordon Goodwin, Harley, Sir Edward (1624–1700), rev. David Whitehead. In *Oxford Dictionary of National Biography*, online ed., edited by Lawrence Goldman. Oxford: OUP. [http://www.oxforddnb.com/view/article/12335. (Accessed 2 May 2016).

[14] To ascertain this, I searched through the ale recipes within the Wellcome Library's digitized collection of 75 seventeenth-century household recipe books. The digitization process included an electronically searchable first line (i.e. title) index. http://wellcomelibrary.org/using-the-library/subject-guides/food-and-medicine/finding-recipe-manuscripts-online/(Accessed 23 June 2015).

so that St. John could make her favored medical salve.[15] Others were part of long-term planning processes to ensure that particular foodstuffs would be available and ready for specific occasions and seasons. For example, St. John lodged early requests to her staff to fatten up turkeys for Christmas and to ripen cheese for her table.[16] The brewing of ale and beer, an essential drink on early modern tables was a central feature of this sort of planning. A number of letters reveal St. John's concern to ensure that the right sort of ale flowed on the various tables in her household.

In one undated letter, St. John warned Hardyman that her husband Sir Walter intended to travel to Lydiard on Lady Day, thus Hardyman should fill the cellar with both strong and small beer.[17] On another occasion, she advised Hardyman to begin brewing beer for the family's stay the next month. Her instructions read as follows:

> … I remember your Ale uses to drink but wel at a qtr old therfor sure you may Brew it now tho we should stay a month beyond Easter when yu bottle it put into every bottle 8 leaves of Garden Scirvie Gras here after I doe intend to Brew our strong Bere by it selfe and small Beer by it selfe then we shal have always Good Beer for sinc we come hither some complaine of smal Beer we use to have at Lidiard so that I would have the strong beer brued now and by it selfe if the celer be not ful alredy fill it and fill the smal Beer celer Also….[18]

St. John's specific directives betray not only the important managerial work which gentlewomen took on as part of their social roles but they also indicate the central place held by ale and beer on early modern tables. Indeed, if the beer was not up to scratch, 'some complaine[d].'

By the mid-seventeenth century, a number of different kinds of drink accompanied food on English tables including ale, beer, perry and cider (Markham 1683, 244; Thirsk 2007, 304–12). Ale, a drink brewed only with malt, was the traditional drink and, as suggested by St. John and Conway's letters, continued to be popular. Ordinary beer, however, was recommended for the tables of noblemen, gentlemen, yeomen and husbandmen by writers such as Markham.[19] The practice of brewing beer—a drink made with both malt and hop—came to England in the fifteenth century (Sambrook 1996, 108). Significantly, the addition of hops afforded the drink a longer shelf-life which made it particularly popular with commercial or common brewers. Both of these drinks were offered at different strengths achieved by different malt and hops to water ratios. Markham advised makers to use one quarter of good malt to brew three hogsheads of beer.[20] Brewed in March or April, March beer was much stronger and, in this case, one quarter of malt would only yield one hogshead of beer. March beer also required aging of around a year but once aged, the beer would last 2–4 years if it kept closed in a cool place (Markham 1631, 245). There were also two different strengths of ale. For strong ale, Markham recommended four bushels or half a quarter to 14 gallons. A weaker ale, which Markham termed 'bottle ale,' used the same amount of malt to brew 20 gallons. The small ale mentioned by St. John was produced by adding barme (a froth taken from the top of fermenting malt liquors) to the drink on

[15]Brian Carne, *Some St. John Family Papers*. Reproduced, with corrections and additions, from the *Reports* of the Friends of Lydiard Tregoz, 27, 28 and 29 (1994–1996), 47 (venison pasty) and 54 (butter). The St. John family papers are now in the Wiltshire and Swindon History Centre. The main file for the letters of Sir Walter and Lady Johanna St. John are found under catalogue number 3430/22. These are all loose leaves and are not foliated. For ease of information retrieval, all references in this article are to Carne's transcriptions in *Some St. John Family Papers*. Copies of the text can be obtained directly from the Friends of Lydiard Tregoz at Lydiard Park.

[16]Carne, *Some St. John Family Papers,* 47–8 (turkeys) and 54, 101–2 (cheese).

[17]Carne, *Some St. John Family Papers*, 68.

[18]Carne, *Some St. John Family Papers,* 95.

[19]Markham suggests that for ordinary beer, 'a quarter of good malt' gives three hogshead of beer' (Markham 1631, 244).

[20]Within country households, the hogshead was used as the unit of measurement for ale and beer. There seems to be a great deal of variation as what exactly constituted a hogshead. According to Pamela Sambrook, in the seventeenth century, this could refer to 54, 56 or 63 gal. A 'quarter' of malt is 8 bushels or 64 gal. Sambrook is careful to remind us that the measurements in the early modern period are anything by standardized (Sambrook 1996, 18 and 273–4).

the morning of drinking. According to Markham, this produced a drink which was ‘a great deale the fresher and be much more lively in taste’ (Markham 1631, 248). While Markham adopted a tone of confident authority in his discussion of ale and beer, the sources indicate that a great amount of variation existed between the different recipes and methods used to brew ale and beer. Somewhat confusingly, the historical actors themselves also used the terms ‘ale’ and ‘beer’ to refer to a range of different drinks. By the seventeenth century, it was not uncommon for ale to be hopped. Thus, in some cases, ‘beer’ was used to designate a weaker brew (Sambrook 1996, 109–11).

The basic process for brewing ale or beer in the early modern country house was fairly simple and consisted of six steps. First, the barley had to go through a malting process. This involved steeping the barley in water to start the germination process and then drying the just germinated barley in a kiln. While some country houses did their own malting, many bought in ready-made malt. Reflecting this practice, Markham's description of brewing in his *English House-wife* begins with malt rather than barley. The second step, called the ‘mashing of the malt,’ involves grinding the malt and mixing it with hot water termed the ‘liquor.’ Here, Markham recommends the maker put the boiling water onto the malt ‘little and little with scoopes or pailes,’ stir well and to let it stand for an hour or so. After which, the maker should run the liquor into a clean trough or vessel, add hops and boil the mixture together for another hour or so. At this point, Markham recommends the maker take up a ‘dishfull’ of the wort and if the hops ‘shrinke into the bottome of the dish,’ they may strain the wort into the cooler and into a ‘great bowle’ containing the yeast. This mixture was then left to cool overnight. The next morning, the work began by taking off any black scum on the top of the barme and stirring the beer. After leaving the mixture to stand for another hour, the maker should give the mixture one final mixing and then pour it into clean and scalded casks (hogsheads) and leaving it to settle. (Markham 1631, 244–5).

Within this multi-stepped production process, householders and brewers faced a number of difficulties. Following Valleriani's terminology, these might involve the workflow (or procedure) or the technique. The first was perhaps logistical and involved the tasks of ensuring that there was a supply of large quantities of water, and of heating and moving gallons of hot water through the system. Water was required not only for preparing the liquor but also for washing and soaking all the different kinds of vessels utilized in the brewing process. Concern to ensure that the hot liquids could be somewhat easily moved around the brew house can be seen in the architecture of contemporary structures and in written documents. Surviving early modern brew houses indicate that buildings were constructed to optimize gravitational flow from one vessel to another. One way to ensure this was to build the structure on the hillside, as suggested by Cressy Dymock in his ‘Propositions concerning brewing of a more excellent sort of Ale and beare’ now stored in the Hartlib papers (Greengrass et al. 2013) (Fig. 1).[21]

The second difficulty concerned the temperature of the liquor and of the wort at points where the malt and the yeast were added to the liquid. This has attracted the attention of both our historical actors and historians of science. In both cases, the temperature of the liquid was crucial. If the temperature of the liquor was too hot, it would cause the malt grains to set or as contemporaries termed it, ‘mak[e] a pudding.’ If the temperature was too cold, the mash would not extract sufficient sugars (Sambrook 1996, 93). This also, of course, affected the taste of the beer. Given our earlier encounters with the kinds of instructions which Lady Johanna St. John sent her steward, it is not surprising that she also had something to say on both these points. She wrote in an undated letter:

> … the Celer had need be fild wth beer and I would have 2 sorts of strong Beer one a good deal smaler than the other and for the Ale I would not have it so strong as that you sent up nothing neer and I would have it boyld for

[21] Published by HRI Online Publications, Sheffield [available at: http://www.hrionline.ac.uk/hartlib (Accessed 23 June 2015)]: 62/39A–39B.

Fig. 1 Image showing the interior of the early eighteenth-century brew house at Charlecote Park, Warwickshire. In the front, we can see the mash tun where the heated water and prepared malt would be mixed. In the back, we can see the brick-encased copper underworks. The brew house at Charlecote Park is unusual in that it had two separate coppers for heating up the liquor and the wort. From (Sambrook 1996, 95)

> 4 or 5 Howers at the least and the water must not boyl but only be redy to boyl when he powers it on the mault that he make the Ale with Brew it so as it may be a fortnight [in] the vessel and as long bottled before we [come] down.[22]

St. John's first instruction, requiring Hardyman and his helpers to boil the wort for 4 or 5 h, was a common practice at the time.[23] As this step was most probably performed in an open vessel, this also affected the taste of the finished drink. On the other hand, St. John's description that the water must not be boiling but only 'redy to boyl when he powers it on the mault' clearly referred to the water temperature at the mashing step. While St. John merely instructed her brewers to ensure that the water was only 'redy to boyl' later printed brewing manuals offer techniques to determine the water temperature. For example, in his 1802 manual *The Compleat Family Brewer*, Thomas Threale advised the brewer to scatter a handful of bran or malt over the surface of the water. This allowed the brewer to better gauge when the water was on the verge of boiling, as this was the point when it was ready for the mashing step. Another method was to test by putting one's finger in the water or to wait until the steam had cleared to the point where the brewer could see his own reflection in the water (Sambrook 1996, 94).

The mashing step in beer brewing is not unfamiliar to historians of science who have studied eighteenth and nineteenth century debates on heating the liquor and the various practices used to determine, control and standardize this step.[24] As Otto Sibum and James Sumner have demonstrated, beer brewing provides an interesting case to study the intersections between practical and theoretical knowledge. Sibum, in particular, has used the various ways of testing the liquor temperature to demonstrate the importance of tacit or gestural knowledge within brewing and, by extension, scientific practices (Sibum 1995). More recently, Sumner has convincingly shown that practitioners, amateurs and natural philosophers alike participated in eighteenth- and nineteenth-century debates about the

[22] Carne, *Some St. John Family Papers*, 68.

[23] A number of recipes for the boiling of ale or beer within contemporary household books also recommend a long boil of the wort. This includes, the recipe to brew 'Barkshire Ale' by Sir Paul Neile in the notebook of Elizabeth and Jacobs and Others (Wellcome MS 3009 fol. 241r) which required a 3 h boil or a recipe to make 'strong beare' in an anonymous late seventeenth-century notebook which required the wort to be boiled for at least 2 h (Wellcome MS 4054, 191).

[24] Interestingly, most of the work done on brewing in England pre-1700 appears to concentrate on the social and cultural effects of beer and alcohol. The literature on this is numerous. Recent surveys include (Withington and McShane 2014) and (Hailwood and Toner 2012). One of the most cited social histories on beer in England is (Bennett 1996).

brewing process (Sumner 2013, 2007). In fact, by the eighteenth-century, the world of beer brewing was one in which many of the figures that Ursula Klein calls hybrid experts or artisanal-scientific experts functioned (Chap. 11).[25]

In both seventeenth-century country house brewing and later commercial brewing, the conversations on how to determine water temperature were largely tied to empirical practice and personal experience. In surveying the practices in commercial breweries in nineteenth-century Manchester, Sibum argues that the brewers relied on 'imitative knowledge' or, as he terms it, 'gestural knowledge' to judge mashing heats. This expertise was highly sought after and the Manchurian brewers in Sibum's story closely guarded their know-how on mashing heats and their specialist knowledge enabled them to command high wages. If we read this scenario backwards in time, one assumes that the Petworth brewer, whom Conway consulted over the 'boiling water step,' had, over the years, also developed his his own methods of gauging mashing heats. But that is not why Conway consulted him.

Despite all the interest over the mashing step, by both historical actors and historians of science, it is, in fact, *not* this step that was discussed in such detail by Conway and Harley. Whereas the later writers of brewing manuals and taste-concerned housewives like Johanna St. John were troubled with the act of mixing malt with hot water, Conway and Harley focused on the boiling of the water *before* the addition of malt. At stake here is not the possibility of brewing a foul-tasting pint of ale but rather the healthful quality of the water used in the ale-brewing process. The act of boiling here is not related to gauging or ensuring a particular water temperature but rather to alter or ameliorate the substance of water. In fact, while our historical actors all discussed heating up and boiling of the same pot of water (as it were), their discussions led to very different paths.

3 Brewing Ale and Household Management

Given that ale and beer were the usual drinks of choice and were often used as a base for health-promoting drinks and medicines, it is not surprising that early modern men and women took an interest in searching for and refining their own favorite recipe. As alluded to earlier in this essay, brewing recipes are often found in contemporary household recipe collections, with many households having gathered and recorded several different ways to brew ale and beer.[26] The collection of multiple recipes for the same medicine or foodstuff was a common practice reflecting householders' interests both in novelty and in collating and comparing different ways of producing household goods. Consequently, Harley's passing on of Sir John Tracey's recipe to Conway and the latter's interrogation of the recipe needs to be framed within these larger practices of recipe exchange and transmission and household management activities. As we learned earlier from the letters between Johanna St. John and her steward Hardyman, gentlewomen and gentlemen took an active interest in managing quotidian household production activities. When it comes to brewing, this interest seemed to have extended across the different sections of early modern society. It was not only ordinary householders who experimented with brewing but the virtuosi of the Royal Society also took an interest. The topic was discussed at a December 1662 meeting and plans were made for a member, Sir Robert Moray (1608/9–1673), to study beer brewing in his native Scotland. While these plans were never carried through,

[25] For a recent discussion of this idea, see (Klein 2012a, b).

[26] Beer and ale recipes make frequent appearances in early modern recipe collections. Many of these recipes have specific curative properties. For example, the Townsend family collection include recipes a 'laxative beer' associated with Dr. Butler and Dr. Fryers's beere for ye scurvy. The collection of Elizabeth Jacob and others includes recipes for cock ale, Sir Paul Nele's 'Barkshire Ale', cleare ale and elder beer: Wellcome Western Manuscript 774, 30 and 70 and Manuscript 3009, fols. 233r, 241r, 275r and 295v.

other members, such as Sir Kenelm Digby (1603–1665), most certainly oversaw the production of beer and ale on their estates (Sumner 2013, 11).

Like many complex and large social organizations, the early modern household was fairly rigidly structured. While members of the household might take on several different tasks at any given point, they had clearly defined roles. Contemporary conduct manuals and in-house rules and directions for servants provided clear outlines to both to their assigned roles within the household and to many aspects of their daily lives. As historians such as R.C. Richardson and Steve Hindle have demonstrated, the master-servant relationship in early modern households was one which was hierarchical and ordered (Richardson 2010; Hindle 2011). Hindle's study of Sir Richard Newdigate (1602–1678) demonstrates that some gentlemen, at least, rather adhered to a management style which might be termed as 'close intervention.' The St. John/Hardyman letters suggest that it was not only gentlemen who tended to keep a close eye on their domestic staff but gentlewomen too. Gathering recipes and other forms of know-how was one way that early modern gentlemen and gentlewomen informed themselves on various aspects of housework. Aside from informal knowledge exchange with family and friends, our actors could also rely on household manuals such as the aforementioned work by Markham. These structures of household management and oversight constitute one frame within which to understand Conway and Harley's discussion of the ale recipe and the codification of household practical knowledge.

Elsewhere and in her contribution to this volume, Pamela Long has introduced the concept of the 'trading zone' for the early modern period and it may be useful to make parallels between the country brew house and the kinds of 'trading zones' described by Long (Chap. 8). Like the engineers, miners and shipbuilders outlined in Long's cases, the Petworth brewer not only contributed to the ongoing conversation but also occupied the crucial position of deciding whether actual trials (involving manpower and other resources) could go ahead. Here, the country brew house provided a setting for substantive exchange between the bookish Conway and the experienced brewer. If we see Conway and Harley's epistolary discussions about the ale recipe as a sort of 'trial on paper,' then in this case, the trial could only be conducted with the help and consensus of the Petworth brewers. From the beginning of the epistolary conversation, Conway was clear that he relied on the brewer to help him assess the ale recipe. In his second letter, Conway wrote 'I find that in the receipt you have made a mistake,' intimating that perhaps the whole affair might have arisen from false transcription or some other such mistake. Conway's main source for determining this mistake can be traced back to his conversations with the brewer who asserted that boiling for 3 h not only had few benefits but that if one just let the malt sit in the water, it would not 'be good at all.' Furthermore, the brewer suggested that the liquor needed to be drawn off and boiled separately. In fact, this unnamed brewer was a constant figure throughout Conway and Harley's conversation.[27] In the next letter, Conway reported that 'they say' (here I think that he again means the brewers) that one must not put scalding hot water onto the malt.[28] The Petworth brewers also featured in the last letter where they seemed to have a final say in whether the recipe was viable. After all, Conway wrote 'I cannot persuade any body to make it.'[29]

In reconstructing Conway and the brewer's conversations from the letters, it is clear that both men brought different skill and knowledge sets into the conversation. Despite the fact that conduct book authors like Markham geared their manuals towards country gentlemen and gentlewomen, their

[27]Letter from Edward Conway to Edward Harley, 14 October 1651, London, British Library, Additional MS 70006, fol. 237r.

[28]Letter from Edward Conway to Edward Harley, 21 October 1651, London, British Library, Additional MS 70006, fol. 239r.

[29]Letter from Edward Conway to Edward Harley, 25 November 1651, London, British Library, Additional MS 70006, fol. 247r.

involvement in processes such as beer brewing was most likely supervisory and managerial. As Sibum has shown us, the practice of brewing relied on a great deal of embodied knowledge. It is really only through long experience and personal observation that a skilled brewer can know the correct temperature to add the hot water to the malt and estimate how long it might take a certain amount of water to boil in a closed copper. Knowing how to complete both of these steps is crucial, because water that is too hot would spoil the malt and boiling a copper dry would ruin an expensive piece of equipment. Secondly, brewing beer or ale was not really a one-man job. Even small-scale country house brewing involved the production of at least a hogshead at a time. Ensuring that all the tables within a household were awash with enough beer required a crew of men and women: someone to stoke the fire to heat the copper, another to stir the mixture in the tun and brave and strong souls to move the hot liquids through the brewing process. So, despite Conway's statements: 'I will make trial of it'—it seems unlikely that he really saw himself in the thick of the action in the brew house. After all, from his life trajectory, he was definitely more comfortable in libraries, salons and even on the battlefield than in the brewery. It is, thus, likely that while it may have been conducted within an ordered hierarchical structure, the working relationship between Conway and the brewer appears to have been a collaborative one with each party bringing different bodies of knowledge into the conversation. The brewer offered hands-on knowledge gained through long-lived experience and Conway, in this instance, contributed knowledge gained from his extensive library and his correspondence with Harley.

To oversee, to manage, to supervise—these terms describe, for the most part, the involvement of gentlemen and gentlewomen such as Conway, St. John and Digby in the brewing process. Recall from the quotations above that while Johanna St. John commanded Hardyman to ensure that the ale was made with water which was merely 'redy to boyl' and not boiling, she did not tell her servants how to judge when the water was 'redy to boyl.' One assumes that she left that kind of experiential knowledge and skill to her brewers. Conway and Harley might also have worked along the same lines. The short recipe at the center of this essay clearly assumes that the maker had substantial prior knowledge and experience in brewing. Gentlemen and gentlewomen such as Conway, Harley and St. John, tended to leave hands-on tasks such as gauging mashing heats to the experts. When it came to brewing ale, their interests did not lie in the area of the technicalities but rather the different steps involved in the brewing procedure. Their role here, it seems, was rather to order and to manage what Valleriani terms as workflow—that is selecting their favored recipe. This separation of household roles goes someway to explain the 'thinness' of the ale recipe and perhaps recipes in general. Consequently, perhaps oddly in letters discussing 'how-to' knowledge, Conway and Harley were not so much interested in how to accomplish each of these steps but in 'what' steps should be included. The separation of workflow and technique in this particular narrative is firmly structured by the local context of the early modern country house.

4 Debating Water Quality

As Conway and Harley's discussions about the ale recipe progressed, their focus turned to contemporary debates on water. In doing so, Conway and Harley joined a number of contemporary discussions. Emily Cockayne has vividly described the filthy, smelly and grotty conditions of the early modern city. Amidst the hubbub of early modern English urban life, clean water was definitely a rare commodity (Cockayne 2007, 198–202). Consequently, anxiety and concerns about water quality, both for drinking and for cooking and brewing, was widespread. Debates over the merits of different kinds of potable water can be found in a number of contemporary regimen guides. The *regimen sanitatis* is a long-standing medical genre which first flourished in the medieval period.[30] These guides offered advice on the preservation of health through careful regulation of the six non-naturals which included air, food and drink, sleep and walking, movement and rest, retention and evacuation and the passions of the soul or emotions. Often penned by physicians, vernacular versions of these guides flourished in late sixteenth- and early seventeenth-century England and were, according to Paul Slack and Mary Fissell, one of the most popular medical genres of the period (Fissell 2011 and 2007; Slack 1979).[31]

Many of these texts offered advice on water drinking and water quality. Some discussed the effects on water by surrounding environmental conditions and the various steps that one could perform to ensure that, if one chose to drink water, one drank water with the best possible health benefits. However, just as Conway and Harley presented two viewpoints, the advice proffered by these texts was by no means uniform. Take, for example, discussions on the merits or dangers of drinking water. Some authors, such as Andrew Boord advised that 'water is not holsome sole by itselfe for an englisshe man.' Rather, English men and women were to take in water in the form of drinks like ale, beer, mead and metheglyn (Boord 1542, Sig, F r). Others, such as Thomas Elyot (1490–1546) and, following him, Thomas Cogan (ca. 1545–1607), argued that water is the 'very naturall' and the most ancient drink (Elyot 1534, 33; Cogan 1584, 206). A later writer, James Hart, also presented a strong case for water drinking in this 1633 *The Diet of the Diseased*. Like Elyot and Cogan before him, Hart argued that water is the most ancient drink but he recognized that two qualities deter people from the use of this 'noble and antient' drink: coldness and crudity. On the latter, he argued that crudity is merely the 'abundance of moisture' and, for Hart, boiling the water and mixing it with a little honey or sugar and a little wine vinegar was the best correction.[32] Across the genre of regimen guides, authors of these texts offered a wide variety of views.[33] Physicians could not agree amongst themselves on the merits or dangers of drinking water, the desirability of different water sources or the utility of boiling as a way of ameliorating water quality.[34] These rich and diverse discussions serve as a frame for Conway and Harley's concerns about water quality.

Whether they were advocates of water drinking or not, many regimen authors offered some sort of schema to rank the different water sources available to early modern men and women. For many, rainwater was the most desirable, 'yf it be received pure & cleane, is moste substyll and penetrative, of any other water.' Spring water, which has 'passeth swyftely amonge great stones or rockes' and cleane ryver water which 'rennethe on great harde stones or pebles' followed. (Elyot 1534, 34r).[35] Yet, here, as on other points, there was a great deal of disagreement amongst the regimen authors. In contrast to Elyot, William Bullein (1515–1576) argued that that river, pond, spring and sea water were not

[30] There is now a substantial body of literature on regimen guides. See, for example, (Nicoud 2007; Cavallo and Storey 2013) and, for the English context, (Wear 2000, Chapter 4).

[31] Note: the majority of the regimen guides cited below have long publishing histories. Thus, while the citations refer to sixteenth-century first edition, these texts continue to be printed in the early part of the seventeenth-century and most likely circulated well beyond the year of the final printed edition.

[32] According to Hart, it was customary, particularly amongst female drinkers, to add sugar (considered hot in temperament) to water to correct the latter's coldness. Hart advises against this.

[33] My analysis here of regimen texts is limited to the authors' diverse views on water potability, water quality and the effects of boiling on water. I acknowledge that research is needed to establish the origins of regimen texts in English, the *fortuna* of particular titles, the relationship between the English regimen guides and those written in other European vernaculars and clarification on the diverse views put forward in English regimen texts. However, these discussion lie beyond the scope of this present essay. Sandra Cavallo and Tessa Storey's forthcoming volume of essays tittled *Conserving Health. The Non-Naturals in Early Modern Culture and Society* leads the way in addressing some of these issues.

[34] This was by no means an 'English' concern. For a discussion of water qualities and cold water drinking in late Renaissance Italy, see (Cavallo and Storey 2013, 218–28).

[35] This categorization is fairly common. For the Italian context, see (Cavallo and Storey 2013, 220).

'pure' because they had been 'mingled with sundry ayres, corruptions, grosenesse, and saltnes' (Bullein 1558, fol. cii). The authors seemed to share a general fear of standing or still water. For example, Andrew Boord, our 'anti-water drinker' advised his readers to avoid 'all waters whiche be standynge and be putryfyed with froth, duckemet, and mudde' even for use in baking, brewing and cooking (Boord 1542, sig. F v). If one intends to use water in these tasks, it should be running water and should be left to stand in vessel for 2–3 h and then strain the upper part thoroughly through a thick linen cloth and throw the water near the bottom of the vessel away.

Within these schemas ranking different kinds of water, many authors drew attention to the natural surroundings of the water source and the different kinds of materials over which water flowed. William Bullein, for instance, claimed that water that had run over clay was preferable to that which had flowed over gravel or stones, as clay as long as it was 'pure' and 'voyde of corruption,' had the ability to clean the water. Bullein was especially anxious about water quality and also warned his readers about 'water running neare into cities and townes, or marish ground, wodes, & fennes.' These, he argued, because there is 'so much filthe in them of carion (corpses) and rotten dunge.' It may not come as a surprise that Bullein ended his section on the properties of water with 'to drincke cold water is evyll' (Bullein 1558, fol. cii r–v). From the correspondence, it is clear that Conway shared the idea that flowing over stones, gravel and, in some cases, clay, affected the quality of the water. In the letter dated 28 October 1651, he wrote 'if somme waters are better then others for severall uses it is from the qualities they have from the earth thorough which they passe, comprehending in the worde Earth all that is in the earth thorough which water doth passe….'[36] Thus, for Conway like for many of the regimen writers, it was crucial to ascertain the origin of one's water source.

In that same letter, Conway also chastised his nephew for the latter's mistaken motion that 'water which hat least weight have greatest bounty.' His reasoning for this was that 'it is very rarely found that one water should be lighter than another' and he reminded his nephew that 'all physitians doe holde that water is not to be judged by the balance.'[37] Conway himself believed that the 'highest' water was that which 'doth soonest goe out of the stomach.'[38] That is, for Conway the discussion had more to do with bodily digestion than the weight of water. While Conway might be dismissive of Harley's views, a number of contemporary regimen writers concurred with Harley and judged the quality of water by its weight. The two staunch advocates for water drinking, Thomas Elyot and Thomas Cogan, argued that water which is 'lightest in poyse or weyght is beste.' They helpfully provided 'dyvers meanes to trye out, whiche is the beste water' including the suggestion that their readers put a piece of linen cloth in the water and lay it out to dry. The water that dried the quickest was the best water (Elyot 1534, 34; Cogan 1584, 209). This view was also shared by William Bullein who wrote in his *The Government of Health*: 'Also waters runnynge toward the east, be pure, comyng oute of harde stonye rockes, and a pinte of that water is lighter then a pynte of the standyng water of welles, or pooles. The lighter the water, the better it is' (Bullein 1558, fol. cii r–v). This vivid description suggests that readers might have compared water from different sources by weighing pint bottles filled with water collected in various locations. Reading these authors, it is clear that Harley was merely one of many who thought that the 'highest' water was literally the 'lightest.'

Within the discussions on water drinkability, a number of the regimen authors advised their readers to boil their water before drinking. William Bullein recommended boiling drinking water as this would 'clense it from corruption' (Bullein 1558, fol. cii v). Andrew Boord also suggested that his

[36] Letter from Edward Conway to Edward Harley, 28 October 1651, London, British Library Additional MS 70006, fol. 241r.

[37] Letter from Edward Conway to Edward Harley, 28 October 1651, London, British Library Additional MS 70006, fol. 241r.

[38] Letter from Edward Conway to Edward Harley, 14 October 1651, British Library, Additional Ms 70006, fol. 239r.

readers seethe and cool their water before mixing it with the wine for drinking. For Boord though, it was better still to only mix wine with 'stilled waters,' particularly distilled water of strawberries, bugloss etc. (Boord 1542, sig. Fi r).[39] Another author who recommended boiling as a way to render water potable was James Hart. Writing in the 1630s, Hart acknowledged that the boiling of waters was a somewhat controversial topic—particularly in terms of exactly how much of the water to boil away. He wrote: 'some willing to boile water to the wasting away of the third part, others of the halfe, which others againe thinke too much; and indeed a meane is the best' (Hart 1633, 117). Like Boord, though, Hart felt that even if it is a bit 'costly' and sometimes not so convenient, distillation was the way to go in making water potable. Other measures suggested by Hart include straining the water through a clean linen cloth and then boiling it again, adding coral to correct the bitterness of the water or, quoting Pliny, adding a little wheat flour to the water, letting the mixture stand for 2 h and then drinking it (Hart 1633, 117–8).

By the later sixteenth century, another consideration began to occupy the regimen writers—drinking water chilled by snow or ice. As Sandra Cavallo and Tessa Storey have shown, cold drinking became fashionable (and a concern for physicians) in Italy in the late Renaissance (Cavallo and Storey 2013, 218–28).[40] This fad also appears to have traveled to London. Writing in the 1580s, Cogan recommended his readers drink 'colde water' but omitted to state whether this cold drink could be at a wintery room temperature or whether it should be artificially chilled (Cogan 1584, 208). Writing 50 years later, James Hart left no such ambiguity. He argued that 'cold water doth better further the concoction of the stomacke than warme' and provided six ways of cooling water. The first, adapted from Galen, was cooling by air. According to Galen and Hart, the Egyptians in Alexandria 'well warmed' their water and cooled it in close earthen vessels exposed first to night air and then in shady places surrounded by cooling herbs during the day. Sailors apparently also exposed their water to night air and then surrounded their bottles with cloths to preserve the 'cold quality.' The boiling of water also played a crucial role in these instances as the 'water once boiled receiveth sooner and easilier the impression of the cold aire; as witnesseth the Prince of Philosophers' (Hart 1633, 118).[41] Other ways of cooling water included the injecting of saltpeter or pouring the water into an earthen pot which was then stored in a deep well, in snow, in a deep cellar or in a running stream (Hart 1633, 118–9). Crucial to our discussions here is that even in debates on cold drinking, the boiling of water was of central importance and took on a rather unexpected function.

When read alongside the debates voiced in contemporary regimen guides, Conway and Harley's lengthy discussion over the 'boiling water step' and its merits in improving water quality seems fairly commonplace. Conversations about water quality and directions to use boiled water for drinking, cooking and brewing preoccupied both early modern physicians and laymen alike. In voicing their opinions in their private letters, Conway and Harley were putting their own spin on a longstanding debate held by regimen writers across Europe. Crucially for our understanding of the 'structures of practical knowledge,' Conway and Harley's foray into water quality formed a 'structure' within which the ale recipe needs to be framed.

[39]The 'cooking' of water also concerned Italian regimen writers: (Cavallo and Storey 2013, 220–1).

[40]On a history of cold drinks, see (De Planhol 1995).

[41]Here in the margin Hart helpfully cites 'Arist. [i?] Meteor. cap 12.'

5 Print, Brewing and the 'Boiling Water Step' in the 1690s

For most of the seventeenth century, while recipes for beer and ale circulated widely, readers who wanted to read about the actual brewing process had to turn to Gervase Markham's *The English Houswife*. In the mid-1690s, though, a number of treatises on the 'art of brewing' became available to the English reading public. These included William Yworth's (died 1715) *Cerevisiarii Comes: or the New and True Art of Brewing* (1692), Thomas Tryon's (1634–1703) *A New Art of Brewing* (London 1690) and Jeffrey Boys' *Directions for Brewing Malt Liquors* (1700).[42] The 'boiling water step' or a variant of it, features in all three accounts and Conway and Harley's ideas about water resurface, now retooled, within these new 'structures' of knowledge.

Our first brewing tact from the 1690s was written by William Yworth, distiller and chemical physician who worked for a spell with Isaac Newton (1643–1727) on his alchemical activities. *Cerevisiarii Comes* presents a method of brewing heavily influenced by iatrochemical and, in particular, Helmontian ideas.[43] In Yworth's scheme of brewing, the boiling, or more correctly in Yworth's terminology the 'decoction,' of water took on a central role. Like the regimen writers noted above and like Conway, Yworth argues that water 'take into their Bodies more or less of the Nature of those things which they passed through' resulting in a 'diversity of waters.' (Yworth 1692, 22 and 24). Different waters, thus, each required specific treatment before use in brewing processes and each would have resulted in different tasting beers. Some, Yworth wrote, required little or no boiling and others required a 'considerable time of decoction' (Yworth 1692, 25). To remedy this problem, Yworth encouraged 'English Gentry, Brewers and other House-keepers, that brew their own beer' to 'apply themselves to Experience [and] … by many Tryals, prove the nature of their Water' (Yworth 1692, 26–7).

A 'boiling water step' of sorts, featured prominently in Yworth's brewing method. He advised: 'First, make your Liquor or Water to boyle very suddenly, and when it boyles with the greatest Violence, immediately damp or pull out your Fire … or immediately remove your Liquor into some proper vessel … and cooll till the height of the Steam or Vapour is so gone as you can see your Face in it….' (Yworth 1692, 52). This step, Yworth argued, would cause the water 'violently to evaporate in such as way, that all the parts may be radically opened … and this wild Gass shall only evaporate, and the Water become free and quiet in tis decoction, *and is more powerful for the extracting Tinctures, and also for the making of essential oyles'* (Yworth 1692, 16). Decoction here is the 'separation of the Wild and Unruly *Gass* which is the grand Enemy and fatal Destroyer of the Life of Man' (Yworth 1692, 18). A version of this step, argued Yworth, already existed as part of the common brewing method in the form of the long boiling of the wort. Thus, gentlewomen like Johanna St. John who ordered their beer to be boiled for 4–5 h were also (perhaps inadvertently) causing the removal of these gases. Yworth's method (a quick violent boil) allowed the drinkers to benefit from the step without 'considerable diminution of the Liquor' (Yworth 1692, 20). While he did not recommend boiling the liquor for 3 h, Yworth's arguments that through the action of boiling or decoction, water 'concentrates into Parts more agreeable' seem to echo Tracey and Harley's initial suggestion of the 'boiling water step' where the act of boiling served to improve the quality of water and so the ale. The similarities here may be far more than mere coincidence.

Yworth's tract was heavily influenced by contemporary alchemical and iatrochemical ideas and contained numerous references to the writings of Jan Baptista van Helmont (1580–1644), George Starkey (1628–1665) and Michael Sendigovius (1566–1636). Harley's chosen explanation of the 'boiling water step' may be due to the company he kept during his London sojourn. Throughout the

[42] *Directions for Brewing Malt Liquors* is marketed as an anonymous tract. This identification is put forward by James Sumner (Sumner 2013, 19 and footnote 33).

[43] For Yworth's connection to Newton, see (Figala and Petzold 1993).

series of letters, Conway tantalizingly referred to Harley consulting 'physicians' in London. One of the biggest mysteries of the episode remains the identities (and thus intellectual blueprints) of the physicians who advised Harley on the 'boiling water step.' Significantly, as a keen letter writer and networker, Conway was in touch with physicians such as Theodore de Mayerne (1573–1655) and Walter Charleton (1619–1707) and figures such as Sir Kenelm Digby. He was, after all, also staying on the estate of the late 'wizard earl,' Henry Percy, Earl of Northumberland. It is, thus, not inconceivable that the elusive 'physicians' with whom Harley was conversing about the 'boiling water step' might well have been Charleton and other members of the '1645 group' of physicians who had deep interests in natural philosophy and new iatrochemical ideas (Cook 1986, chapter 3).[44] According to Charleton's accounts of the period, brewing was one of the activities investigated by the group.[45] Harley's explanation of what happens during the 'boiling water step' is frustratingly vague. However, his position that water consists of different parts and that boiling separates these parts might echo early and mid-seventeenth-century English discussions of atomism and corpuscular philosophy. As Antonio Clericuzio has shown, the interpretation of chemical principles in terms of corpuscles was common in England in this period and the iatrochemists' notion of spirit was influential on English medical writers in the 1640s and 1650s. Kenelm Digby and Walter Charleton, both of whom corresponded with Conway, took part in these early conversations on atomism (Clericuzio 2000, chapter 3). Walter Charleton was also responsible for the first translations of Jan Baptista van Helmont's works into English and so there is a possibility that Harley's ideas might have stemmed from informal conversations on atomism, iatrochemistry and Helmontianism (Booth 2005, 11–8). This might go some way to explaining the connections between a recipe for ale passed around amongst friends in the 1650s and a commercially printed tract from the 1690s.

Unlike Yworth, our second author, Thomas Tryon argued vehemently against the 'boiling water step.' In his 1683 tract *The Way to Health Long Life and Happiness*, Tryon wrote extensively on the virtues of water. Like Thomas Cogan and James Hart, Thomas Tryon was an advocate of water drinking. For him, water was the 'alone thing by which all external things are purified, purged and cleansed' and it was 'more strong and sublime than most imagine, for it contains a most ravishing and excellent *Spirituous Balsamick Vertue*.' This 'innate' virtue enabled it to 'digest and purifie all sorts of food' (Tryon 1683, 143). Tryon further contended that water was the root of all nourishment and reminded readers that water was used 'in Religious Mysteries as a sign of Cleansing and Purifying the Soul from Sins and for curing Diseases' (Tryon 1683, 145).

Like the earlier regimen writers, Tryon also ranked the different kinds of water. Rainwater was again recommended as it was considered the lightest of all waters, cooling and heating up quickly and it did not 'oppress' any parts of the body. It possessed these virtues partly because it was 'drawn up into the Clouds by the power and virtue of the sweet influences of the sun beams' and 'carried to and fro with the Clouds by Gods divine power … therby [giving it] the greater benefit of Motion' (Tryon 1683, 145–6). Rainwater was Tryon's choice for cooking, brewing and washing. River or running waters came next in Tryon's mind and he saw them as the 'richest of Waters, draining and sucking into themselves the far *Saline Vertue* of the Earth' (Tryon 1683, 148). Like others, he claimed that standing waters fall at the bottom of the rankings. Not because, as Bullein would have it, they contain filth but because 'they have not the benefit of Motion, nor the like advantage of passing through various soils' (Tryon 1683, 152).

[44] See also (Frank 1979, 80–4).

[45] Charleton provides a description of the kinds of activities discussed by the group in his *The Immortality of the Human Soul, Demonstrated by the Light of Nature* (London, 1657). Charles Webster provides an in-depth discussion of the treatise in (Webster 1967).

Tryon's main intervention in the area of brewing was his adamant claim that one should not boil at all. These views were discussed at length in his *A New Art of Brewing*. Tryon, it seems, found the long boiling of the wort, as requested by Johanna St. John, a most troubling practice (Tryon 1690, 15). The act of boiling 'evaporates or suffocates the pure Spirits, which are the preservers of the sweet friendly Balsamick Body.' Ale or beer which has undergone a long boil proved to be 'very injurious to several constitutions' as it is predominantly 'hot, fulsom, strong and bitter.' This kind of drink was particularly harmful to those whose were prone to suffer from kidney stones (Tryon 1690, 15–6). Additionally, Tryon also advised against the boiling of the liquor or water. For, the boiling of water would 'irritate and evaporate the subtle fine penetrating Spirits' rendering the water or liquor 'more harsh, hard and fixed' and not as able to 'draw forth or extract the sweet virtues of the Mault' (Tryon 1690, 23–4). To test out his theory, Tryon invited his readers to conduct a trial by leaving some boiled and unboiled river water to stand in two separate vessels. After a while, Tryon argued, 'you will find the boyled Water to stink, and never be sweet again' (Tryon 1690, 24).

Tryon's fear and loathing of heat and boiling might be considered within his rather eclectic intellectual persona. Tryon was an autodidact and one-time shepherd. He became a proto-vegetarian in his twenties, traveled and lived in Barbados and became a prolific and popular author at the age of forty-eight. Advocating a diet of vegetables, grains and pulses, Tryon wrote numerous tracts to advise his readers on the 'way to health.' Aside from restricting themselves to a 'clean' diet of vegetables, Tryon also recommended an ascetic, hygienic 'cold regimen' which avoided 'hot' luxuries such as tobacco, and alcohol. Tryon's way of life was very much part of a movement of 'cold hygiene' which was gaining popularity at the time. This included the recommendations of cold beds, cold bathing and pure air (Smith 2007, 212–23). Seen within this context, his advice to brew ale by heating rather than boiling the liquor and wort is merely part of a larger schema for a particular kind of healthy living.

While there is little in the correspondence to indicate that Harley or Conway were followers of the 'cold hygiene' movement, Harley's final word on the ale recipe suggests that by the end of the discussion, he began to see things along the same lines as Tryon. After all, according to his uncle, Harley intended to try brewing Ale with 'water which is but little boyled.' If Tryon saw the boiling of water as to 'evaporate the subtle fine penetrating spirits,' for Harley, boiling separated the 'thinnest and the finest parte' of the water from the 'most grosse, unholsome and earthy parte.'[46] With scant surviving evidence outlining Harley's points of view, it is difficult to ascertain how closely he might have agreed with Tryon. For now, it suffices to note that in both cases, what was left behind after prolonged boiling was thought to be unhealthful to the human body.

The final tract of relevance, *Directions for Brewing Malt Liquors*, was penned by a 'Country Gentleman' Jeffrey Boys (Sumner 2013, 18–9). *Directions for Brewing* not only describes the brewing process but also offers advice on the selection of water, malt and hops. Particularly of interest here, of course, is Boys' descriptions of the best kind of water for brewing. Like the earlier regimen writers, Boys' account covers water from a range of sources—from 'Pond-Water' to 'Standing Waters' to 'Rain-Water' to 'Thames-Water.' From the start of his account, it is clear that Boys was working from a slightly different theoretical framework from our earlier regimen writers. Here, the author seemed to judge the suitability of a water source for brewing by their 'fat' content and their ability to lather suds with soap. For Boys, standing water in 'fat Grounds' made a stronger drink than water drawn from wells and rain-water which 'lathers the best of any' was good for brewing ale to be drunk immediately but not kept. Water from the Thames, 'taken up about *Greenwich* at Low Water, where it is free from all Brackishness of the *Sea*, and has in it all the Fat and Sullage from this great City of *London*, makes a very strong drink' (Boys 1700, 3). Unlike the regimen writers, Boys clearly did not take issue with the filth of London. Rather, the only disadvantage he perceived in beer brewed with

[46]Letter from Edward Conway to Edward Harley, 25 November 1651, British Library, Additional Ms 70006, fol. 247r.

Thames water was that it would not keep as 'it [was] apt on any considerable and sudden change of Weather, to ferment and grow foul' (Boys 1700, 4). That said, he ends his section by advising readers that the 'best Liquor to Brew with, is that which is taken from a small clear Rivulet or Brooke, undisturb'd by Navigation or Fording' and the best water in England is at Castleton in Derbyshire which 'Owzes from a great Rock, covered over with a shallow Earth and short Gras a top' (Boys 1700, 4). And the 'boiling water step'? Boys' opinion was that that while it did no harm, 'tis a needless expence of Fuel and Time, first to make it too hot, and after to stay till 'tis cooler again.' Rather, Boys recommended a gentle simmer of the liquor. (Boys 1700, 11).[47] Boys' dismissal of boiling the liquor echoes the views of the Petworth brewer who, we might recall, said that 'the boyling of water three howers cannot adde any thing to the water' and could not be persuaded to make Tracey's ale recipe.[48]

From Boys to Tryon to Yworth, our three late seventeenth-century authors demonstrate that the boiling of water in ale brewing remained a contested topic of conversation decades after Conway and Harley's weekly correspondence. Yet while the ale recipe or 'workflow' at the center of these discussions remained largely stable, by the 1690s it was framed by a myriad of different contemporary discussions and knowledge frameworks. As with the 1650s the conversation touched upon both skill and technique in brewing and ideas about health and the human body. However, by the 1690s, the frames or 'structures' extended towards new ideas on cold hygiene, iatrochemistry and more. It is perhaps a testament to the flexibility of the recipe that it is able to participate in these diverse conversations.

6 Conclusions: Practical and Theoretical Knowledge and the Recipe

Our story began on a late summer's night, with two men raising tankards of homebrewed ale to toast the health of another man. At the end of their evening of merriment, the kind host shared his prized ale recipe with his guest who in turn sent it to his uncle with a curious mind. Our trek following the various discussions and debates on how to brew that pint of ale has taken us through manuals of early modern housewifery, letters between mistress and steward, contemporary regimen guides and manuals of brewing. On the way, we had brief run-ins with early adopters of atomism in England, Newton's alchemical assistant, a proto-vegetarian and a physician who (sensibly) told us not to drink corpse and dung-infested waters. Yet, in the midst of this colorful and rich journey, the method of brewing ale remains largely constant and the issue of whether to boil or how long to boil the liquor or wort continues to be the topic of debate. *Why* one should boil or not, however, shifts from author to author, reader to reader and brewer to brewer. The very 'thinness' of a recipe, it turns out, enables it to map onto different systems of explanation. In this case, Sir John Tracey's ale recipe and the subsequent discussions over it were framed by the social structure of the early modern household and changing ideas of health and the body in late seventeenth-century England. The connections between the recipe and knowledge frameworks were by no means constructed in a linear fashion but, like much else in the early modern period, were rather messy nodes of contact. In all these cases the flexibility or ambiguity of the recipe enabled it to endure even as it journeyed across knowledge frameworks. Within each of these frameworks, practical knowledge is structured or restructured in a particular way to adapt to its environs.

[47] Fascinatingly, Boys also suggests that brewers strew a handful of bran onto the surface of the liquor. However in this case not as a way to judge when the water reaches its boiling point rather because the bran will 'make it heat quickly, for simple Water alone will be long ere it Boyl.'

[48] Letter from Edward Conway to Edward Harley, 14 October 1651, British Library, Additional Ms 70006, fol. 239r.

References

Manuscript Sources

London, British Library
Additional MS 70006

London, National Archives
State Papers, Domestic, SP18/16

London, Wellcome Library
Western Manuscript 774, Recipe book associated with the Townsend family.
Western Manuscript 3009, Recipe book associated with Elizabeth Jacob and others.
Western Manuscript 3712, Recipe book associated with Elizabeth Okeover and others.
Western Manuscript 4054, Recipe book, 17th and 18th century.
Western Manuscript 4338, Recipe book associated with Johanna St. John.
Western Manuscript 7113, Recipe book associated with Ann Fanshawe.

Printed Sources

Bennett, Judith M. 1996. *Ale, beer, and brewsters in England. Women's work in a changing world 1300–1600*. Oxford: Oxford University Press.
Boord, Andrew. 1542. *Hereafter foloweth a compendyous regyment or a dyetary of helth, made in Mou[n]tpyllier, compyled by Andrew Boorde of Physiycke doctor.* London: Robert Wyer.
Booth, Emily. 2005. *'A subtle and mysterious machine.' The medical world of Walter Charleton (1619–1707)*. Dordrecht: Springer.
Boys, Jeffrey. 1700. *Directions for brewing malt liquors*. London: J. Nutt.
Bullein, William. 1558. *A newe booke entituled the governement of healthe*. London: Ihon Day.
Carne, Brian. 1994–1996. *Some St. John family papers*. Reproduced, with corrections and additions, from the *Reports* of the Friends of Lydiard Tregoz.
Cavallo, Sandra, and Tessa Storey. 2013. *Healthy living in late Renaissance Italy*. Oxford: Oxford University Press.
Charleton, Walter. 1657. *The immortality of the human soul, demonstrated by the light of nature*. London: Henry Herringman.
Clericuzio, Antonio. 2000. *Elements, principles and corpuscles. A study of atomism and chemistry in the seventeenth century*. Dordrecht/Boston/London: Kluwer Academic Publishers.
Cliffe, J.T. 1999. *The world of the country house in seventeenth-century England*. New Haven/London: Yale University Press.
Cockayne, Emily. 2007. *Hubbub: Filth, noise & stench in England*. New Haven/London: Yale University Press.
Cogan, Thomas. 1584. *The haven of health*. London: William Norton.
Cook, Harold. 1986. *The decline of the old medical regime in Stuart London*. Ithaca: Cornell University Press.
Cooper, Alix. 2006. Homes and households. In *The Cambridge history of science volume 3: Early modern science*, ed. Katharine Park and Lorraine Daston, 224–237. Cambridge: Cambridge University Press.
De Planhol, Xavier. 1995. *L'Eau de neige: Le Tièd et le frais: Histoire et géographie des boissons fraîches*. Paris: Fayard.
DiMeo, Michelle, and Sara Pennell. 2013. *Reading and writing recipe books 1550–1800*. Manchester: Manchester University Press.
Eamon, William. 1994. *Science and the secrets of nature: Books of secrets in medieval and early modern culture*. Princeton: Princeton University Press.
Elyot, Thomas. 1534. *The castel of helth*. London: Thomas Berthelet.
Ferguson, John K. 1898. *Bibliographical notes on histories of inventions and books of secrets*. London: Holland Press.
Figala, Karin, and Ulrich Petzod. 1993. Alchemy in the Newtonian circle: Personal acquaintances and the problem of the late phase of Isaac Newton's Alchemy. In *Renaissance and revolution. Humanists, scholars craftsmen and natural philosophers in early modern Europe*, ed. J.V. Feld and Frank A.J.L. James, 173–191. Cambridge: Cambridge University Press.
Fissell, Mary. 2007. The marketplace of print. In *Medicine and the market in England and its Colonies c. 1450– c. 1850*, ed. Mark Jenner and Patrick Wallis, 108–132. Basingstoke/New York: Palgrave Macmillan.

Fissell, Mary. 2011. Popular medical writing. In *The Oxford history of popular print culture. Cheap print in Britain and Ireland to 1660*, ed. Joad Raymond, 417–430. Oxford: Oxford University Press.

Frank Jr., Robert G. 1979. The physician as Virtuoso in seventeenth-century England. In *English scientific Virtuosi in the sixteenth and seventeenth centuries*, ed. Shapiro Barbara and Robert G. Frank Jr., 59–114. Los Angeles: University of California Press.

Greengrass, M., M. Leslie, and M. Hannon. 2013. *The Hartlib papers*. HRI Online Publications. http://www.hrionline.ac.uk/hartlib. Accessed 13 June 2015.

Hailwood, Mark, and Deborah Toner, eds. 2012. Developments in the brewing, retail and consumption of alcohol in early modern England. *Brewery History* 150: 2–5.

Harkness, Deborah E. 1997. Managing an experimental household: The dees of Mortlake and the practice of natural philosophy. *Isis* 88: 247–262.

Hart, James. 1633. *KAINIKH, or the diet of the diseased*. London: John Beale.

Hindle, Steve. 2011. Below stairs at Arbury Hall: Sir Richard Newdigate and his household staff, c. 1670–1710. *Historical Research* 85 (227): 71–88.

Hunter, Lynette, and Sarah Hutton (eds.). 1997. *Women, science and medicine, 1500–1700: Mothers and sisters of the royal society*. Thrupp: Sutton Publishing.

Klein, Ursula. 2012a. Introduction: Artisanal-scientific experts in eighteenth-century France and Germany. *Annals of Science* 69 (3): 303–306.

Klein, Ursula. 2012b. Savant officials in Prussian mining administration. *Annals of Science* 69 (3): 349–374.

Klein, Ursula, and Emma Spary (eds.). 2010. *Materials and expertise in early modern Europe: Between market and laboratory*. Chicago: University of Chicago Press.

Leong, Elaine. 2008. Making medicines in the early modern household. *Bulletin of the History of Medicine* 82 (1): 145–168.

Leong, Elaine, and Alisha Ranking (eds.). 2011. *Secrets and knowledge in medicine and science, 1500–1800*. Aldershot: Ashgate Publishing.

Mandelbrote, Scott, and William Yworth. died 1715. In *Oxford dictionary of national biography*, online ed. edited by Lawrence Goldman. Oxford: OUP. http://www.oxforddnb.com/view/article/40388. Accessed 17 June 2016.

Markham, Gervase. 1631. *The English house-wife*. London: John Harrison.

Nicoud, Marilyn. 2007. *Les régimes de santé au Moyen Âge: naissance et diffusion d'une écriture médicale en Italie et en France (XIIIe–XVe siècle)*. Rome: École française de Rome.

Pennell, Sara. 2004. Perfecting practice? Women, manuscript recipes and knowledge in early modern England. In *Early modern women's manuscript writing: Selected papers from the Trinity/Trent colloquium*, ed. Victoria Burke and Jonathan Gibson, 237–258. Aldershot: Ashgate.

Pomata, Gianna. 2013. The recipe and the case. Epistemic genres and the dynamics of cognitive practices. In *Wissenschaftsgeschichte und Geschichte des Wissens im Dialog – Connecting science and knowledge*, ed. Kaspar von Greyerz, Silvia Flubacher, and Philipp Senn, 131–154. Göttingen: V&R unipress.

Rankin, Alisha. 2007. Becoming an expert practitioner: Court experimentalism and the medical skills of Anna of Saxony (1532–1585). *Isis* 98: 23–53.

Richardson, R.C. 2010. *Household servants in early modern England*. Manchester: Manchester University Press.

Sambrook, Pamela. 1996. *Country house brewing in England 1500–1900*. London/Rio Grande: Bloomsbury Academic.

Sibum, Heinz Otto. 1995. Reworking the mechanical value of heat: Instruments of precision and Gestures of accuracy in early Victorian England. *Studies in the History and Philosophy of Science* 26 (1): 73–106.

Slack, Paul. 1979. Mirrors of health and treasures of poor men: The uses of the vernacular medical literature of Tudor England. In *Health, medicine and mortality in the sixteenth century*, ed. Charles Webster, 237–273. Cambridge: Cambridge University Press.

Smith, Virginia. 2007. *Clean: A history of personal hygiene and purity*. Oxford: Oxford University Press.

Smith, Pamela. 2012. In the workshop of history: Making, writing, and meaning. *West 86th: A Journal of Decorative Arts, Design History, and Material Culture* 19: 4–31.

Smith, Virginia, and Thomas Tryon. 1634–1703. In *Oxford dictionary of national biography*, online ed. Lawrence Goldman. Oxford: OUP. http://www.oxforddnb.com/view/article/27783. Accessed 28 Nov 2014.

Starza Smith, Daniel. 2012. "La conquest du sang real:" Edward, second Viscount Conway's quest for books. In *From compositor to collector: Essays on the book trade history*, ed. John Hinks and Matthew Day, 191–208. London: British Library.

Starza Smith, Daniel. 2014. *John Donne and the Conway Papers. Patronage and manuscript circulation in the early seventeenth century*. Oxford: Oxford University Press.

Sumner, James. 2007. Michael Combrune, Peter Shaw and commercial chemistry: The Boerhaavian chemical origins of brewing thermometry. *Ambix* 54 (1): 5–29.

Sumner, James. 2013. *Brewing science, technology and print, 1700–1880*. London: Pickering and Chatto.

Thirsk, Joan. 2007. *Food in early modern Britain: Phases, fads and fashions 1500–1760*. London/New York: Continuum Books.
Tryon, Thomas. 1683. *The way to health, long life and happiness*. London: Andrew Sowle.
Tryon, Thomas. 1690. *A new art of brewing beer, ale and other liquors*. London: Tho. Salusbury.
Tryon, Thomas. 1696. *A discourse of waters shewing the particular natures, various uses and wonderful operations both in food and Physick*. London: T. Sowle.
Wear, Andrew. 2000. *Knowledge and practice in English medicine, 1550–1680*. Cambridge: Cambridge University Press.
Webster, Charles. 1967. The college of physicians, "Solomon's House" in commonwealth England. *Bulletin of the History of Medicine* 41 (5): 393–412.
Withington, Phil, and Angela McShane. 2014. *Cultures of intoxication, past and present*, Special Supplement 9. Oxford: Oxford University Press.
Yworth, William. 1692. *Cerevisiarii comes: Or the new and true art of brewing*. London: J. Taylor.

"Which Cannot Be Sufficiently Described by My Pen." The Codification of Knowledge in Theater Engineering, 1480–1680

4

Viktoria Tkaczyk

Abstract

This paper is dedicated to theater engineering and related modes of knowledge production and transfer in early modern Europe. It focuses on two case studies: (1) Leonardo da Vinci (1452–1519) and the rise of theater engineering as a lucrative field of activity for "artist-engineers" in fifteenth- and sixteenth-century Italy, and (2) Giulio Parigi's (1571–1635) "school of theater engineering," the first of its kind, in early seventeenth-century Florence and the subsequent spread of engineering knowledge among European court artists. The paper traces how theater engineers started to transfer and codify their knowledge beyond the realm of publication, how specific schools of "theater engineering" emerged, and how the first manuals and series of machine drawings were published. In parallel, it examines the changing function of theater machines in the period under investigation, from representing religious motives to representing natural phenomena. With this shift, theater engineering gained new epistemic values. It was linked to the emerging experimental philosophy not only by a shared mechanical knowledge, but also by the formation of new iconographic programs of nature.

In 1637 and 1638, the Italian architect Nicola Sabbattini (1574–1654) published the first comprehensive manual of early modern theater engineering, entitled *Pratica di fabricar scene, e machine ne' teatri*.[1] The first volume explains how to design the perspective stage in a courtly theater and arrange the auditorium; the second how to construct machines for special effects such as stage flights, the appearance of a ship in a billowing sea, the transformation of a rock fragment into a man, or an approaching thunder-

[1] For the widespread use of "pratica" as a component of book titles in various fields of knowledge dissemination in early modern Italy, see (Chap. 13).

V. Tkaczyk (✉)
Max Planck Institute for the History of Science, Berlin, Germany

Humboldt Universität zu Berlin, Berlin, Germany
e-mail: tkaczyk@mpiwg-berlin.mpg.de

M. Valleriani (ed.), *The Structures of Practical Knowledge*, DOI 10.1007/978-3-319-45671-3_4

storm with lightning, wind, and rain (Sabbattini 1638). The book stands as a remarkable testament to the translation of practical knowledge on theater engineering into more abstract and systematically arranged instructions. Sabbattini thinks of his instructions as a "theory"[2] but, as his title indicates, always ties them back to his practical experience as court architect to the dukes of Urbino.

Taking Sabbattini's *Pratica* as a turning point, this paper explores the production, distribution, and reevaluation of knowledge on theater engineering in the early modern period. By examining notions such as "imitation," "invention," and "imagination" in the writings and drawings of theater engineers, the paper analyzes the changing value attributed to their profession. It asks how the engineers' mechanical knowledge took shape within their local socioeconomic contexts, focusing on two case studies: Leonardo da Vinci's (1452–1519) technological inventiveness and the Italian festival circuit around 1500; and the spread of Giulio Parigi's (1571–1635) first "school of stage design" across seventeenth-century European courts.

These two cases shed light on the earliest systematic approaches to theater engineering both as distinct fields of knowledge production and distribution and as components of broader fields of artistic, mechanical, and natural knowledge. In the European theater and festival culture of the period under investigation, "artist-engineers" encountered a niche of experimentation that was not necessarily constrained by the same principles and rules as the other arts. Accordingly, I place special emphasis on the classification of theater engineering within and across the early modern canon of the arts. In the sixteenth and early seventeenth century, theater engineering was increasingly regaining its association with the mechanical arts, but at the same time it was also linked to emerging natural philosophy by a shared mechanical knowledge and by its "poetic" capacity to produce new modes of representing nature.[3]

1 Leonardo da Vinci and the Italian Festival Circuit Around 1500

Around 1500, the northern Italian city-states were increasingly experiencing a pressure for civic self-representation. This culture of self-representation found a suitable medium in the festival, the mass impact of which recommended it as a large, competitive, and lucrative sphere of activity for local artisans, court artists, and scholars alike. In the festivals held from the late quattrocento to the early seicento, in particular, a variety of political and clerical interests and social energies consolidated to form new artistic "iconographic programs" that did not follow any previously defined rules (Warburg 1932, 74). This was the context for the design of imaginatively constructed theatrical machines, first for religious processions and triumphal parades, later for *sacre rappresentazioni* and the court theaters.

Leonardo da Vinci was fully involved in this festival culture, and the following analysis will focus on the knowledge encapsulated in his designs for theater machines. First, however, a brief look at the philosophical writings on theater that circulated in Leonardo's time will reveal a surprising gap between the practice and philosophy of theater.

1.1 Theater Engineering: Philosophical Knowledge Versus Practical Knowledge

In late fifteenth-century Europe, the art of theater came to be classed more and more regularly among the mechanical arts. We find entries on "theatrics" in encyclopedias such as the *Speculum vitae humanae* by the Spanish bishop Rodrigo Sánchez de Arévalo (1471) and *Margarita philosophica*

[2]"La Teorica non è difficile, ma è più facile la Pratica" (Sabbattini 1638, 165). The second edition of the manual is available in a simplified English translation (Sabbatini 1958). Here and throughout, all translations are my own unless otherwise attributed.

[3]This paper draws on material discussed in (Tkaczyk 2011).

(1489–1496, published in 1503) by the German humanist Gregor Reisch (1467–1525). Both writers were inspired by the treatment of theater in one of the earliest canons of the mechanical arts, drawn up by the Paris theologian Hugh of St. Victor (1096–1141) in his *Didascalicon* (ca. 1120). It is therefore worth taking a closer look first at Hugh's writings on theater, then at earlier classifications of theater within and across the arts.

Hugh describes *philosophia mechanica* as an autonomous branch of philosophy—alongside *theoretica* (mathematics, physics, and theology), *practica* (politics, ethics, and economics), and *logica* (grammar and the theory of argument) (Hugh of St. Victor 1991, 2: 62).[4] He enumerates the following *septem scientiae mechanicae*: fabric making, armament, commerce (navigation), agriculture, hunting, medicine, and theatrics (Hugh of St. Victor 1991, 2: 74–5). According to Hugh, all these crafts produce *artificialia* and so serve "naked man" by restoring the prelapsarian integrity of his nature (Hugh of St. Victor 1991, 2: 52). As he sees it, mechanical philosophy is not concerned with craftsmanship as such, but with the knowledge of how and according to what rules *artificialia* are produced, and how they imitate nature (Hugh of St. Victor 1991, 2: 51, 61). By virtue of their imitative character, Hugh argues, the crafts provide a key to understanding nature, though a far weaker key than theory (Hugh of St. Victor 1991, 2: 56). In analogy to the *quadrivium* and *trivium* of the *artes liberales*, three of the mechanical arts (fabric making, armament, commerce) pertain to the external covering of nature, whereas the other four (agriculture, hunting, medicine, and theatrics) bear on the internal concepts or rules of nature by means of which she "feeds and nourishes herself" (Hugh of St. Victor 1991, 2: 74–5).

It is important to note here that Hugh's term "theatrics" refers to the "science of entertainment," which encompasses historical knowledge on the recital of epics, the acting out of dramatic roles, singing, instrumental music, choral processions, dances, masks or puppets, and, finally, the proper use of chariots. Hugh was aware that his description of *theatrica* referred to the classical theater of the past, "to which the people once used to gather for the performance" (Hugh of St. Victor 1991, 2: 79). Indeed, it is most likely that Hugh's concept of theater derives from the entries under "De spectaculis" in Latin glossaries or *Hermeneumata* (Sternagel 1966, 76–7). In the *Didascalicon*, however, knowledge on how to design chariots or other devices for theater spectacles was classified as part of *armatura*, or weaponry (Hugh of St. Victor 1991, 2: 76, 79).

In counting the fabrication of theater machines under mechanics, Hugh was building on an even longer, though interrupted, tradition. The term "machine" derives from the ancient Greek word *μηχανή*, which originally signified a theater machine or siege engine or, more generally, a rhetorically, dramatically, or technically beguiling tool (Schmidt-Biggemann 1980). Aeschylus's and Sophocles's tragedies still made sparing use of the *μηχανή*, which became standard beginning with Euripides.[5] The

[4]For Hugh of St. Victor's writings on the mechanical arts in the context of late medieval philosophy, see (Whitney 1990, 75–127; Sternagel 1966, 67–77).

[5]Archaeology has yielded only sparse clues to the visual performance praxis (*ópsis*) of ancient tragedy. Surviving poros blocks in the west *paraskenion* of the Theater of Dionysus in Athens and marble slab rails on the curved *skene* of the theater at Eretria may have formed part of a *mechané* like those used from the fourth century BCE and in the Hellenistic period. However, the use of the *mechané* is not attested with certainty until Aristophanes's comedies (*The Clouds*, vv. 218–38). Synonymously with *mechané*, the terms *eorema* and *aiorai* were also used, or parodically (by Aristophanes) the word *géranos*. The term *ekkyklema* originates with Aristophanes (*The Acharnians*, vv. 408–9). It refers to a wheeled platform on which "living pictures" were rolled out onto the *skene* from behind the stage (probably for the first time in Aeschylus's *Oresteia*) and for the rotating platform on the crane. The first indications of a sky chariot are found in Aeschylus's *Prometheus Bound* (Okeanos's gryphon chariot, vv. 128, 284). As becomes clear from the cloud chariot journey of Artemis in Euripides's *Hippolytus* (v. 1282) and Pegasus's flight on a dung beetle in Aristophanes's *Peace*, a gondola was later sometimes attached to the crane. Subsequent technical reconstructions such as Julius Pollux's *Onomasticon* are instructive, if not always historically reliable (Schmidt 1963, 36–66; Melchinger 1974, 191–200; Taplin 1978, 442–7; Schneider 1989; Newiger 1996a, 98–102; 1996b; Weihe 1992, 11–6).

classical discourse on drama was also an opportunity to praise technology for its social relevance (Kahn 1970). However, in classical philosophy, true knowledge was supposed to be theoretical, related to final ends and higher purposes, while productive knowledge was regarded as dealing with physical needs, the material world, and natural forces. Mechanics was therefore thought of either as a physical but inferior component of theoretical geometry (the Platonic/Augustinian tradition) or as a structurally different but still inferior form of productive knowledge (the Aristotelian tradition).[6] In late Roman antiquity and the early Middle Ages, mechanical knowledge further sank in esteem; the term "machine" was mainly used for simple scaffolds and tools (Sternagel 1966, 11–29). As cultures of spectacle declined, the reputation of theater machines suffered,[7] but appreciation for mechanics began to recover in the high Middle Ages; when the discipline came to be associated once again with terms—such as "intricate," "magical," "artful," "ingenious," or "inventive"—that captured the utility of its products and its specific relation to nature.[8] Around the same time, changing socioeconomic structures in the growing European cities led to what has been called the "medieval industrial revolution" (Whitney 1990, 2–5): ecclesiastical and secular authorities started to patronize craft knowledge, while the invention of new devices and the efficient use of older technologies facilitated agriculture, metallurgy, textile-working, medicine, construction, and warfare (Crombie 1961, 1: 191–222). Both elite scholars and trained individual artisans dedicated themselves more seriously to technical issues, material-based and physical practices, and two lines of mechanical knowledge production were reestablished—one through apprenticeships and workshops, the other scholarly and linked to the study of various branches of philosophy.

There are doubtlessly many exceptions to this general narrative. Still, it is striking that Hugh of St. Victor's *Didascalicon*, which contributed to the increasing acceptance of the *artes mechanicae*, also mentioned both the art of theater as part of the seven productive art forms, and theater engineering as part of mechanics. These inclusions drew much criticism, for the Church Fathers' polemics against morally inferior performances (magical shows, legerdemain, musical performances by touring companies) still echoed in twelfth-century Europe, and new forms of liturgical drama were not yet called "theater" (Sternagel 1966, 99–101). Thus, Hugh's successor Godfrey of St. Victor (ca. 1125–ca. 1195), in his *Microcosmus* of 1180, redefined the mechanical arts by removing theater from the canon and complementing *armatura* with *fabricatura* (civil architecture), though without mentioning theater machines explicitly (Godefroy de Saint Victor 1951, 74).[9]

In this setting, the reappearance of theater in the fifteenth-century encyclopedias is remarkable. *Speculum vitae humanae* lists theater as part of the *artes mechanicae*, though only to attack it in a four-page diatribe (Sánchez de Arévolo 1471, 160–64). Reisch's *Margarita philosophica* emphasizes the social significance of the *artes mechanicae* and offers a much more neutral account of what forms of entertainment count as "theatrica." Yet like others before him, Reisch mentions only classical games and gymnastics, circus performances, gladiator fights, military parades, and the recitation of comedies and tragedies (Reisch 1504, 610–1). Similarly, the eighth book of Italian architect Leon Battista Alberti's (1404–1472) *De re aedificatoria* (1443–1452, published in 1485) proposes the erection of a public theater and points out the theater's potential to "cultivate the mind" and

[6] On these distinctions, see (Whitney 1990, 30–2).

[7] In *De civitate Dei* (22.24), St. Augustine mentions man's ability to fabricate machines for theater spectacles, but does not hide his contempt for both technology and spectacle, which distract man from faith, truth, and eternal life. In Isidore of Seville's *Etymologiae* (book XVIII: *De bello et ludis*), chariots are described as being used in triumphal parades in the context of warfare, and theater as a site of prostitution (Whitney 1990, 52–5, 61–3).

[8] On the rise of the mechanical arts, see (Long 2001; Bacher 2000a, 35–49; Bacher 2000b, 519–55).

[9] Theater was also excluded in Robert Kilwardby's *De ortu scientiarum* of 1279 (Kilwardby 1976, 130–1). See (Bacher 2000a, 37; Whitney 1990, 103, 106–9, 123).

"encourage intellectual energy"—yet makes reference solely to classical Roman performance culture, failing even to mention the more prominent forms of courtly and public theater in his own days (Alberti 1988, 268).

Such was the context in which the artist-engineers involved in the booming Italian festival circuit produced and shared their knowledge on the making of theater machines outside the confines of philosophical literature. They transferred and codified this knowledge in apprenticeships and workshops. Beginning in Italy, in the fifteenth century the machines designed for church plays, festival processions, and courtly performances appeared in a growing body of documentation, whether drawings in the engineers' notebooks or descriptions in city chronicles and the biographies of individual artists. Leonardo da Vinci, too, took part in the Italian festival circuit, but like his peers he did not explicitly comment on how his theater machines were made. Tracing the influences on Leonardo's designs, I will rely mainly on visual evidence and comparisons with etchings and descriptions of machines designed by his colleagues. In common with most of his era's artist-engineers, Leonardo contributed to many fields of engineering at the same time, constructing not only theater machines for festivals, but also machines of war, agriculture, and construction. As a result, Leonardo's theater work can also be read as representative of the wider Italian culture of technology around 1500, or, more to the point, as part of a rapidly developing field of technological experimentation and cultural negotiation over what was mechanically feasible.

Of particular interest in this respect is the relationship between Leonardo's stage designs and his involvement in another field of technical experimentation: the design of flying machines. Leonardo left behind more than 500 sketches of such machines, the earliest dating from 1485 and the last from 1515 (before he entered the services of Francis I). These sketches are now widely known, but at the time Leonardo entrusted them only to secret notebooks. Moreover, there is no conclusive evidence that he ever implemented his designs, let alone tried to make them work.[10] The ample scholarly literature on this Renaissance artist-scientist's studies on flight consistently omits all mention of his theatrical work.[11] I argue, however, that Leonardo would not have produced his considerable body of work on flight, or at least would not have pursued the subject with such insistence, had he not also been witness and contributor to a festival culture in which spectacular flight effects played a particularly salient role.

1.2 Shaping Theater Technology: A Short History of the *Mandorla*

In northern Italy, the earliest standing theaters with elaborate stage machineries were designed and built only over the course of the sixteenth century.[12] However, the process by which triumphal parades that once took place in public streets and squares were gradually transformed into events staged behind courtly walls had already begun by 1500. To this end, the court engineers erected provisional stages in ceremonial halls and designed simple theater machineries. These stage designs served as status symbols in courtly self-representation and as flagships for engineers hoping to attract new patrons among the audience. In January 1496, for example, Leonardo designed a *mandorla* for the comedy *Danae* by Baldassarre Taccone, performed in the Milan palace of Giovan Francesco

[10] See also (Hart 1962, 307, 311–3). For a reconstruction of one of Leonardo's unsuccessful designs for a flying machine, see (Galluzzi 1997, 225).

[11] The only study that takes Leonardo's theatrical work even marginally into account is (Laurenza 2004).

[12] The first standing theater was designed by Bramante in the Belvedere courtyard (Vatican Palace 1520). Leonardo may also have planned a theater building. A sketch in Codex I, fol. 110r, shows a reconstruction of the mobile theater of Curius, as described by Pliny the Elder. See (Steinitz 1969, 5–9).

Fig. 1 Leonardo's sketch for the staging of *Danae* (1496), single sheet. From (Bambach 2003, 449)

Sanseverino (Fig. 1). His sketch for this almond-shaped construction bears striking similarity to the *mandorle* that had been in use since the early fifteenth century in the *sacre rappresentazioni* (a form of religious drama). Bonaccorso Ghiberti (1451–1516) illustrated the workings of this kind of *mandorla* in his *Zibaldone* (ca. 1500) (Fig. 2).[13]

Although at first glance these *mandorle* resemble the *deus ex machina* cranes of ancient Greek theater, they do not indicate the rebirth of an older technology.[14] Rather, they represent a new development in theater technology—one that began with the church liturgy of the Italian duecento and then flourished in the context of the *sacre rappresentazioni* and the (partially secular) triumphal parades of the quattrocento and cinquecento. In the duecento, the strictly regulated repertoire of liturgical gestures was gradually supplemented by pictorial and sculptural representations of the Passion that served as a basis of religious instruction and memory (Belting 1981). As the dramatization of the liturgy (*drammi liturgici*), the extra-liturgical elements of the sermon (*uffizi drammatici*), and encomiums emerged in liturgical practice at the beginning of the trecento, the first actors took the stage to embody the images of the Passion in tableaux referred to as *quadri animati* (Helas 1999, 13–28). And with the rise of Italian festival culture in the quattrocento, these *quadri animati* developed into the most spectacular element of religious processions and the *sacre rappresentazioni*.

In processions, actors were carried through the streets on scaffolds decorated to look like clouds (the *nuvole* or *soleri*). Giorgio Vasari attributes the invention of the *nuvole* to the Florentine engineer

[13] On the practice of copy-drawing in Leonardo's notebooks, see (Chap. 2).

[14] While ancient gods reappear in fifteenth-century drama, there is no explicit reference to ancient Greek technology in early modern theater machine books. The only direct tradition seems to have been Pollux's *Onomasticon* (1502) and previously circulating transcriptions of this work. The earliest explicit reference to Greek stage technology is in the entry "Machines de Théâtre" in Diderot and d'Alembert's *Encyclopédie* of 1765. On the rebirth of pagan gods in Renaissance drama, see (Seznec 1953; Koopmans 2006).

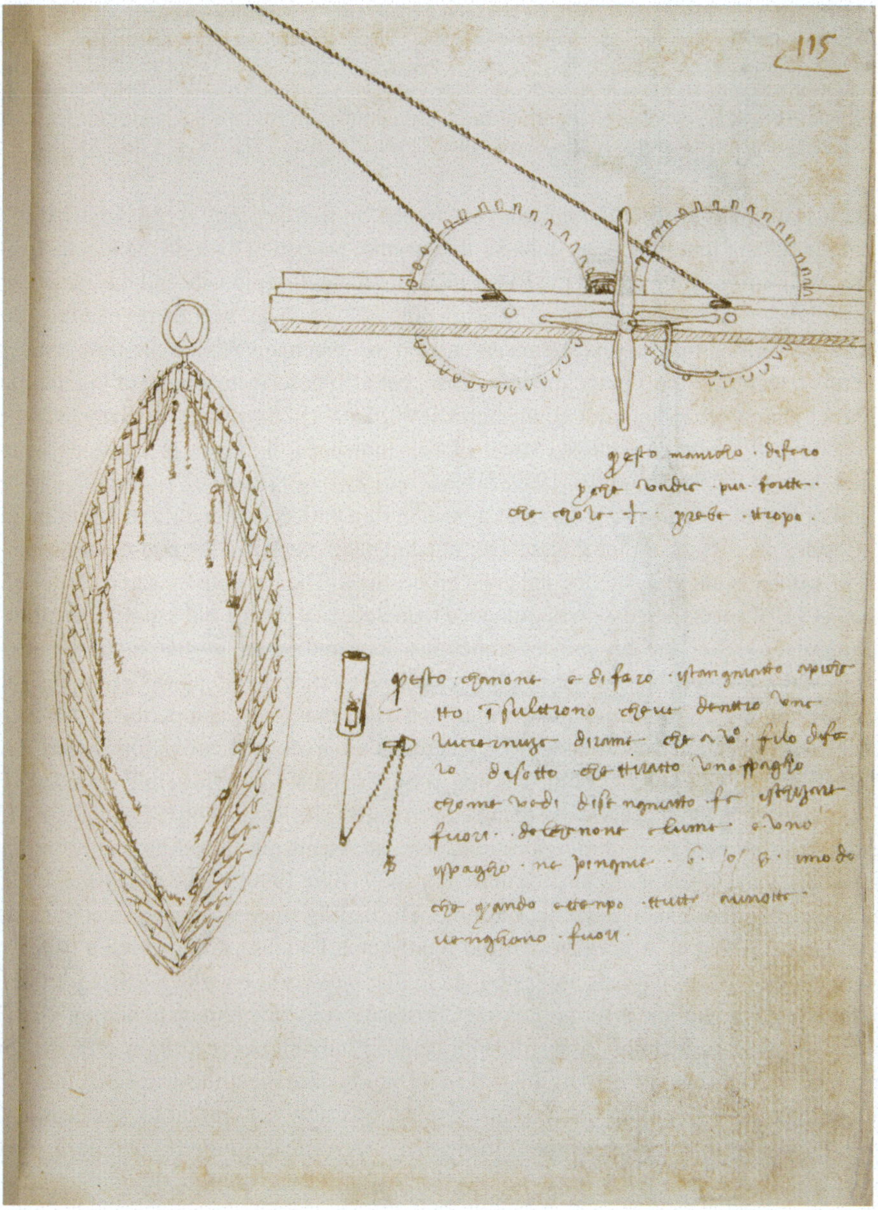

Fig. 2 Sketch of a mandorla in Bonaccorso Ghiberti's *Zibaldone*, Florence, ca. 1500. From (Ghiberti ca. 1500, fol. 115r). Courtesy of *Ministero dei beni e delle attività culturali e del turismo–Biblioteca Nazionale Centrale. Firenze.* Prohibition of further reproduction or duplication by any means

Il Cecca (Francesco d'Angelo, 1446–1488) and provides an extensive description of their construction:

> A square framework was made of planks, about two braccia in height, with four stout legs at the corners.... On this framework two panels were laid crosswise …, in which was fixed a high pole, whereon there was placed a mandorla all covered with cotton-wool, cherubim, lights, and other ornaments, and within this, on a horizontal bar of iron, there sat or stood, according as might be desired, a person representing that Saint whom the particular Company principally honoured as their peculiar patron and protector—to be exact, a Christ, or a Madonna, or a S. John, or some other—and the draperies of this figure covered the iron bar in such a manner that it could not

> be seen.... The whole of this structure, with the pole and the iron bars ... sometimes represented a lily.... Within were porters or peasants, who carried it on their shoulders, placing themselves round the wooden base that we have called the framework, in which, below the places where the weight rested on their shoulders, were fixed cushions of leather stuffed with down, or cotton-wool, or some other soft and yielding material. All the machinery, steps, and other things were covered, as has been said above, with cotton-wool, which made a beautiful effect; and all these contrivances [*macchine*] were called Clouds.[15] (Vasari 1912–1914, 3: 197–8)

It is no accident that Vasari calls the *nuvole* "machines," *macchine*: they were supposed to fool the audience into believing that they floated above the ground, propelled by a divine mechanism hidden within. Thus, the prominent images of the Passion (the archangel's epiphany and the *ascensio Christi*) were not just represented by actors. The impression of a *tableau vivant* was also evoked by the festival floats, whose apparent autokinesis was conceptualized as "mechanical"—at the time, "mechanical" was not merely a technical term in the modern sense, but also described a range of kinetic (animated, or even divine) phenomena (Schmidt-Biggemann 1980, 790–1). According to many chronicles, the mechanical effect of the *nuvole* moved some of the audiences of Easter processions to tears and strengthened them in their faith (Helas 1999, 50–8).

In his description of the *nuvole*, Vasari lists a bewildering variety of figures that might be placed on the floats. Besides processions of the Corpus Domini and the three Magi, the mid-quattrocento saw the emergence of secular triumphal parades and civic processions. The previously sacred cast of characters on the floats—Christ, angels, saints—was expanded to include allegorical and historical figures. Around the end of the century, even the city princes sometimes appeared on the *nuvole*. In these instances, the floating effect was preserved, but was gradually secularized by the new occupants of the scaffolds.

The *sacre rappresentazioni* of the quattrocento borrowed this effect and perfected the technology. The space of representation was rotated 90°: whereas the street processions organized the *nuvole* sequentially to represent the stations of the Passion, the church plays were generally reduced to a single image of the Passion and the scenery was organized vertically. The Russian bishop Abraham of Souzdal gave an exact description of two *sacre rappresentazioni*, an Ascension and a Marian mystery play, that he witnessed during the Council of Florence in March 1439 (Barr 1990).[16] The *Ascension of Christ* was performed at the church of Santa Maria del Carmine. The earthly stage was represented by a stone dais in the church nave. Above it, a wooden scaffold represented Paradise, from which a flying machine descended at the moment of Christ's ascension: a circular *mandorla* on which a group of children in angels' robes circled around the actor portraying Christ to accompany him up to heaven (Fig. 3).[17]

The evolution of stage technology in fifteenth-century Italy shows that flight effects were genuinely reinvented, rather than being rediscovered in the study of ancient theatrical practice. The *mandorla*, part of the *nuvole* in earlier festival processions, was now set upright to represent a *Christus* or *angelus ex machina*. It is no surprise, then, that the vertically moving *mandorla* bears little

[15]"Si faceva un telaio quadro di tavole, alto braccia due in circa, che in su le teste aveva quattro gagliardi piedi ...; sopra questo telaio erano in croce due tavole ..., nella quale era uno stile alto sopra cui si accomodava una mandorla, dentro la quale, che era tutta coperta di bambagia, di cherubini e di lumi et altri ornamenti, era in un ferro al traverso, posta a sedere o ritta, secondo che altri voleva, una persona che rappresentava quel Santo, il quale principalmente da quella compagnia, come proprio avvocato e protettore si onorava; o vero un Cristo, una Madonna, un S. Giovanni o altro; i panni della quale figura coprivano il ferro in modo che non si vedeva.... E tutta questa machina e lo stile et i ferri che tallora faceva un giglio.... E dentro erano facchini o villani, che la portavano sopra le spalle, i quali si mettevano intorno intorno a quella tavola, che noi abbiam chiamato telaio, nella quale erano confitti, sotto dove il peso posava sopra le spalle loro, guanciali di cuoio, pieni o di piuma o di bambagia o d'altra cosa simile, che acconsentisse e fusse morbida. E tutti gl'ingegni e le salite et altre cose erano coperte come si è detto di sopra con bambagia, che faceva bel vedere, e si chiamavano tutte queste machine, nuvole" (Vasari 1768, 2: 81).

[16]For a German translation of Souzdal's report, see (Wesselofsky 1877).

[17]On the history of the *mandorla*, see also (Beijer 1956; Cohen 1926, esp. 152–5; McKinven 1995, 1–51).

Fig. 3 Photograph of model reconstruction (anonymous) of an Ascension play, Florence 1439. From (Zorzi and Sperenzi 2001, 125)

resemblance to the technology of the ancient Greek stage; its components derived primarily from quattrocento cathedral and fortress construction and shipbuilding techniques.[18]

Coming back to Leonardo's design for the performance of *Danae* (1498), it is interesting to note that in this case the *mandorla* featured not biblical figures but pagan deities (according to Taccone's directions: Jupiter, Mercury, and Apollo), an allegorical figure (Eternity), and a mythical but secular character (Danaë). Just as the *nuvole,* originally invented for religious processions, were adapted for triumphal parades, the *mandorle* flying effects of the *sacre rappresentazioni* survived in courtly performances, secularized through the new "cast" of the machines.

This act of secularization corresponds to Leonardo's research on flight and his attempts to break the long-standing taboo surrounding the scientific study and technical realization of flight. At the turn of the fifteenth to the sixteenth century, ancient myths such as that of Icarus and the Christian legend of the fall of Simon Magus still warned scholars away from flights both physical and intellectual (Ginzburg 1976). Nevertheless, Leonardo asserted that human beings would one day be able to build a flying machine that would imitate the movements of avian flight. In the Codex Atlanticus, he wrote that such an instrument would lack "nothing except the life of the bird [*l'anima dell'uccello*], and this life must needs be imitated by the life of man."[19] Aware of the Aristotelian dogma that any externally propelled instrument was inferior to natural movement, Leonardo still believed in the knowledge (*cognizione*) obtained through the careful observation of the flight of birds and practical experience acquired in imitating the bird's preservation of balance, or emulating it by analyzing and repurposing natural forces.[20] Consider Leonardo's notebook *Sul volo degli uccelli*, in which he compares human to avian equilibrium (Leonardo da Vinci 1978, fol. 5r): there, he describes the human body as capable of conquering the skies, in contrast with the predominantly Christian culture of his time, which reserved the physical capacity for flight to animals, supernatural creatures (angels, saints, demons), and extra-natural creatures (witches, monsters).[21]

Leonardo's curiosity drew him to study the subject of flight from many different angles. He gained access to this knowledge by observing the flow of water,[22] but also by comparing fish to birds and swimming to flying.[23] Maritime travel, too, gave him insights into the principles of movement, in turn suggesting the idea of an airship (Heydenreich 1954, 148). Military technology was another area of inquiry: a study of ballistics led Leonardo to his theory of impetus, which became a central component of his flight research (Dibner 2005, 181). Where these areas of research failed to provide Leonardo with answers, he looked for them in unusual places. But especially in his late studies of flight, moments of resignation are frequent. In a page from the Windsor collection, for example, Leonardo writes that neither sea nor air travel by human beings will be able to achieve the perfection with which animals move in these elements (Windsor, 12666r [c. 1515]).[24] Perhaps this explains why Leonardo expressed his admiration for the hitherto inimitable art of animal movement not by conducting aviation experiments, which were doomed to failure, but by retreating into the theater as a safer realm of experimentation.

[18]The example of the *mandorle* of Filippo Brunelleschi, who likely also orchestrated the church plays described by Souzdal, shows that he probably used the technique of attaching levers to the cathedral structure. See (Prager and Scaglia 2004, vol. 2; Scaglia 1991; Vasari 1912–1914, 2: 193–237).

[19]"...[c]he tale strumento composto per l'omo non li manca se non l'anima dell'uccello, la quale anima bisogna che sia contraffatta dall'anima dell'omo" Leonardo da Vinci, Codex Atlanticus, fol. 161a, in (Leonardo da Vinci 1970, II: 228–9).

[20]On Leonardo's notion of mimesis, see (Blumenberg 1999, 60–1).

[21]On early modern flight taboos, see (Hart and Stevenson 1995, 109–27; Tazi 1989; Ginzburg 1991, 122–52).

[22]MS E (1513–1514), fol. 54r, in (Leonardo da Vinci 1955, 471).

[23]See also (Leonardo da Vinci 1955, 430–1).

[24]Quoted in (Gombrich 1976, 143n71; see also the chapter at 39–56).

Fig. 4 Leonardo's bird theater machine. Leonardo da Vinci, *Codex Atlanticus*, 629bv. From (Leonardo da Vinci 1974, 629bv)

1.3 Drawings of Bird Machines

For the play *Orfeo* by Angelo Poliziano (1454–1494), which was performed between 1506 and 1508 in the villa of the French governor Charles d'Amboise (1473–1511), Leonardo designed several special-effects machines, one of which was a device that probably represented the flight of the boy-turned-bird Perdix (Fig. 4).

Before designing this "ocel de la comedia," Leonardo had already presented bird machines on several occasions.[25] These machines most probably imitated the hydraulic and pneumatic devices of classical automata that were documented in the writings of Hero of Alexandria, to which Leonardo refers repeatedly in the Codex Atlanticus (96v-a; 219v-a).[26] The model for Leonardo's "ocel de la comedia," however, is to be found elsewhere: in the effect machines constructed for Whitsun celebrations since the late trecento, during which wooden doves were lowered on strings from church steeples (Helas 1999, 16–7). These doves had symbolized the epiphany of the *spiritus sanctus*; Leonardo's "ocel de la comedia" once again secularized a religious praxis.

At first glance, the "ocel de la comedia" does not look like a theater machine. Instead, the sketch stylistically resembles the movement studies that Leonardo made in his notebook *Sul volo degli uccelli* in 1505, when he was intensely occupied with questions of avian and human flight. There, too, one finds instructions for experiments relating to the shifting of balance, in which, as in the "ocel de la comedia," birds are suspended from trapezes (Figs. 5a–b).

[25] See (Mazzocchi Doglio 1983, 60; Kemp 1981, 170). According to Vasari, in Rome in 1513 Leonardo, "forming a paste of a certain kind of wax, … shaped animals very thin and full of wind, and, by blowing into them, made them fly through the air, but when the wind ceased they fell to the ground" (Vasari 1912–1914, 4: 104). See also Aulus Gellius's *Noctes Atticae* (Gellius 1927, X: 12), on Architas's dove.

[26] On imitations and transformations of Hero's bird automata in the Italian Renaissance, see (Valleriani 2014, 137, 145, 149, 157; Franke 2003, 250).

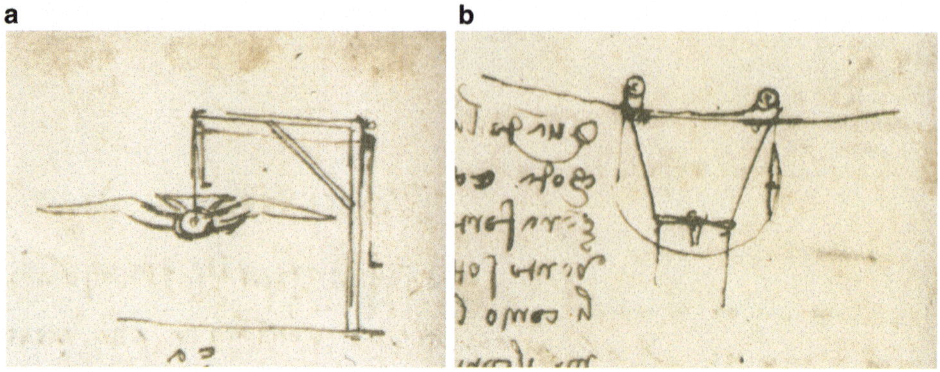

Fig. 5 (**a–b**) Leonardo's instructions for bird flight experiments. Leonardo da Vinci, *Sul volo degli uccelli*, 1505, fols. 15v, 17v. From (Leonardo da Vinci (1505) 1978, 15v and 17v)

Leonardo's illustrated notebooks, in which text and image are correlated in remarkable ways, established a new form of scientific experimentation on paper. For Leonardo, flight was an epistemological object that he could not physically describe but only paraphrase. The stratagem of a provisional conceptual apparatus let him limn the elusive conception of the flying machine by maneuvering with agility between the concepts of "volatili," "strume(n)to," and "uccello."[27] In addition, the drawings in this notebook lend visible form to what the researcher can neither see with the naked eye nor describe in words (Figs. 6a–e).[28]

Through his "cinematographic" sketches (fol. 6r), Leonardo brought processes of continual movement to life on paper. The series appear open, not self-contained; there are empty spaces in between, unrepresented movements that challenge the observer of the drawings to fill in the gaps. Sometimes only thin contours and circles serve to indicate or anticipate the traces of the birds' movements (fol. 10r). Elsewhere, sketchy lines, as if drawn by a ghostly hand, indicate invisible air resistances that surround the body in flight (fol. 8r). In some cases, lateral, frontal, and overhead views of a movement are represented simultaneously, offering the observer multiple perspectives (fol. 7v and 15v).[29]

Leonardo constantly sought to generate a double form of evidence through image and text, yet a degree of inexactitude is inherent to this double medium. Articulated in the inter-media space between image and text is the non-self-contained movement of a quest for the soul of flight (*l'anima dell'uccello*). From a contemporary point of view, the partial sketches of a flight machine in *Sul volo degli uccelli* look extraordinarily advanced (Figs. 7a–c). At no point in the notebook, however, do these partial sketches appear together. They always evince an unfinished quality, the *nonfinito* of an unrepresentable idea.

In this light, we may also read Leonardo's sketch of the "ocel de la comedia" as a "paper experiment," part of his research on flight as documented in *Sul volo degli uccelli*. The visual similarity between these sketches may be regarded as indicating that in the context of his work for the theater, Leonardo performed mechanical experiments that directly corresponded to his flight research. He may have found a space of experimentation in the theater that was denied him (at least officially) in

[27] Leonardo uses the terms "vccello" (fol. 7r), "oeto" (fol. 8r) and "lucel" (fol. 8r), multiple times to refer to flying machines; only once does he speak of "l'uomo ne volatili" (fol. 5r), whereas he speaks more often of the "strumeto" (fol. 15v) (Leonardo da Vinci 1955, 403–92).

[28] On Leonardo's use of graphical methods (sketches and diagrams) to analyze the processes of movement, see (Kemp 2002).

[29] On visual strategies of knowledge production in Leonardo's notebooks, see (Chap. 2).

Fig. 6 (**a–e**) Sketches of bird flight. Leonardo da Vinci, *Sul volo degli uccelli*, 1505, fols. 6r, 10r, 8r, 7v, 15v. From (Leonardo da Vinci (1505) 1978, 6r, 10r, 8r, 7v, 15v)

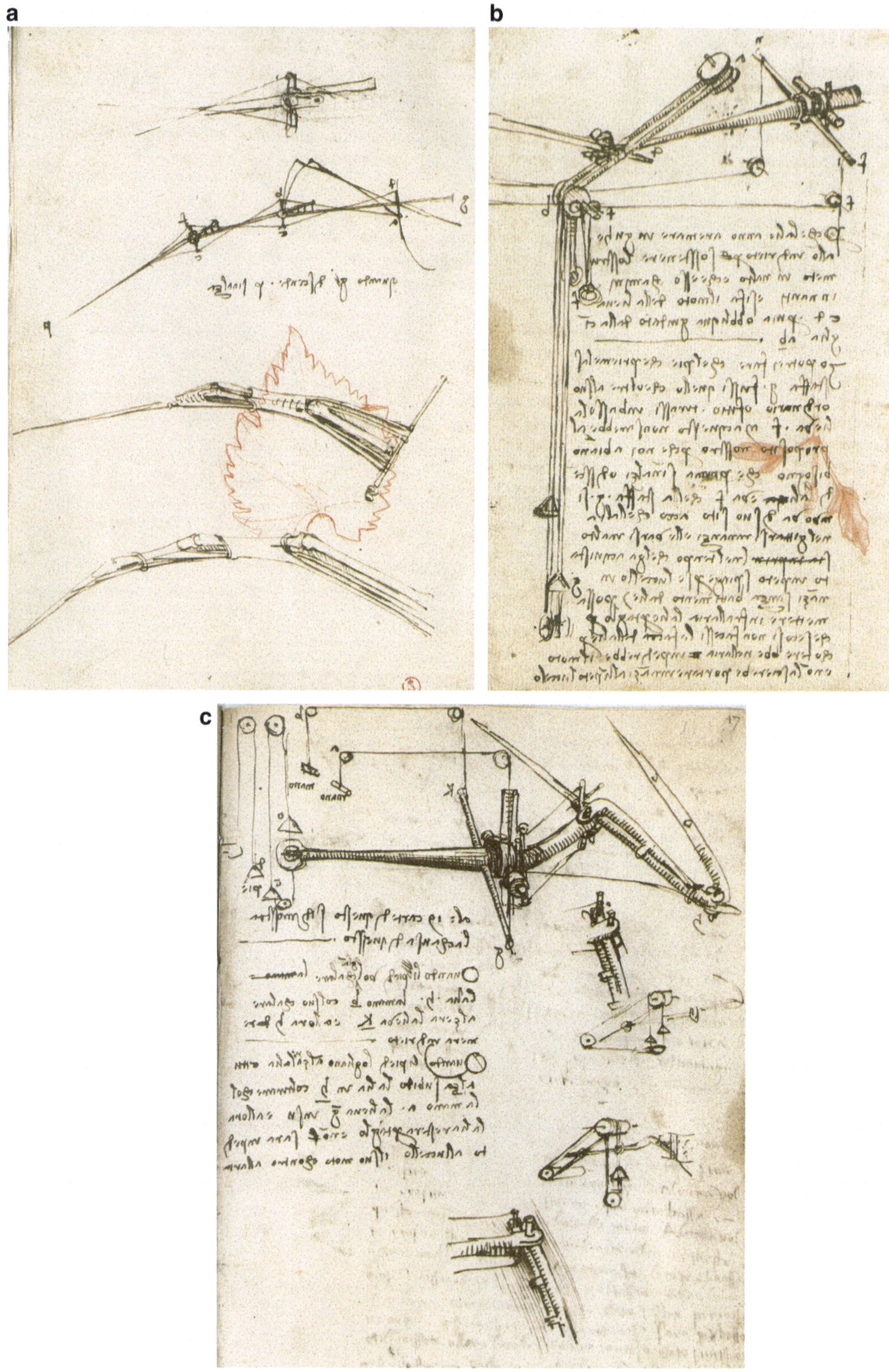

Fig. 7 (**a–c**) Sketches of a flying machine. Leonardo da Vinci, *Sul volo degli uccelli*, 1505, fols. 11v, 16v, 17r. From (Leonardo da Vinci (1505) 1978, 11v, 16v, 17r)

his research on flight.[30] This nexus certainly comes to mind when we turn the sketch for the "ocel de la comedia" over: on the reverse is the design for a bird-watching aviary that was supposed to be constructed in the garden of Charles d'Amboise's villa in Milan (Codex Atlanticus, fols. 271 v-a).

1.4 Shared Fascination

Leonardo's stage designs came without any explicit theoretical reflection. However, his drawings of theater machines—*nuvole*, *mandorle*, bird machines, and so on—show parallels with the general style of special-effect machines in late fifteenth-century northern Italian festival culture. By copying and modifying these model machines, Leonardo participated in the shaping of his era's knowledge on stage design. In addition, his work as a theater engineer seems to be linked to another arena of interest, human flight. The trace that leads us from one arena to the other is Leonardo's fascination with as yet unrealizable machines. The "mac(c)hina" is the overarching term for all the theater machines produced by the experimental and inventive spirit of the Italian engineers in the quattrocento and cinquecento. The machines gradually lost their religious meaning in this process of artistic and technical development. The *nuvole* and *mandorle* were increasingly occupied by secular figures, and the still taboo concept of human flight entered the picture—*ex machina*. Leonardo, too, drafted designs for such machines for the Italian festival circuit, while still taking care to designate the flight machines in his scientific work as "uccelli" or "strume(n)ti." Leonardo's flight instruments were "instruments of thought"[31] more than anything else—they were the motor of a strand of research in technology that extended across multiple domains (anatomy, ballistics, mechanics) to enable new forms of inquiry into the mystery of flight.

In order to make sense of the structural affinities between apparently disparate fields of activity such as stage design and the drafting of flight instruments, one might speak of a "shared fascination." A common feature of Leonardo's theater machines and his flight instruments is that both kinds of devices yoke together different passions—religious admiration (*admiratio*); technical, experimental curiosity (*curiositas*); and discovery or even ingenuity (*inventio*). The conjunction of these three separate passions formed the basis for his efforts to give ever clearer form and material expression to his fascination with flight.

From a modern point of view, Leonardo's theater machines, as contributions to Italian quattrocento and cinquecento festival culture, thus appear in a new light in two specific ways. First, they should be understood less as a rediscovery of ancient Greek theater praxis than as part of a process of technological reinvention that developed out of images of the Passion, religious processions, and triumphal parades, *sacre rappresentazioni*, and early court theater; second, Leonardo's theater machines grew out of a project of flight research that took place behind closed doors and on the pages of notebooks, but in this project Leonardo shared the theater's fascination with flight.

[30] This would contradict the argument of art historian Domenico Laurenza, the only scholar to have recognized the parallels between Leonardo's theater work and his flight studies. Laurenza supports his hypothesis with reference to Leonardo's work from his apprentice years in Florence, and interprets the theater machines of Leonardo's final productive years solely as expressions of a move away from serious research into the fantastical. I argue, however, that Leonardo's bird machine for *Orfeo* shows very clear technical affiliations with his flight research (Laurenza 2004, 10–5, 46, 49–55, 117–8).

[31] This allows us to revise Hankins and Silverman's thesis that only the "elaborate instruments" of the late sixteenth century (such as the telescope or microscope) led to the development of "instruments of the mind"—thought instruments in the sense I use here. In the case of Leonardo's flight studies, one can find no technical *a priori*; the "strume(n)to" as a thought instrument preceded the material construction of the airplane (Hankins and Silverman 1995).

2 The Spread of Giulio Parigi's School of Theater Design in Seventeenth-Century Europe

In sixteenth- and early seventeenth-century Italy, the theater served to represent the fabulous wealth of the courts to a broad international audience. As a result, theater engineering became one of the most prestigious and prolific fields of engineering, though it left no traces in books on machines. Some stage designs are nevertheless well documented, thanks to courtly *descrizioni* (festival souvenir books), chronicles, and account books, as well as letters by audience members and fellow engineers. Among these designs are the ones created by the Medici court architect Giulio Parigi (1571–1635). In this section, I inquire how Parigi's work was described and evaluated by his contemporaries, and trace his wide influence on theater engineering in seventeenth-century Europe with particular attention to the reception of his 1608 designs for the intermezzi of the pastoral *Il Giudizio di Paride* (The Judgment of Paris). Parigi's stage designs figure prominently in Sabbattini's manual *Pratica di fabricar scene, e machine ne' teatri,* which provides instructions for building stage machines to represent natural phenomena. Examining these instructions on theater engineering, I show their significance for seventeenth-century natural philosophy, a connection that is further illuminated through the influence of the British architect Inigo Jones (1573–1652), another student of Parigi's, on Francis Bacon's (1561–1626) view of the epistemic value of theater machines.

2.1 All Theater Engineers' Roads Led to Florence: Parigi's Designs for *La Nave di Amerigo Vespucci*

In 1608, Florence celebrated the wedding of Cosimo II de' Medici (1590–1621) and the Archduchess Maria Maddalena of Austria (1589–1631). On this occasion, the pastoral *Il Giudizio di Paride* by Michelangelo Buonarroti il Giovane (1568–1646) was performed in the Florentine Teatro Mediceo on October 25, 1608.[32] The five acts of the pastoral alternated with intermezzi. As the pastoral itself did not require any machine effects, the intermezzi presented an opportunity to make the whole performance more spectacular and to pay homage to the bride and groom.[33] Departing from the tradition of Italian court theater, these intermezzi were no longer linked by an overarching theme. Each was written by a different author, but they remained in the hands of a single stage designer: Giulio Parigi, formerly the artistic tutor to the young Cosimo II, who held his expertise in high esteem.

The fourth intermezzo, *La Nave di Amerigo Vespucci* (Amerigo Vespucci's Ship), made history with the most technically innovative stage designs of the early seventeenth century.[34] Written by Giovan Battista Strozzi (1551–1634), it recounted an episode from the well-known letter that Amerigo Vespucci (1454–1512) had sent from the New World to his Florentine patron Lorenzo di Pierfrancesco de' Medici (1463–1503). As can be seen in an etching by Parigi's pupil Remigio Cantagallina (1582–1656) (Fig. 8), the stage was turned into a coastal scene, surrounded by a rocky palm-lined landscape and inhabited by exotic animals.

The background shows a calm sea, while water seems to flow in cascades over the front edge of the stage into the orchestra. The small letters in the etching indicate that at least three theater machines

[32] The Teatro Mediceo accommodated 5000 spectators; it was located on the second floor of the Uffizi Gallery and restored in 1586 by Bernardo Buontalenti in the antique style (Solerti 1989, 43; Blumenthal 1980, 38–55, 111).

[33] "Ma perchè l'azione della fauola non ricercaua marauiglie di machine, furono aggiunti gl'intermedi, per render lo spettacolo in tutto, e per tutto mirabile" (Rinuccini 1608, 34).

[34] This section draws on several studies of the Medici festivals: (Blumenthal 1980; 1992; Tintelnot 1939; Nagler 1964; Molinari 1968; Strong 1973; Fabbri, Zorzi, and Tofani 1975; Gaeta Bertelà and Tofani 1969).

Fig. 8 Remigio Cantagallina's etching of the intermezzo *La Nave di Amerigo,* Florence, 1608. From (Zorzi and Sperenzi 2001, 188)

were put into action successively. (A) refers to a ship carrying the conqueror to the New World. The ship was moved from back right to the center of the stage. (B) marks a rock that raised the figure of Tranquillity and a group of music-making nymphs from below the sea-cum-stage. (C) is a spectacular celestial machine crossing the stage heaven (Rinuccini 1608, 47).

In his letter, Amerigo Vespucci had described how the clouds parted when he and his companions reached the New World, illuminating the ship's crew with "great clemency" (Vespucci 2002, 15). But by designing an airship that accompanied the voyage of conquest, Parigi clearly went beyond his source. His celestial machine carried Eternity at the wheel, equipped with an astrolabe and accompanied by a twenty-character chorus of Apollo, the nine muses, and ten European poets—including Dante Alighieri (1265–1321) and Francesco Petrarca (1304–1374). The airship legitimized the imperial aspirations of the Medici dynasty. Its exact design is unrecorded, but it may have been similar to the celestial machine, with space for around twenty performers, that is shown in an anonymous drawing from late seventeenth-century Parma (Figs. 9a–b).[35]

The technology behind the intermezzo is further documented in *Architectura recreationis* (1640) by the German architect Joseph Furttenbach (1591–1667). According to Furttenbach, the ship slid from the background to the forestage on wave-like rails (Furttenbach 1640, 63–4, 68–9).[36] The airship moved on rails high up along the stage heaven.[37]

With such inventions, the wedding celebrations of 1608 no doubt outdid an already long-standing tradition of spectacular festivities, established by the Medici dynasty in the sixteenth century. The revels were postponed twice because of the enormous preparations involved, and a fortune was spent

[35] On this drawing, see (Küster 2003, 132–3).

[36] Similar instructions for constructing machines to represent ships on a billowing sea can be found in (Sabbattini 1638, 109, 119).

[37] Furttenbach put his observations into practice with instructions on how to reconstruct the Florentine stage (Furttenbach 1640, 70).

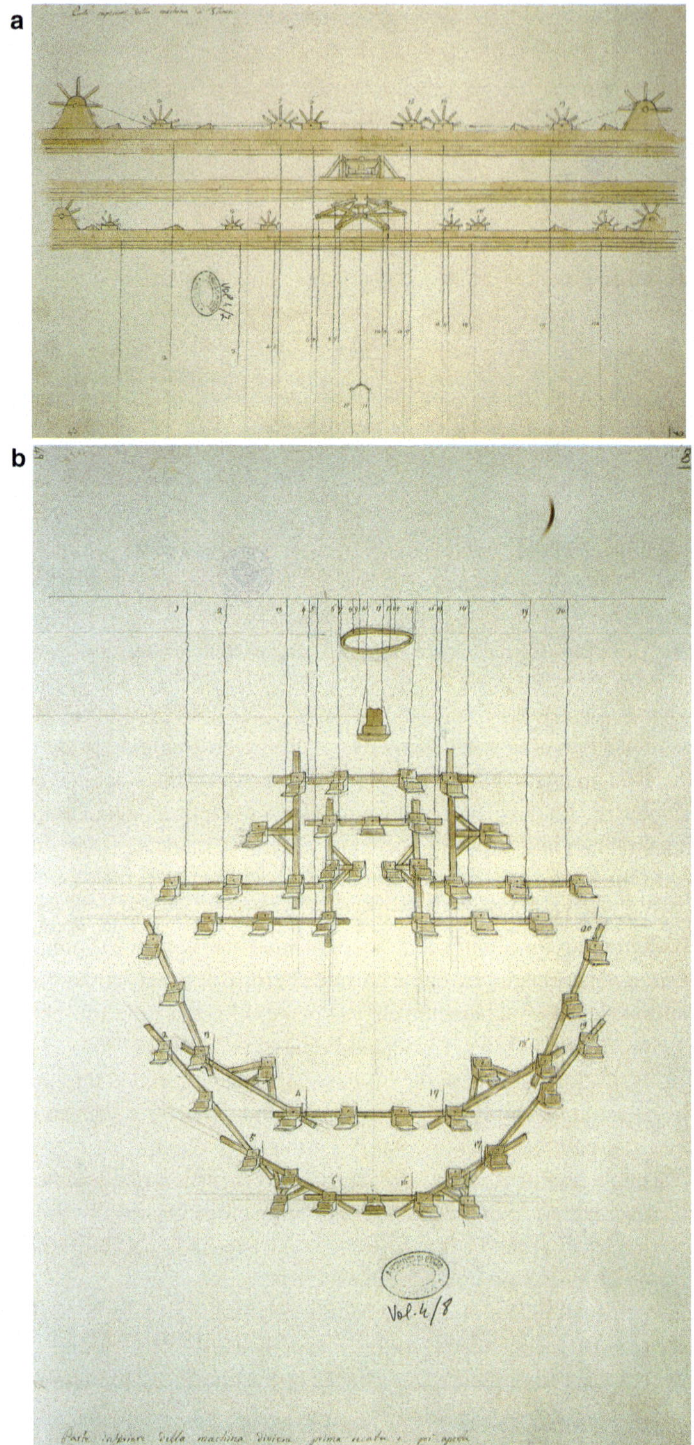

Fig. 9 (a–b) Anonymous sketch of a flying machine (late seventeenth century, Parma). From (Küster 2003, 133)

on naval battles, dance events, fireworks, and theater performances such as *Il Giudizio di Paride*, prompting Furttenbach to criticize the pastoral's elitist character and extravagance.[38] Such criticism did not, however, prevent the Florentines from giving their all. In a letter to the Duke's grand secretary Curzio Picchena (1553–1626), the poet Giovanni de' Bardi (1534–1612) reports that the groom's mother, Christina of Lorraine (1565–1637), even intervened in the preparations of the intermezzi for *Il Giudizio di Paride* with the complaint that the celestial machines were too simple and lacking in variety. The Grand Duchess called for improvements: the most spectacular machine effects must not be shown before the third intermezzo and must not be repeated.[39] The painter and poet Michelangelo also notes that all the stage machineries had to be tested before the performance took place, as there was little practical experience in operating the new technologies.[40]

All these efforts paid off. As the court chronicler Camillo Rinuccini (1564–1649) emphasized in his detailed festival report, the audience's admiration of Parigi's designs was drawn to everything that was new, uncommon ("incognite à noi"), and opulent.[41] Gabriele Bertazzuolo (1570–1626), the engineer of the Gonzagas in Mantua, also wrote bombastically of the fifth intermezzo, which represented an eruption of Mount Etna. He had never witnessed anything like it, Bertazzuolo said, and believed that neither the ancients nor any stage designer in his own days had produced comparable stage effects. He would have loved to draw all the machines, but they moved too fast and the drawings would inevitably be disappointing.[42] The machines were probably invisible to the audience during the performance, but Remigio Cantagallina's etching of the fifth intermezzo further documents Parigi's technological wizardry by revealing the stage technology in a view from beneath the stage (Blumenthal 1992, I: 143) (Fig. 10).

The librettist Giovanni Villifranchi described Giulio Parigi as the "Daedalus of our times" and an "admirable imitator of nature."[43] Parigi's designs were thus celebrated as innovative (in the field of theater engineering) while still compliant with the requirement of not surpassing nature in their mechanical functioning. This assessment needs to be understood in the context of Florentine court culture around 1600. Parigi had been apprenticed to his father, the architect Alfonso Parigi (died 1590), his uncle Bartolomeo di Antonio Ammannati (1511–1592), and the courtly theater engineer Bernardo Buontalenti (1535–1608), who also held an informal academy at his workshop on Via Maggio (Povoledo 1960; Blumenthal 1992, I: 27–8; Fanelli 1973, 1: 301). Parigi started his career as a landscape painter and was initially hired by the Medici as a painter, before succeeding Bernardo

[38]"Das in ein/zwar siben mal verwandeltes Theatrum, oder Scena di Comedie wol 1/2 Thonnen Goldes ist gespendirt worden/welche Comoedi doch nur ein einiges mahl bey einem Fürstlichen Beylager gehalten/dieselbige auch nur die anwesende grosse Herren zusehen/bekommen/die andere vil 1000. Cittadini und Innwohner aber/wie sehr sie sich auch darnach gesöhnet haben/dannoch die Genade zubesichtigen/sie nicht gehaben mögen" (Furttenbach 1663, 111).

[39]In a letter to Curzio Picchena, Giovanni Bardi writes that the the duchess "non vuole del/terzo intermedio ..., s'apra il Cielo, e par mossa da/credenza che il Cielo s'apra più volte." Archivio Mediceo, filza 6068, fol. 386r, in (Blumenthal 1992, I: appendix, 372).

[40]"[N]on mi par da indugiare a esercitar le macchine, e le musiche in sul luogo perche como ho detto ogni magistero vuol lunga, e diligente pratica, altrimente le cose vanno per male via ... si che non si viene presto in cognizione di quelle difficultà che si posson dar le macchine di questa con quella di quella invenzione ... le dificulta s'accrescessero." Letter from Michelangelo Buonarroti il Giovane to Curzio Picchena, undated, probably July or August 1608. Archivio Mediceo, filza 6068, fols. 387r–388v, in (Blumenthal 1992, I: 373–74).

[41]Rinuccini praised the flight effects of the final intermezzo as a "gran merauilia degli spetatori" (Rinuccini 1608, 40).

[42]"In questa [scena] ho visto spezzarsi il palco istesso dalla parte dinanzi sino a terra, cosa che non solo ho mai più visto, ma non ho manco inteso che dalli antichi, nè da moderni sia stato fatto. Queste scene sono state da me designate insieme con li movimenti delle machine, ma per la carestia di tempo non li posso mandare per essere solo schizzati in modo che V. A. non v'havrebbe ponto di gusto." Letter of October 28, 1608, Archivio di Stato, Mantua, in (Solerti 1989, 54–7, quotation 56).

[43]"Dedalo de' nostri tempi ... ammirabile imitatore della Natura" (Villifranchi 1613, 9).

Fig. 10 Remigio Cantagallina's etching of the fifth intermezzo of *Il Giudizo di Paride*, Florence 1608. From (Blumenthal 1980, 53)

Buontalenti in the post of court architect.[44] The intermezzi of *Il Giudizio di Paride* were Parigi's first designs for the Medici court. Buontalenti's influence informed Parigi's creation of perspectival depth in the scenes and his organization of the scene changes (using *periaktoi*), the lighting effects (using mirror reflectors), and additional special machine effects. In particular, Buontalenti's seascape for the fifth intermezzo of *La Pellegrina* (Florence, 1589) shows parallels with *La Nave di Amerigo Vespucci*.[45] There is also good reason to believe that Parigi was inspired by the intermezzo *Il Ratto d'Europa*, performed in connection with the tragedy *L'Idropica* in Mantua in 1608. According to Mantuan chronicler Federico Follino, the engineer Antonio Maria Vianini also designed a seascape on this occasion, but no ship appeared in his intermezzo; instead there was Zeus in the guise of a bull, with Europa on his back. Above this scene appeared a celestial machine carrying Cupid, the god of love (Nagler 1964, 178–82; Solerti 1989, 218–9). Inspired by his fellow engineers, Parigi promoted technological advances in stage design. He improved the playability of the stage heavens, introducing ever larger and heavier flying machines, which could carry up to twenty performers. Moreover, he was the first to have several theater machines working on stage at the same time (Blumenthal 1992, I: 150).

Parigi probably also saw parallels between stage design and other branches of technology such as nautical engineering, which flourished in northern Italy under Ferdinando I de' Medici (1549–1609) and his son Cosimo II and from which he may have borrowed rope-and-knot techniques for arranging the machines. The fulcrum shaft frequently used below stage seems to have been adapted from the

[44]In 1598, Parigi is listed in the Deliberazioni of the Accademia del Disegno in Florence (Blumenthal 1992, I: 32).

[45]Buontalenti initially directed the preparations for the 1608 festival, but he died in June 1608. On the fifth intermezzo for *La Pellegrina*, "The Rescue of Arion," see (Nagler 1964, 87–8, illustration 57). Parigi may also have drawn inspiration for the airship in *La Nave* from *La Pellegrina*'s first intermezzo ("Harmony of the Spheres"). See (Blumenthal 1992, I: 10, 74–8, 125, 140–1, 148–51).

capstan of sailing ships.[46] As court architect to the Medici, Parigi may also have benefited from the equipment used on construction sites, such as lifting devices and cranks.[47]

Parigi's inclusion of theater engineering in a broader study of the arts is reflected in the curriculum of the art academy he founded in 1597 at his house at Via Maggio 36 in Florence, where students from all over Europe were taught Euclidean mathematics, mechanical engineering, painting, and civil and military architecture.[48] Among these students were Cosimo II de' Medici and a number of other European noblemen. In addition, Parigi's academy was a talent pool for theater engineers, including his son Alfonso Parigi (1606–1656), who had assisted his father in the designs for the Medici festivals and later assumed his position as the Medici court architect.[49] This title then devolved on his pupil Fernando Tacca (1619–1685), another "new Daedalus,"[50] who opened the first public theater of his time on Via della Pergola in Florence and expended special effort on designing the effect machines.[51] Other students were Italian architects such as Stefano della Bella (1610–1664), Cosimo Lotti (ca. 1590–1650), and Baccio del Bianco (1604–1657), the French architect Jacques Callot (1592–1635), the German architect Joseph Furttenbach, and the British architect Inigo Jones.[52]

These direct teacher-student relationships were not the only way that Parigi's "school of theater design" spread. His stage designs were exhibited on many occasions: the intermezzi of *Il Giudizio di Paride* were performed again twice, on November 19, 1608, in honor of a visit by the Duke of Mantua, and then on November 27, 1610 (Blumenthal 1992, I: 53). According to the Florentine court chronicler Cesare Tinghi, the cloud machine (*nuvola*) of the fourth intermezzo was subsequently transferred to the Uffizi Gallery, where it was displayed in the Vasari corridors and shown to visitors including Don Ferrante Borgia, the Spanish ambassador to Tuscany.[53] Parigi's designs circulated in etchings by Parigi, Remigio Cantagallina, Jacques Callot, and Stefano della Bella, printed in large editions to propagate the magnificence of the Medici in courtly circles, for example in (Parigi and Cantagallina 1608). The etchings were produced before the performances took place; most of them were poorly printed and intended as illustrations for the *descrizioni* rather than as works of art. Despite not having studied directly with Parigi, numerous architects in seventeenth-century Europe were thus familiar with his stage designs and shared the knowledge produced in the "Parigi school." It is fair to regard Parigi's knowledge on theater engineering as a nodal point in a first regional, then supra-personal network of theater engineering.[54]

---

[46] On the use of early modern navigation technology in theater, see (Edström 1995, 69–104).

[47] On Parigi's architectural projects, see (Berti 1951; Linnenkamp 1958).

[48] On the academy's curriculum, see (Baldinucci 1812, 11: 6–7).

[49] Alfonso Parigi copied the designs for *La Nave di Amerigo Vespucci* in his design for *Le Nozze degli Dèi* (1637) (Blumenthal 1992, I: 232–65).

[50] The epithet "nouello Dedalo" is mentioned in Moniglia's *Descrizione* of Tacca's *L'Hipermestra* (1653). See (Moniglia 1658, 6–7).

[51] The theater is now called Teatro della Pergola. For Tacca's "Garden of Venus" in the production of *Ercole in Tebe* (1661), likewise, the flight machines were deployed in parallel to a nautical scenery. See (Blumenthal 1992, I: 191, 202–6).

[52] Parigi's academy thus contributed to the erudite specialization in "true artistic architecture." See (Chap. 9).

[53] "Ed adì 27 di novembre mangiò con l'ambasciatore [di Spagna, don Ferrante Borgia] il sig. Paolo Giordano et il giorno andò per il corridore a vedere le nugole et intermedi della Comedia grande [Certamente ciò che si era usato per la rappentazione del Guidizio di Paride due anni innanzi]." (On 27 November Paolo Giordano [Orsini] had dinner with the Spanish ambassador Don Ferrante Borgia and took him to the corridors to see the clouds and [other machines of] the intermezzi of the grand comedy [that were certainly used for the performance of the Judgment of Paris from two years ago]). Quoted in (Solerti 1989, 59).

[54] On knowledge networks and the associated media and practices of dissemination, see (Chap. 16).

2.2 Nicola Sabbattini's *Pratica*: All the Stage Is a Book

Some of the instructions in Nicola Sabbattini's *Pratica di fabricar scene e machine ne' teatri*, too, may be read as references to Parigi's work. Sabbattini (1574–1654) studied with the mathematician Guidobaldo del Monte (1545–1607) and served as court architect to the dukes of Urbino from 1610 to 1631.[55] When preparing the *Pratica* for publication in 1637–1638, he had difficulty finding another patron and dedicated the book to Cardinal Girolamo Grimaldi, the vice-legate of Pesaro.

In his preface to the second edition of the *Pratica* (1638), the printer Pietro dei Paolo praises Sabbattini as an author who has "illustrated and written about all fields of mathematics"[56] and has described the machines in the simplest images and words for maximum comprehensibility.[57] Those wishing to learn more about "the finest theory for this practice" are advised to consult Guidobaldo del Monte's *Perspectivae Libri VI* (1600) (Sabbattini 1638, 11). Dei Paolo goes on to commend Sabbattini "for his experience [in theater engineering], but even more for his sustained dedication to his theoretical studies,"[58] and ranks Sabbattini's knowledge of theater engineering as equal to his learning in other fields of civil and military architecture (Sabbattini 1638, 11). Yet Sabbattini regarded his book not merely as a practical application of his theoretical knowledge in geometry. More importantly, he saw it as a translation of experiential and orally transmitted knowledge into written knowledge.[59] This becomes apparent when we look at the 1637 first edition, in which Sabbattini claims to set out in print the practical art of theater, something that has hitherto been possible to admire only at the moment of performance.[60] This understanding of the printed book as a different form of "theater"—one that provides the reader with an imperishable survey of technical artifacts—corresponds to a more general tendency in seventeenth-century literature on machines. Examples are Jacques Besson's (1540–1573) *Theatrum instrumentorium et machinarum* (1578), Heinrich Zeising and Hieronymus Megiser's (ca. 1554–1618) *Theatrum machinarum* (1607–1614), or Vittorio Zonca's (died 1613) *Novo teatro di machine et edificii* (1607).[61] In the final chapter of his second book, entitled "Della facilità della Pratica" (On the ease of practice), however, Sabbattini argues that theoretical writings are a more cumbersome form of communication than observation of the practitioner at work.[62] He points to the "great admiration and delight"[63] that stage effects elicit during the performances; by contrast, he is obviously unsure of the theatrical quality of his writings and drawings.

Sabbattini goes on to refer to his stage designs for a performance of the tragedy *Asmondo* by the highly esteemed poet Giovanni Hondedei that took place at the Teatro del Sole in Pesaro. According

[55] On Sabbattini, see (Eiche 1996; Flemming 1929).

[56] "Chi sa, che doppo queste Pratiche Teatrali non si disponesse l'Autore die participarti l'altre sue Pratiche d'Archittetura Civile, e Militari? Sappi che la sua penna per non starsene oziosa, hà delineato, e scritto sopra tutte le materie di Matematica" (Sabbattini 1638, 11).

[57] However, this brevity may indicate a deliberate withholding of details: Sabbattini may have wanted to avoid precise specifications in order to prevent others copying him. On similar publication strategies in early modern engineering, see (Popplow 1998, 65–98).

[58] "Nicola Sabbattini da Pesaro Ingegnero, c'hà ingegno, stimatissimo per l'esperienza, e molto piu per le teroica acquistata con lungo studio" (Sabbattini 1638, 3).

[59] On the formation of "descriptive geometry" in early modern architectural treatises, see (Chap. 9).

[60] "Dedico questa mia fatica per i Teatri: accioche sotto l'Illustrissimo suo Nome oggi si legga impressa la Prattica Teatrale, conforme sotto i suoi felicissimi auspicij s'ammirò in un istante eretto in Pesaro Teatro si illustre" (Sabbattini 1637, 2).

[61] On alternative meanings of the term *theatrum* in the early modern period, see (Schramm 1996; Bredekamp 2004, 23–44; Schramm et al. 2008; Roßbach 2013).

[62] "La Teorica non è difficile, ma è più facile la Pratica" (Sabbattini 1638, 165).

[63] "amiratione, e diletto" (Sabbattini 1638, 165).

to Sabbattini, the design was worthy of Hondedei's work: "It enhanced the glory of the poet, who, without infringing the rules of art, knew how to facilitate the flight of ingenuity. Therefore, experience teaches us that we can realize with ease and speed what has been explained about the appearance of the theater decorations and machines [in this book]."[64] This final remark is interesting for two reasons: because Sabbattini once again stresses that he believes knowledge of stage design to be more easily imparted in action; and because he counts stage design as equal in artistry to poetry, all the more striking given that poetry was a firmly established component of Renaissance *studia humanitatis* (Kristeller 1951, 510–27). It was in this context that the topos of the poet's wings of ingenuity reentered the discourse (see, for example, the poet flying on the eagle's wings in Dante's *Divina Commedia*). Within poetry, the poet's *inventio* was increasingly conceived of as an act of ingenuity rather than, following the rhetorical tradition, as a mere recomposition of given contents and forms (Kemp 1977, 349). Clearly, Sabbattini claimed the same "flight of ingenuity" for theater engineering.

2.3 Designing Nature in Theater Engineering

Indeed, a closer look at Sabbattini's *Pratica* shows that the engineer did not only copy and recompose the accomplishments of his predecessors, such as Giulio Parigi. The *Pratica* is innovative, for example, in Sabbattini's vigorous efforts to make stage design resemble nature as closely as possible.[65] To this end, he strictly subordinates all scenic elements such as the set, backdrop, and shutters to the laws of central perspective. For the stage ceiling, too, Sabbattini suggests dividing the space into hatches and sections that follow the same "tendency" as the rest of the space. Angled frames in the shape and color of clouds are lowered through openings in the ceiling and incorporated into the unified perspective of the stage (Sabbattini 1638, 128–9). Even the cloud scenery is now subservient to the overall perspective impression.

As well as his instructions regarding cloud frames, Sabbattini provides a long list of instructions on how to make theater machines that simulate thunder and lightning, rainbows, and above all cloud formations (Figs. 11a–c). We are told, for instance, "how part of the sky can gradually become overclouded by starting with a small cloud that becomes larger, and constantly changes color" (Sabbattini 1638, 133–5).

Next comes a description of a cylindrical mechanism that is fitted into the stage sky such that it can be moved on gear wheels from the back to the front of the stage while changing shape and size in relation to the perspectival geometry of the space (Fig. 12a). Sabbattini also explains "how the sky can be made to suddenly cloud over" or "how a cloud can be made to float across the stage" (Sabbattini 1638, 130–2, 149–51) (Fig. 12b).

These directions are a novelty in the history of stage design. Flight effects had previously been reserved primarily for the appearance *ex machina* of the figures of Christian belief or legend or, later, for historical and allegorical figures, as seen in Leonardo's design for *Danae* or Parigi's for *La Nave di Amerigo Vespucci*. In all these cases, cloud scenery served simply to conceal the secret mechanisms, whereas Sabbattini placed them at the center of attention. Without gods, without being illuminated by an apparently supernatural radiance, cloud structures were supposed to become visible in the theater as both spectacular and secular facts of nature.

---

[64] "Opera degna dell'Autore, che stimatissìma' da gl'intendenti, si nel tutto, come nelle parti, porta i vanti del Poeta, che senza traccollare dai confini de i precetti dell'arte hà saputo inalzare i voli' dell'ingegno. Si che l'esperienza insegna, che con molta ageuolezza si mette in opera, quanto si è mostrato intorno all'apparenza delle Scene, e delle Machine" (Sabbattini 1638, 165).

[65] "immitare il piû, che si possa il naturale, & il vero" (Sabbattini 1638, 2).

Thus, the architect not only explains how cloud formations can be staged *ex machina* over the whole stage, he also groups the necessary machines into eight different cloud types. In addition to the directions mentioned above, he describes "how a small cloud can be made to sink while growing in size," or "how a cloud can be made to descend, separating into three parts and subsequently gathering together again as it ascends" (Sabbattini 1638, 148–54) (Figs. 12c–d).

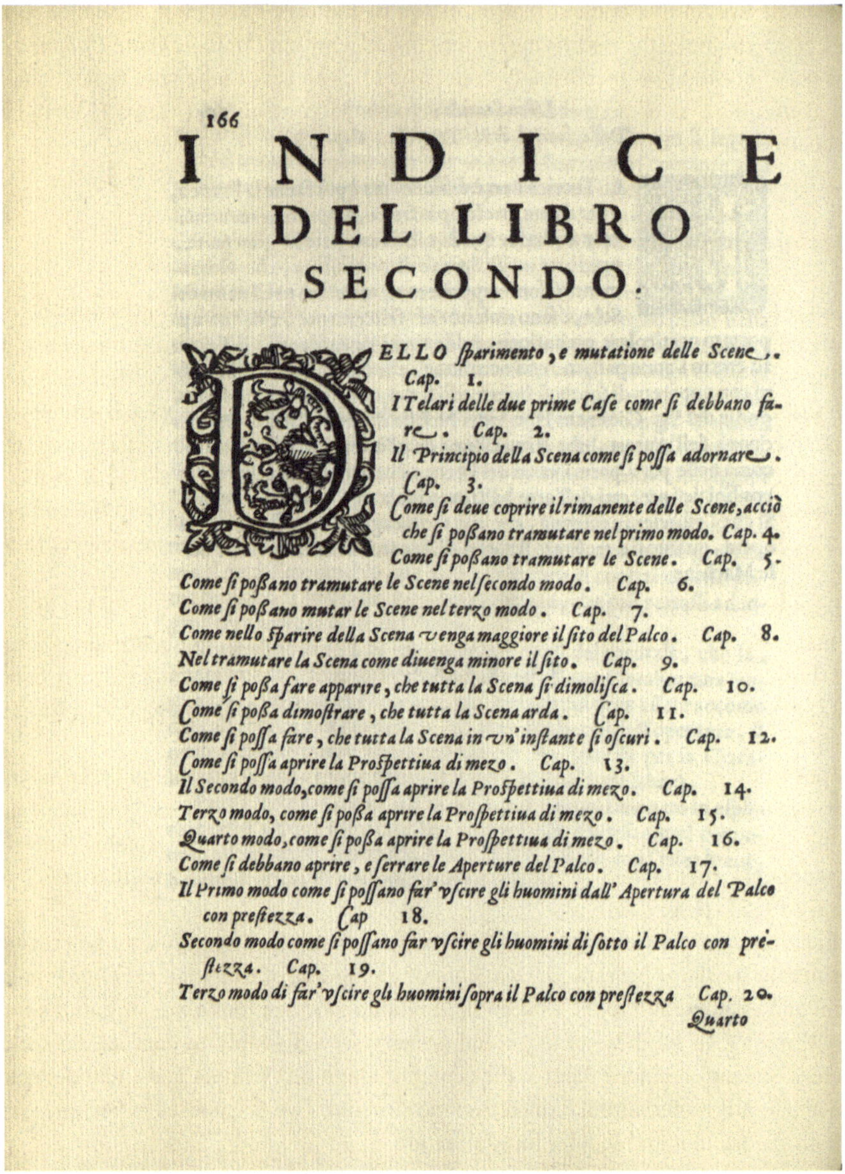
166

INDICE

DEL LIBRO

SECONDO.

Quarto

Fig. 11 (a–c) Table of contents, in Nicola Sabbattini, *Pratica* (1638), Book II. From (Sabbattini 1926, 166–8)

167

Come

Fig. 11 (continued)

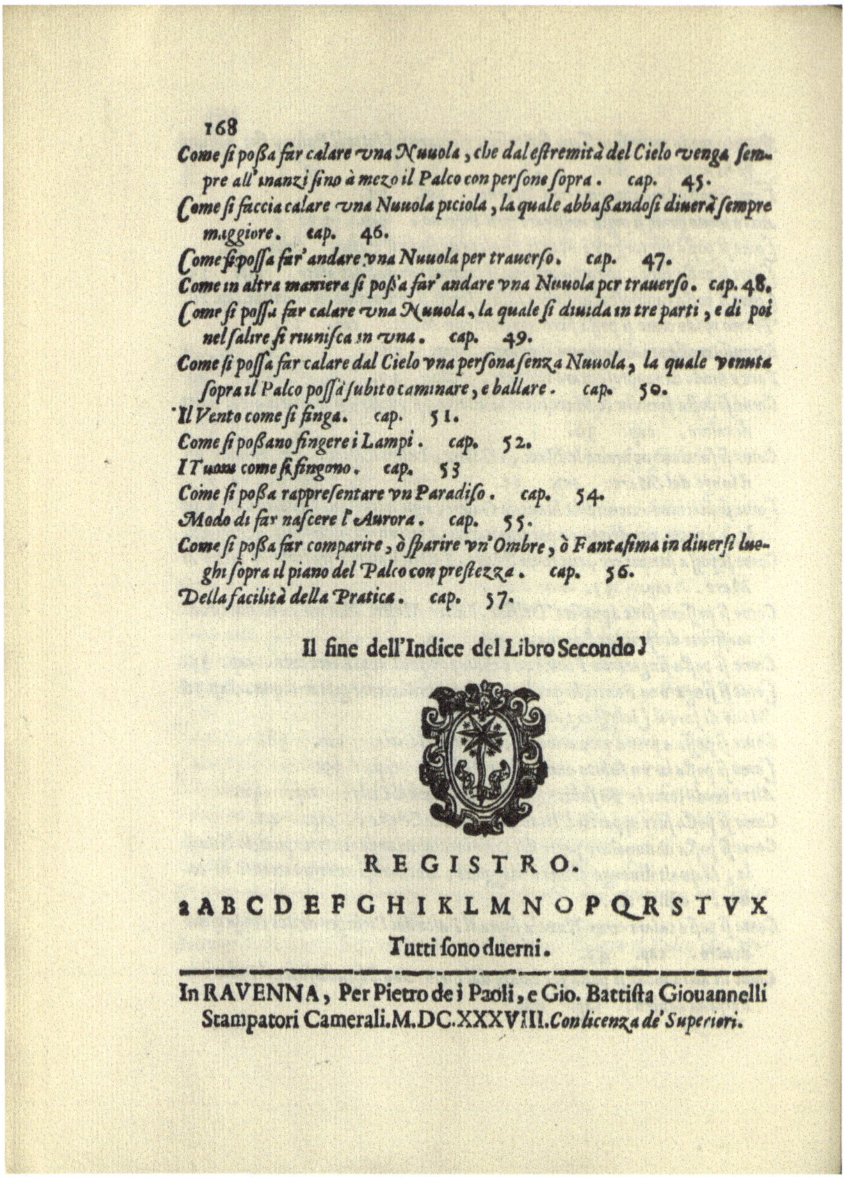

168

Come ſi poßa far calare vna Nuuola, che dal eſtremità del Cielo venga ſempre all'inanzi ſino à mezo il Palco con perſone ſopra. cap. 45.

Come ſi faccia calare vna Nuuola piciola, la quale abbaßandoſi diuerà ſempre maggiore. cap. 46.

Come ſi poſſa far' andare vna Nuuola per trauerſo. cap. 47.

Come in altra maniera ſi poß'a far' andare vna Nuuola per trauerſo. cap. 48.

Come ſi poſſa far calare vna Nuuola, la quale ſi diuida in tre parti, e di poi nel ſalire ſi riuniſca in vna. cap. 49.

Come ſi poſſa far calare dal Cielo vna perſona ſenza Nuuola, la quale venuta ſopra il Palco poſſa ſubito caminare, e ballare. cap. 50.

Il Vento come ſi finga. cap. 51.

Come ſi poßano fingere i Lampi. cap. 52.

I Tuoni come ſi fingono. cap. 53

Come ſi poßa rappreſentare vn Paradiſo. cap. 54.

Modo di far naſcere l'Aurora. cap. 55.

Come ſi poßa far comparire, ò ſparire vn' Ombre, ò Fantaſima in diuerſi luoghi ſopra il piano del Palco con preſtezza. cap. 56.

Della facilità della Pratica. cap. 57.

Il fine dell'Indice del Libro Secondo.

REGISTRO.

a A B C D E F G H I K L M N O P Q R S T V X

Tutti ſono duerni.

In RAVENNA, Per Pietro de i Paoli, e Gio. Battiſta Giouannelli Stampatori Camerali. M.DC.XXXVIII. *Con licenza de' Superiori.*

Fig. 11 (continued)

In the same year that Sabbattini's *Pratica* proposed to "demystify" cloud phenomena on the stage, the French philosopher René Descartes (1596–1650) published *Les météores*, the essay that appeared alongside the *Discours de la méthode* in 1637. In this text, Descartes denies that clouds are sites of divine presence, as the gaze turned reverentially to the sky in early modern Europe had assumed. The sky's opaque whiteness, argues Descartes, is neither a marvel of nature nor the outer bound of human research and knowledge. Instead, clouds (and all other weather phenomena) can be regarded as

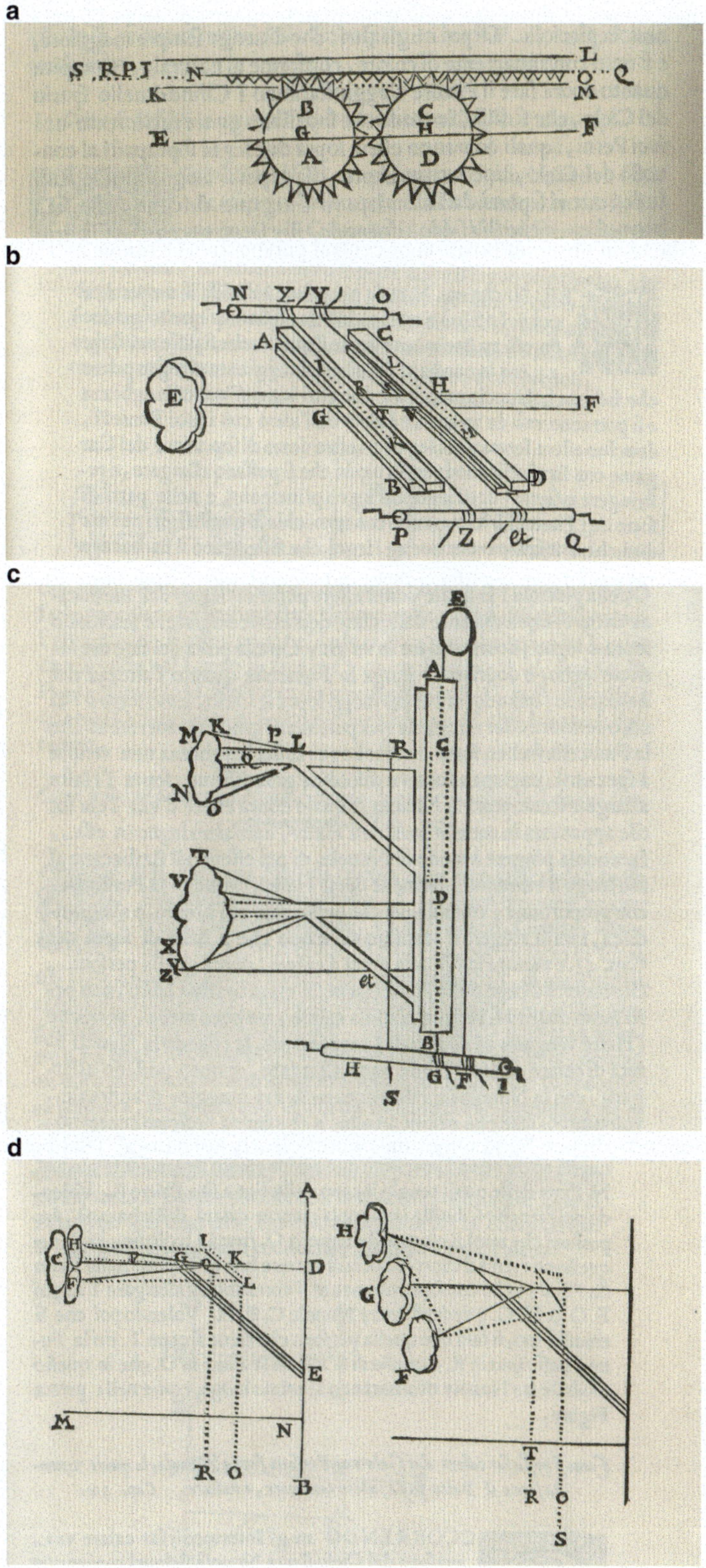

Fig. 12 (**a–d**) Sketches of theater cloud machines. From (Sabbattini 1926, 134, 151, 148, 153)

touchstones of thought—which ultimately means as facts of the Cartesian production of evidence.[66] In the *Discours*, Descartes uses a notion of evidence borrowed from mathematics both metaphysically and physically. He declares "that which presented itself to my mind so clearly and so distinctly" to be the model for approaching all the epistemic problems of other sciences (Descartes 2011, 16, 18). What is revealed to the observer in sensible nature is grounded (deductively) in self-certainty. It has frequently been suggested that Descartes's method of constructing self-evidence was influenced by the perspective structures of early modern painting, with the final point of Cartesian certainty linked to the ideal observer position of central perspective (Krämer 2003).[67] A similar correspondence might exist between Descartes's perspective on meteorology and the early seventeenth-century stage designs that subordinate all cloud décor to the geometric space, as described by Sabbattini.

The weather phenomena investigated by Descartes in *Les météores* also appear in manifestly theatrical explanations, as he deploys machines made by mechanical engineers as thought models for natural processes. Clouds, Descartes writes, are formed through the condensation of "vapors," water particles that condense into small units under the influence of the cold and the wind (Descartes 2011, 269–74, 298–306). To illustrate how these "vapors" float in the air, Descartes uses a mechanical model: if one observes two sticks placed crosswise and rotating on a shared axis, they seem to fill the entire motion radius. According to Descartes, water particles warmed by the sun move just as quickly. They take on the shape of a sphere and ascend due to their large surface and minimal weight (Descartes 2011, 271). Descartes also compares evaporation processes and the origin of wind to a machine that was later used in theater to create mist effects: the "aeolipile," a heatable spherical apparatus with a small opening that was filled with water (Descartes 2011, 287–97).

At first sight, it might seem that Descartes's meteorology and Sabbattini's stage practice are results of the same paradigm shift. That argument has been made, albeit from two opposing perspectives. On the one hand, weather phenomena are assumed to have been made intelligible by the invention of optical instruments such as the telescope, which not only forced natural philosophers like Descartes to craft a new image of nature, but also created a need in the theater for stage designs that would represent these new ways of looking at nature (Nelle 2001); on the other, it is argued that early modern books on machines by Sabbattini and other engineers paved the way for discoveries in natural philosophy because engineering provided science with mechanical models showing the hidden mechanisms of nature in a new light (Werrett 2001). Both perspectives are based on the assumption that theater engineers and natural philosophers alike were seeking new media strategies for the representation of nature.

However, it is worth noting that Sabbattini's stage designs sometimes followed other rules of representation than those of natural philosophy. His stage effects were designed to operate so as to "correspond to nature," and sometimes were "clearly not natural" (Sabbattini 1638, 254, 257). Paradoxically, his attempt to surpass nature's spectacular phenomena was precisely what led Sabbattini to develop a representative cloud typology that was far ahead of contemporary natural philosophy and went to the heart of later meteorological discoveries. Although Descartes strove to make weather phenomena evident as explainable facts, his description of cloud formations and cloud heights remained extremely

[66]"It is our nature to have more admiration for the things above us than for those that are on our level, or below. And although the clouds are hardly any higher than the summits of some mountains, and often we even see some that are lower than the pinnacles of our steeples, nevertheless, because we must turn our eyes toward the sky to look at them, we fancy them to be so high that poets and painters even fashion them into God's throne, and picture Him there, using His own hands to open and close the doors of the winds.... This leads me to hope that if I here explain the nature of clouds, in such a way that we will no longer have occasion to wonder at anything that can be seen of them, or anything that descends from them, we will easily believe that it is similarly possible to find the causes of everything that is most admirable above the earth" (Descartes 2011, 263).

[67]On central perspective and the reader of Descartes's work, see (Horn 2000).

undifferentiated (Descartes 2011, 298–306).[68] The first writer to bring greater precision to the subject was the chemist and amateur meteorologist Luke Howard (1772–1864). In his essay "On the Modification of Clouds," Howard laid out the Latin nomenclature to distinguish between cloud formations that is still in use today: *cirrus, cumulus, stratus*, and so on. Transient changes in the cloud formations could be described using compounds such as *cirrocumulus* or *cirrostratus* (Howard 1803, 5–11).[69] In fact, Sabbattini had carried out comparable work on the "cloud myth" in his *Pratica* as early as 1638. In the series of machine types mentioned above, which differentiated the representation of clouds according to their form and modification, *cumulus* and *cirrocumulus* had already achieved visibility as products of stage design. However, Sabbattini's book was written neither for the theater-going public, nor as a technical manual for colleagues, nor as a contribution to natural philosophy. It was addressed to the titular bishop Honorato Visconti, from whom Sabbattini hoped to gain further commissions. The dramatic and detailed description is subjugated to the strategic objective of winning over and enthralling a single reader. If the *Pratica* nonetheless also had repercussions for meteorology, these must be considered marginal phenomena, incidental to the colorful prospectus.

2.4 Francis Bacon's Theater of Nature

In the theater of the early seventeenth century, the emergence of new ways of seeing nature went almost unnoticed on the margins of the mise-en-scène, but Francis Bacon, in the utopian fragment *New Atlantis* (published in 1627), was already developing far more spectacular methods for the reinvention and representation of nature. If human understanding and the senses gave rise to misleading theories ("idols"), Bacon aimed to make the facts of nature experimentally evident. Theoretical knowledge should no longer be derived from the contemplative observation of the sky, as in the *contemplator coeli*, a symbol of the ancient thinker. Bacon's ideal figure is the scholar of the House of Solomon in the *New Atlantis*, who, in his laboratory-like towers and houses, gathers theoretical evidence through active observation and the experimental imitation of nature: "We use these towers, according to their several heights and situations, for isolation, refrigeration, conservation; and for the view of divers meteors; as winds, rain, snow, hail; and some of the fiery meteors also.... We have also great and spacious houses, where we imitate and demonstrate meteors; as snow, hail, rain, some artificial rains of bodies and not of water, thunders, lightnings" (Bacon 1859b, 157–8).

As Bacon saw it, new empirical evidence, at least in his utopia, resulted from artfully designed instruments, machines, and experiments that could help in gaining "the knowledge of Causes, and secret motions of things" and "enlarging of the bounds of Human Empire" (Bacon 1859a, 156). The mechanical arts were regarded as a way of revealing the extraordinary, like the wondrous phenomena of nature.[70] Even if thought experiments such as these could not be realized in the seventeenth century, Bacon's utopian text sought to inspire the reader to devise new, playful practices of evidence production in natural philosophy.

Such utopian writings contain a creative potential that can, at least to a certain extent, be compared to engineering in the court theater. The comparison with theater is perhaps surprising, in that Bacon took grave offence at the automata exhibited in the cabinets of curiosity of his time. These artifacts, he believed, imitated nature in a virtuous fashion but did not facilitate any deeper insight into her

[68] The measuring of cloud heights was first undertaken by Johannes Kepler. Distinctions between cloud heights are first found in Immanuel Kant's geography lectures (1756). The first, impracticable, cloud nomenclature was developed by Robert Hooke in 1663. See (Middleton 1969, 265–70; Körber 1987, 138, 168).

[69] On the description of changes in the cloud formation, see (Hamblyn 2001).

[70] On the ambivalence of the concept of wonder in Bacon's meteorology, see (Janković 2000, 44–50).

secret motions. He spoke of the danger that they "may stun the intellect, bind it, and cast their own particular spell on it so that it becomes incapable of further knowledge, because it thinks that nothing of that kind can be done except in the same way as those things were done" (Bacon 2000, 150).[71] By contrast, Bacon had only the highest praise for the courtly theater of his time. He had tried his hand at dramatic writing as a young man, and was familiar with the masques realized by the British stage designer Inigo Jones at the courts of James I (1566–1625) and Charles I (1600–1649)—which is all the more interesting given that Jones had studied in northern Italy between around 1598–1601 and 1613. As mentioned, Jones was a student of Giulio Parigi's and later copied Parigi in his designs for the British court (Strong 1984, 70–1).[72]

In his essay "Of Masques and Triumphs" (1625), Bacon emphasized that flying machines, descending from the stage heavens, evoked in the viewer the desire to look behind the scenes, and the

> … alterations of scenes, so it be quietly and without noise, are things of great beauty and pleasure; for they feed and relieve the eye, before it be full of the same object. Let the scenes abound with light, specially coloured and varied; and let the masquers, or any other, that are to come down from the scene, have some motions upon the scene itself before their coming down; for it draws the eye strangely, and makes it with great pleasure to desire to see that it cannot perfectly discern. (Bacon 1859a, 468)

The aesthetics of the masque may, then, have been what paved the way for Bacon's experimental approach to nature and his advocacy of playful experimentation with the inner processes and outward appearances of nature for the better comprehension of both.

The correspondence between the Renaissance theater and Bacon's natural philosophy can be described in terms of perspective: like the theater audience observing the raging sea on stage from a safe place, Bacon was ensconced in his London armchair when he described the shipwrecked crew in the *New Atlantis*, which is simultaneously a departure into a New World of science.[73] However, it was not the safe perspective of the audience per se, but rather the inventive energy of stage design that attracted Bacon to the theater—the play of hide-and-seek, of illusions and the technical knowledge of their making. Bacon took the theater machine not as a model, but as an inspiration for his program in natural philosophy, which proposed to reveal the obscure and dark sides of nature.

2.5 Departing from the Stage into New Worlds of Thought

In early seventeenth-century court theater, the degree of spectacle was still largely circumscribed by the resemblance between the stage wonder and nature.[74] Machines were background media. Beginning in the mid-seventeenth century, however, the mechanical arts began to move into the limelight. This development was driven by competition among the stage designers of the European court theaters, and led to an atmosphere of extravagance. In *Apparati scenici* (1644), for instance, the Venetian architect Giacomo Torelli (1608–1678), who became chief theater engineer at the court of Versailles, thanks Giulio Parigi for his inspiration but claims to outdo him in the domain of technology (Torelli 1644).[75] The paradigmatic example is Torelli's stage design for Pierre Corneille's (1606–1684) *Andromède*, which premiered at the Petit Bourbon in Paris in 1650. The stage direction for the scene

[71] On Bacon's critique of wonder chambers, see (Bredekamp 1995, 63–80). On the significance of the marvels of art and nature in Bacon's reform of the philosophy of the supernatural, see (Daston and Park 1998, 220–31).

[72] On Bacon's experience of the theater, see (Campbell 1960, 182–3; Blumenthal 1992, 152, 287–8, 334).

[73] On the comparison between shipwreck narratives in philosophy and on the theatrical stage, see (Blumenberg 1997).

[74] The aim was "as far as possible, to imitate the natural and the true" (Sabbattini 1638, 175).

[75] The influence of Parigi's designs for *La Nave di Amerigo Vespucci* is most visible in Torelli's design for the prologue of *Bellerofonte* (Venice, 1642). See (Bjurström 1961, 20, 60, 101).

in which Andromeda is carried off by Aeolus clearly shows which way the wind of this new aesthetic was blowing:

> There is a loud sound of rolling thunder accompanied by multiple flashes of lightning, executed with such exactness that the illusion's likeness to nature provokes horror and wonder. Meanwhile, Aeolus is seen ascending with eight winds, on each side are four winds…. The two winds that floated in the air at his sides now flee, one to the left, the other to the right…. The two other winds, in the wings of the theater to the left of Aeolus, now move forwards in the airy heights and, after they have made an artful tour like two whirlwinds, float to the right of the theater.[76] (Corneille 1651, 37–8)

Corneille still argues for the theater's resemblance to nature, but the playful machine spectacle at the center of his dramaturgy belies that pretext. He also stresses that what is important is not the eloquence of the drama; instead, the spectator's "spirit" should be moved by the effects of Torelli's machines.[77]

A similar development is evident in the work of the German architect Joseph Furttenbach, another student of Parigi's. Furttenbach's instructions for theater stage designs read as direct references to what he observed during his studies in Florence. Yet whereas Sabbattini's *Pratica* is an essentially neutral construction manual, Furttenbach expatiates on the anticipated aesthetic effects. Rather than giving a simple and clear account of theater machines, he exalts them in poetic words. This also explains the direction for a scene change that must take place so swiftly that "the onlookers, astounded by such a hasty transformation, may hardly know what is happening to them so that they sit there with almost bewildered senses. This transformation then brings a special refreshment of the human spirit."[78]

In his instructions for the "first cloud," Furttenbach describes how the machines disappear before the eyes of the onlookers, but "the sagacious viewer, upon further reflection, will derive much more pleasure from this machine."[79] Only with the spectator's "reflection," then, does the mechanical aesthetic unfold its effect. Not unlike Sabbattini, Furttenbach concludes his writings on theater engineering with the remark that what occurs in the process of viewing cannot be put into words:

> Contemplation of the scenery makes for a blissful and profoundly reflective experience that moreover quickens the human eye and heart, which cannot be sufficiently described by my pen; still less can I express all the other delights contained in it. The sights that go into it charm not only one inexperienced in these things but even the master himself, and indeed leave him so baffled, that man in his imperfection is, one might say, dumbfounded by them and his senses ravished. Whence his reason wanders in a different and new world so lovely that a melancholy soul is refreshed and strengthened and a man's life extended.[80]

---

[76]The set for *Andromède* reused Torelli's designs for the *Orphée* opera. See (Bjurström 1961, 151).

[77]Corneille justified the lack of beautiful verses by remarking that the function of the spectacle is to awaken visual sensation, "not to move the spirit through the power of the thoughts [of literature], or to touch the heart through the delicacy of the passions" (Corneille 1651, 13).

[78]"[D]ie Zuseher in solcher so eylfertigen Verwandlung bestürzt werden/auch kaum wissen mögen/wie ihnen geschihet/dannenhero sie gleichsam verzucktes Sinnes da sitzen/welche transmutation dann/deß Menschen Geist sonderbare Erquickung bringet" (Furttenbach 1663, 119).

[79]"Der verständige wird im weitern nachsinnen noch vil mehr Ergötzligkeit bey dieser machina finden" (Furttenbach 1663, 125).

[80]"Was für ein holdseelig/tieff nachsinndendes/beneben des Menschen Auge und Herz erquickendes Anschawen umb die Prospectiva … seye/das mag mit eigener Feder nit zur genüge beschrieben/noch vil weniger deroselben vil insich habende/so wol erfrewliche Ding/ausgesprochen werden. Sintemahlen der also hinein lauffenden Augenpuncten/nicht allein den hierinn unerfahrenen/sondern auch den Meister selbsten/…/dermassen verführet/ja solcher massen bestürtzet/das der so unvollkommene Mensch gleichsam darüber erstummet/und sein Sinnlichkeit entzucket wird. Dannerhero er mit seiner Vernunft in einer anderen newen liebreichen Welt umbherschweiffen thut/so gar/das hierdurch manichmal die Melancolische Gemüther erfrischet/gestärcket/und zu längerem Leben angereizet wird" (Furttenbach 1663, 111).

In Furttenbach's view, even the architect's practical knowledge is thus part of an aesthetics of effect that cannot easily be translated into the language of an architectural treatise.[81]

This link between court theater aesthetics and philosophical inspiration also appears in seventeenth-century scientific utopias. The German physician and collector Johann Daniel Mayor (1634–1693), for example, had studied in Padua in the 1650s and was familiar with northern Italian theater. In Mayor's 1670 essay "See-Farth nach der Neuen Welt ohne Schiff und Segel" (A Sea Voyage to the New World, Without Ship and Sails), the protagonist Daedalus is traveling to a utopian place called "Cosmosophorum" that, unlike Francis Bacon's *New Atlantis*, cannot be reached by sail. Instead, Daedalus departs from a Venetian opera stage, on a flying machine designed by theater engineers in Venice (Mayor 1670, 5–6). With this machine, Daedalus reaches the new and peaceful society, which is busy establishing wonder cabinets, art collections, and laboratories of the experimental sciences. While in Old Europe inventions such as the compass, print, or gunpowder are in danger of being misused for military purposes, in "Cosmosophorum" all these inventions exist in peaceful contexts (Mayor 1670, 4–5). This is why even flying machines can be invented and used there, whereas Mayor advises his readers in the Old World to exercise flights of the mind instead (Mayor 1670, 81). At the end of his essay, Mayor provides a recipe for all those willing to try such imaginary flights. Among the ingredients are a good appetite for wisdom, much ingenuity (or an inventive energy), and finally Fortuna—luck (Mayor 1670, 81–4). But where is this luck to be found? Mayor has an answer ready: in the New World, in utopia (Mayor 1670, 84). This is where the essay comes full circle, as the luck required for imaginary flights can only be found at the flight's destination. The only way for the mind to take off, then, seems to be attending an opera performance in Venice.

I will conclude this chapter with a final example linking theater machines with the new spirit of natural philosophy. In the essay "Drôle de pensée" (1675), the German philosopher Johann Gottfried Leibniz (1646–1716) sets out his plans for an "Académie des Jeux"[82] dedicated to natural philosophy and the arts, which was to follow the model of the Berlin Academy of Science but on a much larger scale.[83] The academy would be spread across the city of Paris and house all sorts of spectacular experiments, performances, and collections. Among its many attractions, it would accommodate shadow plays and the "marionettes of the Marais or the Pygmies" (Leibniz 1983, 567), so that visitors would be able to see "endless metamorphoses, salti mortali and flights, and the sorcerer Circe" (Leibniz 1983, 568). By mentioning the Théâtre du Marais, Leibniz invoked the most prominent machine theater of late seventeenth-century Paris—at the Marais, the spectacular stage machines were often even presented without any human actors appearing on stage. The theater engineer Denis Buffequin had designed the stage for Claude Boyer's (1618–1698) *Ulysse dans l'Isle de Circé ou Euriloche foudroyé*, which became famous for its numerous stage flight effects, for the Marais in 1648.[84] As many of Buffequin's stage designs were revived at the Marais in the subsequent years, Leibniz could have seen *Ulysse* during his stay in Paris.[85] In the same year that Leibniz published "Drôle de pensée," the entrepreneur Alexandre de Rieux (known as the Marquis de Sourdéac) (1620–1695) also staged the machine opera *Circé* by Thomas Corneille (1625–1709) and Jean Donneau de

[81] Here, Furttenbach connects with the discourse on the inevitability of errors in the translation of practical knowledge into written records, as described in (Chap. 6).

[82] Leibniz later added the sub-title "plus tost Academie des Ieux" ("or rather, Academy of Games") (Leibniz 1983, 562).

[83] The projects that Leibniz mentions in "Drôle de Pensée" can for the most part be traced back to the experimental and theatrical culture that he witnessed during his stay in Paris from 1672 to 1676. See (Bredekamp 2004, 45–63, 237–46; Daston and Park 1998, 215–20; Lazardzig 2007, 221, 228–37). The literature on Leibniz's plans for a learned society is extensive; see, for example, (Brather 1993; Böger 1997; Bredekamp 2004, 23–9, 170–89).

[84] During the performance of *Ulysse*, the Théâtre du Marais was already equipped with a spectacular upper and lower stage. Reviewers spoke of the flying machines of Aeolus and Circe (Dietrich 1958a, 202, 207).

[85] On the revivals of Buffequin's productions at the Marais, see (Dietrich 1958b, 331).

Visé (1638–1710) at the Comédiens du Faubourg Saint-Germaine.[86] The *Gazette d'Amsterdam* reviewer writes of this performance that the "stage flights and other machine effects are so spectacular that they surpass all imagination."[87] A comparable game with the observer's imagination was Leibniz's objective with his Paris academy.[88] Yet the academy's attractions were neither to overwhelm the visitor, like the performance of *Circé* as described in the *Gazette d'Amsterdam*, nor did Leibniz intend to give human imagination free rein—at least, this is how he argued in a similar, earlier proposal for a "Theatrum naturae et artis."[89] Rather, he hoped that the academy's attractions would "open the visitor's eyes, inspire him to new inventions, and provide him with beautiful views, and familiarize the whole world with an endless series of useful and ingenious novelties" (Leibniz 1983, 565).

3 Conclusion

This paper has looked at the generation, transfer, and codification of knowledge on theater engineering between around 1480 and 1680. Examining Leonardo da Vinci's contributions to the festival and theater culture of his time, I traced the transformation of theater engineering from a rather lowly form of craftsmanship into a lucrative and prolific branch of engineering. In the late fifteenth century, the art of theater reentered the canon of the mechanical arts. However, philosophical and encyclopedic writings on the theater omit all reference to contemporary festival culture, where theater engineering flourished. The only surviving records of the theater machines designed in this period are found in biographies of artists, city chronicles, account books of churches and courts, and the notebooks of artist-engineers.

What makes Leonardo's theater machines an interesting case for the study of the structures of practical knowledge is that they were, firstly, deeply embedded in northern Italian festival culture, as can be seen in the evolution of the *mandorla*, which was invented as part of the cloud machine used for religious processions and secular parades, then adopted first in the designs of celestial machines in the *sacre rappresentazioni* and later as part of the flight-effect machines of courtly performances. A second source of inspiration for Leonardo's theater machines may have been his research on human flight, including his numerous designs for flying machines. Striking here is that in Leonardo's work, the parallels between stage design and other fields of engineering are already more than just technological: both theater machines and flight machines served him as "instruments of thought" to develop an experimental approach to nature that transcended traditional lines of thinking.

This attribution of epistemic value to theater engineering became increasingly explicit in the early seventeenth century. Architects such as Giulio Parigi were admired for their technical skill in designing ever more advanced stage effects, but they were careful to emphasize that the technical function of their designs did not exceed the bounds of nature. Parigi linked stage design to other fields of civil architecture and included it in the curriculum of his art academy. In Parigi's day, training in theater engineering took on the aspect of regular schooling, with master-pupil relationships at court theaters and architects from all over Europe studying in northern Italy or learning about Italian stage design from festival descriptions and etchings.

[86] This conjunction is pointed out by (Lazardzig 2007, 202–5).

[87] "[L]es mouuements des vols et des machines de cette dernière sont si extraordinaires qu'ils passent l'imagination," *Gazette d'Amsterdam* (February 14, 1675). Quoted in (Lazardzig 2007, 205).

[88] On Leibniz's concept of games, see (Schramm 2003, 194).

[89] Leibniz's notion of the *theatrum naturae et artis* was shaped by the desire to hone the mind using encylopedic arrangements of natural and artificial objects (Leibniz 1986, 898). See also (Bredekamp 2004, 27).

Among those students was Nicola Sabbattini, the first to publish a manual on theater engineering, the *Pratica*, and the first to reflect on the different character of knowledge transfer via actual performances and via books—the book here seen as a *theatrum machinarum*. These theater machine books by Sabbattini, Torelli, or Furttenbach are part of a whole series of publications that appeared in the realm of European court theater starting in the mid-seventeenth century.[90] In them, the architects not only reproduced their technological knowledge on stage design, but reflected more generally on how the theater audience's imagination can be fired by the inventive spirit of engineering. These books' emphasis on the machine's effect on the audience may partly be explained by the fact that they were intended less as textbooks than as advertising brochures, which the authors hoped would attract potential employers. Most of the writers were apprenticed in Italy and belonged to a community of court architects in Florence, Venice, Versailles, or Vienna that shared practical and theoretical knowledge. Their publications may thus be seen as representative of the dominant discourse on theater engineering in seventeenth-century Europe. Yet outside this discourse, in many places, such as smaller courts or schools, theater engineering was still developing rather idiosyncratically. Rather than professional engineers trained in Italian academic circles, it was local carpenters, painters, rope makers, sculptors, and dressmakers who were commissioned to produce theater performances there. As I have shown elsewhere,[91] surviving account books and performance reports illustrate the somewhat unsystematic training of these craftsmen and the plethora of local distinctions with regard to the purpose and form of the theater machines, construction materials, and the technological terminology used by local craftsmen.

Nevertheless, this paper's focus on the grand court theaters of Europe—with its rather linear narrative of knowledge accumulation in theater engineering—has allowed us to trace the transformation of practical engineering knowledge into a multilayered philosophical knowledge about both mechanics and nature. Sabbattini, for example, compared the inventive spirit of theater engineering to that of poetry, but was careful to ensure that the visible effects of his machines did not go too far beyond what appeared natural. In its attempt to assimilate stage effects to natural effects, Sabbattini's exacting typology of cloud machines evinces parallels with Descartes's essay on meteorology and even with Luke Howard's later typology of clouds. The effect was not so much an immediate contribution to a new understanding of nature. Rather, the playful hide-and-seek, the oscillation between natural appearance and technical skill, was what may have prompted philosophers such as Francis Bacon, or later Daniel Mayor and Gottfried Wilhelm Leibniz, to codify theater machines as a suitable inspiration for natural philosophy and so to re-structure this area of mechanics.

References

Alberti, Leon Battista. (1443–1452) 1988. *On the art of building in ten books*. Trans: Neil Leach, Joseph Rykwert, and Robert Tavernor. Cambridge, MA: MIT Press.

Bacher, Jutta. 2000a. Artes mechanicae. In *Erkenntnis, Erfindung, Konstruktion. Studien zur Bildgeschichte von Naturwissenschaften und Technik vom 16. bis zum 19. Jahrhundert*, ed. Hans Holländer, 35–49. Berlin: Gebr. Mann Verlag.

Bacher, Jutta. 2000b. "Ingenium vires superat." Die Emanzipation der Mechanik und ihr Verhältnis zur Ars Scientia und Philosophia. In *Erkenntnis, Erfindung, Konstruktion. Studien zur Bildgeschichte von Naturwissenschaften und Technik vom 16. bis zum 19. Jahrhundert*, ed. Hans Holländer, 519–555. Berlin: Gebr. Mann Verlag.

Bacon, Francis. (1620) 2000. *The New Organon*. Trans: Michael Silverthorne, ed. Lisa Jardine and Michael Silverthorne. Cambridge: Cambridge University Press.

[90] For partial English translations of these machine books, see (Carini Motta 1987; Hewitt 1958).

[91] See the case of the theater at Schloss Friedenstein in Gotha, examined in (Tkaczyk 2011, 267–83).

Bacon, Francis. (1625) 1859a. Of masques and triumphs. In *The works of Francis Bacon*, vol. 6, ed. James Spedding, Robert L. Ellis, and Douglas D. Heath, 467–468. London: Longman.

Bacon, Francis. (1627) 1859b. New Atlantis. In *The works of Francis Bacon*, vol. 3, ed. James Spedding, Robert L. Ellis, and Douglas D. Heath, 119–166. London: Longman.

Baldinucci, Filippo. (1682) 1812. *Notizie dei professori del disegno da Cimabue in qua....* Milan: Società Tipografica de'Classici Italiani.

Bambach, Carmen C. (ed.). 2003. *Leonardo da Vinci master draftsman*. New Haven: Yale University Press.

Barr, Cyrilla. 1990. Music and spectacle in confraternity drama of fifteenth-century Florence: The reconstruction of a theatrical event. In *Christianity and the Renaissance: Image and religious imagination in the quattrocento*, ed. Timothy Verdon and John Henderson, 376–404. Syracuse: Syracuse University Press.

Beijer, Agne. 1956. Visions célestes et infernales dans le théâtre du Moyen-Âge et de la Renaissance. In *Les fêtes de la Renaissance*, vol. 1, ed. Jean Jacquot, 405–418. Paris: CNRS Éditions.

Belting, Hans. 1981. *Das Bild und sein Publikum im Mittelalter: Form und Funktion früher Bildtafeln der Passion*. Berlin: Gebr. Mann Verlag.

Berti, Luciano. 1951. Giulio and Alfonso Parigi. *Palladio: rivista di storia dell'architettura* 4: 160–164.

Bjurström, Per. 1961. *Giacomo Torelli and Baroque stage design*. Stockholm: Almquist & Wiksell.

Blumenberg, Hans. (1979) 1997. *Schiffbruch mit Zuschauer. Paradigma einer Daseinsmetapher*. Frankfurt am Main: Suhrkamp.

Blumenberg, Hans. 1999. "Nachahmung der Natur." Zur Vorgeschichte der Idee des schöpferischen Menschen. In *Wirklichkeiten, in denen wir leben*, 55–103. Stuttgart: Reclam.

Blumenthal, Arthur B. 1980. *Theatre art of the Medici*. Exhibition catalog. Hanover: University Press of New England.

Blumenthal, Arthur B. (1986/1984) 1992. *Giulio Parigi's stage designs: Florence and the early Baroque spectacle*. New York: Garland.

Böger, Ines. 1997. *"Ein seculum ... da man zu Societäten Lust hat." Darstellung und Analyse der Leibnitzschen Sozietätspläne vor dem Hintergrund der europäischen Akademiebewegung im 17. und frühen 18. Jahrhundert*, 2 vols. Munich: Utz.

Brather, Hans-Stephan (ed.). 1993. *Leibnitz und seine Akademie. Ausgewählte Quellen zur Geschichte der Berliner Sozietät der Wissenschaften 1697–1716*. Berlin: Akademie Verlag.

Bredekamp, Horst. 1995. *The lure of antiquity and the cult of the machine: The Kunstkammer and the evolution of nature, art, and technology*. Trans: Allison Brown. Princeton: M. Wiener.

Bredekamp, Horst. 2004. *Die Fenster der Monade: Gottfried Wilhelm Leibniz' Theater der Natur und Kunst*. Berlin: Akademie Verlag.

Campbell, Lily. (1923) 1960. *Scenes and machines on the English stage during the Renaissance*. Cambridge: Cambridge University Press.

Carini Motta, Fabricio. 1987. *The Theatrical writings of Fabrizio Carini Motta: Translation of Trattato sopra la struttura de' theatri e scene, 1676 and Costruzione de teatri e machine teatrali, 1688*. Trans: Orville K. Larson. Carbondale: Southern Illinois University Press.

Cohen, Gustave. (1906) 1926. *Histoire de la mise en scène dans le théâtre religieux français du Moyen Âge*. Paris: Champion.

Corneille, Pierre. 1651. *Andromède. Tragédie, représentée avec les machines*. Paris: Charles de Sercy.

Crombie, Alistair C. 1961. *Medieval and early modern science*. Cambridge, MA: Harvard University Press.

Daston, Lorraine, and Katharine Park. 1998. *Wonders and the order of nature, 1150–1750*. New York: Zone Books.

Descartes, René. (1637) 2011. *Discourse on method, optics, geometry, and meteorology*. Rev. edition (trans: Paul J. Olscamp). Indianapolis: Hackett.

Dibner, Bernd. 2005. Maschinen und Waffen. In *Leonardo: Künstler, Forscher, Magier*, ed. Silvio A. Bedini, 166–189. Munich: Bassermann.

Dietrich, Margret. 1958a. Der barocke Corneille. Ein Beitrag zum Maschinentheater des 17. Jahrhunderts. Teil I. *Maske und Kothurn* 4 (2–3): 199–219.

Dietrich, Margret. 1958b. Der barocke Corneille. Ein Beitrag zum Maschinentheater des 17. Jahrhunderts. Teil II. *Maske und Kothurn* 4 (4): 316–345.

Edström, Per Simon. 1995. *Liten handbok i gammal teaterteknik*. Varmdö: Arena Teater Institutet.

Eiche, Sabine. 1996. Niccolò Sabbatini. In *The dictionary of art*, vol. 27, ed. Jane Turner, 482. New York: Grove.

Fabbri, Mario, Elvira G. Zorzi, and Anna Maria P. Tofani (eds.). 1975. *Il luogo teatrale a Firenze*. Milan: Electa.

Fanelli, Giovanni. 1973. *Firenze: architettura e città*. Florence: Mandragora.

Flemming, Willi. 1929. Afterword to Nicola Sabbattini, *Anleitung Dekorationen und Theatermaschinen herzustellen*, ed. and trans. Willi Flemming, 281–91. Weimar: Gesellschaft der Bibliophilen.

Franke, Birgit. 2003. Automaten in höfischen Lustgärten der Frühen Neuzeit. In *Automaten in Kunst und Literatur des Mittelalters und der Frühen Neuzeit*, ed. Klaus Grubmüller and Markus Stock, 247–268. Wiesbaden: Harrassowitz.

Furttenbach, Joseph. 1640. *Architectura Recreationis, Das ist: Von Allerhand Nutzlich und Erfrewlichen Civilischen Gebäwen....* Augsburg: Schultes.
Furttenbach, Joseph. 1663. *Mannhaffter Kunst = Spiegel/Oder Continuatio, und Fortsetzung allerhand Mathematisch = und Mechanisch = hochnutzlich = so wol auch sehr erfrölich delectionen, und respective im Werck selbsten experimentirten freyen Künsten....* Augsburg: Schulte.
Gaeta Bertelà, Giovanni, and Annamaria Petrioli Tofani (eds.). 1969. *Feste e apparati medicei da Cosimo I a Cosimo II*. Florence: Olschki.
Galluzzi, Paolo. 1997. *Mechanical marvels: Invention in the age of Leonardo*. Florence: Giunti Editore.
Gellius, Aulus. 1927. *The attic Nights of Aulus Gellius*. Trans: John C. Rolfe. Cambridge, MA: Harvard University Press.
Ghiberti, Bonaccorso. Ca. 1500. *Zibaldone*, Codex Banco Rari 228. BNCF. Florence.
Ginzburg, Carlo. 1976. High and low: The theme of forbidden knowledge in the sixteenth and seventeenth centuries. *Past & Present: A Journal of Historical Studies* 73: 28–41.
Ginzburg, Carlo. 1991. *Ecstasies: Deciphering the witches' Sabbath*. Trans: Raymond Rosenthal. New York: Pantheon.
Godefroy de Saint Victor. (1180) 1951. *Microcosmus*, ed. Philippe Delhaye. Lille: Facultés Catholiques.
Gombrich, Ernst H. 1976. *The heritage of Apelles*. Oxford: Phaidon.
Hamblyn, Richard. 2001. *The invention of clouds: How an amateur meteorologist forged the language of the skies*. London: Picador.
Hankins, Thomas L., and Robert J. Silverman. 1995. *Instruments and the imagination*. Princeton: Princeton University Press.
Hart, Ivor B. 1962. *The world of Leonardo da Vinci: Man of science, engineer and dreamer of flight*. New York: Viking Press.
Hart, Clive, and Kay Gilliland Stevenson. 1995. *Heaven and the flesh: Imagery of desire from the Renaissance to the Rococo*. Cambridge: Cambridge University Press.
Helas, Philine. 1999. *Lebende Bilder in der italienischen Festkultur des 15. Jahrhunderts*. Berlin: Akademie-Verlag.
Hewitt, Barnard (ed.). 1958. *The Renaissance stage: Documents of Serlio, Sabbattini and Furttenbach*. Trans: Allardyce Nicoll, John H. McDowell, and George R. Kernodle. Coral Gables: University of Miami Press.
Heydenreich, Ludwig Heinrich. 1954. *Leonardo da Vinci*. Basel: Holbein.
Horn, Angelica. 2000. Das Experiment mit der Zentralperspektive: Filippo Brunelleschi und René Descartes. In *Descartes im Diskurs der Neuzeit*, ed. Angelica Horn, Wilhelm Friedrich Niebel, and Herbert Schnädelbach, 9–32. Frankfurt am Main: Suhrkamp.
Howard, Luke. 1803. On the modifications of clouds, and on the principles of their production, suspension, and destruction: Being the substance of an essay read before the Askesian Society in the session 1802–1803. *Philosophical Magazine* 16 (62): 97–107; 16 (64): 344–57; 17 (65): 5–11.
Hugh of St. Victor. (1176–1177) 1991. *The Didascalicon of Hugh of Saint Victor: A Medieval guide to the arts*. Trans: Jerome Taylor. New York: Columbia University Press.
Janković, Vladimir. 2000. *Reading the skies: A cultural history of the English weather, 1650–1820*. Manchester: Manchester University Press.
Kahn, Arthur D. 1970. Every art possessed by man comes from Prometheus: The Greek tragedians and science and technology. *Technology and Culture* 11: 133–162.
Kemp, Martin. 1977. From "mimesis" to "fantasia." The quattrocento vocabulary of creation, inspiration and genius in the visual arts. *Viator. Medieval and Renaissance Studies* 8: 349–397.
Kemp, Martin. 1981. *Leonardo da Vinci: The marvellous works of nature and man*. London: J. M. Dent.
Kemp, Martin. 2002. Die Zeichen lesen. Zur graphischen Darstellung von physischer und mentaler Bewegung in den Manuskripten Leonardos. In *Leonardo da Vinci: Natur im Übergang*, ed. Frank Fehrenbach, 207–228. Munich: Fink.
Kilwardby, Robert. (1279) 1976. *De ortu scientiarum*, ed. Albert G. Judy. Oxford: Oxford University Press.
Koopmans, Jelle. 2006. De la survivance des dieux antiques à la survie de l'humanité. In *Dieu et les dieux dans le théâtre de la Renaissance*, ed. Jean-Pierre Bordier and André Lascombes, 235–247. Tours: Brepols.
Körber, Hans-Günther. 1987. *Vom Wetteraberglauben zur Wetterforschung. Aus Geschichte und Kulturgeschichte der Meteorologie*. Innsbruck: Pinguin-Verlag.
Krämer, Sybille. 2003. Über die Rationalisierung der Visualität und die Visualisierung der Ratio: Zentralperspektive und Kalkül als Kulturtechniken des "geistigen Auges." In *Bühnen des Wissens: Interferenzen zwischen Wissenschaft und Kunst*, ed. Helmar Schramm, Hans-Christian von Herrmann, Florian Nelle, Wolfgang Schäffner, Henning Schmidgen, and Bernhard Siegert, 50–67. Berlin: Dahlem University Press.
Kristeller, Paul Oskar. 1951. The modern system of the arts: A study in the history of aesthetics, Part 1. *Journal of the History of Ideas* 12 (4): 496–527.
Küster, Ulf (ed.). 2003. *Theatrum mundi. Die Welt als Bühne*. Exhibition catalog. Munich: Haus der Kulturen.
Laurenza, Domenico. 2004. *Leonardo on flight*. Florence: Giunti Editore.

Lazardzig, Jan. 2007. *Theatermaschine und Festungsbau. Paradoxien der Wissensproduktion im 17. Jahrhundert*. Berlin: Akademie Verlag.

Leibniz, Gottfried Wilhelm. (1675) 1983. Drôle de pensée, touchant une nouvelle sorte de REPRESENTATIONS. In *Sämtliche Schriften und Briefe*, vol. 1 part 4, ed. Akademie der Wissenschaften der DDR, IV, 562–568. Berlin: Akademie-Verlag.

Leibniz, Gottfried Wilhelm. (1679) 1986. Agenda. In *Sämtliche Schriften und Briefe*, vol. 3 part 4, ed. Akademie der Wissenschaften der DDR, IV, 894–902. Berlin: Akademie-Verlag.

Leonardo da Vinci. 1955. *The notebooks of Leonardo da Vinci* (ed. and trans: Edward MacCurdy). New York: George Brazillier.

Leonardo da Vinci. 1970. *The literary works of Leonardo da Vinci* (ed. and trans: Jean-Paul, Richter). 3rd edition. London: Phaidon.

Leonardo da Vinci. 1974. *Il codice Atlantico (Edizione in facsimile dopo il restauro dell'original conservato nella Biblioteca Ambrosiana di Milano)*, vol. 7. Florence: Giunti-Barbèra.

Leonardo da Vinci. (1505) 1978. *Il codice sul volo degli uccelli*. Reprinted with translation and notes in *Traktat über den Flug der Vögel. In der königlichen Bibliothek von Turin*, ed. Augusto Marioni and Sigrid Braunfels. Würzburg: Georg Popp.

Linnenkamp, Rolf. 1958. Giulio Parigi architetto. *Rivista d'arte* 33 (7): 51–63.

Long, Pamela O. 2001. *Openness, secrecy, authorship: Technical arts and the culture of knowledge from antiquity to the Renaissance*. Baltimore: Johns Hopkins University Press.

Mazzocchi Doglio, Mariangela. 1983. Leonardo "apparatore" di spettacoli a Milano per la corte degli Sforza. In *Leonardo e gli spettacoli del suo tempo*, ed. Mariangela Mazzocchi Doglio, 40–76. Milan: Electa.

Mayor, Johann Daniel. 1670. *See-Farth nach der Neuen Welt/ohne Schiff und Segel*. Kiel: Joachim Reuman.

McKinven, John A. 1995. *Stage flying: 431 B.C. to modern times*. Glenwood: David Meyer.

Melchinger, Siegfried. 1974. *Das Theater der Tragödie. Aischylos, Sophokles, Euripides auf der Bühne ihrer Zeit*. Munich: C.H. Beck.

Middleton, W.E. Knowles. 1969. *Invention of the meteorological instruments*. Baltimore: Johns Hopkins University Press.

Molinari, Cesare. 1968. *Le nozze degli dèi. Un saggio sul grande spettacolo italiano nel seicento*. Rome: Madio Bulzoni Editore.

Moniglia, Giovanni Andrea. 1658. *Descrizione della presa d'Argo e de gli amori di Linceo con Hipermestra*.... Florence: Stamperia di S.A.S.

Nagler, Alois. 1964. *Theatre festivals of the Medici, 1539–1637*. New Haven: Yale University Press.

Nelle, Florian. 2001. Von der beobachteten zur inszenierten Natur: Descartes und der Regenbogen im Wasserglas. In *Theatralität und die Krisen der Repräsentation*, ed. Erika Fischer-Lichte, 392–410. Stuttgart: Metzler.

Newiger, Hans-Joachim. 1996a. Ekykklema und Mechané in der Inszenierung des griechischen Dramas [1990]. In *Drama und Theater. Ausgewählte Schriften zum griechischen Drama*, 96–106. Stuttgart: M & P Verlag für Wissenschaft und Forschung.

Newiger, Hans-Joachim. 1996b. Zum Theater der Tragödie [1985]. In *Drama und Theater. Ausgewählte Schriften zum griechischen Drama*, 86–95. Stuttgart: M & P Verlag für Wissenschaft und Forschung.

Parigi, Giulio, and Remigio Cantagallina. 1608. *Come: rapp. Nelle nozze del Sermo Principe di Toscana l'an 1608*. Florence.

Popplow, Marcus. 1998. *Neu, nützlich und erfindungsreich: Die Idealisierung von Technik in der frühen Neuzeit*. Münster: Waxmann.

Povoledo, Elena. 1960. Parigi, Giulio; Parigi, Alfonso. In *Enciclopedia dello spettacolo*, vol. 7, 1675–1678. Rome: Casa editrice Le Maschere.

Prager, Frank D., and Gustina Scaglia. 2004. *Brunelleschi: Studies of his technology and inventions*. Mineola: Dover Publications.

Reisch, Gregor. 1504. *Margarita philosophica*. Freiburg: Johann Schott.

Rinuccini, Camillo. 1608. *Descrizione delle festa fatte nelle reali nozze de' Serenissimi principi di Toscana d. Cosimo de' Medici, e Maria Maddalena arcidvchessa d'Avstria*. Florence: Appresso i Giunti.

Roßbach, Nikola. 2013. *Poiesis der Maschine: Barocke Konfigurationen von Technik, Literatur und Theater*. Berlin: Akademie Verlag.

Sabbattini, Nicola. 1637. *Pratica di fabricar scene, e machine ne'teatri*. Pesaro: Flaminio Concordia.

Sabbattini, Nicola. (1637) 1638. *Pratica di fabricar scene, e machine ne' teatri*. 2nd edition. Ravenna: Pietro de' Paolini, e Gio. Battista Giovannelli.

Sabbattini, Nicola. (1638) 1926. *Pratica di fabricar scene e machine ne' teatri*. Facsimile. Trans: Willy Fleming. Weimar: Gesellschaft der Bibliophilen.

Sabbattini, Nicola. (1638) 1958. Manual for constructing theatrical scenes and machines. Trans: John H. Dowell. In *The Renaissance stage: Documents of Serlio, Sabbattini and Furttenbach*, ed. Barnard Hewitt, 37–177. Coral Gables : University of Miami Press.
Sánchez de Arévalo, Rodrigo. 1471. *Speculum vitae humanae*. Augsburg: Gunther Zainer.
Scaglia, Gustina. 1991. Building the cathedral in Florence. *Scientific American* 264: 66–72.
Schmidt, Wieland. 1963. *Der deus ex machina bei Euripides*. Eiserfeld: Fotodruck EDWE.
Schmidt-Biggemann, Wilhelm. 1980. Maschine. In *Historisches Wörterbuch der Philosophie*, vol. 5, ed. Joachim Ritter and Karlfried Gründer, 790–802. Darmstadt: Wissenschaftliche Buchgesellschaft.
Schneider, Helmut. 1989. *Das griechische Technikverständnis: Von den Epen Homers bis zu den Anfängen der technologischen Fachliteratur*. Darmstadt: Wissenschaftliche Buchgesellschaft.
Schramm, Helmar. 1996. *Karneval des Denkens: Theatralität im Spiegel philosophischer Texte des 16. und 17. Jahrhunderts*. Berlin: Akademie Verlag.
Schramm, Helmar. 2003. Feuerwerk und Raketentechnik um 1700. In *Bühnen des Wissens. Interferenzen zwischen Wissenschaft und Kunst*, ed. Helmar Schramm, Hans-Christian von Herrmann, Florian Nelle, Wolfgang Schäffner, Henning Schmidgen, and Bernhard Siegert, 183–213. Berlin: Dahlem University Press.
Schramm, Helmar, Ludger Schwarte, and Jan Lazardzig (eds.). 2008. *Spuren der Avantgarde: Theatrum machinarum. Frühe Neuzeit und Moderne im Kulturvergleich*. Berlin: De Gruyter.
Seznec, Jean. 1953. *The survival of the pagan gods: The mythological tradition and its place in Renaissance humanism and art*. Trans: Barbara Sessions. New York: Pantheon Books.
Solerti, Angelo. (1905) 1989. *Musica, ballo e drammatica alla Corte Medicea dal 1600 al 1637. Notizie tratte da un diario con appendice di testi inediti e rari*. Bologna: Arnaldo Forni Editore.
Steinitz, Kate T. 1969. *Leonardo architetto teatrale e organizzatore di feste*. Florence: G. Barbèra.
Sternagel, Peter. 1966. *Die artes mechanicae im Mittelalter: Begriffs- und Bedeutungsgeschichte bis zum Ende des 13. Jahrhunderts*. Kallmünz: Lassleben.
Strong, Roy. 1973. *Splendor at court: Renaissance spectacle and the art of power*. Boston: Houghton Mifflin.
Strong, Roy. 1984. *Art and power: Renaissance festivals 1450–1650*. Berkeley: University of California Press.
Taplin, Oliver. 1978. *Greek tragedy in action*. Oxford: Oxford University Press.
Tazi, Nadia. 1989. Celestial bodies: A few stops on the way to heaven. In *Fragments for a history of the human body*, vol. 2, ed. Michel Feher, Ramona Naddaff, and Nadia Tazi, 518–552. New York: Zone Books.
Tintelnot, Hans. 1939. *Barocktheater und barocke Kunst: Die Entwicklungsgeschichte der Fest- und Theaterdekorationen in ihrem Verhältnis zur barocken Kunst*. Berlin: Gebr. Mann Verlag.
Tkaczyk, Viktoria. 2011. *Himmels-Falten. Zur Theatralität des Fliegens in der Frühen Neuzeit*. Munich: Fink.
Torelli, Giacomo. 1644. *Apparati scenici per lo teatro nouissimo di Venetia*. Venice: Vecellio.
Valleriani, Matteo. 2014. Ancient pneumatics transformed during the early modern period. *Nuncius* 29: 127–173.
Vasari, Giorgio. (1550) 1768. Vite del Cecca ingegnere fiorentino. In *Le vite de' più eccellenti architetti, pittori et scultori italiani, da Cimabue infino a'tempi nostri*, part 2. Florence: L. Torrentino.
Vasari, Giorgio. 1912–1914. *The lives of the most eminent painters, sculptors and architects*. Trans: Gaston du C. de Vere. 4 vols. London: Macmillan.
Vespucci, Amerigo. (1504) 2002. *Der Mundus Novus des Amerigo Vespucci*, ed. Robert Wallisch. Vienna: Verlag der österreichischen Akademie der Wissenschaften.
Villifranchi, Giovanni. 1613. *Descrizzione della barriera e della mascherata, fatte in Firenze a' XVII. & a' XIX. di febbraio MDCXII*. Florence: Appresso Bartolommeo Sermartelli e fratelli.
Warburg, Aby. 1932. Die Bildchronik eines florentinischen Goldschmiedes. In *Die Erneuerung der heidnischen Antike: Kulturwissenschaftliche Beiträge zur Geschichte der europäischen Renaissance*, vol. 1, ed. Gertrud Bing and Fritz Rougement, 69–77. Leipzig: Teubner.
Weihe, Richard Emanuel. 1992. *Die Theater-Maschine. Motion und Emotion*. Zurich: Zentralstelle der Studentenschaft.
Werrett, Simon. 2001. Wonders never cease: Descartes's "Météores" and the rainbow fountain. *British Journal for the History of Science* 34 (3): 129–147.
Wesselofsky, Alexander. 1877. Italienische Mysterien in einem Reisebericht des XV. Jahrhunderts. *Russische Revue: Monatszeitschrift für die Kunde Russlands* 10: 425–441.
Whitney, Elspeth. 1990. *Paradise restored: The mechanical arts from antiquity through the thirteenth century*. Philadelphia: American Philosophical Society.
Zorzi, Elvira Garbero, and Mario Sperenzi (eds.). 2001. *Teatro e spettacolo nella Firenze dei Medici: modelli dei luoghi teatrali* exhibition catalogue. Firenze: Olschki.

Shooting with Ink

5

Jochen Büttner

Abstract

The sixteenth century saw the emergence of preclassical external ballistics. It was engendered by a new representational means—the trace of the path of a shot outlined on paper. This resulted in the concept of a trajectory in which practitioners's knowledge of the behavior of gunshots was fused with knowledge from other domains. In the practical literature of the sixteenth and seventeenth centuries, preclassical external ballistics increasingly became a self-contained subject. The different accounts given were abstracted and segregated from the practical knowledge on which they were based and often focused on aspects that had little or no relevance for the actual practice of shooting. Preclassical external ballistics had virtually no impact on the practice of gunnery in the sixteenth and seventeenth centuries and only marginal improvements were made due to its application to gun shooting in the field. Preclassical external ballistics reestablished contact with the practice of shooting from the second half of the seventeenth century onwards when learned men with close ties to practice, who relied on the help of gunners and their infrastructure, transformed guns into experimental systems in order to probe the claims of external ballistics. This was mainly done because of the precarious status it had earned due to its divergence from practical knowledge and gunnery practice.

J. Büttner (✉)
Max Planck Institute for the History of Science, Berlin, Germany
e-mail: buettner@mpiwg-berlin.mpg.de

M. Valleriani (ed.), *The Structures of Practical Knowledge*, DOI 10.1007/978-3-319-45671-3_5

Not long after the practice of shooting gunpowder artillery had become firmly established in the Latin West in the fourteenth century, a second, closely linked tradition was born—writing about shooting.[1] Just as the use of cannons proliferated over the next centuries, so did the amount of writing about it. The sixteenth and in particular the seventeenth century saw an ever increasing output of military text books, many of them pertinent to questions of gunnery.[2] This proliferation is certainly related to the introduction of the printing press in the previous century. However, it can be argued that it was caused predominantly by changes within the field itself. In particular, this new type of literature advanced to meet a need to train gunners on a grand scale, but also a growing need for technological knowledge transfer, due to political leaders' awareness of the military importance of shooting. In short, just as the practice of gunnery changed dramatically in the period from the fourteenth to the seventeenth century, so did the writing about it. Notwithstanding the increasing diversification of the sources conveying information about shooting gunpowder artillery, in the following, these sources will be referred to collectively by the umbrella term "the practical literature of gunnery."

Writing about shooting, or "shooting with ink" as it is referred to in the title, is not shooting, and the knowledge codified in the practical literature of gunnery should not, it will be argued, be accepted too hastily as a faithful rendition of the practical knowledge of gunners. Nor should it be taken for granted that this literature was aimed at gunners in an attempt to improve practice, despite frequent assertions to this effect in the works in question. With the aim of analyzing how shooting and shooting with ink were actually related, and how this engendered what has been referred to in hindsight as a "science of shooting," this article surveys the practical literature of gunnery from its beginnings up until roughly the end of the seventeenth century.

The use of gunpowder artillery incorporated an extremely wide range of activities, from the production of powder via the fabrication of guns and their carriages, to the testing and finally the shooting of guns in the field, to mention just a few. Equally we must include numerous other activities which might appear to the modern reader to be only incidentally related, but fell under the responsibility of gunners during this period, such as the preparation of fireworks, for instance.[3] Each of these activities was, in turn, subdivided into a large number of subordinate tasks, all of which are treated in the practical literature of gunnery.[4] This article focuses on just one aspect of these activities, that is, the shooting itself, and even more specifically, on what today would be called external or exterior ballistics.

External ballistics is an anachronistic category, defined as the (theoretical) study of the flight of a shot after it has left the mouth of a gun and before it hits the target. Applying the notion to the early

---

[1]Exterior ballistics in the early modern period has been a long standing interest of mine. I am hence particularly grateful to Matteo Valleriani for bringing together the working group *The Structures of Practical Knowledge* and thus giving me the occasion to finally put my thoughts on the topic on paper. The feedback from the participants in the discussions during the two workshop events in which the present volume was prepared, but also that from the participants of the 2011 Mellon Conference *The Temper of Evidence*, at which I presented an earlier version of this paper under the title *Bullet proofs: Practical knowledge of gunnery as a source of evidence in early modern theories of ballistics*, has greatly influenced and hopefully also improved the thoughts I have tried to express here.

[2]Military bibliographies from the seventeenth and eighteenth century, such as (Rumpf 1825), even though far from complete, give a good impression of the impressive amount of military books printed in the period.

[3]The terminology used throughout this work does not reflect actors' categories. Indeed, this would hardly be possible, as the relevant technical terminology was still nascent during the period in question, such that terminology varied over time and across cultural backgrounds. Thus, for instance, the term artillery was initially reserved for small firearms, at least in England, and only gradually came to refer to great ordnance. I will use "gun" as a generic term for large firearms and further differentiate between "cannons"—guns for shooting at low angles—and "mortars"—guns used for shooting at high angles. I further will use the word "projectile" for anything that was shot from a gun, be it a cast iron or a stone ball, a bomb, a grenade or even an animal cadaver.

[4]In preparing the cannon for a shot, William Eldred in his *The Gunners Glass*, for instance, distinguishes 11 different steps. See (Eldred 1647).

modern period, I will subsume all information pertinent to the flight of a shot under the term external ballistics. In this sense, a statement such as "a shot will deviate from its expected course in the presence of side wind" falls under the category of external ballistics. However, a statement like "a shot from a hot cannon differs, all other things being equal, from that from the same cannon when cold" refers to the category of internal ballistics. These categories are similar but more broadly interpreted than modern use of the terms would imply.

With regard to the study of the flight of projectiles, a development can be identified that only in hindsight was qualified as theoretical or scientific. The works of Nicolò Tartaglia (ca. 1499–1517) from the mid-sixteenth century, Galileo's (1564–1642) theory of projectile motion from the mid-seventeenth century, and Robins's studies on external ballistics from the mid-eighteenth century are commonly considered milestones in this development. Owing not least to Rupert Hall's seminal study *Ballistics in the Seventeenth Century*, this science of gunnery has often been perceived of as virtually independent from a practice of gunnery whose progress is understood as hinging on the integration of theoretical insights into practice.[5]

Recently historians of science have come to question such dichotomous views of practice and theory. For the concrete case of ballistics, Cathrin Ann France has done just this in her excellent, recently published dissertation "Gunnery and the Struggle for the New Science (1537–1687)," in which she argues that "from Tartaglia on, all endeavoured to mediate between practical demands and the theoretical tools available, recognising and attempting to overcome the limitations of a particular theory. Thus there was an iterative process over time as both technology/practice and theory both developed and moved closer together." France, while skeptical concerning their separation, maintains the basic dichotomy between theory and practice.

It will be argued here that, in order to understand the role of practical knowledge for the formation of thought on ballistics, it is more instructive to focus on the difference between shooting and shooting with ink, i.e., between the practical knowledge and the representations of practical knowledge. This perspective allows us to discern a continuous transition from the first emergence of the very question of external ballistics in a literature then with very strong ties to practice, to the subsequent establishment of external ballistics as a self-contained topic within that very literature. It will be shown that the different approaches to external ballistics contained in the practical literature all share the same basic structure, a structure referred to here as preclassical external ballistics. Preclassical external ballistics, it will be argued, was for the most part a necessary consequence of the codification of practical knowledge based on a particular representational means—the trajectory.

Preclassical external ballistics on paper, for reasons to be discussed, abstracted practical knowledge, giving rise to tensions and an increasing separation between such abstractions and the practical knowledge on which they were based. It was not so much the promised application of preclassical external ballistics to gunnery practice, as the realization of these tensions that caused engineer-scientists to collaborate with gunners in performing systematic trials with guns and, thus, to eventually reconnect preclassical external ballistics with practice.

[5] Cathrin A. France has summarized Hall's position as follows: "Hall's thesis is based on a division between scientists, who had little or no contact with practice, and writers of gunnery manuals, who incorporated the theory into their manuals, usually when it was out of date, to no apparent purpose" (France 2014, 288).

1 Practical Literature vs. Practical Knowledge

The information conveyed by the practical literature of gunnery concerning external ballistics was rooted, as will be demonstrated, in the practical knowledge of gunners and, with some qualifications, was meant to be of use in the practice of gunnery. At the same time, the information codified in the practical literature could not straightforwardly represent the practical knowledge of gunnery and its structures.

The argument to be made hinges on a particular understanding of the difference between practical knowledge in the sense of "knowledge-how," as it is often referred to on the one hand, and explicit representations of such knowledge on the other hand. Thus some general remarks on practical knowledge and its codification are in order (Fantl 2014; Wiggins 2012). Knowledge-how is understood here as knowledge directing actions: it is knowing how to X to obtain Y. Such knowledge becomes manifest primarily in the corresponding action, i.e. in doing X. It is at any rate the individual knowledge of the person performing the action, i.e. an agent's knowledge, or agential knowledge. It is not usually put into words, let alone written down or otherwise symbolically codified. Practical actions may encompass external statements or acts involving symbols, such as oral commands or instructions in collaborative actions, the recording of length measures, the notation of construction marks, and the like. Although part of the act, such statements are neither to be equated with, nor do they represent, the practical knowledge in any strong sense.

Making practical knowledge explicit, for instance in sharing processes, necessarily entails representing it externally. Speaking or writing about, or referring in any other way to doing X, however, is not equivalent either to doing X or usually even to doing part of X. In fact, many people can do X and I believe we are justified in saying that they know how to X, even if they are not able to tell us the least bit about how they do.

Explanations of how to X, it may be argued, are necessarily part of any artisanal practice, particularly when it comes to sharing and instructing. However, when an artisan provides an apprentice with information about woodworking, for example, he is not working wood, he is explaining how to work wood. Hence what manifests itself in the artisan's instruction and what he conveys to the apprentice is first and foremost not knowledge-how, but knowledge about knowledge-how, that is, knowledge about knowledge.

Explicated knowledge-how may, in turn, become part of practical knowledge. Practical rules formulated in the context of instruction may be internalized by apprentices, supplemented by personal experience and again made explicit when training a successive generation. However, this need not necessarily be true for any explication of knowledge-how. In fact, much of the knowledge about external ballistics that we find codified in the practical literature of gunnery was, as will be argued, not reintegrated into practice or practical knowledge, until at least the end of the seventeenth century.

As indicated above, any explication of knowledge-how entails that the knowledge be externally represented. Codification necessarily involves selection, abstraction and, more often than not, the emendation of practical knowledge with knowledge from other domains.

In external representations of actual practice, real objects are substituted by symbols along with certain syntactical rules about how to transform these symbols. By means of such transformations, actions can be performed with the symbolically represented objects, just as the real-world counterparts can be acted with and manipulated in the physical world. Symbolic objects and actions are clearly more abstract than their real-world counterparts; symbolic actions, in particular, may sometimes have no real-world counterpart at all.[6]

[6]For an instructive example of how symbolic actions can generate further abstractions and thus new knowledge, see (Damerow 1996, 1999).

The symbolic system or systems used in the external representations of practical knowledge—language spoken or written, drawing, numbers arranged in tables and or combined with symbols representing mathematical operation, any combination of the aforementioned, etc.—were not, as a rule, created for this purpose. They come with their own semantics and syntax and often serve as the conduit for background knowledge from different domains of knowledge attached to the symbolic system. As a consequence, representations of practical knowledge, indeed, typically exceed the scope of the knowledge itself.

The discrepancy between the practical knowledge of the agents and the information codified in practical literature is thus, it can be argued, to a certain extent characteristic of any attempt to make practical knowledge explicit. As will be seen in the following, this discrepancy is particularly obvious in the case of external ballistics.

2 Preclassical External Ballistics vs. Practical Knowledge of Gunnery

The development of external ballistics paradigmatically exemplifies the general remarks on the relation between external representations of practical knowledge and practical knowledge itself. As will be argued, the trajectory which was to become the principal object of preclassical external ballistics, first only emerged, in the realm of shooting with ink, as the result of a particular way of codifying the practical knowledge of gunnery in a symbolic object. Thus, rather than a theory of shooting, preclassical external ballistics should more appropriately be regarded as shooting on paper, i.e. symbolical shooting, with a convoluted relation to shooting.

3 Practical Knowledge of External Ballistics

Information pertinent to external ballistics is sparse in the earliest practical literature on gunnery.[7] This is hardly surprising in view of the practice of shooting at the time. The huge iron bombards of the fifteenth century were battering pieces which were brought as close to enemy walls as possible in order to breach them.[8] Knowledge about the flight of the shot obviously had very little relevance for such practices. The little that did matter, such as, for instance, the fact that a shot leaves the gun in the direction of the axis of the bore—a trivial yet indispensable precondition of shooting—was not usually mentioned by the authors, most likely because it was deemed self-evident.[9]

[7] The number of works from the fifteenth century, in which shooting gunpowder weapons is treated in some detail, is small. Most famous among them is the so-called *Feuerwerksbuch*, presumably from the first quarter of the fifteenth century, of which numerous manuscript versions have survived. The first print was made in 1529. Numerous reductions and translations followed until way into the seventeenth century (Schmitt 1972, 41).

[8] Before and, indeed, well after the onset of gunpowder artillery, trebuchets were used as siege engines throwing wall breaching projectiles (Chevedden 2000). Shooting a trebuchet is substantially different from shooting a cannon. The launch angle of a trebuchet in particular depends on the release mechanism of the sling. These mechanisms were certainly adjusted when building and test firing an individual trebuchet. Whether the release angle was altered in shooting is unclear and in fact seems unlikely. Otherwise the shots could be altered by adjusting the weight of the shot, the counterweight, and in particular the sling length (Vemming 1993). This required a great amount of expertise, expertise which, however, could not simply be transferred to shooting gunpowder artillery.

[9] Information which is not made explicit usually pertains to knowledge that is not specific to shooting, but hinges on more universal experiences that can be made with heavy bodies in motion. Jürgen Renn and Peter Damerow distinguished this type of knowledge from practical knowledge, which is generated in a specific context and depends on the availability of means that are not universally given. They refer to it as "intuitive knowledge." For a comprehensive discussion of their concept of layers of knowledge and their relations, see (Damerow et al. 2006).

However, ordnance changed markedly beginning in the fifteenth century, and with it the practice of shooting. In the mid-fifteenth century, high-angled fire with large-bore, short-barreled mortars throwing stones and often incendiary shells over the walls of fortresses and towns became more common. By the end of the fifteenth century, cast bronze and later, iron guns replaced those of forged iron, and a trend toward producing lighter guns is discernible. Boring machines came to be used to correct the bores of guns after manufacture, which increased the precision of shooting. Furthermore, trunnions integrated with the guns were cast, and wheeled gun carriages were introduced. As a result, artillery became more versatile, and began to see more widespread use (Manucy 1985). Moreover, artillery also became increasingly used in field operations as well as in sieges. Together with artillery, fortification architecture developed, or more precisely, coevolved. In reaction to the increasing use of guns, the towers of medieval fortifications were replaced by bastions, effectively increasing the distance from which enemy artillery were forced to operate (Duffy 1996). This, in turn, called for guns with increased range. In consequence, the period saw "long sieges with little to show at the end for the huge cost and loss of life involved" (France 2014, 225).

This brief outline of the development of gunpowder artillery does injustice to the complex historical processes it summarizes. A more detailed account of the development of shooting gunpowder, however, lies beyond the scope of this article, so the reader is referred to the extensive literature on the

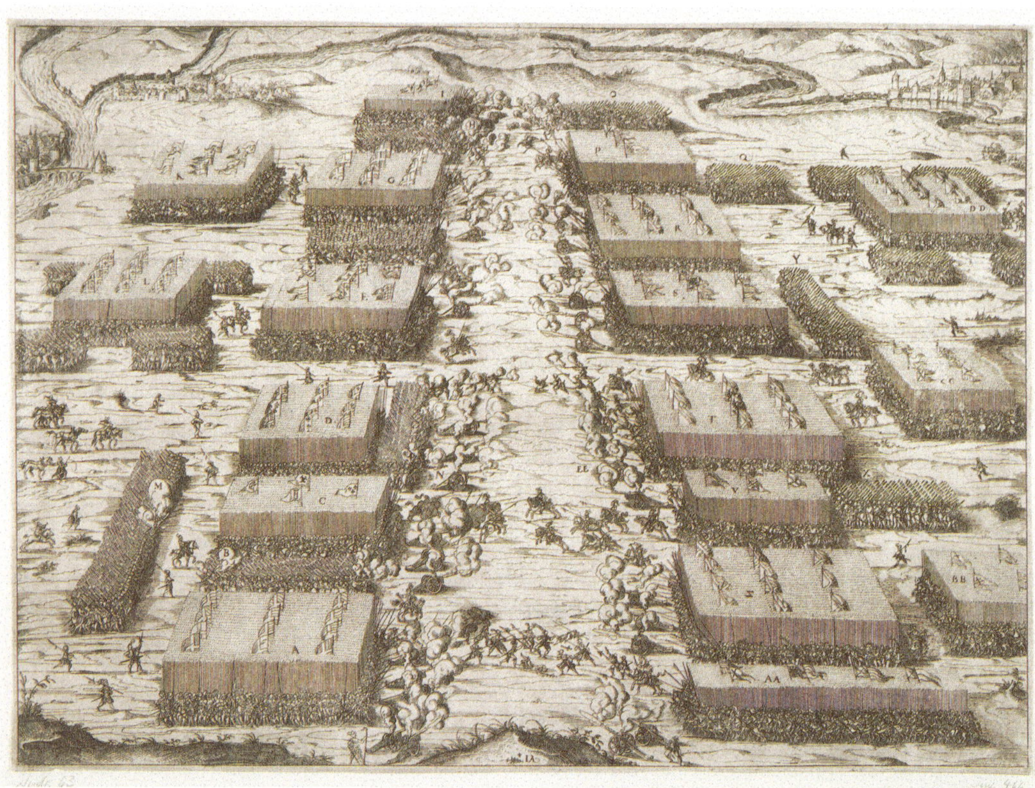

Fig. 1a Field battle between the French and Austrians in an etching by Jost Amman from the second half of the sixteenth century. The cannons shoot at targets within an extremely close range, as was common in field combat situations. Etching by Jost Amman, 1560–1573, JAmman WB 3.17, Herzog Anton Ulrich-Museum. Courtesy of Herzog August Bibliothek

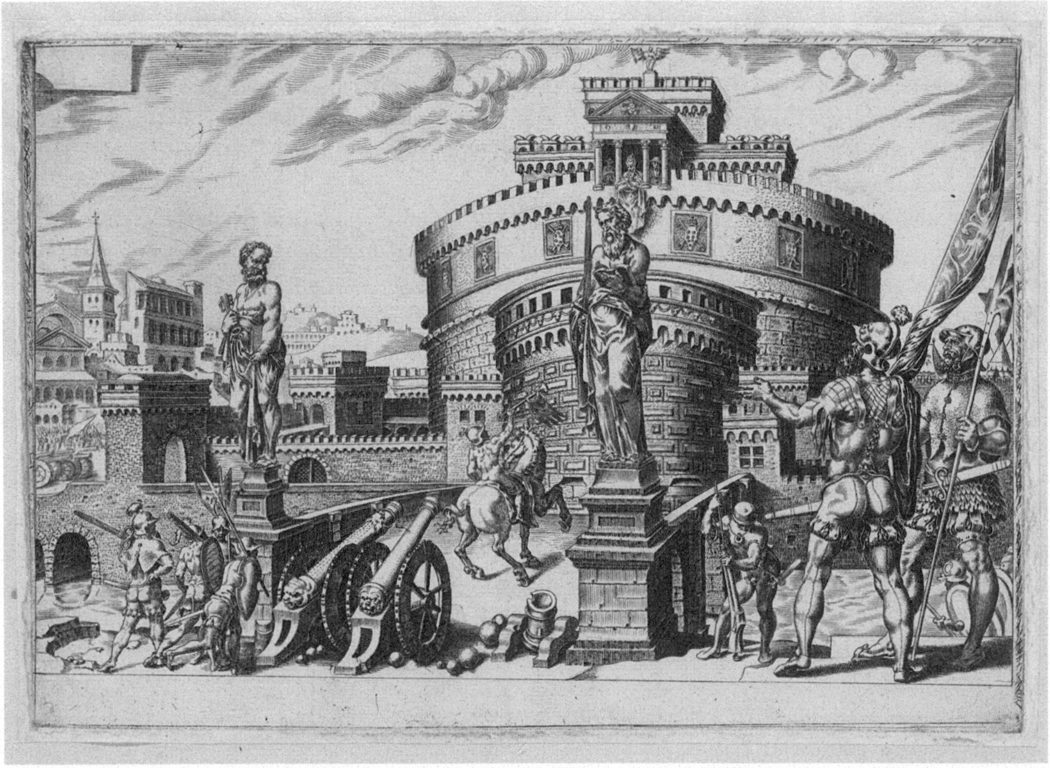

Fig. 1b Castel Sant'Angelo under siege during the sack of Rome in 1527 in an engraving by Maarten van Heemskerck. The distance from the outer wall of the castle to the facing bank of the Tiber is approximately 150 m, showing that cannons were often shot within point blank range even during sieges. Maarten van Heemskerck, 1555–1556. Object no. RP-P-BI-6609. Courtesy of the Rijksmuseum. The Netherlands

topic.[10] With regard to the present argument, in any case, it is less the details but the general trend observable in the significant change of technology, and thus, also the practice of shooting gunpowder artillery that is interesting. Even though most fire was and remained direct, in field operations, indirect fire with mortars was increasingly attempted in siege operations (Figs. 1a–b).[11]

Direct fire refers to shooting situations in which the target is hit without the need to substantially adjust the elevation of the gun with respect to the line connecting the target and the origin of the shot. This is also referred to as shooting within point blank range.[12] Indirect fire, i.e. shooting in situations

[10] For a detailed account of the early development of gunpowder artillery, in particular, the reader is referred to (Schmidtchen 1977).

[11] When cannons were used in sieges at the time, it seems they were typically used to breach the enemy's walls.

[12] Targets that can be hit without adjusting the elevation of the gun against the direct line connecting gun and target are said to be within point blank range. This range depends not only on the characteristics of the gun, but also on the size of the target. The idea of a point blank range is implicit in the earlier practical literature of gunnery in the assumption of an initial straight part. The relevance to practice is sometimes indicated by the terms used to refer to the straight line, such as, for instance, "Scheibenschuss." Shooting within point blank range is not to be confused with shooting at point blank, which refers to shooting in a horizontal direction. A cannon aimed at point blank was also said to lie "right with the mark" or "right with the metal" (Bailey 1726). The term is said to derive from the white bull's-eye of archery targets. An interesting discussion on the relation between shooting at point blank and shooting at targets at close distances is contained in Capobianco's *Corona e palma militare di artiglieria* first published in 1598 (Capobianco 1598). See (Capobianco 1647, 31–43).

in which the flight of the shot significantly deviates from a straight line, will in the following be referred to as "aimed shooting." Aimed shooting, referred to in the sources as "shooting at random" or "Bogenschießen," is where the practical knowledge pertinent to external ballistics had its predominant origin and place.[13] Consequently, the increasing use of mortar bombardments in the seventeenth century may indeed have contributed to the proliferation of external ballistics in the practical literature of gunnery in that period.[14]

Information pertinent to external ballistics does not initially come directly in the form of statements about the flight of a shot. Rather, the practical literature abounds with statements that can be interpreted from the modern perspective as based on knowledge pertinent to, if not the flight of the shot as such, then certain features thereof, such as its range. Such statements often come in the form of instructions about how to act under certain given conditions and/or goals of action. In his *Archiley-Kriegskunst*, first published in 1615, Johann Jacob von Wallhausen (1580–1627), for instance, writes:

> To shoot from a fortress/the higher the cannon is planted/the farther a shot can be made with it. (Wallhausen 1617, 59)[15]

The knowledge concerning the flight of the shot intrinsic to this statement is rather obvious and could be reformulated in propositional form as: "All other things being equal, the higher the place from which a shot is being fired, the greater its range."

In some cases, the relation between the statements in the literature and the knowledge they convey concerning the flight of the shot is less obvious. Thus, for instance, on the next page Wallhausen writes:

> If you lie in front of a fortress/and the fortress is placed up high/the piece at the bottom as well/do not take the point to low/but well above/the place you desire to strike. (Wallhausen 1617, 60)[16]

Again, from a modern perspective, we can reformulate this into a propositional statement representing the underlying knowledge, namely: "All other things being equal, the deviation of the point of impact from the line of sight is greater for the higher of two shots at targets located at the same horizontal distance." Although not immediately obvious, this does indeed happen to be a true statement from a modern perspective.

Not all such statements are immediately commensurate with our modern understanding of external ballistics. On the very same page Wallhausen remarks:

[13] The expression "Bogenschießen" is used for instance by Wallhausen: "Aus einer Festung auff ein Volck oder in ein Quartier zu schießen/so hinder einem Berg/oder hoch verschanzt liegt/muß solches mit Bogenschützen geschehen. Muß in acht genommen werden die Distantia oder Weite/danach das Pulver in seiner Quantitet" (Wallhausen 1617, 60). Shooting at random, sometimes also "randon," refers to shooting at an angle. The difference between direct shooting with cannons and indirect shooting with mortars is perfectly illustrated by Zubler's *Nova Geometrica Pyrobolia* of 1608, where the discussion of shooting cannons only treats shooting within point blank range (Zubler 1608).

[14] According to Cathrin A. France, the change in the frontispiece from the first to the second edition of Blondel's *L'Art dejetter des Bombes* bears witness to the increasing "importance of siege warfare in the military conquests of this period … and in particular of the contemporary interest in the use of bombs and mortars in this type of warfare" (France 2014, 235).

[15] "Aus einer Festung zu schiesen/je höher das Geschütz gepflanzet/je weiteren Schuß man damit thun kann." Author's translation here and in the following, unless otherwise noted.

[16] "So du vor einer Festung ligst/und die Festung in der Höhe ligt/das Stück auch in der Nidere/… muss man den Punkten nicht zu nieder/sondern wol oben/den Ort so du begerest zutreffen/ergreifen."

> Shooting over grassland or dust/and over water too/[the amount of] powder and elevation have to be increased/ otherwise [the shot will be] much too short/because the vapors of water and grassland much tire out the shot/and the earth as it were pulls it down toward it and the strong air or humidity impedes the shot much. (Wallhausen 1617, 60)[17]

This remark, which can also be found in a similar form in other works, can be reformulated into knowledge-that concerning the flight of the shot, for instance as: "All other things being equal, flight is such that the range is shorter if the shot is made over grass or water."[18] In this case it is entirely unclear, however, whether and how truth value can and should be attributed to this proposition from a modern perspective.

A specific aspect of the knowledge-how of gunners received particular attention in the practical literature of gunnery, namely, how the range of the shot could be varied by altering the angle of elevation of the piece. This knowledge will be referred to in the following as knowledge about the range behavior of guns, i.e. knowledge of how the range reached by the shot varies qualitatively with the variation in the elevation of the gun's barrel. This knowledge is regularly portrayed in the practical literature as an important prerequisite, if not the essence, of aimed distance shooting.

A clear idea of the practice of aimed distance shooting can be gleaned, for instance, from William Bourne's (ca.1535–1582) *The art of shooting great ordinance*. In Chap. 5, entitled "How to give levell with any peece of Ordnance, to make a shotte, as the most sortes of Gunners use to doe, although there be no Arte in it," the author describes the common practice of shooting at a mark as follows:

> First by their judgementes they doe give that so manye ynches advantage as they suppose will reach the marke, and then by the first lighting or falling of the shot, hee doth see whether it be shorte or gone over the marke, and if it be shorte, then at the next shooting hee will give the peece more advantage and if it bee over, then he will give the peece lesse advantage and so by divers times shooting off the peece at a marke, they will finde howe many ynches and partes will keepe the length of the marke. (Bourne 1587, 19)

Thus, in essence, depending on whether a previous shot fell short or exceeded the mark, gunners knew how to give more or less advantage to the gun, that is, they knew how to appropriately alter the elevation angle. With a cannon shot at low angles, if the shot fell short, they increased the elevation; vice versa, if the shot was too long, they aimed lower. Shooting with a mortar piece at high angles, the opposite was the case. If the shot fell short of the mark, one had to lower the gun to make a longer shot.[19] Such descriptions of aimed distance shooting appear frequently in the literature, and one is led to believe that gunners inevitably had to alter the elevation of their piece to correct the range of the shot.

However, a second possibility existed to vary the range of a successive shot, namely varying the amount of powder used.[20] Variation of the amount of powder is less frequently reported as a correction method than altering the elevation angle, yet existing accounts testify to the fact that it was practiced.

---

[17] "So man uber ein Wiesengrundt oder Wasen/wie auch uber Wasser schiest/muß man an Pulver und auch anrichtung deß Geschütz zugeben/sonst viel zu kurz/dieweil der Dunst von dem Wasser und der Wiesen die Kugel sehr ermattet/ unnd zur Erden gleichsam nieder an sich ziehet unnd die starcke Lufft oder Feuchtigkeit die Kugel sehr zuruck helt."

[18] For a similar statement concerning shots made over water see for instance (Brechtel 1591, chapter 11). For the distinction between knowledge-how and knowledge-that, see (Ryle 1945).

[19] It has been claimed in the literature that gunners used quantitative rules independent of external ballistics in adjusting the elevation of their guns. Such rules are indeed presented in the practical literature, sometimes explicitly as practitioners's rules. As will be argued here, this attribution is questionable and it is far from clear whether practitioners used such rules before and independently from their formulation on paper.

[20] There were, of course, rules for the upper limits of the amount of powder that would allow a safe operation of the gun. For a study of the rules according to which powder loads were determined, see (Walton 2008).

So, for instance, Leonhard Fronsperger (ca.1520–1575), referring to shooting with mortars, writes in his *Vom Geschütz, Feuerwerk und von Festungen*:

> This is why a shot that is too high or too short can be improved by taking away or adding powder alone. (Fronsperger 1557, 22)[21]

Altering the elevation of a gun is somewhat more complicated and time-consuming than altering the powder load, suggesting that the latter practice was more common than it would appear by judging only the sparse references made to it in the practical literature of gunnery. As will be argued below in more detail, authors promoting the idea of preclassical external ballistics had a particular interest in overemphasizing the variation of the angle of elevation over that of the amount of powder as being the common practice. Thus the image of the practice of shooting they imparted should be approached with caution.[22]

Generally, the practice of aimed shooting and, consequently, knowledge pertinent to the flight of the shot, played only a minor role in the earlier practical literature. Often the works contain only a few scattered remarks on aimed shooting, amounting to a couple of pages in books with page counts ranging in the hundreds, whereas other topics related to shooting are treated in much greater depth. This scheme is maintained in many of the practical works even after preclassical external ballistics had firmly taken hold in the course of the sixteenth century. On the one hand, this certainly reflects the fact that shooting situations requiring a more detailed command of the flight of the shot were still the exception well into the seventeenth century. As can be gleaned from warfare and field engagement accounts, the effective range in the sixteenth century was usually between 150 and 220 m, a range at which the deviation of a shot from a straight line was indeed of little significance.

On the other hand, the scarcity of information pertinent to aimed distant shooting can be understood to reflect a general difficulty in codifying the knowledge which allowed gunners to reasonably master the manifold erratic factors influencing the outcome of a single shot and, at least from time to time, shoot successfully. Especially in the case of high angle fire with mortars, controlling shooting was apparently difficult. Fronsperger, for instance, states about such mortar shots:

> Thus it has now seriously to be noted that one cannot actually make a certain shot as from other pieces … because they have a wide barrel and are short. (Fronsperger 1557, 22)[23]

---

[21] "darumb mus man einen zu hohen oder zu kurzen schuss allein mit abnehmung oder zugebung des pulvers … bessern." By mentioning the height reached by a shot, this quote by Fronsperger draws attention to an aspect of the flight of the shot besides range, that had practical relevance in certain situations, for instance, when attempting to hit a target behind an obstacle like a town wall.

[22] In 1695 Edmond Halley issued a proposal in the *Philosophical Transactions* regarding mortar shots, to always shoot at a 45-degree angle instead of altering the elevation, and to adjust the shot by varying the amount of powder. He claims that this procedure was implied by "later Experience." However, his proposal was apparently motivated less by experience than by the fact that, according to his understanding of Galilean theory of projectile motion, a shot at 45 degrees is the most efficient in the sense that it reaches the greatest range with the least muzzle velocity and thus the least amount of powder: "But a third and no less considerable Advantage is, in saving the King's Powder which in so great and many discharges, as we have lately seen, must needs [sic] amount to considerable value" (Halley 1695, 68). Varying the load was not a recent practice and, as will be argued, shooting at an elevation in the vicinity of the maximum range of elevation has considerable disadvantages, implying that Halley was not particularly well acquainted with the constraints of actual practice and had made his proposal based primarily on a theoretical inference from external ballistics.

[23] "So ist nun ernstlich zu merken das man keinen eigentlich gewissen schuß wie aus anderen Stücken … thun kann denn sie haben eine weit rohr und sind kurz."

Fig. 2a Annunciation scene in a painting by Guido da Siena from 1262 to 1279. Light rays indicate the path of the dove as the manifestation of the Holy Spirit. Guido da Siena, *Annunciation*, 1262–1279. Tempera on wood panel. Caroline G. Mather Fund. y144. Princeton University Art Museum. Public Domain. Wikipedia Commons

In battle, mortars were sometimes shot with whatever ammunition was at hand, such as stones and occasionally even animal cadavers. The experience, which allowed gunners to develop a judgment of how to act in such situations, could not easily be put into words and jotted down as instructions in recipe-like form, as had been done with descriptions of the production of gunpowder, for instance, or for taxonomies of guns.

To summarize, one can state that in the earlier practical literature of gunnery information pertinent to external ballistics is sparse. For the most part, the little that can be found refers to direct shooting and is predominantly formulated with reference to concrete actions and their goals, i.e. leaning closely on knowledge-how. This information can be reformulated as, or better translated into, knowledge-that concerning certain features of the flight of a shot, i.e. knowledge concerning external ballistics. Such a translation presupposes, however, that the flight of a shot is perceived as an epistemic object in the first place, i.e. as an entity in itself endowed with features that can be described and studied.

It will be argued in the following that preclassical external ballistics, which emerged and unfolded in the sixteenth century, is essentially characterized by just that, i.e. by the perception of the trajectory of gunshots as an epistemic object. The trajectory is pertinent to, and could emerge only in, the realm of shooting with ink, yet it was based on and inextricably related to actual shooting.

4 The Trajectory

The modern reader is all too familiar with seeing motion represented by a line rendering the path a travelling object has taken. Such trajectories, or trajectory traces, integrate the past positions of an object over time, and make them visible on paper as a line. With regard to shooting, such trajectories can be found not only in the practical literature of gunnery, they are also present in artistic renditions, for instance those of battle scenes. As natural as the idea of symbolically representing motion by

Fig. 2b Muzzle flash indicating the path of a cannonball shot from an early gun. Miniature in a manuscript from the beginning of the fifteenth century of Konrad Kyeser's *Bellifortis*. From (Keyser 1967, vol. 1)

trajectory traces may seem to us, it was apparently not seen as such the fifteenth century, when writing about shooting became more prolific (Figs. 2a–b).

Predecessors of trajectory traces can be discerned in medieval iconography. In Annunciation scenes, for instance, beams of light are occasionally depicted as emanating from the heavens towards Mary. The dove as the manifestation of the Holy Spirit appears to descend from the heavens on those beams, and as such, they indicate the path the dove is taking.[24] Yet rather than being mere lines representing motion, these beams represent physically perceivable objects, namely light rays. Similarly, with regard to the representation of projectile motion, Konrad Kyeser's (1336–after 1405) *Belifortis* contains an illustration of a shot with an early gun, where in some of the manuscripts the cannonball appears to be frozen in mid-air, connected to the mouth of the gun by a straight line rendered as a muzzle flash; i.e. again, not as a mere representation of the path, but as a physically perceivable phenomenon.[25] Moreover, the insinuated path is straight, as in the case of the Annunciation scenes (Fig. 3).

[24]The earliest Annunciation scene I have been able to identify that follows the iconography described in the text, is a painting by Guido Da Siena of 1270. For an interesting case of trajectory traces in art that was apparently inspired by external ballistics, see (Topper 2007).

[25]Illustrations of guns with muzzle flashes and cannonballs suspended in the air are to be found in the manuscripts (MS Philos. 63, University of Göttingen Library, fol. 105v and MS.1360, Bibliothèque municipale de Besançon, fol. 121r).

Fig. 3 Study by Leonardo of mortar fire with trajectories falling beyond the walls of a fortress. From *Codex Atlanticus*, fol. 72r (Leonardo da Vinci 1973, vol. 1, fol. 72r)

The very first non-trivial cases I have been able to identify of the drawn trajectory traces of gunshots, i.e. cases in which the curved motion of a projectile is represented by a line that does not symbolize a physical carrier, come from Leonardo da Vinci's (1452–1519) notebooks. A sketch in *Codex Atlanticus* shows a number of mortars fired over the walls of a fortress. A huge number of trajectories is drawn for each of the mortars, thus the image not only represents the time of one shot, but depicts many successive shots to illustrate a fortress being covered by enemy fire (Fig. 4).[26]

Leonardo employed trajectories not only in illustrational contexts, however, but also in more explorative contexts, as is exemplified here by a diagram from the *Codex Madrid*. Leonardo drew this particular diagram in the context of dynamical considerations related to the observation that the trajectories of gunshots are asymmetrical.[27]

Trajectory traces of motions other than gunshots can be found in Leonardo's manuscripts as well, for instance in diagrams pertaining to his study of bird flight (Chap. 4). As I have not been able to identify trajectory traces before Leonardo, he may have been the first to avail himself of this particular representational tool.[28] Ultimately, however, this is a question for future research. He was, however, quite definitely the first to use the trajectory trace to represent the flight of a shot, and it is

[26] Another rendition of mortars shooting fragmentation bombs contained in *Codex Atlanticus* does not show trajectories, but represents motion by integrating the different positions of the bombs at different times in the trajectories of the shots (Leonardo, *Codex Atlanticus*, Bibliotheca Ambrosian Milan, fols. 31r and 33r).

[27] A detailed discussion of the context in which Leonardo drew this diagram is contained in (Büttner et al. 2003).

[28] The issue of trajectory traces is touched upon in (Gal and Chen-Morris 2012). The authors are interested in trajectories not so much as a representational tool in general, but in their role as mediators between the physical world and its mathematical description by early modern thinkers.

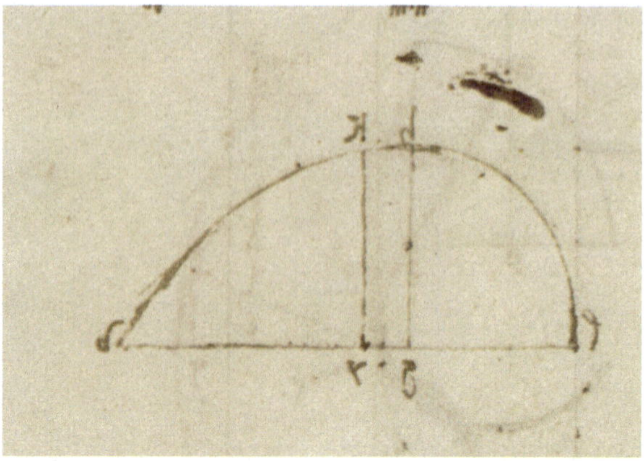

Fig. 4 Drawing of a trajectory in the context of dynamical considerations regarding projectile motion. From Leonardo da Vinci, *Codex Madrid*, fol. 147r (Büttner et al. 2003, 8)

thus entirely possible that this representation of gunshots played a crucial role in the emergence of the very concept of a trajectory trace.[29]

The introduction of the trajectory trace of the flight of a shot, henceforth referred to as simply the trajectory, was clearly not engendered by theoretical speculation. The trajectory was rather a representational tool, whose introduction enlarged the repertoire of what could be rendered using the traditional representational means of ink and paper. The introduction of the trajectory, which was to become the foundation of preclassical external ballistics, had, as will be demonstrated, striking consequences. With the trajectory, not only a new symbolic object originated, but also a new epistemic object to which one could refer mentally and/or explicitly and which furthermore could be both the subject and also a means of intellectual activity.

It is the trajectory which first provided the opportunity to synthesize practical knowledge concerning the different specific features of the flight of the shot in a single object. At the same time, however, the trajectory bestowed early modern thinkers with the challenge of providing information that was available only in part, if at all, through practical experience, as the drawing of or constructing of a trajectory entails concrete choices that leave no part of the trajectory undetermined. How certain parts of the flight of a shot look precisely, escapes the attention of the observer, who by verbalizing his or her practical experiences, may thus be able to state that the trajectory is curved, without, however, being able to specify more concretely the form of the curve. Yet drawing trajectories makes concrete choices necessary, and in making such choices, as is argued above, it was necessary to draw upon resources other than practical knowledge to fill in the missing information. To trace trajectories, the authors thus integrated practical knowledge while at the same time amending it in various ways.

Representation of the trajectory on paper thus served as a conduit for knowledge, abstractions, and representations from other domains of thought, most notably geometry. Indeed, in drawing trajectories, early modern thinkers typically resorted to well-known geometrical elements, predominately lines and arcs, to construe certain parts of their trajectories. Non-specifiable curvature based on practical experience was thus usually specified as the arc of a circle. Moreover, they used geometrical constructions and thus geometrical knowledge to specify the properties of, and rela-

[29] I have not been able to identify any trajectory traces representing non-straight motions in art before the end of the fifteenth century. Certain forms of representation are akin to trajectory motions, yet arguably not quite the same. In geometry, certain curves which René Descartes later termed mechanical, like for instance the Archimedean spiral line, were conceived as being composed of different motions. These curves could be generated using mechanical means, but did not in any strong sense represent the motions used to generate them. Astronomical diagrams showing planetary orbits, rather than making motions which are in principle visible on paper, arguably represent or illustrate astronomical models.

tions between, the lines and arcs from which they composed their trajectories. A faithful rendition of the perceivable reality was thereby usually not the main incentive for the concrete choices made, rather preference was given to other aspects such as calculability or constructability over achieving a realistic depiction.[30]

Furthermore, trajectories constructed geometrically of straight and curved parts lent themselves particularly well to a dynamical interpretation of projectile motion in the framework of Aristotelian natural philosophy in the tradition of Albert of Saxony and Johannes Buridan (1205–1363).[31] Thus the straight parts of the trajectory were usually identified with a violent motion along the direction of the shot, and the downward part with a natural motion. The curved part was often considered as a mixed motion. Thus besides the practical knowledge of gunnery and geometrical knowledge, yet another body of knowledge was thus drawn on to specify features of the trajectory. The majority of trajectories drawn up to the mid-seventeenth century, for instance, end in a straight line vertically downward, as was implied by Aristotelian dynamics, but this did not necessarily correspond to the knowledge of gunners at the time.

5 Tensions Between Shooting and Shooting with Ink

Since early modern authors of the practical literature of gunnery needed to resort to additional knowledge resources when drawing or constructing trajectories, tensions between their representation and the practical knowledge on which these representations were ultimately based, emerged almost necessarily. This will be illustrated by an example based on a diagram from Diego Ufano's (died 1613) widespread treatise on artillery, first published in 1613 (Ufano 1613). The diagram illustrates the basic fact that shots fired from a higher position go farther, a fact that was well known to practitioners, as we have already seen. In the illustration of the two shots, one from the foot and one from the top of the tower, the shot from the higher place indeed hits the small castle on the left, which is farther away from the tower (Fig. 5).

The act of shooting cannons is symbolically represented and embedded in an idealized scenery. The two cannons placed at the foot and on top of the tower are not realistic depictions, but rather abstractly rendered icons of a generic cannon, signaling that the expressed dependence of the range of a shot on the elevation from which it is fired holds true not only for a particular cannon or a particular type of cannon, but indeed universally for all types of guns.

The statement that of two shots fired, the one fired from a higher position goes farther, applies only *ceteris paribus*, of course, a condition which never quite holds in the physical world but can be presupposed to hold in the abstract world on paper. The author of the drawing obviously took for granted that the readers would understand the two shots as being ideally made under equal conditions. He merely indicated, by placing a quadrant in both cannons, that they are indeed elevated at the same angle.

To convey the information that the shot fired from higher up will go farther, marking the points of impact and relating them in some way uniquely to one of the two cannons, would have sufficed. However, the trajectories of the two shots are both drawn out fully, introducing a subtle inconsistency in Ufano's diagram. Small differences in the height of the places from which a cannon is fired should neither affect its behavior, nor the trajectory of the shot. Both trajectories should be of identical shape,

[30] That the aim of drawing or constructing trajectories was usually not to produce faithful renditions of perceived reality is illustrated vividly by the difference between the trajectory shown on the frontispiece of *Nova Scientia* and the trajectories Tartaglia constructed in his book. For a more detailed discussion, see (Büttner et al. 2003).

[31] For Buridan's interpretation of projectile motion, see (Maier 1940).

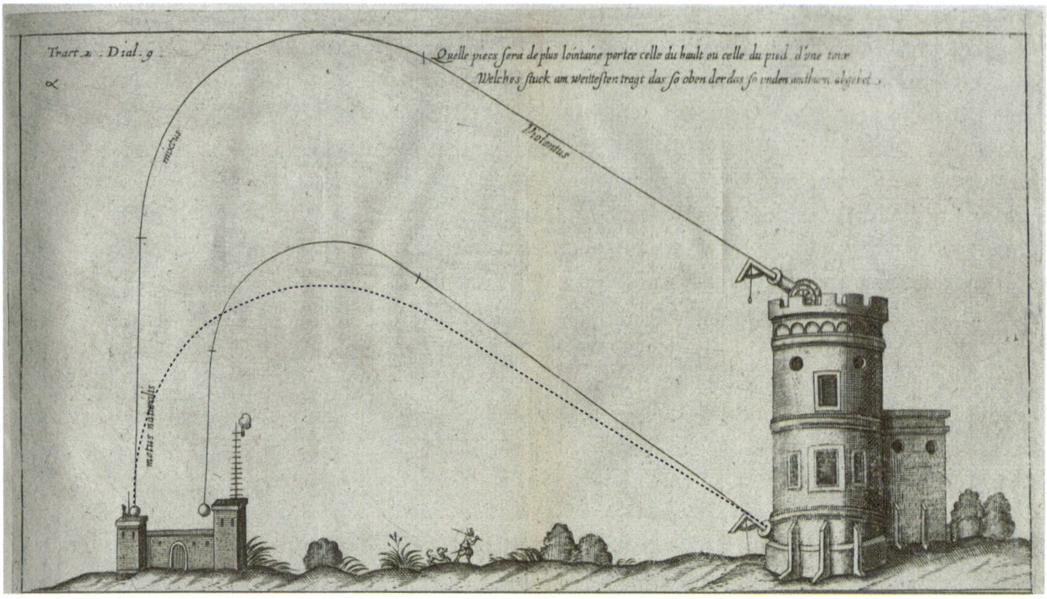

Fig. 5 Trajectories of two shots. The diagram is a reproduction from the 1628 French edition of Diego Ufano's widespread treatise on artillery, first published in Spanish in 1613. The dotted trajectory for the cannon at the foot of the tower has been added. Its shape is identical to the shape of the trajectory of the shot fired from the top of the tower. From (Ufano 1628, 55). Courtesy of the Library of the Max Planck Institute for the History of Science, Berlin

but merely be transposed in parallel. However, they are not. Indeed, if both trajectories had the same shape, the projectiles would impact at exactly the same spot, as I have indicated by adding a third trajectory drawn as a dotted line on the original diagram.

In essence, in order to correctly represent practical knowledge regarding the dependence of the range on the height from which a shot is fired, Ufano violated another aspect of practical knowledge. What ultimately forced him to do so is the fact that, due to their vertical downward parts, the ranges of his trajectories do not increase from a certain point onwards. A symbolic act on paper, the direct comparison of two shots, differing solely with regard to the height from which they are fired, thus revealed that the particular construction of the trajectory Ufano used cannot be brought into accordance with all aspects of practical knowledge. To cover this incompatibility, Ufano subtly manipulated his diagram and drew two slightly different trajectories. Since his trajectories are not rigidly constructed, this does not immediately strike the eye and probably went unnoticed by most of his readers.

Ufano's trajectories, in particular their vertical downward part which gives rise to the incompatibility in first place, seem to indicate his commitment to Aristotelian dynamics. One may thus be inclined to read the drawing as a tension between assumptions based on theoretical, natural philosophical speculation and practical empirical knowledge. However, other trajectories depicted in Ufano's book, for instance those shown in Fig. 10 here, do not show the familiar scheme, and in particular do not have a vertical downward part. Ufano apparently adjusted his trajectories depending on the particular information he wanted to convey in a diagram. Ascribing to Ufano a strong commitment to Aristotelian dynamics is thus inappropriate.

The tension in his diagram is to be addressed more adequately as a tension between practical knowledge and the representation of this knowledge on paper by means of the trajectory. In tracing his trajectories Ufano was forced to make choices where practical knowledge could not guide him, and in making these choices he resorted to solutions he was familiar with. He apparently chose those solutions not so much because of their theoretical underpinnings, but because they allowed him to adopt the representational tool of the trajectories to illustrate facets of practical knowledge.

6 Preclassical External Ballistics

The fundamental idea of the trajectory enabled two further conceptions. Together with the trajectory itself, they constitute the core of what will be referred to as preclassical external ballistics.

Approaches subsumed under this term in the following are considerably different from each other. Yet a common structure can be recognized, which unifies these otherwise widely varying approaches and which delineates the scope of, and defines, preclassical external ballistics. First, the concept of an angle-dependent trajectory, i.e. the understanding that the trajectory is essentially characterized by the variation of its shape, depending on the elevation of the shot over the full range of angles from zero (or even less) to ninety degrees. Second, the conception of the trajectory as form-invariant, i.e. the idea that the trajectories of shots at a particular elevation always have the same shape independent of the particular gun, shot or load used. For a concrete shot, these latter factors are understood to provide a scale to the form-invariant trajectory. Less than the result of theoretical speculation, as will be seen, both conceptions are an extension of practical knowledge, facilitated by the symbolic representation of that knowledge in the form of the trajectory on paper.

The idea that the trajectory can be characterized by the variation of its shape depending on the elevation of the shot over the full range of angles seems hardly remarkable to the modern reader, not least because he is all too familiar with the mathematical formalism of functions. Historically seen, however, it constitutes a daring and consequential abstraction rooted in practical knowledge, but only possible, as will be argued, in the realm of shooting with ink.

The idea of a full-angle range trajectory implies or presupposes a gun capable of shooting at any angle. Such guns, however, did not exist at that time. Cannons were shot at low angles and generally could not be elevated above a certain angle. Mortars were shot predominantly at angles above 45°, apparently usually at angles between 60 and 80°. For a gunner of the day, cannons and mortars were entirely different things, a difference reflected even in the terminology used to describe their actions—cannons shot, mortars threw. The two types of guns were used in different situations for completely different purposes (Fig. 6).

On paper, however, it was possible to conceive of guns which, overcoming practical limitations, were indeed capable of firing over the full range of angles. The earliest example of such a full-angle gun, and thus of a full-range trajectory, is to be found in Leonardo's manuscripts. A diagram from

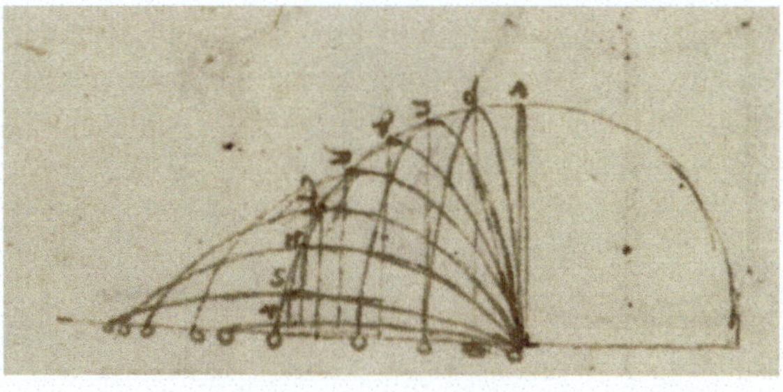

Fig. 6 The full angle-dependent trajectory in a diagram by Leonardo da Vinci. From *Codex Madrid*, I. fol. 147r (Leonardo da Vinci 1974, I, fol. 147r)

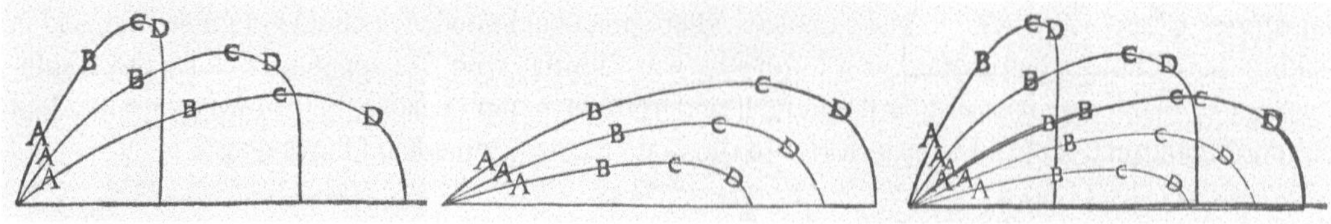

Fig. 7 Trajectories of mortar shots (*left*) and of cannon shots (*middle*). From (Bourne 1587, 39–40)

Codex Atlanticus shows the trajectories of nine shots made from the same gun at different elevations. Clearly, the diagram is not meant to illustrate nine different, specific shots. The set of trajectories rather illustrates the characteristic variation of the shape of the trajectory over the full angle range from the horizontal to the vertical shot. The nine trajectories thus instantiate and illustrate a more general constructive principle employed, namely that the culmination points of the trajectories all lie on a circle around the origin of the shot, where the trajectories are nearly symmetrical with respect to this point. Only for the flattest of the nine shots did Leonardo violate this rule, for reasons to be discussed below.

Constructions of full-range trajectories, such as the one by Leonardo, allowed mortar shots to be perceived as the extension of cannon shots at steeper angles, or vice versa, to perceive cannon shots as the extension of mortar shots at lower angles. In the realm of practice, not only would there have been very little reason to establish such a link, but it would have made little sense for gunners. On paper, however, the physical limitations could be overcome, and experience with and knowledge about low and high angle shots could be combined in one object, the angle-dependent trajectory.[32] While enriching the representation of practical knowledge, this integration, as will be demonstrated, gave rise to questions and conclusions mostly alien to the realm of practice. In particular, the question of the angle at which a gun reaches its maximum range arguably first emerged, as will be detailed below, as a consequence not of shooting but of constructing full-range trajectories (Fig. 7).

The disparity between the abstract idea of a full-range trajectory and actual practice is vividly displayed in William Bourne's *The arte of shooting in great ordnance*, published in London in 1587. In the chapter headed "What manner of course the shot flyeth in the ayre," Bourne presented two diagrams, one showing the trajectories of three low shots (15, 30 and 45°) and the other showing the trajectories of three high shots (45, 60 and 75°). These diagrams are remarkable in at least two respects. On the one hand, they provide one of the few notable exceptions known from the sixteenth century of trajectories that are not perceived as made up from lines and arcs.

Their second remarkable feature manifests itself only when both diagrams are scaled appropriately and combined into a single diagram, as has been done in Fig. 7. It thus revealed that the 45-degree shot is, within the limits of a freehand drawing or of a woodcut, identical in both diagrams. Thus rather than as two different, separate diagrams—one representing the characteristic variation of the shape of cannon shots, and the other the characteristic variation of shape mortar shots with a variation of the elevation—the two diagrams together can be regarded as a composite diagram showing a set of five shots from the same abstract gun at different angles, thus embracing the idea of a full-range trajectory. Issuing two separate diagrams, Bourne, it appears, made a concession to practice and practitioners, paying tribute to the practical differences between cannon and mortar shots.

[32] Tartaglia was quite aware that the concept of the full-angle trajectory was not native to practical knowledge, claiming, with regard to its consequences drawn on paper, "I found with a very evident argument that a piece of artillery can hit one place along two different paths (or at two different elevations) and I found the method of how to execute this in reality (a subject never heard or conceived by anyone else, ancient or modern)" (Valleriani 2013, 75).

By its very nature, the full-range trajectory cannot be specific to a particular gun or type of gun. It generically represents shots at different angles with one and the same abstract or symbolic gun, and thus implicitly entails the assumption that the only factor that has an influence on the shape of the trajectory is the angle of elevation. The full-range trajectory thus immediately implies the second concept characterizing preclassical external ballistics, namely the understanding that all the other factors besides elevation that influence the trajectory—such as powder, length of the cannon, weight of the ball, perfection of the bore, etc.—can be subsumed under a single second factor, discernible, from a modern perspective, as the muzzle velocity of the shot. Whereas the angle of elevation essentially determines the form of the trajectory, all other factors taken together provide the trajectory with its scale.

It is this feature that, at least according to its adherents, potentially tied preclassical external ballistics to practice and, by making it potentially applicable, endowed the project with meaning and legitimacy. If the idea of a form-invariant, full-range trajectory indeed concurred with physical reality, then preclassical external ballistics would have potentially been useful in exactly the way its advocates claimed. One trial shot with a given cannon would then suffice to provide scale to the form-invariant, full-range trajectory for that particular cannon. All successive shots with that cannon, made at whatever elevation but loaded in the exact same manner, could then, if potential complicating factors could be controlled, be fully anticipated in every sense. Shooting, the claim and promise, would thus be perfected. Thomas Digges (1546–1595), for instance, announced this program as follows:

> … the perfect Arte to shoot at all Randons from one grade of altitude to 90 with all kinds of peeces, with Rules infallible to find out their severall Ranges, & that not onely on level grounds equidistant from the *Horizon* but also upon plaines, enclining or declining whatsoever the angle of inclination be. (Digges and Digges 1590, 361)

The idea of the form-invariant trajectory is already present in Leonardo's manuscripts as well, as is illustrated here, by a diagram taken from manuscript M showing mortar shots fired at the same elevation, using different amounts of powder (Figs. 8a–b). Patently the trajectories have the same form, they are merely scaled with respect to each other.

Not only does the idea of the form-invariant trajectory represent yet another bold abstraction, it is also false from a modern perspective. The form of the trajectory for a given elevation does indeed depend on the muzzle speed, and is thus different for two different cannons shot at the same elevation. Yet the very idea of a form-invariant full-range trajectory was not only implied on paper but, as can be argued, instructed by practice.

In 1579 Thomas Digges, known mostly for his comments on Nicolaus Copernicus (1473–1543), published his *Stratioticos*. As he stated in the foreword, the book had been started by his father and finished by him. Appended as a final chapter is a discussion of ballistics entitled "Certeine Questions in the Art of Artillery by Mathematical science." The chapter opens with the following general considerations:

> The diversitie of the force and violence of great Artillerie, and of the farre or nigh shooting and variable randge of their Bullets, doth chiefely arise & grow of these 4 principall causes: Powder, Peece, Bullet, and Randon. In Powder is considered Quantitie and Qualitie. In the Peece the length and proportion of the Cylinder. In the Bullet his waight and Quantitie. In Randon the degrees of Altitude from the Horizon or plain wher the Peece must play. These may be called the Prime, substantiall or effectuall causes.
>
> There are also other causes and circumstances which may likewise worke some alteration: as the Raritie or Densitie of the ayre. The Winde. The diuerse wadding and ramming in chardging. The inequalitie of the ground or Plat-forme, and wheeles of the Carriadge reculing. The regular or irregular Boring. The distempering of the Peece, this is an important point, hotte or colde, and well or ill fitting of the Bullet. These may be termed secondarie or accidentall causes of alteration, which by Practize and use will best learned, and their errours by discretion reformed, albeit for some of them certain Rules may also be prescribed. (Digges 1579, 181)

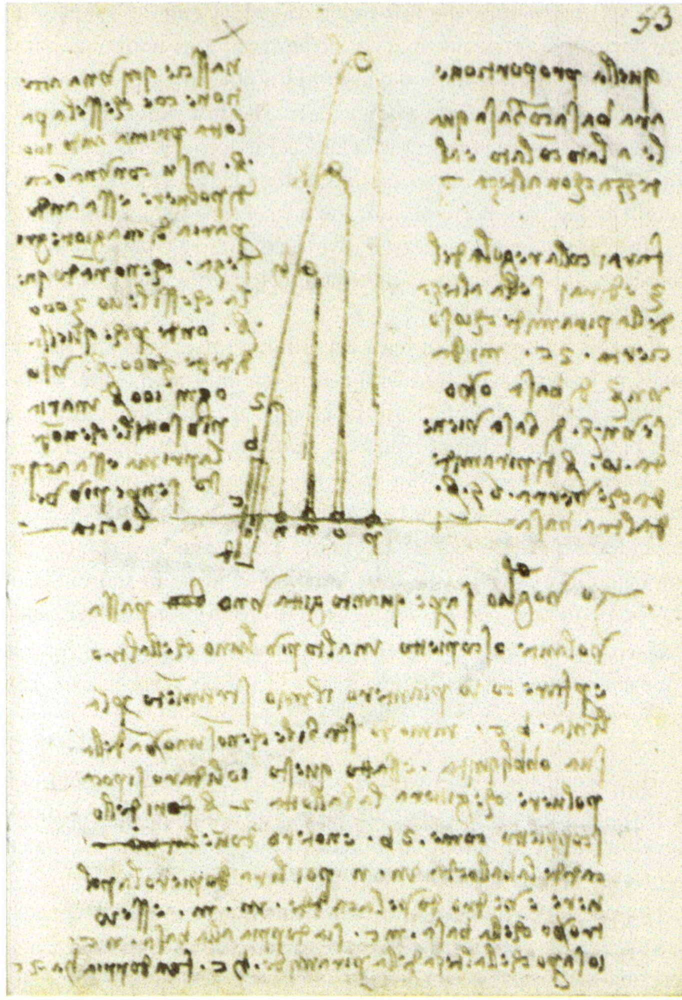

Fig. 8a The concept of a form-invariant trajectory illustrated by a diagram. From Leonardo's Ms. M (Leonardo da Vinci 1986, vol. 12)

Digges had a clear awareness that the ability to discriminate among the relevant parameters in terms of their effects on the shot, was essentially achieved from practice. A regularity in the relation between a given parameter and its effect—say the amount of powder used, judged by the range reached by the shot—is discernible only to the extent that the unavoidable alterations in other parameters, primary or circumstantial, can be controlled so their influence on the effect will not completely mask the regularity. We have no reason to doubt that, just as Digges claimed, practitioners were able to do so, and thus not only knew, for instance, that using more powder would regularly result in a greater range, but that the different parameters were largely independent with regard to their effect—that using a different load would not qualitatively alter the elevation-range behavior of a cannon, thus indeed suggesting a form-invariance of the trajectory. That causal relations between the relevant factors and the effect are recognizable and knowable—despite not being evident in each individual shot for a multitude of reasons, most of which the gunners were perfectly aware of—is characteristic of practical knowledge in general.

Practical knowledge concerning the parameters and a qualitative understanding of their effect on the outcome of the shot, thus directly informed the basic conceptions of preclassical external ballistics.

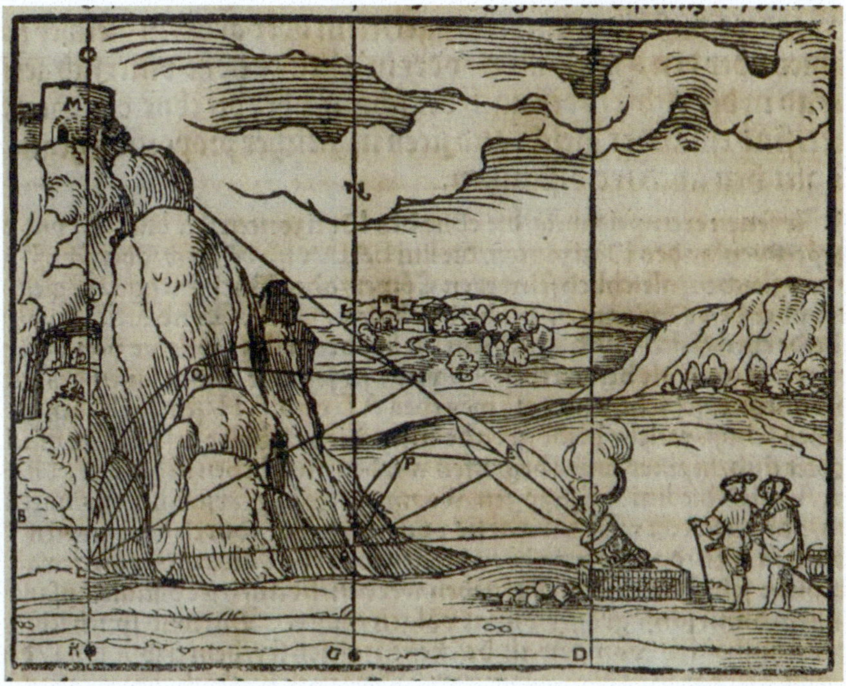

Fig. 8b The concept of a form-invariant trajectory illustrated by a diagram from Walther Hermann Ryff, (1558), plagiarizing Tartaglia's *Nova Scientia*. From (Ryff 1558, 14v). Courtesy of the Library of the Max Planck Institute for the History of Science, Berlin

Knowledge about the effects of varying the elevation of cannon and mortar shots, respectively, was synthesized in the full-range trajectory. As far as the effect of a variation in the other principal parameters on the trajectory is concerned, the assumption that the shape of the trajectory is not affected by altering these parameters seems to be almost inescapably implied by practical knowledge. Preclassical external ballistic can thus essentially be characterized as an abstraction of practical knowledge on paper, mediated by the representational tool of the trajectory, an abstraction that would neither have been possible nor telling without the skill, knowledge, and experience of practitioners.

The complete structure of preclassical external ballistic is, as has been seen, already present in Leonardo's works. His manuscripts remained unpublished in the period under investigation, however, and the question of what influence, if any, they exerted on later thinkers, remains open to debate. The first author to expound preclassical external ballistics in a published work, was Nicolò Tartaglia with his *Nova Scientia*, first issued in 1537.[33]

In the incipit of the *Nova Scientia,* Tartaglia gave an almost programmatic summary of preclassical external ballistics:

> In the second [book], the qualities, similarities and proportions … of the distances, of such bodies are shown and demonstrated (geometrically) according to the various ways in which the bodies can be ejected, that is, thrown violently through the air. (Valleriani 2013, 63)

As the comparison to Leonardo's pondering on the matter shows, Tartaglia did not create preclassical external ballistics. Indeed, as argued above, the basic structure of preclassical external ballistics

[33]For a survey of Tartaglia's *Nova Scientia*, see the foreword by Matto Valleriani to the new edition (Valleriani 2013), which also contains a comprehensive bibliography.

emerged as an almost necessary consequence of combining the practical knowledge of gunners with the idea of representing gunshots by trajectories. For once, it thus does not seem overly daring to engage in an exercise in counterfactual history and claim that, had it not been for Tartaglia, someone else would have expounded the idea of preclassical external ballistics in published form in the sixteenth century.

The book was extremely influential in promoting preclassical external ballistics. Tartaglia marked his approach as a new science and, despite clear reverberations of a practical background, the work is more abstract and omits most of the other topics discussed in the practical literature of gunnery from about the same period. Thus, even if Tartaglia was not the creator of preclassical ballistics, he was certainly one of its most influential promoters. Many successive authors more or less faithfully reproduced Tartaglia's version of preclassical external ballistics, occasionally without acknowledging their source. More often than not, however, authors gave preclassical external ballistics their own interpretations, different from Tartaglia's but retaining the basic structure.

In the late sixteenth, and then particularly in the seventeenth century, more and more books were published which, like the *Nova Scientia*, focused more or less exclusively on external ballistics, and in which canonical topics of the earlier literature were omitted. But even in works which otherwise remained more directly and concretely committed to practice and practical knowledge, and which retained the canon of topics from practical gunnery literature, external ballistics increasingly became a topic of its own, which following the publication of the *Nova Scientia* was often treated in one or more separate chapters.

Even authors who avoided an explicit treatment of preclassical external ballistics usually at least acknowledged its existence, variably referring to it, for instance, as "geometrische Büchsenmeisterei," or "the art of shooting arithmetically," and apparently often felt obliged to justify why they did not discuss it. Thus, for instance, Fronsperger writes in his *Vonn Gesschütz unnd Fewrwerck* of 1557:

> … and even though for such prediction/more beautiful arts of geometry are needed/but others have written of it/…/I will not report anything thereof in this my book. (Fronsperger 1557, 22)[34]

Despite its grounding in practical knowledge, preclassical external ballistics gave rise to mismatches between the theoretical accounts and practical knowledge, which became ever more apparent, as it became more established in the practical literature. This will be elucidated in the following example, which received considerable attention in the period in question, examining the question of the elevation at which a cannon shoots farthest.

7 Maximum Range Elevation

Tartaglia in the *Nova Scientia* was the first to claim, at least in written form, that cannons will shoot farthest onto a horizontal plane if they are inclined at an angle of 45°. In the dedication letter to his book Tartaglia gave a quite detailed account of how he came to make this claim in first place:

> During the year 1531 … I was asked by a close and kind friend of mine, a very skilled bombardier … about the method to set up a piece of artillery in such a way that it shoots the farthest. Although I did not have any experience in such an art (… I have never discharged any artillery …), I nevertheless (as I wished to serve a friend) promised to quickly provide him with a resolute answer. After I had carefully contemplated and ruminated this subject, I concluded and demonstrated to him by means of natural and geometrical arguments that the mouth of the piece had to be elevated so that it addresses straightly 45 degrees above the horizon. (Valleriani 2013, 67)

According to Tartaglia, his friend was not entirely convinced of the solution, as being a gunner, he believed that the maximum range would be reached at an angle below 45°. As it happened, so

[34] "Und wiewohl zu derselbigen absehung/viel schöner künst aus der geometri gebraucht werden/und aber andere davon geschrieben haben/…/so wil ich in diesem meinen Buch weiter davon nichts melden/."

Tartaglia's story continues, the next year two bombardiers had an argument about the very same question, one of them claiming that shots go farthest at 45°, whereas the other claimed that the barrel had to be elevated at a lower angle in order for the shot to go farthest. As Tartaglia relates, the two practitioners went ahead to settle their dispute by means of a test and:

> A certain quantity of money was bet … and they finally arrived at the experiment. A culverin of twenty [libra] was brought to the countryside around Santa Lucia. Each of them shot according to what they suggested and without any difference in reference to the ball and to the charge. The one who shot according to our conclusion reached … 1972 perches…. The other, who shot two points lower [at 30 degrees], reached only 1872 perches. For this reason, all the bombardiers and the other persons recognized the truth of our conclusion, although before the experiment they had doubted it and most of them had been of the contrary opinion as it seemed to them that the piece was elevated too much. (Valleriani 2013, 71)

Tartaglia's narrative is undeniably indebted to an attempt to accentuate his own achievements. Indeed, as will be argued, it may be seriously doubted that the events he described even took place, or at least that they took place in the way he depicted. Quite independent of its historical reliability, Tartaglia's account, however, offers a good starting point for the analysis, as it raises a number of crucial questions.

To begin with, Tartaglia allocates the origin of the question regarding the maximum range elevation to the domain of practice. It is a gunner who first posed him the question, and independently of this, two other master gunners disputed about it. It is thus implied that the question of the maximum range elevation arose in practice.[35] Indeed knowledge that cannons at low angles shoot further with increasing elevations, and that mortars at high angles shoot further with decreasing elevation, seems to inevitably imply the existence of a maximum range elevation and could at least potentially have resulted in the question as to the concrete angle.

As already argued, however, gunners had no reason to link cannon and mortar shots in such an abstract manner. Rather, the link was established in the external representations of shots on paper with the construction of full-range trajectories. How this inevitably raised the question of the maximum range angle is illustrated perfectly by the diagram from Leonardo da Vinci's *Codex Madrid* already discussed. The construction underlying the diagram basically assimilates the range behavior known to practitioners from high-angle shots with mortars and extends it to low angles. Thus a conflict with practical knowledge is created. For shots at low angles the range should, as every gunner knew, decrease with a decreasing angle of elevation. According to Leonardo's construction, however, shots would go ever farther the lower the elevation of the gun, so that the horizontal shot would in fact be the one to reach maximum range. It is probably Leonardo's own realization of this conflict that induced him to violate his construction rule for the flattest shot in his diagram, and to give it a smaller range. According to Leonardo's construction, gunshots would hence reach the maximum range at an elevation far below 45°.

Do we have reason to assume that the question of the maximum range angle raised in the constructions of trajectories on paper posed itself in the domain of practice as well, as Tartaglia would have us believe, and if so, how was it answered and what relevance did it have there?

To begin with, no practical knowledge concerning the maximum range elevation could have existed if cannons had not at least occasionally been shot at angles close to 45°. Yet the Dutch Gunner Trolis Nielson Brink, for instance, states in his book of 1681, from which I translate from the German translation *Gründlicher Unterricht der Attilerie* published in 1699:

[35] In the dedication, Tartaglia claims to have "concluded and demonstrated by means of natural and geometrical arguments" that the maximum range is reached at an elevation of 45° (Valleriani 2013, 67). The eighth definition of the second book of the *Nova Scientia* states that the initial straight parts of a shot end "farther away from its beginning and above the plane of the horizon than [if it were] elevated in any other way" (Valleriani 2013, 163). This insinuates that the angle of maximum range is 45°, but a demonstration is missing.

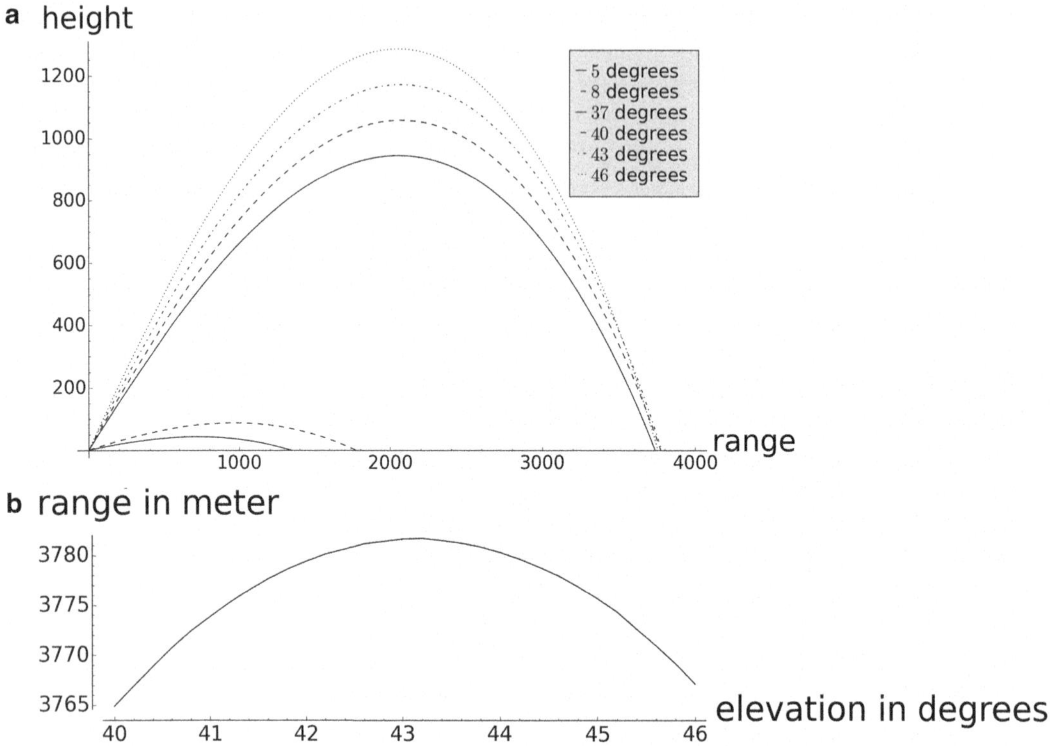

Fig. 9 (**a**) Trajectories for shots at different elevations. The trajectories have been calculated using parameters as they would have been typical for a seventeenth century culverin. (**b**) Range of the shot depending on the gun's elevation for angles closes to the maximum range elevation (approximately 43 degrees). Graph by © Jochen Büttner

> Thus quite a few make believe/one could shoot with a piece towards 45 degrees/where I have up to now/wherever I have been/found not one/that could be placed that high on its carriage on a plane field/or one would have to dig a hole in the earth/and hang the carriage's tail in it/to lift up the front of the piece. And if an advantage could be had from such shooting/one could have made the carriage accordingly/so that the piece could be elevated to 45 degrees in the field. (Brink 1699, 70)[36]

Shooting at angles close to 45° would indeed have made very little sense in a siege or fieldcombat situations. The reason for this can be gleaned from the two graphs given in Figs. 9a–b. Both graphs are based on the calculation of trajectories for iron round shot of 5.2 inch (13.2 cm) calibre weighing 18 pounds (ca. 8.2 kg), as would have been typical for a sixteenth-century culverin (Manucy 1985, 35). No reliable data for the muzzle velocity of such a cannon is available. Based on the point blank ranges given in the literature a muzzle velocity close to the speed of sound, i.e. of 330 m/s, has been assumed for the calculations.[37]

The first graph shows the trajectories of shots at 5- and 8-, as well as of 37-, 40-, 43- and finally 46-degree elevations. Whereas in the case of the first two trajectories an elevation difference of 3°

[36]German translation of (Brink 1681). "So geben auch etliche vor/man könne mit einem Stücke nach 45 Graden schiessen/da ich doch bis dato/wo ich auch gewesen/nicht eins gefunden habe/das man so hoch in der Affüte auff ebenem Felde stellen konnte/oder man müsst ein Loch in die Erde graben/und der Affüten Schwantz darein hengen/umb das Stuck forn in die Höhe zu bringen. Und wenn man von solchem Schiessen einigen Vortheil haben könnte/solte man die Affüte darnach können lassen machen/damit man ein Stück wohl solte auff 45 Grad auf ebenem Felde stellen können."

[37]For a height of the muzzle of 1 m the calculations based on a muzzle velocity of 330 m/s result in a point blank range of 143 m which compares well with the point blank range of 180 paces (ca. 137 m) which is listed in a military dictionary of 1702 for the "*Culverin* of the least size" (Viollet-le-Duc 2005, 207).

results in a range difference of more than 400 m, the latter four shots all fall within less than 50 m difference.[38] The second graph shows the dependence of the range of a shot on its elevation in the vicinity of the maximum range elevation, which in the case considered is approximately 43°.[39]

The angle of elevation is the only parameter that changes in the calculation and it causes the varying ranges plotted in the graphs. However, an ideal *ceteris paribus* is never achievable in practice. Practitioners knew, for instance, that successive shots made with the same cannon systematically increased in range even if all other factors were kept fixed, as expressed, for instance, by Ernst Braun in his *Novissimum Fundamentum & Praxis Artilleriae* of 1687:

No doubt many gunners (Büchsenmeister) and constables, inexperienced in this geometric reason/will, following common understanding, put/that such two shots/in so quick succession/from one cannon to the same target/in the same direction and with equal load without any advantage [of one over the other]/will be equal to each other/but in truth it is devised much different/because the other shot/will go much father/than the first/as is fairly proven. (Braun 1687, 44)[40]

Other factors influencing the shot, such as wind, are entirely beyond human control.

As a consequence, the ranges of real shots made with real cannons at different elevations would scatter around the ranges in a graph representing an idealized situation. The virtual cannon associated with the graph reaches its maximum range at an elevation of 43°. The graph is extremely flat around this angle. In other words, the range differences of shots fired in an angle range of plus or minus 5° around this angle are very small. For any limited number of shots in this angle regime, the range variations will thus completely mask the ideal range, due to the small but unavoidable variations in the influence parameters from shot to shot This fact only becomes discernible when comparing a very large number of shots. In direct shots, in contrast, the unavoidable variations in the influence parameters play virtually no role at all. When shooting at a target over short distances, differences in muzzle velocity and external conditions such as wind make almost no difference, providing another reason why this was the predominant way of shooting.[41]

Aimed distance shooting based on knowledge of the range behavior apparently worked well at low angles, where the graph is steep. However, this method breaks down completely around the maximum range angle. If a cannon was originally aimed at 38°, for instance, and the first shot fell short of the

[38] Standard ballistics software cannot be used for calculating trajectories of early modern artillery as these programs are restricted to small calibres and mostly to cylindrical projectiles. I have therefore adopted my calculations of trajectories from Anthony Collins (http://arc.id.au/CannonBallistics.html Accessed 6 January 2015). The external ballistics of sixteenth- and seventeenth-century cannons turns out to be particularly interesting because the parameters (diameter of shot, velocity) happen to be in a region where the flow behaviour of the air around the projectile changes from laminar to turbulent. Furthermore, the muzzle velocities are close to the speed of sound such that additional forces, caused by the pressure, increase due to the building up of shock waves ahead of a projectile and cause a significant increase in the drag coefficient.

[39] The angle of maximum range is smaller for the same projectile fired with a higher velocity as well as for a lighter round shot of the same calibre, i.e. for instance, a stone shot.

[40] "Das wohl etliche Büchsenmeister und Constabel die dieses geometrischen Grundes unerfahren seind/dem gemeinen Verstand nach setzen werden/das solche beyde Schüsse/so stracks nacheinander/aus einer Canon zu gleiche Ziel/in gleicher Richtung und gleicher Ladung ohn allen vorteil/geschehn/einander gleich sein werden/erfindet es sich doch in Wahrheit viel anders/dann der andere Schuß/viel weiter gehn wird/dann der Erste/wie solches genugsam zu erweisen."

[41] The almost complete indifference of shots within point blank range on changing external conditions is very clearly expressed in Vannocci Biringuccio's *Pirotechnica*, published posthumously in 1540: "all guns both ancient and modern ones, except the mortar shoot in a straight line at the desired thing if they are made accurately…. It cannot help causing the ball to strike the exact place you have designated unless the ball is hindered by something in issuing … [or] from the gun's not being straight, or not well bored and clean inside, or from … the powder. I also do not wish to omit telling you as a warning that, as long as the gun is straight, every error made in its operation is yours and not that of the gun." (Biringuccio 1540), cited according to (Walton 2008, 69).

aim, there would have been no way for a gunner to foresee whether or not a successive shot made at a greater elevation, say, 40°, would go further.[42]

That the shots near the maximum range elevation would have had little to no relevance for practice in the period under investigation, is strikingly confirmed by an artillery instruction manual, in which it is stated as late as the mid-nineteenth century regarding the maximum range that:

> … it is supposed in the case of mortars to be somewhat below 45° but great importance need not to be attached to the absolute determination of the true angle, inasmuch as the deviation of three or four degrees from the angle of 45 will not sensibly impair the range.[43]

Two additional reasons made such shots near the maximum range elevation undesirable in practice. First, striking power fell off rapidly over long ranges, and, moreover, upon a certain range it became difficult to discern the target.[44]

This shows that we should approach Tartaglia's account of the bet and trial shooting with great caution. A test based on just two shots is a quite meaningless endeavor from a practical perspective. If such a test was nevertheless conducted in the way Tartaglia described, it would not have been a particularly wise choice for any gunner of the time to bet money on its outcome, as the odds of such a bet must have been very close to fifty-fifty.

Cannons were, however, occasionally shot at angles close to 45°, even though such shots made little sense in siege or field combat situations. Rather, they were made, if only occasionally, outside combat, mainly on newly emerging artillery grounds in the context of artillery training, but also for testing guns and making trial shots.[45] In such a controlled environment the multiple factors influencing each shot could be much better controlled, and the results observed much better and under much safer circumstances than in combat.[46]

Statements like the following by Wallhausen, show that gunners did indeed possess knowledge pertinent to the question of maximum range elevation:

> According to the highest elevation as is the fortieth and forty-fifth degree/approximately 6000 passes. (Wallhausen 1617, 15)[47]

In agreement with what was said above about the range behavior in the vicinity of the maximum range elevation, Wallhausen, rather than specifying a concrete angle, specifies an angle range between 40 and 45° as corresponding to the maximum range of a cannon. This more faithfully represents gunners' knowledge than the specification of a concrete angle. Moreover, it makes it plausible that sixteenth-century gunners confronted with the claim that cannon shots reach a maximum range at 45°, as in Tartaglia's story, would have answered that a maximum range angle should be somewhat lower.

[42] Halley saw an advantage of shooting mortars close to the maximum range elevation, stating that gunners at such an elevation "may shoot with much more certainty, for that a small Error committed in the Elevation of the Piece, will produce no sensible difference in the fall of the Shot" (Halley 1686, cited according to France 2014, 278). He affirmed that this practice was applied by "*French* Engineers in their late Sieges."

[43] Cited from the 1848 English translation of a French manual of 1841 (Anonymous 1848).

[44] Cathrin A. France has rightly emphasized that the striking power or effect on the target played an important role, in particular for Tartaglia. The practical relevance of and knowledge about striking power awaits further study. Based on the assumption that striking power fell off over the course of the shot, allegedly rooted in practical knowledge, Newton devised an argument against the parabolic shape of the trajectory of gunshots in a letter to Collins in 1674 (France 2014, 265).

[45] Steven Walton discusses the establishment of shooting grounds and artillery gunnery schools in England in (Walton 1999).

[46] Before being used in the field, guns were test fired according to standardized routines. After a number of successive shots using increasing loads they were inspected for signs of structural instability (Hall 1952, 13).

[47] "Nach der höchsten elevation, als des 40. und 45. Grads/an die 6000 Schridt."

Thus, gunners may have had an answer to the question about maximum range elevation, but the question itself was mostly irrelevant for the practice of shooting. Why, then, did it take on such a central role for shooting with ink? Moreover, why and how did Tartaglia's answer become almost unequivocally accepted, despite it being at odds with practical knowledge?

To comply with the practical knowledge regarding the qualitative range behavior of shots, any construction of a trajectory needed to yield, as we have seen, a monotonously increasing range with an increasing angle for low angle shots and a monotonously decreasing range for high angle shots. The angle at which range behavior changes is then necessarily the maximum range elevation. Finding a construction that has this characteristic and which is not based more or less directly on the a priori assumption of a particular maximum range angle as a determining element of the construction, is not entirely trivial.[48]

A maximum range elevation of 45° is, moreover, a particularly convenient assumption, as it allows symmetry arguments to be exploited when specifying the construction rules for the trajectory. Indeed, Tartaglia's claim that the maximum range is reached at an elevation of 45° can similarly be exposed as an a priori assumption made for the sake of being able to construct the trajectories in the first place. Here, rather than using Tartaglia's example, the role of the maximum range elevation as a defining element of constructions of full range trajectories will be elucidated based on an example from Ufano's book (Fig. 10).

The respective diagram from Ufano's book may at first glance look like a geometrical construction of trajectories, but it is not. Rather, it must be considered an unusually rendered and illustrated range table. What is important is thus, less the drawing, than the numbers given at the bottom, which represent the ranges of shots at different angles, varying alternately in steps of 7 and 8°. These numbers immediately impart that the drawing is not a construction, but rather quite a resourceful way of providing the table rows for the different elevations of the shots. Whereas the range differences between the different trajectories are equal in the drawing, they decrease downwards toward 45° according to the ranges given, just as one would expect.

About 70 years later, in 1683, François Blondel (ca. 1618–1686) published a slightly corrected version of Ufano's drawing in his book on ballistics, which appeared 2 years later in German translation

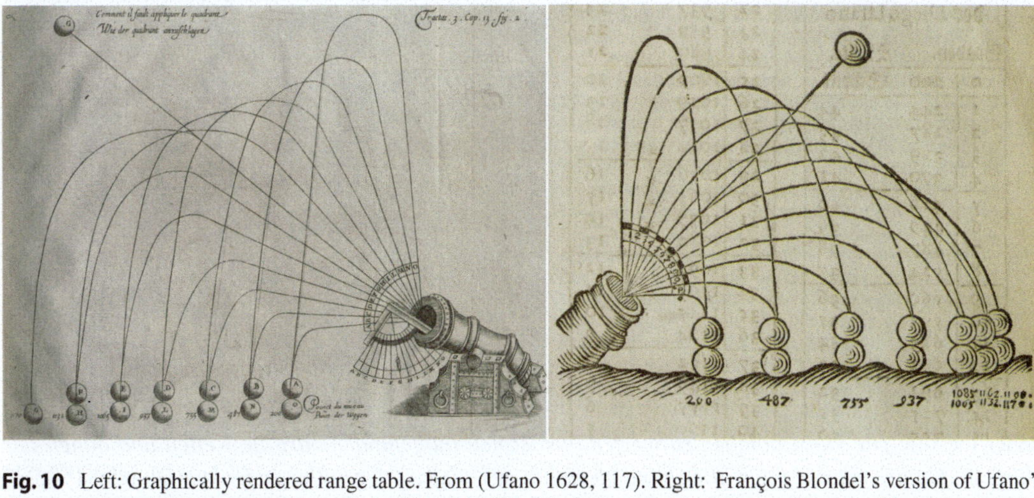

Fig. 10 Left: Graphically rendered range table. From (Ufano 1628, 117). Right: François Blondel's version of Ufano's diagram, using slightly corrected data. From (Blondel 1686, 21). Courtesy of the Library of the Max Planck Institute for the History of Science, Berlin

[48] Some theories of external ballistics indeed produced monotonously increasing range with decreasing angles of elevation, with the horizontal shot being the one to reach the maximum range. An example is provided by Flurance Rivault's *Les Elemens de l'Artillerie* of 1608, which was criticized by Blondel for exactly this reason; see (Rivault 1608). For Blondel's critique see (Blondel 1686, 36–9).

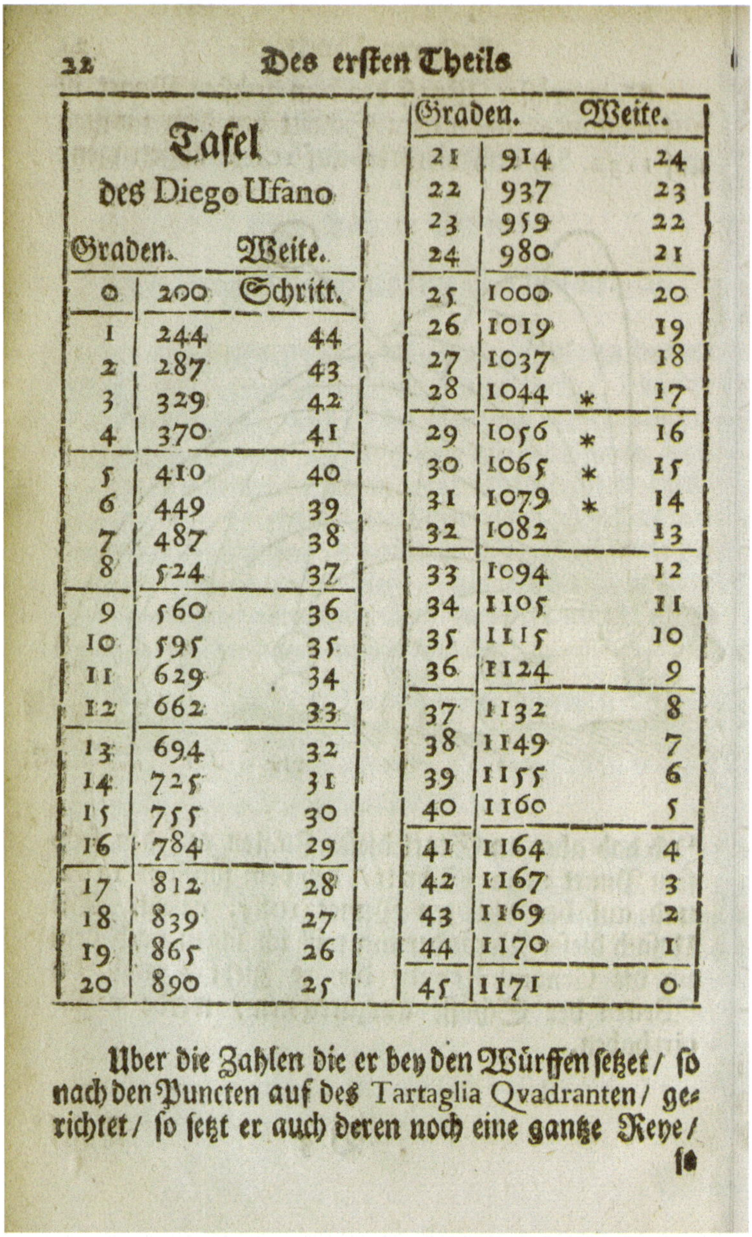
22 Des ersten Theils

Tafel des Diego Ufano

Graden.	Weite.		Graden.	Weite.	
0	200	Schritt.	21	914	24
1	244	44	22	937	23
2	287	43	23	959	22
3	329	42	24	980	21
4	370	41	25	1000	20
5	410	40	26	1019	19
6	449	39	27	1037	18
7	487	38	28	1044 *	17
8	524	37	29	1056 *	16
9	560	36	30	1065 *	15
10	595	35	31	1079 *	14
11	629	34	32	1082	13
12	662	33	33	1094	12
13	694	32	34	1105	11
14	725	31	35	1115	10
15	755	30	36	1124	9
16	784	29	37	1132	8
17	812	28	38	1149	7
18	839	27	39	1155	6
19	865	26	40	1160	5
20	890	25	41	1164	4
			42	1167	3
			43	1169	2
			44	1170	1
			45	1171	0

Uber die Zahlen die er bey den Würffen setzet / so nach den Puncten auf des Tartaglia Qvadranten / gerichtet / so setzt er auch deren noch eine gantze Reye / so

Fig. 11 François Blondel's interpretation of the range table by Ufano. From (Blondel 1686, 22). Courtesy of the Library of the Max Planck Institute for the History of Science, Berlin

under the title *Die kunst bomben zu werfen das ist neu aufgefundene art die weiten und höhen der würfe und bogenschüsse nach aller hand elevation der stücke oder böller zu finden* (Blondel 1683).[49] In his version of Ufano's diagram, Blondel has the ranges of the shots fired at 30, 37 and 45° falling very close to each other, and his drawing thus somewhat corresponds to the numbers more faithfully than Ufano's original (Fig. 11).

[49] Blondel's work was originally written about 10 years earlier, as can be gleaned from the printer's address to the reader. For Blondel's discussion of Ufano, see (Blondel 1686, 20–31); the diagram discussed here can be found on page 21.

If the given ranges are not the result of a geometrical construction, it might seem that they must represent empirically recorded data, which, however, they do not. As Blondel has already remarked, the numbers follow a simple numerical rule. In fact, Blondel's book comprises a completed and corrected version of Ufano's range data rendered as a real table, listing the ranges of shots for each degree between 0 and 45°. In addition, Blondel added a third column in which the range differences between any two successive shots, differing in elevation by one degree, are listed—to illustrate the rule which Ufano had applied in generating his range data.

Blondel's representation perfectly illustrates that Ufano simply assumed a linear decrease in the range difference with an increase of the angle. To make things even simpler, he chose the initial range difference between the shots fired at 0 and one degree to be 44 distance units, so that the range difference between the 43 and the 44° shot would amount to exactly one. As the 200 paces listed as the point blank range is a very unlikely empirical value, it can safely be stated that not a single one of Ufano's numbers represents empirically recorded data. At the same time, these numbers are not totally detached from an empirical knowledge based on practical experience. They are the results of a numerical rule, which itself was formulated based on generalizing empirical knowledge of the qualitative range behavior. In order to be able to formulate such a rule, Ufano directly assumed a maximum range elevation of 45° as a determinative element for the specification of his rule.[50]

As an aside, the symmetry assumption invoked created a direct tension with practical knowledge when rendered as a diagram by means of different trajectories, which, again, Ufano subtly concealed in the diagram. The symmetry assumption implies that a shot at 90°, i.e. a shot made vertically upwards, would have had a range equal to that of the point blank shot. This is not only implausible to practicing gunners, but contradicts universal experience. In order to conceal this contradiction in the diagram, everything is slightly tilted so that the steepest shot, which goes vertically upwards according to the underlying scheme, seems to have a small angle with respect to the vertical, and thus could plausibly also result in a non-zero range.

John Roberts of Weston, probably realizing the contradiction entailed in assuming, as the majority of authors did, a symmetry in the range behavior between point blank and vertical shot, turned the argument around to argue for a maximum range at an angle below 45°:

> Therefore it stands with more reason, and experience also hath found, that the utmost randon of any peece should be (being duely charged & discharged) at 42. degrees. (Roberts 1639, 38)

According to him, shots at 84° have the same range as point blank shots and hence he argues that the maximum range angle must be 42°, implicitly presupposing symmetry in range behavior as well. Moreover, as no gun was capable of shooting at point blank and then being elevated to 84° and shot again, Weston's assumption that these two shots reach the same range can be unmasked not as practical, empirical knowledge but as a conventional assumption, since his argument would be sound only if the ranges compared are reached by shots from one and the same cannon.

Yet the assumption of a maximum range angle of 45° quickly became so dominant that one finds hardly any alternatives to it, such as in the case of Weston, in the practical literature of the late sixteenth and seventeenth centuries. It should not be forgotten, however, that this assumption was, after all, in conflict with practical knowledge. Some authors of the time indeed remained aware of the fact that most of the contemporary accounts of ballistics deviated from practical knowledge in this respect. Thus, for instance, Thomas Digges writes:

[50] The rules used to generate the data in the range tables of the practical literature are often even simpler than Ufano's rule. Thus Eldred, for instance, whose tables list the ranges for every degree of elevation up to only 10°, simply assumed a linear increase in the range with elevation (Eldred 1646, 69 (irregular numbering: after 72, before 56)).

> Even Tartalea the Italian, albeit he were an excellent Geometer, taking upon him to deliver sundrie Demonstrations in this newe Science, yet for wante of Practise, and Experience, hath erred even in the first Principles, and so consequentlye in the whole substaunce of his discourse. This Tartalea averreth the Angle of 45 grades, as meane between the Horizontal and Vertical mountes of Eleuations, to be the Angle of the most Randons: an Errour knowen even to the first Practicioners. (Digges 1579, 189)

Digges here resorts precisely the example of the maximum range elevation to illustrate Tartaglia's ignorance of practical knowledge. He himself seems to have regarded the fact that this angle is not 45° as a kind of king's road to an improved account of external ballistics. In his book he made some very interesting remarks to this end, but did not work out a full solution.[51]

Indeed, there is at least one example of a sixteenth-century approach to external ballistics that is not based on more or less directly making an assumption about the maximum range elevation as a determinative element. It is to be found in an undated Italian text, written presumably not long after 1565 and most likely authored by a certain Giusto Aquilone (Aquilone after 1565).[52] The author, following the general scheme of preclassical external ballistics, provides a relatively simple construction that complies with Euclidean geometry and results in trajectories satisfying an astounding number of elements of practical knowledge (Fig. 12).

Aquilone's construction leads, in particular, to the conclusion that the elevation angle for the maximum range must be about 42.5°. The author is perfectly aware that this sets his construction apart from other accounts of his time, emphasizing:

> There is a dispute among the populace and a certain Tartaglia who has initiated it and they say that shooting a ball whatever piece of artillery at 45 degree that it will go farther from the center than in any other elevation. This is false. (Aquilone after 1565, 16)

He even mirrors Tartaglia's report of a dispute between two gunners and a successive trial shooting. According to Aquilone, such a trial took place in Ibiza in 1565. Of course, in his version of the story, the gunner triumphs who holds that the maximum range elevation should be lower than 45°.

These last examples show that the tension or conflict that existed between the theoretical claim that the maximum range is reached at an elevation of 45° and practical knowledge was recognized, if only occasionally, and that alternative approaches existed. Yet this recognition would not make any real impact on shooting with ink and what is more, it can be claimed, on shooting, for a long time.

More generally, the above brief survey of approaches to and discussions of the question of the maximum range elevation show that, far from being a time of stasis as it has occasionally been portrayed, the period between Tartaglia and Galileo is characterized by an intense elaboration of preclassical external ballistics, its internal consequences, and the new questions engendered. This resulted in a great number of different accounts, filling and interpreting the basic structure of preclassical external ballistics differently. In turn, most of these accounts were recognized by other authors who in turn discussed them in their own practical books on gunnery.[53]

---

[51]Digges elaborated his approach to external ballistics in his *Pantometria* of 1591. A more detailed discussion is contained in (France 2014, 82–92).

[52]A somewhat more detailed account of Aquilone's theory is contained in (Büttner et al. 2003).

[53]Once the structure of external ballistics had been established, drawing or constructing trajectories was no longer mandatory. Once established as an epistemic object, they could be invoked, for instance, by means of verbal description of the underlying geometrical constructions. An example is provided in Henry Phillippes's chapter on gunnery in his *A mathematical manual*, first printed in 1669 (Phillippes 1693).

Fig. 12 A full solution for the geometrical construction of the trajectory by G. Aquilone. Prat. n. 438 mss. Biblioteca Nazionale Centrale di Firenze (Büttner et al. 2003, 20)

Galileo's considerations concerning the flight of projectiles differ from earlier preclassical external ballistics in that he and later his followers—much more than earlier authors had done, if at all—related external ballistics to other phenomena such as the fall of heavy bodies or the shape of water-jets. They attempted to exploit the relations they perceived when drawing up explanations, which they embedded in a wider natural philosophical context. Indeed, the trajectory thus became one of the most challenging objects of early modern mechanics.[54] In this sense Galileo's account of the parabolic shape of the trajectory can be understood as being theoretical. After the publication of his *Discorsi*, Galileo's theory became established and provided the foundation for most accounts and further studies

[54]For the concept of challenging objects of early modern mechanics, see (Renn 2001), further developed in (Büttner 2008a) and (Büttner 2008b).

of the flight of the shot. Yet at its core, the theory can be exposed as an instance of preclassical external ballistics.[55]

Galileo's theory thus served as a conduit for the practical knowledge of gunnery in studies of the flight of the shot way beyond the seventeenth century. In this way, preclassical external ballistics mediated through Galileo's theory, continued to act as a scaffolding in the works of authors who had no direct contact with practical knowledge, but mainly consulted earlier works on the subject as a source of information. Although some of these authors essentially pursued a natural philosophical endeavor, the majority continued to produce practical literature in which they often stripped Galileo's theory of its wider implications and treated it as a mere account of the flight of the shot for the most advantageous shooting practice.

Blondel who propagated Galileo's approach in this manner was aware that it remained rooted in practice and stated:

> Because for once nothing is more certain than this being well conceived and implemented practice. (Blondel 1686, 372)[56]

At the same time, however, the embedding of the Galilean theory in a wider explanatory framework, which doubtlessly contributed to its success, arguably resulted in a solution that was a less adequate account of the actual flight of a shot than earlier accounts had been. Thus, the increasing dominance of the Galilean solution in practical literature made the implied reapplication of external ballistics to the practice of shooting artillery even more precarious, at least from a modern perspective.

8 Applying Preclassical External Ballistics

That gunnery theory was largely independent from and found no application in gunnery practice in the sixteenth and seventeenth centuries has been common knowledge since it was influentially argued in the early 1950s by Bert Hall. In his *Ballistics in the Seventeenth Century* he stated that:

> The standard of engineering technology was not merely insufficient to make scientific gunnery possible, it deprived ballistics of all experimental foundation and almost of the status of an applied science, since there was no technique to which it could, in fact, be applied. (Hall 1952, 16)[57]

For Hall, the reasons why preclassical external ballistics was not and indeed could not be applied, were to be sought entirely on the side of practice, where, according to his understanding, practitioners were unable to control their war technology sufficiently to allow for the successful application of ballistic theories. Hall's opinion is noticeable in its contrast to the position expressed by the majority of

---

[55] Thomas Harriot's attempts to derive the shape of the projectile trajectory lend themselves to comparison. As Matthias Schemmel has shown, Harriot shares a number of basic assumptions with Galileo, such as, for instance the decomposition of projectile motion or a times-squared law of fall. Unlike Galileo, however, Harriot invoked range data from the practical literature of gunnery in his considerations and thus ended up with trajectories whose shape is described by a parabola, tilted with respect to the vertical. For details, see (Schemmel 2008).

[56] "Denn einmal ist nichts gewisser als das dies wohl ausgesonnene und ins werk gesetzte praxis."

[57] In "Gunnery, Science and the Royal Society" Hall expressed his position even more distinctly: "... existing military art was incapable of adopting a mathematical theory of projectile flight and applying it to practice, nor [so far as one can tell] did it ever attempt to do so, at least before the death of Newton. Therefore, if the learned men developed the theory out of a desire to solve useful problems, they were mistaken" (Hall 1983, 116).

contemporary authors of practical literature, who more often than not conceded that gunners were able to achieve their goals precisely because they were accomplished in controlling the manifold parameters that affected the result of shots, technological as well as external. As the remark by Thomas Digges quoted above shows, this was furthermore understood as the precondition of a potential application of preclassical external ballistics.

As exemplified by the case of the maximum range elevation, the relation between theories of external ballistics and gunnery practice was convoluted. Indeed, the question of the maximum range elevation arose, as has been argued, on paper—and no practice existed to which the answers that were given could meaningfully be applied. This suggests submitting to reanalysis of Hall's claim that gunnery theory was not brought to bear on practice and his broad-brush-stroke explanation as to why this was the case. To this end I will distinguish the question of empirical from that of practical applicability. First it will briefly be discussed whether preclassical external ballistics adequately described the physical phenomenon at hand. It will then be asked what consequences an application would have entailed for practice, and finally, how this could have served practice, if at all.

9 The Empirical Adequacy of Preclassical External Ballistics

If Hall saw the reason for the non-applicability of "a mathematical theory of projectile flight" to be situated in in the practitioners' inability to sufficiently control their technology, this apparently held true for him only after Galileo's theory of projectile motion, as the earlier approaches:

> … reveal the fundamental uninventiveness of the mind for during two generations at the height of the Renaissance and of the rapid expansion of the use of artillery scarcely a useful footnote was added to the ballistical writings of Tartaglia, and make plain the barrenness in the field of technique, equal to the philosophical inadequacy of pre-Galilean mechanics, of the early manuals of gunnery. Although the art of shooting at long ranges was confessed to be the flower of gunnery, it was quite divorced from the physical speculations of the schools and there was no bridge between them until Galileo founded a new ballistics which derived nothing from the Tartaglian doctrines expounded by the professional experts. Yet ballistics as a science could only begin when the theory of projectiles emerged on to the field of the battle from the studies of the philosophic commentators, to introduce a new and permanent characteristic to society. (Hall 1952, 48)

If prior to Galileo accounts of external ballistics were inadequate, then the question of their applicability to a capricious technology need not be discussed at all. Not all historians have been as dismissive with regard to the pre-Galilean works engaging in preclassical external ballistics as Hall. A tendency to assume that Galileo's theory of projectile motion, judged in hindsight as a scientific achievement, is also by and large empirically more adequate, and that it should thus offer a more suitable basis for practical application, nevertheless seems to prevail.

Modern physics, however, instructs us how disturbingly wide off the mark, to use a fitting metaphor, the predictions of the Galilean theory of projectile motion in fact are from the empirically observable phenomena. This had already been realized in the mid-eighteenth century by Robins, who gave a drastic example of the discrepancy between the predictions of Galilean/Newtonian ballistic and empirical data:

> A Musketball 3/4 of an Inch [ca. 1.9 cm] in Diameter, fired with half its Weight of Powder from a Piece 45 Inches long [ca. 1.14 m], moves with a Velocity of near Feet in 1 00 [1´´]. Now if this Ball flew in the Curve of a Parabola, its horizontal range at 45° would be … about 17 Miles. Now all the practical Writers assure us, that this Range is really short of half a Mile.[58]

[58](Robins 1742), cited according to (Rickey 2008, 8).

Robins's understanding of the enormous difference between theoretical prediction and empirically observable fact was based on knowledge of the muzzle velocity of the gun, knowledge which had first become accessible with his invention of the ballistics pendulum. Before that, resorting to Galileo's, or in fact any other theory of external ballistics, to predict the behavior of gunshots would have required, as has been seen, assimilating or gauging the form-invariant trajectories of theory to the particular gun used via a trial shot.[59]

From a modern perspective, trajectories of projectiles result from the behavior of bodies in motion under idealized conditions altered by effects mostly arising from the fact that those motions are made through the medium of air. The latter effects are subject to a separate theory which has shown them to be much greater than anyone could or would have expected at the time. Such a perspective could emerge, however, only as a long-term consequence of the historical process under investigation here. It would be thus whiggish to judge theories of external ballistics, in particular Galileo's, based on such a perspective.

From Tartaglia onwards, authors conceded that air, but also other factors, may have given rise to differences between prediction and empirically observable fact. Yet Galileo's account, like earlier accounts, was thought of as applicable to practice, at least in principle, and thus essentially treated as a genuine account of the path shots take through the air in the real world rather than in an abstract, idealized one, i.e. as in external ballistics.[60]

Taking the early modern accounts of external ballistics for what they in fact are, it can and indeed has been demonstrated that Galileo's theory does not necessarily give an empirically more adequate account of real trajectories than some of the earlier approaches, as the shape of trajectories of sixteenth- and seventeenth-century cannons indeed closely resembled those trajectories as constructed by the Tartaglias or Aquilones who came before.[61]

---

[59]Galileo's ideal parabolas have a scale with respect to their initial velocity, i.e. given the initial velocity and elevation the horizontal range is determined. From a modern perspective, gauging Galileo's trajectories to the range a cannon reached in a trial shot, as application of the theory in the seventeenth century required, amounts to accounting for the shots from this cannon, using Galilean trajectories calculated for cannon of a muzzle velocity that was actually much lower. For one particular elevation, the effect of air resistance on the range thus disappears entirely, and the difference between the theoretical prediction of ranges for other elevations and the ideally observable empirical ranges, is thus comparatively small.

[60]Galileo is aware of some of the abstractions of his account of projectile motion with regard to the observable physical phenomena. He discusses them at length, arguing that that his theory indeed models physical reality well and is thus applicable to shooting. In conclusion, he states: "The enormous momentum [impeto] of these violent shots may cause some deformation of the trajectory, making the beginning of the parabola flatter and less curved than the end; but, so far as our Author is concerned, this is a matter of small consequence in practical operations, the main one of which is the preparation of a table of ranges for shots of high elevation, giving the distance attained by the ball as a function of the angle of elevation; and since shots of this kind are fired from mortars [mortari] using small charges and imparting no supernatural momentum [impeto sopranaturale] they follow their prescribed paths very exactly" (Galilei et al. 2010, 256). "Questo soverchio impeto di simili tiri sforzati può cagionar qualche deformità nella linea del proietto, facendo 'l principio della parabola meno inclinato e curvo del fine; ma questo, poco o niente può esser di progiudizio al nostro Autore nelle praticali operazioni: tra le quali principale è la composizione d'una tavola per i tiri che dicono di volata, la quale contenga le lontananze delle cadute delle palle tirate secondo tutte le diverse elevazioni; e perchè tali proiezzioni si fanno con mortari, e con non molta carica, in questi non essendo sopranaturale l'impeto, i tiri segnano le lor linee assai esattamente" (Galileo 1890–1909, 279). In addition to providing tables, Galileo proposed how his theory could be applied by gunners using a simple hanging chain (Damerow et al. 2001). Thus, on the flip-side of Galileo's achievement in theoretically accounting for the idealized case, resides a serious misjudgment on his part about the capacity of his theory to model external ballistics.

[61]In applying the model of a linear resisting medium, (Stewart 2012) shows that the trajectories of shots in a highly resistive medium approach shapes closely approximated by right-angled triangles. A more realistic model results in skewed trajectories, as shown in Fig. 9a. The higher the muzzle velocity and the lighter the shot for a given caliber (for instance in the case of a stone ball), the more the trajectories deviate from a symmetrical shape.

Closer overall resemblance in shape, i.e. greater empirical fit from a modern perspective, does not automatically entail greater practical adequacy, as practitioners, as detailed above, were less interested in the shape than in the individual characteristics of the trajectory, like its range or the maximum height reached.[62] Thus the question of the extent to which different external ballistics formulated in the sixteenth and seventeenth centuries were capable of reproducing, and thus potentially also anticipating, practically relevant data, depends on what actually mattered in practice. Moreover, answering the question presupposes knowledge about the real trajectories of shots fired with sixteenth- and seventeenth-century guns. Modern physics can furnish this information based on information about the characteristics of contemporary guns, such as muzzle velocity, or the weight, size and form of projectiles. Experimental research on these characteristics has only just begun and the question as to what aspects of the shot practically mattered at that time requires further study.[63]

Even with such questions left unanswered, it can be argued, however, that the stumbling block for attempts to make the theory bear fruit in practice, lie not so much in the theories' capability or lack thereof to make passable practically relevant predictions, but in the concrete problems and consequences their application entailed.

10 The Practical Applicability of Preclassical External Ballistics

The practical literature of gunnery portrays the knowledge it imparts as applicable, and this, by and large, also holds true with regard to the accounts of external ballistics given therein. However, the authors generally show a clear awareness of the contentious status of preclassical external ballistics in relation to the practice of gunnery and the knowledge of gunners. In particular, they acknowledge that putting preclassical external ballistics to practice would necessitate altering established practice and pose demands on abilities of practitioners—abilities which could not necessarily be presupposed.

Indeed, the geometrical constructions or calculations that would be required in order to anticipate the result of a gunshot based on preclassical external ballistics would have called for a rather high degree of mathematical skill from the gunners, and even the possession of such skills would not guarantee them being readily applicable in combat situations.[64] Two trends can be observed in the practical literature which, arguably, are based on an aspiration to overcome this limitation. The first is the tendency to embody external ballistics in instruments that would allow the gunners to achieve the desired results by mechanical manipulation alone, or which would at least greatly simplify and diminish the amount of calculations required.

[62] The difference between better empirical fit and practical adequacy alluded to in the text, becomes evident if it is brought to mind that rather bizarre constructions of trajectories can be devised which, however, reproduce the correct range behavior. Such trajectories would not describe the flight of the shot through the air well, but could at least potentially be used to elevate the gun for aimed shooting. A historical example is provided by the triangular trajectories Daniel Santbech presented in his *Problematum astronomicorum et geometricorum sectiones septem* of 1561 (Santbech 1561), which are discussed in more detail in (Büttner et al. 2003).

[63] For a summary of the results of test firings with small arms of the Landeszeughaus Graz, see (Krenn et al. 1995). Modern external ballistics takes effects other than air drag into account to infer sixteenth- and seventeenth-century trajectories.

[64] In his 1647 book *The Art of Gunnery,* Nathaniel Nye stated with regard to the mathematical qualifications required of a gunner that "He ought to have skill in Arithmetick, to Adde, Subtract, Multiply, Divide, to work any conclusion by the single or double Rule of Three; to abstract both the Square and Cube Roots and to be perfect in the Art of Decimal Arithmetick" (Nye 1670, 41). Moore gave a similar list in his *Practices in Arithmetic, Geometry, and Astronomy* of 1674, but further qualified "and though ordinary Gunners may be excused from all this Knowledge, yet Master-Gunners, and those that desire to be knowing in this Profession, must not hereof be ignorant" (Moore 1674, 84).

Tartaglia allotted the gunner's quadrant a central place in his *Nova Scientia*, and even insinuated that it was his own invention despite the fact that, as Valleriani has shown, it had already been in use more than 100 years earlier in the Husseit wars.[65] Tartaglia's quadrant is basically an angle measuring device and, as such, its use in the practice of gunnery was quite independent of external ballistics. Arguably, it was used primarily to alter the elevation after a failed shot at an angle of elevation previously determined, or simply to recreate the conditions of previous, successful shots after which the guns had to be readjusted due to recoil.

Yet Tartaglia insinuated that the quadrant could have a use exceeding the customary application:

> This instrument should also be easily usable to investigate the variety of shots of each piece of artillery and, similarly, of each mortar. (Valleriani 2013, 195)

Later instruments did indeed incorporate the constructions underlying specific accounts of external ballistics, and thus allowed the direct reading of ranges when applied to the elevated gun. The recorded ranges merely had to be applied to one's own gun based on range data obtained in a trial shot and straightforward calculation.

Yet this does not seem to have completely solved the problem. William Eldred (ca. 1563–ca. 1646), in his *The Gunners Glass* of 1646, for instance, writes:

> The ground of all shooting is by the Quadrant; but it is troublesome and unfit to be used in time of service or in haste;

An alternative to embodying the theories of external ballistics in instruments was to boil their results down to more readily usable tables, giving the ranges of shots at different elevations, usually from zero to 45°. Famously, Tartaglia promised to provide such a range table in the *Nova Scientia* but failed to do so. Using a range table required some mathematics, too, but the mathematical operations to be carried out entailed no more than calculating ranges by means of the rule of three.

Employing tables was even further simplified by having tables that listed ranges for not only one, but multiple different types of cannons. If a gunner found his own cannon among the types listed, or if his cannon had at least reached a range reasonably close to that of one of the cannon types listed in a trial shot, he could simply use the data in the corresponding column to elevate his piece depending on his distance from the target.

That such tables were indeed understood to reduce and substitute for the more complete theories of external ballistics to readily applicable data is illustrated perfectly by another passage from Eldred's *The Gunners Glass:*

> I will endeavor to prove the truth of the same as briefly and truly as I can referring my self to others more learned and practiced in the Mathematiks to take some pains to find the Arch or way of a shot, being in my judgement hard to find, but of great use when it is found, in the mean time accept of this The Table of Randoms. (Eldred 1646, 53)[66]

Shooting by instrument or table added a new task to the scope of activities of gunners, that is, the determination of the distance and height difference between their gun and the target. Tartaglia saw that shooting by instrument or table entailed changes in common practice, and that this, in turn, required new skills on the part of the gunner:

> But then I realized that all these subjects (Magnificent Lord) are not really useful to the bombardier if he does not know the distance to the place he needs to strike. (Valleriani 2013, 75)

[65] For Tartaglia's quadrant, see Valleriani's discussion in (Valleriani 2013).

[66] The pagination is irregular; the table referred to is contained on the next two pages bearing the page numbers 72 and 69.

Accordingly, a good part of his *Nova Scientia* is devoted to providing methods of measuring distance and height.

The implementation of preclassical external ballistic in practice thus represented, or would have represented, a rather drastic intervention within existing practice. As will be detailed below, contemporaries could have had profound reasons to doubt that the complications thus introduced by applying theory, would pay off in the form of improved shooting results.

Increasingly, preclassical external ballistics came to be looked upon and referred to as theory apart from practice, which could inform the latter in ways other than through direct application in shooting. Often authors alluded to this distinction prominently in the titles of their works, such as Johann Sigmund Buchner, for instance, who entitled his 1682 book *Theoria Et Praxis Artilleriæ* (Buchner 1682). Theory, even if not directly applicable or applied, could, as claim some authors, have the benefit of helping gunners to better understand what it is that they do and act as a shortcut in acquiring the necessary skills.

While they perceived the relation between preclassical external ballistics and the practice of gunnery, and in particular the applicability of the former to the latter, as a vexing question, none of the authors whose works have been consulted resorted to a position that looks upon external ballistics as pure theory, for the time being inapplicable to practice. Such a position has been attributed in the past to Evangelista Torricelli (1608–1647), who in his *De motu gravium* wrote in 1644:

> … and if by the impediment of the medium the parabolas become pretty deformed, or if by many other accidents the shots turn out most inconsistent, it is enough to most indubitably have satisfied the school of the mathematicians if not that of the gunners. (Torricelli 1644, 229)[67]

Yet Cathrin Ann France has convincingly shown that this "retreat to the safe haven of mathematical certainty" does not represent Torricelli's actual stance, as "letters show that he wanted to investigate how the difficulties of gunnery practice could be overcome; he had no doubt that the theory was true in the material world" (France 2014, 154–82).

11 Application to What and What For?

Could practice have reasonably been improved, given that theories of external ballistics were in principle sufficiently adequate empirically, and that gunners had the means to take advantage of them? As will be argued, a biased, incomplete and, occasionally, simply false image of shooting gunpowder artillery was coined by those advocating preclassical external ballistics in order to emphasize its applicability and the expected benefit thereof. This image needs to be rectified before our question can be answered.

As elucidated above, the authors of the practical literature insinuated that shooting close to the maximum range elevation was relevant in practice, which it demonstrably was not. The distortions between the image of shooting and actual practice, however, far exceed this individual aspect and are much more pervasive (Fig. 13).

[67]In context the passage reads: "Per cauar dunque qualche regola dall'esperienze, era necessario il farle esattamente, a tutti gradi della eleuazione, in tutte le forti de i pezzi, con tutte le varietà delle polueri, e le diuerse materie delle palle, e forsi anco direi che era necessario che le facesse ogni Bombardiere da se stesso. Cose quasi impossibili a ridursi sotto regole e cauarne certezza alcuna, se la Teorica, e la Geometria non ce ne daua manifesta scienza mediante proposizione del Galileo, nell quale primo di tutti egli hà auuertito, & insegnato a noi, che i proietti camminano tutti per vna linea parabolica. Su questa supposizione fonderemo lo strumento promeso: se poi per l'impedemento del mezzo le parabole venghino troppo deformarte, o per moti altri accidenti i tiri riescono incostantissimi, ci bastera auer sodisfatto indubitatamente alla scuola de Matematici, se non a quella de Bombardieri."

Fig. 13 Exploding gun. From (Aneau 1552, 34). Courtesy of the Staatliche Bibliothek Regensburg

Pointedly speaking, external ballistics as an applied theory, is a theory of elevating the gun to an aim. Gunnery practices, however, encompass much more than shooting, and shooting much more than aiming the gun. With an increasing focus on external ballistics, many of the aspects of gunnery practice, as they had been discussed in detail in earlier works, were eclipsed in later books in favor of an emphasis on shooting itself. With regard to shooting, hitting their aim was certainly not the only objective for gunners. Thus, for instance, guaranteeing their own and their crew's well-being by regularly servicing their cannons and paying attention to basic safety measures was an important part of gunners' practice.

Against this, shooting at a target was increasingly allotted the central place in the presentations of gunnery practice, and aimed distance shooting was overemphasized—although, as has been argued, it must have been much less frequent and important than shooting within point blank range, where the latter practice, by definition, can hardly be improved by theory (Fig. 14).[68]

This polarity is illustrated, in almost comical fashion, by a diagram extracted from Wallhausen's *Archiley Kriegskunst* of 1617 (Wallhausen 1617).[69] The diagram shows the familiar symbolic cannon capable of shooting at elevations at least up to 45°. From the cannon a set of trajectories extend, corresponding to elevation angles of 30, 20 and 10° below and 0, 10, 20, 30, 40 and 45 above the horizontal plane, and thus the diagram seems to embrace preclassical external ballistics.

Yet the shots below the plane of the horizon leave the diagram, thus ending essentially nowhere. As regards the point blank shot and the shots at higher elevations, the author placed a mountain in the scene, which gets in the way of the shots before they even start to curve. Thus, while superficially engaging in preclassical external ballistics, the author has effectually reduced the full trajectories to straight shots within point blank range, i.e. exactly the type of shots predominant in practice and in no need of theory.

Only the shot at 45° goes over the mountain, and has a curved path. However, it clearly misses what was obviously its target, the fortress on top, and eventually lands behind the mountain so that the shot

[68]Often the authors of the practical literature of gunnery asserted that the relevance of external ballistics was restricted to shooting at long range, but did not indicate how germane such shooting was to gunnery. Thus Tartaglia, for instance, in the epistle of the *Nova Scientia*, acknowledged that the improvement his new science could bring to the art of shooting "concerns long shots" (Valleriani 2013, 77).

[69]The diagram discussed in the text is contained between (Wallhausen 1617, 58–9).

Fig. 14 Impeded shots in diagram. From (Wallhausen 1617, 58). Courtesy of Sächsische Landesbibliothek – Staats- und Universitätsbibliothek Dresden (SLUB). Deutsche Fotothek, 01054 Dresden/Milit.A.60, misc.2

is wasted. One cannot help but read the diagram as a comment on the futility of applying preclassical external ballistics to practice.

Other contexts overemphasized in the practical literature are situations in which gunners constantly readjusted their guns to aim at new targets, and likewise situations where the target was positioned much higher or lower than the gun. Thus Bourne, for instance, while acknowledging the fact that gunners were indeed able to perfectly master the situations they were usually confronted with based on their experience, advocated external ballistics by asserting that it could improve the results of shooting in exactly those two types of circumstances in which gunners otherwise, according to him, would tend to err.

> And to bring that peece onto any other place, they must doe as at the first, to prove what will reach the marke: and therefore this kind of shooting is to no purpose, but onely in a Town of Castell, in the time of service, for him that hath the charge of Ordnaunce, to prove what the peece will doe at every marke, as touching the keeping of the length of the marke, wherby they may the better shoote at their enemies when they doe serve, otherwise it will be to no great purpose: for as often as you doe alter or change your peece, or take the peece away to serve in another place, so oft you are to seeke, and to prove thinge newe againe…. And if you shoote by the Degrees in a Quadrant, then the highnesse or lownesse of the ground shall cause them to erre, as I doe shew in the thirteenth Chapter. Therefore, if I were worthy to give counsell, I could shew them how to use the matter, that they might atteyne to know the length of the marke at the first shotte, but I never saw it so handled, whereby they should atteine it. (Bourne 1587, 20)

This is to be contrasted with a practice of shooting, in particular in siege situations, where constant re-aiming was not called for. In an account of the Siege of Elfsborg in Sweden by the Danish and English in 1612, for instance, one reads:

> King [James I] began to playe with 7 peeces upon one of the towers of the castell contynually tyll ten a clocke, at which tyme he had beaten downe parte of the tower, having spent 200 shot.[70]

Thus until the tower had finally come down, each of the seven cannons had fired an average of about 30 shots at essentially the same spot. More than initially hitting the target, the problem was thus to recreate and sometimes create the conditions that would allow one to repeatedly hit a target after a successful hit with successive shots.[71]

Moreover, shooting up and down appears to have been more complicated than shooting on a horizontal plane, at least as far as aiming the gun is concerned, only from the perspective of preclassical external ballistics—almost all of its accounts attribute a special role to the vertical direction in some form or another, thus giving it preference. At least at short distances, however, the gunners' model of correcting the elevation against the line connecting cannon and target does not generally distinguish with regard to differences in the height between the gun and the target.

Thus, Eldred in *The Gunners Glass* advanced the example of shooting at a ship from a platform 20 yards above sea level. Information about the height of the platform is used, based on sighting, to determine the range difference to the ship but does not affect the shooting itself, which is treated as if the shot over this distance were made on the horizontal plane. Eldred describes some pages later as follows (Eldred 1647, 79):[72]

> Then I say the ship is 58 score off; and I must give a Demi-Culverin almost 4 degrees to reach the mark. (Eldred 1646, 94)[73]

As already stated, the application of preclassical external ballistics to practice essentially amounts to providing the gunners with a means of aiming their cannon, i.e. determining the angle of elevation of their piece in a concrete situation. The angle of elevation was, however, only one of a multitude of factors influencing the outcome of each shot. As part of the act of shooting their guns, gunners deliberately adjusted a number of different factors to control the shot and hit their target. Other factors preferably had to be held constant, which, in turn, required special actions and knowledge on the part of the gunners as well.

With regard to powder, for instance, its composition and amount had to be chosen depending on the gun, the projectile, and sometimes also, as we have seen, on the position of the target; different rules existed according to which this was done.[74] The dampness and granularity of the powder had to be monitored and the other factors adjusted accordingly. In a chapter headed "Die Ungleichheit der Materie verhindert in der Praxi, daß die Reguln von the Theorie ihre rechte Würckung nicht haben können," Blondel stated:

> … because many kinds of powder/so in their stuff and in their effect/being very much different/are being used. Who does not know/that one and the same powder/that is being made from one and the same stuff/will show its strength/depending on wether it is small or coarse grained/moister or dryer/older or fresher/and who can ensure

[70] Cited according to (Wilson 1945, 726).

[71] Humphrey Barwick, in his *A Breef Discourse Concerning the Force of All Manual weapons of Fire*, mentions, for instance, that cannons sometimes recoiled from six to eight feet (Barwick 1592).

[72] Once the distance to the target is known "I know," Eldred continues, "by the Table what allowance to give my Peece."

[73] In this second example of shooting at a ship that Eldred gives, the distance to the barque was determined from the knowledge of its length.

[74] Steven Walton, who has compiled data for powder loads from five English gunnery books from the late fifteenth to the late sixteenth century, concluded that "it is clear that there was no single rule, and even within one author a simple rule (for powder charge) does not exist" (Walton 2008, 74).

> that in every shot as much powder went into the piece or in the chamber as did the other time; evidently there is more powder in a weight/when it is dry than when it is wet/and in a [volume] measure of the charge more small than coarse grained powder. (Blondel 1686, 369)[75]

Gunners, moreover, had to be able to account for factors outside of their direct control, such as imperfectly bored cannons and changing external conditions like wind, and had to react suitably. Variations in each of these factors and in the way gunners handled them was a potential source of inaccurate results. Indeed, in almost any of the works of practical literature one finds a list of sources of inaccuracy which needed be controlled by the gunner. Some even conceded that there were probably some factors with a potential influence on the shot not yet known.

Halley, commenting on the difference between theoretically expected and observed behavior in experiments made with a mortar, was keen to emphasize that those differences:

> … were usually less than the Errours committed in ordinary Practice, by the unequal Goodness and Dryness of the same sort of Pouder, by the Unfitness of the Shott to the Bore, and by the Loosness of the Carriage. (Halley 1686, 20)

While clearly intended as an argument for the correctness of theory, and thus indirectly for its applicability, Halley concomitantly highlights the problem of its application. The potential improvement that could be gained by applying external ballistics would mostly be masked and thus annulled by the other "Errours commited." Indeed, elsewhere Halley concedes that gunners:

> … loose all the geometrical accuracy of their art from ye unfitness of ye bore to ye ball, and ye uncertain reverse of ye gun, which is indeed very hard to overcome.[76]

Application of external ballistics may thus have been capable, in principle, of giving good or even perfect guidance to gunners in elevating their pieces. Yet, we have good reason to believe that what Halley describes would regularly have been the case, and that applying external ballistics would have only altered the results of shooting a little. As much as depending on the elevation, these results also depended on other factors that had to be controlled but for which no theoretical guidance could be given, as well as on unavoidable imperfections and changing external influences.

A quote from Robert Norton (died 1635) in *The Gunners Dialogue Vvith the Art of Great Artillery* conjures up an image of shooting quite different from the one given by those championing preclassical external ballsitics:

> To faile at the first shott, if he be not acquainted with the Peece and Marke is passable; and at the second to fayle is pardonable, but to faile of a faire shott at the third time, is too much, and argues but little judgement and discression in such a Gunner. (Norton 1628, 115–6)

If we accept Norton's to be the more realistic image of the practice of shooting, it is indeed difficult to see how shooting aided by the theories of external ballistics could have substantially altered the state of affairs.

The extent to which external ballistics nevertheless began to exert some influence on practice is, according to my understanding, a completely open question which awaits further study. There is some

---

[75] "da man sich doch so vielerlen Pulvers/so an seinem Zeug und an seiner Würckung/so gar sehr unterschieden ist/bedienet. Wer weiß nicht/daß einerlen Pulver/das ist daß so aus einerlen Zeug gemacht ist/seine Krafft erweise/nach dem es kleiner oder gröber gekörnet/feuchter oder truckner/älter oder frischer ist/und wer kann versichern daß am vermittelst des Gewichts oder der Maas allzeit gleiche Ladung bekomme/und daß eben bei jedem Schuß grad einmal soviel als das andere mal vom Pulver in das Stück oder die Kammer hinein gekommen seie; Angesehen viel mehr Pulver auf das Gweicht geht/wenn es trucken als wenn es feucht ist/und in eine Maas von der Ladung mehr klein als grob gekörntes Pulver."

[76] (Halley 1932, 52), cited according to (France 2014, 272).

evidence to suggest that gunners did indeed start to avail themselves of range tables. It is doubtful, however that the tables used, at least initially, were the ones that can now be found in the practical literature, almost all of which, without exception, provide data generated based on external ballistics. Nathaniel Nye (1624–after 1647), in his *The art of gunnery,* appealed instead to gunners to do what he convincingly claims to have done himself, namely to produce their own range tables from systematic trial shots with their own guns:

> Every one that will learn to shoot at Random, must draw his Peece into a levell ground, where first shooting level, he must observe that distance in feet or paces, then mount his Peece to one Degree, and mark where that shall graze: thus find the distance of every Degree, from the level to the 10th degree, and by these distances make a Table, to which annex the Degrees against the distances.... (Nye 1670, 98)

It thus seems entirely plausible that though the very idea of range tables was engendered by external ballistics, the way they came to be used was essentially decoupled from the range predictions resulting from the different accounts.

To sum up, in the sixteenth century, and even more so in the seventeenth, an image of gunnery practice began to emerge within its practical literature that set apart and overemphasized the shooting practice of elevating the gun to aim in certain situations that required aimed distance shooting, and for which external ballistics promised to provide guidance. In a sense, it could be argued, Hall fell victim to the very image that began to be coined in the period under investigation. If one accepts elevating the gun for shooting at long ranges to be the core of gunnery practice, then it is necessarily implied that external ballistics is gunnery theory, and that the reason for the failure to apply this theory can be sought in the practice peripheral to this core activity.

A more balanced picture of gunnery practice, however, reveals elevating the gun for shooting at long ranges to be a less pivotal task among the wide spectrum of gunners' pursuits. For those engaged in shooting, the little appeal that the conclusions of preclassical external ballistics had, which held in the realm of shooting on paper, can thus arguably be attributed to the fact that external ballistics as an applicable theory was much too narrowly conceived.

External ballistics may, nevertheless, have come to be applied to a limited and possibly increasing extent. Yet the application of external ballistics to shooting in battle could not have given rise to knowledge accountable for the further development of theory, which then necessarily seems to have been essentially an inner-theoretical one detached from the empirical data provided by practical knowledge. Nor could the benefit of the application have been anywhere close to what was frequently promised so confidently in the practical literature.[77] So if preclassical external ballistics could not live up to its promise, why did it persist and, moreover, how could it have proliferated and turned into an independent branch of study?

12 Shooting with Ink for Whom and to What End?

The realization of the inefficacy of preclassical external ballistics in practice did not result in its expulsion from gunnery's practical literature. Rather the opposite was the case. The fundamental idea of preclassical external ballistics proliferated and took on an ever more central role in books on gunnery, often at the expense of topics that had canonically been treated in the earlier works as part of gunnery practice. To understand this trend, it first needs to be acknowledged that the practical literature of

[77] Ken Adler made the point that the situation did not essentially change in the eighteenth century, despite the seemingly great advances in external ballistics. "The functional contribution of mixed mathematics to the practical success of the late-eighteenth century artillery engineers, therefore was largely illusory...." (Adler 1999, 116).

gunnery was not quite what it appears and often explicitly pretends to be—information laid down by practitioners for practitioners.

A comprehensive study of authorship and readerships of the practical literature and their changes over time and national backgrounds is far beyond the scope of this article. Based on existing studies, however, some general remarks are permissible. Some authors, particularly of the earlier works, were indeed men who shot guns as part of their professional occupation.

A good part of the authors of the practical literature, however, certainly had not. Tartaglia famously prided himself in never having shot a cannon. Indeed, the professional background of the authors of the works considered covers a wide spectrum, among them higher-ranking military officers in close contact with gunners, but not necessarily practicing gunners themselves; technical advisers on civil and military projects best described as military engineers; architects and practical mathematicians with or without a strong background in scholarly culture; and engineer-scientists. In particular, later authors were thus often observers of the practice they described rather than actors taking part.

In writing their books on gunnery practice these authors unquestionably followed a wider cultural trend. At the same time, their individual motives to write their books were certainly vastly different. We should avoid reducing their motives to a competition for patronage, although it is surely true that, as France states, theories of external ballistics were "a form of knowledge capital that was passed on to future military writers and theorists, who nevertheless endeavoured to draw off the maximum in interest that they could gain along the way" (France 2014, 286).

Which audience the practical literature of gunnery was intended for and which audience it eventually found is not easily answered. Especially in the earlier works, the presentational style tends to vary within one and the same book—some parts indeed seem to speak directly to the practicing gunner, whereas other parts seem to be intended more for persons outside the field with a general interest in the subject. Notably, the works on gunnery are written almost without exception in the vernacular. Many of the books went to more than one edition and were frequently and rather quickly translated into other languages, suggesting a fairly wide audience.

In his illuminating study of the practical literature of gunnery in England in the Elizabethan era, Steven Walton distinguished "how-to manuals which instruct readers how to use artillery from textbooks which analyze and attempt to systematize the study of artillery." He suggests two sorts of audiences, "didactic for the books, and mnemonic for the manuals," where according to him, manuals mostly came in manuscript form. This difference between pen and print reminds us that printed books were, if not luxury items, still objects of substantial value, affordable primarily to those in privileged situations. Writing in France in the second half of the seventeenth century, in a society dominated by military culture, Blondel stated:

> Following this example even today the most noble persons exercise this branch of mathematics, which mainly pertains to the things of war…. (Blondel 1686, 373)[78]

This corresponds well to what Walton concludes based on his analysis of the relevant English literature:

> The printed practical literature was aimed at the educated reader, but not one necessarily a reader employed militarily. It was in particular not primarily aimed at the practicing gunner.
>
> While a practicing gunner might pick one of these books up, the profession clearly had more to do with craft knowledge than even the best "textbook" … could offer. Mathematicians who engaged the subject, despite the best insight and training available, found dead-ends. … the military men using artillery were less interested in analyzing its performance than those on the periphery who thought that analysis might be their entry ticket into that profession. (Walton 1999, 332)

[78] "Diesem Exempel nach üben sich auch heut zu Tag die allervohrnehmsten Personen in diesem teil von der Mathematic, welcher hauptsächlich zun Kriegssachen gehört…."

Thus master gunners, in particular, may occasionally have been among the readers; the average gunner most certainly was not. Indeed, it is usually explicitly indicated in the practical literature that the books can neither exhaustively cover the art of shooting nor replace its teaching, as the art cannot be learned from the book or at least not from the book alone. Humphrey Barwick, in his *A Brief Discourse concerning the Force and Effect of All Manual Weapons of Fire*, states:

> Thinking this sufficient, to proove my argumentes manifest enough onto all that dooth knowe the use of weapons of war: and for such as are ignorant, let them learne of them that dooth truely know them. (Barwick 1592)

The practical literature of gunnery, whose emergence and success thus has to be understood against the background of a context extending far beyond gunnery and its immediate periphery, creating a niche in which the increasing abstraction of preclassical external ballistics, with regard to practice, was able to flourish and be developed further largely undisturbed by its increasing tension with practical knowledge and the untenable nature of its promise to improve practice. Pointedly speaking, preclassical external ballistics, with its hybrid nature between practice and theory, parasitically hinged on the genre of practical literature from which it gradually emancipated itself, thereby transforming the very genre itself.[79]

That this transformation was accompanied by an increasing decoupling from practice was occasionally realized, if virtually ignored, as a marginal note in a copy of Blondel's book shows:

> It is obvious that since the divided charge of powder alone produces all of those effects, goodbye to all the square-edges, angles, parabolas, conic functions and to this entire hodgepodge of mathematics, which is useful only for displaying the profound knowledge of the author, but which could never be understood by any gunner.... Where the devil was this author's head when these gunners taught him that and why did he not take this opening that their experience gave him?[80]

Did preclassical external ballistics thus lose all contact with practice and continue to evolve as pure speculative theory?

13 External Ballistics Meets Practice

External ballistics reestablished contact with practice much later in the second half of the seventeenth century, in a way different from the one authors of the practical literature proposed, i.e. its application to shooting in the field. Rather, some of the claims made by preclassical external ballistics were increasingly called into question and submitted to systematic experimentation. These experiments did not take place in the field, but in controlled situations. Stimulus often came from those firmly situated in the world of learning but much closer to the practical world of gunners than the natural philosophers at universities, i.e. from men who have been referred to as engineer-scientists. Although they were close to practice, these men more often than not lacked the expertise to shoot a gun themselves usually resorting to enlisting the help of gunners to carry out their experiments. Moreover, those experiments relied on the infrastructure that had been built up around gunnery, such as shooting grounds, as guns could not be fired just anywhere.

---

[79] Cathrin A. France likewise resorted to the concept of a niche in her study on the emergence of external ballistics: "Tartaglia seems to have found a niche or gap in which he could construct a distinct category for his talents similar to those artist engineers such as Brunelleschi and Leon Battista Alberti who had successfully introduced the theory of proportions and perspective to create a 'scientific conception'" (France 2014, 42). I agree, but would emphasize that Tartaglia neither created the niche nor did he occupy it alone.

[80] The quote is taken from a marginal note that was added to Blondel's book, presumably in the late seventeenth or early eighteenth century. (Gerbino and Blondel 2010) here quoted according to (France 2014, 217).

The scheme outlined above is perfectly illustrated by an episode that has been reconstructed based on the correspondence between Giovan Battista Renieri and Torricelli. In a letter written to Torricelli on 2 August 1647, Renieri complained that the results of a systematic experiment conducted with a cannon, made in the presence of gunners, deviated from the predictions of Galileo's theory, which Torricelli had further elaborated and promoted, including a proposal for an instrument that would allow shooting according to this theory. Torricelli took Renieri's concerns seriously and even made proposals as to what could have gone wrong and how to improve further trials, which were apparently made, albeit without conclusive results. Torricelli's untimely death in 1647 put an end to this particular episode. As this episode has been subject to a number of detailed historical studies, another example of how the results of external ballistics came under scrutiny and were subjected to systematic experimentation will be presented here (Segre 1983, 492).[81]

The example is taken from the context of the Royal Society, and although it focuses on shooting, it first leads back to shooting on paper. Indeed, in the second half of the seventeenth century authors of the practical literature began to appeal explicitly to practitioners to put the results expounded to the test and report on their experiences. Blondel, for instance, wrote:

> These experiments having been carried out properly & repeated several times, could highlight the reliability of the application of the rules of the Theory, & by this means produce marvelous effects for the Art of throwing Bombs…. (Blondel 1685, 519)[82]

He had in fact criticized that the range data he had found in the literature, which were supposedly empirical, were the result of inferences based on external ballistics. Commenting on the tables prepared by a company of French royal gunners at the order of Louis XIV (1638–1715), he showed that the ranges listed follow a simple scheme, and stated:

> … [that] they made those their tables more according to their opinion, than they should have taken the trouble to carry out precise and certain *Experimenta* (Blondel 1686, 53)[83]

Somewhat earlier, in 1667, Sir Robert Moray (ca. 1609–1673) in England had issued a similar call for experimentation with guns in the *Philosophical Transactions* under the title "Experiments for Improving the Art of Gunnery." The goals were, first, "finding the point-blank distance; secondly, finding the quantity of powder for the just charge of any piece; and thirdly, what gun (for size, bore, weight, metal etc) shoots the farthest."[84]

Of the three goals, only the first is directly pertinent to external ballistics. Indeed, most systematic tests with guns were largely aimed at technical problems, with the solution being judged based on the trajectory, and predominantly the resulting range. Such systematic trials concerning technical aspects of the gun go back even further and appear to have withdrawn from practice more naturally, as documented, for instance, by Eldred's attempts at a systematic exploration of the effect of recoil:

---

[81] The episode described by this letter is discussed in detail, for instance, in (Segre 1991; Meli 2006; France 2014); it is summarized in (Segre 1983; 1994).

[82] "Ces expériences êtant bien faites & répétées plusieurs fois, pourroient donner de grandes lumiéres pour la sûreté de l'application des régles de la Théorie, & produire par ce moien des effets merveilleux pour l'Art de jetter les Bombes…."

[83] "sie diese ihre Tafeln vielmehr nach ihrer Meinung eingerichtet/als das sie sich sollten die Mühe genommen haben genau und sichere *Experimenta* zu machen."

[84] (Steven Walton 2008) has emphasized that technical "trials are reported as early as the mid-sixteenth century which were designed to determine whether longer cannon did indeed shoot further. Extraordinarily long guns were cast, fired, and then a few inches were cut off at a time. Luis Collado reported that a 48-calibre long piece had been cast and was then cut off in 6-calibre long segments at each trial."

… for it is most certain, that a Peece that doth not reverse will much out-shoot a Peece that doth reverse; as I have tryed often times; but in the yeere 1610 I was to shoot 3 Demi-Culverins … the peeces had 3 degrees by the Quadrant; and two of these peeces were upon Wheels, the other lay in her Carriage upon the ground, and could not reverse, but out shot the other two above 40 yards…. I have often proved to this effect. (Eldred 1646, 36)[85]

Some years after Moray's appeal, in 1674, Robert Anderson (before 1668–after 1696) published his *The Genuine Use and Effects of the Gunne* (Anderson 1674). He had undertaken "a series of experiments in ballistics with a gun cast at his own expense," the results of which he submitted to Galilean theory. Anderson's book provoked quite a controversy.[86]

According to Domenico Meli, it was most likely this controversy that prompted a group of men, many of them members of the Royal Society, to come together in September 1674 to perform experiments with guns at Blackheath, as is documented by entries in Robert Hooke's (1635–1703) diary. Among these men were the Master of the Ordnance Sir Thomas Chicheley (1614–1699), the Surveyor of the Ordnance Sir Jonas Moore (1617–1679), Lord William Brounker (1620–1684), Hooke and others, including gunners (Meli 2006). Details are not given, but Hooke noted that:

Tryd Experiment of bullet with good [results] found it very neer a parabola. (Hooke 2005)[87]

Moore seems to have continued experiments, as in a meeting of the Royal society on 18 April 1678 he:

… alledged, that in shooting Granados he had found, that the greatest random was below forty-five degrees of inclination. And that shooting at twenty degrees would fly much farther than shooting at seventy. (Birch 1757, 399)

Experiments of the kind Moore was referring to indeed crop up in the correspondence of John Flamsteed (1646–1719), to whom Moore was an important patron.

In September 1681, William Molyneux (1656–1698) wrote a long letter to John Flamsteed, who had become the King's Astronomical Observator in 1675 and a member of Royal Society in the following year. After remarks concerning astronomical matters, Molyneux leaves "the Businesse of Astronomy" and asks Flamsteed to enlight him "in a Particular of an other Nature, viz, relating to the Gun and Mortar piece:"

There is One Mr Wotton here with us who shewed me a Table of Horizontall Ranges calculated by you (as he told me) therein you make 42 degrees 1/2 the mounter for the utmost Randon whereas I think it is demonstrable that 45 degrees must needs be the utmost randon, For Toricellio has demonstrated that Randons are in Proportion to each other as Duble the sines of Elevation, if so then certainly 45° must be the greatest Randon…. But that perhaps which has deceived many in this case is theire relying on experiments made upon the mountures of 45.44.43.42.41. etc which differ but little among themselves in comparison to the Higher degrees of 70.75.80. etc. so as to determine by experiment the utmist randon is very difficult especially where such an unruly impetus as is that of gunpowder is used, which tho all circumstances be observed the same will seldome give randons the same…. This, Sir, I propose to you because a friend of mine has been at the charge of casting a small mortar-peice, and I am dayly making experiments and observations thereby, and am therefore desirous to hear what any one can say concerning it…. (Forbes 1995, 815)

Flamsteed had his answer ready. Indeed, 4 years earlier, in May 1677, he had sent a letter to Collins, who was one of the central figures in the Anderson controversy, informing him that an experiment had been made to determine the angle of maximum range. After a detailed description of the setup, and after providing a table with the ranges achieved, Flamsteed summarizes:

[85] Eldred's book contains numerous similar accounts of trials made by him or others, often even listing the precise date when those trials were performed.

[86] The controversy triggered by the publication of Anderson's book is discussed in some detail in (Meli 2006, 204–6).

[87] Cited according to (France 2014, 266.)

> They conclude the longest shot to be made at 42 degrees and what they find by the shot from the bow the find also in the Arrow shot from it and in the bullet discharged from the Mortar peece. The longest range in all being about 42 degrees. And the rest of the ranges in one proportionalle to the like of the same elevations in the other. (Forbes 1995, 558)

In his answer to Molyneux, Flamsteed refers to the experiment of 1677, and in so doing conveys information on who conducted it:

> As for the Table of Ranges which hee shewed you I assure you it agrees admirably with the experiments here lately mad(e) by one Mr Colsons a diligent man and Mr Lindgard, whereby we found that the Ranges above 45 went not so far from the peece, as those under, considerably. (Forbes 1995, 815)

The Mr. Lindgard referred to is most likely John Lindgard, who is listed as an Ordnance Office fireworker from 1676 to 1679 and thus a professional of gunnery practice (Forbes et al. 1997, 32, note 10). Molyneux finally built and test-fired his own mortar and a number of letters were sent back and forth between him and Flamsteed over the course of the next 2 years, in which the two men discussed the results of the tests and their implications. Above all, the discussion shows how difficult it was to obtain consistent results even under ideally controlled conditions. For instance, Molyneux wrote on 1 August 1682 about his own experiments with mortar shots:

> But truly Sir I must needs say that what I have performed herein was not with any great successe, notwithstanding all manner of Caution used … we can hardly make shotts anything nigh each other, tho all things be observed the same, if we make 12 or 14 shotts at the same Elevation … perhaps 8 or 10 of these will come together in an Area of about 30 to 400 foot square, but the others will be so extravagant, that there is no relying on them, some perhaps being 20 yards short and others 20 yards over…. (Forbes 1997, 27)

These historical events traced above, are symptomatic of a development that took hold in the second half of the seventeenth century. Learned men familiar with gunnery practice and often directly engaged in the administrative, coordinating, and planning aspects thereof, turned war machines into experimental systems with which preclassical external ballistics was scrutinized. They were able to do so only by involving gunners, their knowledge and expertise, and the infrastructure of gunnery. Rather than an independent branch of scientific activity, these experiments thus remain embedded in practice. The way in which these experiments prepared what has been referred to as the ballistics revolution of the eighteenth century remains a topic for future study.

14 Summary

I have argued that the practical knowledge of gunnery and the external ballistics that emerged in the sixteenth and seventeenth centuries are not two independent epistemic approaches with entirely different roots, related merely in that they partially pertain to the same phenomenon occasioned by gunshots. Rather, theory, or more precisely the abstraction of practical knowledge, began when early modern authors of practical literature started to codify or represent practical knowledge with a new representational tool—the trajectory, understood as the path of a shot traced out on paper. This particular form of representation mediated the transformation of practical knowledge into what has been referred to here as preclassical external ballistics. This preclassical external ballistics was marked by two conceptions, namely that the trajectory can be characterized by the variation of its shape, depending on the elevation of the shot, and that the trajectories of shots at a given angle, but under otherwise varying conditions, will retain their shape and merely be scaled differently. The elaborations of the basic structure of preclassical external mechanics led to different accounts of external ballistics, in which the practical knowledge of gunners that had given rise to preclassical external mechanics continued to act as a scaffold.

At the same time, the proposed accounts necessarily introduced abstractions and thus began to segregate from the practical knowledge upon which they were based. A detailed study of the discussion of the problem of the maximum range elevation has revealed two principal reasons for this. On the one hand, the trajectory was underdetermined by practical knowledge, which provided information about only some of its features, such as, for instance, the qualitative dependence of range on the angle of elevation in specific angle regimes, for specific types of guns. Authors of preclassical external ballistics thus needed to emend missing information and thereby resorted to drawing on knowledge from other domains such as geometry or Aristotelian dynamics. On the other hand, on paper, the consequences of shooting under conditions not realizable with real guns for practical reasons, or at least not relevant to shooting them in battles, could be explored. Shooting at, or close to, the maximum range elevation, despite being attributed to practice, was thus revealed to pertain mostly to shooting on paper. The main reason for the divergence between preclassical external and practice, usually understood as the result of theoretical choices such as failing to account for drag effects, is in contrast understood here as residing in the step from practical knowledge to codified knowledge.

Moreover, it has been argued that the years of failing to apply external ballistics to the practice of gunnery in the sixteenth and seventeenth centuries cannot be attributed to the fact that the accounts given reproduced the empirical phenomenon insufficiently. Rather application would have entailed drastic changes to established practice, changes which had to be weighed against the realization that even a perfect theory could not have offered much improvement to gunnery practice—primarily because all external ballistics had to offer as an applied theory, was guidance on how to elevate a gun to aim at a target where distance was known. Not only was this problem less central to gunnery than it is portrayed in the literature, but moreover, the outcome of a single shot depended on a great number of factors over which gunners apparently had reasonable control. A theory giving perfect command over just one factor was not capable of improving shooting to any great extent.

It has been argued that, in order to accentuate the potential advantage of applying external ballistics, its protagonists coined a biased image of the practice of gunnery in the practical literature. That this was possible, and that external ballistics proliferated despite its tension with practice, has been traced back to the fact that the practical literature addressed not only practitioners, but primarily educated readers, and that thus a niche was created in which external ballistics was able to develop further separate from the practice itself, it built a new structure of knowledge (Chap. 1). It has finally been argued that contrary to the authors intentions that external ballistics be applied in gunnery practice, this is not the way in which this theory reestablished contact with practice, although it may have influenced practice to a limited extent. Rather, learned men with close ties to practice, relying on the help of gunners and their infrastructure, turned guns into experimental systems to probe the claims of external ballistics precisely because of its precarious status with regard to practical knowledge.

References

Adler, Ken. 1999. French engineers become professionals; or, how meritocracy made knowledge Objective. In *The sciences in enlightened Europe*, ed. William Clark, Jan Golinski, and Simon Schaffer, 94–125. Chicago: University of Chicago Press.

Anderson, Robert. 1674. *The genuine use and effects of the gunne as well experimentally as mathematically demonstrated/by Robert Anderson; with tables of projection, &c., Exactly calculated and their use exemplified, by Thomas Streete*.... London: Printed by J. Darby for William Berry and Robert Morden.

Aneau, Barthélemy. 1552. *Picta poesis*. Lugduni: Apud Ludovicum & Carolum Pesnot.

Anonymous. 1848. *Instruction upon the art of pointing cannon*. Washington, DC: J. and G. S. Gideon.

Aquilone, Giusto. ca. 1565. Ms. II, IV, 337 – *Della formatione de tre cerchi, ne quali si fonda tutto il transito, che fa la palla tirata per aria in qual si voglia elevatione di artigleria*. Biblioteca Nazionale Centrale di Firenze.

Bailey, Nathan. 1726. *An universal etymological English dictionary, comprehending the derivations of the generality of words in the English tongue ... with additions*. London: Darby.
Barwick, Humfrey. 1592. *A breefe discourse, concerning the force and effect of all manuall weapons of fire and the disability of the long bowe or archery, in respect of others of greater force now in vse. With sundrye probable reasons for the verrifying therof: the which i haue doone of dutye towards my soueraigne and country, and for the better satisfaction of all such as are doubtfull of the same*.... London: Printed [by E. Allde] for Richard Oliffe, and are to be solde in Paules Churchyard at the signe of the Crane.
Birch, Thomas. 1757. *The history of the Royal Society of London for improving of natural knowledge from its first rise, in which the most considerable of those papers communicated to the society, which have Hitherto not been published, are inserted as a supplement to the philosophical transactions*. London: A. Millar in the Strand.
Biringucci, Vannoccio. 1540. *De la pirotechnia libri x: dove ampiamente si tratta non solo di ogni sorte & diversita di miniere, ma anchora quanto si ricerca intorno à la prattica di quelle cose di quel che si appartiene à l'arte de la fusione over gitto de metalli come d'ogni altra cosa simile à questa*. Veneto: V. Rossinello.
Blondel, François. 1683. *L'art de jetter les bombes/par M. Blondel*. Paris: l'auteur et N. Langlois.
Blondel, François. 1685. *L'art de jetter les bombes*. La Haye: Arnout Leers.
Blondel, François. 1686. *Die Kunst Bomben zu werffen, Das ist: Neu-ausgefundene Art die Weiten und Höhen der Würffe und Bogen-Schüsse, nach allerhand Elevationen der Stücke oder Böller zu finden: Sammt beygefügten Mathematischen Demonstratiionen von der Natur und Eigenschafft aller Würffe und deren Bewegung*.... Sultzbach: J. Holst für J. Hoffmann.
Bourne, William, d. 1587. *The arte of shooting in great ordnaunce contayning very necessary matters for all sortes of seruitoures eyther by sea or by lande. Written by William Bourne*. London: [By Thomas Dawson] for Thomas Woodcocke.
Braun, Ernst. 1687. *Novissimum Fundamentum & Praxis Artilleriae Oder Nach itziger besten Manier Neuvermehrter und gantz Gründlicher Unterricht*. [Danzig]: Author, Beckenstein.
Brechtel, Franz Joachim. 1591. *Büchsenmeisterey, das ist kurtze doch eigentliche Erklerung deren Ding, so einem Büchsenmeister fürnemlich zu wissen von nöten: also der rechte Gebrauch des grossen Geschützes, was damit außzurichten ...: samt getreulicher Unterweisung mancherley Feuerwerck*. Nürmberg: Kath. Gerlachin.
Brink, Trolis Nielson. 1699. *Gründlicher Unterricht von der Artillerie: Darinnen alles, was zu dieser curieusen und nützlichen Wissenschafft nöhtig ist, deutlich angewiesen, und durch viele beygefügte Figuren vor Augen geleget wird*. Hamburg: bei Zacharias Härtel zu finden.
Brink, Trolis Nielson. 1681. *Beschryvinge van de artillerye, leerende in 't korte alle 't geene dat tot de beginselen en fondamenten van nooden is: Seer dienstig voor alle leerlingen ende jonge practisijns, die haer in dese wetenschap soecken te oeffenen: In ordre gestelt en met figuren versien, en seer klaer en kort beschreven voor alle lief-hebbers*. Graven-hage: Gerrit Rammazeyn.
Buchner, Johann Siegmund. 1682. *Theoria et praxis artilleriæ*, 1st ed. Nürnberg: Hoffmann.
Büttner, Jochen. 2008a. Big wheel keep on Turning. *Galilaeana* 5: 33–62.
Büttner, Jochen. 2008b. The pendulum as a challenging object in early-modern mechanics. In *Mechanics and natural philosophy before the scientific revolution*, ed. Roy Laird and Sophie Roux, 223–237. Dordrecht: Springer.
Büttner, Jochen, Damerow Peter, Renn Jürgen, and Schemmel Matthias. 2003. The challenging images of artillery: Practical knowledge at the roots of the scientific revolution. In *The power of images in early modern science*, ed. Wolfgang Lefèvre and Jürgen Renn, 3–27. Basel: Birkhäuser.
Capobianco, Alessandro. 1598. *Corona, e palma militare di artiglieria, nella quale si tratta dell'inuentione di essa, e dell'operare nelle fattioni da terra, e mare, fuochi artificiati da giuoco, e guerra; & d'vn nuouo instrumento per misurare distanze: con vn aggiunta della fortificatione moderna, e delli errori scoperti nelle fortezze antiche, tutto a proposito per detto essercitio dell'artiglieria, con dissegni apparenti, & assai intelleggibili*. Veneto: Francesco Bariletti.
Capobianco, Alesandro. 1647. *Corona e palma militare di artiglieria, nella quale si tratta dell'inuentione di essa, e dell'operare nelle fattioni da terra, e mare, fuochi artificiati da giuoco, e guerra; & d'vn nuouo instrumento per misurare distanze: con vn aggiunta della Fort*. Veneto: Antonio Bariletti.
Chevedden, Paul E. 2000. The invention of the counterweight trebuchet: A study in cultural diffusion. *Dumbarton Oaks Papers* 54: 71–116.
da Vinci, Leonardo. 1973. *Il codice Atlantico*. Firenze: Giunti-Barbèra.
da Vinci, Leonardo. 1974. *Der Kodex Madrid*. Frankfurt am Main: Fischer.
da Vinci, Leonardo. 1986. *I manoscritti dell'Institut de France. Il manoscritto M*. Firenze: Giunti-Barbèra.
Damerow, Peter. 1996. *Abstraction and representation: Essays on the cultural revolution of thinking*. Dordrecht: Kluwer.
Damerow, Peter. 1999. *The material culture of calculation. A conceptual framework for an historical epistemology of the concept of number*. Berlin: Max-Planck-Institut für Wissenschaftsgeschichte.
Damerow, Peter, Jürgen Renn, and Simone Rieger. 2001. Hunting the white elephant: When and how did Galileo discover the law of fall? In *Galileo in context*, ed. Jürgen Renn, 29–149. Cambridge: Cambridge University Press.

Damerow, Peter, Malcolm Hyman and Jürgen Renn. 2006. *Weight, motion and force: Conceptual structural changes in ancient knowledge as a result of its transmission*. Berlin: Max Planck Institute for the History of Science.

Digges, Leonard, d. 1579. *An arithmeticall militare treatise, named stratioticos compendiously teaching the science of nu[m]bers, as vvell in fractions as integers, and so much of the rules and aequations algebraicall and arte of numbers cossicall, as are requisite for the profession of a soldiour. Together with the moderne militare discipline, offices, lawes and dueties in euery wel gouerned campe and armie to be obserued: long since atte[m]pted by Leonard Digges gentleman, augmented, digested, and lately finished, by Thomas Digges, his sonne*. London: Henrie Bynneman.

Digges, Leonard, and Thomas Digges. 1590. *An arithmeticall vvarlike treatise named Stratioticos compendiously teaching the science of nombers as well in fractions as integers, and so much of the rules and æquations algebraicall, and art of nombers cossicall, as are requisite for the profession of a soldier. Together with the moderne militare discipline, offices, lawes and orders in euery well gouerned campe and armie inuiolably to be obserued. First published by Thomas Digges Esquire anno salutis 1579. And dedicated vnto the right Honorable Earle of Leicester, lately reuiewed and corrected by the author him selfe, and also augmented with sundry additions. Aswell concerning the science or art of great artillerie, as the offices of the Sergeant Maior Generall.* London: Imprinted by Richard Field.

Duffy, Christopher. 1996. *The fortress in the early modern world, 1494–1660*, Siege warfare, vol. 1. London/New York: Routledge.

Eldred, William. 1646. *The gvnners glasse vvherein the diligent practicioner may see his defects, and may from point to point reforme and amend all errours that are commonly incident to unskilfull gunners. Set forth by way of a dialogue: Betweene an experienced gunner and a scholler, with tables of randons. Approved by the authors own practice, with certain easie wayes and rules for taking of distances: Never before this time published. Whereunto is annexed part of the excellent worke, published in Spanish by Diago Uffano, containing a dialogue between an experienced captaine, and a new generall of the artillery. Shewing the office and duty of all officers belonging to the traine of the artillery, with the manner of batteries and sieges, and many other things now discovered for the benefit and good of the common-wealth. By William Eldred, sometimes Mr. Gunner of Dover-Castle.* London: Printed by T. Forcet for Robert Boydel.

Fantl, Jeremy. 2014. Knowledge how. In *The Stanford encyclopedia of philosophy*, ed. Edward N. Zalta. http://plato.stanford.edu/archives/fall2014/entries/knowledge-how. Accessed 30 May 2016.

Forbes, Eric Gray, Lesley Murdin, and Frances Wilmoth. 1995. *The correspondence of John Flamsteed, the first Astronomer Royal*. Boca Raton: CRC Press.

Forbes, Eric Gray, Lesley Murdin, and Frances Wilmoth. 1997. *The correspondence of John Flamsteed, the first Astronomer Royal*. Boca Raton: CRC Press.

France, Catherine Ann. 2014. *Gunnery and the struggle for the new science (1537–1687)*. Phd, University of Leeds. http://etheses.whiterose.ac.uk/7912. Accessed 30 May 2016.

Fronsperger, Leonhart. 1557. *Vonn Geschütz vnnd Fewrwerck, wie dasselb zuwerffen vnd schiessen, Auch von gründlicher zuberaitung allerley gezeugs, vnd rechtem gebrauch der Fewrwerck, wie man die in Schimpff oder Ernst, von der hand, aus Fewrbüchsen oder Böllern, zu lust oder gegen den Feinden sol werffen, schiessen, oder in stürmen, in vnd aus den Besatzungen zu brauchen ; das ander Buch: Vonn erbawung, erhaltung, besatzung vnd profantierung der wehrlichen Beuestungen*. Frankfurt am Main: Zephelius.

Gal, Ofer, and Raz Chen-Morris. 2012. Nature's drawing: Problems and resolutions in the mathematization of motion. *Synthese* 185 (3): 429–466.

Galilei, Galileo. 1890. *Le opere di Galileo Galilei: edizione nazionale sotto gli auspicii di sua maesta il Re d'Italia*, ed., Antonio Favaro, Isidoro del Lungo, V. Cerruti, G. Govi, G. V. Schiaparelli, Umberto Marchesini. 20 vols. Florence: G. Barbera.

Galilei, Galileo. 2010. *Dialogues Concerning Two New Sciences*. Trans: H.C.A. de Salvio. New York: Cosimo Classics.

Gerbino, Anthony. 2010. *François Blondel: Architecture, erudition, and the scientific revolution*. London/New York: Routledge.

Hall, Alfred Rupert. 1952. *Ballistics in the seventeenth century: A study in the relations of science and war with reference principally to England*. Cambridge: Cambridge University Press.

Hall, A. Rupert. 1983. Gunnery, science, and the Royal Society. In *The uses of science in the age of Newton*, ed. John G. Burke, 111–41. Berkeley: University of California Press.

Halley, Edmond. 1686. *A discourse concerning gravity and its properties: Wherein the descent of heavy bodies and the motion of projects is briefly, but fully handled: Together with the solution of a problem of great use in gunnery*. London: Royal Society.

Halley, Edmond. 1695. A proposition of general use in the art of gunnery, shewing the rule for laying a mortar to pass, in order to strike any object above or below the horizon. *Philosophical Transactions* 19: 68–72.

Halley, Edmond. 1932. *Correspondence and papers of Edmond Halley, preceded by an unpublished memoir of his life by one of his contemporaries and the "Éloge" by D'Ortous de Mairan*. Oxford: The Clarendon Press.

Hooke, Robert. 2005. *The diary of Robert Hooke, M.A., M.D., F.R.S., 1672–1680, transcribed from the original in the possession of the corporation of the city of London (Guildhall Library) Edited by Henry W. Robinson ... and Walter Adams ... with a foreword by Sir Frederick Gowland Hopkins.* Ann Arbor: University of Michigan Library.

Keyser, Konrad. 1967. *Bellifortis. Facsimile-Druck: der Pergament-Handschrift Cod. Ms. philos. 63 der Niedersächsischen Staats- und Universitätsbibliothek Göttingen.* Düsseldorf: VDI.

Krenn, Peter, Paul Kalaus, and Bert Hall. 1995. Material culture and military history: Test-firing early modern small arms. *Material Culture Review/Revue de La Culture Matérielle* https://journals.lib.unb.ca/index.php/MCR/article/view/17669/22312. Accessed 30 May 2016.

Maier, Anneliese. 1940. *Die Impetustheorie der Scholastik.* Wien: A. Schroll & Co.

Manucy, Albert. 1985. *Artillery through the ages: A short illustrated history of the cannon, emphasizing types used in America.* Washington, DC: US Department of the Interior, Division of Publications.

Meli, Domenico Bertoloni. 2006. *Thinking with objects: The transformation of mechanics in the seventeenth century.* Baltimore: The Johns Hopkins University Press.

Moore, Jonas, Sir. 1674. *A mathematical compendium; Or, Useful practices in arithmetick, geometry, and astronomy, geography and navigation, embattelling, and quartering of armies, fortification and gunnery, gauging and dyalling explaining the logarithms, with new indices; Nepair's rods or bones; making of movements, and the application of pendulums: With the projection of the sphere for an universal dyal, &c. Collected out of the notes and papers of Sir Jonas Moore. By Nicholas Stephenson,* ed. N. Stephenson. London: Printed and sold by Nathaniel Brooke at the Angel in Cornhill.

Norton, Robert. 1628. *The gvnner shewing the whole practise of artillery: With all the appurtenances thereunto belonging. Together with the making of extraordinary [sic] artificiall fire-workes, as well for pleasure and triumphes, as for warre and seruice.* London: A.M.

Nye, Nathaniel. 1670. *The art of gunnery wherein is described the true way to make all sorts of gunpowder, gun-match, the art of shooting in great and small ordnance, excellent ways to take heights, depths, distances, accessible or inaccessible, either single or divers distances at one operation to draw the map or plot of any city, town, castle, or other fortified place, to make divers sorts of artificiall fire-works, both for war and recreation, also to cure all such wounds that are curable, which may chance to happen by gunpowder or fire-works : This treatise is composed for the help of all such gunners and others, that have charge of artillary, and are not well versed in arithmetic and geometry : All the rules and directions in this book being framed both with and without the help of arithmetick/by Nathanael Nye....* London: Printed for William Leak.

Phillippes, Henry. 1693. *A mathematical manual; containing tables of logarithms for numbers, sines and tangents: With the manifold use thereof, briefly explained and applied, in arithmetick, geometry, astronomy, geography, surveying, navigation, dyalling, gunnery and gauging.* London: Printed for Richard Mount, at the Postern on Tower-Hill.

Renn, Jürgen. 2001. Editor's introduction. An engineer-scientist, artist, and courtier at the origins of classical science. In *Galileo in context*, ed. Jürgen Renn, 1–8. Cambridge: Cambridge University Press.

Rickey, Fred. 2008. The impact of ballistics on mathematics. The work of Robins and Euler in the eighteenth-century." In *Proceedings of the 16th ARL/USMA Technical Symposium.* Atlantic City, NJ.

Rivault, David Flurance. 1608. *Les Elemens de l'Artillerie.* Paris: Chez Adrien Beys.

Roberts, John. 1639. *The compleat cannoniere: Or, The gunners guide wherein are set forth exactly the chiefe grounds and principals of the whole art, in a very briefe and compendious forme, never by any set forth in the like nature before. With divers excellent conclusions, both arithmeticall and geometricall belonging thereunto: As also sundry serviceable fireworkes, both for sea and land service. a study delightfull and very usefull for men of the best quality, and imbrac'd by the greatest princes. Written by iohn roberts of weston neere bathe, gentleman, practitioner and professor in the arte thereof.* London: Printed by J. Okes, and are to be sold by George Hurlock at his shop in Thames-street neere S. Magnus Church.

Robins, Benjamin. 1742. *New principles of gunnery: Containing the determination of the force of gun-powder, and an investigation of the difference in the resisting power of the air to swift and slow motions.* London: J. Nourse.

Rumpf, H. F. 1825. *Allgemeine Literatur der Kriegswissenschaften: Versuch eines systematisch-chronologischen Verzeichnisses aller seit der Erfindung der Buchdruckerkunst, in den vornehmsten europäischen Sprachen, erschienenen Bücher über sämmtliche Kriegswissenschaften:(duitsch en fransch).* Berlin: G. Reimer.

Ryff, Walther Hermann. 1558. *Der Architektur fürnembsten, notwendigsten, angehörigen Mathematischen vnd Mechanischen künst, eygentlicher bericht, vnd verstendliche vnterrichtung, Zu rechtem vertandt der lehr Vitruuii: in drey fürneme Bücher abgetheilet Von rechtem verstandt, Wag und Gewicht, etliche Büchlein zu sonderlichem verstandt, künstlicher Mechanischer invention, der Geometrischen messung angehenckt [u.a.]. Durch Gualtherum H. Rivium.* Nürmberg: Gabriel Heyn.

Ryle, Gilbert. 1945. Knowing how and knowing that: The presidential address. *Proceedings of the Aristotelian Society, New Series* 46: 1–16.

Santbech, Daniel. 1561. *Problematum astronomicarum et geometricorum sectiones septem.... Omnia nunc recens conscripta in communem gratiam studiosorum Matheseos & in lucem edita*. Basileae: Per Henrichum Petri, et Petrum Pernam.

Schemmel, Matthias. 2008. *The English Galileo: Thomas Harriot's work on motion as an example of preclassical mechanics*, Boston studies in the philosophy of science, vol. 268. Dordrecht: Springer.

Schmidtchen, Volker. 1977. *Bombarden, Befestigungen, Büchsenmeister: Von D. Ersten Mauerbrechern D. Spätmittelalters Zur Belagerungsartillerie D. Renaissance: E. Studie Zur Entwicklung D. Militärtechnik*. Düsseldorf: Droste.

Schmitt, Wolfram. 1972. *Deutsche Fachprosa des Mittelalters: ausgewählte Texte*. Berlin/New York: Walter de Gruyter.

Segre, Michael. 1983. Torricell's correspondence on ballistics. *Annals of Science* 40 (5): 489–499.

Segre, Michael. 1991. *In the wake of Galileo*. New Brunswick: Rutgers University Press.

Segre, Michael. 1994. Schuß – Und kein Treffer – Ballistik von Tartaglia bis Torricelli. *Kultur & Technik* 1: 30–33.

Stewart, Seán M. 2012. On the trajectories of projectiles depicted in early ballistic woodcuts. *European Journal of Physics* 33(1): 149.

Topper, David R. 2007. Art illustrates science: Galileo, a Blemished Moon, and a parabola of blood. In *Quirky sides of scientists*, 109–21. New York: Springer.

Torricelli, Evangelista. 1644. *Opera geometrica Evangelistae Torricellii*. Florentiae: typis Amatoris Masse & Laurentii de Landis.

Ufano, Diego. 1613. *Tratado de la artillería y uso della platicado por el capitán Diego Ufano en las guerras de Flandes*. Brussels: Ivan Momarte.

Ufano, Diego. 1628. *Artillerie ou vraye instruction de l'artillerie et de ses appartenances aller Zugehör*. Rouen: Jean Berthelin.

Valleriani, Matteo (ed.). 2013. *Metallurgy, ballistics and epistemic instruments the nova scientia of Nicolò Tartaglia: A new edition*. Berlin: Edition Open Access.

Vemming, Hansen P. 1993. Experimental reconstruction of a medieval trebuchet. *Acta Archaeologica* 63: 189–208.

Viollet-le-Duc, Eugène-Emmanuel. 2005. *Castles and warfare in the middle ages*. Dover Pictorial Archive Series. Mineola: Dover Publications.

Wallhausen, Johann Jacob von. 1617. *Archiley Kriegskunst, Darinnen Gelehrt Und Fürgetragen Werden Die Initia Und Fundamenta Dieser Edlen Kriegkunst/Vor Diesem Niemals so Compediosè, Methodicè, Diludicè Und Rectò an Tag Gegeben. Mit Schönen Kupfferstöcken Perfecte Angewiesen Und Beschrieben*. Hanaw [Siegen]: in verlegung des Authoris.

Walton, Steven A. 2008. Perceptions of the performance of Cannon Shot before 1700. *Journal of the Ordnance Society* 20: 69–80.

Walton, Steven Ashton. 1999. *The art of gunnery in renaissance England [microform]*. Thesis (Ph.D.), University of Toronto.

Wiggins, David. 2012. Practical knowledge: Knowing how to and knowing that. *Mind* 121 (481): 97–130.

Wilson, A. W. 1945. The story of the gun. Part I: 1242 to 1650. *The Field Artillery Journal* 12: 12–35.

Zubler, Leonhard. 1608. *Nova Geometrica Pyrobolia Neuwe Geometrische Büchsenmeisterey: Das ist: Grundtlicher Bericht, wie man durch ein neüw Geometrisch Instrument mit sonderer behendigkeit jedes Geschütz klein oder groß, bey tag oder nacht nit allein richten, sonder zugleych auch desselben höhe vnd weite messen sol ; Mit schönen kunstreychen Kupfferstucken geziert vnd an jetzo erstlich allen Kunstliebenden, sonderlich aber Zeüg- vnd Büchsenmeisteren zu gutem an tag geben*. Zürich: Jonas Gessner.

Doing It Wrong: The Translation of Artisanal Knowledge and the Codification of Error

6

Sven Dupré

Abstract

An important aspect of structuring practical knowledge is the codification of error. Rather than writing down how to proceed, authors write down what not to do. Writing down how to do something right, is probably as old as mankind's writing abilities. Recipes on Babylonian clay tablets suggest as much. However, I suggest that writing about doing it wrong seems to first emerge in the early modern period. Before the seventeenth century the recipe literature shows evidence of silent changes. During processes of transmission authors changed recipes without explicit notification to adapt them to new local and material conditions. Tried in new contexts, recipes seemed to no longer work, or the results were considered unsuccessful. Occasionally, readers of recipes even jotted down that a procedure did not work in the margin of a recipe book. Equally common in the recipe literature is the listing of variations on recipes offering the reader many different ways to do something right. However, seemingly new in the seventeenth century is the process of writing how-to, as found in earlier named or anonymous sources, followed by the explicit signal that a recipe does not work and suggestions for ways to change it to make it work. This is what is called 'the codification of error' in this essay. Faced with the flood of practical knowledge in the early modern period, it seems to be an equally powerful means to reorganize practical knowledge than bringing it together under a limited number of postulates. This essay explores the problem of the limits of language inherent to the codification of practical knowledge, and the codification of error as a strategy to overcome the problem.

This project has received funding from the European Research Council (ERC) under the European Union's Horizon 2020 research and innovation programme (grant agreement No 648718)

S. Dupré (✉)
Utrecht University, Utrecht, The Netherlands

University of Amsterdam, Amsterdam, The Netherlands
e-mail: S.G.M.Dupre@uu.nl

M. Valleriani (ed.), *The Structures of Practical Knowledge*, DOI 10.1007/978-3-319-45671-3_6

Fostered by Protestantism and the revival of Augustinian anthropology, the self-image of seventeenth-century natural philosophy, especially in England following the work of Francis Bacon (1561–1626), was that of a restoration of Adamic knowledge (Harrison 2007).[1] Adam was attributed with an encyclopaedic knowledge. His epistemic superiority was thought to reside in his capacity to name all things. The Fall meant the 'confusion of tongues,' and the beginning of the problem of translation. Since sin was equated with error, Adam's Fall was considered the source of ignorance and epistemic error. It was also the moment where sensory knowledge, which until the Fall had only been a distraction from the direct perception of knowledge with the eye of the soul, attained a certain value. It was only through externally imposed methodological constraints that 'fallen' human minds could avoid error and be rightfully guided. Experimental natural philosophy as it emerged in the seventeenth century aimed at the restoration of Adamic knowledge, a state which could be reached by developing methods such as experimental testing to avoid and erase error. This essay is concerned with the codification of error as a response to the limits of language. The limits of language are problems inherent to the codification of practical knowledge, as Viktoria Tkaczyk also proposes (Chap. 4). I argue that the codification of error emerging in the early modern period is one particular strategy to overcome these linguistic limits.

The massive translation migration of classical texts from antiquity into Latin and the vernaculars during the Middle Ages and Renaissance, often via Arabic and Hebrew, has traditionally received much attention from intellectual historians and historians of science and medicine (Montgomery 2000; Goyens et al. 2008). In contrast, in this essay the focus is on contexts involving the confrontation of erudition with local knowledge. This entailed problems of how to express information and knowledge for which no equivalent in learned circles and the languages they used was available. The most typical context in the early modern period is that of the encounter with local plants and plant names, in particular outside Europe. Travellers were confronted with a plethora of unknown names, terms and concepts. However, the definition and naming of both natural phenomena and the categories to which they belonged could sometimes prove controversial even in the same dialect. One of the major problems was the lack of completeness: what was included in the book of nature kept expanding, with a consequent expansion of vocabulary. Structurally similar to the processes of naming in the encounter with indigenous, non-European plants is the invention of vocabulary in the context of the artisanal workshop and the appropriation of this terminology in print and translation when no equivalent in Latin or the vernaculars existed.

The English word 'translation' derives from the Latin *translatio*, meaning to transfer or transport from one place or state to another, a handing over, or even, in horticulture, a grafting of one living stem onto another. When referring to writing practices, *translatio* might mean not only translation in the modern English sense of conveying the meaning of one linguistic expression into another, but copying, or even the employment of a word in an unusual way in order to convey a figurative or metaphorical meaning. Put otherwise, problems about whether (or how) situated knowledge can be moved to other places, involve questions of translation. The word 'translation' in other languages also suggests a 'rendering,' with connections to 'change,' 'transformation' and 'handing over.'

Translation is the process by which information and knowledge is transferred from one place to another, often being altered in the process. Change is not incidental to translation but rather its very essence, even when it is not deliberate. Wherever people of different languages, or even people of the same language but different origins, interact and speak with one another, transformations are at hand. Such instances deserve attention from historians of science and of ideas. Scientific communication does not necessarily imply a universal or underlying meaning to which words and other signifiers point, nor does it necessarily imply incommensurability. Translation is a process integral

[1] See especially Chaps. 1, 4.

to human communication. However, beyond this philosophical point, underlining how the problem of translation is integral to the human condition, it is worth studying the practice of translation historically because over time translation practices have changed. In this essay I argue that the translation—also in the sense 'of rendering' or 'verbalization'—of artisanal knowledge changed in the early modern period.

This goes to the heart of one of the major issues in the history of early modern science, which has been considered and re-considered by many since the generation of Leonardo Olschki and Edgar Zilsel. The study of the translation of artisanal knowledge is a way to approach the question of the origins of early modern science in artisanal cultures. In her most recent book and elsewhere in this volume, Pamela Long argues that in the fifteenth and sixteenth centuries "trading zones" (a term she borrows from Peter Galison's analysis of modern physics) emerged creating the conditions for the exchange of knowledge and empirical values between artisans and scholars (Long 2011, 94–126). Early modern trading zones consisted of "arenas in which the learned taught the skilled, and the skilled taught the learned, and in which the knowledge involved in each arena was valued by both kinds of 'traders'" (Long 2011, 95). In the concept of the 'trading zone,' language is emphasized. Galison describes such 'zones' as "characterized by the development of pidgin languages and, at times, full-scale Creole languages as a means of communication between people from two different cultures" (Long 2011, 125). Although less emphasized in her contribution to this volume, language is nevertheless a central concern for Long (Chap. 8). Most appropriately, she suggests that the invention of fire made possible a whole range of crafts, but above all, according to Vitruvius, prompted 'early humans to gather around fires, and this new proximity led them to begin to communicate and acquire language' (Long 2011, 64). However, while Long is primarily focused on detecting interactions between vernacular artisans and learned humanists who knew Latin and had received a university education, it is important to emphasize translations of artisanal knowledge, or the processes of re-structuring practical knowledge in writings on the arts.

When shifting the focus of translation studies to the confrontation of the translator with local knowledge and objects, multilingualism is the key issue that demands attention. For example, Florike Egmond points out that Latin and vernacular languages and names co-existed in the practice and publication of plant knowledge from the 1550s (Egmond 2013, 131–62). Although Latin names were used to facilitate international communication, Latin did not have a special conceptual status (in developing a classification) nor did it have a higher status than the vernacular languages or local names for indigenous plants. This offers an important corrective to the dominant historiography, which has almost entirely focused on the translation movement from Latin to the vernaculars and the so-called rise of the vernacular (Burke and Po-chia Hsia 2007; Weitemeier 2004, 1280–1295). Instead, Egmond argues, multilingualism was the norm. As we will see below, multilingualism is typical for communication in a trading zone in which knowledge is exchanged between artisanal and scholarly cultures.

1 Recipes and Errors

Writing down how to do something correctly is a practice that is probably as old as humankind's ability to write in the first instance. There is a long history of writing about technology, taking the form of recipes, reaching back to Antiquity. We have clay tablets inscribed with Babylonian glass recipes, and the so-called Leiden and Stockholm Papyri contain recipes for several arts and crafts with a long history reaching well into the early modern period—a longevity facilitated by the 'thinness' of recipes, as Elaine Leong argues (Chap. 3) (Figs. 1 and 2). In fact, one of the most common formats used to reproduce artisanal experience is the 'recipe' telling the reader 'how-to' (Clarke et al. 2012;

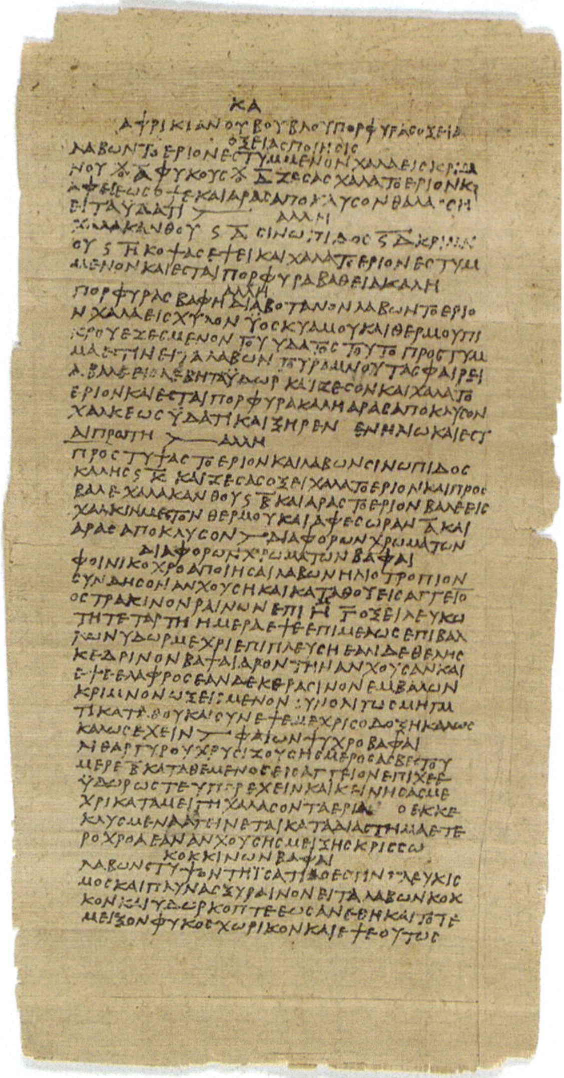

Fig. 1 *Stockholm Papyrus*, ca. 300 AD. (*Papyrus Graecus Holmiensis*, acc 2013/75). Courtesy of the National Library of Sweden

Leong and Rankin 2011; Smith 2010). By the early modern period recipes were ubiquitous. The correspondence between how-to instructions and practice is often problematic. Words typically fall short of unambiguously describing the materials and processes leading to the desired result. Historians of technology have only attributed a limited (if any) role to recipes in the learning of a trade within the context of master-apprentice relationships in the workshop (Wallert 2012). Learning by doing was the standard. Recipes are translations: they moved artisanal knowledge beyond the workshop.

An important aspect of structuring practical knowledge is the codification of error. Rather than writing down how to proceed, authors write down what not to do. Writing about doing it wrong seems to first emerge in the early modern period. Before the seventeenth century the recipe literature shows evidence of silent changes. During processes of transmission authors changed recipes without explicit notification, adapting them to new local and material conditions, because tried in new contexts, recipes seemed to no longer work, or the results were considered unsuccessful. This is not to say that, occasionally, readers of recipes did not jot down that a procedure did not work in the margin of a

Fig. 2 *Apparato della Fonderia dell'Illustrissimo et Eccellentissimo Signor Don Antonio. Nel quale si contiene tutta l'Arte spagirica di Teofrasto Paracelso*. 1604, vol. 2, 308–9. (Magli. XVI, 63, vols. I–IV). Courtesy of Ministero dei beni e delle attività culturali e del turismo–Biblioteca Nazionale Centrale di Firenze. Prohibition of further reproduction or duplication by any means

recipe book. Here is just one example that that was indeed the case: Wolfgang Seidel (1492–1562), prior but also copyist at Tegernsee monastery, wrote two *Kunstbücher* (today in Munich at the Bayerische Staatsbibliothek) between 1540 and 1550 (Neven 2014, 23–52, 30–6). Seidel collected recipes from the libraries of Tegernsee and from the libraries of neighbouring cloisters during his stay at St. Ulrich's and St. Afra's Abbey in Augsburg. Seidel also relied on exchanges with artisans and scholars, whom he cites as his sources in his *Kunstbücher*. Seidel's marginal annotations offer his comments on the recipes. For example, on folio 53r, Seidel notes down in the margin of a recipe for the melting of crystal that the recipe was not of use to him and that a better way to melt crystal is found on another folio of the same manuscript (Fig. 3).

In the early modern period such evaluative notes moved from the margin of the text to the body of the text. What emerges is the process of writing how-to, as found in earlier named or anonymous sources, followed by the explicit signal that a recipe does not work and suggestions for ways to change it to make it work. This is termed 'the codification of error.' In this way writing down artisanal experience begins to reflect reading. Faced with the flood of practical knowledge in the early modern period it was a powerful means to reorganise practical knowledge. An important factor in the emergence of codification of error was the conception of natural knowledge as an enterprise devoted to the restoration of Adamic knowledge as described above. Francis Bacon was one of the most prominent spokesmen of this self-understanding of natural knowledge as the undoing of the errors consequent of the

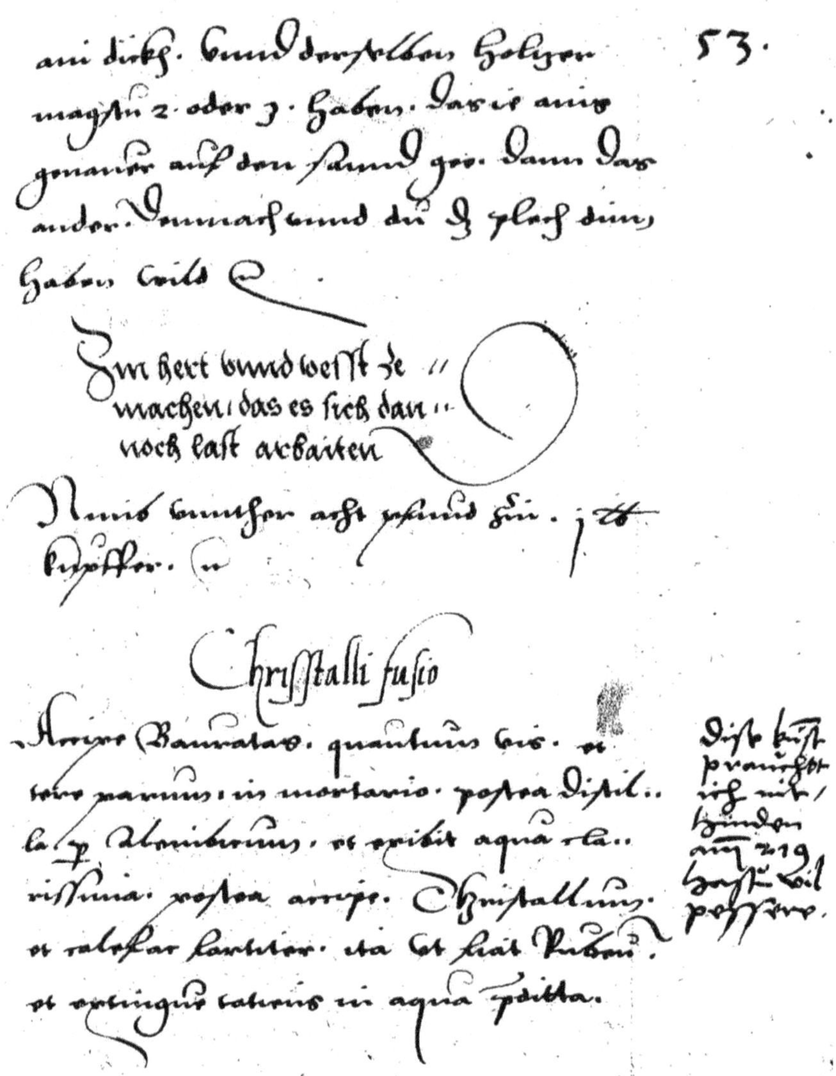

Fig. 3 Wolfgang Seidel, *Kunstbuch oder von manigerlai Handwerchskünsten,* 1550–1506, fol. 53r. Courtesy of Bayerische Staatsbibliothek München, Cgm 4117, fol. 53r

Fall (Harrison 2007, 172–85). Bacon described the sources of error—the 'idols of the mind'—which included sensory errors as well as errors of the internal senses and the intellect. Bacon also believed that the errors could be rectified by natural means, and he suggested the use of optical instruments to circumvent the failures of the human senses and practices of note-taking to undo the errors of memory. However, as we will see below, although Bacon's significance should not be underestimated, the emergence of the codification of error cannot be simply attributed to Bacon's conception of the project of experimental natural philosophy. Another important factor in the emergence of the codification of error was the movement of recipes into print and the specificities of early modern print culture.

Fig. 4 Johannes Kunckel's 1679 Frontespice. From (Kunckel 1679). Courtesy of Herzog August Bibliothek Wolfenbüttel: [M: Od 215]

Codification of error is a sign of the fixity of print.[2] This does not entail the erasure of all typographical errors, as Eisenstein has pointed out.

> In the hands of ignorant craftsmen, the printing of texts led to the multiplying of error (all errors being 'compounded by pirated editions').... But under the guidance of technically proficient masters, printing provided a means of transcending the limits imposed by scribal procedures. (Eisenstein 2002, 92)

The codification of error would be much more difficult to achieve without the stabilization and preservation of texts characteristic for print.

The focus in this essay is on Johannes Kunckel's (1630–1703) translation of Antonio Neri's (1576–1614) *L'arte vetraria*, the first printed book on glass-making, published in 1612: *Ars vitraria experimentalis oder Vollkommene Glasmacher-Kunst* (1679) (Fig. 4).

---

[2]For the sense given to 'fixity of print,' see (Eisenstein 2002). In response to criticism by Adrian Johns and others, see (Johns 1998).

2 Johannes Kunckel and Gold Ruby Glass

Both Neri and Kunckel were alchemists, and their work on glass-making is best conceived within their broader alchemical pursuits. In the early modern period alchemy was a diverse field (Principe 2013). Metallic transmutation of base metals into more noble ones, especially the making of gold, is probably the most ready association of alchemy. However, not only was there considerable divergence of opinion in how to make the philosopher's stone to bring about metallic transmutation, *chrysopoeia* (as this field was called) was merely one aspect of early modern alchemy. A second sub-field was medical alchemy. Made a cornerstone of alchemical practice by John of Rupescissa (1310–1362) in the fourteenth century, the importance of chemically produced or enhanced medical preparations immensely expanded following the writings of Paracelsus (1493–1541) in the sixteenth century. A third subfield of early modern alchemy was chemical manufacture. This is the field we are concerned with in this essay. Continuing a recipe literature going back to Antiquity (the best known examples are the Stockholm and Leiden Papyri) early modern artisanal workshops increasingly adopted chemical methods to make all kinds of goods and materials: pigments, dyes, alloys, perfumes, among others. The entire range of ideas and practices dealing with the production and manipulation of material substances and their properties—whether the making of gold and silver, or the making of medicines, dyes, pigments, glass, and so on was referred to as alchemy in the early modern period. In short, far from being limited to the making of gold, alchemy was the practice and theory of material transformation (Dupré et al. 2014).

It was possible for an early modern alchemist to be involved in one or two of these fields, but he did not need to be involved in all three. For example, Paracelsus gave the field of medical alchemy a boost, but he did not dabble in gold-making. Interestingly, although it is not possible to fully develop this here, given that these three fields are separate domains of alchemy the language belonging to the field of metallic transmutation was sometimes used to describe the processes of glass-making as if the production of glass equalled finding the philosopher's stone. Thus, glass-making had a special status as alchemical technology (Dupré 2014a; Beretta 2009).

Born into a family of glass-makers, and trained as an apothecary, in 1667 Kunckel entered the service of the Elector of Saxony, Johann Georg II (1613–1680) (Triotzsch 1982; Kuhnert 2008). His task was to work through the alchemical writings in the possession of the Dresden court. It seems that in the course of his labors Kunckel acquired considerable knowledge of the secrets of alchemy. It is also clear that Kunckel was committed to a chrysopoeiaic project in this period. After his move to Brandenburg (where his main activity was the making of glass) he continued his transmutation experiments on the Pfaueninsel, the island in the river Havel between Berlin and Potsdam that he received as a gift from the Elector of Brandenburg Friedrich Wilhelm (1620–1688) (Rau 1972) (Fig. 5). Kunckel first came to Brandenburg in 1678 through the brokerage of Christian Mentzel (1622–1701), the court physician, who was also responsible for the Elector's collection of Chinese books. Shortly after his arrival in Brandenburg, Kunckel was appointed to direct the glassworks in Potsdam, turning the production of local forest glass into luxury Venetian-style crystal glass. He built his laboratory on the Pfaueninsel for the production of crystal glass in 1686. Kunckel was granted two privileges—one for glass beads, the other for gold ruby glass. The development of gold ruby glass was Kunckel's claim to fame (Kerssenbrock-Krosigk 2001). This glass of a deep yet translucent ruby color was highly fashionable around 1700 and valued as highly as hard-paste porcelain would be only a few decades later (Fig. 6).

This short biography should be sufficient to show that Kunckel moved between artisanal and scholarly cultures, between the laboratory and the library, and between reading and doing. Published in 1679, his *Ars vitraria experimentalis* came at the moment he established himself as expert in glass-making and chymical knowledge at the court in Brandenburg. An important factor in the emergence

Fig. 5 Pfaueninsel (Kaninchen Werder). Detail from H. Klockhoff's topographic map of Berlin, Spandau and Potsdam (Amsterdam, 1780). Courtesy of SLUB Dresden/Deutsche Fotothek

of codification of error is the rise of experts at court in connection with the luxury industry, a social 'structure' of practical knowledge discussed by Ursula Klein (Chap. 11; Dupré 2014b, "Introduction," vii–xix). These courts were 'trading zones,' and the experts connected scholarly and artisanal cultures. Especially Kunckel's work on alchemy at the Dresden court brought Kunckel to acquire an interest in the written word serving his work on glass. Typical for the hybrid expert moving between artisanal and scholarly worlds, Kunckel's *Ars vitraria experimentalis* organises artisanal knowledge according to scholarly formats (Klein 2012, 2013) (Fig. 7).

In *Ars vitraria experimentalis oder Vollkommene Glasmacher-Kunst*, first published in 1679, Kunckel did not give the secret of gold ruby glass away. Commenting on Neri's recipe using gold as a red colorant, he wrote:

> Because I have spent so much effort and labour and because it is such a peculiar rarity, no one will blame me for not sharing it this time.[3]

By not telling the secret, this book served as a way for Kunckel to promote himself.[4] Given that the book was first printed precisely when Kunckel moved to the Brandenburg court, the book served to

[3]"weil es mir aber gar viel Zeit Müh und Arbeit gekostet und eine sehr rare Sache ist als wird mich niemand verdencken daß ichs vor dißmahl nicht gemein mache." Johannes Kunckel, *Ars vitraria experimentalis oder vollkommene Glasmacher-Kunst* (Frankfurt 1689), 89.

[4]Kunckel used the rhetoric of books of secrets. For books of secrets, see (Eamon 1994; Leong and Rankin 2011; Wheeler 2009).

Fig. 6 Gold Ruby Glass, Potsdam Glass Workshop, covered goblet, 1715. Courtesy of Staatliche Museen zu Berlin, Kunstgewerbemuseum

establish his expertise more than anything else, although Kunckel did imagine a readership of the book among glass-makers.

If Kunckel's book did not serve to make public his most important secret, what then was in the book? The first part consists of a German translation of Neri's *L'arte vetraria*. The first edition of Neri's book appeared in 1612 in Italian. Neri was a Florentine alchemist (Grazzini 2010; Galluzzi 1982). The recipes on glass he gathered in *L'arte vetraria* were based on his experiences and readings in the glass workshop at the Casino San Marco in Florence and in the house of the Portuguese merchant-banker Emmanuel Ximenez (1564–1632) and the glass workshop of Filippo Gridolfi in Antwerp (Dupré 2010) (Fig. 8).

What did Neri's book contain? Neri's *L'arte vetraria* consists of seven books, of which only the first is devoted to the making of *cristallo* and thus colorless glass. The other six books discuss several recipes for colored glass, in which the Medici, Neri and Ximenez were all interested, especially those that imitated precious stones (Dupré 2014a, 146–7). Neri's book on glass is partly based on recipe collections for glass circulating in manuscript, such as the Montpellier manuscript *Ricette per far vetri colorati e smalti d'ogni sorte havute in Murano* (1536) and various other anonymous collections circulating in Florence (Moretti and Toninato 2001). Moreover, a very important source, especially for book 5 on the imitation of precious stones, was Isaac Hollandus. The Antwerp painter Otto Van Veen singled Hollandus out as his most important authority in alchemy. Hollandus was a post-Paracelsian, probably fictive character to whom numerous manuscripts, in Dutch and translated, were attributed, and who reached mythic proportions at the time, sometimes even as an alleged predecessor and source of Paracelsus (Van Gijsen 2010, 310–30). Neri read Hollandus's work in the library of

Fig. 7 Title Page of Johannes Kunckel. From (Kunckel 1679). Courtesy of Herzog August Bibliothek Wolfenbüttel: [M: Od 215]

Ximenez (Dupré 2014a, S. 155–6). Beyond single recipes for the making of artificial gemstones, reading the *Opera Mineralia* of Isaac Hollandus reinforced for Neri the similarities between glass and minerals and confronted him with a natural philosophy of vitrification. However, while partly a product of reading, experience was also important to Neri's book. For example, book two of *L'arte vetraria* on the production of chalcedony glass shows silent but interesting variations with respect to earlier written sources related to Neri's experiments. It is important to understand in what ways Kunckel's translation differed from Neri's book. The difference has to do with the fact that in Kunckel's book his *Anmerckungen* (comments) follow each book of *L'arte vetraria* in translation.

Following Kunckel's translation and comments on Neri's *L'arte vetraria*, the *Ars vitraria experimentalis* contains, firstly, Kunckel's German translation of Christopher Merrett's (1614–1695) notes on Neri's *L'arte vetraria* and Kunckel's *Observationen und curieusen Erinnerungen* on

Fig. 8 Filippo Gridolfi, Serpent stem goblet, 1590–1599. Courtesy of MAS | collectie Vleeshuis, Serpent stem goblet AV.1659 (Photo Bart Huysmans, Michel Wuyts)

Merrett's notes. Merrett was a practicing physician in London, a fellow of the Royal College of Physicians and a founding member of the Royal Society in London (as Kunckel also advertised on the title-page) (Allen 2004). As part of the Royal Society's history of trades programme he translated Neri's *L'arte vetraria* into English as *The Art of Glass* in 1662. This was a considerably expanded translation, not just a rendering in English, but with the addition of Merrett's "Observations." This separate section of the book discussed Merrett's views on the nature, antiquity and use of glass, followed by notes on the different recipes of Neri, such as new experimental findings, gleaned from workmen's anecdotes, practical experience and materials of his own production. However, the bulk of these notes are references to other writings of Boyle, Glauber and, above all, Della Porta. Merrett's translation of Neri is also preceded by a short list of general rules and observations on procedures, which (Merrett argues) he includes to avoid the repetitious nature of Neri's recipes.[5] In sum, Merrett's translation shows scholarly interventions, which partly foreshadow Kunckel's as we will see, but he also takes other directions than Kunckel. Kunckel does not issue a list of generalizations to circumvent Neri's repetitions.

Kunckel's title page signals in various ways his use of learned genres to organise artisanal knowledge. While the book is in German, Kunckel uses a Latin title, *Ars vitraria experimentalis* of his own making, but clearly echoing the format of Baconian experimental history behind the Royal Society's history of trades programme and Merrett's *The Art of Glass* (Klein 2003; Pastorino 2011). On the

[5] "To avoid our Authors Repetitions," (Merrett 1662, n.p.).

same title page he also advertises his book as a '*commentario*' referencing the commentary tradition in natural philosophy and other theory-dominant fields of knowledge. However, throughout the book and on the title page, Kunckel emphasizes that he offers "*Anmerckungen*" and "*Observationen*," terminology already used by Merrett to refer to his comments on Neri. Merrett and Kunckel might have had the epistemic genre of medical *observationes* in mind (Pomata 2010, 2011). This genre consisted of collections of case histories, and emerged in the sixteenth century from another typical humanist genre, the *curationes*. These epistemic genres did not give a blueprint for Kunckel's codification of error. Although medical failures were reported in the *curationes* and *observationes*, they were less prominent in the *observationes*, in which authors did not focus on the efficacy of cures as was typical for the *curationes*. However, already in fifteenth-century philology, *observationes* indicated a work that focused on separate parts of a text, a sense of the term, which applies well to Kunckel's use of it (Pomata 2010, 200). Kunckel used learned formats—without any obvious preference—to organise artisanal knowledge.

While from the fifteenth century more and more artisans claimed authorship and wrote down how to go about creating their works in a variety of formats, they were also well aware of the limits of their abilities to transmit their knowledge through writings. For example, the French potter Bernard Palissy (1510–1589) made 'Practice' reluctant to tell 'Theory' the secret of white enamel, considered the source of all other colors. However, Palissy had 'Practice' say, this was not a refusal for economic reasons, but simply because words were an ineffective way to learn a craft. "Even if I used a thousand reams of paper to write down all the accidents that have happened to me in learning this art," Practice responds,

> You must be assured that, however good a brain you may have, you will still make a thousand mistakes, which cannot be learned from writings, and even if you had them in writing you wouldn't believe them until practice has given you a thousand afflictions. (Palissy 1957, 192)[6]

Only long and sustained experience, including the making of mistakes inherent to the apprenticeship, leads to the acquisition of knowledge, to properly know how to create colorful glazes. Although artisans recognised the epistemic significance of making mistakes, the codification of error was not essential to the codification of artisanal knowledge. Precisely, because (as Palissy noted) it was important to personally experience the mistakes, the errors themselves were not put down in writing. Simply reading them would not make you believe them.

In sixteenth-century art theory error became an important concept (Ostrow 2006, 267–91, 278–9). The cause of the occurrence of error in the arts was considered to be deficiency in judgement, or *giudizio dell'occhio*, a notion of the highest significance in early modern art theory. A deficiency in judgement meant that artists failed to adhere to the rules of the profession. Errors were thus considered to be the consequence of breaking these rules. These rules were known as 'misure' or 'measures' and could often be formulated mathematically. In sixteenth-century writings on the arts, such as the work of Paolo Pino and Giovanni Paolo Lomazzo (1538–1592), errors were often associated with ignorance of perspective and the absence of good proportions. For example, Prospero Bresciano's statue of Moses was criticized for not using the proper working methods of the sculptor, and precisely in terms of 'Moses' being ill proportioned (Fig. 9). However, as we will see below, the rules were certainly not limited to those that could be mathematically formulated. If the eye lacked judgement, the artist's hand could not be properly guided. Thus, in the sixteenth century, error became an important category in art criticism. However, while the rules were codified, and deviations from the rules noted in criticisms of the arts, the errors were not codified, although making mistakes was essential to learning.

[6]For Palissy and experience, see also (Shell 2004).

Fig. 9 Prospero Bresciano and Leonard Sormani, *Moses*, 1587–1588. Fontana dell'Acqua Felice. © Marie-Lan Nguyen/ Wikimedia Commons

How does Kunckel's translation compare to the Baconian experimental history Kunckel referenced on the title page? For Bacon it was important that the researcher did not immediately jump to conclusions, but in a first stage of his investigations, gets lost in the "woods of experience" (Giglioni 2014). This represented a sort of pre-linguistic phase of research. However, to obtain the right material for experimental histories, on which natural philosophy was built, it was necessary for experience to become 'literate.' Experience could only obtain this literate stage—that is, transform into '*experientia literata*'—if written down in reports:

> When all the experiments of all the arts have been collected and arranged, and come with one man's knowledge and judgement, many new things, useful to our life and condition, can be discovered by means of that very translation of experiments from one art to others, i.e. by that experience which I have called literate. (*Novum Organum*, Book I, aphorism 103 in *The Instauratio Magna Part II*, 161)

For Bacon, the recording of experience had to follow strict rules so that the resulting experimental history was not any collection of experiments, but a collection of relevant experiments generated in a controlled manner, according to the true order of experience and digested according to the rules of *experientia literata* (Jalobeanu 2014). The most important rule is that of the experimental variation of parameters. To some extent Bacon followed Renaissance authors, such as Desiderius Erasmus (1466–1536), who considered *varietas* a constitutive aspect of nature. However, Bacon was also convinced that beneath the bewildering variety of appearances, there was a 'secret order' of things. The codification of error had its place in Bacon's literate experience. He advised that the errors that the researcher committed during his enquiries and discoveries be included in experimental histories (Pastorino 2011, 545). That said, Bacon's emphasis was very much on variety rather than error.

3 Kunckel's Translation of Neri's *L'arte vetraria*

Kunckel's language in *Ars vitraria experimentalis* further illustrates that the book was written in a 'trading zone' by a figure impossible to categorize in the mutually exclusive categories of artisan and scholar. Kunckel's translation of the first section of the book of Neri's *L'arte vetraria* begins as follows:

> How to distract the salt from the powder Rochetta and the Soda Hispanica in order to prepare the Fritta Cristalli which the Italians call Bollito. This is the foundation of the art of glassmaking following a completely new and secret way.[7]

This is about the making of crystalline glass (*cristallo*, named after rock crystal because the Venetian glassmakers of Murano succeeded in making glass that was as clear and transparent as natural rock crystal), which is the basic material of which all other types of colored glass are produced, according to Neri. The interesting point is that Kunckel did not translate several terms from Neri's Italian. One of the most striking of those terms is *fritta cristalli*. Why did Kunckel not translate those terms?

One might argue that Kunckel simply opted for a certain regime of translation. 'Foreignization' was one of the options open to translators, as we know (Hermans 2007, vol. 2, 1420–1428). Keeping the exotic terms and providing a glossary was Merrett's choice. We might suspect that Kunckel leaves the term untranslated simply because he is forced to do so for lack of an appropriate term in German, employed in contemporary glass workshops in Brandenburg. According to Paul Taylor, Karel van Mander only left Italian terms untranslated if he could not use a Dutch term employed in contemporary painters's workshops in the Netherlands (Taylor 2013). However, this is not the case for *fritta*, as Kunckel tells us in his *Anmerckungen* on this section of Neri's book. The term in German existed: '*Gemenge.*' Neri writes:

> I have often heard from you glassmakers that you would really like to know what is meant by the word *Fritta* which is used many times in this book. So here I tell you that it means no more or less than when you say: *Gemenge*; and one *Gemenge* is made of so much ash and sand and so much salt and the other *Gemenge* of so much. Thus what you call *Gemenge*, is here named *Fritta*. So now you also understand this word.[8]

While the German equivalent for *fritta* was used in German glass workshops, glassmakers, apparently hearing or reading the term '*fritta*' according to Kunckel wondered what it meant. Instructing German glassmakers, Kunckel does not create a monolingual translation aimed at teaching them how to do something (such as in the example here, the secret of the making of crystalline glass), but offers them a foreign language vocabulary to speak about what it is they do.

[7] "Wie das Salz aus dem Pülverlein Rochetta und aus der Soda Hispanica auszuziehen vermittelst welches die Fritta Cristalli von den Italiänern Bollito genannt zubereitet wird als darinnen des fundament der ganzen Glassmacher-Kunst bestehet mit einer ganz neuen und geheimen Manier" (Kunckel 1689, 2).

[8] "Auch habe ich zum öfftern von euch Glaßmachern gehöret daß ihr gern wissen möchtet was doch mit dem Wort *Fritta* angedeutet werde welches zu vielen malen in diesem Buch erwehnet ist. So vermelde ich euch hiermit daß es allemal so viel bedeutet als nach eurer Art zu reden wenn ihr sprecht: *Gemenge*; als das *Gemenge* ist gemacht von so viel Asche und Sand und so viel Salz und das andere *Gemenge* von so viel: Allso das was ihr *Gemenge* heisset, wird hier *Fritta* genennet ausdaß ihr dieses Wort auch verstehet" (Kunckel 1689, 35).

Historians of science have primarily focused on two processes of translation: Latinization of vernacular artisanal knowledge and vernacularization of Latin scholarly knowledge. As already mentioned in the introduction, even a recent and sophisticated book on the origins of the new sciences in artisanal cultures, such as Pamela Long's, couches this in terms of exchange and communication between a Latin learned culture and vernacular artisanal knowledge (Long 2011, 95). In contrast, although the book is mostly in German, Kunckel consciously created a multi-lingual environment by leaving terms un-translated.

It was also a multilingual context, the availability of Neri's book in more than one language that shaped the writing and translating of Kunckel's *Ars vitraria*. The first edition of Neri's *L'arte vetraria* in 1612 was not only followed by more Italian editions in 1661, 1663 and 1678, and Merrett's English translation, but also by a Latin translation of Neri, which for the first time appeared in 1668. Kunckel explicitly mentioned on the title page of his book that this was a translation of Neri's publication from the Italian and the Latin. In a comment on Neri's recipe on golden colored glass in the third book, Kunckel writes:

> This process was the most tiresome and caused me more trouble than any other recipe in this book: This is because the quantity described by the Author is not correct but false. I had always thought that this was a mistake in the Latin version. But after I had laid my hands on the Italian version (in which it is described by the Author himself) I have found out that it mentions the same dose. I do not know if this is a mistake in the Italian print or by the author himself.[9]

This tells us several things: that Kunckel's first access to Neri's book was through its later Latin edition; that Kunckel considered the Italian edition, by the author himself (as he writes) to be the most authoritative one; that his translation was the outcome of comparing and evaluating editions in more than one language. Multilingualism was a source for translation as much as it was its outcome.

The multilingual outcome of Kunckel's translation effort is not limited to the use of Italian terms even when a German equivalent is available. His preface to the book is sprinkled with terms such as *observiret, demonstrationen, operiret, experimentiret, determinirte dosin*—all of which are highlighted, in the preface and also elsewhere in the book, because they are printed in a different letter-type. Most of these terms carry epistemic significance. Just one example illustrating this point would be: "It will please the diligent artisan that I only speak of those things which I have not borrowed from others but *operiret, experimentiret* and learned with my own hands" (Fig. 10).[10]

Another example, directly referring to Kunckel's claim that he tried every single one of Neri's recipes himself, would be that "I do not speak here of things which I have not myself investigated and *geapprobiret*."[11] Again, this epistemic category is highlighted in another letter-type. Kunckel does not just pride himself on first-hand experimentation, he also warns his reader that the recipes found in the book might not work upon the reader's first trial. This does not mean that the recipes are wrong; it is more likely that the reader made a mistake. If the reader finds himself in such a situation, Kunckel's advice is to try again, and again and again. After all, he reminds his reader, it requires practice to be a master.

[9] "Es hat mir dieser hierinn beschriebene Proceß die allergröste Müh und Verdruß über alle andere so in diesem Buch enthalten verursacht: Indeme die von dem Autor beschriebene Dosis ganz unrecht und falsch ist. Ich habe immer zu vermeint es sein in der Lateinischen Version ein Fehler gewesen; nachdem mir aber das Italiänische (als worinnen es der Autor selbst beschrieben) zu händen kommen habe ich befunden daß die Version mit demselben hierinnen ganz zutrifft: Ob es nun in dem Italiänischen Druck oder von dem Autor selbst versehen kann ich nicht wissen…." (Kunckel 1689, 85–6).

[10] "Es wird auch der fleissige Künstler in diesen und allen andern Dingen was ich hier tractire, befinden daß ich auch die jenige Warheit welche ich nicht von einem andern entlehnet oder erlanget sondern mit diesen meinen händen *operiret, experimentiret* und gelernet allhier aufgeschrieben und bewiesen habe" (Kunckel 1689, Vorrede).

[11] "Letzlich füge ich noch dieses hinzu … daß … ich nichts allhier mittheile welches ich nicht selbst untersuchet und *geapprobiret* habe" (Kunckel 1689, Vorrede).

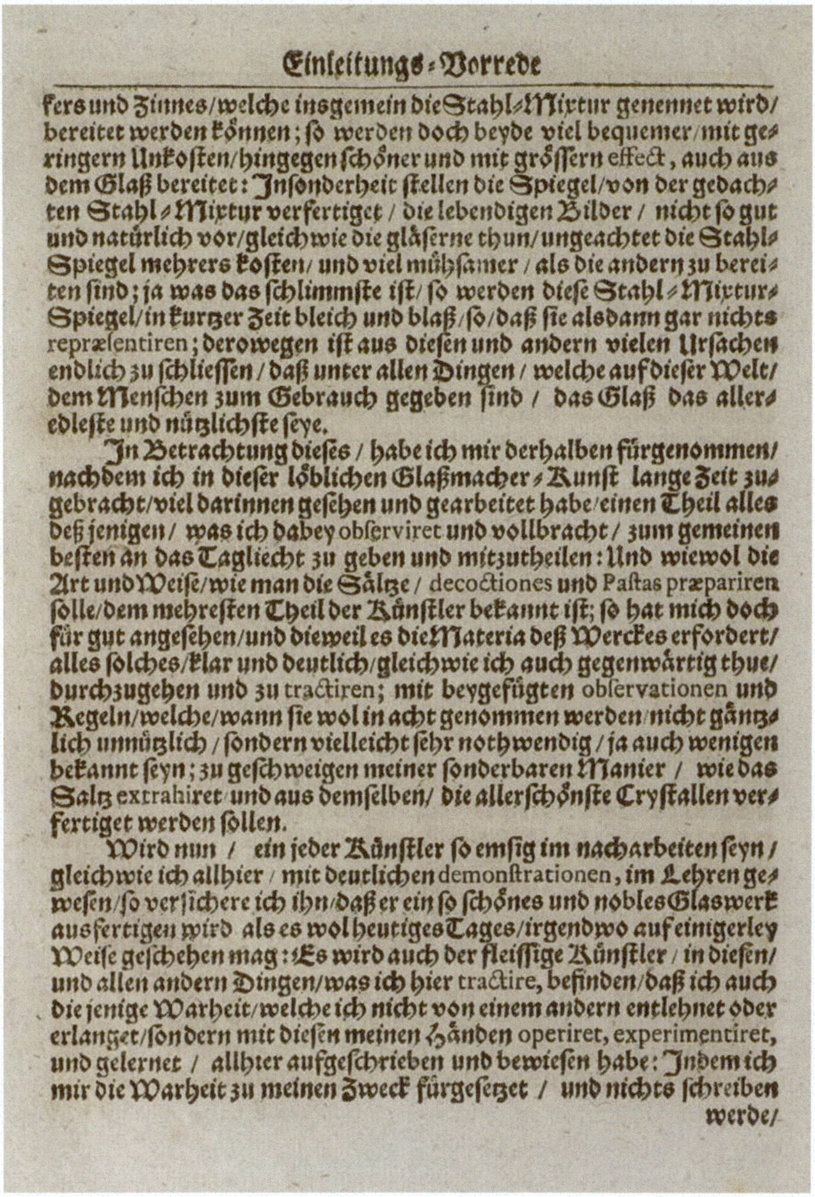

Einleitungs-Vorrede

fers und Zinnes/welche insgemein die Stahl-Mixtur genennet wird/ bereitet werden können; ſo werden doch beyde viel bequemer/mit geringern Unkoſten/hingegen ſchöner und mit gröſſern effect, auch aus dem Glaß bereitet: Inſonderheit ſtellen die Spiegel/von der gedachten Stahl-Mixtur verfertiget/ die lebendigen Bilder / nicht ſo gut und natürlich vor/gleichwie die gläſerne thun/ungeachtet die Stahl-Spiegel mehrers koſten/ und viel mühſamer / als die andern zu bereiten ſind; ja was das ſchlimmſte iſt/ ſo werden dieſe Stahl-Mixtur-Spiegel/in kurtzer Zeit bleich und blaß/ſo/daß ſie alsdann gar nichts repræſentiren; derowegen iſt aus dieſen und andern vielen Urſachen endlich zu ſchlieſſen / daß unter allen Dingen / welche auf dieſer Welt/ dem Menſchen zum Gebrauch gegeben ſind / das Glaß das alleredleſte und nützlichſte ſeye.

In Betrachtung dieſes / habe ich mir derhalben fürgenommen/ nachdem ich in dieſer löblichen Glaßmacher-Kunſt lange Zeit zugebracht/viel darinnen geſehen und gearbeitet habe/einen Theil alles deß jenigen/ was ich dabey obſerviret und vollbracht / zum gemeinen beſten an das Tagliecht zu geben und mitzutheilen: Und wiewol die Art und Weiſe/wie man die Sältze / decoctiones und Paſtas præpariren ſolle/dem mehreſten Theil der Künſtler bekannt iſt; ſo hat mich doch für gut angeſehen/und dieweil es die Materia deß Werckes erfordert/ alles ſolches/klar und deutlich/gleichwie ich auch gegenwärtig thue/ durchzugehen und zu tractiren; mit beygefügten obſervationen und Regeln/welche/wann ſie wol in acht genommen werden/nicht gäntzlich unnützlich / ſondern vielleicht ſehr nothwendig / ja auch wenigen bekannt ſeyn; zu geſchweigen meiner ſonderbaren Manier / wie das Saltz extrahiret/und aus demſelben/ die allerſchönſte Cryſtallen verfertiget werden ſollen.

Wird nun / ein jeder Künſtler ſo emſig im nacharbeiten ſeyn / gleichwie ich allhier / mit deutlichen demonſtrationen, im Lehren geweſen/ſo verſichere ich ihn/daß er ein ſo ſchönes und nobles Glaswerk ausfertigen wird als es wol heutiges Tages/irgendwo auf einigerley Weiſe geſchehen mag: Es wird auch der fleiſſige Künſtler / in dieſen/ und allen andern Dingen/was ich hier tractire, befinden/daß ich auch die jenige Warheit/welche ich nicht von einem andern entlehnet oder erlanget/ſondern mit dieſen meinen Händen operiret, experimentiret, und gelernet / allhier aufgeſchrieben und bewieſen habe: Indem ich mir die Warheit zu meinen Zweck fürgeſetzet / und nichts ſchreiben werde/

Fig. 10 Excerpt from the preface of Johannes Kunckel. From (Kunckel 1689). Courtesy of Herzog August Bibliothek Wolfenbüttel: [M: Od 216]

Pamela Smith has argued that recipes and how-to books are not always, or not only, about telling readers how to get something done, but about how to do things right (Smith 2012). They are not—or not only—about the transmission of knowledge, but about the transmission of epistemic values which were very important in artisanal workshops. One example is continuous attention, and indeed, the repetition of practices on which Kunckel also puts so much emphasis. The transmission of epistemic values is one of the primary aims of Kunckel's translation of Neri. The foreignization of terminology, catching epistemic categories (observation and experimentation), offers the reader a richer vocabulary when discussing how to do things right.

Experimentation led Kunckel to evaluate Neri's recipes. Sometimes his assessment was wholeheartedly positive. More often however, experimentation led him to adapt recipes specifically to local circumstances, practices and raw materials. An ingredient used by Neri could be too expensive, or just impossible to obtain in Brandenburg. One of the most common changes was because of different furnace conditions. For example, in his comments on a *lattimo* color in book 3, chapter 55, Kunckel writes:

> The author teaches that the composition should stay in the furnace for 18 days and nights, which, however, is entirely unnecessary, especially in our German glass furnaces, where it should not stay longer than three days and nights.[12]

However, the changes introduced following experimentation and adaptation to local materials and knowledge did not suffice to teach the craft of glassmaking. Kunckel was very conscious of the limits of writing and reading. It was a matter of practice. The glassmaker needed to be equipped with a good eye, as Kunckel called it, *Augenmaß* (Kunckel 1689, Vorrede). Perhaps those limits were felt more strongly in glass-making than any other craft, because it was a practice that relied so heavily on the visual element. *Augenmaß* was called for when judging the quantities of ingredients crucial to obtain colors: small differences of quantities could result in big differences of color. "None of this can be taught on paper," Kunckel concluded, proclaiming the limits of language.[13]

It is to characterize this unspeakable property of artisanal knowledge that Albrecht Dürer (1471–1528) already evoked the term Kunckel uses for the same reason, *Augenmaß* (eye measure), a term that like *giudizio dell'occhio* connects visual perception and judgment. In the *Ästhetische Exkurs* (1528), Dürer wrote:

> But if you have learned to measure well, and added reason to practice so that you can make a thing with free assuredness and know how to do justice to a thing, then it is not necessary to measure it all the time, for your accomplished art endows you with a good eye measure and your practiced hand obeys.[14]

Related to period notions of *ingenio* or *ingenium* (ingenuity) as innate talent or natural ability without instruction, Dürer's *Augenmaß* was *Wissen* (knowledge) partly acquired through practice (Doorly 2004, 272). *Augenmaß* guided the hand of the skilled artisan and allowed him to avoid *yrthumb* (error) and *falscheit* (falseness). *Augenmaß* appears to generally play more of a role in Dürer's epistemology and his understanding of imagination, a concept very difficult to define in Dürer's writings and for which he avoided the use of the available terminology (such as *phantasie*) (Parshall 2013; Panofsky 1943, 279). Following Dürer, Kunckel used the idea of *Augenmaß* to express that essential aspects of the making of art were a matter impossible to fix in words, relating the awareness of the limits of language to avoiding errors.

In response to these limits Kunckel introduced the codification of error. He used his translation of Neri as a vehicle not just for making changes, adapting to local circumstances and procedures whenever those differed from those found in Neri, but to write down whenever Neri went wrong. Long neglected in the history of science, Giora Hon, Jutta Schickore and Friederich Steinle have argued for the importance of learning from error and for 'going amiss' in experimental research (Hon et al.

[12]"Lehret der Autor daß die Composition, 18. Tage und Nacht im Ofen stehen soll welches ganz unnötig sonderlich in unsern Deutschen Glaß-Oefen da es nicht 3. Tag und Nacht stehen darff" (Kunckel 1689, 88).

[13]"Dieses alles aber kann wie gedacht auf dem Papier allhier nicht gelernet warden" (Kunckel 1689, Vorrede).

[14]"Aber so du wol messen hast gelernt, und den verstandt mit sambt dem brauch uber kumen, also das du ein ding auß freyer gwißheyt kanst machen und weyst einem yetlichen ding recht zu thon, als dann ist nit alweg not, ein ydlich ding alweg zu messen, dan dein uberkumne kunst macht dir ein guten augen maß, als dann ist die geübt hand gehorsam" (Dürer 1969, 297).

Fig. 11 Stemmed bowl, chalcedony glass, Venice, ca. 1510. Courtesy of Staatliche Museen zu Berlin, Kunstgewerbemuseum

2009).[15] Schickore has for example shown how microscopy in the eighteenth century was primarily occupied with detecting and documenting sources of observational error (Schickore 2007). Against this background, Kunckel's identification of errors is even more important than his occasional suggestions of alternatives to Neri's recipes and procedures to make them work. For Kunckel *Ars vitraria* was not just a book about what to do but also what not to do. As Kunckel himself wrote in the preface to the 1679 edition: "*Was sich thun and nicht thun lasset*" (Kunckel 1679, Vorrede).

The full impact of Kunckel's codification of error only becomes clear in comparison with earlier collections of recipes. Recipes changed but this was often unnoticed. The standard was silent change. Authors of recipes corrected a recipe, sometimes because workshop practice demanded such, but they did not note that they had corrected their source recipe, or how exactly the source recipe was lacking. Moreover, authors of recipes seemed more interested in collecting various ways to arrive at the same or a similar result, than in sorting out the right ways from the wrong. One of the manuscript sources of Neri's *L'arte vetraria* listed no less than seventeen different ways to make chalcedony glass—just one after the other, without distinction (Moretti and Toninato 2001, 43). Neri himself still lists three. His emphasis is on the source of experimentation and the networks of material and personal connections in which the recipes were caught. He mentions he practiced the first recipe in Florence at the Casino under the direction of Niccolo Landi in 1601; the second recipe served him when making chalcedony glass in Flanders, and was admired by Grand Duke Ferdinando de' Medici (1549–1609); the third recipe was the outcome of experiments in the glass workshop of Gridolfi in Antwerp in January 1609, and Neri added that he presented two pieces of chalcedony glass made according to this technique to the Prince of Orange (Neri 1612, 39–48). While as we have seen Neri read recipes on glassmaking and occasionally named his sources, he did not write down the errors he found in his source recipes. Kunckel, on the other hand, did (Fig. 11).

Sometimes Kunckel completely rejects a recipe of Neri's in his *Anmerckungen*. For example, on a recipe for imitating lapis lazuli, he writes: "The author shows, I am absolutely certain, that he did not try it himself or else he would have regarded it differently and abstained from including it in the book because it does not work at all."[16] A recipe could be wrong in different ways. As we have already seen,

[15] See also (Kaiser and Creager 2012).

[16] "Der Autor lehret bin ich ganz gewiß versichert daß er es nicht gemacht er würde es sonst anders befunden und nicht also hier gefeßet haben denn es gehet im geringsten nicht an" (Kunckel 1689, 100).

one of the more common errors concerned the quantities of ingredients. At other times, procedures followed by Neri were wrong—not always because they led to the wrong result but because they contained operations conducted in vain, which were not necessary to reach the goal or simply too cumbersome. For example, on one of Neri's recipes for chalcedony glass, Kunckel noted: "The author makes this recipe difficult and expensive, while it can be done with much less effort and expense."[17] There was more than one way to be wrong.

4 Conclusion

The case of Kunckel's *Ars vitraria experimentalis* shows that translation is an epistemic process. Kunckel's translation of artisanal knowledge is really two translations: one between workshop practice and words and another between two languages. Kunckel was conscious of the limits of language to render workshop practice, and harked back to the concept of '*Augenmaß*,' originally used by Dürer and closely associated with 'judgment by eye' as used in art theory, to express those limits. Translating recipes between two languages heightens the awareness of these limits. Questions on the relation between texts and practices become acute in the case of translations between languages of recipes and writings on technology. One wonders whether translation between languages, when confronted with locally specific names and terms, would be even possible without recourse to practice—that is, without trying out the recipe? Kunckel did try out Neri's recipes, and in the long history of artisanal recipes, this made him an exceptional reader. As I have argued, more typically, recipes are translations of workshop practice moving artisanal knowledge to readerships outside the physical boundaries of the artisanal workshop.

Kunckel's translation is part of a transformative process in which the conception of artisanal knowledge itself changed. The codification of practical knowledge resulted in a reflection of the nature of artisanal knowledge. The most marked difference between Neri's *L'arte vetreria* and preceding collections of recipes is the book's naming of people and places, not only of patrons and cities, but also of recent authorities, even if they are as elusive as the nonexistent Isaac Hollandus. For Neri, artisanal knowledge became the product of experience and reading. Kunckel's codification of error pushed this one step further. The example of his translation shows how writing down artisanal knowledge came to reflect the process of reading. As I have argued, this process took place in the context of 'trading zones,' such as the courts Neri and Kunckel worked for, and in the hands of 'learned artisans.' This social structure of practical knowledge connects to a re-definition of this knowledge.

Translation is an epistemic process, and following Bacon, making experience literate (that is, writing it down following specific rules), an essential aspect of the process of knowledge acquisition. For Bacon the most important rule was that of variety; the strategy of varying experimental parameters so as to create strings of experiences resulting in a certain experimental history. Error only played a less important part in Baconian experimental history than it did in other epistemic genres such as the medical observationes. Following the recognition of the epistemic significance of errors, and the importance of learning by making mistakes, in artisanal writings and art theory from the sixteenth century, Kunckel's translation emphasized the codification of error. For Kunckel, the codification of error was a response to the limits of language. While Kunckel did not consider the codification of error to lead to more abstract levels of knowledge, as Matteo Valleriani points out in the introduction, it is indicative of a development towards a more elaborate knowledge system.

[17] "Der Autor macht diese Composition nur sehr schwer und kostbar da sie doch mit viel geringer Mühe und Kosten könnte gemacht warden" (Kunckel 1689, 75).

References

Allen, D.E. 2004. Merret, Christopher (1614–1695). In *Oxford dictionary of national biography*. Oxford: University Press; Online Edition. May 2013. http://www.oxforddnb.com/view/article/18599. Accessed 1 June 2015.
Beretta, Marco. 2009. *The alchemy of glass: Counterfeit, imitation and transmutation in ancient glassmaking*. Sagamore Beach: Science History Publications.
Burke, Peter, and R. Po-chia Hsia (eds.). 2007. *Cultural translation in early modern Europe*. Cambridge: Cambridge University Press.
Clarke, Mark, Bert De Munck, and Sven Dupré. 2012. *Transmission of artists' knowledge*. Brussels: KVAB.
Doorly, Patrick. 2004. Dürer's 'Melencolia I:' Plato's abandoned search for the beautiful. *The Art Bulletin* 86: 255–276.
Dupré, Sven. 2010. Trading luxury glass, picturing collections and consuming objects of knowledge in early seventeenth-century Antwerp. *Intellectual History Review* 20 (1): 53–78.
Dupré, Sven. 2014a. The value of glass and the translation of artisanal knowledge in early modern Antwerp. In *Trading values in early modern Antwerp*, Netherlands yearbook for art history, ed. Bart Ramakers, Christine Göttler, and Joanna Woodall, 138–161. Leiden: Brill.
Dupré, Sven (ed.). 2014b. *Laboratories of art: Alchemy and art technology from antiquity to the 18th century*. Heidelberg: Springer.
Dupré, Sven, Dedo von Kerssenbrock-Krosigk, and Wismer Beat. 2014. *Kunst und Alchemie. Das Geheimnis der Verwandlung*. Munich: Hirmer Verlag.
Dürer, Albrecht. 1969. In *Schriftlicher Nachlass*, vol. 3, ed. Hans Rupprich. Berlin: Deutscher Verlag für Kunstwissenschaft.
Eamon, William. 1994. *Science and the secrets of nature: Books of secrets in medieval and early modern culture*. Princeton: Princeton University Press.
Egmond, Florike. 2013. Names of naturalia in the early modern period: Between the vernacular and Latin, identification and classification. In *Translating knowledge in the early modern Low Countries*, ed. Harold J. Cook and Sven Dupré, 131–162. Zürich/Berlin: LIT.
Eisenstein, Elizabeth. 2002. An unacknowledged revolution revisited. *The American Historical Review* 107: 87–105.
Galluzzi, Paolo. 1982. Motivi paracelsiani nella Toscana di Cosimo II e di Don Antonio dei Medici: alchimia, medicina chimica e riforma del sapere. In *Scienze, credenze occulte, livelli di cultura*, 31–62. Florence: Olschki.
Giglioni, Guido. 2014. From the woods of experience to the open fields of metaphysics: Bacon's notion of silva. *Renaissance Studies* 28 (2): 242–261.
Goyens, Michèle, Pieter De Leemans, and An Smets (eds.). 2008. *Science translated: Latin and vernacular translations of scientific treatises in medieval Europe*. Leuven: Leuven University Press.
Grazzini, M.G. 2010. Discorso sopra la Chimica. The Paracelsian Philosophy of Antonio Neri. *Nuncius* 27: 411–467.
Harrison, Peter. 2007. *The fall of man and the foundations of science*. New York: Cambridge University Press.
Hermans, Theo. 2007. Concepts and theories of translation in the European Renaissance. In *Übersetzung: Ein Internationales Handbuch zur Übersetzungsforschung*, vol. 2, ed. Kittel Harald, 1420–1428. Berlin: W. de Gruyter.
Hon, Giora, Jutta Schickore, and Friedrich Steinle (eds.). 2009. *Going amiss in experimental research*. Dordrecht: Springer.
Jalobeanu, Dana. 2014. Constructing natural historical facts: Baconian natural history in Newton's first paper on light and colors. In *Newton and empiricism*, ed. Eric Schliesser and Zvi Biener, 39–65. Oxford: Oxford University Press.
Johns, Adrian. 1998. *The nature of the book: Print and knowledge in the making*. Chicago: University of Chicago Press.
Kaiser, David, and Angela Creager. 2012. The right way to get it wrong. *Scientific American* 306: 72–75.
Klein, Ursula. 2003. Experimental history and Herman Boerhaave's chemistry of plants. *Studies in History and Philosophical of Biological and Biomedical Sciences* 34 (3): 533–567.
Klein, Ursula. 2012. Introduction: Artisanal-scientific experts in eighteenth-century France and German. *Special Issue of Annals of Science* 69: 303–306.
Klein, Ursula. 2013. Chemical experts at the Royal Prussian Porcelain Manufactory. *Ambix* 60 (2): 99–121.
Kuhnert, Lothar. 2008. *Johann Kunckel, Ritter von Löwenstern: Die Erfindung der Nanotechnologie in Berlin: Ein Bericht*. Berlin: L. Kuhnert.
Kunckel, Johannes. 1679. *Ars vitraria experimentalis oder vollkommene Glasmacher-Kunst*. Frankfurt und Leipzig: Kunckel.
Kunckel, Johannes. 1689. *Ars vitraria experimentalis oder vollkommene Glasmacher-Kunst*. Frankfurt und Leipzig: Kunckel.
Leong, Elaine, and Alisha Rankin (eds.). 2011. *Secrets and knowledge in medicine and science, 1500–1800*. Farnham/Burlington: Ashgate.
Long, Pamela O. 2011. *Artisan/Practitioners and the rise of the new sciences, 1400–1600*. Corvallis: Oregon State University Press.

Merrett, Christopher. 1662. *The art of glass*. London: Octavian Pulleyn.
Montgomery, Scott L. 2000. *Science in translation: Movements of knowledge through cultures and time*. Chicago: University of Chicago Press.
Moretti, C., and T. Toninato. 2001. *Ricettario vetrario del Rinascimento. Trascrizione da un manoscritto anonimo veneziano*. Venice: Marsilio.
Neven, Sylvie. 2014. Transmission of alchemical and artistic knowledge in German mediaeval and premodern recipe books. In *Laboratories of art: Alchemy and art technology from antiquity to the 18th century*, ed. Sven Dupré, 23–52. Heidelberg: Springer.
Ostrow, Steven F. 2006. The discourse of failure in seventeenth-century Rome: Prospero's Bresciano's Moses. *The Art Bulletin* 88: 267–291.
Palissy, Bernard. 1957. *The admirable discourses of Bernard Palissy*. Urbana: University of Illinois Press.
Panofsky, Erwin. 1943. *The life and art of Albrecht Dürer*. Princeton: Princeton University Press.
Parshall, Peter. 2013. Graphic knowledge: Albrecht Dürer and the imagination. *The Art Bulletin* 95: 393–410.
Pastorino, Cesare. 2011. Weighing experience: Experimental histories and Francis Bacon's quantitative program. *Early Science and Medicine* 16: 542–570.
Pomata, Gianna. 2010. Sharing cases: The observations in early modern medicine. *Early Science and Medicine* 15: 193–236.
Pomata, Gianna. 2011. Observation rising: Birth of an epistemic genre, 1500–1650. In *Histories of scientific observation*, ed. Lorraine Daston and Elizabeth Lunbeck, 45–80. Chicago/London: University of Chicago Press.
Principe, Lawrence M. 2013. *The secrets of alchemy*. Chicago/London: University of Chicago Press.
Rau, H. G. 1972. Das Glaslaboratorium Johann Kunckels auf der Pfaueninsel. In *Ausgrabungen in Berlin* 3: 148–171.
Schickore, Jutta. 2007. *The microscope and the eye: A history of reflections, 1740–1870*. Chicago: University of Chicago Press.
Shell, Hanna Rose. 2004. Casting life, recasting experience: Bernard Palissy's occupation between maker and nature. *Configurations* 12: 1–40.
Smith, Pamela H. 2010. Why write a book? From lived experience to the written word in early modern Europe. *Bulletin of the German Historical Institute* 47: 25–50.
Smith, Pamela H. 2012. Craft techniques and how-to books. In *Transmission of artists' knowledge*, ed. Mark Clarke, Bert De Munck, and Sven Dupré, 75–84. Brussels: KVAB.
Taylor, Paul. 2013. Italian artistic terms in Karel van Mander's Schilder-Boeck. In *Translating knowledge in the early modern low countries*, ed. Harold J. Cook and Sven Dupré, 197–216. Zürich/Berlin: LIT.
Triotzsch, Ulrich. 1982. Kunckel von Löwenstern, Johann. In *Neue Deutsche Biographie*. Berlin: Duncker & Humblot. 13, 287 f. http://www.deutsche-biographie.de/pnd11872536X. Accessed 1 June 2015.
Van Gijsen, A. 2010. Isaac Hollandus revisited. In *Chymia. Science and nature in medieval and early modern Europe*, ed. Mario López Pérez, Didier Kahn, and Mar Rey Bueno, 310–330. Newcastle: Cambridge Scholars.
von Kerssenbrock-Krosigk, Dedo. 2001. *Rubinglas des ausgehenden 17. und des 18. Jahrhunderts*. Mainz: Verlag Phillip von Zabern.
Wallert, A. 2012. Standards and procedures for art technology and mediaeval encyclopedias. In *Transmission of artists' knowledge*, ed. Mark Clarke, Bert De Munck, and Sven Dupré, 69–74. Brussels: KVAB.
Weitemeier, Bernd. 2004. Translation and the role of vernacular languages in medieval Europe. In *Übersetzung: ein internationales Handbuch zur Übersetzungsforschung*, vol. 2, ed. Kittel Harald et al., 1280–1295. Berlin: W. de Gruyter.
Wheeler, Jo. 2009. *Renaissance secrets: Recipes and formulas*. London: V&A Publishing.

The Chiaroscuro Woodcut in the Early Modern Period

7

Eileen Reeves

Abstract

This essay investigates the way in which technical, historical, and aesthetic information about the chiaroscuro woodcut circulated in the early modern period. It argues that specific directives concerning the design, carving, and printing of these colored, multi-block woodcuts were relatively limited in scope and detail, and that when they were transmitted, such information was presented alongside a double origin story favoring an Italian version over an early, unsung German prototype. The real circulation of knowledge concerning the chiaroscuro woodcut was therefore transmitted with the artifact itself, and it is thus significant that prints frequently gesture to the unseen and often idealized types of wood from which the image was carved. The absence of a single, canonical aesthetic regulating such prints, the simultaneous presence of several unpredictable elements—inks, paper quality, and especially alignment of the blocks—and the constant reuse of older designs all militated against the uniform and linear development of the genre, and contributed to its ongoing association with artistic experimentation. This variety aside, however, the chiaroscuro woodcut does increasingly depend upon a set of strategies designed to conceal the genre's relative inadequacy as portraiture: such, for example, are the numerous profiles, downcast heads, shielding gestures, hooded visages, and rear-facing protagonists. These compensatory techniques, and the extraordinary challenge posed by the frontal portait of the living subject, also explain the genre's frequent inclusion of severed heads, scenes of beheading, and skulls. Put differently, the thematic content of the chiaroscuro woodcut is generally legible as an explicit comment upon both the possibilities and the limits of this medium.

E. Reeves (✉)
Department of Comparative Literature, Princeton University, Princeton, NJ, USA
e-mail: ereeves@princeton.edu

M. Valleriani (ed.), *The Structures of Practical Knowledge*, DOI 10.1007/978-3-319-45671-3_7

Fig. 1 Chiaroscuro woodcut frontispiece of Pietro Aretino's Stanzas in *Praise of My Lady Angela Serena*. Francesco. DG2002/542. Courtesy of Albertina, Vienna

Book-buyers and browsers in the Veneto encountered an odd volume in the winter of 1537. The strangest feature of Pietro Aretino's (1492–1556) *Stanzas in Praise of My Lady Angela Sirena* was not, as its publisher Francesco Marcolini (1500–1559) seemed to suggest in his dedication, that these chaste verses followed closely on the author's depictions of decadent courtiers and frollicking prostitutes, but rather that the work was accompanied by a novel print, the chiaroscuro woodcut (Fig. 1).

Though two-block woodcuts in printed books were an extraordinary rarity, and though Marcolini himself was occasionally identified as a designer, carver, or engraver, he did not gesture to the image. A passing allusion to Titian (ca. 1488–1576) led some to suppose that the celebrated Venetian artist had provided the design (Aretino 1537, frontispiece).[1] The print itself was a curious variant on the emerging authorial portrait: Aretino figured in the unlikely guise as a lovestruck shepherd sojourning at the seashore, a muscular cousin to Niccolò Vicentino's slightly earlier chiaroscuro woodcut of Saturn. The poet included a reference to Titian in a letter dedicating the stanzas to the Empress Isabella (1503–1539); Marcolini quickly published these and other missives describing Aretino's paternal love for Angela Sirena, and the labor involved in this lyrical composition (Aretino 1542, 440–41, 453–54).

[1]On Marcolini as a printmaker, see also (Aretino 1537, III: vii; Doni 1552a, 84–6; Karpinski 2004, 106–7).

The tribute, less warmly received by Angela, proved entirely offensive to her husband. He and her kinsmen objected; she suffered and soon died; a line-block version of the original chiaroscuro woodcut appeared in 1552 in another work from Marcolini's press, where Aretino's friend Anton Francesco Doni (1513–1574) made the bold claim that the rustic verses had offered the late lucky lady immortality (Doni 1552a; 1552b, fols. 70v–71). Or so the story went, though some suggested instead that there was neither an angelic siren nor an aggrieved spouse, and that the entire episode was an Aretine publicity stunt.

The tangential tale of the chiaroscuro woodcut, the subject of this essay, is by comparison the more lucid narrative. At stake in the story of this medium are some of the same issues that confronted readers of Aretino's *Stanzas*: the identity of the image-makers, the circulation of their work, and the tension between the mimetic expectations of some consumers and the aesthetic autonomy of these artifacts. More generally, this essay concerns the codification of the knowledge needed to produce what were relatively expensive and complicated prints. While an explicit description of the process was available in print in Italian from the mid-sixteenth century, much of the relevant knowledge must have been inferred from the chiaroscuro woodcuts themselves: their creation depended upon attention to prevailing trends, reverse engineering, and experimentation. This information, moreover, was increasingly complemented by explicit allusions to the technical limits of such prints, such that chiaroscuro woodcuts encoded both material indices of how they were made, and aesthetic references to representational strategies beyond their makers' capabilities.[2]

1 Origins

Like many other processes and technologies of the period, this sort of woodcut has two origin stories.[3] The earlier and slightly more primitive version of the print emerged, as a series of experiments and more or less without fanfare, in Southern Germany before 1500.[4] In this setting, the eventual prints were an outgrowth of efforts to find a printed alternative to the stencils often used to color woodcuts by hand; flat patches of opaque hues were simply added to the overall design, and no adjustments were made for the contours of the object or for the interplay of light and shadow. In this preliminary phase, moreover, the color of the paper support played no particular role in the composition of the printed image. The genuine chiaroscuro woodcut only appeared when the entire image had been conceived in terms of interlocking areas of dark lines, one or more mid-tones, and the highlighted contrast provided by the paper. In general, these early German instances featured an independent dark line-block from which the entire design could be read, even without the structuring detail provided by the midtone and highlight.

The second and more celebrated birth, trumpeted by Giorgio Vasari (1511–1574) as the true origin of the chiaroscuro woodcut, dates to 1516. It is associated with Ugo da Carpi (1480–1523), perhaps via German influence in Venice, the city where woodblock printing already formed an alternative, if not a kind of resistance, to copperplate engraving. Originally a "mediocre painter" of friezes and other external decorations in or near his native Carpi, once in Venice Ugo became an established master of conventional woodcuts, enjoying a very successful collaboration with Titian (Vasari 1568, 303). More crucially for the intermedial struggle between relief and intaglio prints,

[2] On the resistance of quattrocentro engraving to reverse engineering, however, see (Roberts 2013).

[3] On other disputes over German versus Italian technological origins, see (Wood 2008, 63–71).

[4] On the overall development of color printing and the early German chiaroscuro print see (De Simone 2004, 54–6; Jecmen and Spira 2012, 32–4, 57–61, 66–101). For other instances of color printing, see (Hind 1963, 2: 442, 444, 496, 725).

Ugo had reproduced woodcut versions of Marcantonio Raimondi's (1480–1534) engravings, as if to reverse the directionality of the latter's well-known engraved imitations of Albrecht Dürer's (1471–1528) woodcuts.[5] Ugo sought and received a privilege for his chiaroscuro woodcuts from the Venetian Senate shortly before departing for Rome. He presented himself in this petition of July 1516 as both more and less than an established artisan:

> I, Ugo da Carpi, as a cutter of images in wood, have been a long time in this excellent city of Venice, and have consumed my youthful years here, and having now reached a considerable age, would like to live on my accustomed profession, rather than depending on others. Nor do I want my talent sullied or my livelihood taken away. Having found a new way of printing in *chiaro et scuro,* something new and never yet undertaken, a beautiful thing, and of great utility to those many people who delight in drawing, and moreover, since I have already carved and will carve things never yet done nor even imagined by others, on account of these labors I humbly ask and beg that your Lordships concede that no one be permitted or even dare to imitate any design or woodblock of mine.[6]

While the purpose of the request was to announce a novel technique and to secure protection against those who would be tempted to copy his work, Ugo's point of departure suggests an imperiled livelihood; both the allusion to the unhappy possibility of dependence on others and the not entirely accurate reference to a frail and aged body raise the specter of his penurious early years as a painter of friezes and façades. The gesture to sullied talent—literally, "violated genius"—recalls Titian's slightly earlier complaint about those imitators who imperilled "his honor," while both the insistence on prints made and yet to be made, and the depiction of copyists as sexual predators appear vernacular versions of a Latin colophon published by Dürer in 1511 and addressed to "you who lie in wait and who ravage the labor and genius of others."[7] Particulars of Ugo's claims thus seem inaccurate or overstated, but the general narrative would endure. It is the story of an art so vulnerable that only its most experienced practitioners, seizing on unimagined novelty, could survive.

The technique itself was not described in print until 1550 when, in the final paragraphs of the Proem of the first edition of his *Lives of the Most Celebrated Artists,* Vasari introduced the chiaroscuro woodcut as a recent invention. In keeping both with the overall schema of the progressive development of the arts and with his preference for design over color, he placed no great emphasis on the striking hues most noted by modern viewers, but described these woodcuts instead as an inspired imitation of engraving, a generalization which was appropriate for some of Ugo's works.[8] Vasari's initial account of the process by which these prints were made presupposed an image readily divided into four discrete visual elements:

> The original inventor of three-block woodcuts was Ugo da Carpi who, in replicating copperplate engraving in order to show shadows, midtones, and highlights in addition to the design, discovered this sort of print, carving them in boxwood or pear, which are superior to other woods. He made them in three blocks, placing the lines and contours of everything on the first block, all colored elements near those outlines on the second block, and on the third [both] the highlights and the ground, leaving the white of the paper for the lights, and coloring the rest for the ground.
>
> This last block is made in this way: take a piece of paper freshly printed from the first block, the one with the lines and contours, and place it along the plank of the pearwood, and weigh it down with other paper that isn't

[5]See (Servolini 1937; Karpinski 1976; Johnson 1982). On another façade painter with multiple and transferable skills, see (Chap. 8). For recent correctives to Vasari's account of the conflict between Dürer and Raimondi, see (Pon 2004; Rinaldi 2009).

[6]For the text of the petition, see (Pon 2004, 187 n. 39; Rinaldi 2009, 282).

[7]See (Servolini 1937, 107–8; Rinaldi 2009, 281–2).

[8]On Vasari's complicated adherence to a model of organic growth, development, and decline, see (Payne 2001); for instances where Ugo seems to have elaborated or even surpassed an engraving with a chiaroscuro woodcut, (Johnson 1982, 20–2, 39–46, 61–2, 64). For evidence, however, that woodblock carvers might be valued more in fiscal terms than designers, see (Chap. 15).

> damp, pressing it so that it leaves an impression of these outlines on the [third] block. Then the painter should use white lead mixed with gum to mark the highlights on this block; once this is done, the carver should cut out the areas marked in this way. This is the block that is used first, because once it is prepared with oil paint, it will print the ground everywhere except for where the highlights have been carved out, as that part of the paper will remain white.
>
> The second block is the one with the shadows. It should be quite flat and covered with watercolor, except where there should be no shadows, and here it is hollowed out. The third block, which was the first to be carved, is the one where the outlined part is hollowed out all over, except where there are no outlines touched by the black of the paint. These are printed in a press, and are run through three times, that is, once for each block, in such a way that they are all aligned. And this was certainly a most beautiful invention (Vasari 1568, 65–6).

Vasari's description has several striking features, the most significant of which is its constant oscillation in address and focus. Though he begins and concludes with an emphasis on the historical nature of Ugo's innovation, Vasari also includes directives seemingly aimed at any reader, remarks for the connoisseur of the notional print, and depictions of several quite specialized tasks. These shifts typify an artifact distinguished both by multiple origin stories, by experimentation with aesthetic and economic limits, and by repeated episodes of appropriation and re-invention, and their contribution is not simply to the expected education of the consumer, but also to the apprenticeship of the eventual creators of more such woodcuts. As if to underscore the somewhat fluid nature of the labor involved, Vasari does little more than gesture to these various artisans. But the painter who daubed the different blocks with the appropriate colors and highlights might have relied on a design sketch produced by an invisible (and perhaps unknowing) confederate; the woodcarver in his turn was likely aided by an unseen printer who managed the difficult alignment of the images, coped with the often unpredictable quality of inks and paper, and with the uneven patterns of wear on the blocks. Inexperience with or inattention to this last part of the process could and evidently did result in the common problem of poor registration; more harmful still was the work of printers so unfamiliar with the technique that they used inks of insufficient contrast, or even of the same color, for different blocks.[9]

The second and more puzzling feature of Vasari's description is his final allusion to the line-block, characterized in both the original edition of 1550 and in the expanded second edition of 1568 as "the one where the outlined part is hollowed out all over, except where there are *no* outlines touched by the black of the paint." The presence of the negative here must have proved confusing to some readers, especially those interested in learning whether this block could be used independently to produce a conventional print, or if the function of the lines was rather to provide an additional shade through an abundance of hatching. For readers aware of other experiments in color printing, Vasari's curious addition of the negative, in effect a reversion to intaglio rather than relief carving, would have recalled a hybrid such as Domenico Beccafumi's (1486–1551) *Two Nude Men in a Landscape*, where linear engraving was printed, with some difficulty, on top of a two-block chiaroscuro woodcut, or Hubert Goltzius' (1526–1583) *Lively Images of Almost All the Emperors,* a single volume of 1557–1561 featuring over a 100 etched medaillons enhanced with tone blocks.[10]

Whatever the origin of this error, Vasari returned briefly to the chiaroscuro woodcut in a subsequent biographical sketch of Ugo, but the additional details did not clarify matters. Here he depicted such prints as imitations of engravings that were themselves efforts to copy chiaroscuro drawings.[11] More crucially, he offered a slightly different explanation of the process. After stating that Ugo's initial attempt had been a two-block woodcut based on Raphael's (1483–1520) drawing of a Sibyl

[9] For one such instance, see (Johnson 1982, 85). The variability in materials and instruments—gunpowder and bores—likewise contributed to somewhat unpredictable results in ballistics (Chap. 5).

[10] On this experimental print and its overall relationship to Beccafumi's work see (Lincoln 2000, 58–62, 107–9); for the medaillons, as well as other hybrids later in the sixteenth century, see (Bialler 1993, 30–4, 56–63, 69–76).

[11] For a rapid overview of the range of early modern chiaroscuro drawings, see (Petherbridge 2010, 52–7).

reading by candlelight—a subject of particular suitability to print and to a tenebrous manner—Vasari alleged that the emboldened artist then undertook a three-block configuration. "The first [block] rendered the shadows; the second, stained with a lighter shade, the midtone; the third, hollowed out, produced the lightest shade and the highlights of the white paper" (Vasari 1568, 304).

This formulation appears to discard both the line-block and the troubling negative in the original description: here, the print is constituted by a combination of shadows, midtones, light tones, and highlights. In practical terms, the implicit elimination of an independent line-block suggested that the same print could not be produced and sold in both monochrome and chiaroscuro versions, though this dual-purpose product was and would continue to be familiar throughout the sixteenth century, culminating in the work of Hendrick Goltzius (1558–1617), where tone blocks might be added decades later to enhance a conventional print (Bialler 1993, 78–9, 81, 87, 90, 95, 174, 210). Vasari's silent passage from a mode more closely associated with the German origins of the chiaroscuro woodcut to one more typical of Ugo and his Italian successors seems to be the first step in a developmental narrative.[12] In the wake of the invention, Vasari relates, the Sienese Baldessare Peruzzi (1481–1537) distinguished himself with a single print as an extraordinarily adept master of the emerging genre, and Francesco Mazzola (1503–1540), better known as "il Parmigianino," created a large chiaroscuro woodcut of Diogenes "that was a more beautiful print than anything Ugo ever made." "But the most beautiful of all," Vasari concludes, "were those made by the Sienese Domenico Beccafumi after Parmigianino's death" (Vasari 1568, 304).

While his copious biographies of the latter two artists added little more in the way of technical information about the chiaroscuro woodcut, Vasari offered several important details. Parmigianino was robbed of his drawings, engraving plates, and woodblocks by his assistant Antonio da Trento (1508–1550), Vasari related; he was able to recover the latter items from the friend with whom his former collaborator had left them, but the permanent loss of the drawings drove him to despair.[13] Not surprisingly, this anecdote privileges the artist's drawings—possibly designs for prints—over the materials used to imitate and reproduce these unique creations, and makes what was Parmigianino's alone more valuable than the blocks and plates collaboratively produced by him and his carver Antonio. Vasari's assertion that Parmigianino's large chiaroscuro woodcut of Diogenes was more beautiful than anything Ugo had ever made is similiarly motivated, as it suppresses the fact that Ugo carved this work after Parmigianino's design. And the laconic statement in the biography of Parmigianino that the artist "had certain chiaroscuro prints carved, among them … a sizeable *Diogenes,*" likewise effaces the effort of the carver, despite the fact that the inscription sometimes included on this work recorded the names of both men.[14]

Vasari's mistaken assumption that Baldessare Peruzzi, taking up Ugo's technique, cut rather than designed a chiaroscuro woodcut positions him early in the developmental narrative; by the same logic, the gesture to the chiaroscuro woodcuts both designed and cut by Beccafumi presents them as the pinnacle of the genre.[15] Beccafumi, Vasari recounted, had planned to cast a series of 12 Apostles in bronze for decoration in the cathedral of Siena, but had died in the spring of 1549 before executing much of this work. Vasari went on to offer the persuasive suggestion that Beccafumi's chiaroscuro woodcuts of some of Christ's disciples date from this late period. These prints were superbly executed,

---

[12]As Johnson notes, however, about half of Ugo's chiaroscuro woodcuts are reliant on a key-block; see (Johnson 1982, 4).

[13](Gregory 2012, 36; Franklin 2003, 214) both point out that *stampe di rame e di legno* must refer not to the prints but to the plates and blocks used to make them. The eighteenth-century art critic, connoisseur, and printmaker Anton Maria Zanetti allegedly discovered the designs in the Arundel Collection. See (Affò 1784, 66–7).

[14]On the skill involved in this low-contrast inscription see (Lincoln 2000, 83–4).

[15]On the mistake involving Peruzzi, see (Gregory 2012, 287–8); on Peruzzi's likely acquaintance with Ugo, see (Johnson 1982, 32–4).

Vasari noted, adding that one was now in his own *Book of Drawings*. Since the first edition of the *Lives* was published in 1550, this prized possession would have figured as the best and most recent instance of chiaroscuro woodcuts.

2 Quiet and Unpretentious in Appearance

That cursory allusion to planks of boxwood and pear as the artist's material of choice, likely drawn from a celebrated and well illustrated early fourteenth-century treatise on agriculture, found quick elaboration in Anton Francesco Doni's *Mondi,* first published in 1552, and adorned with a woodcut designed by Vasari himself.[16] An astute observer of all that concerned the world of publishing, Doni provided an energetic account of the invention of the printing press—a drunken German's bout of experimentation with viscous plant resin—and concluded with a knowing reference to woodblocks made "of pear, of boxwood, and of sorb." Later that same year, Doni included in his *Marmi* a woodcut allegedly designed by his publisher Marcolini, rather familiarly described in the same text as Vasari's sidekick, in preparation for a forthcoming book of poetic emblems devoted to the theme of love. Doni reported that he had arranged for the images of this phantom emblem book by contacting an artist in Rome, "to see whether he wanted to draw them on boxwood, in order to make prints" (Doni 1552a 84–6, 92; Rizzarelli 2008; Gregory 2012, 63–131, especially 66–8, 117 n 34).

The trio of desirable woods soon reappeared in Pietro Cataneo's (1510–ca. 1569) *First Four Books on Architecture,* originally published in 1554, in a discussion of the strengths of various materials: "pear, sorb, and boxwood, in addition to their many other uses, are especially suited to woodcarving for prints. The best of these is the boxwood tree, but because large ones are not found, in sizeable prints pearwood is commonly used, since it is softer, and less knotty than the sorb" (Cataneo 1554, fol. 32v; Chap. 2). Influential then as now for his treatises on mathematics and architecture, the Sienese Cataneo perhaps owed this bit of specialized knowledge about woodblocks to his teacher Baldessare Peruzzi, or to Domenico Beccafumi, in whose circle he worked and to whom he may have been related through marriage, or to a contemporary and kinsman in Siena who worked in the book trade.[17]

The picayune attention to the particular sorts of hardwoods best suited for prints is striking in that no other aspect of Vasari's description of the chiaroscuro woodcut underwent further elaboration. Stipulations about materials, unlike descriptions of the process itself, were relatively brief, relevant to both conventional and chiaroscuro prints, and rich in cultural antecedents. Given that the blocks remained, for the most part, inaccessible to the consumer, we might expect a less than perfect correlation between the sorts of wood suggested by certain prints and the actual material of extant specimens; the fact that apple, walnut and cherry, for instance, are rarely mentioned as suitable for printmaking, but seem to have been used for some early conventional woodcuts is perhaps a significant finding.[18] An additional variable, at issue in foreign editions, was the availability and relative size of those trees

[16] Pietro de' Crescenzi's *Ruralia commoda* of ca. 1304 was repeatedly published in Latin and various vernaculars in the sixteenth century; on the suitability of pear and boxwood for carving, see Book Five, chapters 20 and 36; for discussion and reproduction of woodcuts taken from Crescenzi's work, see (Hind 1963, 2: 480, 483, 607, 640, 641, 655–6, 658 n. 2; Doni 1552a, b, fols. 27r and v; Gregory 2012, 74, 76). On Vasari's own early interest in botanical illustration, see (Tongiorgi 2001, 181).

[17] On Cataneo see (Bruschi 1979; Lincoln 2000, 85, and 175 n72, n73; Payne 2012, 95, 96, 97, 99, 110, 127, 142; Merrill 2013, 14–5).

[18] On the tentative identification of early woodblocks made of other woods, see (Landau and Parshall 1994, 22; Parshall and Schoch 2005, 72–3). The mutability of botanical names from one region to another, and particularly in translation, was an additional complication (Chap. 6).

in particular climates: Gabriel Chappuis' French translation of Doni's *Mondi,* for example, passed in silence over the boxwood, perhaps better known for its prominence in topiary, and alluded only to the pear and the sorb (Doni 1578, 62).

The expense in producing the blocks, the long life expected of them, the perceived need for increased detail and thus the finest possible grain, and the wood's vulnerability to damage from insects, all ensured a sustained interest in the material support of the design. Woodblocks were routinely transferred from one publisher to another, being as valuable as printing equipment and paper stock; many had a shelflife of well over a century.[19] They could be sold or rented out, used as collateral for loans, and they constituted, in most cases, the printmaker's most significant source of wealth (Grivel 1994, 33–57, especially 42–50). In 1558, for instance, the sculptor Alessandro Vittoria (1525–1608) would purchase from the engraver and miniaturist Giambattista Pittoni a book of drawings by Parmigianino and what he described as "a woodblock of pear, with the design for the Cumaean Sibyl and the Emperor Octavian, from the hand of the same Francesco Parmigianino" (Pittoni 1907, 20; Goldfarb 1981, on 315, 330 n. 34).

Even the banal registers of the bookseller's inventory convey the efforts to preserve such materials: while no details emerge about the composition of the 300 different blocks listed in 1570 in a notarized catalog of a Milanese printer's wares, the fact that 200 of them were stored in a special fruitwood box was recorded (Stevens 1995, 646, 659). Woodblocks could also be chased down by authorities and repossessed in the interest of civic pride, as happened with the unusual blocks of white lines on black, many two-sided and variously said to be of pear or of boxwood, with which the painter Sebastiano Aragonese had reproduced ancient Brescian tombstones in 1564. Rediscovered in the hands of one "Marco Polo, a horseman of this city," in 1611, they were immediately purchased for 15 *scudi,* and awarded to an antiquarian; some survive to this day (Lang 2006, 98, 100).

But the insistence on materials is not to suggest that knowledge of the appropriate types of wood and of the desirability of a cut along the plank were dependent upon Vasari; such considerations circulated both before and after his *Lives* and independently of it. The issue was of special concern to botanists, whose books increasingly included woodcuts, often borrowed, of the specimens under discussion, and whose texts sometimes offered brief synopses of the uses of particular woods.[20] In his *Herball* of 1597 John Gerard (1545–1612) noted, in the course of his description of the pear tree, its utility in printmaking: "the timber or wood it self is very tractable or easie to be wrought upon, exceeding fit to make moulds or prints to be graven on, of color tending to yellowness." He returned, four pages and eight images later, with a slightly more emphatic reference in his discussion of the wild pear, whose wood "likewise serveth to be cut into many kindes of moulds, not only such prints as these figures are made of, but also many sorts of pretie toies" (Gerard 1597, 1267, 1271).[21] Like much else in the volume, these passages are paraphrases of remarks made in Latin in 1587 by the Flemish botanist Remert Dodoens (1517–1585) (Dodoens 1583, 787; Chap. 15). The unobtrusive, even coy, tenor of such allusions doubtless has to do with the constant appropriation of woodblocks—whatever they were actually made of—in copiously illustrated works of this sort. Dodoens' own dependence on pearwood is signalled in later editions of his vernacular work, where the pear figures prominently on

[19] On the transfer of woodblocks, and for references to those preserved today in museums, see (Hind 1963, 2: 417, 432, 436, 438, 446, 448, 584, 609, 635–6, 640, 643, 647, 650–1, 659, 671, 731, 741). On the long-term significance of the exception to this pattern in the case of Ugo da Carpi, see (Johnson 1982, 10). On the extraordinary longevity of some woodblocks, see (Chap. 15).

[20] More generally, on the desire of botanists to portray not just distant but also vanished specimens, see the discussion of "numismatic botany" in (Keller 2014 432–6). For an explicit reference to woodblocks' origin and survival as a form of plant life, see (Chap. 15).

[21] On Gerard's unacknowedged dependence on the work of others, particularly a community of foreigners in London, see (Harkness 2007, 15–56).

the title pages (Dodoens 1608, 1644). While his French translation of Dodoens' earlier Flemish-language work did not include this allusion, his compatriot Charles de l'Ecluse (1526–1609), likewise a botanist, took great pains to secure both a "carpenter who is the only one who knows how to cut proper blocks of well-seasoned pearwood," as well as "the best carver in town, the son of Virgilus Solis" (Roze 1899, 82).[22]

Though it would have needed little elaboration among carvers themselves, some variant on this kind of information, destined for the consumer and designed to enhance aesthetic appreciation, also appeared in the woodcuts themselves, where trees or wooden objects were often a focal point. The frequent gestures to boxwood in both prints and texts were to some extent a function of Pliny's somewhat ambiguous description of that tree in his *Natural History,* and the lacunae and errors in various editions and translations of the work. All readers would have learned, for instance, that "boxwood is considered a first-rate material, as it is rarely knotted, except at the root;" and that the wood "is prized for its hardness and pallor, while the tree is valuable in ornamental gardening." Some would have also encountered the curious phrase, "it is quiet and unpretentious in appearance," an addition so bland and uninformative as to encourage and even to authorize the repeated emphasis on the unseen material from which prints emerged (Pliny 1535, 273).[23]

What appears to be an odd insistence on topiary, for instance, in works such as a woodcut for the Cologne Bible of 1478 featuring Saint Luke writing, or in several passages and woodcuts concerning gardening in the original version and subsequent translations of the *Hypnerotomachia Poliphili* of 1499 may have been a way of calling attention to the *other* use of boxwood; put differently, the gratuitous detail of those carefully shaped trees stands in for the unseen carved blocks on which these well-illustrated books depend. In 1562, Antoine du Pinet (ca. 1510–ca. 1584), the influential French translator of Pliny's *Natural History,* rendered the Latin author's description of this tree in a manner that strongly implied an actual connection between topiary and print making: after noting the hardness and pallor of the wood, he added *on se sert de sa plante à vigneter & historier en verdure.* In context, of course, the phrase might be translated "this tree is used for ornamental vines and figures," but because *vignettes* referred even then to woodcut borders, and woodcarvers themselves were identified as *tailleurs d'histoires,* readers would have been inclined to see such an association, despite the obvious anachronism, as a feature of Pliny's text. It comes as no surprise to discover that the translator was based in Lyon and would publish numerous woodcut images in an atlas of 1564 (Pliny 1562, 592).[24] Allusions to the suitability of boxwood for producing woodcuts of plants continue to appear, if only in passing, when that sort of print had fallen into irreversible decline: John Johnston's (1603–1675) *Dendrographia,* published in 1662 and accompanied by engravings, is a typical instance. His reference to the utility of boxwood in image production gives way to cautionary words about the difficulty and danger in splitting the tree's bark, reinforcing the medium's association with the rough and rustic arts of yesteryear (Johnston 1662, 247–8).

The striking number of such prints featuring musical instruments likewise seems at once to have been a general indicator of elite tastes and a specific gesture to the many items which, like the woodblock, were made with these same materials (Chap. 11).[25] Because *buxus* could refer not just to

[22] On Clusius' deteriorating relationship with the woodcarver, however, see also (Roze 1899, 86, 90).

[23] The phrase *lenis quies materiae silentio quodam* does not appear in all editions or translations. On the carver's effort to avoid knots, see (Chap. 15).

[24] More generally on du Pinet's translation, see (Tomlinson 2012).

[25] Examples of chiaroscuro print featuring lutes, often made of pearwood, include an Orpheus-like figure designed by Parmigianino, and Antonio Campi's mid-sixteenth century *Allegory of Music;* for boxwood flutes, see Parmigianino's *Pan* and Ludolph Büsinck's *Flute Player;* for a viol, components of which could include pearwood and sorb, see Parmigianino's *Marsyas;* for a dulcimer, which likewise could involve pearwood, see Parmigianino's *Saint Cecilia.* For conventional prints see for example (Mather 1975).

boxwood trees and to flutes, trumpets, and other pipes made of this wood, but also to the wax-covered writing tablets of Antiquity, these last items likewise enjoy special prominence in woodcuts. Such tablets appear in Hans Baldung's (1480–1545) *Fall of Man* of 1511, in Hendrick Goltzius' chiaroscuro woodcut *Demogorgon in the Cave of Eternity* of 1588, in three different chiaroscuro woodcuts of the Sibyl designed in the 1640s by Guido Reni (1575–1642) and cut by Bartolomeo Coriolano (ca. 1590–1676), and in a chiaroscuro woodcut thesis print which they produced in this same period. A celebrated literary precedent, moreover, would have made the very idea of prints made of boxwood blocks of particular appeal: one of the elegies of the Roman poet Propertius is a playful complaint about a stolen tablet of such material, ending with the suggestion that the thug who took it would return the boxwood for a cash reward (Sextus Propertius 1996, 113–4). Though the point in these verses is that the poet's best work has been lost, having been overwritten by the thief's vulgar accounts, within the ambit of the woodcut this literary referent attempted to mask the fact that blocks were stolen, and prints were copied through slavishly recarved blocks, precisely because of the information they contained. It is noteworthy that Ugo da Carpi, who experimented constantly with ways of inserting his name into his prints so as to retain some degree of ownership, and who saw his work plagiarized with some frequency, moved from a small, easily overlooked wooden tablet in his early *Lamentation,* to a signature along a fragile border, and finally to durable inscriptions incised on conveniently placed stones, as if to minimize this particular connection with boxwoods.[26]

Individual prints might also refer to the importance of certain types of wood to their overall appearance. Consider in this connection *Goats at the Foot of a Tree,* a chiaroscuro woodcut of ca. 1530–1535, Venetian in origin, and perhaps based on a design by Titian (Gnann et al. 2014, 119) (Fig. 2).

Close to 20 in. in height and 8½ in width (50.8 by 21.5 cm), the image is striking not just for its size, but also for the curious framing of its apparent subject. These features are related. The height and width of the print appear dictated by the tree it represents, or better, by the size of the woodblock itself, given that the image does not include what we can only assume are the uppermost branches. The approximate matching of the contours of the upright tree with what must have been those of the block is a way of showing that the plank, rather than the endgrain, was the relevant dimension for this sort of print. While the clusters of dense, rounded, paired leaves have a generic air, and the bark is fanciful and unidentifiable, Peter Breughel (1525–1569) would imitate the twisted trunk, aged but adorned with new shoots, in reverse, in a pen and ink drawing in the 1550s (Orenstein 2001, 88–9). Both the foliage and the trunk bear a strong resemblance to the wild boxwood, which adapts to rocky soils, requires little cultivation, and can reach 30 ft in height, though only after centuries of growth.[27] As Cataneo's allusion to the rarity of large boxwood trees and Chappuis' silence suggest, specimens of such size were difficult to obtain; the printmaker publicizes, or perhaps idealizes, the material at his disposal. Given that the bark bears no resemblance to that of the boxwood, the image seems based on a general description such as Pliny's text or a visual sketch of the size, shape, and locale of such trees, rather than on close acquaintance with living specimens.

The strong association of the evergreen boxwood with color, moreover, is clearly subordinated here to the rather literal connection between the dimensions of the tree and the block. Put differently, the pigments in this print are no attempt to approximate either its verdant boughs or the proverbial pallor of the wood itself: they convey the object without recourse to these clumsier representational

[26]On his efforts to claim ownership, first of the process, and then of particular prints, see (Johnson 1982, 20–1, 30, 52, 64–6, 70).

[27]The most celebrated wild boxwoods in Europe today are in the forest of Coudrée in the Haute Savoie and at Birr Castle in County Offaly, Ireland; both are about 400 years old. Seventeenth and eighteenth century writers mention large boxwoods in Normandy, Picardy, Ile de France, the Vivarais, Franche-Comté, near the Spanish Pyrenees, in Surrey, and in mountainous regions generally, while references to the disappearance of old growth boxwood became a refrain by the late nineteenth century.

Fig. 2 Oversized chiaroscuro woodcut, perhaps after design by Titian, *Goats at the Foot of a Tree*. 1530–1535. Courtesy of the British Museum

strategies.[28] That the boxwood was a traditional icon of longevity would not have been irrelevant to this ambitious Venetian print, but the most curious touch is in the presence of the goats that seek to destroy the tree, as these same animals would have once provided the parchment surface on which the most valuable images were painted and preserved. Given that miniaturists frequently doubled as early woodblock designers, and eventually abandoned the old craft of painting individual masterpieces on parchment for the newer one of mass produced images, the implicit struggle of the goats and the boxwood appears to capture the shift from one medium to another.[29]

We might even take this gesture to such a shift as a corrective to several curious details in Albrecht Dürer's *Adam and Eve,* where the artistic utility of different sorts of wood is as important as their symbolic meanings. In this celebrated engraving, Adam bears aloft the severed branch of a sorb tree, from which a tablet, presumably boxwood, is suspended. It is inscribed with a Plinian signature whose verb tense indicates the ongoing, open-ended work of the greatest artists—"Albrecht Dürer of Nuremberg was making this in 1504"—and stands in rough symmetry with the hybrid bough of apple and fig held by Eve.[30] Not surprisingly, there is a sharp gradient from the uplifted sorb and boxwood down to the evil fruit and fig: the relatively soft and splintering branches of the latter had been famously identified in one of Horace's *Satires* as a "useless wood," disastrously carved into a statue of Priapus by an inept artisan.[31] But the fact that *all* the types of wood depicted here are conveyed through a copperplate engraving is the more crucial feature: it is as if their utility to the printmaker had been temporarily interrupted, if not definitively discarded, in the face of a dynamic artistic career. *Goats at the Foot of a Tree,* by contrast, would insist on the vitality of the woodcut, or at least of the chiaroscuro woodcut, but at the expense of the parchment-based miniature.

Extremely large chiaroscuro prints might insist in yet another way on the relevance of the wood from which the blocks were fashioned. A prime instance is *The Fall of the Giants,* a collaborative effort of Reni and Coriolano from around 1638–1641, and loosely based on Reni's recent painting of the same subject (Fig. 3).

While the painting appears to predict, in its pronounced emphasis on hatching, its conversion to another medium, we might also argue that the eventual print anticipates and minimizes the inherent difficulties of the transition from single canvas to multiple paper copies (Spear 1997, 311). Measuring about 34 ½ by 24 ½ inches (87.6 by 62.2 cm), this three-block print is composed of four joined sections, and thus involves 12 pearwood blocks in all. The blocks, inevitably, are slightly misaligned and out of register; despite their considerable size, they are necessarily dwarfed by the subject. These technical flaws and material constraints achieve a brilliant form of aesthetic recognition and containment in the print. Wood seems initally to figure here as the pathetic index of defeat, in that it appears only in the three cudgels, unobtrusive, inadequate, and combustible, with which the victims seek to defend themselves against Jove's thunderbolts. But it is in fact the source of all that we see, as our ability to recognize the general effect—colossal stones falling upon upside-down giants—depends upon the relative stability and correct orientation of the imperishable or renewable resource of the woodblocks.

[28]The expression *buxo palliodiora,* "paler than boxwood," appears in Ovid's *Metamorphoses* IV. 134 and was a cliché from Antiquity through the early modern period.

[29]On the evolution of miniaturists to early woodcut designers, see (Hind 1963, 2: 421, 574, 597–8, 623, 626, 632, 643; De Simone 2004, 27–8, 33–5, 37–9, 92–4; Armstrong 1993); on the left-right reversals suggesting relative inexperience with the printed image, see (Armstrong 1993, 343, 346, 353), as well as (De Simone 2004, 31).

[30]The inscription is a variant on the story Pliny related about Apelles in the preface of the *Natural History,* and generally associated with the urgent dynamism of grand projects. On the motif, see (Barkan 1999, 89, 116–7, and 366 n. 90).

[31]"Once I was the trunk of a fig tree, a useless wood,//When the craftsman, uncertain whether to make a stool or Priapus,//Chose to make me a god." Horace, *Satire* 1.8: 1–3. Giorgio Vasari refers in his *Life of Leonardo* to a peasant who had made himself an ugly and distorted buckler out of figwood; the young artist refashioned this humble object into a splendid shield with a Medusan motif.

Fig. 3 Oversized, multi-block, multi-part chiaroscuro woodcut of *The Fall of the Giants* by Guido Reni and Bartolomeo Coriolano. 1638. 574904i. Courtesy of the Metropolitan Museum of Art, Purchase, Joseph Pulitzer Bequest, 1917, www.metmusem.org

3 Legacy

While it is clear that information about the optimal material and size of the woodblocks circulated before and independently of Vasari, could be addressed to artisans, consumers, and to a general public, could take several different verbal or graphic forms, could range from factual specifications to allusive imagery, and could even gesture to the very gradual evolution of the medium as it distinguished itself from rivals, there is no denying Vasari's influence on the subsequent development of the chiaroscuro woodcut. On the contrary, the importance of his account must have been considerable, as it remains by far the most explicit discussion of the art for two centuries. To summarize the main points made in the *Lives,* the chiaroscuro woodcut appeared to have less to do with woodcuts than with engravings and highlighted drawings, its origins were Italian, and its development moved silently from independent prints, where a line-block conveyed the entire design, to the more sophisticated interdependent sort. The standard three-block arrangement could be elaborated; the darker blocks seem to have been carved first, but printed last; designers and carvers both coveted the blocks, but desired the drawings still more. Material details such as paper quality and ink color never entered into the discussion, but the occasional print was worth collecting.

Vasari's legacy is easily detected within the context of historigraphy and connoisseurship. Predictably, his choice of *primus inventor* was disputed; the Dutch Carel van Mander (1548–1606) and the German Joachim von Sandrart (1606–1688), among others, argued for the Germanic origins of the chiaroscuro woodcut. The latter, writing in the 1680s, noted first that the art had already existed "for a long time" in Germany when Ugo created it for Italians, and elsewhere dated it to 1503, pointing out that "three-block combinations such as yellow and white, or blue and white, or black and white with the shadows and highlights added" were common in Germany at that point (Sandrart 1683, 195, 206–7). Sandrart's description of the process, soon repeated verbatim by popularizers such as Christoph Weigel (1654–1725), was brisk: one block conveyed the main lines and the shadows, another the color, an optional third a paler hue, and the shade of the paper provided highlights (Weigel 1698, 212–3).

The perfunctory allusions to these prints also lacked a stable nomenclature. Vasari's own shift from *tre pezzi* to *tre stampe* to *tre forme* to indicate the number of blocks involved was almost certainly an effort to distinguish the chiaroscuro woodcut from large prints composed of sections aligned side by side, for the latter were commonly described by their title and either the size of the paper, or "in two pieces," "in three pieces," and so forth (Hind 1963, 2: 446, 448). This wavering terminology and Vasari's passing allusion to chiaroscuro drawings both meant that his successors relied on cumbersome and relatively uninformative names. In his *Lives* of 1642, Giovanni Baglione (1566–1643) used the expression *gl'intagli a tre facce, finte, di chiaro oscuro in legno,* "prints in three blocks imitating chiaroscuro in wood," while Carlo Ridolfi in a rival publication of 1648 referred in passing to "many careful prints in copper and in wood that appear touched with watercolor and white highlights" (Baglione 1642, 395; Ridolfi 1648, 72). In his *Felsina Pittrice,* published in 1678 and focused on Bolognese artists, Carlo Cesare Malvasia (1616–1693) made frequent and approving mention of prints "cut in wood with two blocks" (Malvasia 1678, I: 116, 117, 118, 119, 130, 255; II: 595). Even in his otherwise elegant Latin Sandrart drew on *Xylographia ... duobus pluribusve trunculis,* "woodcuts ... with two or more blocks," the latter being, literally, "little branches" (Sandrart 1683, 206).

English discussions of the chiaroscuro woodcut, while inflected by a readiness to adopt Vasari's account and Italian nomenclature in general, were not more informative. Few would have been better qualified to explain the technique than John Evelyn (1620–1706): an authority on trees and wood, with training as a draftsman and experience in etching, he had visited important art collections in Italy in the 1640s. His *Sculptura* of 1662, while focused on the endpoint of the mezzotint engraving, betrays genuine interest in the chiaroscuro woodcut from its origins until 1600, which suggests that either Evelyn or his Italian acquaintances were unfamiliar with the many significant prints of this sort in the first four decades of the seventeenth century. But if Evelyn slightly outdistances Vasari, it is above all in the concision and obscurity of his account:

> Ugo da Carpi did things in stamp, which appeared as tender as any drawings, and in a new way of *chiaro oscuro,* or *mezzo tinto,* by the help of two plates exactly counter-calked[32]; one serving for the shadow, the other for the heightening. And of this he published a *Sibyl* after Raphael, which succeeded so rarely well, that he improved the curiosity to three colours.... This occasioned many to imitate him (Evelyn 1769, 52).[33]

As woodcuts in general and the chiaroscuro woodcut in particular declined in popularity, the emphasis on their color, already slight, seems further to have diminished. While the shift from two to three blocks had earlier been treated as a form of refinement, or as a marker of Ugo da Carpi's growing confidence in his invention, in the 1670s Malvasia gestured most often to two-block prints made collaboratively in the 1630s by Reni and Coriolano, reserving his sole hints of criticism for the two such

[32] "Counter-calked" is an English neologism based on the technical term *contraccalcare* in Italian, to produce a mirror-image copy of a design in order to insure its correct orientation in printing.

[33] See also (Evelyn 1769, 128). The latest reference to chiaroscuro woodcuts is to the work of Andrea Andreani prior to 1600; see (Evelyn 1769, 56). For Evelyn's ambiguous assessment of the work of Hubert Goltzius, "in two colours," see (Evelyn 1769, 81); for mention of Christoffel Jegher's conventional woodcut of Peter Paul Rubens' *Temptation of Christ,* made without reference to the chiaroscuro woodcuts, see (Evelyn 1769, 73). On Evelyn's great interest in the maintenance of forests, see (Chap. 15).

works that happened to involve three blocks.[34] A series of French writers subsequently outdid each other in effacing the color elsewhere associated with these prints. In his *Dictionnaire* of 1690, the influential Antoine Furetière (1619–1688) offered a restrictive and somewhat confusing definition, the wording perhaps taken from Sandrart: "chiaroscuro, in pictorial terms, is a print made of two blocks, either black and white, or black and yellow." In 1698 his compatriot Monnier, following Vasari in a historical account of the arts of design, described the chiaroscuro woodcut as an invention "that made prints look as if they had been highlighted with a brush." A year later, the engraver and historian Florent le Comte (1655–1712), cataloguing a large collection of art objects, alluded to the prolific late sixteenth-century Italian printmaker Andrea Andreani (1558–1629) as one who "made many things in wood, in the highlighted manner," identified Beccafumi's woodcuts, so prized by Vasari, and so seemingly modern even today, as "most Gothic in content," and distinguished between conventional and chiaroscuro prints only with the discreet addition of the word *rehaussé,* "highlighted," for the latter (Le Comte 1699, 165, 170–1, 186).[35]

The connoisseurship that induced Vasari to keep Beccafumi's chiaroscuro woodcut in an album of prized drawings had evidently proved persuasive to some readers. Among the most significant collections was that of the French translator Michel de Marolles (1600–1681), assembled over the course of 40 years. Louis XIV's (1638–1715) minister Jean-Baptiste Colbert (1619–1683) bought the entirety of these 123,400 woodcuts, engravings, and etchings; this rather thrifty purchase would become the basis of the Royal Library's *Cabinet des Estampes,* and Marolles' own descriptive catalogue of the collection, first published in 1666, a reference point for subsequent scholars and connoisseurs (Félibien 1666, 70, 270; Le Comte 1699, (1) 181–6, (2) 128ff; Baillet 1685, 2: 269–70; Brice 1685, 1: 55). In that work, Marolles treated the chiaroscuro woodcut as a separate genre, and estimated that he owned 500 such prints (Marolles 1666, 21–2). His descriptions tended either toward generic expressions of approval—"excellent works" "of singular beauty"—or quick references to easily recognizable subjects; of the "3150 different images of the Virgin, some with the Christ Child, some alone," 39 were chiaroscuro woodcuts (Marolles 1666, 15). Classified simply as "Chiaroscuro," these prints were defined as "works of various masters, expressed in three colors, that is to say, by means of three different blocks" (Marolles 1666, 21). Because the catalogue was intended to convey both aesthetic and commercial value, Marolles' audience would have learned nothing of the historians' disputes over the origins of these prints, nor of the techniques deployed and wood involved, nor of the relative successes and failures of the different works in the collection.

4 Signs of Rivalry

As Vasari's influence did not result in any elaborations or revisions of his brief technical description of the chiaroscuro woodcut, we must assume that such knowledge was already familiar to some fraction of his successors, and a matter of indifference to most others. Because this sort of print underwent two subsequent waves of popularity—the first from the late 1580s through 1610, the second from the late 1630s through 1640—we might also infer that artisans relied upon other sources of aesthetic and technical information. This expertise will certainly have been transmitted orally, and perhaps in manuscript form as well. Both circuits of information would have been by design limited to small audiences, but the prints themselves would have disclosed other particulars to *cognoscenti.* It appears that woodblock designers and cutters, whose professional opportunities were steadily diminishing as

[34] Johnson notes that with his experimental *Archimedes* Ugo eventually contrived a print of five or six blocks; see (Johnson 1982, 72; Malvasia 1678, I: 116, 119).

[35] On the uncanny "modern" look of Beccafumi's work, see (Lincoln 2000, 13, 96).

copperplate engraving and etching grew in importance, appropriated the specialized genre of the chiaroscuro woodcut as a means of survival.

The decline of the woodblock print had long been anticipated. Though Vasari himself had provided designs for woodcuts for many works, his emphasis on the connection between engraving and the chiaroscuro woodcut in the Proem of his *Lives* was in part an effort to explain growth in a mode of image production many expected to wither away (Gregory 2012, 68–114). The rise of the engraved book illustration and the fall of the woodblock, clearly related phenomena, are associated with the mid- to late- sixteenth century, though woodcarvers had begun to imitate intaglio techniques, particularly types of hatching, decades earlier, as if in recognition of this competition.[36] The fact that woodblocks could be made the same height as type meant that they, unlike engravings, could be printed on the same press as books, but even this convenient feature became increasingly less significant (Bury 2001, 9–10, 40).[37] While woodblock images continued to be used for a wide array of paratextual elements—printer's marks, vignettes, headpieces, tailpieces, historiated capitals, diagrams printed alongside the text—authors, book sellers, and readers increasingly preferred intaglio rather than relief methods for the more elaborate elements of a book.

Such desires were in part aesthetic: engraving and etching generally provided a sharper, more refined, and more nuanced image; engraving, in particular, seemed to offer a greater range in shadow and light, as Vasari's gesture to the common denominator of chiaroscuro would suggest. Fiscal considerations were also significant; in general, copperplate engravings were quicker to make, easier to refresh and to amend, and invulnerable to insects. Woodblocks, on the other hand, could generate greater numbers of copies than plates; profits depended on higher print runs and on the preservation of older blocks. This difference may have increased the association of relief prints with lower production standards, with popular taste, and at times with a dated aesthetic.[38] Woodcuts themselves also encoded information concerning this struggle, and in ways that would become pertinent to the eventual chiaroscuro prints. In this section I will examine two such tendencies in conventional early modern woodcuts.

Though many early woodcuts—particularly those prized by collectors and overrepresented in museums and libraries today—exhibited a high degree of finish, the norm, especially in printed books, was lower. Because the relief line is almost invariably thicker than its incised counterpart, it is generally assumed that woodcuts offered less in the way of pictorial information, or suffered initially from crowding, and over time, from illegibility. While woodblock artisans sometimes imitated the overall appearance of engraving by relying on very fine-grained wood, thin and swelling lines, and complex hatching, ink or paper of inferior quality could and did compromise these costly, time-consuming effects. Poorly inked surfaces, for instance, which were difficult to avoid in all but the smallest print runs, would produce illegible images. The chain lines in the paper could introduce a striped and not always appropriate pattern, and because paper was often of higher quality on one side than on the other, prints on adjacent pages could likewise vary considerably in appearance. And even where these problems were avoided, very close inspection of the finest lines can reveal puddled and unevenly

[36]On the particular importance of Flemish engravers in France in the 1570s, see (Grivel 1994, 50 and 57 n. 109); on the general tendency of designers and carvers to imitate engraving in woodcuts, see (Hind 1963, 2: 433, 434, 451, 455, 469, 483, 485, 502, 526, 544, 568, 570, 580, 594, 604, 669, 742, 752; Ivins 1969, 27, 47–9, 50; De Simone 2004, 53, 59–61; Leutrat 2007, 35).

[37]The first illustrated books in Milan, Florence, and Bologna, dating to the 1470s, involved engravings rather than woodcuts; see (Hind 1963, 2: 521, 526, 527–8).

[38]On the longevity or occasional reemergence of woodcuts in texts designed for travelers, tourists, and antiquarians, see (Zorach 2008, 14, 66, 105, 123, 132–4, 137–9, 145–8); more generally on woodcuts in seventeenth-century illustrated books see (Lehmann-Haupt 1977, 171–227); on the relative costs of woodcuts versus engraving see (Bury 2001, 40–1).

distributed ink, an effect that did not necessarily work to the print's advantage (Ivins 1969, 47; Ivins 1987, 6, 15, 21, 22; De Simone 2004, 166).[39]

Cross-hatching, especially if it involved a shimmering distribution of light and dark, rather than a simple grid-like pattern, was of particular appeal to some woodblock designers and carvers through the sixteenth and seventeenth century (Bury 2001, 40, 42–3; Lehmann-Haupt 1977, 37, 48, 57, 68, 122, 124–6; Lejeune 2014, 147–80: 159–61; De Simone 2004, 167–70, 194–7). While other printmakers and connoisseurs, if not every potential consumer of the work, would be quick to recognize that such virtuoso effects were much harder to achieve in relief than in intaglio, artisans' gestures to the practical limits of cross-hatching in conventional woodcuts are the more relevant issue in the reconstruction of practical knowledge. These references, incorporated within the woodblock print itself, are part of an ongoing conversation about the evolution of the printed image; like those allusions to the desirability and dimensions of certain types of wood, they offer a commentary on what could and could not be attempted in the medium.

Consider in this connection the first French-language book of emblems, Guillaume de la Perrière's (1503–1565) *Théâtre des bons engins,* originally published by Denys Janot in Paris in 1540 and republished with slightly altered woodcuts in Lyons by Jean de Tournes (1504–1564) 15 years later (Saunders 1993, 63–79; Rawles 1993, 49–61). This small work was to undergo numerous reprints, pirated editions, and translations to Dutch and English over the course of the next few decades, and seems singularly well suited, in its moralizing emphasis on thrift and clarity, to the economical and workmanlike format of short text and woodcut image. The book's value cannot be ascribed to the ethical lessons it imparts, as these are, by and large, the expected admonitions about the wiles of women and flatterers, the swift passage of time, the value of effort and prudence, and so forth. Despite its opening references to emblems of Antiquity and the early modern period, its frequent allusions to "good literature," to "Greek, Latin, and Hebrew writers," and its knowing gestures to various dicta of Pythagoras, Homer, and Giovanni Boccaccio (1313–1375), de la Perrière's text itself is undistinguished, or distinguished above all by its watchful and slightly defensive tone (De la Perrière 1539, Preface and Emblems, 7, 8, 9, 10, 57, 62, 63, 69, 87, 88). Its real interest is in what the work discloses about the graphic medium on which its message depends; it is especially significant that the author implies, rather unusually, that he himself was among the artists laboring over the emblems.[40]

Some of what de la Perrière has to convey emphasizes the hybrid nature of the printed image, since the finished product resembles painting, but the method of production is, of course, a form of low-relief sculpture. Thus, for instance, the fifteenth emblem explicitly compares the scholar who seeks to know too much to the painter whose minute attention to his canvas imperils the work. This tacit gesture to the spare representational strategies of the woodblock carver is reinforced both by the woodcut image itself and by two other emblems whose focus is sculpture. One of these criticizes the man who attempts to split stones with a small knife, a figure of inadequacy and the counterpart to the overly meticulous painter. The other claims that the ancient Greek sculptor Lysippus, rather than de la Perrière or an artisan guided by him, made the image at hand, an affirmation that foregrounds the woodcut even as it appears to efface it (De la Perrière 1539, emblems 15, 33, 63).

But the most striking feature of the emblems lies in the careful allusions to those areas where the woodblock was already beginning to lose ground to engraving. The *Théâtre des bons engins* repeatedly presents fishing nets, tennis rackets, tiled floors, checked fabrics, chain mail, a spider's web, chessboards, barred windows, and fine mesh, for these items all mimic on an enlarged and thus manageable scale the cross-hatching characteristic of intaglio surfaces (De la Perrière 1539, emblems 2, 5, 8, 10, 15, 23, 26, 29, 37, 38, 40, 41, 49, 54, 59, 70, 76, 83, 87, 91, 95, 96). The gratuitous nature of

[39] On paper production for books in the early modern period see (Needham 2011, 21–4).

[40] "Epistre … à Madame Marguerite de France," in (De la Perrière 1539. A3, A5); the title page also refers to the author as one who has *limé,* meaning either "cut away" or more generally "polished" these emblems. See also (Sharratt 2005, 68).

Fig. 4 Detail from the School of Franz Floris, *Wedding at Cana* (1560–1570), showing both etching and chiaroscuro woodcut. *The Meals of Jesus Christ.* Detail. Courtesy of the British Museum

many of these grids, lozenges, and the like suggests that the real agenda of the work is less a set of lessons for some wary courtier than a staged conflict between two print media. The Parisian publisher Denys Janot anticipated that this work would be pirated by others as soon as his 4-year privilege expired, but the faux cross-hatching guaranteed that the appropriation would lose much of its meaning in an engraving, where surfaces were routinely rendered in this pattern.[41]

While this stategy constitutes a kind of magnification of cross-hatching, a scaling up and reification that result in an assemblage of nets, rackets, tiled floors, and so forth, the same work also rehearses the conflict with engraving through effacement. The *Théâtre* addresses the increasing dominance of engravings over woodblocks in portraiture by foregrounding the limits of individualized facial representation. That several emblems involve disguise is of no great surprise in early modernity, though readers would likely have been puzzled by the image of a man holding an oversized mask in order to confront a lion. But the truly curious image of a headless woman is even less easily assimilated; the fact that the emblem states that a woman without a head would be the true marvel shifts the focus away from the normative (De la Perrière 1539, emblems 6, 16, 60). While the texts routinely offer the banal suggestion that appearances are misleading, the tacit emphasis of the images falls upon the shortcomings of woodcut portraits in particular.

The specificity of this message is worth attention, as it suggests that this book of emblems offers, under the cover of its shopworn ethical messages, a commentary on the medium itself. Apart from frequent sylvan backdrops, six of the 100 images feature trees in prominent fashion; two involve the pear, much favored by woodblock cutters (De la Perrière 1539, emblems 65, 72, 80, 81, 82, 87). More crucially, de la Perrière's work ends with an emblem of *Diligence* bearing a cornucopia—the icon of a sort of wealth embodied in plentitude, rather than generated by scarcity—as if to make explicit the rewards an industrious printmaker might expect.

But it is the first emblem that is perhaps the most significant; it features the rather unattractive double-faced god Janus, and claims that wisdom consists in both lucid recall of the past, and an attempted vision of the future. It is clear that the slackness of this formulation, its seeming adaptability to any audience, and the smug suggestion that the antiquarian flavor of many of the emblems would somehow help such readers forge their own individual fortunes, function as the obverse face of a specific set of lessons directed exclusively to those who designed, cut, and published woodblock images. Those latter admonitions were simple, and they can be reduced to a preoccupation with the growing competition posed by engraving.

In foregrounding the technical challenges of cross-hatching and of portraiture, the *Théâtre des bons engins* offers a surprisingly accurate vision of the future of the woodcut. It anticipates by about 30 years the rather peculiar contrast of cross-hatching in the hybrid chiaroscuro print *Wedding at Cana*, where two media compete in the foreground, even the incised white lines of the woodblock being notably thicker than the darker ones of the etching (Bialler 1993, 69) (Fig. 4).

And it precedes by more than 200 years the defeated formulation offered in the *Dissertation on the Origin and Progress of the Art of the Woodblock,* where the author, a typefounder and engraver, tabulated the labor involved in a square of cross-hatching produced by 20 horizontal and 20 vertical lines:

[41] "A Monsieur le Prévot de Paris," in (De la Perrière 1539, first two unnumbered pages). He was pirated by the Lyonnais Denis de Harsy, who produced an unillustrated edition often mistaken as the original; see (Rawles 1993).

> If one were to etch 40 lines of cross-hatching in order to produce shadows or the like, this would involve 40 cuts, or 40 strokes of the burin if the copper plate were to be engraved, and it is easy to execute. These cross-hatches, so crucial to prints, sorely try the patience of the woodblock carver, and often drive him to despair. In order for him to produce 40 cross-hatches in relief on the block, he must make at least 1444 strokes to remove 361 [i.e., 19 by 19] small square or diamond-shaped pieces of wood in 4 cuts. The only advantage that this art holds is that one can print 50,000 to 60,000 fine copies from a single block—about 1200 to 1500 in one day—while one can produce only around 2000 from a copper plate, running about 150 to 200 a day. This benefit, along with its antiquity, meant that the woodblock could retain, for a little while, some of the most celebrated artists, some of whom practiced etching as well (Fournier 1758, 62–4).

5 Enhanced with a Bit of Color

I've argued that by the mid-sixteenth century both artisans and interested onlookers would have known something of the competition between woodcuts and engravings and of the disadvantages traditionally associated with the former medium; those who had access to Vasari's *Lives*, first published in 1550, would also be familiar with his version of the development and production of the chiaroscuro woodcut, but even those entirely unaware of that text might well have encountered such prints. It is worth noting that while many chiaroscuro woodcuts deploy crosshatching that appears to mimic that of engraving, two important variants emerge. The first is a pronounced over-reliance on that technique, normally associated with shading and contouring. Such touches are typical of extremely familiar subjects, and serve as a means of distinction, as in *The Madonna of the Roses,* carved by Antonio da Trento after a design by Parmigianino around 1527, where the net-like pattern nearly effaces the Christ child (Fig. 5).

Fig. 5 Heavily crosshatched faces in Francesco Parmigianino's chiaroscuro woodcut *Madonna of the Roses*. The Virgin and Child with the infant St John the Baptist; an oval composition. Chiaroscuro woodcut from two blocks in green. 1527. Francesco Parmigianino, *Madonna of the Roses*. 1527. Courtesy of the British Museum

Fig. 6 Ugo da Carpi's chiaroscuro woodcut, *Hercules Chasing Avarice from the Temple of the Muses*. 1516–1517. Courtesy of the Metropolitan Museum of Art, Rogers Fund, 1920, www.metmusem.org

A similar though slightly less disturbing tendency is apparent in Ugo da Carpi's early *Hercules Chasing Avarice from the Temple of the Muses,* where dense cross-hatching covers the face of one of the Muses, and lends a curious air to the hero's right side and uplifted arm, as if the visual index of his strength (Fig. 6).

The second and more interesting development entails a magnification of this technical issue; just as in the conventional woodcuts of the *Théâtre,* one encounters nets, barred windows, checkerboard floors and the like. The effect here, however, is notably less apologetic, for the transfer of these patterns toward particular objects and away from shade and contour suggests that the latter are fully conveyed through other means, typically one or more color blocks. Some mid-sixteenth century instances would include Cornelius Anthonisz's (1505–1553) *Last Supper*, Franz Floris' (1517–1570) *David Playing the Harp before Saul* and his *Bull and Bear Hunts,* while Giuseppe Bianchino di Cosimo's scaly *Sea Monster,* only slightly later, might also fall under this rubric. The most notable examples of this particular graphic touch, however, are in the rough squares on the tree trunk in Ugo's *Hercules Chasing Avarice,* in the celebrated *Diogenes* of 1527, where the lozenge-shaped marks on the rooster's back are not crosshatching but rather a realistic representation of its skin (Fig. 7), and in the profusion of chain mail, woven surfaces, and barred windows in Andrea Andreani's archaizing *Triumphs of Caesar* of 1599.

Fig. 7 Detail from Ugo da Carpi's chiaroscuro woodcut Diogenes featuring rooster's skin. Diogenes, seated in front of his barrel, reading from a book, with a plucked chicken. Chiaroscuro woodcut from three blocks. Ugo da Carpi, *Diogenes*, 1527, Detail. Courtesy of the British Museum

It is likely that the addition of printed color was initially imagined as both a solution to the particular technical issues confronting designers and carvers of woodcuts, and as a feature of significant appeal to consumers. Hand-colored and stencil-colored woodcuts—the latter much more common—had long been a familiar commodity. Some readers sought to enliven de la Perrière's emblem book with the application of various pigments—with questionable results—while others trusted the matter to a *brieffmaler,* an artisan producing hundreds, if not thousands, of similarly colored images (Fig. 8).

These varied widely in the value of materials and execution; while Jost Amman's (1539–1591) verse claims that gold or silver were used to enhance prints, such embellishments were obviously not the norm.

The mechanical application of color to portraits, particularly to series with relatively little in the way of individuation, was somewhere between a fantasy and a fully realized process in the mid-sixteenth century. When the erudite historian Paolo Giovio (1483–1552) sought to publish images as well as biographical sketches of celebrated men, he alluded, somewhat inconclusively, to the appeal that color would add. Writing to Anton Francesco Doni in the fall of 1548, Giovio professed that he "wouldn't desire anything other than for these images [of military men] to be a little larger than those of ancient

Fig. 8 Artisan coloring printed images with stencil in Jost Amman's *On All the Mechanical Arts*. From (Amman 1574, *Brieffmaler*). Courtesy of the Library of the Max Planck Institute for the History of Science, Berlin

medals, and perhaps enhanced a bit with color, for more dignity" (Doni 1551, 47–8). Such wishes were expressed about a decade before the emergence of Hubert Goltzius' immense volume of chiaroscuro prints, *Lively Images of Almost All the Emperors From Julius Caesar to Charles V,* the basis of which were not purported images of the sitters themselves, but coins or medals cast during their lifetime. Color and size clearly enhanced these imperial icons, as did Goltzius' gambit of occasionally including an empty frame to suggest that his images derived from authentic sources (Bialler 1993, 30–4; Gaylard 2013, 178–96, especially 180–2). While Giovio might have known of Goltzius' forthcoming work, given that it was 12 years in preparation and required wide-ranging research, the model with which he was likely concerned, and the one he hoped to surpass with the addition of color was probably Andrea Fulvio's (ca. 1470–1527) *Images of the Famous,* an influential volume published in Rome in November 1517 and featuring more than 200 conventional woodcuts of celebrated figures of antiquity, or rather of coins and medallions made in their likeness. It is especially interesting, therefore, that those black and white images are occasionally ascribed to Ugo da Carpi, newly arrived in that city; such a chronology would suggest that the chiaroscuro woodcuts alone did not provide sufficient livelihood. Independently of the medallion-like *Images of the Famous,* it is worth noting that the heads of Aeneas and Anchises in Ugo's 1518 chiaroscuro print of this pair and the child Ascanius ultimately derive from ancient Roman coinage (Johnson 1982, 68, 71; Lincoln 2000, 90, 176 n. 77) (Fig. 9).

Fig. 9 Coin-based facial images in Ugo da Carpi's chiaroscuro woodcut *Aeneas Carrying Anchises*, 1518. Courtesy of the British Museum

The notion that color might somehow enhance likenesses was a persistent one, animated at once by aesthetic, technical, and fiscal concerns. Carefully designed and well-executed chiaroscuro woodcuts could, in theory, improve on both the quantity and quality of prints colored by hand and by stencil. More importantly, the broad swatches of pigment could eliminate some of the most time-consuming and difficult effects undertaken by the woodcarver. Thus, for example, the mid-tone in a multiple-block print might be used to convey either shadow or intensity of hue, both of which could have involved dense hatching or cross-hatching in a conventional print. Alternatively, a single tone block, hollowed out in small areas intended to show highlights, and combined with a second block of either the interdependent or independent variety, would need less in the way of cutting to convey contours, depths, and the illusion of different colors. Such approaches, less reliant on the precision and compactness generally associated with hatching and cross-hatching, achieved a more painterly effect, often in prints of somewhat larger dimensions. Given that paper constituted a significant expense for printsellers, and that sizes were one index of price, this ability to scale up would have been crucial to some designers and carvers.

There is a certain convergence of fiscal, technical, and aesthetic considerations in the use of color in the backdrops of many chiaroscuro woodcuts; these tendencies, to a degree, seem to have determined the subject matter of such prints. Distant, lightly sketched landscapes, conveyed by a midtone and minimally complemented by the darker line block and highlight, are typical of numerous Italian chiaroscuro woodcuts of the 1520–1540s, many of which were reissued by Andrea Andreani between the 1580s and 1610. Whether these impressionistic backgrounds featured buildings, countryside, or some panoramic combination, the approach meant that about one third of the image involved neither dense hatching or cross-hatching, nor especially exacting alignments of midtone, keyblock and highlight. Another strategy for diminishing the labor-intensive and unpredictable nature of the artisan's work lay in the designer's choice of subject matter: scenes featuring cloudy, windswept, or tenebrous skies are also common in mid- to late-sixteenth century Italian chiaroscuro woodcuts, and constitute a practical way of concentrating attention on a single and often dramatic moment. Such approaches find further development in the three-block prints of Hendrick Goltzius, dating to the 1580–1590s, where the backdrop might consist of swirling masses of fire, smoke, clouds, or waves, usually conveyed by a tone-block and highlights (Fig. 10).

That said, to judge from extant chiaroscuro woodcuts—almost certainly skewed toward works preserved for their aesthetic prestige, rather than representative of all sixteenth century production—printmakers often sought to avoid any obvious reductions in labor. The dissociation of the chiaroscuro woodcut from cost-cutting maneuvers is made most graphically, to use the obvious adverb, in Ugo's *Hercules Chasing Avarice from the Temple of the Muses,* a large print of 1516–1517, soon plagiarized by another less talented and energetic woodcarver and copied by an engraver (Johnson 1982, 30–5, 85–6). Avarice, bearing a pouch of coins, a metal urn, dish, and chains, is driven by Hercules' club from the impassive or rather approving company of Apollo, Minerva, and eight muses. The rather primitive lesson here—that stinginess is inimical to the arts—seems no more than a violent variant on the usual association of creativity with liberality. What is more interesting than the implicit plea for patronage, however, is the pantomime of an intermedial struggle: a rough wooden implement triumphs over refined metal. This is not simply to say that the woodcut was the more forceful medium, but rather to suggest that its means involved expenditure, not hoarding or economizing. The dense hatching mentioned above, radiating along the back and uplifted arm of Hercules, as well as on the spine of Apollo, and elsewhere throughout this two-block print convey a medium that yields nothing to engraving.

Ugo further insisted on the crucial role of the material itself to this print by inserting the wedge-like trunk on which one of the muses rests into the middle foreground. The squarish pattern of its rough trunk is typical of the boxwood, sorb, and some species of pear, but the presence of a lyre and a flared

Fig. 10 A typical background in Hendrick Goltzius' chiaroscuro woodcut *Pluto*. 1580–1590. Los Angeles County Museum of Art, Public Domain. Wikipedia Commons. © Vincent Steenberg

pipe makes the first of these woods the most likely referent. Ugo's manner of handling the attributes of the eight muses here is itself worth consideration in connection with his evolving aesthetic of the chiaroscuro woodcut. The muse seated in the right foreground is clearly Thalia, identified by the comic mask she holds. Just behind her is Melpomene, whose attributes are normally a tragic mask, a leafy crown, and a sword or Hercules' club; she wears such a crown, while Hercules himself wields the cudgel. Polyhymnia, typically distinguished by a pensive look, leans against the tree trunk; Clio, holding a woodwind, is at her side.

Four other muses encircle Clio, and theirs is a less obvious but not impossible identification. The maiden next to Minerva's staff, her shadowed visage covered with cross-hatching, is Urania, whose typical attributes include a globe, a short staff, an uplifted face, and a crown of stars. Just as Hercules' club hovers above Melpomene, so a single star, inscribed in the pediment on which Apollo is seated, appears in the foreground below Urania; her small staff seems sufficiently indicated by the presence of Minerva's larger one. This striking economy of means appears likewise to characterize the trio clustered in the background, all in profile and virtually identical. From left to right, they are Terpsichore, whose attribute is the lyre, Euterpe, associated with the flute, and Erato, identified with both the flute and the lyre. What would normally figure as their distinguishing features, in other words, finds approximations elsewhere in the scene by Apollo's lyre and the woodwind, something between a trumpet and a flute, positioned between Clio and Euterpe. The fact that these figures enjoy such minimal identification has less to do with Ugo's esteem for the particular discipline that they represent, though Erato's role as muse of mimic imitation is an especially appropriate touch.[42]

It is above all significant, of course, that one of the muses is missing, or rather invisible, in this configuration. This can only be Calliope, whose attribute is the boxwood tablet. Not surprisingly, the print itself is a version of that tablet, but it is more striking that Calliope figures as the double of Avarice; the latter, burdened by metal and on the verge of expulsion, is clearly less powerful than the unseen presence of that muse. While the print functions as an allegory of the intermedial struggle, or rather as a fantasy of that conflict, its most interesting aspect concerns the relationship between the aesthetic goal of recognizable pictorial protagonists and the technical features of this medium. Given that the muses are normally identified through their particular attributes and by their membership in a group of nine, Ugo appears to have deliberately undercut such strategies in this work: their attributes, distributed about the scene, shared by several, or delineated with ambiguity, have little to do with individual characteristics, and there are nine young women here only because Avarice substitutes temporarily for Calliope. The eight muses depicted in the print, moreover, bear strong facial resemblances to each other, as if a single individual had simply been portrayed in various poses. As if to reinforce his emphasis on the dissociation of the individualizing portrait with the chiaroscuro woodcut, Ugo presents the heavily shadowed face of Urania and the blank stare of the comic mask, equally unfathomable, as icons of the illegible visage.

Such a gesture was prescient, and constitutes an enduring codification of the technical limitations of the chiaroscuro woodcut (Chap. 6). Successful individualized frontal portraits are extremely rare in this genre: examples would be limited to Hans Burgkmair's (1473–1531) three-block print of Hans Paumgartner (1512), Hendrick Goltzius' three-block print of Gillis van Breen (1588), the three-block *Head of Christ* by the late sixteenth-century Monogrammist N.B., a reedition of Albrecht Dürer's conventional woodcut of Ulrich Varnbühler of 1522 as a chiaroscuro print in 1620, Peter Paul Rubens' (1577–1640) and Christoph Jegher's (ca. 1578–ca. 1652) four-block image of Doge Giovanni Cornaro (1632–1636), and François Dieussart (ca. 1600–1661) and Jan Lievens' (1607–1674) two-block bust of an old man (1630–1640). The very low incidence of such works has to do with our scant tolerance for misalignment in facial features, particularly in the eyes, and to a lesser degree, around the mouth. Until the individual tone blocks could be properly and reliably registered, facial images in chiaroscuro woodcuts are characterized either by recognizable flaws—misplaced lines and highlights, assymetrical eyes and skewed mouths, or oversized, heavily shadowed blobs reminiscent of the comic mask—or by a set of strategies designed to avoid this technical hurdle (Figs. 11 and 12).[43]

Individuals in many prints, regardless of the subject, appear in profile, which reduces the chances of misalignment and does not involve the direct gaze, or have downcast eyes and obliquely held heads, or shield their faces, or in numerous cases are depicted only from the rear.[44]

---

[42] On the Muses and their attributes, see (Waters 1881, 468–9).

[43] For a small sample of prints with heavily shadowed eyes, see especially *Death and the Courtesan,* after Hans Sebald Beham; Antonio da Trento after Parmigianino, *Circe, The Martyrdom of Peter and Paul,* and *Saint Cecilia;* Niccolò Vicentino after Parmigianino, *Augustus and the Tiburtine Sibyl,* and *Adoration of the Magi;* the anonymous Italian master, *Jesus Healing the Lame Man;* Andrea Andreani after Giambologna, *Christ Led Away;* Andrea Andreani after Alessandro Casolani, *Christ Carrying the Cross.* For an extant example of the effect of poor registration on facial features, see the Metropolitan Museum of Art's version of the mid-sixteenth century Master ND's *Holy Family,* and the related study of (Jenkins 2013, especially 138); see also Antonio Campi's *Mystic Marriage of Saint Catherine*, a two-block print of 1547 held at the Fine Arts Museum of San Francisco.

[44] Among the many examples featuring rear-facing protagonists, see Antonio da Trento after Parmigianino, *The Martyrdom of Peter and Paul*, and *Narcissus*; anonymous carver after Parmigianino, *Sacrifice of Abraham*; Hendrick Goltzius, *Pluto*. For instances of shielded or blindfolded faces, see Domenico Beccafumi, *Nude Men in a Landscape*; Niccolò Vicentino after Polidoro da Caravaggio, *Ajax*; Tobias Stimmer, *The Synagogue*; Giovanni Gallo after Marco Pino, *Cain Killing Abel*; Andrea Andreani after Giambologna, *The Rape of the Sabine Women*; Andrea Andreani after Iacopo Ligozzi, *Allegory of Virtue*.

Fig. 11 Typical instance of heavily shadowed eyes in a detail from Francesco Parmigianino's chiaroscuro woodcut *The Tiburtine Sibyl and Augustus*. Mid 1520s. Library of Congress, USA. Public Domain. Wikipedia Commons. © Adam Cuerden

The fact that portraiture was so difficult clearly conditioned the chiaroscuro woodcut, but we might argue that this high technical threshold was eventually codified as an aesthetic tendency. In other words, while characteristics such as averted or profiled faces seem transparent references to a short-coming of the medium, the issue itself—the very elusive association of individual identity with chiaroscuro woodcut—begins to inform these prints, appearing as a bold thematic choice rather than as a mere defensive strategy. Such options are readily apparent in works devoted to disguise and metamorphosis, as for example Ugo da Carpi's *Circe and the Companions of Ulysses*, where the goddess contemplates an indistinguishable mass of human and animal faces, but the aesthetic issue soon resulted in a much broader pattern (Fig. 13).

The rarity of the engaged frontal gaze of a recognizable living individual in chiaroscuro woodcuts led to an abundance of prints featuring sleepers, moribund sitters, decapitations, skulls, and masks, for such scenes, however unnatural, naturalized the enduring technical difficulty.[45]

6 Codification

We might draw several conclusions about the way in which the technical knowledge necessary to produce the chiaroscuro woodcut emerged, circulated, grew, and underwent organization and abstraction in the early modern period. That these developments often departed from linear progress toward

[45]For instances of beheadings, see Ugo da Carpi after Raphael, *David Beheading Goliath*; Antonio da Trento after Parmigianino, *The Martyrdom of Peter and Paul*; anonymous carver after Parmigianino, *Sacrifice of Abraham*; Ludolph Büsinck, *Judith*; Giovanni Gallo after Marco Pino, *Perseus*; Bartolomeo Coriolano after Guido Reni, *Salome and Herodias*. For emphasis on skulls, see Hans Burgkmair, *Lovers Surprised by Death*; Hans Wechtlin, *Skull*; Anonymous Master, *Child Sleeping on a Skull*; Master YHS after Raphael, *Saint Jerome*; Hendrick Goltzius, *Hercules and Cacus*; Andrea Andreani after Alessandro Casolani, *Woman Contemplating Skull, Saint Francis of Assisi*; Andrea Andreani after Giovanni Fortuna, *Allegory of Death*; Bartolomeo Coriolano after Guido Reni, *Christ Child with Skull.*

Fig. 12 Detail from the Master ND's chiaroscuro woodcut after Rosso Fiorentino *Holy Family* showing misaligned blocks. Master ND, *Holy Family*, 1544. Detail. Courtesy of the Metropolitan Museum of Art, Harris Brisbane Dick Fund, 1928, www.metmusem.org

a single and increasingly uniform aesthetic or technical goal is certain. Such departures have several causes: the double origin of the chiaroscuro print, the widespread diffusion of woodcutters and artisans themselves, the frequent appropriation of designs and woodblocks, the reissue of numerous chiaroscuro woodcuts, particularly in Northern Italy, 50–80 years after their emergence, and above all, the inclination of artists, consumers, and the genre as a whole toward experimentation.

We might attribute this sometime resistance to aesthetic codification to several factors. Some artisans and consumers would have found Vasari's account of the chiaroscuro woodcut, emerging several decades after the print itself, somewhat confusing in its particulars, or actively misleading in its chronology and in its teleological tendencies. Many others, to be sure, would not have encountered the text at all; their knowledge of the chiaroscuro woodcut would emerge either from familiarity with printmaking, or from the artifact itself, or both. As engraving flourished, and the professional opportunities of designers and woodcarvers diminished, the chiaroscuro woodcut represented a refuge of sorts.

Fig. 13 Unrecognizable faces thematized in Francesco Parmigianino's chiaroscuro woodcut *Circe and the Companions of Ulysses*, 1520–1550. Courtesy of the British Museum

Much of what can be inferred from the artifact—its extraordinarily self-referential tendencies—derives from its place within an intermedial struggle, and from its creators' constant efforts to distinguish their work from the conventional woodcut.

Thus, for instance, the early woodcut's attempt to mimic the fine cross-hatching of engraving, and its sometime inclusion of larger, more manageable squares and lozenges in the form of nets, barred windows, tiles, and the like finds an echo in the chiaroscuro woodcut, characterized by the same labor-intensive techniques for shading, and more interestingly, by the more sophisticated representations of the boxwood's bark, a chicken's skin, and by chain mail. The chiaroscuro woodcut, more costly than its conventional counterpart, appears to insist more on the material of the largely invisible blocks from which it was made. Such impulses answered to several needs. Allusions to other valuable commodities made of boxwood—a range of musical instruments, for instance—served to distinguish both artisan and consumer from those trafficking in the lowly conventional print. Such references also involved, on occasion, classical precedents, offering a pedigree of sorts to the wooden blocks, and a robust cultural corrective to the usual economistic assumption that the more valuable the material of the plate or blocks, the more precious the print produced by it.[46] But the emphasis on the particular

[46]On the occasional experiments with plates composed of metals more precious than copper, see (Viljoen 2011).

sort of wood used for the blocks might also serve to bolster weak or unverifiable proprietary claims: given that designs were often plagiarized from existing prints, that blocks were sold, stolen, and appropriated as collateral, and that they could be made of woods other than the canonical boxwood, pear, and sorb, allusions to those coveted substances mask a fragile professional context.

While the presence of colored inks, in combination with the variable quality of paper, seems to have contributed to what is invariably described as the "experimental" nature of the chiaroscuro woodcut, it correlates poorly with realistic or naturalistic impulses, and only weakly with any labor-saving techniques.[47] Color seems to have been put to opportunistic uses in larger mid-sixteenth century prints featuring stormy, smoky, or darkened skies, or panoramic backdrops; it also served to enliven older prints, or to mask worn areas of blocks (Gnann 2014, 50–3; Bialler 1993, 82, 87, 90, 95, 174). The tension between masculine, austere Design and feminized, flimsy Color, so often the subject of rather arid academic discourses and occasionally of painting, finds fresh ground here; it seems possible that the several chiaroscuro woodcuts of *Unequal Marriage,* featuring ill-assorted couples and more popular in tone than most such prints, refer instead to these graphic components and their uneven relationship.[48]

The fact that the individual portrait was, by and large, a goal that eluded the chiaroscuro portrait may be the most significant among the self-referential features that characterize these prints. Here the codification of knowledge crucial to the production and appreciation of the chiaroscuro woodcut reaches an apogee. Until the viscosity of the inks, the quality of the paper, and above all registration of the various blocks could be controlled, such portraits—already, perhaps, a riskier investment in terms of potential consumers than a mythological scene or landscape—were rare, but the technical impediment gave rise to a rich range of strategies and to an impressive host of variants on the living image. That such a self-referential art, so relentless in its allusions to its material basis, to its vanishing connection with conventional woodcuts, to its long-term struggle with engraving and etching, to its desired cultural pedigree, and to its costliness, should have so little to do with portraiture is interesting, for such gestures constitute a kind of self-portrait.

We might in fact close with an image that functions as the genre's self-portrait, *Narcissus,* a two-block chiaroscuro woodcut of 1527–1531 by Antonio da Trento after Parmigianino (Fig. 14).

It features the protagonists of the third book of Ovid's *Metamorphoses,* the young man whom the luckless Echo first encountered as he caught timid deer in nets.[49] Those animals have vanished, the nets have been transferred to an abstract pattern strewn across the landscape and his body, and he is intent on a facial image that we cannot see and that he does not recognize. Despite his averted head, Narcissus, his trappings as a one-time hunter, his strange activity in these dark woods, and his eventual fate are identifiable above all by the temporary presence of Echo in the lower left-hand corner of the print.

The moment must be just after their disastrous encounter, where he has proclaimed, "I'll die before all that I have might be yours," and she has replied, "all that I have might be yours," before hiding her face among the leaves. The original exchange—*Ante emoriar quam sit tibi copia nostri* and *sit tibi copia nostri*—conveys at once the failed romantic encounter and more crucially, the particular suitability of that story to the chiaroscuro woodcut (Ovid, *Metamorphoses* III: 391–2). We understand what we see and even what must remain invisible to us simply because of Echo, a sylvan figure destined to melt back into the landscape, the very icon of repetition, whose act of copying offers us, like the print itself, both clarity and obscurity.

[47] On experimentation, see (Bialler 1993, 115–6, 133; Lincoln 2000, 50, 52, 58, 78; Raftery 2005, 6, 8; Takahatake 2010a, 112, 118; Takahatake 2010b, 50). On the relatively weak links of woodcuts with naturalism and mimesis, see (Chap. 15).

[48] For two instances, see (Gnann 2014, 53, 60).

[49] Ovid, *Metamorphoses* III: 356. On the importance of Ovid's *Metamorphoses* throughout early modernity, see (Chap. 14).

Fig. 14 Antonio da Trento after Francesco Parmigianino's chiaroscuro woodcut *Narcissus*. 1527–1531. Courtesy of the Metropolitan Museum of Art, Rogers Fund, 1922, www.metmusem.org

References

Affò, Ireneo. 1784. *Vita di Parmigianino*. Parma: Filippo Carmignani.

Amman, Jost. 1574. *De omnibus illiberalibus sive* mechanicis. Frankfurt: Hartmann Schopper.

Aretino, Pietro. 1537. *Stanze in lode di Madonna Angela Serena*. Venice: Francesco Marcolini.

Aretino, Pietro. 1542. *Del Primo libro de lettere*. Venice: Francesco Marcolini.

Armstrong, Lilian. 1993. The master of the Rimini Ovid: A Miniaturist and woodcut designer in Renaissance Venice. *Print Quarterly* 10(4): 327–363.

Baglione, Giovanni. 1642. *Vite*. Rome: Andrea Fei.

Baillet, André. 1685. *Jugemens des scavans*. Paris: Antoine Dezallier.

Barkan, Leonard. 1999. *Unearthing the past*. New Haven: Yale University Press.

Bialler, Nancy. 1993. *Chiaroscuro woodcuts. Hendrick Goltzius (1558–1617) and his time*. Amsterdam: Rijksmuseum.

Brice, Germain. 1685. *Description nouvelle de ce qu'il y a de plus remarquable dans la ville de Paris*. The Hague: Abraham Arondeus.

Bruschi, Arnaldo. 1979. Cataneo, Pietro. In *Dizionario biografico degli italiani*, 22.

Bury, Michael. 2001. *The print in Italy: 1550–1620*. London: British Museum Press.
Cataneo, Pietro. 1554. *Quattro primi libri di Architettura*. Venice: Aldine.
De la Perrière, Guillaume. 1539. *Théâtre des bons engins*. Paris: Denys Janot.
de Marolles, Michel. 1666. *Catalogue de livres d'estampes*. Paris: Fréderic Léonard.
De Simone, Daniel (ed.). 2004. *A heavenly craft: The woodcut in early printed books*. New York: George Braziller.
Dodoens, Remmert. 1583. *Stirpium*. Antwerp: Plantin.
Dodoens, Remmert. 1608. *Cruydt-boeck*. Leyden: François van Ravelingen.
Dodoens, Remmert. 1644. *Cruydt-boeck*. Antwerp: Balthasar Moretus.
Doni, Anton Francesco. 1551. *La Zucca*. Venice: Francesco Marcolini.
Doni, Anton Francesco. 1552a. *I marmi*. Venice: Francesco Marcolini.
Doni, Anton Francesco. 1552b. *I mondi libro primo*. Venice: Francesco Marcolini.
Doni, Anton Francesco. 1578. *Mondes célestes, terrestres, et infernaux*. Lyon: Barthelemy Honorati.
Evelyn, John. 1769. *Sculptura*, 2nd ed. London: J. Murray.
Félibien, André. 1666. *Entretiens sur les vies et sur les ouvrages des plus excellens peintres*. Paris: Pierre le Petit.
Fournier, Pierre Simon. 1758. *Dissertation sur l'origine et les progrès de l'art de graver en* bois. Paris: J. Barbou.
Franklin, David. 2003. *The art of Parmigianino*. New Haven: Yale University Press.
Gaylard, Susan. 2013. *Hollow men: Writing, objects, and public image in Renaissance Italy*. New York: Fordham University Press.
Gerard, John. 1597. *Herball*. London: John Norton.
Gnann, Achim, David Ekserdjian, and Michael Foster. 2014. *Chiaroscuro. Renaissance woodcuts from the collections of Georg Baselitz and the Albertina, Vienna*. London: Royal Academy of the Arts.
Goldfarb, Hilliard. 1981. Chiaroscuro woodcut technique and Andrea Andreani. *The Bulletin of the Cleveland Museum of Art* 68 (9): 307–330.
Gregory, Sharon. 2012. *Vasari and the Renaissance print*. Farnham: Ashgate.
Grivel, Marianne. 1994. Printmakers in sixteenth-century France. In *The French Renaissance in prints from the Bibliotheque Nationale de France*, ed. Jacobson Karen. Los Angeles: Grunwald Center for the Graphic Arts.
Harkness, Deborah. 2007. *The jewel house: Elizabethan London and the scientific revolution*. New Haven: Yale University Press.
Hind, Arthur M. 1963. *An introduction to the history of the woodcut*, 2 vols. New York: Dover.
Ivins, William M. 1987. *How prints look*. Boston: Beacon.
Ivins, William M. 1969. Jr. *Prints and visual communication*. Cambridge, MA: MIT Press.
Jecmen, Gregory, and Freyda Spira (eds.). 2012. *Imperial Augsburg: Renaissance prints and drawings, 1475–1540*. Washington, DC: National Gallery of Art.
Jenkins, Catherine. 2013. The chiaroscuro woodcuts of the Master ND at Fontainebleau. *Print Quarterly* 30 (2): 131–143.
Johnson, Jan. 1982. Ugo da Carpi's chiaroscuro woodcuts. *Print Collector* 57–58: 2–87.
Johnston, John. 1662. *Dendrographia*. Frankfurt: Heirs of Matthew Merian.
Karpinski, Caroline. 1976. Some woodcuts after early designs of Titian. *Zeitshrift für Kunstgeschichte* 39 (4): 258–274.
Karpinski, Caroline. 2004. Light always at play with shadows: Prints in Titian's service. In *The Cambridge companion to Titian*, ed. Patricia Meilman, 95–118. Cambridge: Cambridge University Press.
Keller, Vera. 2014. Nero and the last Stalk of *Silphion:* Collecting extinct nature in early modern Europe. *Early Science and Medicine* 19: 424–447.
Landau, David, and Peter Parshall. 1994. *The Renaissance print 1470–1550*. New Haven: Yale University Press.
Lang, Giancarlo. 2006. Un protagonista del Rinascimento bresciano: Sebastiano Aragonese. In *Produzione e circolazione del libro a Brescia*, ed. Valentina Grohovaz, 95–114. Milan: Vita e Pensiero.
Le Comte, Florent. 1699. *Cabinet des singularitez*. Paris: Nicolas Le Clerc.
Lehmann-Haupt, Hellmann. 1977. *Introduction to the seventeenth-century woodcut*. New York: Abaris Books.
Lejeune, Maud. 2014. Preparatory drawings for woodcuts by Renaissance Lyonnais artist Bernard Salomon. *Master Drawings* 52 (2): 147–180.
Leutrat, Estelle. 2007. *Les débuts de la gravure sur cuivre en France: Lyon 1520–1565*. Geneva: Droz.
Lincoln, Evelyn. 2000. *The invention of the Italian Renaissance printmaker*. New Haven: Yale University Press.
Malvasia, Carlo Cesare. 1678. *Felsina Pittrice*. Bologna: Erede di Domenico Barbieri.
Mather, Christine K. 1975. Maximilian I and his instruments. *Early Music* 3 (1): 42–46.
Merrill, Elizabeth. 2013. The 'Trattato' as textbook: Francesco di Giorgio's vision for the Renaissance architect. *Architectural Histories* 1 (1): 20: 1–19.
Needham, Paul. 2011. *Galileo makes a book*. Berlin: Akademie Verlag.
Orenstein, Nadine (ed.). 2001. *Pieter Breughel the elder: Drawings and prints*. New York: Metropolitan Museum of Art.

Parshall, Peter, and Rainer Schoch. 2005. *Origins of European printmaking*. New Haven: Yale University Press.
Payne, Alina. 2001. Vasari, architecture, and the origins of historicizing art. *Res* 40: 51–76.
Payne, Alina. 2012. *The telescope and compass*. Florence: Olschki.
Petherbridge, Deanna. 2010. *The primacy of drawing*. New Haven: Yale University Press.
Pittoni, Laura. 1907. *Dei Pittoni, artisti veneti.* Bergamo: Istituto Italiano d'Arti Grafiche.
Pliny. 1535. *Historia mundi denuo emendata non paucis* locis. Basil: Frobenius.
Pliny. 1562. *L'Histoire du monde*. Lyon: Claude Senneton.
Pon, Lisa. 2004. *Raphael, Dürer and Marcantonio Raimondi*. New Haven: Yale University Press.
Raftery, Andrew. 2005. Collaborating through a medium: A new chiaroscuro woodcut after Parmigianino. *California Printmaker* 2–9.
Rawles, Stephen. 1993. The Daedalus affair. In *Intellectual life in Renaissance Lyon*, ed. Philip Ford and Gillian Jondorf, 49–61. Cambridge: Cambridge University Press.
Ridolfi, Carlo. 1648. *Le Maraviglie dell'arte*. Venice: Giambattista Sgava.
Rinaldi, Stefano. 2009. Marcantonio Raimondi e la firma di Dürer. *Opera Nomina Historiae* 1: 263–306.
Rizzarelli, Giovanna. 2008. Doni e Ariosto: Illustrazioni per il *Furioso* riusate nei *Marmi. Italianistica* 37 (3): 103–117.
Roberts, Sean. 2013. Tricks of the trade: The technical secrets of early engraving. In *Visual cultures of secrecy in early modern Europe*, ed. Timothy McCall, Sean Roberts, and Giancarlo Fiorenza, 182–207. Kirksville: Truman State University Press.
Roze, Ernest. 1899. *Carolus Clusius. Biographie et Correspondence.* Paris: J. Rothschild.
Saunders, Alison. 1993. Paris to Lyon and back again. In *Intellectual life in Renaissance Lyon*, ed. Philip Ford and Gillian Jondorf, 63–79. Cambridge: Cambridge University Press.
Servolini, Luigi. 1937. Il maestro della xilografia a chiaroscuro. Ugo da Carpi. *Gutenberg Jahrbuch* 107–114: 39–66.
Sextus Propertius. 1996. *Elegies,* ed. Harold Edgeworth Butler and Eric Arthur Barber. Hildesheim: Georg Olms.
Sharratt, Peter. 2005. *Bernard Salomon: Illustrateur Lyonnais*. Geneva: Droz.
Spear, Richard. 1997. *The "Divine" Guido*. New Haven: Yale University Press.
Stevens, Kevin. 1995. Vincenzo Girardone and the popular press in counter-reformation Milan: A case study (1570). *Sixteenth Century Journal* 26 (3): 639–659.
Takahatake, Naoko. 2010a. Coriolano. *Print Quarterly* 27 (2): 103–130.
Takahatake, Naoko. 2010b. VGO. Ugo da Carpi, the engravings: Woodcuts and chiaroscuro by Titian, Raphael and Parmigianino, 317–321.
Tomasi Tongiorgi, Lucia. 2001. The study of the natural sciences and botanical and zoological illustration in Tuscany. *Archives of Natural History* 28 (2): 179–193.
Tomlinson, Rowan. 2012. 'Plusieurs choses qu'il n'avoit veues:' Antoine du Pinet's Translation of Pliny the Elder (1562). *Translation and Literature* 21: 145–161.
Vasari, Giorgio. 1568. *Le vite de' più eccellenti pittori, scultori e architettori.* Florence: Giunti.
Viljoen, Madeleine. 2011. 'To Print or Not to Print: Hendrick Goltzius' 1595 *Sine Baccho et Cerere Friget Venus* and Engraving with Precious Metals. *Zeitschrift für Kunstgeschichte* 74 (1): 45–76.
von Sandrart, Joachim. 1683. *Academia nobilissimae artis pictoriae*. Nuremberg: C. S. Froberg.
Waters, Clara Erskine. 1881. *Legendary and mythological art*. Boston: Houghton Mifflin.
Weigel, Christoph. 1698. *Abbildung der Gemein-Nützlichen Haupt-Stände*. Regensburg: Weigel.
Wood, Christopher S. 2008. *Forgery replica fiction*. Chicago: University of Chicago Press.
Zorach, Rebecca (ed.). 2008. *The virtual tourist in Renaissance Rome*. Chicago: University of Chicago Press.

8 Multi-tasking "Pre-professional" Architect/Engineers and Other Bricolagic Practitioners as Key Figures in the Elision of Boundaries Between Practice and Learning in Sixteenth-Century Europe: Some Roman Examples

Pamela O. Long

Abstract

This essay focuses on the multi-variant careers of five individuals, most from practical and technical backgrounds, all of whom worked in a variety of occupations in the city of Rome in the late sixteenth century. Its goal is to investigate the working lives of these men and their friendships and the communicative networks within which they participated. This paper explores their practices as they were involved in trading zones—arenas of substantive communication between individuals from skilled (apprenticeship trained) and learned (university trained) backgrounds. It suggests that such trading zones between the skilled and the learned were also characterized by a certain fluidity of occupation and self-identification and that this fluidity contributed to the elision of the boundaries between the two groups. It thereby contributes to the long-standing and on-going discussion concerning the relationships of artisanal and learned cultures in the sixteenth and seventeenth centuries and the influence of those relationships on the development of new empirical methodologies.

In a recent publication, I have used the concept of "trading zones" to suggest that in the sixteenth century, there was increasing substantive communication between learned people who knew Latin and had studied in a university, and practitioners who through apprenticeship or on-the-job training possessed practical skills such as surveying, and other skills involving practical mathematics, the making and use of instruments, building construction, hydraulic engineering, and the like.

P.O. Long (✉)
Independent Scholar, Washington, DC, USA
e-mail: pamlong@pamelaolong.com

M. Valleriani (ed.), *The Structures of Practical Knowledge*, DOI 10.1007/978-3-319-45671-3_8

I suggested that the number and range of trading zones between the learned and the skilled increased dramatically especially in the sixteenth century. Further, within such trading zones, the people "trading" tended to become more like one another. The learned became more skilled and the skilled more learned, or, to put it in another way, learned men might acquire skills, such as measuring and surveying while skilled men might study Latin and discussed and/or read ancient Greek and Latin texts (usually in translation) relevant to understanding ancient history and Roman ruins (Long 2011, esp. 94–126; Long 2015).

In the history of science, the phrase "trading zones" was first used by the historian of science Peter Galison, who in part was inspired by the work of anthropologist Michael T. Taussig. Galison uses "trading zones" to describe arenas in which people from different groups engage in exchanges and at the same time can have completely different ideas about what they are doing. He points to an example given by Taussig concerning monetary exchange between the peasants and the landowning classes in the Cauco Valley of Columbia. The trading zone involves exchanges such as the payment of wages, the purchase of goods, and the payment of rent. Whereas the two groups interacted in perfectly functional ways, they understood the transaction differently. The landlords consider money as neutral, and having certain properties such as the ability to accumulate and become capital. For the peasants, money acquired in certain ways possessed moral properties. For example, in baptisms of babies, sometimes there was a secret baptism of money—the godparent hid a peso note in his or her hand while the priest baptized the infant. According to local belief, the bill is baptized, rather than the child. After being put into circulation, the godparent, now the godparent of the bill, calls it by name three times and the faithful pesos will return to the owner along with its kin.[1]

Galison has taken The Cauco Valley Market in Columbia as an example of a "trading zone," that is, a zone in which the peasants and land owners view money (and commodity exchange) in entirely different ways and yet complete successful transactions of buying and selling (It should be noted that Taussig himself does not use the term and has a different set of interests). Galison developed the notion of "trading zones" into a powerful metaphor for the exchanges that occur among specialists working on various aspects of quantum physics in the twentieth century. To simplify greatly, one of Galison's underlying concerns was to reject the unity of science embraced by the positivists of the early to mid-twentieth century—a unity based on the notion of a "protocol" language of observation that functioned across all theory. Antipositivists rejected the notion that such a language was possible and thought of theory and experiment as inextricably linked. In contrast to both, Galison and other historians and philosophers of science wanted to think of theory and experiment as separate on the one hand, but "intercalated" on the other, involving a partial autonomy of experiment from theory, but also the ability of the two entities to interact. However, Galison found the separate practitioners of theory, experiment, and instrumentation as possessing entirely distinct communities and incommensurable beliefs. How, he wondered, could such distinct cultures do anything but pass each other by with no interaction whatsoever? (Galison 1997, 782–3)

His answer, focusing on "local coordination, rather than global meaning" invokes anthropological linguistics in the form of pidginization and creolization. These refer to linguistic forms through which quite disparate cultures (say, for Galison, theoretical physics, engineering, and experimental physics) communicate at the overlapping boundaries of their respective disciplines. A pidgin is a "contact" language constructed with at least two languages, a language not used by anyone as a native tongue, but rather to negotiate particular local exigencies. A creole is also created from two or more languages, but it is complex and well developed enough to function as a native language among some groups. Pidgins can evolve out of their context-specific use and become used more generally and thus evolve into creoles (Galison 1997, esp. 830–5: 783; Foley 1988, 162–83).

[1]See (Galison 1997, 781–844). I have summarized Galison's description. Taussig's own discussion—(Taussig 2010)—does not mention "trading zones" and deals with quite different problems.

Turning to fifteenth and sixteenth century Europe, I use the concept of "trading zones" in some ways quite differently from Galison's usage. Nevertheless, the concept has been useful because it helps to frame and analyze arenas in which groups that had been to a large degree separate increasingly engaged in substantive communication. These groups consisted primarily of artisans and other practitioners (trained as apprentices in workshops or in hands-on instruction at, say, construction sites) and learned men (trained in Latin at universities). These two kinds of locales were usually quite separate in the medieval and late medieval eras. Medieval university instruction was presented entirely in Latin, a language not accessible to people being trained as stonecutters, goldsmiths, or shoemakers. And guild apprenticeships entailed long years of hands-on training from a young age, not accessible to the more socially elite university-bound (male) individual. In the later period, after the eighteenth century, professionalization of both engineering and architecture brought more formal educational demands and dictated requirements for licensing. No longer was it likely that learned, university-educated people who were not engineers would put out suggestions regarding the practicalities of hydraulic engineering, or, say, bridge repair.

I suggest that multiple trading zones developed in Europe especially from the sixteenth century. In certain arenas, artisans and practically trained individuals and those who were university trained increasingly engaged in substantive reciprocal communication and shared their respective expertise. Such trading zones do not include, for example, instructions by a patron for a specific design of a painting or building. Rather, it suggests the development of a mutual inquiry in which learned men and women desire to know something about a practical skill, such as measurement, and skilled people want to know something accessible to the learned, such as the content or meaning of an ancient Latin or Greek text (Long 2011, 94–126).[2]

In a later essay, Galison raised the issue of the extent to which the respective parties to any particular trading zone understood the other discipline. That is, how much of the substance of each of the respective sub-disciplines was in the trading zone and how much was outside of it? His answer is that this could range from a miniscule amount to quite substantial (Galison 2010, 25–52). This issue is much more difficult to resolve in the sixteenth century than in the twentieth, because the available evidence is much less extensive. Moreover, in one way at least, the situation is radically different. Rather than hyper-specialized sub-disciplines characteristic of modern physics and related areas, the sixteenth-century figures that I discuss in this paper lived before the development of modern professionalism. It is precisely this pre-professionalism, the lack of standard training, educational, and licensing requirements for architects, engineers, and the like, that gave these men multiple tasks, but also a kind of flexibility that facilitated the participation in trading zones. The "traders" that I discuss in the sixteenth century shared at least some interests with their modern counterparts, but came from a variety of backgrounds and possessed very different skill sets.

The application of trading zones to the earlier period must also be qualified by the fact that the categories themselves of "artisanal" and "learned" often became blurred as the learned acquired skill and the skilled acquired learning.[3] This paper specifically focuses on this blurring of categories in sixteenth-century Italy, and on the variety of identities and tasks that sixteenth-century "architects" and other workers took on. It should be underscored that one of the many differences between this context and that

[2]I say "men and women," although as far as I know, there were no female architect/engineers working in Rome in the sixteenth century. However, in other arenas, such as medicine, there were numerous female practitioners in various parts of Europe, and some of them were involved in trading zones as I have described them. See esp. (Rankin 2013). See also (Chap. 6), which usefully complicates the idea of trading zones with attention to translation.

[3]Centered on the eighteenth century, Ursula Klein's "Hybrid Experts" (Chap. 11), discusses individuals who were both learned and practiced so to speak, and the important roles they played in early modern economies. I would suggest that such experts were at least in part an outcome of the earlier development of trading zones (as I have described them). However, the sixteenth century is not a seamless unity with the eighteenth and more work is needed on these issues focused on the transition from the earlier to the later period.

occupied by twentieth-century physics centers on specialization. Twentieth century physics entails diverse fields of hyper-specialized professions. In the areas that are the focus of my own study—architecture, building construction, printing, engraving, engineering, and the like, we are in a pre-professional period. That is not to say that people that designed and built buildings or aqueducts in the sixteenth-century lacked skill. It is rather to note that the standard educational or training or licensing requirements that characterize modern professions, whether of architecture or engineering, did not exist. As Anthony Gerbino, focusing on the eighteenth-century figure François Blondel, has rightly pointed out, modern historians of architecture particularly have tended to anachronistically project their modern profession onto their early modern counterparts and therefore have ignored the varied tasks and occupations undertaken by these sixteenth century men that were outside of the purview of modern architecture, which is centered on design. In this paper I emphasize the transformation of identities that sixteenth-century practitioners could accomplish in this ambiguous context (Gerbino 2010, 1–7 and *passim*).

1 The Roman Context

This paper centers on one particular context—mid-late sixteenth-century Rome. It focuses individuals who mostly were practitioners, many of whom were called architects (*architetti*) at one time or another, but also a variety of other names. I investigate the great variety of tasks undertaken by such practitioners and their simultaneous participation in the culture of learning by which they acquired new identities and sometimes higher social status. I suggest that their "pre-professionalism," that is—their ability to work without specific training or licensing requirements; their ability to respond to the exigency of patronage and contracts in multiple and diverse ways; and to carry out highly diverse tasks—was a key condition that facilitated increasingly porous boundaries between learning and practice and the two-way traffic back and forth across such boundaries.

At the center of this paper are five case studies—five individuals who spent at least some of their working lives in the city of Rome. They came from varied, mostly practical, brackgrounds and operated under rubrics often not covered by existing guild structures—architecture/engineering, cartography, surveying, practical mathematics, engraving, printing, and publishing. The "bricolagic" flexibility of such practitioners, I suggest, had significance for the creation of trading zones between the skilled and the learned, and helped to break down the boundaries between them.

In the second half of the sixteenth century, the city of Rome drew numerous workers and practitioners of various kinds and, was a particularly conducive locale for the development of trading zones. I suggest that three aspects of the Roman context contributed to this environment. First, during this time, Rome experienced a huge building and construction boom that included major hydraulic projects and resulted in the significant reshaping of the city.[4] At the same time, a rage for antiquarian investigation developed, with ancient ruins and topography, and the collection of ancient coins, inscriptions, statues, and fragments becoming an obsessive interest to humanists and practitioners alike.[5] Finally a flourishing print industry developed consisting of numerous fiercely competitive shops, many specializing in single-sheet copperplate engravings of images—often featuring maps of the city as a whole or images of Roman monuments and ruins.[6] This paper focuses on a few individuals as examples—Leonardo Bufalini (died 1552), Pirro Ligorio (1514–1583), Stefano (Étienne)

[4]Still one of the most comprehensive introductions to Rome in the second half of the sixteenth century is (Delumeau 1957–1959). There is a large literature focused on various aspects of building construction and hydraulic engineering in this period. See as a start, (Long 2008; Marconi 2004; Rinne 2010; Simoncini 2008; Simoncini 2011). For a more specialized study of one aspect of construction (wooden roof structures in churches), (Valeriani 2006).

[5]There is a large and growing literature on antiquarianism. See esp. (Curran 2007; Jacks 1993; Miller 2007; Rowland 1996; Schnapp 1997; Stenhouse 2005; Weiss 1988).

[6]See esp. (Infelise 2002; Witcombe 2008; Zannini 1980).

Duperac (died 1620), Mario Cartaro (1540–1620), and finally, Egnazio Danti (1536–1586) and his extended family. These individuals were quite different one from the other. Nevertheless, their careers and interests exemplify the great diversity of their practical activities, their access to the cultures of both practice and learning, and the rising social status that most of them enjoyed.

2 Leonardo Bufalini: Woodcutter, Architect, Military Engineer, Carpenter, Cartographer, Antiquarian

Leonardo Bufalini was born in Udine, a city that was then part of the Republic of Venice in northeastern Italy. He arrived in Rome the early 1530s. His fame is based on his wood-cut map of 1551, a large, free-standing ichnographic printed map of Rome made almost entirely by orthographic projection, the first free-standing plan of Rome since antiquity to be drawn true to scale throughout (Fig. 1).

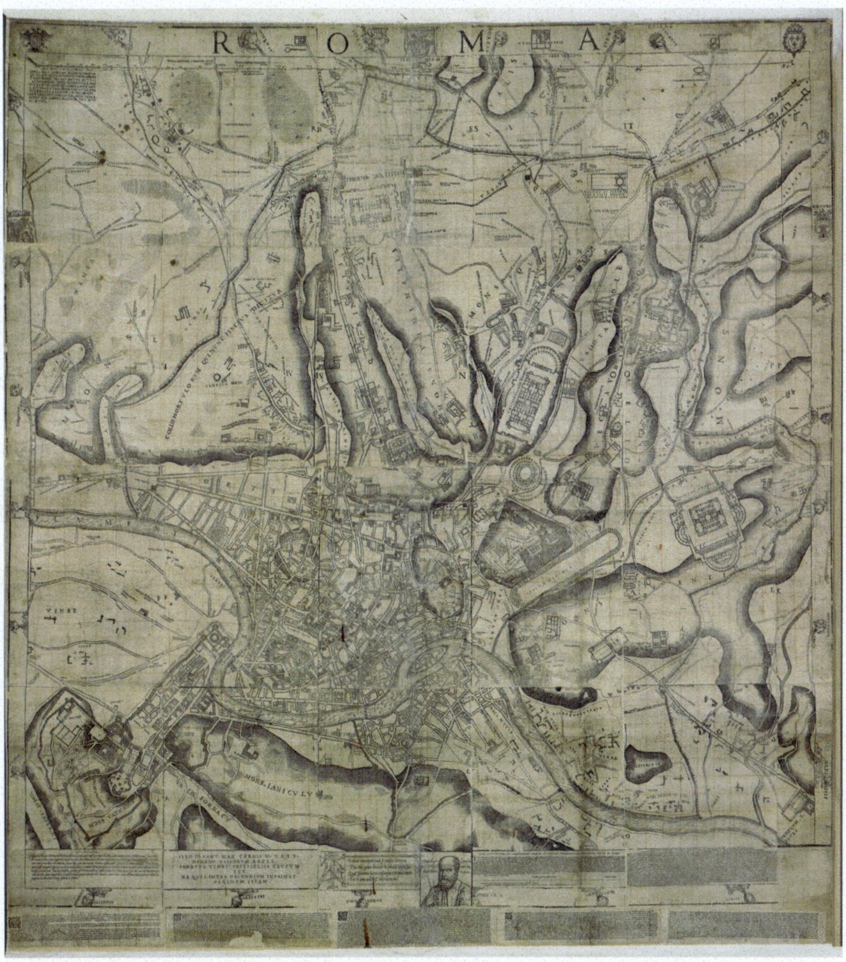

Fig. 1 Leonardo Bufalini, *Roma*, London: British Library. Woodcut on 24 sheets (200×190 cm). First published in 1551 (no copies extant), this is one of three extant copies published by Antonio Treviso in 1560. Copyright © The British Library Board, All Rights Reserved, Maps S.T.R. (175.)

Each viewpoint is represented as though from directly above in strict orthogonal projective. The map displays measured relationships between interrelated parts, preserving dimensions according to an established scale.[7]

Sparse but telling documentary sources show that aside from making his map, Bufalini also worked as a military engineer. Some of that evidence comes from the treatise on military engineering by another military engineer, Francesco de Marchi (1504–1576), who was involved in the refortification of Rome in the 1530s and 1540s under Pope Paul III (1468–1549). The Farnese pope's effort, which was led by Antonio da Sangallo the Younger (1484–1546), entailed a massive plan to fortify the city and projected the construction of eighteen bastions, which would greatly strengthen the third-century Aurelian walls (a project never completed). De Marchi mentions Bufalini ("*il cap.* [i.e., *il capitano*] *Leonardo da Udene*") as one of the participants in a debate in front of Paul III on these fortifications. De Marchi also says Bufalini went with him to investigate the ancient ships at the bottom of Lake Nemi near Rome (this was on July 15, 1535), a topic of ardent humanist interest ever since Leon Battista Alberti (1404–1472) had investigated these ships a 100 years before. De Marchi describes Bufalini as "a talented architect, the one who measured all of Rome inside and out, and had it printed with hills, theaters, temples, streets, and other things indicated, in which I helped perhaps for 6 months for my pleasure and to learn more" (Marchi 1599, fol. 78r (Bk. 3, ch. 34)).[8] Other documents show Bufalini, again described as a "fabro lignario" (a carpenter), as part of a team that constructed a bastion at the Vatican. And in his will, he is called a *faber lignarius* (Ehrle 1911, 16, 55).[9]

A list of workers in Antonio Blado's (1490–1567) print shop where his map was printed, lists a "Leonardus Venetus Intagliator," which probably refers to Bufalini and which suggests that he himself cut the woodblocks for his map (Ehrle 1911, 17). On that map, addressing the reader, Bufalini mentions that he created it "not only with the geometer's square and compass, but also with the mariner's [i.e., a magnetic] compass." The bottom tier of the map contains his self portrait and images of instruments (Fig. 2).

The instrument at the top is a surveying instrument, a *bussola*, that is, a primitive theodolite with a magnetic compass at the center of a graduated circle, divided into degrees, and a pivoting arm to take sightings. The other instruments are not surveying instruments, but carpenter's tools—a mason's level, and geometer's compasses—suggesting perhaps that Bufalini identified himself as much a carpenter as a surveyor and mapmaker.[10]

Although Bufalini's map has often been seen as having been made purely in the interests of military defense and fortification, Jessica Maier, in her definitive study, has argued persuasively that it was

[7]No copies of the 1551 map are extant. However, three copies (two complete and one partial) of a 1560 edition published by the military engineer Antonio Trevisi are extant, with the addition (on the bottom of the map) of letters by Trevisi that concern flood prevention of the Tiber River. The extant maps are: London, British Library, Maps S.T.R. (175.); and Vatican City, BAV, Stampe Geogr. I. 620. Riserva, which are complete, but differ in the letters by Antonio Trevisi, and BAV, Barb. Lat.4432, which is missing four of the main sheets and four of the smaller side sheets. There is also a facsimile edition: (Ehrle 1911). For recent studies, see (Ceen 2012; Friedman and Schlapobersky 2003; Maier 2012a; Maier 2015).

[8]For the debate in front of Paul III, and 42r–v (Bk. 2, ch. 82) on the exploration of the boat at the bottom of Lake Nemi, "valente Architetto, il quale misurò tutta Roma dentro e fuori, e la pose in istampa con tutti li monti e Theatri, et Tempij, strade e altre cose segnalate, al qual io aiutai forse sei mesi per mio piacere, e per più imperare." For De Marchi's life, see esp. (Lamberini 1990, 447–54). For Alberti's investigation of the ships, see (Grafton 2000, 248–52). For Antonio da Sangallo's work on bastion fortification, see esp. (Adams and Pepper 1994, 61–74; Fiore 1989, 175–80).

[9]See (Ehrle 1911, esp. 13–22) for a detailed discussion of the extant documents concerning Bufalini and (Ehrle 1911, 55–6) for a transcription of his will.

[10](Bufalini 1551:Vatican City, BAV, Stampe Geogr. I. 620. Riserva), and (Ehrle 1911): "Ad Lectorem, "non ad normam, olum [solum], et circinum sed ad pyxidem etiam nauticam." See also (Frutaz 1962, 1: 168–70 and 2: Pianta CIX, Tav. 189–209; Huelsen 1969, 38–41; Scaccia Scarafoni 1939, 78–82. For Bufalini's life, see esp. the foundational study which accompanies a full-scale reproduction of BAV Stampe Geogr. I.620. Riserva: (Ehrle 1911, Introduzione; Palagiano 1972; Maier 2007; Maier 2015). See also (Insolera 1985, 112–22) for a detailed, descriptive discussion of the map. I would like to thank Wolfgang Lefèvre and Jim Bennett for discussing the instruments depicted on the map with me.

Fig. 2 Leonardo Bufalini, *Roma* (detail), London: British Library. Self-portrait of Bufalini and his instruments, with a Latin inscription addressed to Bufalini: "If one wishes to assess the immense effort [of this undertaking], the Romulan land owes more to you [i.e. Bufalini] than to its own Lord, which [land] has been consumed by fire and arms and years, through your agency now, in no age is it about to perish." Copyright © The British Library Board, All Rights Reserved, Maps S.T.R. (175.)

also made to present Rome as a magnificent city, worthy of the ancient Romans. This can be shown by the fact that the map displays not only contemporary buildings but also ancient structures that no longer existed. In his letter to the reader at the lower margin, Bufalini confirms that his map displays the whole city through time from ancient to contemporary: "Whoever you are, Leonardo Bufalini from Friuli asks that you not think vilely, because he believes that he gives to you now the most beautiful of all things, which is to say, Rome, and this her twin; nor indeed did he consider it satisfactory to you to have placed before your eyes the one Rome which is inhabited today, unless he also added the ancient one, once mistress of the whole world, almost having been called back from the grave with very much labor, great expense and long-enduring vigilance."[11] Bufalini here is explicit that he is presenting the city as a whole, timelessly—the ancient city along with the modern, the measured walls, but also no-longer-existing ancient buildings. His map reveals engineering practice in its accurately measured walls (which walls, he may have measured in conjunction with Paul III's fortification team). But the map also displays structures of the ancient city that were not extant in Bufalini's lifetime, but were studied by means of ancient texts, by learned humanists, antiquarians, and clearly by Bufalini himself. Bufalini was much admired in these circles. For example, the renowned humanist and learned student of antiquities, Onofrio Panvinio (1529–1568), in the preface of his *De Republica Romana* of 1558 wrote that Bufalini's plan was "the fruit of incredible labor and persistent study of twenty years" (Ehrle 1911, Introduzione, 20).[12]

Bufalini's assessment of his own achievement (or perhaps that of one or more of his contemporaries) can perhaps be surmised from the inscription in the cartouche to the left of his portrait: "If one wishes to consider the immense effort, the Romulan land [i.e., Rome] owes more to you [i.e., Bufalini] than to its own Lord [i.e., the pope, Julius III (1487–1555)], [a land] which has been consumed by blazing fire and armies and the years. Through you now in no age is it about to perish."[13] Rome in this declaration has been saved by an image of the city, namely the map itself, which, as has been noted, portrays no-longer existing ancient as well as contemporary structures. Bufalini has thus done more than the popes (or the sitting pope, Julius III), to save the capitol city from the ravages of fire, armies, and time itself.

Bufalini's biographer, Francesco Ehrle noted that Bufalini's Latin was "rough and incorrect" (Ehrle 1911, 21).[14] Whether or not, he did inscribe his map in Latin. He was esteemed among learned humanists such as Panvinio, and accompanied De Marchi and others to Lake Nemi on an antiquarian quest to investigate ancient ships. On his map, as Maier has particularly underscored, he displays an interest in ancient Rome as well as the contemporary city. At the same time, the references to him as a carpenter and an engraver, as well as an "*architetto*" (De Marchi's word), combined with the instruments both of surveying and of woodworking on the map itself, suggest a kind of fluidity of identity, and a familiarity with both the technical skills of carpentry, engraving, surveying and cartography, and the researches of antiquarian scholars whose admiration he seems to have enjoyed.

3 Pirro Ligorio: House Facade Painter, Fresco Painter, Architect, Antiquarian, Designer of Sculpture, Garden Designer, Hydraulic Engineer

Pirro Ligorio (ca 1513–1583) arrived in Rome from Naples around 1534. Although he claimed to have come from a noble family, nothing is known about his early education and training. That it must have been practical is suggested by his earliest known employment in Rome. This work involved painting

[11] (Bufalini 1551, Ad Lectorem): "Quisquis es, Rogat te Leonardus Bufalinus Foroiulien ne tu vile existimes, quod ille tibi iam dat, omnium rerum pulcherrimam se dare credit/Romam scilicet et hanc geminam: neque enim satis tibi factum duxit, redivivam istam unam quae hodie colitur ante oculus posuisse: nisi vete/rem etiam, totius olim orbis dominam, summo labore, maxima impensa, diuturna vigilia, quasi e sepulchro excitatam addidisset;" and see (Maier 2007; Maier 2012b; Maier 2015).

[12] The manuscript is BAV Vat. Lat 6783. See (Ferrary 1996, 56), "Leonardus quidam Burfalinus Foroiuliensis faber lignarius incredibili labore et pertinaci XX annorum studio totam urbem commensuravit et ligneis typis eius ichnographiam expressit."

[13] Bufalini, Roma, "Debeat immensum vult si pensare laborem/Plus tibi quam domino Romula terra suo. Quae flamma fuerat consumpta et armis et annis./Per te iam nullo est interitura die." I would like to thank Kenneth Gowens for discussing this passage with me. Gowens noted that dispite Ehrle's assessment (see below), there is only one error in the passage—"fuerat consumpta" for "erat consumpta," a very common mistake.

[14] "molto ruvido e scorretto."

the facades of Roman houses and palaces (in this Roman tradition, numerous house facades displayed decorative historical or mythical scenes, of which only a very few examples are still extant). Ligorio also is recorded as carrying out fresco painting and in 1546 making a *gonfalone*—a religious banner for processions. During the 1540s he turned himself into an antiquarian—avidly collecting ancient inscriptions and coins, and studying, excavating and drawing ancient ruins and sites. He was helped by a number of learned humanist friends, most importantly the scholar of ancient Greek, Benedetto Egio. In 1549 he entered the service of Ippolito II D'Este (1509–1572) as an "*antiquoro*."[15]

In the 1550s Ligorio received his first architectural commissions. He also published a short book on the antiquities of Rome (the only book published during his lifetime despite his thousands of pages of writing and drawing). At the same time, he created several copperplate engraved single-sheet images of reconstructed Roman sites, such as the Circus Maximus that were published,[16] and in addition two maps, one of Rome with both ancient and contemporary structures (1552)[17] (Fig. 3) and one of the ancient city (1553).[18]

In 1561, he published a large wall map of ancient Rome which in some ways was dependent on Bufalini's, but was also based on his extensive knowledge of Roman coins and other antiquities (Ligorio 1561).[19]

Pirro's obsession with antiquities, including coins and statues, as well as his knowledge of ancient ruins (including their excavation and measurement) is evident in his writings, often illustrated, which have remained in manuscript form until recent years. He carried on sometimes bitter arguments about the details of ancient Rome, most notably one about the location of the ancient Roman forum. The extent to which he knew ancient Greek and Latin is a matter of disagreement among scholars. Certainly at the very least, he knew (or taught himself) some Latin. We also know that he had ongoing associations and friendships with learned humanists who helped him in linguistic matters, as he helped them in archeological.[20]

Soon after Giovanni Angelo de Medici (1499–1565) became Pius IV, Ligorio was made chief architect of the Vatican Palace. He designed the building known as the Casino, an architectural gem in the Vatican garden, and decorated it with numerous ancient busts and statues, many of which he had excavated himself. He was also in charge of decorating the so called *Terza Loggia*, or the *Loggia di*

[15]See esp. (Anon. 2005; Coffin 2004, including 145–6, Appendix I: House Facades at Rome Painted by Ligorio; Gaston 1988; Palma Venetucci 2003; Schreurs 2000).

[16](Ligorio 1553). I have used the online edition with commentary by Margaret Daly Davis, http://archiv.ub.uni-heidelberg.de/artdok/volltexte/2008/562 (accessed 22 January 2016).

[17]British Library, Maps* 155 (5), "URBIS ROMAE SITUS CUM IIS QUAE ADHUC CONSPICIUNTUR VETER. MONUMENT. RELIQUIIS PYRRHO LIGORIO NEAP. INVENT." See also (Frutaz 1962, vol. 1, 170–1 and vol. 2, Pianta CXI, Tav. 222), who reproduces the example in the Biblioteca Nazionale di Roma, P.R. 13; (Huelsen 1969, 41–3; Karrow 1993, 349–58; Scaccia Scarafoni 1939, 82, no. 143).

[18]British Library, Maps* 155. (35), See also (Frutaz 1962, vol. 1, 60–1 and vol. 2, Pianta XVI, Tav. 25); Huelsen 1969, 43–4; Scaccia Scarafoni 1939, 23–4).

[19]British Library, C.25.d.9(1.). See (Frutaz 1962, vol. 1, 61–3, vol. 2, Pianta XVII (Tav. 26–32), vol. 3, Pianta XVII a and b 7, Tav. 671–84). On this map, see the classic study by (Burns 1988).

[20]Ligorio is the focus of a growing body of scholarship, including a national edition of his works (*Edizione Nazionale delle Opere di Pirro Ligorio*). For a list of his archeological manuscripts, see (Mandowsky and Mitchell 1963, 130–40). Most of Pirro's manuscript writings are located in Naples, Biblioteca Nazionale, Cod. XIII.B.1–10 (10 vols.); Oxford, Bodleian Library, Cod. Canon. Ital. 138 (1 vol.); Paris, Bibliothèque Nazionale, cod. Ital. 1129=St. Germ. No. 86 (1 vol.); and Turin, Archivio di Stato di Torino ms. a.III.3.1- ms. a.II.5, an alphabetical dictionary of antiquities (18 vols.); and ms. a.II.6. (XIX) – ms. a.II.17 (30)—12 volumes on topics such as coins, medals and, the earthquake of November 16, 1570 that devastated Ferrara. Fourteen of the eighteen volumes of the alphabetical encyclopedia in Turin were copied for Queen Christina of Sweden in the seventeenth century, and are now in the Vatican Library—BAV Ottob. lat. 3364- Ottob. lat. 3377. For the argument about the location of the ancient Roman Forum, see (Laureys and Schreurs 1996).

Fig. 3 Pirro Ligorio, *Urbis Romae situs cum iis quae adhuc conspiciuntur veter. monument. Reliquiis Pyrrho Ligorio Neap. Invent.* (Venice: Michaelis Tramezini, 1552). London: British Library. Etching and Engraving, (40.2×54.5 cm.). Pirro Ligorio's first map showing both contemporary and ancient structures, including those, such as the Circus Neronis on the lower left, which no longer existed in Pirro's time. Copyright © The British Library Board, All Rights Reserved, Maps* 155. (5)

Cosmographia which featured landscapes painted by and maps designed by Stefano Du Perac (to be discussed below). In the early 1560s, Ligorio also seems to have surveyed the barely functioning ancient aqueduct, the *Acqua Vergine* that ended at the Trevi Fountain. He had undoubtedly hoped to receive the contract to repair and reconstruct it. His large, beautiful map of the aqueduct is extant in the Uffizi (Coffin 2004, 83–105).[21]

However, at the death of Pius IV, Pirro lost his position as chief architect of the Vatican. Pius IV's successor, Pius V (1504–1572) was primarily interested in the spiritual reform of the church and the moral rectitude of the city, and was deeply unsympathetic to humanist antiquarianism. As soon as he became pope, he began to get rid of the Vatican collection of ancient busts and statues that Pirro had so assiduously collected. Pirro himself, after a brief imprisonment, found new employment outside of Rome. In the late 1560s he returned to the patronage of Cardinal d'Este and was occupied in helping to build and decorate the Villa d'Este in Tivoli, including its famous garden, entailing an elaborate design and the construction of aqueducts and extensive hydraulic works.[22]

In November 1569, Pirro left Tivoli for Ferrara, entering the service of Alfonso II d'Este (1533–1597), duke of Ferrara. He spent the first years there working on antiquarian collecting and study and became a member of a new academy, and he designed triumphal arches for the entry of Henry III

[21] For the Casino, see (Fagiolo and Madonna 1972; Friedländer 1912; Losito 2000; and Smith 1977). For Pirro's work on the aqueduct, see (Long 2008, esp. 1123–29); and for the map itself, Florence, Uffizi, no. 4236A.

[22] For Pius V, see esp. (Pastor 1951–52; Lemaitre 1994, esp. 201–40). For an introduction to Pirro's garden at the Villa d'Este, see esp. (Coffin 2004, 83–105).

(1551–1589) in 1569. On Nov. 16, 1570, a terrible earthquake destroyed most of the city. As a result, Ligorio wrote a history of earthquakes and designed an earthquake resistant house, a project said to be the first of its kind in Europe. After the earthquake, Ligorio returned to architecture, redesigning part of the duke's damaged palace for the display of antiquities, a new library and many other architectural projects. In Ferrara, Pirro also wrote most of the hundreds of pages of his alphabetic encyclopedia of antiquities and other manuscript books now in Turin.[23]

Whatever his specific early training, in Rome Pirro moved from being a painter to one of the most knowledgeable students of ancient ruins and antiquities in Rome. He was called an *antiquaro*, but also involved in bitter disputes over ancient Roman ruins and antiquities. His eclectic career reached great heights as he became architect of the Vatican, designer of gardens, a hydraulic engineer, and finally, investigator of natural phenomena such as earthquakes. Undoubtedly he was substantially an autodidact. Yet his writings consistently emphasize not his practical knowledge, but rather his learning. The above-mentioned alphabetic encyclopedia in Turin is filled with references to ancient Greek and Roman texts. It also contains numerous personal observations and drawings of structures that at least initially seem to have been made on site. The encyclopedia also contains numerous transcriptions of inscriptions found in various locations, and drawings of hundreds of coins, usually both sides. Indeed his famous large map of ancient Rome published in 1561 is crowded with depictions of ancient buildings, largely derived, as Howard Burns has shown, from ancient coins (Burns 1988, 27–31).[24]

4 Stefano Duperac: Painter, Engraver, Cartographer, Antiquarian, Garden Designer, Architect

Étienne Dupérac (in Italian, Stefano Duperac) was a painter, designer and engraver who came from Paris. In the late 1550s he traveled to Italy and, after a fruitful stay in Venice, worked in Rome for 20 years between 1559 and 1578. While he lived in Rome he engraved numerous images of ancient monuments and other sites in the city, many of them published by the printer, Antonio Lafreri (1512–1577) (Ehrle 1909, Introduzione; Witcombe 2008, 107–221). Between 1561 and 1563, he designed and painted the 13 landscapes in the frieze of Pius IV's loggia in the Vatican, and he designed the maps as well. During his work on the loggia he would have met Pirro Ligorio who at that time was architect of the Vatican Palace and creator of the complex iconography of the loggia decoration. Both Duperac and Ligorio, who were probably friends or at least acquaintances, traveled in the circle of antiquarian scholars anchored by the humanist Augustinian monk Onofrio Panvino (1530–1568) and by Fulvio Orsini (1529–1600).[25]

The extensive collaboration between the humanist Onofrio Panvinio and Duperac was based on reciprocal interests and needs. Panvinio had developed an extensive knowledge of ancient Roman and ecclesiastical history, which he shared with Duperac. Panvinio was a great admirer of Benedetto Egio and Pirro Ligorio. As Emmanuel Lurin has argued, Duperac's skill at image making was crucial to Panvinio's antiquarian studies. Duperac, on the other hand, used his skill as a painter and an engraver, but also the antiquarian learning that he acquired from Panvinio to ascend the social and professional

[23]The tract on the earthquake is Turin, Archivio di Stato di Torino, Cod. a.II.15.J.28, "Libro, o Trattato di diversi Terremoti raccolti da diversi Autori per Pirro Liborio mentre la Città di Ferrara à stata percossa, et ha tremato per un simile acciolente del moto della terra;" and (Ligorio 2005). See also (Martin 2011, 74–6). For an overview of Ligorio's career in Ferrara, see (Coffin 2004, 107–37).

[24]See also (Campbell 1988).

[25]The definitive study of Duperac including a complete catalog of his prints and drawings is (Lurin 2006, 1: 13–5) for a summary biography, and (Lurin 2006, 1: 39–237) for his extensive collaboration with Onofrio Panvinio. See also (Bragaglia Venuti 2002 [2003]; Bragaglia Venuti 2006; Ferrary 1996).

ladder. As will be discussed, he ending up in the exalted position of architect to King Henry IV of France (1553–1610).[26]

Duperac created three maps of Rome, two of the ancient city and one of the modern. The first, *Specimen, seu perfecta urbis antiquae imago* (Form or Perfect Image of the Ancient City) is a small (60×46 cm.) image of the city depicting only its ancient features. On the left is an address "studioso Lectori" (to the studious reader) by "Stefanus Duperac architect[us]," which emphasizes that the image of the ancient city was created from information from ancient writers and from remains that have survived to the present, which have been "very accurately drawn."[27] Seemingly having gained confidence, Duperac next created a large plan of the ancient city, which was first published in 1574, perhaps in anticipation of the Jubilee that pope Gregory XIII (1502–1585) had planned for the next year.[28]

From this image of ancient Rome, Duperac turned to the creation of a plan of the modern city. The first edition of Duperac's panoramic view of contemporary Rome, *Nova Urbis Romae Descriptio* (New Description of the City of Rome) was published by Antonio Lafreri or his shop in 1577 (Fig. 4). The spectacular view of the contemporary city, composed on four sheets and dedicated to the French King Henry III, is slightly more than a meter wide and 820 mm high. It shows a birds-eye view of the city nestled in a sweeping landscape of the surrounding countryside with roads leading to and through the Bufalini-inspired walls, which stand out prominently. As Franz Ehrle points out, many buildings, churches, ancient monuments, but also houses, are displayed individually as they appeared, rather than in stereotypical, identical forms. There is a precision of detail and individualization that is notable, and by far the greatest number of buildings that are precisely drawn and labeled are churches and basilicas, which are labeled as Temples.[29]

Emmanuel Lurin has cogently tied the 1577 plan to Duperac's collection of views of ancient monuments, *Vestigi dell'antichità di Roma* (Vestiges of the Antiquities of Rome), first published in 1575. The creation of both plan and collection of views was guided by two seemingly contradictory principles: the careful analytical description of urban space as viewed, but also an evocative living portrait of the city (Lurin 2008, esp. 51–7).

[26](Lurin 2006, 1: 39–237), for his extensive collaboration with Onofrio Panvinio. For Panvinio's ideas about images, see (Lurin 2006, 1: 77–89; 2007). See also (Ferrary 1996, 26–38) for Panvinio and Duperac, and (Ferrary 1996, 55) for Panvinio's friendship with Ligorio.

[27](Frutaz 1962, vol. 1: 66–7 and vol. 2: Pianta XXI, Tav. 36,"accuratissime delineate;" Huelsen 1969, 58–60; Lurin 2008; Scaccia Scarafoni 1939). The map can be found in the Biblioteca dell'Istituto di Archeologia e Storia dell'Arte in Rome (Roma XI.16.3.10), as well as in several Speculum Collections as listed in Huelsen.

[28]London, British Library, 23810 (4) for the fragment with the privilege from Gregory XIII; and for the complete map, Maps 155 (7), *Urbis Romae Sciographia ex Antiquis Monumentis Accuratiss. Delineata*." (Huelsen 1969, 60–1) identifies this example in the British Library as the unique example of the first 1574 printing. However, as (Frutaz 1962, vol. 1: 67–8 and vol. 2: Pianta XXII, Tav. 37–45) pointed out (1: 67), the phrase "Romae excudebat Franciscus Villamena" on the upper right hand corner of this example indicates that it is a reprint, since Francesco Villamena (or Villamoena) (1565–1624) did not arrive in Rome until the late 1580s during the reign of Sixtus V. Frutaz reproduces a second reprint by Giovanni Giacomo De Rossi from the mid-late seventeenth century. See also (Aragozzini and Nocca 1993, 38–9 and Tav. 8), who identify the original publisher as Vaccheria. It can be assumed that no example of the first printing is extant. The plates are conserved at the Calcografia Nazionale di Roma (now the Istituto Nazionale per la Grafica). See (Petrucci 1953, 126, n. 1439).

[29]London, British Library, 23805. (8.). See also (Bury 2001, 136–7; Ehrle 1909) for a facsimile reprint with an introduction (25–33) that includes a complete list of labeled items that are on the map; (Frutaz 1962, vol. 1: 186 and vol. 2: Pianta CXXVII, Tav. 247–55; Huelsen 1969, 66–8, including the reprint editions). For a summary, see (Aragozzini and Nocca 1993, 50–1 and Tav. 7). (Lurin 2008, 50–1) suggests that the Latin dedication to Henry III was written by the humanist Giovanni Francesco Peranda, secretary to the Caetani family for whom Duperac worked regularly in the 1570s and that the printing was delayed by waiting for Perenda to compose the Latin dedication and for the French court to approve it.

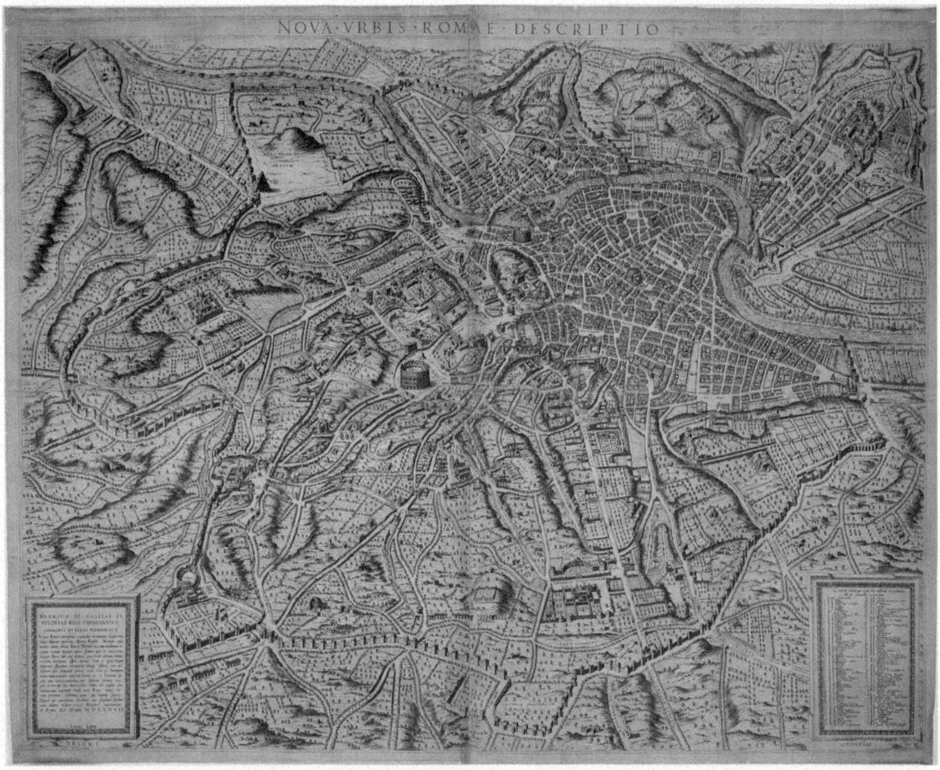

Fig. 4 Étienne Duperac, *Nova Urbis Romae Descriptio* (Rome: Antonio Lafreri, 1577). London British Library Etching (79.4 × 100.7 cm.), a depiction that carefully depicts contemporary buildings and the Roman countryside, including the roads leading into the city, relevant to pilgrims. Copyright © The British Library Board, All Rights Reserved, Maps *23805

Duperac dedicated several of his maps to French sovereigns, perhaps indicating that he wanted to return to his home country. As Lurin points out, in these dedications, he presents himself as the creator of images and as a specialist in ancient architecture. He finally did return to France and in 1578 is documented as being in Caen in northwestern France where he was employed by the municipality to inspect and survey a recently constructed wall. He moved with his family to Paris and his subsequent employment is notable for the high social status of his patron/employers. He designed a number of elaborate tombs, hired by various French dukes—one for cardinal Charles de Vaudémont, for which he was hired by Duke Philippe-Emmanuel de Lorraine (1558–1602) for a church in Nancy in north-east France. He also designed and built a tomb for René du Rochechouat (1528–1587) at Poitiers in the central-western part of the country, and finally the tomb of Marie de Billy at Goussainville, now a suburb of Paris (Lurin 2006, 1: 602–3 and 606–17).

Primarily however, Duperac worked in France as an architect and designer of gardens. In 1582 he was working for the Duke of Aumale, Charles de Guise (1555–1631) in charge of the architecture of the duke's houses and chateaux, including the Châteaux d'Anet in Normandy, for which he helped to design the gardens. In the late 1580s, he was employed by Catherine de Medici (1519–1589) where for her property (the "Hermitage" in the village of Chaillot near Paris), he designed a villa with a hippodrome in antique style. His design, which was never executed, was influenced by the investigations of Pirro Ligorio and Panvinio on ancient circuses. Duperac also worked as an architect and garden designer at the Chateau and gardens at Tuileries. Finally he became the "architect of the king" and in this capacity became the primary creator of the gardens built by Henry IV at Châteax-Neuf de Saint-Germain-en-Laye, now in the western suburbs of Paris (Lurin 2006, 1: 618–89).

Étienne Duperac, one of the most prolific engravers of late sixteenth-century Rome, completely abandoned his work of engraving when he returned to France. As Lurin notes, this renunciation of the most artisanal part of his work was certainly voluntary, and was a sign of his social rise and intellectual ambitions (Lurin 2006, 1: 603). While in France he is called a painter, an architect—and on two occasions, an engineer (*ingenieur*)—but never an engraver (Lurin 2006, 1: 614 and 638). His social rise is confirmed by the marriage of his daughter Arthémise (an unusual name referring to an ancient Greek ruler—Artemisia of Caria, the Queen of Halicarnassus) to a member of "*la petite noblesse parisienne*," Jean Bourdin (Lurin 2006, 1: 694–5). What is clear is that Duperac achieved social elevation by means of his Roman experience of becoming an antiquarian with the help of Pirro Ligorio and Ottavio Panvinio and combining his knowledge of antiquity with the skilled craft of painting and engraving, and later in Paris, the design of buildings and gardens.

5 Mario Cartaro: Engraver, Printer, Publisher, Cartographer, Architect, Engineer

Far less is known about Mario Cartaro than about his contemporary Duperac. Cartaro was from the city of Viterbo not far from Rome. He was an engraver who arrived in Rome around 1560, when he would have been about 20 years old. V. Federici first suggested that the name Cartaro was not a true surname, but referred (with a Roman dialect ending denoting emphasis) to his skill as a cartographer. When he arrived as a young man, the Roman print business was not only flourishing, but engendered bitter competition and rivalries. Cartaro became an engraver and printer, working with (among others), Antoine Lafréry, the most powerful and prolific printer in Rome. Lafréry, who italianized his name to Antonio Lafreri after he arrived from the Franche-Comté in eastern France, died intestate under suspicious circumstances in 1577. During the conflict that ensued over the inheritance of the hundreds of valuable plates and prints in his shop, a young engraver in the shop, Gerolamo Modenese was murdered. Cartaro was among those initially suspected of being an accomplice in the group around Lafreri's rival printer, Lorenzo Vaccari (fl. 1574–1608). Although the murder was never solved, the investigative committee, the *Commissario della Camera*, gave Cartaro the task of dividing Lafreri's estate with its large number of prints and plates, indicating perhaps, his good standing in the wider community, and/or his intimate knowledge of the print business.[30]

Cartaro's career in Rome involved many aspects of the flourishing print culture, including engraving, printing, and publishing. He himself was a prolific printmaker. Over a hundred of his prints are known, not only of maps but of other kinds of images—copies of well-known paintings, including religious images, an elaborate framed portrait of Pius V, and two views of the Vatican palace and gardens. He seems to have run a flourishing print business from his own shop and hired other printers and engravers, including his younger brother Cristofano. He also created large maps of ancient and contemporary Rome. His other cartographical works include matching terrestrial and celestial globes.[31]

Although more work is needed on Cartaro and his career, his large map of ancient Rome, published in 1579 offers a window into some of his interests (Fig. 5).

[30]For Cartaro, see (Borroni 1977, vol. 20, 796–99; Federici 1898; Maier 2015, 143–52). For the murder, see especially, (Witcombe 2008, 166–9).

[31]For Cartaro's career in Rome, see (Almagià 1913; Cattaneo 1998, 2000). A signed celestial globe is in the Museo Galileo in Florence (inventory no. 123), and the terrestrial at the Observatory of Monte Mario in Rome. See http://www.mhs.ox.ac.uk/epact/catalogue.php?ENumber=87909 (Accessed 23 January 2016).

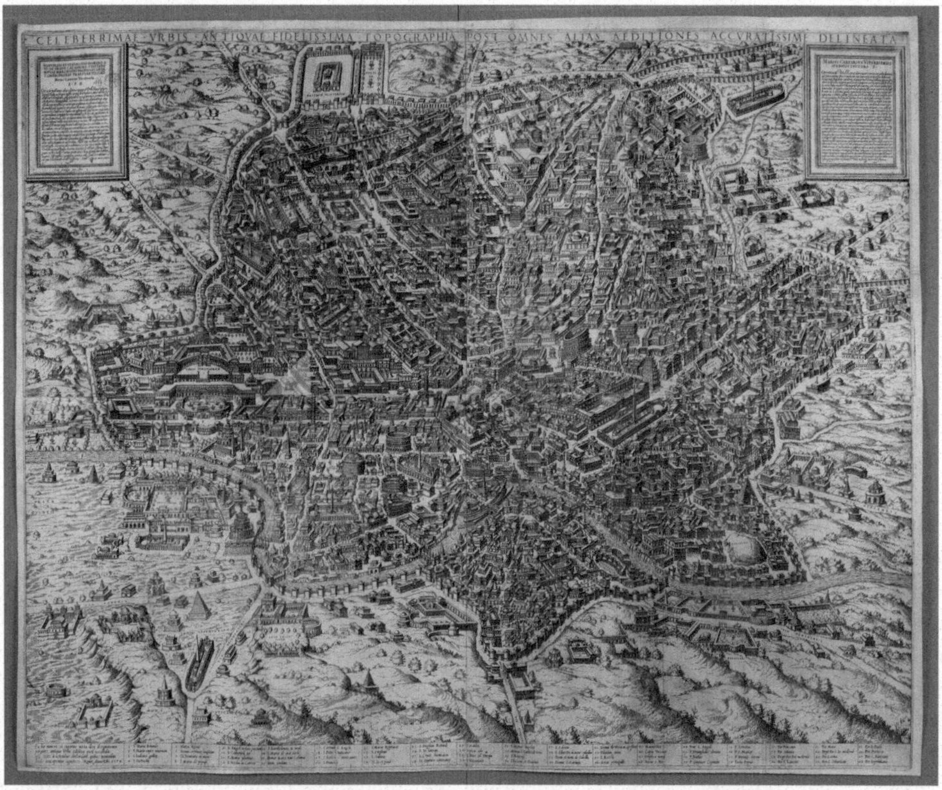

Fig. 5 Mario Cartaro, *Celeberrimae Urbis Antiquae Fidelissima Topographia Post Omnes Alias Aeditiones Accuratissime Delineata* (Rome: Mario Cartaro, 1579). London: British Library Etching (91 × 113 cm.). Cartaro shows the ancient city built out to the Aurelian walls with beautifully intact buildings. Copyright © The British Library Board, All Rights Reserved, Maps* 155. (10.)

It is an ichnographic plan, following Bufalini, which also includes the elevation of buildings. It is oriented in the usual way with the north to the left and Vatican Hill at the lower left. While most of the city is packed with ancient buildings, major roads into the city are particularly visible. Sixty-eight numbers on the lower edge of the map indicate buildings—often the sites of modern churches that are not actually depicted—e. g., 18, a number found inside the Circus of Gaius and Nero indicating "S. Petri in Vaticano," and as Cartaro explains on the far left of the numbers, they are there to tell us where the modern buildings are when their ancient counterparts had diverse names. The plan of ancient Rome is crowded with intact ancient buildings inside of and (in contrast to Cartaro's 1576 plan of contemporary Rome) right up to the Aurelian Walls with only a small amount of relatively empty countryside outside of the walls. In a plaque on the upper left, the plan is dedicated to "Giovanni Giorgio Caesariano" (Giovangiorgio I Cesarini, 1550–1585), marchese of the town of Civitanova in the Marches on the Adriatic Sea, and gonfalonier of the Roman commune. Cesariano, the dedicatee, was a member of an upwardly mobile family who had ascended to Roman nobility at the end of the fifteenth century, by claiming ancient Roman roots, exploiting their office of standard-bearer, building a Roman palace and collecting antiquities and paintings. Cartaro, who was attempting to hang on his coat-tails, tells Caesariano that the plan has been made "with high study and not small expense by our men." Yet, it would "be viewed unfavorably by the bites of detractors and the envious." In order to be defended "from the perverseness of these times," it needed the protection of the patronage of a powerful person, namely Caesariano.[32]

On a plaque in upper right corner of the plan, Cartaro addresses the "Studious Reader." Here he notes that since the boundaries of the ancient city are uncertain, he has enclosed it within the extant walls, although he believed it had been larger before the destructions of the barbarians. He emphasizes that: "we will have situated ancient buildings in this manner with very true intervals, as much as was it was possible to gather from the appearances of the vestiges today." He continues, "you will have therefore Rome both Old and New [i.e., his large plan of contemporary Rome published in 1576], with two plates having been fashioned by us of the same magnitude." Thus with the same measures being observed in both, one can compare the two cities, ancient and modern, at a glance. Finally, Cartaro points to the care with which the plan has been made, "having been supported with the study of very many learned men, [and] having been strengthened with the diligent inspection of all ancient fragments which are extant with the use of mathematical instruments," and not only buildings on their sites but also clearings, forums, plazas, streets and alleys.[33] Cartaro here emphasizes the importance of antiquarian research as a collaborative effort entailing personal observation among learned men (including, by implication, himself). The work of the map also involved the use of mathematical instruments, suggesting the practices of mathematical measurement and surveying.

Unlike Duperac who turned away from engraving (and the cartography that accompanied it), Cartaro increasingly concentrated on mapmaking, as he also turned away from Rome in favor of the south. He made maps of places such as the Island of Ischia (1586) and the Pozzuoli region in Campania (1588). Around 1589, he moved to Naples and there was made *ingegnere delineatore* (engineer draftsman) and cartographer to the *Giunta dei Regi Lagni*, that is, the council in charge of the canal system of Naples. He held this post from 1589 until his death in 1620, and made a series of drawings of the entire Royal canal (no longer extant). In addition, he joined the mathematician and cartographer Nicola Antonio Stigliola (1546–1623) who had been occupied from at least 1583 with a chorographical survey of the entire Kingdom of Naples. When Cartaro joined this ambitious cartographical project he was named the *tabulario* (surveyor) to the Royal Chamber. A set of colored manuscript maps of the kingdom of Naples as a whole and then each of its 12 provinces from this project (drawn between 1590 and 1594) are housed in the Biblioteca Nazionale di Napoli. In the interests of state secrecy, they were not printed or published (in contrast to Cartaro's Roman maps). After 1593, Cartaro was named "*Ingegnere della Regia Corte*" (Engineer of the Royal Court), where he was sent on various missions (concerned with fortification, for example) along with others such as the architect/engineer Domenico Fontana (1543–1607). It is notable that Mario Cartaro's son, Paolo followed his footsteps as a cartographer and draughtsman in the employ of the kingdom of Naples.[34]

[32] London, British Library, Maps* 155. (10.), *Celeberrimae Urbis Antiquae Fidelissima Topographia Post Omnes Alias Aeditiones Accuratissime Delineata* (Rome: Mario Cartaro, 1579), upper left plaque: "summo studio ac suptu nostris viribus non mediocri," "non bene tutam detractorum atque invidorum morsibus," "a temporum horum privitate;" See (Frutaz 1962, 1: 68–9 and 2:Pianta XXIII, Tav. 51–5; Huelsen 1969, 68–70; Maier 2012b, esp. 742–7; Maier 2011, esp. 46–50, 2015, 143–52; Rocchi 1902, vol. 1: 121–72 and vol. 2: Tav. XXIII). For the Cesarini family in Rome, see (Sickel 2007).

[33] London, British Library, Maps* 155. (10.), *Celeberrimae Urbis Antiquae Fidelissima Topographia*, Upper right plaque: "Studioso Lectori," "eadem [i.e., aedificia] suis propriis locis eo pacto situaverimus, ut verissimis intervallis quantum ex apparentibus hodie vestigiis colligi potuit," "Habes itaque Romam et Antiquam et Novam duobus tabellis eiusdem magnitudinis," "plurimorum doctorum hominum studio suffulti et diligenti inspectione omnium que extant antiquorum fragmentorum roborati, mathematicorum instrumentorum usu."

[34] For Cartaro's activity in Naples and his work with Stigliola, see esp. (Almagià 1972, 1: 87–100, and 2: XI–XXIII), for facsimiles of the maps—the originals of which are in Naples, Biblioteca Nazionale di Napoli, Ms. xii.D.100)—(Marino 1992, esp. 12–3; Palagiano 1974; Valerio 1982, 1993, esp. 50–8, 2007, esp. 962–70).

Like Duperac, Cartaro moved from the highly competitive and vibrant print culture of Rome to the employ of noble patrons elsewhere—in Cartaro's case that of the rulers of the Kingdom of Naples. Unlike Duperac, he did not actually give up engraving. But he did seem to expand his activities of surveying and cartography, as he also gained official positions in the royal bureaucracy and expanded his activities to such matters as the inspection of canals and fortification.

6 Egnazio Danti: Painter, Mathematician, Instrument Maker, Cartographer

Dominican mathematician and cartographer, Egnazio Danti came from an eclectically distinguished family of Perugia who collectively combined artisanal skills such as painting, sculpture, and goldsmithing with mathematics, instrument making, and writings of various kinds. It is useful, I think, to see him in the context of his upwardly mobile, artisanal family.[35] The family was originally called the Rinaldi family. However, Egnazio's grandfather, Piervincenzo, changed the family name to Danti out of admiration for the great Italian poet. A goldsmith with great family ambitions, Piervincenzo translated into Italian John Sacrobosco's *On the Sphere*, a widely-used introduction to geocentric cosmology. Egnazio's aunt Theodora (fl. early sixteenth century) was a painter who had studied under Perugino (circa 1446–1523) and who also made instruments and was known for her love of astronomy and mathematics. His father Giulio Danti (1500–1575) was an architect, goldsmith, and metal caster, who had worked on the fortifications of Perugia under Pope Paul III, and was reported to have written on the flooding of the Tiber River and on architecture. His older brother Vincenzo Danti (1530–1576) was a leading sculptor of the Florentine mannerist style who is considered one of the most distinguished sculptors of the sixteenth century. Vincenzo wrote an (unfinished) treatise on anatomy and human proportions for which he claimed to have carried out 83 human dissections. He was also a highly skilled poet, and a member of Florence's literary academy, the Accademia Fiorentina. Egnazio's younger brother Girolamo (1547–1580) was a goldsmith and painter.[36]

In 1555 after attending the *studio* in Perugia, Egnazio entered the Dominican order at the age of 19. Unlike the other men treated here, he acquired a Latin university education. He spent 7 years in the Dominican *studio* of Perugia, which was the center of Dominican study in the Roman province, and contained a library that included many manuscripts on perspective, optics, and astronomy, which he studied and absorbed. Egnazio followed the traditional course of study which included 3 years of philosophy (dialectic, philosophy, and metaphysics), and 4 years of theology centered on the *Summa* of Thomas Aquinas. He had previously learned painting from his father and had become a skilled mathematician. In 1562 he joined his brother in Florence where he was commissioned by duke Cosimo I to create and paint a set of maps of the world on the cabinet doors of the Guardaroba of the Palazzo Vecchio, and to create a large terrestrial globe for the center of the room. During his time in Florence, Danti also constructed astronomical and other kinds of instruments, such as his anemoscope used to measure the direction of the winds, and a quadrant and armillary sphere mounted on the church of Santa Maria Novella. The instruments provided data such as the equinox and the length of

[35] For Egnazio Danti, see (Fiorani 2005, esp. 42–59, 100–2, 135–6, 155–7, 171–202) and for the *Terza Loggia*, (Fiorani 2005, 29–30, 141–2, 236–41, 250–1); and see also esp. (Dubourg Glatigny 2002; Fiore 1986a; Tabarroni 1996).

[36] For Sacrobosco's *Sphere*, see (Chap. 16). For Giulio Danti and Piervincenzo Danti, see (Fiore 1986b, vol. 32: 664–7 (Giulio), 667 (Piervincenzo)). For Girolamo Danti, see (Sapori 1986). For Vincenzo Danti, see esp. (Fiore 1986c; Summers 1979, 1996; Cole 2011; and Cole and Gamberini 2016). Other details about the family can be found in the thorough account of (Dubourg Glatigny 2002, esp. 547–55). For Giulio Danti's writings, see (Oldoini 1678, 198, Cum edidisset Opus de Alluvione Tyberis, et Annotationes in ornamenta Architecturae.)

the year and were meant to assist Cosimo in his (unfulfilled) ambition to reform the Julian calendar. Danti also taught mathematics and cosmography to the children of his patrons.[37]

Cosimo's death and the dislike that Cosimo's son and successor, Francesco I de' Medici (1541–1587), had for Egnazio forced him to leave Florence. In 1576 he moved to Bologna where he became professor of mathematics and continued his prolific activity of writing and published books on practical mathematics.[38] He continued his cartographical efforts as well. He surveyed Bologna and the surrounding countryside and created maps. He also returned to visit his native city of Perugia several times, creating a map of the city while also undertaking extensive land surveys of areas of Campania, Lazio, and Urbino. While in Bologna he resumed the manufacture of anemoscopes, as well as other instruments (Dubourg Glatigny 2002, 577–96).

Gregory XIII called Egnazio to Rome, not specifically for cartographical reasons, but to assist in calendar reform, which he was well equipped to do. The pope also enlisted him in cartographical activities, beginning with the assignment of completing the third Loggia that had been undertaken by Pius IV. Danti did so by designing a world map divided into two hemispheres, and creating ten maps of Africa, Asia, and America. In addition, the wall decoration included city views to compliment the world maps—which have not survived. He also designed the pope's Tower of the Winds, a tower created in imitation of the ancient tower of the winds in the agora of Athens. Danti's sixteenth-century version measured both time and the direction of the winds (the later with an instrument invented by Danti himself, the anemoscope).[39]

The pope also ordered the creation of the *Sala Bologna*, which served as his dining room and which included a monumental perspective plan of Bologna on the south wall and a huge map of the Bolognese countryside on the western wall. The maps were created by Danti, who had surveyed the Papal States, including Bologna in 1578. The maps of the *Sala Bologna* worked in tandem with accompanying paintings that formed a complex painted iconography that served as, to cite Francesca Fiorani, a "visual synthesis of the mixed sciences of map-making, perspective, and astronomy, the so-called *scientiae mediae* that made it possible to transfer the abstract principle of Euclidean geometry into practice" (Ceccarelli and Aksamija 2012; Fiorani 2005, 140–70: 149).

The second great map cycle of the Vatican palace was the famous Gallery of Maps, also commissioned by Gregory XIII and also created by Danti. He made the cartoons for the maps, and supervised a large number of painters. The gallery, which was largely completed in 1581–1582, is constructed so that the rectilinear axis of the corridor stands for the central Apennine ridge of the Italian peninsula. It is constructed so that the viewer walking down the center of the hall begins in northern Italy and walks to the south. The forty maps, each about three by four meters, display the regions of Italy, four Mediterranean Islands (Malta, Corfu, the Tremiti, and Elba), four ports (Genoa, Venice, Ancona, and Civitavecchia), and the papal territory of Avignon in southern France. Each map was provided with a scale, a wind rose, coordinates of latitude and longitude and one or more legends explaining the peculiarities and history of the particular territory (Almagià 1944–1955, vol. 3; Fiorani 2005, esp. 171–207).[40]

[37]For the Guardaroba of the Palazzo Vecchio, see esp. (Fiorani 2005, esp. 16–59) and for his mathematical activities, (Dubourg Glatigny 2002, 555–77); and see also (Heilbron 1999, esp. 62–8) for the instruments.

[38]For the broader context of practical mathematics in this period from the point of view of a particular genre, see (Chaps. 13 and 16).

[39]For the maps of the *Terza Loggia*, see (Almagià 1944–1955, vol. 4; Fiorani 2007, esp. 816–8, 2005, esp. 237–44; Schulz 1987, esp. 101–7). For this and Danti's other activities in Rome, see (Dubourg Glatigny 2002, 596–605). For Danti and calendar reform, see especially (Casanovas 1983) and for the Tower of the Winds with much material on Danti, (Courtright 2003).

[40]For color reproductions of the maps, and accompanying essays, see (Gambi and Pinelli 2008). Many of the maps were later touched up and some were completely replaced in the seventeenth century, including the map of Lazio with its inserted image of Rome with the Sistine obelisks in place (painted by Luc Holste between 1636 and 1637).

It is evident that Danti in one important way can be distinguished from the previous four individuals in that he had received a formal Latin education. Yet he was also in the most fundamental ways, a mathematical practitioner who could and did survey territory and make maps as well as mathematical instruments. And he was also a trained painter. Danti's many skills and activities can be attributed to his own extensive studies and abilities in mathematics and instrument making, but also to his eclectic family background that provided him, so to speak, with practical and artisanal knowledge that he was able to use in a variety of ways.

7 Conclusion

In his fine study of the culture of the professions in the late Renaissance (centered on Tomaso Garzoni's (1549–1589) *Universal Piazza of all the Professions of the World*), George McClure documents a growing fluidity in matters of professional identity during the fifteenth and sixteenth centuries. This development is accompanied, he shows, by a trend favoring occupational choice, rather than the assumption of a predetermined craft or profession (traditionally determined by the family). He also documents the growing importance of artisanal occupations within a variety of popular discourses such as books of jokes and games, and encyclopedias. These developments, he suggests, are associated with other trends such as the growing commercialization of the book market.[41] In one sense then, the professional fluidity of the five men discussed in this paper can be seen as part of a more general trend, especially noticeable in the late sixteenth century.

This fluidity of professional identity and the rising status of at least some artisanal crafts surely facilitated the varied activities of the five men discussed in this paper—and there is no reason to think that their mobility and the variety of tasks that they undertook were highly unusual. Still, diversity of tasks and mobility were probably facilitated by the circumstance that many of the occupations discussed here—engineering, architecture, printing, engraving, surveying, and the like, were either relatively new and/or not usually encompassed by traditional guilds.

It is well to keep in mind in addition that in Rome itself and in many other areas of Europe, the skills possessed by those working in such occupations were vitally needed in essential ways in many of the flourishing businesses and industries of the late sixteenth century. They include building construction, driven by an expanding population and an increased desire for conspicuous consumption and luxury; the great development of theater, as Viktoria Traczyk has discussed in this volume (Chap. 4); the flourishing business of printing of both books and single-sheet engravings including maps; the thriving practice of cartography, which fed the demand for maps of all kinds—stimulated in part by the discovery of new lands; the industries of mining and the concomitant metallurgical crafts, including minting and goldsmithing; the production of luxury fabrics; and the great expansion of gunpowder artillery, discussed by Jochen Büttner also in this volume (Chap. 5). In a recent essay, Marcus Popplow has pointed to the importance of an expanding and increasingly intricate technological culture of early modern Europe, characterized by both formalization and interaction, as key both to the material culture of the period and to the culture of knowledge (Popplow 2015). Certainly the expanding and increasingly intertwined and intricate realms of technological and practical culture contributed to the flexibility and mobility of certain kinds of practitioners as it also tended to stimulate the development of trading zones.

Returning to Rome itself, an important characteristic of the sixteenth-century city was the presence of avid antiquarianism entailing the investigation of ruins and other antiquities. Antiquarian investigations by humanists and others made Rome a particularly important "trading zone" between practitio-

[41] (McClure 2004); and the book's main focus, (Garzoni 1587).

ners and learned men. The study of antiquities was every bit as empirical as was the developing field of natural history and its sub-disciplines, such as the study of plants.[42] Antiquarian investigation required searching for and digging for "specimens," such as coins, medals, statues, and the ruins of ancient buildings. It required careful observation and measurement, including techniques of surveying and other mathematical techniques such as those needed for cartography. And it required a thorough knowledge of ancient texts, such as those of Pliny, Frontinus, and Vitruvius as well as of ancient histories and other texts that mention events that took place in particular locales. It required the ability to read and transcribe inscriptions, even partially mutilated ones, and to think of ruins in terms of the whole structure of which they were the remains. It also demanded an understanding of topography and an ability to compare ancient and contemporary topographical features. Learned humanists who were avid antiquarians often did not have all the requisite skills to carry out their investigations to the full and the same was true of painters, builders, and other practitioners with similar interests.

Another factor crucially important to the Roman context (and beyond as well) was the growing importance of graphic representation. If one was investigating a coin, a metal, a statue or part of a statue, a fragment of a column, or an inscription, a crucial part of its study was to describe it carefully and to draw it. The craft of drawing and graphic design was possessed by each of the five protagonists of this paper. They used this skill both in the study of antiquity and in the transmission of information and discussion of the discoveries either in books or in single sheet engravings. Not only were individual objects depicted but entire structures (such as the Column of Trajan or the Baths of Diocletian). The importance of graphic description extended to topography as we have seen. Cartographical depictions of the whole city—the ancient city as it was thought to have been, the contemporary city, and a combination of the two—proliferated in the second half of the sixteenth century. All of the individuals who are the focus of this paper were engaged in making maps at one time or another. Such maps were almost always collaborative projects.

A second form of graphic practice was writing. Pirro Ligorio and Egnazio Danti wrote extensive tracts and treatises. That Pirro's writings for the most part failed to be published certainly was through no lack of desire on his part. The other three—Bufalini, Duperac, and Cartaro, did not write treatises or other printed books. But they communicated forceful messages on the cartouches of their maps as we have seen.

In conclusion one can ask whether bricolagic practices such as those described in this paper led to the development of the modern professions such as architecture, engineering, or even archeology. I would say, only in a very indirect way. The development of these modern professions belongs more to the late-eighteenth and nineteenth centuries than to the sixteenth. If anything, in addition to acquiring educational, training, and licensing requirements, the modern professions lost some of the occupational fluidity that has been depicted in this paper. In the sixteenth century, this fluidity allowed for occupational change and rising social status among certain practitioners, and it facilitated communication between practical and learned cultures.

And what about the development of knowledge? Diverse as they were, these men and others like them produced genuinely new knowledge. Their maps contained a wealth of newly synthesized information and were created by the deployment of a variety of technologies involving surveying and measurement. The archeological information that they produced would serve as the basis for the development of modern archeology, no matter how much of it was revised and corrected. Cartography developed apace from the sixteenth century, in Rome but of course, far beyond Rome as well. And Danti invented new instruments and wrote numerous tracts explaining both old and new instrumental and mathematical methods.

[42](Chap. 15) is also relevant to antiquities, especially in its description of the complex mix of individuals involved and the varied cultural uses to which stable material objects (in this case woodblocks) were put.

As important as such substantive contributions however, was the development of approaches which integrated traditional learning with empirical methodologies. What in late sixteenth century Rome made a particular archeological fact true, or a map a realistic representation? The answer involved a combination of learned textual references and observational descriptions, including measurements and mathematical surveys. Knowledge came to be based on both learned argumentation often based on ancient Greek and Roman texts, and empirical methodologies of investigation. This combination of approaches facilitated bricolagic practices and fluid occupations in sixteenth century Rome. It is also true that the fluidity of occupational categories encouraged participation in trading zones between learned and practically trained men.

References

Adams, N., and S. Pepper. 1994. *The architectural drawings of Antonio Da Sangallo the younger and his circle, Fortifications, machines, and festival architecture* ed. N. Admas. vol. 1, 61–74. New York: Architectural History Foundation and Cambridge, Mass.: MIT Press.

Almagià, R. 1913. Intorno a un cartografo italiano del secolo XVI. *Rivista Geografica Italiana* 20 (1): 99–112.

Almagià, R. 1944–1955. *Monumenta cartographica vaticana iussu Pii XII P. M. consilio et opera procuratorum Bibliotecae Apostolicae Vaticanae*. 4 vols. Vatican City: Biblioteca Apostolica Vaticana.

Almagià, R. 1972. Studi storici di cartografia napoletana. Ch. 3: Una carta del Regno di Napoli per iniziative del Governo. In *Cartografia generale del Mezzogiorno e della Sicilia*, ed. E. Mazzetti, vol. 1, 87–100, and 2, XI–XXIII. [Naples]: Edizioni Scientifiche Italiane.

Anonymous. 2005. Ligorio, Pirro. In *Dizionario biografico degli italiani* vol. 65: 109–114.

Aragozzini, G., and M. Nocca. 1993. *Le piante di Roma dal Cinquecento all'Ottocento*. Rome: Dino Audino Editore.

Borroni, F. 1977. Cartaro, Mario. In *Dizionario biografico degli italiani* vol. 20: 796–799.

Bragaglia Venuti, C. 2002 (2003). Étienne Dupérac e i paesaggi della loggia di Pio IV. *Rivista dell'Istituto Nazionale d'Archeologia e Storia dell'Arte*, III ser., 25: 279–310.

Bragaglia Venuti, C. 2006. Étienne Dupérac and Pirro Ligorio. *Print Quarterly* 23 (4): 408–413.

Bufalini, L. 1560. *Roma*. Rome: Antonio Blado.

Burns, H. 1988. Pirro Ligorio's reconstruction of ancient Rome: The *Anteiquae Urbis Imago* of 1561. In *Pirro Ligorio: Artist and antiquarian*, ed. R.W. Gaston, 19–91. Milan: Silvana Editoriale.

Bury, M. 2001. *The print in Italy, 1550–1620*. London: British Museum Press.

Campbell, I. 1988. Pirro Ligorio and the temples of Rome on coins. In *Pirro Ligorio: Artist and antiquarian*, ed. R.W. Gaston, 93–120. Milan: Silvana Editoriale.

Casanovas, J. 1983. The Vatican tower of the winds and the calendar reform. In *Gregorian reform of the calendar*. Proceedings of the Vatican conference to Commemorate its 400th anniversary, 1582–1982, eds. G.V. Coyne, M.A. Hoskin and O. Pedersen, 189–198. Vatican City: Specola Vaticano.

Cattaneo, A. 1998 and 2000. Mario Cartaro, incisore viterbese del XVI secolo. *Grafica d'arte: Rivista di Storia dell'Incisione Antica e Moderna e Storia del Disegno*. 9 (35): 2–9; 11 (41): 6–14; and 11 (42): 3–11.

Ceccarelli, F., and N. Aksamija. 2012. *La Sala Bologna nei Palazzi Vaticani: Architettura, cartografia e potere nell'età di Gregorio* XIII. Venice: Marsilio Editori.

Ceen, A. 2012. Bufalini 1551: Distortion and Rectification. In *Piante di Roma dal Rinascimento ai Catasti*, ed. M. Bevelacqua and M. Fagiolo, 128–133. Rome: Artemide Edizioni.

Coffin, D.R. 2004. *Pirro Ligorio: The Renaissance artist, architect, and antiquarian*. University Park: Pennsylvania State University.

Cole, M.W. 2011. *Ambitious form: Giambologna, Ammanati, and Danti in Florence*. Princeton: Princeton University Press.

Cole, M., and D. Gamberini. 2016: Vincenzo Danti's Deceits. *Renaissance Quarterly* 69 (4): 1296–1342.

Courtright, N. 2003. *The Papacy and the art of reform in sixteenth-century Rome: Gregory XIII's tower of the winds in the Vatican*. Cambridge: Cambridge University Press.

Curran, B. 2007. *The Egyptian Renaissance: The afterlife of ancient Egypt in early modern Italy*. Chicago: University of Chicago Press.

de Marchi, F. 1599. *Della architettura militare libre tre*. Brescia: Comino Presegni.

Delumeau, J. 1957–1959. *Vie économique et sociale de Rome dans la seconde moitié du XVIe siècle*. 2 vols, Bibliothèque des Écoles Française D'Athènes et de Rome, 184. Paris: E. de Boccard,

Dubourg Glatigny, P. 2002. Egnatio Danti O.P. (1536–1586): Itinéraire d'un mathématicien parmi les artistes. *Mélanges de l'École Française de Rome: Italie et Méditerranée* 114: 543–605.
Ehrle, F. 1909. *Roma Prima di Sisto V: La Pianta di Roma du Pérac-Lafréry del 1577*. Vatican City: Biblioteca Apostolica Vaticana.
Ehrle, F. 1911. *Roma al tempo di Giulio III: La pianta di Roma di Leonardo Bufalini del 1551 riprodotta dall'esemplare esistente nella Biblioteca Vaticana*. Vatican City: Biblioteca Apostolica Vaticana.
Fagiolo, M., and M.L. Madonna. 1972. La Roma di Pio IV: La 'Civitas Pia', la 'Salus Medica', la custodia 'angelica.' *Arte illustrate* 5 (4): 383–402.
Federici, V. 1898. Di Mario Cartaro, incisore viterbese del secolo XVI. *Archivio della R. Società Romana di Storia Patria*. 21: 535–552.
Ferrary, J.L. 1996. *Onofrio Panvinio et les antiquité romaines*. Rome: École Française di Rome.
Fiorani, F. 2005. *The marvel of maps: Art, cartography and politics in Renaissance Italy*. New Haven: Yale University Press.
Fiorani, F. 2007. Cycles of painted maps in the Renaissance. In *The history of cartography*. Vol. 3: *Cartography in the European Renaissance*, ed. D. Woodward, Pt. 1, 804–829. Chicago: University of Chicago Press.
Fiore, F.P. 1986a. Danti, Egnazio (al secolo Carlo Pellegrino). In *Dizionario Biografico degli Italiani*, vol. 32, 659–663. Rome: Istituto della Enciclopedia Italiana
Fiore, F.P. 1986b. Danti, Piervincenzo. In *Dizionario Biografico degli Italiani*, vol. 32, 664–667. Rome: Istituto della Enciclopedia Italiana
Fiore, F.P. 1986c. Danti, Vincenzo. In *Dizionario Biografico degli Italiani*, vol. 32: 667–673. Rome: Istituto della Enciclopedia Italiana
Fiore, F.P. 1989. Rilievo topografico e architettura a grande scala nei disegni di Antonio da Sangallo il Giovane per le fortificazioni di Roma al tempo di papa Paolo III. In *Il disegno di architettura. Atti del Convegno Milano 15–18 febbraio 1988*, ed. Paolo Carpeggiani and Luciano Patetta, 175–180. Milan: Guerini Associati.
Foley, W.A. 1988. Language birth: The processes of Pidginization and Creolization. In *Language: The sociocultural context*, ed. F.J. Newmeyer. Linguistics: The Cambridge Survey, 4. Cambridge: Cambridge University Press.
Friedländer, W.F. 1912. *Das Kasino Pius des Vierten*. Leipzig: K. W. Hiersmann.
Friedman, D., and P. Schlapobersky. 2003. Leonardo Bufalini's Orthagonal *Roma* (1551). *Thresholds* 28: 10–16.
Frutaz, A.P. 1962. *Le piante di Roma*. 3 vols. Rome: Istituto di Studi Romani.
Galison, P. 1997. *Image and logic: A material culture of microphysics*. Chicago: University of Chicago Press.
Galison, P. 2010. Trading with the enemy. In *Trading zones and interactional expertise: Creating new kinds of collaboration*, ed. Michael E. Gorman, 25–52. Cambridge, MA: MIT Press.
Gambi, L., and A. Pinelli (eds.). 2008. *La Galleria delle Carte geografiche in Vaticano/The Gallery of Maps in the Vatican*. 3 vols, 3rd ed. Modena: Franco Cosimo Panini Editore.
Garzoni, T. 1587. *Le piazza universal di tutte le professioni del mondo, e nobili et ignobili*. Venice: Gio. Battista Somasco.
Gaston, R.W. (ed.). 1988. *Pirro Ligorio: Artist and antiquarian*. Milan: Silvana Editoriale.
Gerbino, A. 2010. *François Blondel: Architecture, erudition, and the scientific revolution*. London: Routledge.
Grafton, A. 2000. *Leon Battista Alberti: Master builder of the Italian Renaissance*. New York: Hill and Wang.
Heilbron, J.L. 1999. *The Sun in the church: Cathedrals as solar observatories*. Cambridge, MA: Harvard University Press.
Huelsen, C. 1969. *Saggio di bibliografia ragionata delle piante icnografiche e prospettiche di Roma dal 1551 al 1748*. Reprint. Rome: Bardi Editore.
Infelise, M. 2002. *Prima dei giornali: Alle origini della pubblica informazione (secoli XVI e XVII)*. Rome: Editori Laterza.
Insolera, I. 1985. *Roma: Immagini e realtà dal X al XX secolo*, 3rd ed. Rome: Editori Latereza.
Jacks, P.J. 1993. *The antiquarian and the myth of antiquity: The origins of Rome in Renaissance thought*. Cambridge: Cambridge University Press.
Karrow Jr., R.W. 1993. *Mapmakers of the sixteenth century and their maps*. Chicago: Speculum Orbis Press for the Newberry Library.
Lamberini, D. 1990. Francesco De Marchi. In *Dizionario Biografico degli Italiani*. Rome: Istituto della Enciclopedia Italiana, vol. 38: 447–454.
Laureys, M., and A. Schreurs. 1996. Egio, Marliano, Ligorio, and the Forum Romanum in the 16th Century. *Humanistica Lovaniensia* 45: 385–405.
Lemaitre, N. 1994. *Saint Pie V*. Paris: Fayard.
Ligorio, P. 1553. *Libro di M. Pirrho Ligori Napolitano delle antichità di Roma nel quale si tratta de' circi, theatri e anfitheatri, con le Paradosse del medesimo auttore, quai confutano la commune opinione sopra varii luoghi della città di Roma*. Venice: Michele Tramezino. [On line edition: http://archiv.ub.uni-heidelberg.de/artdok/volltexte/2008/562. Edited by M. D. Davis (accessed 22 January 2016)].

Ligorio, P. 1561. *Anteiquae urbis imago accuratissime ex vetusteis monumenteis formata*. Rome: M. and Fr. Tramezini.
Ligorio, P. 2005. *Libro di diversi terremoti,* ed Emanuela Guidoboni. Rome: De Luca.
Long, P.O. 2008. Hydraulic engineering and the study of antiquity: Rome, 1557–1570. *Renaissance Quarterly* 61 (4): 1098–1138.
Long, P.O. 2011. *Artisan/practitioners and the rise of the new sciences, 1400–1600*. Corvallis: Oregon State University Press.
Long, P.O. 2015. Trading zones in early modern Europe. *Isis* 106 (4): 840–847.
Losito, M. 2000. *Pirro Ligorio e il casino di Paolo IV in Vaticano: l'essempio delle cose passate*. Rome: Palombi.
Lurin, E. 2006. *Étienne Duperac, graveur, peintre et architecte (vers 1535–1604): Une artiste-antiquaire entre L'Italie et la France*. 3 vols. These de Docteur de L'Université Paris IV, Sorbonne.
Lurin, E. 2007. Les restitutions de scènes antiques: Onofrio Panvinio iconographe et inventeur d'images. In *Programme et invention dans l'art de la Renaissance*, ed. M. Hichmann, J. Koering, and Ph. Morel, 153–173. Actes Colloque de la Villa Medici, 2005. Rome: Villa Medici.
Lurin, E. 2008. Étienne Dupérac vedutista e cartografo: La costruzione della pianta di Roma del 1577. In *La città dei cartografi. Studi e ricerche di storia urbana*, ed. C. De Seta, B. Marin, and M. Tuliano, 49–59. Naples: Electa Napoli.
Maier, J. 2007. Mapping past and present: Leonardo Bufalini's Plan of Rome (1551). *Imago Mundi* 59 (1): 1–23.
Maier, J. 2011. *Roma Renascens*: Sixteenth-century maps of the eternal city. In *Rome: Continuing encounters between past and present*, ed. D. Caldwell and L. Caldwell, 35–55. Surrey: Ashgate.
Maier, J. 2012a. Leonardo Bufalini e la prima pianta a stampa di Roma, 'la più bella di tutte le cose.' In *Piante di Roma dal Rinascimento ai Catasti*, ed. M. Bevelacqua and M. Fagiolo, 117–127. Rome: Artemide Edizioni.
Maier, J. 2012b. A 'True Likeness:' The renaissance city portrait. *Renaissance Quarterly* 65(3): 711–752.
Maier, J. 2015. *Rome measured and imagined: Early modern maps of the eternal city*. Chicago: University of Chicago Press.
Mandowsky, E. and C. Mitchell (eds.). 1963. *Pirro Ligorio's Roman antiquities: The drawings in MS XIII, B.7 in the National Library in Naples*. London: Warburg Institute, University of London.
Marconi, N. 2004. *Edificando Roma Barocca: Macchine, apparati, maestranze e cantieri tra XVI e XVIII secolo*. Città di Castello: Edimond.
Marino, J. 1992. Administrative mapping in the Italian states. In *Monarchs, ministers, and maps: The emergence of cartography as a tool of government in early modern Europe*, ed. D. Buisseret, 5–25. Chicago: University of Chicago Press.
Martin, C. 2011. *Renaissance meteorology: Pompanazzi to Descartes*. Baltimore: Johns Hopkins University Press.
McClure, G.W. 2004. *The culture of profession in late Renaissance Italy*. Toronto: University of Toronto Press.
Miller, Peter N. (ed.). 2007. *Momigliano and antiquarianism: Foundations of the modern cultural sciences*. Toronto: University of Toronto Press.
Oldoini, A. 1678. *Athenaeaum augustum in quo Persinorum scripta publice exponutur*. Perugia: L. Ciani & F. Desideri.
Palagiano, C. 1972. Bufalini, Leonardo. In *Dizionario biografico degli italiani*, vol. 14: 798–999.
Palagiano, C. 1974. *Gli atlantini manoscritti del Regno di Napoli di Mario e di Paolo Cartaro*. Rome: Istituto di geografia dell'Università.
Palma Venetucci, B. 2003. Pirro Ligorio and the rediscovery of antiquity. In *The rediscovery of antiquity: The role of the artist*, eds. Jane Fejfer, Tobias Fischer-Hansen, and Annette Rathje, 63–68. Copenhagen: Museum Tusculanum Press, University of Copenhagen.
Pastor, L. von. 1951–1952. *History of popes*. Vols. 17–18: Pius V (1566–1572), ed. R. Kerr. Reprinted. London: Routledge & Kegan Paul/St. Louis, Mo.: B. Herder Book Co.
Petrucci, C.A. 1953. *Catalogo generale delle stampe tratte dai rami incise posseduti dalla Calcografia Nazionale*. Rome: La Libreria dello Stato.
Popplow, M. 2015. Formalization and interaction: Toward a comprehensive history of technology-related knowledge in early modern Europe. *Isis* 106 (4): 848–856.
Rankin, A. 2013. *Panaceia's daughters: Noblewomen as healers in early modern Germany*. Chicago: University of Chicago Press.
Rinne, K.W. 2010. *The waters of Rome: Aqueducts, fountains, and the birth of the Baroque City*. New Haven: Yale University Press.
Rocchi, E. 1902. *Le piante icnografiche e prospettiche di Roma del secolo XVI*. 2 vols. Turin and Rome: Roux e Viarengo.
Rowland, I.D. 1996. *The culture of the high Renaissance: Ancients and moderns in sixteenth-century Rome*. Cambridge: Cambridge University Press.
Sapori, G. 1986. Danti, Girolamo. In *Dizionario Biografico degli Italiani*. Rome: Istituto della Enciclopedia Italiana, vol. 32, 663–664.

Scaccia Scarafoni, C. 1939. *Le piante di Roma possedute dalla biblioteca dell'Istituto e dalle altre biblioteche governative della città*. Rome: La Libreria dello Stato.
Schnapp, A. 1997. *The discovery of the past*. Trans. I. Kinnes and G. Varndell. New York: Harry N. Abrams.
Schreurs, A. 2000. *Antikenbild und Kunstanschauungen des neapolitanischen Malers, Architekten und Antiquars Pirro Ligorio (1513–1583)*. Cologne: Walther König.
Schulz, J. 1987. Maps as metaphors: Mural map cycles of the Italian renaissance. In *Art and cartography: Six historical essays*, ed. David Woodward, 97–122. Chicago: University of Chicago Press.
Sickel, L. 2007. Der Romgedanke in der Selbstdarstellung der Familie Cesarini im Rom des 16. Jahrhunderts. http://www.ducatocesarini.it/articoli_e_studi/articolo%20di%20Sickel%20L.pdf. Accessed 24 January 2016.
Simoncini, G. 2008. *Roma: Le trasformazioni urbane nel cinquecento*. I: *Topografia e urbanistica da Giulio II a Clemente VIII*. Florence: Leo S. Olschki.
Simoncini, G. (ed.). 2011. *Roma: Le trasformazioni urbane nel cinquecento*. II: *Dalla città al territorio*. Florence: Leo S. Olschki.
Smith, G. 1977. *The Casino of Pius IV*. Princeton: Princeton University Press.
Stenhouse, W. 2005. *Reading inscriptions and writing ancient history*. Bulletin of the Institute of Classical Studies, Supplement 86. London: Institute of Classical Studies.
Summers, D. 1979. *The sculpture of Vincenzio Danti*. New York: Garland Publishers.
Summers, D. 1996. Danti Family (1) Vincenzo [Vincenzio] Danti. *Dictionary of Art*. 34 vols, ed. Jane Turner, vol. 8, 511–513. New York: Grove Dictionaries.
Tabarroni, G. 1996. Ignazio [Egnazio] Danti. In *Dictionary of* Art, ed. Jane Turner, 34 vols, vol. 8, 513. London: Macmillan.
Taussig, M.T. 2010. *The devil and commodity: Fetishism in South America*. 30th Anniversary Edition. Chapel Hill: University of North Carolina Press.
Valeriani, S. 2006. *Zimmermannskunst und Kirchenbau von der Spätantike bis zur Barockzeit*. Petersberg: Michael Imhof Verlag.
Valerio, V. 1982. The Neapolitan Saxton and his Survey of the Kingdom of Naples. *The map collector* 18 (1): 14–17.
Valerio, V. 1993. *Società uomini e istitutzioni cartografiche nel mezzogiorno d'Italia*. Florence: Istituto Geografico Militare.
Valerio, V. 2007. Cartography in the Kingdom of Naples during the early modern period. In *The history of cartography*. Vol. 3: *Cartography in the European Renaissance*, ed. Woodward, Pt. 1, 940–973. Chicago: University of Chicago Press.
Weiss, R. 1988. *The Renaissance discovery of classical antiquity*. 2d ed. Oxford: Basil Blackwell.
Witcombe, C.L.C.E. 2008. *Print publishing in sixteenth-century Rome: Growth and expansion, rivalry and murder*. London: Harvey Miller Publishers.
Zannini, G.L.M. 1980. *Stampatori e librai a Roma nella seconda metà del cinquecento*. Rome: Fratelli Palombi Editori.

Architectural Knowledge

9

Wolfgang Lefèvre

Abstract

The subject of this chapter is the structure of architectural knowledge in the long sixteenth century. In its first section, the social dimension of the structure of architectural knowledge is addressed, which is not regarded as a feature particular to this field of practical knowledge, but something that can be studied in almost every complex field of such knowledge. In the second section, the focus lies on the central figure of the field of architectural knowledge, namely the architect. Special attention is paid to the development of the scope of his tasks during the Renaissance and to a trend towards specialization among architects. Finally, aspects of the mathematical knowledge of Renaissance architects are the topic of the third and last section. In particular, the knowledge structure of geometrical procedures that were employed and developed by architects over centuries and that cannot be regarded as applications of learned (Euclidean) geometry will be discussed; rather, these procedures eventually constituted the starting point of a new sub-discipline of learned geometry: descriptive geometry.

Bisher nur geprahlt, was die Produktion der Wissenschaft verdankt,
aber die Wissenschaft verdankt der Produktion unendlich mehr

Friedrich Engels

In the last two or three decades, practical knowledge has been made a favored subject of studies by historians of science. Once ignored and underrated, it has become a cherished subject, the significance of which was seen to require vindication against the pretensions of scholars and scientists. Accordingly, no further explanation is required before starting with the statement that there are many fields in

W. Lefèvre (✉)
Max Planck Institute for the History of Science, Berlin, Germany
e-mail: wlef@mpiwg-berlin.mpg.de

M. Valleriani (ed.), *The Structures of Practical Knowledge*, DOI 10.1007/978-3-319-45671-3_9

which the interplay of practical and scientific knowledge has proved to be a decisive factor in the latter's development. However, the ways in which this interplay took place in different fields of practice has still not been sufficiently explored. In this paper, I will focus on the long sixteenth century's construction trade and try to discuss some aspects characteristic of the knowledge of its practitioners.

1 The Social Structure of Architectural Knowledge

The knowledge needed and employed in a field of practice usually consists of a spectrum of different kinds of knowledge. The field of constructing buildings, bridges, fortresses and the like is no exception in this regard. In this field of architectural practice we can generally distinguish three main kinds of knowledge: First, the kind of knowledge pertaining to execution, that is, pertaining to the real process of erecting a building; this kind of knowledge includes that of the craftsmen involved in this process, that is, the knowledge of brick masons, stone masons, carpenters, metalworkers, and so on; second, the kind of knowledge pertaining to designing the building, and planning and supervising its execution; this kind includes the knowledge of principles and rules regarding the design, that of constructing plans and templates, and knowledge of the crafts involved in guiding and supervising the construction process; and third, knowledge pertaining to logistic, economic, and "political" management, that is, knowledge needed for, the procurement and administration of financing, the procurement and transportation of materials, employing personnel and controlling contracts, and, in many cases, the knowledge needed to pilot the project through the tides of its political environment.

If it is about the construction of a simple, traditional house, it may be possible that a craftsman, an experienced mason or carpenter, possesses all the practical skills and also all the different kinds of knowledge needed for this work. But it goes without saying that this is impossible in the case of ambitious building projects such as cathedrals, castles, palaces, and town halls. Such challenging projects can be realized only through the cooperation of various professions. Division of labor is required, and this holds for the past, no less than for the present. If we look at the construction site of a medieval cathedral, for instance, we encounter a highly differentiated and hierarchical ordered structure of different professionals responsible for different tasks.

Stefan Bürger compiled an instructing table (Table 1) representing the division of labor characteristic of large-scale building projects in the German-speaking countries towards the end of the Middle Ages.[1] Number one, the *Bauherr*, is the client; number two, the *Baumeister/Bauverwalter*, is the general administrator; number three and four are governmental agents of a city or a country, who can be ignored for the moment; number 5, the *Werkmeister*, is the architect and number 6 his deputy; number 7 finally, the *werkführende Parlier*, is the foreman. The list is obviously incomplete; for after this foreman follow of course the *Meister* (masters) of the different crafts with their teams of *Gesellen* (journeymen) and workers—master mason, brick as well as stone mason, master carpenter, master smith, and so on.

This structure of the division of labor and the corresponding hierarchy of social categories involved in major building projects is not peculiar to the late Middle Ages or to Germany. More or less the same structure and hierarchy can be found on construction sites of the Italian Renaissance, as can be shown for the construction of the cupola of Florence cathedral (Haines 2011–12; Schlimme et al. 2014, section 2.2.1)[2]

[1] For the organization of large building projects in the Middle Ages in general, see for instance (Binding and Linscheid-Burdich 2002; Binding 1993, sections 3.2, 3.3 and 3.9; Bürger 2009, 29).

[2] For the organization of the building trade in Renaissance Florence in general, see (Goldthwaite 1982, part II).

Table 1 Cast of characters at large-scale building projects in the German-speaking countries towards the end of the Middle Ages, according to (Bürger 2009, 29)

Personenkreis	Funktion	Kompetenzen	Quellen	Identifikation
1. Bauherr/Stifter	Initiator, Auftraggeber; Finanzierung	Evtl. Anteil an Konzeption/Planung; bestellt Baumeister; beauftragt Meister	Stiftungsurkunden, Rechnungen, Verträge etc.	Bildnisse, Inschriften oder Stifterzeichen als Standeszeichen (Wappen)
2. Baumeister/Bauverwalter	Vertreter der Bauherren; Verwaltung der Finanzen	Einfluss auf Konzeption/Planung; Administration/Kontrolle; keine Werkführung	Protokolle, Verträge, Annalen etc.	Bildnisse, Inschriften oder Standeszeichen (Wappen)
3. Landeswerkmeister	Landesweite Bauaufsicht	Große Anteile an Konzeption/Planung; administrative, fachliche Aufsicht/ Kontrolle	Bestallung, Jahressold	Meisterzeichen als Standeszeichen (in Wappenschild)
4. Stadtwerkmeister, Ratsbaumeister	Kommunale Bauaufsicht	Große Anteile an Konzeption/Planung; administrative Aufsicht und Kontrolle aller städtischen Baustellen	Bestallung, Jahressold	Meisterzeichen als Standeszeichen; baugebundenes Bildnis
5. Werkmeister	Bauhüttenleitung; Werkführung	Konzeption/Gesamtplan; Entwurf; keine administrative Bauaufsicht	Verträge, Rechnungen, Gutachten	Meisterzeichen als Signatur
6 Stellvertretender Parlier	Stellvertretende Bauhüttenleitung; keine Werkführung	Vorlagen und Fertigung von Einzelformen; Detailentwurf und Ausführung	Anstellung durch Werkmeister oder Bauhütte	Nennung in Wochenlohnlisten
7. Werkführender Parlier	Stellvertretende Werkführung; Bauhüttenleitung	Ausführer und Planer von Bauabschnitten	Verträge, Rechnungen, Gedinge	Meisterzeichen als Signatur möglich

in the first third of the fifteenth century or for Saint Peter in Rome (Ackerman 1991, 368)[3] in the sixteenth century.

Division of labor implies the division of knowledge, competence, and expertise. Division of labor means specialization, not only with regard to skills but also with regard to experience and knowledge. A carpenter does not need to know how to handle the finances of a construction project. And the manager of finances does not need to know how to construct a roof structure. The various specializations that cooperate in a construction project differ not only in kind according to the various tasks, but also in scope. The scope of competences and knowledge of a master mason or a master carpenter is certainly impressive. But it is narrow compared to that of the supervisor or that of the general manager of the project.

All of these different competences, narrow ones and wider, constitute the body of practical knowledge of the construction trade. Thus this multifaceted body of practical knowledge also has a social dimension. It is the web of narrow and wider specialized competences of different collaborators that imparts a social structure to the knowledge of a practical field.

The structure of knowledge of a field of practice is thus not first and foremost a mental structure; it is also more than the structure built out of shared assumptions, experiences, convictions, and beliefs. It is in fact the structure of fragmented pieces of knowledge of different cooperating players with different social profiles.

Considering the complexity of the body of practical knowledge in such a field, and in particular its essentially fragmented character, it seems advisable not to aim at a comprehensive portrait of its structure, but to focus instead on particular sides or dimensions of this structure. With regard to the history of science, it makes sense to focus on and pay special attention to competences where the interplay of erudite and practical knowledge can be studied. In the case of the construction trade such an interplay can be expected in connection with the knowledge needed to design and plan buildings, that is to say, in connection with the knowledge of architects. The social figure of the architect is therefore our next subject.

2 The Social Figure of the Architect

It is notoriously troublesome to use the term "architect" with respect to the long sixteenth century. Although this designation begins to be used by historical actors in Romanic-speaking countries in this very period, its precise meaning is not at all clear. For instance, sometimes the title "architect" seems to have been reserved for artists who were distinguished from master builders (Ackerman 1991, 362) and thus reflects a significant change that took place in the Renaissance. In the Middle Ages, the practitioners who led and supervised the construction process of a building, and sometimes even designed and planned it, had generally come from the ranks of master builders, that is, master masons or master carpenters[4]; since the fifteenth century, particularly in Italy, these tasks were consigned more and more to artists as well, be they goldsmiths, sculptors or painters. It thus seems that, right from its first

[3]See also (Schlimme et al. 2014, section 2.2.2). For the organization of ecclesiastical as well as communal constructing projects in sixteenth-century Italy, see (Schlimme et al. 2014, section 2.2).

[4]In this article, I use the term "master builder" as a generic term comprising master craftsmen involved in the construction process, namely foremost the master (stone and/or brick) masons and master carpenters. In the Renaissance as well as in the Middle Ages, these masters, particularly master stone masons, often led construction processes and can therefore be regarded as precursors of the modern architect or construction engineer. In Stefan Bürger's table (Table 1), number 5, the *Werkmeister*, could probably be called a "master builder," but not number 2, the *Baumeister*.

appearance, the term "architect" indicates a latent distinction between, or discrimination in favor of architecture as opposed to construction engineering, which still persists today.[5]

But not only the term is problematic. The profession itself is not well demarcated in this age. Whereas clear professional patterns—above all apprenticeship training and guild laws—existed for craftsmen such as masons or carpenters, as well as for artists like goldsmiths, painters, and sculptors, no such institutions existed for the practitioners who became responsible for the execution of major construction projects, that is, construction projects that transgressed in scope and complexity the ordinary construction tasks that master builders had been used to leading and executing.[6] Although some sporadic attempts at establishing institutionalized training for these practitioners can be traced back to the sixteenth century,[7] for institutions like the *École des Ponts et Chaussées* or the Prussian *Bauakademie,* one has to wait another 200 years.[8]

For all of these reasons I want to stress that the designation "architect" is used here as an analytical term geared towards certain tasks and functions, namely towards designing, planning, and supervising. In other words, in this article, instead of using often ambiguous historical designations like *Werkmeister, maître maçon,* and so on, I use the term "architect" to designate practitioners who, whatever their training or title, designed buildings, bridges, or fortresses and/or planned and supervised their execution.[9] A clear methodical focus on this distinct group of practitioners, here designated as "architects," will allow us in the following, to study more precisely the interplay between practical and erudite knowledge.

The social role of the architect becomes immediately visible when we consider the general social structure characteristics of a large construction project. On the one side we have all the professionals and workers who erect the building and on the other side the client and his agents who manage the economical and legal affairs. Between these two sides we find the architect (or the architects) who designs the building according to the wishes and economic potential of the client, carries on negotiations with the management about hiring/firing people, procuring materials, and so on, and instructs and supervises the various master craftsmen. He is the figure in the middle of the social structure of construction projects whose tasks demand a broad, multifaceted spectrum of expertise in matters practical as well as economical, and often diplomatic skills as well.

This complex status and role of the architect did not emerge only with the Renaissance, but was also characteristic of the medieval architect whatever his title—"*magister lapicidae*," "*maitre maçon*," or "*Werkmeister*" (Klein 2009, 14).[10] However, the Renaissance was an age of important changes as regards the profile of the architect.[11] There were four challenges in particular that prompted these changes.

---

[5]For the sometimes latent, sometimes manifest tension between the artistic and the engineering dimensions of architecture, see for instance (Philipp 2012; Pfammatter 2012).

[6]For the uncertain and unsettled professional status of Renaissance architects who worked without any affiliation to a guild-controlled building craft, see (Chaps. 2 and 8).

[7]For a first institutionalization of the profession of the architect-engineer in 1563, the *Collegio degli architetti, ingegneri e agrimensori* in Milan, see (Goldthwaite 1982, 365f. and Scotti 1983).

[8]The *École des Ponts et Chaussées* was founded in 1747 and the Prussian *Bauakademie* in 1799. From the seventeenth century on, architectural design was taught at the academies of fine arts that were founded in several European countries on the model of the academies in Florence and Rome. But training the technical competency of the construction trade remained the purview of the crafts well into the eighteenth century. In this respect not even the famous French *Académie royale d'architecture,* founded by Jean-Baptiste Colbert in 1671, was an exception. For the instruction of architectural design, see (Mai 2012; Johannes 2009).

[9]Often, such "architects" were also responsible for the logistics of materials. The division of labor between the architect and the administrator is generally not clear-cut in this age.

[10]For the use of the term *architect* in the Middle Ages see, for instance, (Pevsner 1942; Binding 1999).

[11]For the development of the profile and profession of architects in the Italian Renaissance, see (Goldthwaite 1982, chap. 7).

First, a new field of architecture had to be mastered, namely a new military architecture. As is well known, in the second half of the sixteenth century a new style of fortification was developed[12] in order to match the developments in artillery.[13] This military architecture required not only a new extent of mathematical work, but also, and particularly, a hitherto unknown extent of precise surveying work. There was also another field where advanced surveying and particularly leveling abilities were expected of an architect, namely the field of hydraulic engineering.[14] This field was not entirely new, but experienced considerable expansion and refinement in the Po Valley and in the Netherlands in this period. Architects engaged in fortification or hydraulic projects should perhaps be called architect-engineers.[15]

Second, during this period new architectural drawing methods were developed and used, which involved new geometrical techniques. Presentational drawings, which had been rather exceptional in the Middle Ages, became standard and were now usually drawn in perspective.[16] The technique of constructing architectural plans also underwent a major change through the invention of the combined views technique, that is, the method of deriving orthogonal plans—ground plan, elevation, and section—from each other (Lefèvre 2004). This method implied techniques of geometrical projecting that were not thoroughly new, but considerably refined. This will be examined in the last part of this paper.

Third, the humanistic rediscovery and appreciation of the Roman antiquity also had an impact on the field of architecture and stimulated architects to study Roman architecture. Many Renaissance architects even undertook archeological investigations of the architectural remains of Rome,[17] and at least elementary familiarity with Vitruvius, the Five Orders, and other shibboleths of the age was generally expected of an architect of any repute.

Fourth, new ways of communicating came into being with the advent of printing—the printing not only of texts, but of images as well. Practicing architects in the Renaissance did not usually write theoretical essays or books—Leon Battista Alberti's (1404–1472) *De re aedificatoria*, which—by the way—does not contain a single drawing, is rather an exception.[18] But we encounter several architects who composed pattern books, that is, books that are laconic as regards theoretical issues but lavish as regards illustrations and plans of exemplary buildings and architectural elements—including columns, capitals, and window sills.[19]

Thus one can state that the sphere of an architect's tasks and functions expanded considerably over the long sixteenth century. And so did the sphere of knowledge and expertise architects had to possess. It is no wonder that not every architect mastered the whole spectrum of the enlarged practical field. One could even ask whether there was any architect at all who was equally competent in all

[12] For the development of military architecture in the early modern period, see for instance (Arnold 2002) or (Duffy 1979–85).

[13] For the development of artillery in the early modern period, see for instance (Hall 1997).

[14] For a recent study on hydraulic engineering in sixteenth-century Milan, see (Maffioli 2013).

[15] Renaissance architects were also frequently active in the field of mechanical engineering; for instance, Filippo Brunelleschi and Francesco di Giorgio Martini, to name just the most prominent ones of the fifteenth century.

[16] Linear perspective owed its (re-)invention and development in this age no less to architects—Brunelleschi and Leon Battista Alberti—than to painters—Piero della Francesca and Albrecht Dürer.

[17] These investigations implied refined methods of surveying and graphically recording ruins.

[18] For treatises on architecture written by Renaissance architects, particularly of the fifteenth century (Leon Battista Alberti, Filarete (Antonio Averlino), and Francesco di Giorgio Martini), see (Thoenes 1995). For such treatises of the long sixteenth century (Pietro Cataneo, Philibert de l'Orme, Vincenzo Scamozzi, and others), see (Kruft 1985, chaps. 3–7).

[19] The most famous and influential of these architectural pattern books are Sebastiano Serlio's *Sette libri dell'architettura* (published from 1537 to 1575), Jacopo Barozzi da Vignola's *Regola delle cinque ordini d'architettura* (1562), and Andrea Palladio's *Quattro libri dell'architettura* (1570), if we take the liberty of regarding the work of the latter as a pattern book. For the impact of books like these on the development of architecture, see (Carpo 2001).

subjects of the architectural trade. In any case a tendency towards specialization is unmistakable among the architects of the age.

There was the traditional civil architect who showed shortcomings with regard to new engineering tasks connected with fortification, water management, and surveying.[20] The well-known polemic of the architect and Vitruvius translator Cesare Cesariano (1475–1543) against architects without a proper Vitruvian education (Cesariano 1521, I, 2v–3r) should not be taken as just an expression of arrogance, but must be seen against the background of his native city Milan's efforts in hydraulic engineering and cadastral mapping, which were seriously compromised by feckless surveying. In the 1550s, the city of Milan introduced a formal examination for people who wanted to be registered as architect-engineers (Scotti 1983; Maffioli 2013, 206ff.).

On the other hand, there were military architects, who were often regarded as merely technical experts without any competence regarding the aesthetic dimension of architecture. Particularly in sixteenth-century Florence, civil and military architects were sharply discriminated against. Bernardo Puccini, for instance, who served Florence as a proficient military architect for many years, was not admitted to this city's *Accademia del Disegno*. The reason: He had not been apprenticed to a painter, a goldsmith, or a sculptor and thus did not count as an artist of *disegno*. In other words, only builders trained in one of the arts of *disegno* were regarded as true artistic architects (Lamberini 1990, 127; Schlimme et al. 2014, 141).[21]

This leads to a third group of specialized architects, to the artist architects, that is, trained artists who were commissioned with architectural tasks because of their artistic excellence (Schlimme et al. 2014, 140f.). Almost all of the famous architects of the Roman Renaissance—Donato Bramante (1444–1514), Raphael Santi (1483–1520), Baldassare Peruzzi (1481–1537), Michelangelo Buonarroti (1475–1564)—belonged to this group. Some of these artist-architects lacked sufficient expertise in structural matters. Bramante's disasters are notorious (Ackerman 1970). Antonio da Sangallo the younger (1484–1546), an experienced architect who helped to prevent some of Bramante's structures from imminent collapse (Ackerman 1970),[22] was regarded as a mere craftsman by the sculptor Benvenuto Cellini (1500–1571), because Antonio had been apprenticed to a carpenter rather than to a painter or sculptor (Cellini 1829, 367).

Finally, a fourth group must be mentioned, which could be called the *studiolo* architects, that is, erudite humanists like Andrea Palladio's (1508–1580) mentor Daniele Barbaro (1513–1570) and the Sebastiano Serlio (1475–ca.1554) publisher Jacopo Strada (1507–1588), who did not construct real buildings themselves but were experienced and competent with respect to architectural theory, that is, with respect to all questions regarding classic architecture, mathematics, and mechanics.

Particularly this fourth group immediately suggests categorization as a group of *hybrid* or *savant experts*, that is, of experts whose expertise rests on learned knowledge as well as on practical experience and knowledge.[23] True, these *studiolo* architects stand out among architects for their familiarity with erudite knowledge (though sometimes lacking full-fledged practical experiences—I will instance some of these *studiolo* architects in the last part of this paper). But some familiarity with erudite knowledge, particularly with geometry in the tradition of Euclid, can also be found among military

[20] Civil and military architecture, which had been merely different tasks of a fifteenth-century architect, began to become different, specialized branches of architecture in the sixteenth century. See (Schlimme et al. 2014, 205).

[21] The Accademia di San Luca in Rome, which began work in 1593, admitted architects only in 1640 (Schlimme et al. 2014, 150). For Vasari's and Puccini's views and contrary assessments of the role of *disegno* (in contrast to geometrical drawing) in architecture, see Puccini's *Proemio* to his treatise on military architecture in the possession of the *Bibliotheca Medicea Laurenziana*, in: (Lamberini 1990, 335; Bredekamp 2007, 79).

[22] According to Vasari, Antonio also saved structures by Raphael from imminent collapse; see (Vasari 1981, 454).

[23] For the concept of *savant experts*, see (Chap. 11).

and artist architects and even among traditional civil architects. Renaissance architects display a broad spectrum of ways how practical knowledge was interlocked with learned knowledge.

Together, these four different types of Renaissance architects represented an expanded and multi-faceted body of knowledge in which an interplay of practical knowledge with two kinds of learned knowledge can be observed, namely with humanistic scholarship regarding Roman architecture and classic or philosophical aesthetics on the one hand and, on the other, with mathematics. The interplay between practical and erudite mathematics will be the subject of the next and final section of this paper. As regards the humanistic dimension of the architectural knowledge of the Renaissance, which I cannot go into here, I will only remark that in this respect, too, an interplay between practical and theoretical knowledge could be studied: Architects did not simply apply rules which humanistic erudition had extracted out of Vitruvius and other sources; their practical knowledge was, in fact, a prerequisite for the interpretation of these sources.

3 Architecture as Mathematical Practice

Anthony Gerbino and Steven Johnston gave their book *Compass and Rule* the subtitle *Architecture as Mathematical Practice* (Gerbino and Johnston 2009). Indeed, architecture became a mathematical practice in the Renaissance, as should by now be obvious, with regard to the new military architecture and to ambitious hydraulic engineering projects, which required highly developed surveying and leveling techniques along with new mathematical instruments. Such techniques were also employed for exact recordings of the remains of classical buildings. And, last but not least, in the context of architectural drawing practice, refined mathematical techniques were required and developed for the construction of orthogonal plans as well as geometrical diagrams from which templates of intricately shaped architectural elements could be derived.

In the last two or three decades it has become fashionable among some historians of science to regard practitioners's claims to mathematical (or other scientific) proficiency as mere rhetoric and attempts to enhance their social status. In the case of Renaissance architects, it cannot reasonably be denied that mathematical abilities and skills served as prerequisite conditions for architects in charge of advanced building projects. However, it makes sense to ask how the practical mathematics of these practitioners relates to theoretical mathematics in the tradition of Euclid.

Instead of a general discussion of this question, which would go far beyond the scope of this paper, I will discuss some specific mathematical techniques used by architects of this age that cannot be understood simply as applied theoretical mathematics.

My first example concerns a very simple geometrical operation, namely the doubling or halving of a square (Figs. 1a–b). This operation can be traced back to Euclid and other classic sources, was never forgotten, and constituted a main technique of Gothic design practice up to and through the fifteenth century (Hahnloser 1972, 111f.; Bork 2011, 5f., 26, 31ff.). Here we have a geometrical construction with a learned or theoretical underpinning, which was used by master builders. But, despite the erudite dimension of this operation, it seems questionable whether it can be regarded as an application of erudite geometry.

First, a geometrical construction like the doubling of a square, even in Euclid's *Elements* itself, is part of learned or theoretical geometry only when it is united with a proof that the area of the resulting square is exactly twice (or half) as large as that of the original square. The procedure of constructing such a square—and there have been various procedures as the documents show—can be performed without any knowledge of how to prove its correctness. It can be learned and transmitted as

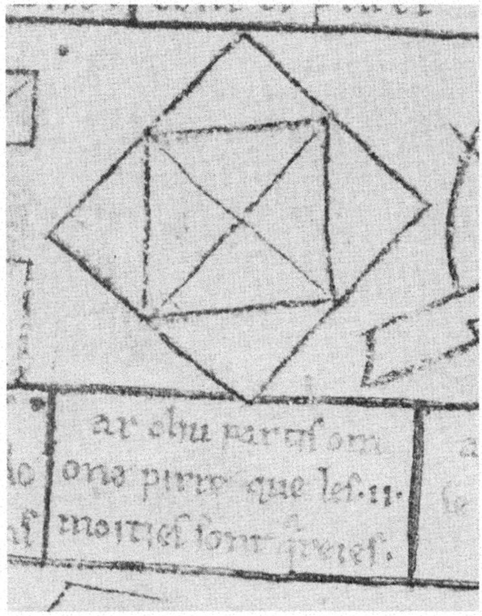

Fig. 1a Doubling/Halving of squares. Villard de Honnecourt, Sketchbook, 13th cent. From (Hahnloser 1972, plate 39 o, image 98)

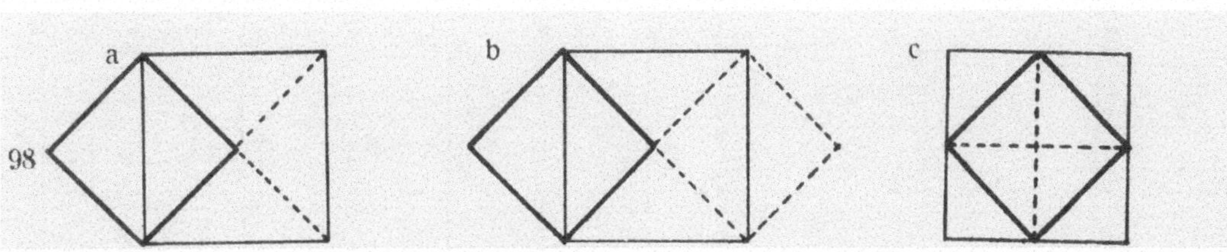

Fig. 1b Doubling/Halving of squares. Tradition of Square Halving (after Hahnloser). Platon, Pythagoras, Euclid/ Villard. From (Hahnloser 1972, image 98)

know-how. And it is not even clear whether it can be considered an indirect application of erudite geometry, since the procedure may be derived from one that predates the Euclidean proof.[24]

Second, it is an open question whether the master builders of the Gothic age—even when acquainted with geometrical issues as transmitted in the *quadrivium* (or in treatises in the tradition of the *agrimensores*) (Victor 1979)[25]—cared for mathematical exactness, namely for the exact geometrical properties of a square or any other geometrical diagram. Their obvious concern and interest were the possibilities of creating complex shapes using procedures like that of square doubling (Figs. 2a–c).

We know from documents of the fifteenth century, for instance, that such procedures of doubling or halving squares were used to design graded pillars and pinnacles, a characteristic decorative element of Gothic churches. This technique was also employed to determine the proportions among the parts of a building, for instance, the proportions of the galleries and the open space of a cloister (Figs. 3a–b).

[24] See, for instance, Jens Høyrup's discussion of the Babylonian tablet YBC 7289 in (Høyrup 2002, 261f.).

[25] See also the literature in (Høyrup 2009, 376f.).

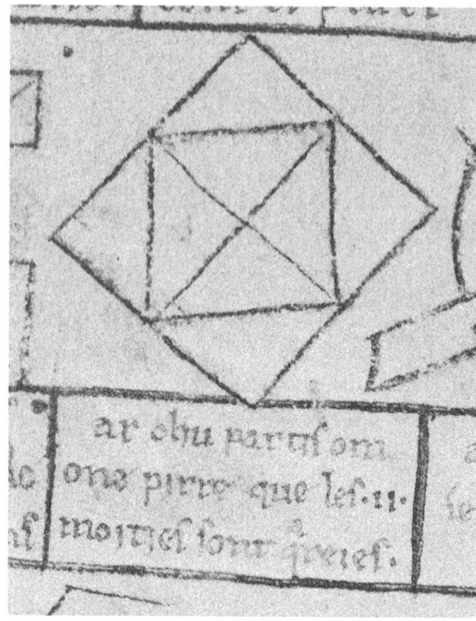

Fig. 2a Designing by square halving. Villard de Honnecourt, Sketchbook, 13th cent. From (Hahnloser 1972, plate 39 o)

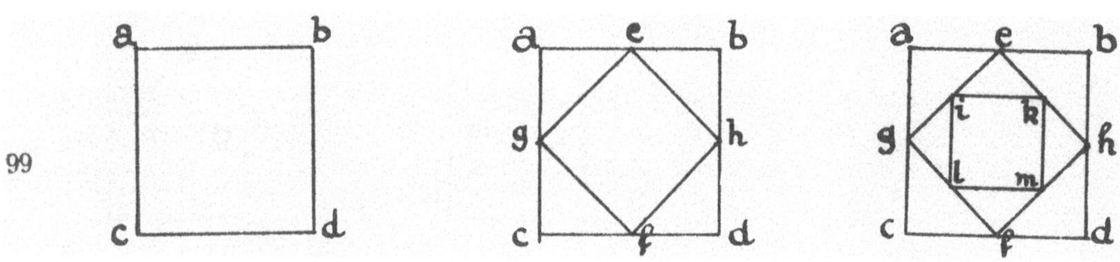

Fig. 2b Designing by square halving (after Hahnloser). From (Roriczer 1486). (Hahnloser 1972, Image 91)

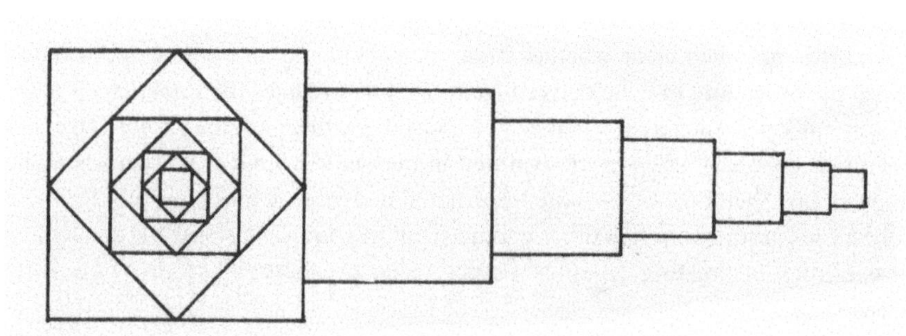

Fig. 2c Designing by square halving (after Hahnloser). From (Lacher 1856). (Hahnloser 1972, image 99)

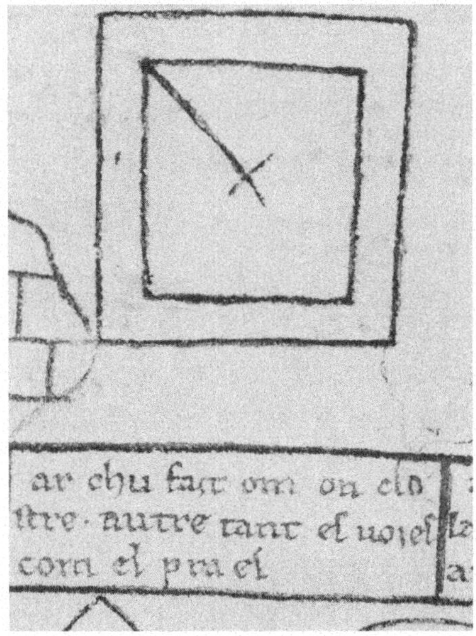

Fig. 3a Designing a cloister by square doubling. Villard de Honnecourt, Sketchbook, 13th cent. From (Hahnloser 1972, plate 39 k)

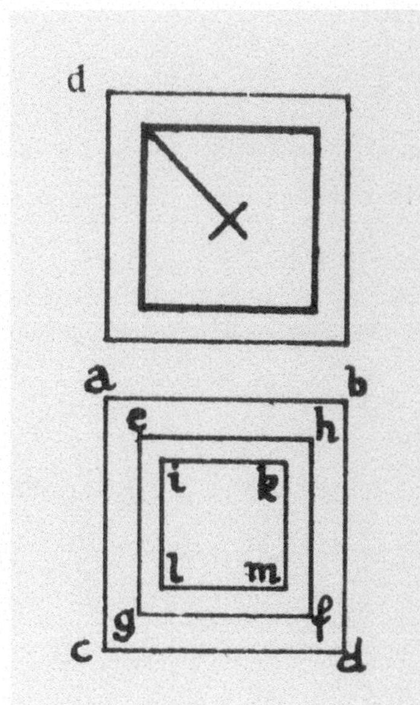

Fig. 3b Top: Designing a cloister by square doubling (after Hahnloser). Villard de Honnecourt; *Bottom*: Designing a cloister by square doubling (after Hahnloser). From (Roriczer 1486). (Hahnloser 1972, images 98, 99)

We encounter in this design practice of Gothic architects an instance of constructive geometry, which does not seem to require any underpinning with theoretical geometry. Lon Shelby, in discussing what is probably the most important document of Gothic design procedures of the high Middle Ages, namely Villard de Honnecourt's (13th cent.) *Sketchbook*,[26] wrote:

[26]More precisely, folios 39 to 41 of this sketchbook can be regarded as the most important document of the practical geometry of Gothic builders. See the editions of this sketchbook by (Hahnloser 1972; Barnes 2007). See also (Bechmann 1993).

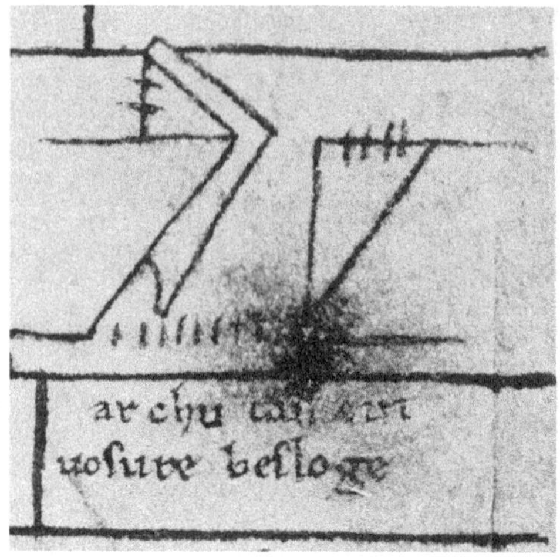

Fig. 4a Measuring angles with a straight edge. *Left*: Villard de Honnecourt, Sketchbook, 13th cent. From (Hahnloser 1972, plate 39 i)

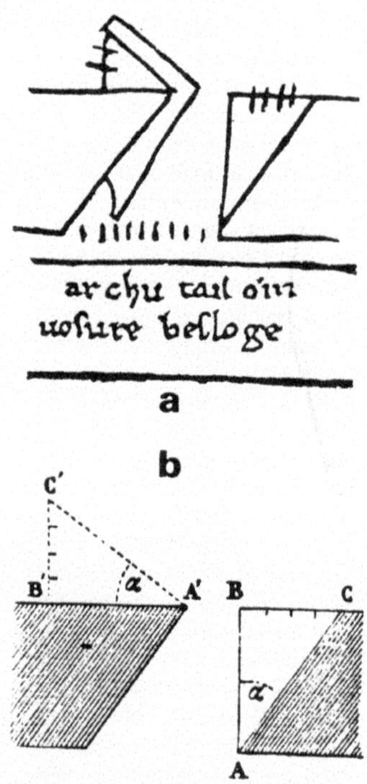

Fig. 4b Measuring angles with a straight edge. Roland Bechmann's reconstruction. From (Bechmann 1993, 174)

> … stereometrical problems were solved by medieval masons primarily through the physical manipulation of geometrical forms by means of instruments and tools available to the masons…. We may thus characterize the practical geometry of Villard's Sketchbook more precisely as constructive geometry, by means of which technical problems of design and building were solved through the construction and physical manipulation of simple geometrical forms—triangles, squares, polygons, and circles. (Shelby 1972, 409)

The physical manipulation of simple geometrical forms included the manipulation of tools such as the straight edge, which allowed substituting the difficult measurement of angles for the far easier measurement of distances (Figs. 4a–b). In this case, too, it is completely irrelevant whether or not the master builder or the stonecutters knew that, by employing the straight edge in this way, they were capitalizing on the geometrical properties of right-angled triangles as explored in Greek antiquity and canonically conceptualized in Euclid's *Elements*.

My second example again concerns an apparently simple, though tricky, geometrical operation, namely the compressing or stretching of circles (Fig. 5). The creation of such deformed circles was needed in the context of vaulting bays with a rectangular ground plan by a groined vault (Figs. 6a–b).

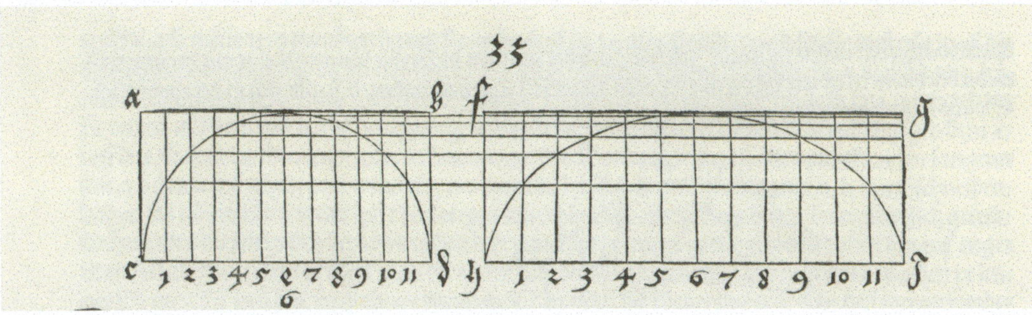

Fig. 5 Creation of non-trivial curved shapes. From (Dürer 1625, I, fig. 33). Courtesy of the Library of the Max Planck Institute for the History of Science

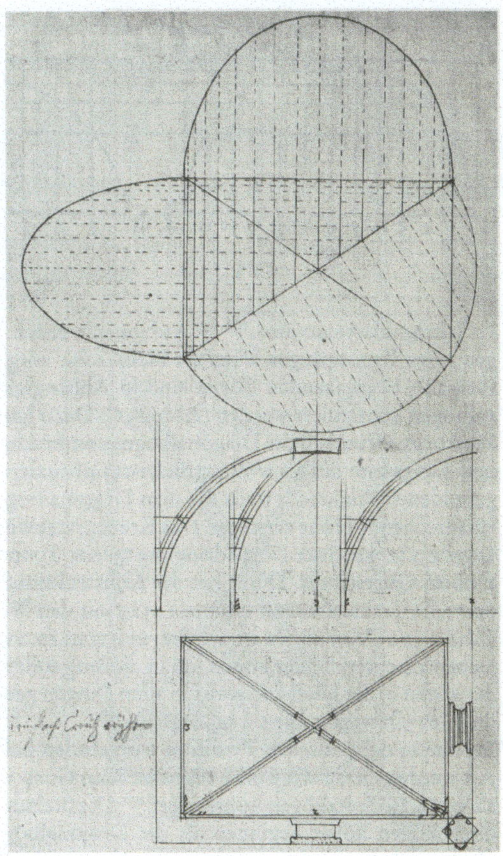

Fig. 6a Vaulting a rectangular bay by a groined vault. The diagonal arch describes a circle segment, whereas the traverse arches describe deformed circles. Orthogonal plans. From (Müller 1990, 145ff)

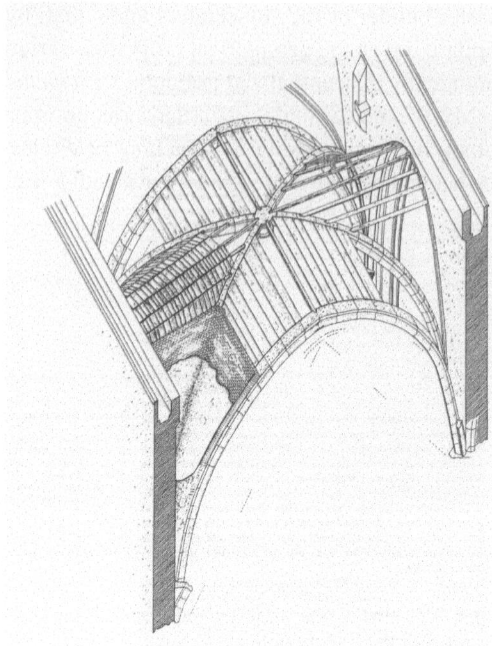

Fig. 6b Vaulting a rectangular bay by a groined vault. The diagonal arch describes a circle segment, whereas the traverse arches describe deformed circles. Isometric view. From (Müller 1990, 145ff)

This geometrical technique or trick can be traced back to antiquity. For instance, the famous *entasis* construction that was used for the columns of the Apollo temple in Didyma consists precisely in the stretching of a circle (Haselberger 1999).[27] This technique, while essentially trivial, deserves attention in our context because the resulting curves transgress the world of geometrical forms that had been the subject of learned geometry up to that time; they are by implication not circles, but neither are they conic sections. Although the point-by-point construction of such curves employs the same tools as Euclid, namely a compass and ruler, it leads into a world of geometrical forms beyond the form arsenal of learned geometry. Clearly, the construction of such forms cannot be regarded as an application of erudite geometry. Moreover, in such forms and the techniques of their creation we encounter a geometry that existed and developed independently of erudite geometry; that is to say, we encounter practical geometry as a second realm of geometry besides theoretical geometry.

My last example concerns techniques of geometrical projection used in the building trade. These techniques constituted a realm of practical geometry that deserves special interest because it stimulated learned mathematicians to study its implications, which eventually led to the origin of a new mathematical discipline, namely descriptive geometry.

First, a few words about the practical context of these techniques. Their original and most important practical context was stone masonry, that is, building with natural stones. In Gothic architecture, except for regions where brick masonry dominated due to a lack of suitable natural stones, architectural elements such as the ribs of groined vaults, pillars, the tracery of windows, pinnacles, and other decorative elements were made out of natural stones which had to be dressed for this purpose. This also holds for Renaissance architecture[28] where mortarless stone masonry flourished, particularly in France, and intricately shaped elements of profane buildings—staircases, entire vaults, vaulted corridors, oriels like the famous trompe of *Chateau d'Anet*—were built out of dressed stones, *ashlars* or *voussoirs*, as they were called by stone masons.

[27] See also (Wilson Jones 2009, 99ff.). For *entasis* constructions in the Renaissance, see (Becchi 2014).

[28] For the principal difference between the Gothic and Renaissance mode of using natural stones, which I cannot go into here, see (Rabasa-Diaz and Calvo-López 2009, particularly sect. 3).

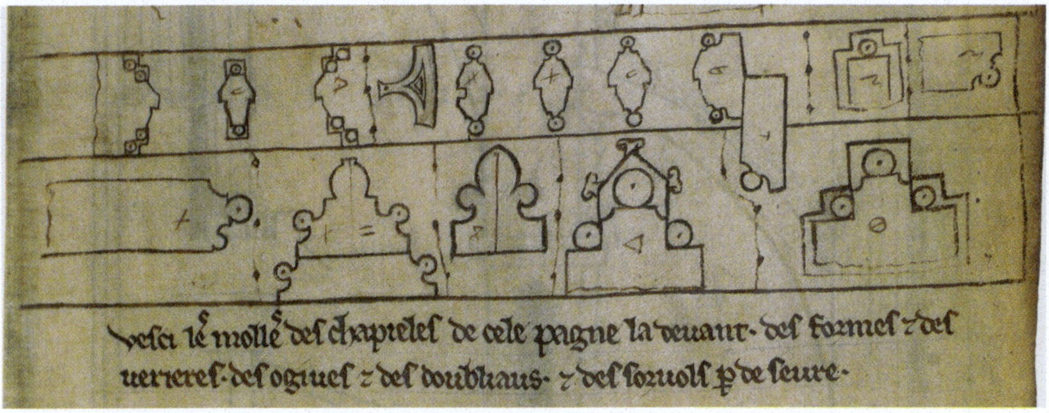

Fig. 7 Villard de Honnencourt's templates for pillars, rib profiles etc. From (Barnes 2007, plate 63)

As a matter of fact, an ashlar (or at least all of its bearing joints) must be dressed before it can be put in its designated place within the architectural element at hand. This trivial requirement implies a peculiar difficulty of the art of stonecutting: The shape of the stone that fits with the shapes of the adjacent stones must be anticipated exactly. This may be no big deal when rectangular ashlars are needed, but when oblique or curved shapes are required, it is a tricky job and, as we will see, can become terribly difficult.

There were many methods and means for accomplishing the exact dressing of ashlars. For instance, complementary angles of oblique shapes could be transferred mechanically from one stone to the next by means of a straight edge, as we saw above (Figs. 4a–b).

Templates were a chief tool in stonecutting. For an ashlar of the rib of a Gothic groined vault, for example, a template was needed only for the profile of the ashlar, namely a section of this profile (Fig. 7).

The alignment, size and curvature of the ashlar were determined by means of combined ground plans and elevations, which allowed geometric projecting from one plan onto the other and vice versa. Since the ribs of a groined vault, in contrast to its traverse arches, almost always described segments of circles, the size of the segment of a particular ashlar could be determined by projecting its horizontal extension, as given in the ground plan, onto the arch as given in the elevation (Figs. 8a–c).

This employment of geometrical techniques for representing the shape of solids, namely stereometric bodies, on plans, that is, on flat planes, as practiced by Gothic master builders, experienced a dramatic development when used by Renaissance architects to develop the templates for the surfaces of ashlars with intricate curvature. The projection techniques developed in this context of advanced stereotomy are far too complicated to be expounded here.[29] I hope, however, to convey an impression of its subtlety and sophistication by showing two concrete examples of combined plans for two types of oriels (Figs. 9a–b, and 10a–b).

We know very little about the development of the geometrical projection techniques employed for the construction of these plans in the first half of the sixteenth century, that is, before the French architect Philibert de l'Orme (1510–1570) published such plans in books III and IV of his *L'Architecture* in 1567 (De l'Orme 1567).[30] Thus, it appears as if these highly accomplished mapping techniques

[29] For the art of stereotomy in the long sixteenth century, see (Sakarovitch 1998, chaps. 1 and 2; Rabasa Díaz 2000, chaps. 2 and 3; Camerota 2004, sect. 5).

[30] His earlier *Nouvelles Inventions pour bien bastir et à petits Fraiz* (1561) is about wooden roof constructions, see (Campa 2006).

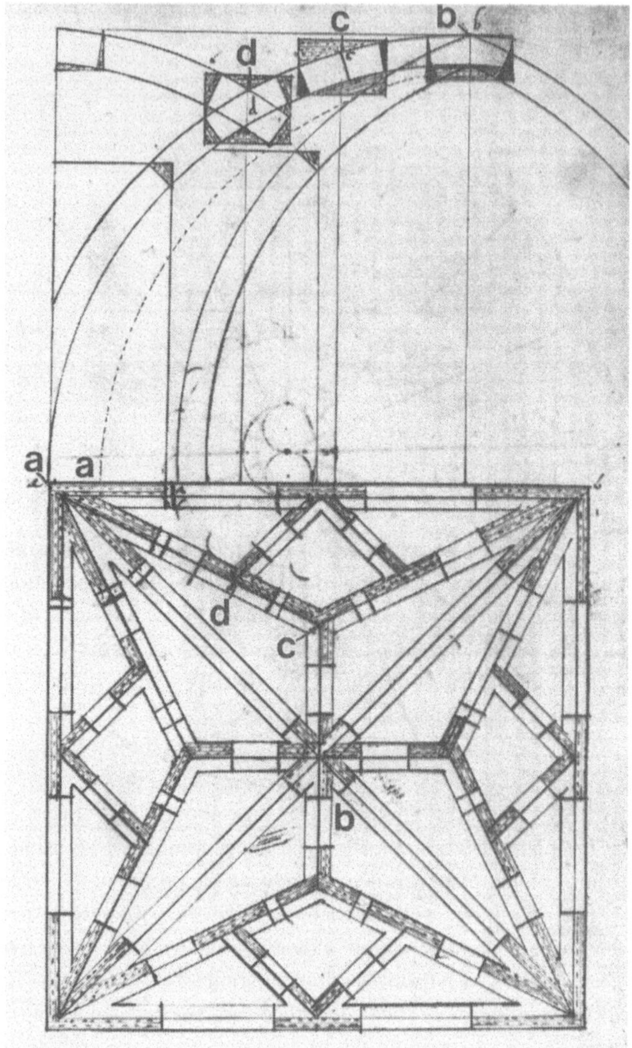

Fig. 8a Designing by combined ground plan and elevation. Contruction of a vault from Fracht von Andernach, Jacob. n.d. [*Musterbuch*]. Archiv der Stadt Köln, Ms. W° 276, [end of the sixteenth century]. Elevation of the ribs and ground plan of the vault. From (Müller 1990, 178)

surfaced suddenly. However, the fact that we have almost no documents of the development before de l'Orme does not necessarily suggest a stroke of genius accomplished by him alone.[31] Rather, this fact indicates that these techniques originated and reached maturity in the practice of stonemasons who wrote no texts and left no marks besides the works they built.[32]

In any case, it goes without saying that these sophisticated techniques of geometrical projection cannot be regarded as applications of learned mathematics. No theory of geometrical projection existed, no descriptive geometry that could be applied. Rather, these techniques represent a highly developed practical geometry that existed independently of and alongside the erudite geometry of the

[31] Actually, de l'Orme laid claim to authorship of one of the two methods of stonecutter's stereometry, namely to the construction "*par panneaux*" which differs from that "*par équarissement*" (Camerota 2004, 203 f.). For these methods, see (Trevisan 1999).

[32] "As proven by the names [sc. *trompe de Montpellier*, *trompe quarrée* etc.] still used to define some of these vaults, they are architectural types coming from the Romanesque and Gothic tradition of Southern France" (Camerota 2004, 203).

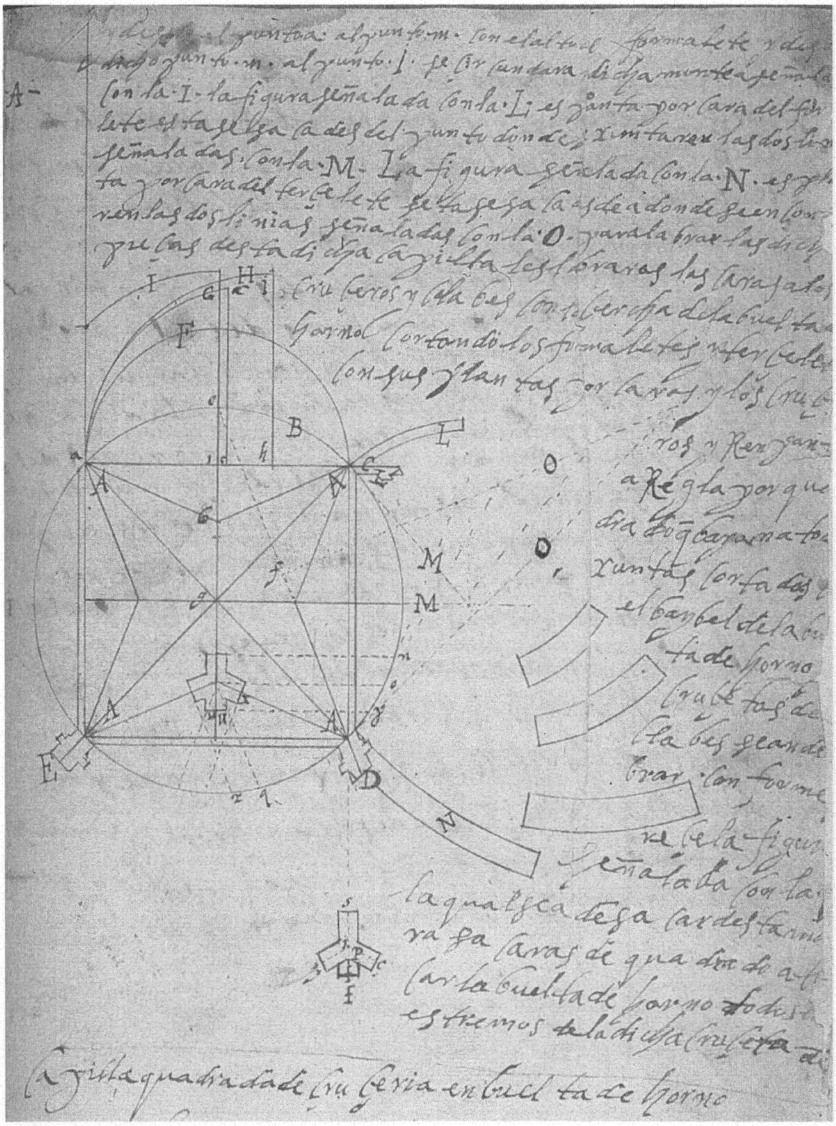

Fig. 8b Designing by combined ground plan and elevation. Construction of a vault from de Guardia, Alonso. *Collección de trazas de montea*. Venetia, 1560. From (Rabasa Diaz 2000, image 165)

age. It is, on the other hand, hardly imaginable that artists and master masons could have developed these techniques without considerable geometrical knowledge.[33] Thus, one can perhaps say, these

[33] Nothing seems to be known about any formal mathematical education of de l'Orme (Blunt 1958, chap. 1; Potié 1996). Being the son of a master builder, he became a master builder himself, who stood out for his uncommon familiarity with classical architecture acquired through journeys and commerce with humanists. He might also have acquired his mathematical knowledge through such commerce. This possibility is suggested by two other pioneers of practical geometry of the age, the artists Piero della Francesca and Albrecht Dürer, who acquired their mathematical knowledge not by formal education but by commerce with educated mathematicians or humanists—Luca Pacioli in the case of Piero and, in Dürer's case, Willibald Pirckheimer and Johann Tscherte. Even less is known about the education of Alonso de Vandelvira, who also stemmed from a master builder's family, than about de l'Orme's (Sanabria 1989, 282).

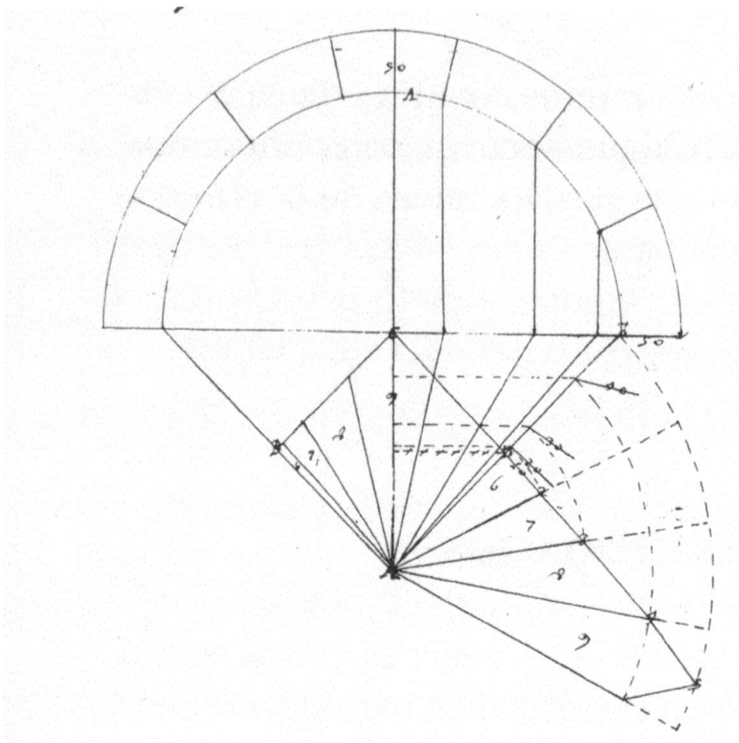

Fig. 8c Designing by combined ground plan and elevation. Construction of an arch from de Vandelvira, Alonso. *Libro de trazas de cortes de Piedras*. Bibl. Nacional de Madrid, Ms. 12.719, [c. 1580]. From (Rabasa Díaz 2000, image 10)

Fig. 9a Three-dimensional shape of a trompe de Montpellier from De la Rue, Jean-Baptist. *Traité de la coupe des pierres*. Paris: De L'Imprimerie Royale, 1728. From (Lenz 2009, image 14.5)

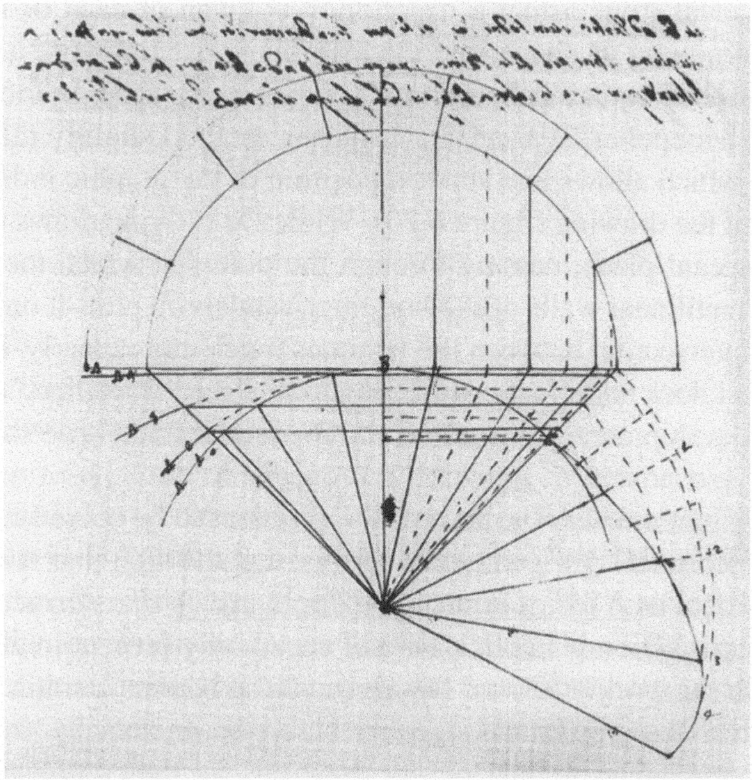

Fig. 9b Constructive drawing for the "trompe de Montpellier" from de Vandelvira, Alonso. *Libro de trazas de cortes de Piedras*. Bibl. Nacional de Madrid, Ms. 12.719, [c. 1580]. From (Camerota 2004, image 6.20)

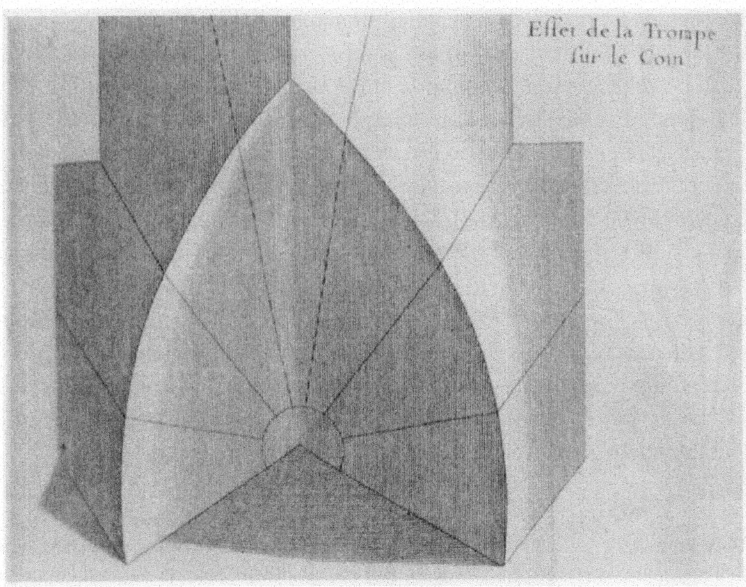

Fig. 10a Three-dimensional shape of a trompe quarrée from De la Rue, Jean-Baptist. *Traité de la coupe des pierres*. Paris: De L'Imprimerie Royale, 1728. From (Lenz 2009, image 14.4)

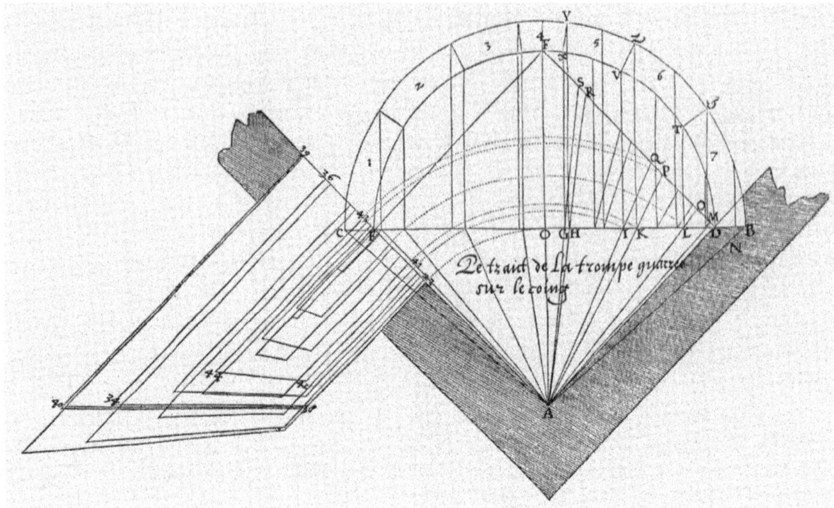

Fig. 10b Philibert de l'Orme's design of a trompe quarrée. From (De l'Orme 1567). (Camerota 2004, image 6.17)

practitioners most likely applied geometrical theories in the tradition of Euclid—however, and this is the decisive point here, they did so not to map solids onto planes, but to invent and refine their own mapping techniques.

Anyway, whatever this sophisticated practitioners's geometry may have owed to learned geometry, we should not finish this section about the interplay of erudite and practical knowledge in the construction trade without at least indicating—by a short forecast of this geometrical projection techniques' further development—what learned geometry reversely owed to this practitioners's geometry.

Turning accordingly toward the known contributors to the further development of these techniques we encounter practitioners—skilled stonemasons like Mathurin Jousse (1607–1650), master mason of La Flèche, who published treatises on stereometric constructions (Jousse 1642) as well as accomplished architect-engineers like Jean Baptiste de la Rue (1697–1743), son of a constructor, who published an influential treatise on stereotomy in 1728 (De la Rue 1728) that gained him admission to the French *Académie royale d'architecture*. And we encounter more and more men with high mathematical competence and even scholarly mathematicians. The Jesuit François Derand (1588–1644), for example, who published an influential treatise on mortarless vaulting in 1643 (Derand 1643), taught mathematics at the Jesuit college at La Flèche and served as one of the order's architects. However, it appears that he belonged to the group of *studiolo* architects, mentioned above, with limited practical competence. In Girard Desargues (1591–1661), finally, we encounter one of the heroes of the history of mathematics who wrote a four-page draft about geometrical projections pertaining to advanced stereotomy (Desargues 1640) that addressed fellow mathematicians rather than stonemasons. Historians of mathematics regard this draft as the birth certificate of descriptive geometry, which Judith Field called "a new, non-Greek way of doing geometry."[34]

[34] Judith Field in a 1995 article on Desargues published on a website of the School of Mathematics and Statistics at the University of St. Andrews (http://www-history.mcs.st-and.ac.uk/Biographies/Desargues.html. Accessed 30 May 2016). She refers there to projective geometry. But she could have said the same about descriptive geometry, since the stories of these two disciplines are strikingly parallel. "While stereotomy, together with carpentry, provides one of the richest examples of the uses of applied geometry, it is also at the root of a branch of erudite geometry, namely descriptive geometry. To sum up the situation, one might say that stereotomy is to descriptive geometry what perspective is to

Indeed, this new geometry was not rooted in the erudite geometry of the Greeks but in the practical geometry of stonemasons, which experienced an unprecedented perfection in the sixteenth century.

Desargues' draft is also of particular interest as a document that marks the point where a theoretical descriptive geometry detached itself from the practical geometry of the stonemasons. It took another century and a half before this new, non-Greek, geometry assumed its classical theoretical form in Gaspard Monge's (1746–1818) *Géométrie descriptive* published shortly before 1800 (Monge 1798–1799).[35]

4 Conclusion

In concluding I want to emphasize once more the three points I intended to make, above all, in this paper.

First, the structure of knowledge of any practical field with division of labor cannot be conceived of as a mental structure. Rather, the knowledge employed in such a field consists of the various pieces of knowledge of the collaborating actors and, thus, possesses a social structure. These pieces do not form a whole according to an epistemological order, but are fragmented elements of a fabric that complement one another due to the nature of the division of labor at hand.

Second, with regard to the interplay of practical and erudite knowledge within the knowledge fabric of the construction trade in the long sixteenth century, the architect, that is, the practitioner who designs and plans the building and/or supervises the construction process, deserves special attention. Many of the architects of this age can be regarded as hybrid or savant experts whose expertise rested on practical knowledge *and*—in a full spectrum of degrees—on erudite knowledge, first of all mathematics and also scholarly knowledge about classic Roman architecture.

Third, with respect to the mathematics of Renaissance architects, the interplay of practical geometry and learned geometry in the tradition of Euclid cannot be conceived of as an application of the latter by the former. Rather, architects developed, in the context of stereotomy, a sophisticated practical geometry besides learned geometry, a practical geometry that transgressed the canonic arsenal of geometric entities and eventually gave birth to a new mathematical discipline—descriptive geometry.

References

Ackerman, James S. 1970. Notes on Bramante's bad reputation. In *Studi Bramanteschi – Atti del Congresso Internazionale*, 339–349. Milan: De Luca Editore.

Ackerman, James S. 1991. Architectural practice in the Italian Renaissance. In *Distance points – Essays in theory and Renaissance art and architecture*, 361–384. Cambridge, MA: MIT Press.

Arnold, Thomas F. 2002. *The Renaissance at war*. London: Cassell.

Barnes Jr., Carl F. 2007. *The portfolio of Villard de Honnecourt. A new critical edition and color facsimile*. Burlington: Ashgate.

projective geometry. The parallel between the evolution of stereotomy and perspective is indeed striking. Both practices developed during the Gothic period—whether on stone cutting work sites or in painters' workshops. The first treatises were edited during the Renaissance and the mathematicians of the 'Monge School' explicitly theorized stereotomy and perspective at the end of the 18th and beginning of the 19th century" (Sakarovitch 2003, 72). It remains to add that stereotomic techniques did not originate with Gothic architecture but can be traced back to times before classical antiquity; see, for instance, (Semper 1863, 10. Hauptstück).

[35] See also (Dhombres 1992).

Becchi, Antonio. 2014. Fokus: "Die Gestalt der Säule." In *Wissensgeschichte der Architektur*, vol. 3, ed. Jürgen Renn, Wilhelm Osthues, and Hermann Schlimme, 369–396. Berlin: Edition Open Access.
Bechmann, Roland. 1993. *Villard de Honnecourt. La pensée technique au XIIIe siècle et sa communication*. Paris: Picard.
Binding, Günther. 1993. *Baubetrieb im Mittelalter*. Darmstadt: Wissenschaftliche Buchgesellschaft.
Binding, Günther. 1999. Architectus, magister operis, wercmeistere: Baumeister oder Bauverwalter im Mittelalter. *Mittellateinisches Jahrbuch* 34 (1): 7–28.
Binding, Günther, and Susanne Linscheid-Burdich. 2002. *Planen und Bauen im frühen und hohen Mittelalter*. Darmstadt: Wissenschaftliche Buchgesellschaft.
Blunt, A. 1958. *Philibert de l'Orme*. London: Zwemmer.
Bork, Robert. 2011. *The geometry of creation. Architectural drawing and the dynamics of gothic design*. Farnham: Ashgate.
Bredekamp, Horst. 2007. *Galilei – Der Künstler*. Berlin: Akademie Verlag.
Bürger, Stefan. 2009. Werkmeister – Ein methodisches Problem der Spätgothikforschung. In *Werkmeister der Spätgothik*, ed. Stefan Bürger and Bruno Klein, 18–36. Darmstadt: Wissenschaftliche Buchgesellschaft.
Camerota, Filippo. 2004. Renaissance descriptive geometry. In *Picturing machines. 1400–1700*, ed. Wolfgang Lefèvre, 175–208. Cambridge, MA: The MIT Press.
Campa, Maria Rita. 2006. "Le nouvelles inventions pour bien bastir et a petits fraiz" by Philibert de l'Orme: A new way to conceive wood roof covering. In *Proceedings of the second international congress on construction history*, vol. 1, 525–541. Newcastle: University of Newcastle upon Tyne.
Carpo, Mario. 2001. *Architecture in the age of printing – Orality, writing, typography, and printed images in the history of architectural theory*. Cambridge, MA: The MIT Press.
Cellini, Benvenuto. 1829. Discorso dell' archittetura. In *Ricordi Prose e Poesi di Benvenuto Cellini*, ed. Francesco Tassi, vol. 3, 364–373. Florence.
[Cesariano, Cesare]. 1521. *Di Lucio Vitruvio Pollione de architectura libri dece*, Como: G. da Ponte.
de Guardia, Alonso. 1560. *Collección de trazas de montea*. Venetia
De l'Orme, Philibert. 1567. *Le premier tome de l'architecture de Philibert de l'Orme*. Paris: Federic Morel.
De La Rue, Jean-Baptist. 1728. *Traité de la coupe des pierres*. Paris: De L'Imprimerie Royale.
de Vandelvira, Alonso. c. 1580. *Libro de trazas de cortes de Piedras*. Bibl. Nacional de Madrid, Ms. 12.719.
Derand, François. 1643. *L'Architecture de voûtes ou l'Art des traits et coupe des voûtes*. Paris: Chez Sebastian Cramoisy.
Desargues, Girard. 1640. *Brouillon project d'exemple d'une manière universelle du S.G.D.L. touchant la practique du trait à preuves pour la coupe des pierres en l'architecture*. Paris: by the author.
Dhombres, Jean (ed.). 1992. *L'Ècole normale de l'an III. Leçons de mathematiques. Laplace, Lagrange, Monge*. Paris: Dunod.
Duffy, Christopher. 1979–1985. *Siege warfare – The fortress in the early modern world*. 2 vols. London: Routledge.
Dürer, Albrecht. 1625. *Underweysung der Messung*. Nurmberg: [n.p.].
Fracht von Andernach, Jacob. n.d. End of the sixteenth century. [Musterbuch]. Archiv der Stadt Köln, Ms. W° 276.
Gerbino, Anthony, and Stephen Johnston. 2009. *Compass and rule – Architecture as mathematical practice in England*. New Haven, CT: Yale U.P.
Goldthwaite, Richard A. 1982. *The building of Renaissance Florence. An economic and social history*. Baltimore: The Johns Hopkins U.P.
Hahnloser, Hans R. (ed.). 1972. *Villard de Honnecourt: Kritische Gesamtausgabe des Bauhüttenbuches ms. Fr. 19093 der Pariser Nationalbibliothek*. Graz: Akademische Druck- u. Verlagsanstalt.
Haines, Margaret. 2011–2012. Myth and management in the construction of Brunelleschi's Cupola. *I Tatti Studies: Essays in the Renaissance*, 14–15: 47–101.
Hall, Bert S. 1997. *Weapons and warfare in Renaissance Europe – Gunpowder, technology, and tactics*. Baltimore: The Johns Hopkins U.P.
Haselberger, Lothar. 1999. *Essence and appearance. Refinements in classical architecture*. Philadelphia: University of Pennsylvania Museum Publications.
Høyrup, Jens. 2002. *Length, width, surfaces: A portrait of old Babylonian algebra and its kin*. New York: Springer.
Høyrup, Jens. 2009. Constructural procedures in "practical geometries." In *Creating shapes in civil and naval architecture. A cross-disciplinary comparison*, ed. Horst Nowacki and Wolfgang Lefèvre, 367–377. Leiden: Brill.
Johannes, Ralph (ed.). 2009. *Entwerfen. Architekturausbildung von Vitruv bis Mitte des 20. Jahrhunderts. Geschichte – Theorie – Praxis*. Hamburg: Junius.
Jousse, Mathourin. 1642. *Le Secret d'Architecture decovvrant fidelement les traits geometriques, covppes*. La Fleche: George Griveau.
Klein, Bruno. 2009. Werkmeister oder Architekt? In *Werkmeister der Spätgothik*, ed. Stefan Bürger and Bruno Klein, 13–17. Darmstadt: Wissenschaftliche Buchgesellschaft.
Kruft, Hanno-Walter. 1985. *Geschichte der Architekturheorie – Von der Antike bis zur Gegenwart*. München: Beck.

Lacher, Lorenz. 1856. *Unterweisung [1516]*. Leipzig: T. O. Weigel.
Lamberini, Daniela. 1990. *Il principe difeso. Vita e opere di Bernardo Puccini*. Florence: Editrice La Giuntina.
Lefèvre, Wolfgang. 2004. The emergence of combined orthographic projections. In *Picturing machines – 1400–1700*, ed. Wolfgang Lefèvre, 209–244. Cambridge, MA: The MIT Press.
Lenz, Martina. 2009. The Squinch of Anet. In *Creating shapes in civil and naval architecture. A cross-disciplinary comparison*, ed. Horst Nowacki and Wolfgang Lefèvre, 321–341. Leiden: Brill.
Maffioli, Cesare S. 2013. A fruitful exchange/conflict – Engineers and mathematicians in early modern Italy. *Annals of Science* 70 (2): 197–228.
Mai, Ekkehard. 2012. Kunstakademien und Architektenausbildung. In *Der Architekt – Geschichte und Gegenwart eines Berufsstandes*, vol. 2, ed. Winfried Nerdinger, 537–548. Munich: Prestel.
Monge, Gaspard. 1798–1799. *Géométrie descriptive: leçons données aux écoles normales, l' an 3 de la république*. Paris: Baudouin, an 7.
Müller, Werner. 1990. *Grundlagen gotischer Bautechnik*. Munich: Deutscher Kunstverlag.
Pevsner, Nicolaus. 1942. The term "architect" in the middle ages. *Speculum* 17: 549–562.
Pfammatter, Ulrich. 2012. Architekt und Ingenieur – Aspekte der Entwicklung und Zukunft der Baudisziplinen. In *Der Architekt – Geschichte und Gegenwart eines Berufsstandes*, vol. 2, ed. Winfried Nerdinger, 559–576. Munich: Prestel.
Philipp, Klaus Jan. 2012. Euphorie und Ernüchterung – Architektur und Kunst. In *Der Architekt – Geschichte und Gegenwart eines Berufsstandes*, vol. 2, ed. Winfried Nerdinger, 549–558. Munich: Prestel.
Potié, Philippe. 1996. *Philibert de l'Orme. Figures de la pensée constructive*. Marseille: Parenthèses.
Rabasa Díaz, Enrique. 2000. *Forma y Construcción en piedra. De la cantería medievale a la estereotomía del siglo XIX*. Madrid: Ediciones Akal.
Rabasa-Díaz, Enrique, and José Calvo-López. 2009. Gothic and Renaissance design strategies in stonecutting. In *Creating shapes in civil and naval architecture. A cross-disciplinary comparison*, ed. Horst Nowacki and Wolfgang Lefèvre, 167–192. Leiden: Brill.
Roriczer, Matthäus. 1486. *Büchlein von der Fialen Gerechtigkeit*. Regensburg: by the author.
Sakarovitch, Joel. 1998. *Épures d'architecture. De la Coupe des pierres à la géométrie descriptive – XVIe–XIXe siècles*. Basel: Birkhäuser.
Sakarovitch, Joel. 2003. Stereotomy, a multifaceted technique. In *Proceedings of the first international congress on construction history, Madrid, 20th–24th January 2003*, ed. S. Huerta, 69–79. Madrid: F. Dragados.
Sanabria, Sergio Luis. 1989. From gothic to Renaissance stereotomy. *Technology and Culture* XXX (2): 266–299.
Schlimme, Hermann, Dagmar Holste, and Jens Niebaum. 2014. Bauwissen im Italien der frühen Neuzeit. In *Wissensgeschichte der Architektur*, vol. 3, ed. Jürgen Renn, Wilhelm Osthues, and Hermann Schlimme, 97–367. Berlin: Edition Open Access.
Scotti, Auror. 1983. Il collegio degli architetti, ingegneri e agrimensori tra il XVI e il XVIII secolo. In *Construire in Lombardia – Aspetti e problemi di storia edilizia*, ed. Aldo Castellano and Ornella Selvafolta, 92–108. Milan: Electa.
Semper, Gottfried. 1863. *Keramik, Tektonik, Stereotomie, Metallotechnik (= vol. 2 of idem: Der Stil in den technischen oder tektonischen Künsten.)*. Munich: Friedrich Bruckmann.
Shelby, Lon R. 1972. The geometrical knowledge of mediaeval master masons. *Speculum* 4: 395–421.
Thoenes, Christof. 1995. Anmerkungen zur Architekturtheorie. In *Architekturmodelle der Renaissance – Die Harmonie der Bauten von Alberti bis Michelangelo*, ed. Bernd Evers, 8–39. Munich: Prestel.
Trevisan, Camillo. 1999. The hidden proportions in the *trait* of the *trompe* of Anet. *Disegnare idee e immagini* 16: 59–66.
Vasari, Giorgio. 1981. Antonio da Sangallo. In *Le Opere di Giorgio Vasari*, vol. V. Florence: Sansoni.
Victor, Stephen K. 1979. *Practical geometry in the high middle ages*. Philadelphia: The American Philosophical Society.
Wilson Jones, Mark. 2009. Greek and Roman architects' approach to curvature. In *Creating shapes in civil and naval architecture. A cross-disciplinary comparison*, ed. Horst Nowacki and Wolfgang Lefèvre, 93–118. Leiden: Brill.

Knowledge by Design—Architecture and Jade Models During the Qianlong 乾隆 Reign (1735–1796)

10

Dagmar Schäfer

Abstract

Archival household records and artifactual evidence suggest that the Qing emperors and 'court officials' use of models was part and parcel of a systematic appropriation of in-between communication devices in technical and artistic production. This essay analyses when and how officials and craftsmen employed models with a view on architecture and jade production. What made such sampling fashionable and what function did actors of the period assign to such devices for the circulation of aesthetic and technical concerns? Starting with a highly personal case of communication between the emperor and jade craftsmen the chapter shows how models became part of the standardization of methods of circulation across trade boundaries. Bureaucratic handling promoted the unification of descriptive and prescriptive modes that included both sketching and three-dimensional modeling. Clerks further explained material, size and color schemes on yellow strips attached to the model. In such bureaucratic procedures actors considered the aesthetic and technical information given in models conclusive and valid. Imperial workshops then functioned more as design bureaus than as producing sites per se.

A well-known story originating during Qianlong's 乾隆 (1711–1799) reign that anticipated the downfall of the Qing dynasty 清 (1644–1911) concerns Niohuru Heshen 钮祜禄·和珅 (1750–1799). At the height of his career, Heshen smuggled his henchman, Hushitu 胡什圖 (n.d.), into the palace

This essay was inspired by a collaborative project between the Palace Museum Research Institute, Beijing and Research Group I at the MPIWG, Berlin that ran from 2008 to 2011. Much of the research and ground work on jade was done by Guo Fuxiang who, even though initially extremely skeptical, went out of his way—and time and again returned to the archives—to find materials that would answer my many questions. Further thanks go to Zhang Shuqian, Zhang Qiong, Song Lingping, Wang Guangyao and Luo Wenhua for their inspiring research work and our many intellectually challenging debates.

D. Schäfer
Max Planck Institute for the History of Science, Berlin, Germany

Technical University, Berlin, Germany
e-mail: dschaefer@mpiwg-berlin.mpg.de

M. Valleriani (ed.), *The Structures of Practical Knowledge*, DOI 10.1007/978-3-319-45671-3_10

to steal models of Qianlong's retirement home, the Palace of Tranquil Longevity (*Ningshou gong* 寧壽宮). In the historical account, the model's theft has received less attention than the whopping 80 million silver taels that Heshen also embezzled. And yet, when the Jiaqing Emperor 嘉慶 (1760–1820), (reign 1796–1820), took over after Qianlong's death, he made sure that Heshen's breach of confidence in acquiring the model made it on the public list together with the embezzlement (Liu et al. 2000, 9859–60, ch. 242, *jiaqing* 4, month 1).[1] With the stolen architectural model Heshen was able to replicate the palace's design, only slightly scaled down, for his own garden and residence. In a world in which material display exemplified status and ideals, Heshen had managed the impossible. He had arrogated imperial rights with impunity, until Jiaqing's ascension.[2]

Long before the recently revived interest in the heritage of the imperial architect clan of Lei 雷, the story of Heshen pinpoints the Qing court's systematic use of material models in imperial architectural planning. Klaas Ruitenbeek describes the process as follows: "In designing a building, use was made of models, and of drawings and plans. Miniature models were made by architects of large projects to show their patrons what they intended to build" (Ruitenbeek 1996, 49–50). Historian of architecture Jing Qingsheng situates the use of models and sketches within a long tradition of state approaches to knowledge circulation in architecture (Jing 1985, 10–4).[3] Liu Dunzhen traces the use of scaled models back to textual sources of the Sui Dynasty 隋 (581–618), suggesting that architects at that time sent models of local buildings to the central government for approval and also dispatched models from the central government to local districts to ensure unified styles (Liu et al. 1984). Despite such acknowledgements, the practical and cognitive function of models, sketches and artifacts has hitherto been paid little heed in the historical study of China's artistic, technical and scientific developments.

Archival household records and artifactual evidence suggest that the Qing emperors' and court officials' use of models in architecture was part and parcel of a larger, much more systematic appropriation and application of communicative devices of technical and artistic production. Sketches, models, test pieces, patterns and styles (*shiyang* 式樣) were collected for jade, textile and enamel production, embroidery, interior design, bronze casting, ship building, and astronomy, to name only a few examples. Chen Kaijun has recently shown that the Qing often used models within fiscal processes (Chen 2014, 138–78, mainly ch. 4). The model was a materialized accounting tool. Focusing on architecture and jade, this essay raises two additional points: (1) when officials and craftsmen employed such means in different crafts for bureaucratic purposes, they considered the aesthetic and technical information given in models as conclusive and valid. (2) Within state bureaucracy, methods of circulation were standardized across different trade boundaries.

Artifactual and archival evidence substantiates that, towards the end of the Qing era, emperors and court actors became obsessed with details. In the matter of communication between the emperor and production sites, only a one-to-one material copy would do. The boundary between aesthetic and communicative function—items used for religious, social, or display purposes or models produced for the circulation of technical information—was thus easily crossed. Technical documentation on colors, patterns and forms proliferated because the court was unable to force skilled craftsmen from the South to move to the new capital in Beijing. In this case the increase of bureaucracy in the technical sector indicates that the court and emperor were *not* in control.

[1] For He Shen's biography see (Zhao 2012, vol. 8, 3199–202, ch. 319, *liezhuan* 106). Jiaqing officially ascended the throne in 1796 but Qianlong still continued to have much power in the last 4 years as an ancestor-emperor.

[2] (Yoon 2012) has lately revised the story and shown that Heshen only began to abuse his power after Qianlong's abdication. In (Akcetin 2007), the introduction suggests that in a similar case related to the province of Gansu the term corruption needs to be reconsidered, and that material goods were in fact part of a gift economy before they were utilized for political prosecution by Jiaqing.

[3] See also (Jing 1988, 15; Feng 2007).

Along these lines, I suggest that the imperial court employed models as a way to intercede in a field of knowledge to which they otherwise had no access. From this viewpoint, the growing proliferation of technical records, the increase of written information, illustrations and modeling, is expressive of an imperial wish for participation rather than simply a case of the authoritarian Qing court exercising control of major representational trades. The elite's desire to participate in craft production spurred modeling and an increased interest in accurate, true-to-life imagery and depiction in the production of arts and crafts.

1 Copying and Imitation: Jade Production

The Manchu Qing court was concerned about jade because of its emblematic importance to the legitimacy of rulership in the Chinese world. Jade was mostly found at the margins of the Qing empire: Yunnan and Gansu. During the Ming era 明 (1368–1644), Suzhou in Jiangnan had become the center of jade production. Establishing the capital in Beijing had hampered court control over this sector considerably. In the year 1773 Qianlong was at the height of his rule. This was the year that also saw the imperial commissioning of the largest collection of Chinese literature, the encyclopaedic *Complete Library in Four Sections* (*Siku quanshu* 四庫全書). The 38th year of Qianlong's reign thus stands clearly under the spell of texts and a momentous Manchu-Qing imperial attempt to put its mark on Chinese culture and intellectual debates. For the Imperial Household Agency (*Neiwufu* 內務府), the 1770s were a busy time. Requests and demands came in thick and fast, challenging the Court Imperial Workshop's (*Zaoban chu* 造辦處) capacity and expertise. Amidst intellectual debates about the validity of classical knowledge, the need to nominate a new heir to the throne, and receptions of foreign tributary wares, Qianlong, however, spared a thought for the production of a jade model, probably even engaged with it himself, although he may well have relied on the well-oiled machinery that ensured that his view would be implemented.

On record is the case of a wooden model for a jade carving presented to the Qianlong Emperor in the year 1771 to instigate the production of a larger jade piece depicting the classical Chinese scene of "Zhou Chu Beheading the Flood Dragon" (*Zhou Chu zhan jiao* 周處斬蛟) (Fig. 1). Whereas the models of the palace were archived by the Lei clan, Heshen's story suggests that in the era of Qianlong the palace models were still owned by the palace, or at least, stored in the palace itself. Documents recording a wooden model (*muyang* 木樣) used in jade production suggest that the court systematically collected and archived samples in the imperial workshops for use and re-use. Models used for the production of this piece are remarkable on two levels: firstly, the rather large number of models that were used with some illustrating the various steps to take when cutting the stone. Secondly, the final model itself embodies every sophistication of the final design. Yellow paper strips are glued to the object explicating materials, color and size.

A growing obsession with detail is central to the court's involvement with jades. And it meant that any perceived line between the model as a tool or object d'art was indeed blurred. Qianlong, fascinated by the delicate carvings of the later model, ordered that, "the wooden mock-up should be hollowed-out inside and decorated with patterns on the outside. Let the artisans of the Palace of Inspired Luck (*Qixiang gong* 啟祥宮) finish this off properly, and when ready it should be gilded and lacquered in the same way as a gilt-lacquered jar."[4]

[4]eds. Zhongguo diyi lishi dang'an guan, Xianggang zhongwen daxue wenwuguan 2005 [hereafter *Zonghui*], vol. 34, 263, Qianlong 36, month 11, "Xingwen" 行文 (Official Communication Documents).

Fig. 1 Wooden model for a jade stamp, Guangxu era, attached note reads: "carve white (baiyu 白玉) or greenish (qingyu 青玉) jade following this model (zhao ci yang ke 照此様刻)." 10.95 × 12.5 cm; photographer: Ma, Xiaoxuan; copyright: The Palace Museum

Guo Fuxiang traced the procedure of production in detail following the archival paper trail:

- On 3rd of [?] month the green jade object "Zhou Chu Beheading the Flood Dragon" delivered from Suzhou was submitted for consideration to the throne together with its wooden mock-up.
- The imperial response called for the jade object to be "temporarily taken out pending a time when there would be a [suitable] place and then installed there. The wooden mock-up should be hollowed-out inside and decorated with patterns on the outside. Let Qixiang Gong finish this off properly, and when ready, it should be gilded and lacquered in the same way as a gilt-lacquered jar."
- On the 28th of the same month an order came down from the throne that "the display object 'Zhou Chu Beheading the Flood Dragon' currently in store should be displayed in the Hall of the Scholar (*Xueshi Tang* 學士堂) in the Palace of Yang-Scenery (*Jingyang Gong* 景陽宮)."
- On the date Qianlong 38/9/26 [10th of November 1773] the wooden mock-up of "Zhou Chu Beheading the Flood Dragon" now gilded and lacquered, was submitted for consideration to the throne. The imperial response called for it to be "matched with a wooden-core sand-coated stand, and when ready, be put to use for Buddha-offerings" (*Zonghui*, vol. 10, 419–20; Guo Fuxiang 2010).

The jade object and its model had distinct careers after completion of the project. The jade 'display object' was kept in store for 3–4 weeks, then put on display in the Hall of the Scholar in the Palace of Yang-Scenery. At a later date the wooden mock-up, duly modified, gilded and lacquered, was finally provided with a stand and destined for ritual use in Buddha-offerings.

This case reveals more than whether technology and art in this era were disjunctive. It shows that aesthetic choices and technical considerations were interrelated—an issue the Qing rulers acknowledged and used for their own purposes. What was perceived as art and what as a technical tool or auxiliary was at the individual's discretion and in the eye of the beholder (*Zonghui,* vol. 36, 419–20, "Suzhou" 蘇州).

2 Jade Under Control: Ruling Skills

When discussing jade, Qianlong often elaborated technical details and processes illustrating that the contemporary evaluation of copying skills were high. Not originality but skill was relevant for Qianlong's connoisseur mind. When the emperor clarified his ideas on the design of a stone and the artisans in response elaborated on material quality and capabilities, Qianlong appears as a ruler to have been concerned with the details of the jade trade, interested in adopting the technical terminology, and learning about specific methods such as the coloring of stones.

In the 18th year of his rule (1753), for instance, Qianlong had appreciated a jade goblet of a dark purplish-red color and lustre with two baby-shaped handles, which he thought resembled unearthed antiques of the pre-Han era. However, after passing his hands over the goblet, he noticed its unusually uneven surface and, upon further scrutiny, realized that the purplish-red luster appeared to be a separate layer rather than the inner natural color of the jade. Yao Zongren 姚宗仁 (n.d.), a jade craftsman from a renowned Suzhou jade carving family, who had served the Inner Court for over 20 years, was called in. At first glance Yao, with a smile on his face, identified the piece as an imitation of an antique made by his own grandfather.

Such incidences were recorded to enhance the image of the emperor as an attentive ruler and art expert. The document implies that Qianlong had suspected the connection even before Yao came in. Qianlong further advertised his understanding of the technique in an essay: 'On Jade Goblet' (*Yubei ji* 玉杯記) in which he emphasized Yao's expertise and reports that the jade was "perforated with a diamond drill. Small pits emerged that are then stained with molten amber. Then the jade piece is tempered on a low fire." The emperor also pondered the time-consuming process, pointing out that the jade had to soak in the molten amber for several hours or even days.[5]

The Qianlong Emperor's aim in poeticizing these details might have been aesthetic, since identifying antiques was a useful way of demonstrating expertise in the Chinese arts. But his imagery also stressed that connoisseurship required a thorough understanding of basic techniques. In several other incidences Qianlong instructed artisans on methods of color impregnation and on how to make good use of the natural color of jades to achieve the requested effect.[6] Archivists included such reports to enhance the image of an all-knowing and all-caring ruler, but they could not have done so if the emperor had not been genuinely interested in the technical processes of copying.

Continuity is another issue that marks Qianlong's interest in jade techniques. Knowledge that he had once gained would not be forgotten. By the 50th year of his reign (1785), and thus 30 years after his first reported encounter with craftsman Yao, Qianlong had become a real expert in the theory of jade coloring. Qianlong scrutinized an octagonal jade piece with a round hole in the center (*Yuwangtou* 玉輞頭) at his Imperial Summer Resort (*Bishu shanzhuang* 避暑山庄) in Chengde city, and, identifying it as Han Dynastic 漢 (206 BC–220 AD) in style, reminisced upon its gorgeous ancient color and the smoothness of Yao's technique: "Today's imitations of ancient jade have the color of oil. The carving patterns are shallow, and the texture is coarse and clumsy. They can only stand up to a superficial scrutiny and no (real) expert is deceived for longer than a first glance" (Hong 1993, vol. 8, 485, Fifth collection 五集, ch. 17, 7a–b, "Chant about the Han Yuwangtou," author's translation).

Five months later, in 1786, Qianlong references craftsman Yao again when bemoaning the decline of contemporary jade craftsmanship in the poem 'Chant about the Han Dynastic Yuwangtou Bottle' (*Yong Han Yuwangtou ping* 咏漢玉輞頭瓶). Qianlong suspected that the Han era *Yuwangtou* bottle had to be an imitation of an earlier one: "The jade, though hard and solid, shows some blemishes. Who added these joins and flaws? The color was achieved through soaking, tempering and dyeing. This imitation, done hundreds of years ago, is far better than the oily staining achieved by contemporary

[5]All quotes in the above paragraph are from (Hong 1993, vol. 10, 357, ch. 5, 12a–b). The jade staining technique is known as "Amber Burn" according to (Yang 2006, 399–407).

[6]Qianlong, for instance, asked Yao Zongren to examine the quality of a Han jade to see if there was any similarly naturally colored ancient jades for the production of a ring. (*Zonghui* 2005, vol. 18, 708, Qianlong 17, month 11, "Chronicle"). In another example, Qianlong asked the jade artisan Wen Dan 文旦 (n.d.) to produce a green jade flared-rim bowl, but then to bury it for 5 or 6 years to achieve the right color (*Zonghui,* vol. 18, 808, Qianlong 21, month 12, "Chronicle").

methods." His criticism was harsh: "Staining jades has its *dao*. Contemporary craftsmen, however, stain without care, performing it like plasterers."[7]

In sum, events around jade production imply that issues of authenticity and the delicacy of execution informed Qianlong's assessment of jades. Having realized that jades could be stained to make them look antique, Qianlong ensured that his officials were aware of the technique, ordering the full text of *On Jade Goblet* to be carved on the boxes of jade goblets (which are now preserved in The Palace Museums in Beijing and Taipei) (Zhang 1986). Such actions, however, do not imply that Qianlong appreciated a well-made imitation less than an original antique. Knowing the difference did not imply a value judgment per se.

3 An End to Models: Training Craftsmanship

In their uniqueness, the records on jade production and the use of models for production purposes disclose the emperor's interest in technological know-how as much as they show the limitation of the imperial courts' influence. Craftsmen resisted imperial interference and did not move to the capital or delivered low quality work to exempt themselves from service.

In jade design, the ruling class was forced to work around such resistance by changing the flow of information. They enforced their authority by telling the craftsman how to cut the stone, inscribing their interpretation of the cutting task on the stone. Not every stone could be cut into every shape. Thus every piece had to be examined. Unskilled cutting could easily destroy excellent pieces. The Qianlong Emperor acknowledged the precious material by making an effort to know individual jade carvers by name, and in due course he also claimed to recognize their specific work. Officials may have served as the actual mediators. But the records suggest intimate contact between the emperor and ordinary craftsmen. Artisans could thus rise to the status of imperial advisors through meeting with the emperor or his staff each day. Gu Jichen 顾继臣 (n.d.), a jade craftsman from Suzhou, was on stand-by to identify the types and quality of jade materials. The already mentioned Yao Zongren was an acknowledged expert on jade furnishings from the Han Dynasty. It seems that these craftsmen communicated manufacturing techniques and knowledge directly to the emperor (Yang 2006, 401).

Qianlong not only reported on techniques in writing, but also regularly inquired about the number of apprentice jade craftsmen (*xueshou yujiang* 學手玉匠). Apprentices were chosen after a practical test: "Pick some quality jade materials and choose several apprentice jade craftsmen to serve in Qixiang Palace" (*Zonghui*, vol. 24, 693, Qianlong 24, month 9, "Ruyi guan" 如意館); "Choose five apprentices from the imperial workshop office to serve in Qixiang Palace" (*Zonghui*, vol. 27, 213, Qianlong 27, month 10, "Ruyi Guan"); "Ask apprentices to do the job" (*Zonghui,* vol. 41, 748, Qianlong 43, month 3, "Ruyi Guan"). Sometimes, Qianlong himself directly passed on the order to select apprentices from the Warehouse Management Section (*Guangchu si* 廣儲司) of the imperial workshops and assigned them to learn jade ware manufacture in Qixiang Palace" (*Zonghui,* vol. 29, 508, Qianlong 30, month 1 "Ruyi Guan").

Attempting to build a pool of experts for his own use, Qianlong often attracted masters to the imperial palace workshop with the promise of a temporary assignment that would end with the successful training of apprentices. In the year 1779, Qianlong allowed the master of jade containers Ping Qi 平七 (n.d.) "to return to the South, as the apprentice jade craftsmen in Qixiang Palace have sufficiently grasped the technique for jade container manufacture." The workshop officials opposed Qianlong's order though, arguing that "while the apprentices are now in fact capable of executing the inside-cavity

[7] For further elaboration, see (Guo Fuxiang 2010; Hong 1993, vol. 8, 485, Fifth collection 五集, ch. 17, 7a–b, "Chant for the Han Yuwangtou").

carving, they have not yet mastered the methods and size of the various profiles. We hereby ask for Your Majesty's permission to retain Ping Qi for another 2 or 3 months until the apprentices have grasped the external forms." Qianlong then "asked Ping Qi to return to Suzhou the next spring" (*Zonghui*, vol. 42, 712, Qianlong 44, month 10, "Ruyi Guan"). Ping Qi left Beijing after a year while his students commenced the large-scale manufacture of jade bowls, plates and trays that were used at banquets until the end of the Qing era. Another significant aspect of Ping Qi's appointment is that this jade craftsman found the tools provided by the Studio of the Ornamental Cepter (*Ruyi guan* 如意館) wanting and, one month after arriving in Beijing, requested to have two sets of tools delivered from Suzhou directly to the Ruyi Guan, including:

> Eight benches, four sets of lathes, four *shankous* [a special tool used for jade ware processing], four direction tablets, four long sand bands, four *jin* [a unit of weight equal to half a kilogram] of thin hemp rope, eight iron salt trays, four Chinese pounds (*jin* 斤=ca. 1/2 pound) of iron wire, forty iron nails, four copper basins, four small shovels, four pounds of rosin, two pounds of bentonite, four sandalwood shafts, four large and small shovels, two anvils, a pair of scale rules, one large set of pincers and one small set of pincers, one large pair of iron shears and one small pair of iron shears, four blocks of three *cun*-square and three *chi*-deep Chinese fir wood. (*Zonghui*, vol. 42, 666, Qianlong 44, month 2, "Ruyi Guan").[8]
>
> 坐凳八個, 榆木鏇床四副, 山口四個, 向牌四個, 長砂圈四個, 線廠繩四斤, 鐵鹽盤八個, 細鐵絲四斤, 鐵釘四十個, 銅盆四個, 小鐵鏟四個, 鬆香四斤, 土粉二斤, 檀木軸四根, 大小鐵鏟四個, 大小鐵砧子二個, 矩尺一副, 鐵鉗子大小二把, 鐵剪大小二個, 三寸見方杉木長三尺四塊.

This set of tools was delivered to Ping Qi directly. The second set was delivered via the Suzhou Silk Weaving and Dyeing Bureau (*Suzhou zhiran ju* 蘇州織染局) and transported to the capital by two of its officers. It consisted of

> 50 pieces of iron sheet, one-*chi* and two-*cun* long and nine *cun*-wide; 150 bundles of regular thin iron wire, each with a weight of ten *liang* [a unit of weight equal to fifty grams]; 50 bundles of hollowed iron wire, each with a weight of ten *liang*; 50 large sand drills and 50 small sand drills, adaptive to the model; 20 *qiantas*[9] one-*chi* in diameter and two-*cun* in length; four *qianta* wood shafts; eight three-*cun* by six-*cun* paring shafts with wooden bowls, with a complete set of tail nail sandalwood shafts; and 2000 *jin* of emery (*Zonghui*, vol. 42, 619, Qianlong 44, month 2, "Xingwen."
>
> 長一尺二寸寬九寸鏇條鐵皮五十張;原條鐵絲一百五十把, 每把重十兩; 過眼鐵絲五十把, 每把重十兩; 大小砂鑽五十個, 隨樣; 徑一尺二寸千它二十個; 千它木軸四根; 徑三寸至六寸帶木碗鏇軸八根, 尾釘要齊全; 喇叭頭木軸六根, 要檀木; 寶砂二千斤.

Ping Qi hence had tools custom-made for at least four apprentices, establishing a jade container studio with segmented labor organization. Ping Qi believed that Suzhou's jade carving tools were superior to those that could be obtained in Beijing and requested different tools from two different units corresponding to the tasks that had to be performed. The Studio of the Ornamental Ceptre (*Ruyi guan*) was responsible for the manufacture of general tools and equipment such as benches, lathes, direction tablets, iron trays, copper basins, shovels, iron nails, and anvils, while the Suzhou weaving and dyeing bureau produced specialized tools and materials with higher technical requirements. This was as much a matter of shared responsibility as one of accountability.

Clearly the court's household masters and the Qianlong Emperor considered that Ping Qi's expertise earned him concessions, such as letting the artisan decide when he wanted to arrive in Beijing and train apprentices. The officials emphasized that his trainees should be given enough time to learn the necessary skills. Having a good command of the handiwork required practical experience of the operational steps. This was a process that could not be fully replaced by reading and viewing transcripts and drawings.

---

[8]The *cun* was a unit of length equal to one-third of a decimeter, and *chi* was a unit of length equal to one-third of a meter.

[9]*Qianta* was a special tool used for jade ware processing.

4 Models in Chinese History and During the Qing Dynasty

Evidently the Qing state's use of models concerned power schemes as much as knowledge flows. Financing, labor allocation, and aesthetic control are popular themes in historical discussions of the Qing period's material culture along with the organizational intervention of the state and scholar officials into the practical arts. What is new to this discourse is to inquire into the impact mediating bodies—people or objects—actually had. Design has long been a major theme in art studies. It raises questions within the history of science and technology on materials, objects and how they were seen as contributing to knowledge flow.

As the example of Qing jade demonstrates, actors in China embraced design as a crucial step in knowledge-making, giving room to combined considerations of technical, aesthetic, scientific and material concerns.[10] It is in this sense that modeling features high in the history of Early Modern European practical knowledge circulation, its methods and ideals of visualizing and objectifying nature, man, and the universe or examining, testing and representing scientific and technological developments or ideas. Research has tackled, for instance, the use of models in eighteenth century scientific academies and institutions in countries such as Britain, Italy or France. Apart from experimentation, and the training of apprentices, models were also used to pledge legal rights of ownership on innovative products and ideas to the state or patrons.[11] Alongside a growing interest in the role of craftsmen-practitioners, artisans and artists in the making of early modern European sciences, recent years have furthermore seen publications on models as a feature at the intersection of arts and science. These studies have shown how both scholars and artisans used real-sized or scaled models to consider genealogies of problems, and to form their approaches to their environment beyond visualization and by means of touch and taste (Rheinberger 2010, 8; Daston 2006, 60–7).

In narrations on transfers any material object could become a model and means of knowledge transfer. Just as the European courts featured their (since much studied) Chinoiseries, the Qing Emperors indulged in the exoticism of foreign worlds: Indian, South East Asian and European cultures alike (Kahn 1985).[12] In around 1671 the Italian Jesuit Claudio Filippo Grimaldi (1638–1712)

> … organized an elaborate scientific conversazione for the young Kangxi Emperor, and among the … curiosities which were exhibited … there figured both a model steam-carriage and a model steam-boat.… [T]he experimental models had been begun by Fr Verbiest already in about 1665. (Needham et al. 1965, 225–6)[13]

Conversely not much is known about Chinese and Manchu intellectual approaches to models as a way to grasp the world and circulate knowhow. They have hardly been touched upon outside the arts. One reason for this might be that Joseph Needham's grand endeavor on China's *Science and Civilisation* overemphasized the role of models for research on China's technical and scientific knowledge circulation. Later generations responded to Needham's often exclusive focus on technical content by tuning out the artifact's technical and informative function altogether.

At times miniature modeling was a respected trades in the arts, best known in garden architecture and the art of Bonsai (*penjing* 盆景).[14] Archaeological excavations have yielded house models, figurines of humans, animals and natural environs or miniatures of farming implements, looms, tools and

[10]For an exemplary discussion of the role of design in technological developments and its historiography see (Fallan 2010, 7–9).

[11]For an overview to this literature see (Popplow 2002).

[12]Much of Qing exoticism concerned Central Asia (Chayet 2004, 46; Hay 1999, 201).

[13]Grimaldi was very interested in optics. He also introduced the magic lantern to China (Mannoni 2000, 72).

[14](Siu 2013, xxii) suggests that Qianlong's garden of Perfect Brightness in the Forbidden City Palace "was a microcosm of his vast realm.… Hence to study this one imperial garden is to study the Qing empire in miniature." For an overview of miniaturizing in the arts see (Cammann 1982).

everyday items aplenty from ancient settlements or burial sites. Models thus simultaneously and consecutively functioned as funeral objects (*mingqi* 冥器) and objects d'art. Literature also attests to a combined ritual and functional use of models and artifacts in state bureaucracy. For centuries scholars have pondered the *Mingtang* 明堂 model that represented a cosmological view as an architectural form (Zhang 2005).

Historians identify the Qing dynasty's approach towards artifacts, sketches and models as "exceptional" (*wuqian* 無前) or "advanced" (*fada* 發達) in the context of a lack of research and sources on the multiple and changing historical purposes of three-dimensional models in prior dynasties (Wang et al. 2006, 5). However, understanding ritual, technical and artistic developments in previous dynasties is critical for a contextualization of the Qing court's bureaucratic handlings of models and their attitude towards three-dimensional modeling as sources of information and knowledge production. In a broader historical context, the Qing dynasty's specific approach to models poses questions on the role of different dynastic managerial practices in settling and changing cultural and epistemic values in the Chinese world, and even more so on the many ways Chinese actors circulated practical knowledge and information about materials, forms and processes.

5 Bannermen and Household Records

Bannermen (*shanbao* 善保, *baoyi* 包衣) such as Heshen contributed significantly to the changing role and nature of communication in the arts and crafts. Like the eunuchs of the Ming dynasty, the more military inclined bannermen were the invisible middlemen, that is the extended eyes, ears and hands for all mundane and delicate matters at the court and in the state. Despite their relatively low status the emperor met them regularly. The actual producers of most of our historical accounts, that is, officials ranging from clerk to minister, were envious of the bannermen's influence. But, unlike the Ming eunuchs, bannermen built networks and formed what Chen Kaijun has coined a "technocratic" class of their own (Chen 2013). Men like Nian Xiyao 年希堯 (1671–1738) who organized porcelain production or Cao Yin 曹寅 (1658–1712) from a silk-savvy Han-Chinese bannermen clan, gained influence in the court.[15] They networked and bonded while trading information across the empire and accounted for aesthetic and technical information by way of sketches, models and the migration of artifacts and men. Heshen represents the corrupt remains of a group of highly trained literate bannermen experts who were able to negotiate between many worlds: Chinese and Manchu tribes, craftsmen and members of the imperial clan. Bureaucracy was their arena and models were one of their weapons.

The household records (*zaoban chu huoji dang'an* 造辦處活計檔案) of the Qing court reveal that the use of models and sketches in the organization of labor and materials was the result of the incessant tension between the court's wish to ensure a smooth circulation of practical knowhow, efforts to control developments in the arts and crafts, and a desire to restrict access to information to protect imperial or personal concerns.[16] Unlike published or even contemporary public literature (manuscripts or books), this preliminary level of recording and accounting is more reflective of how things actually worked. Archives reveal experimentation with materials, failures and problems. Bottom-up developments become apparent that were later sold as top-down. Records affirm that craftsmen resisted

[15]Nian Xiyao's standing as a bannerman poised him for a career in Qing bureaucracy (Kleutghen 2015, 61). Nian had also served in the Ministry of Works. Cao Yin moved from the Judicial Ministry to the post of a textile commissioner (Spence 1988, 87).

[16]Often circulation proved more difficult than keeping secrets, as Pamela O. Long has also argued for early Modern European cultures (Long 2001).

imperial power and officials struggled with the realities of production, providing stop-gap solutions as well as innovative experimental methods that eventually failed when implemented. In short, such documents reveal the subject of much intellectual discussion during this period: the tension between knowledge and action.

6 Knowledge and Action

The relation between knowledge and action was a major theme of contemporary scholars such as the Ming loyalists Wang Fuzhi 王夫之 (1619–1692), Huang Zongxi 黃宗羲 (1610–1695), and the son of a cloth merchant and prominent scholar of evidential learning (*kaozheng* 考證) Dai Zhen 戴震 (1724–1777), as they pondered the role of academic and political pursuit in their own lives. Morality rather than the tension between practice and theory was a central theme. James Watson describes this trend in the context of statecraft as 'orthopraxy,' emphasizing that knowledge was not only codified in terms of how to do things. It was about how to do the 'right' things (Watson 1985, 292–324).[17] By contrast, most of these men had no doubt about practice and theory as indicators of social status. Dirtying one's hands was for commoners while literacy lay at the heart of true understanding. These men's philosophical debates and philological inquiries circled around the proper combination of how, why and how things happened, was related and in which way wording or material culture could properly reflect such concerns.[18] Knowing and doing were intertwined.

This rhetoric of morals also served to authorize the need for regulative canons for the production and use of ritual utensils, clothing or architectural features, indicating that Qing scholars subscribed to the eleventh century philosopher and politician Zhu Xi's 朱熹 (1130–1200) notion of material orders as powerful instruments and representations of moral order (Bray 1998). The 'ten thousand things' (*wanwu* 萬物) embodied metaphysical understanding and the recognition of a higher, feasible, efficient order, and thus required careful consideration and control. Items of ritual, social, and political purpose such as clothing, tributary wares or buildings, represented emblems of systematic choices and were thus discussed and constructed as the nucleus and enactment of true understanding. As scholars, Qing literati often repudiated such concerns. True scholarship turned away from pragmatic political concerns in disgust. Officials, the same scholars established regulations, rules or guidelines (i.e. formative knowledge), and developed plans and templates of action in various media: illustrations, objects and texts. They were the mediators, the brokers in charge of collecting, rationalizing and circulating knowledge fit for action. Their efforts influenced their thoughts, whether they defended and justified their role or came to dread it.

The effects can be seen both on the large and small scale. This era's scholars created a range of texts on (1) lineages of thought, (2) dispositions for action (i.e. on customs and habits, *fengsu* 風俗, rites, *li* 禮 etc.) and (3) accounts of labor, materials and ideas (that is administrative literature). This threesome is distinguishable also in the structure of this era's literature that features administrative guides, texts on governance and scholarly debates on knowledge, policies or ritual practice (*li*). Dai Zhen, for instance, discussed the *Six Books* (*Liu shu lun* 六書論, publ. 1745) and also noted on the *Illustrations of the Artificers Record* (*Kaogong ji tu zhu* 考工記圖注 publ. 1746). In his early years he elaborated on the *Erya Dictionary* (*Erya wenzi kao* 爾雅文字考, ca. 1748). During his forties Dai

[17]See also (Sutton 2007) and Watson's response in the same journal (2007). I think though that here the authorization scheme, i.e. that which makes people do things, is given too much credit.

[18]Duan's philological work is closely connected to hero figures of the late history of science in China such as Ruan Yuan 阮元 (1764–1849). Ruan Yuan published a biographical survey of astronomers and mathematicians. Both worked jointly on a review of the thirteenth classics (Duan 1988).

Zhen engaged in waterways and canals (compiling the *Zhili Hequ shu* 直隶河渠书, 1768), he contributed to the revision of the Local Gazetteer of Fenyang Circuit (*Fenyang xianzhi* 汾陽縣志 1771, printed 1776), studied phonetics (*Shenglei biao* 聲類表 Table of Phonetic Categories 1776) and rites (*Yili jishi* 儀禮集释 Explanatory Collection of the Yili). Mathematical issues interested Dai Zhen throughout his life.[19] True scholarship, these intellectuals presumed, had to embrace it all.

We see a recurring set of classical texts, such as the *Artificers Record* (*Kaogong ji* 考工記), the *Discourses of the State* (*Guoyu* 國語) and the *Rites of Zhou* (*Zhou li* 周禮) referenced in such debates, promoting material culture as a way to establish identity and status, shape society and state and grounding rules of production and use. A huge body of commentary literature reflects minor and substantial shifts in planning ideologies, methods of organization and management. The practical implications of such public debates became apparent in the catalogues/codes of rules and regulations that each dynastic state installed for the production of ritual implements. Scholars and officials meanwhile wrote down their thoughts on management and taxing of everyday material goods. This rather sober literature of official state historiography and the detailed, matter-of-fact archival recording is only preserved from the Qing era. It was the Song dynasty, though that provided the literary basis for the Qing effort of building knowledge which Willard Peterson has identified in the context of the great encyclopedic efforts such as the *Collection of Books and Illustrations, Old and New* (Gujin tushu jicheng 古今圖書集成) as an itemized way to understand the world (Peterson 1975). In this context we can see that an ideal became an exemplar, the exemplar turned into a method and the method turned into a legitimizing tool for systemic choices. In the case of the Qing dynasty, court officials used Song texts to authorize their control over arts and crafts and issue standards of validity in a way that enabled emperors to change structures of power and notions of creativity. Design, moved from the realm of ritual into the remit of the imperial court and modeling, became an administrative tool. An institute for the production of models (*zaoyang chu* 造樣處) was created and a model of the final product became an indispensable step of production.

Charlotte Furth has suggested that early works of Qing scholars, such as Huang Zongxi's *Mingru xue'an* 明儒學案 (Cases on the Ming Dynastic *Ru*-School) reflect a particular approach to organizing knowledge and practice: the case study (*an* 案) (Furth 2007, 3–4). Huang Zongxi herein represents a scholar of the early Qing dynasty who resurrected the culture of the late Ming era, while the Manchu Qing Kangxi (1662–1722) (reign 1662–1722), Yongzheng (1678–1735) (reign 1723–1735), and Qianlong Emperors mobilized the exemplars of a much more remote and idealized past, 'investigating into antiquity and upholding culture' (*jigu youwen* 稽古右文) (Cheng 1992). Here we see the intellectual side of building knowledge on individual instances and items again. In this period scholars also shifted towards empiricism in textual studies known as 'evidential studies' (*kaozheng xue* 考證學) (Elman 1984), in which scholars such as Dai Zhen further pushed towards an understanding of the actual practice. From Huang Zongxi to Dai Zhen the nature and impact of discourse on case studies changed, evolving into a text genre and a methodological approach across fields of action and reasoning such as the legal, pharmaceutical or political. Once evaluated as efficient, successful or effective, a case became a norm. An exemplar or a general practice, a particular instance, turned into a general truth. Situational ethics were turned into exemplary conduct and individual intuition made collective knowledge.

Imperial politics reflect such notions when assuming that, if people were innately capable of learning from models, the ideal rulership would be to govern by moral example and not by educative norms. The norm, however, turned into the measurement and means of producing a common standard. This could also be an exemplary norm of ritual bearing and its material representations—which we see reflected in the increasing number of scholarly debates about clothing orders, ritual utensils or

[19] For a chronological overview of Dai Zhen's publications see (Zhai 1992).

imperial tributary wares. On the level of implementation, scholars within administrative structures multiplied the media of description, accounting meticulously for color, form, format and patterns, sizes, materials and methods of production.

On a smaller scale we can see such views also affecting intellectual attitude. When philologist Duan Yucai 段玉裁 (1735–1815) explained in his Annotated Explanations of Characters and Texts (*Shuowen jiezizhu* 說文解字注) that “a mold is a law” (*mo fa ye* 模法也), he equally acknowledged the central normative function that such in-between devices of technical production had long held in Chinese minting and metal production.[20] And as Chinese statecraft time and again invested heavily into material production, models and sampling also attested to a bureaucratic means of communication, verifying and authorizing material practices as Duan Yucai’s interpretation of the word ‘mold’ indeed suggests. In that regard the role of three-dimensional models, miniatures and artifacts in Chinese approaches to knowledge communication in and across the arts and crafts is indeed open to debate.

7 Architecture: Models of Involvement, Models in Use

Among the many models that the imperial architects Lei used, Zhu Qingzheng’s study highlights the novel technique of the ‘ironed paper model’ (*tangyang* 燙樣) (Zhu 2004, 4). This minimized model was produced with grass paper, sorghum stalk, oil wax (*youla* 油蜡) and wood formed by ironing implements. One of the still extant models depicts the Palace of Prolonging Happiness (*Yanxi gong* 延禧宮) (the first of the six palaces in the east) established in the Yongle 永樂 (1360–1424), (reign 1403–1424) era around 1420 and repaired twice in 1686, during the Kangxi reign.[21] Models of the Qing period visualized the building and contextualized it in many regards. The Yanxi Palace model, for instance, also featured the first interior wall behind the screen wall and showed that the facing gates inside the courtyard (*yingbi* 影壁) were made of wood, that it consisted of two courtyards with a five roomed main hall and yellow tiles, that east and west halls had each three rooms and a front courtyard (Fig. 2).

A model of the Yanxi palace seems to have existed for centuries. When between 1845 and 1855 a fire destroyed part of the palace twice, construction models of all palaces were used as guidance for renovations and repairs.[22] Such uses of models continued into the late Qing dynasty. In the year 1872 (tongzhi 同治 11, month 11), for instance, the servant Zhang Lingxi 張伶喜 (n.d.) requested access to a model to ensure that: “Repairs on the Yanxi palace project are pursued according to the regulations.” The archival record further notes that imperial household officials were to scrutinize the draft and estimate costs based on the extant documentation of the palace. The case was approved with the remark that repairs could start based on the model.[23] The costing of repairs, so this case suggests, was based primarily on precedents and viewing the model. On-site surveying seems not to have taken place.

[20] Duan Yucai notes additionally a different emphasis with regard to the materials used: “for wood, it is called a *heng* 橫 (horizontal stencil), for metals it is called a *rong* 鎔 (mold), for earth/soil it is *xing* 型 (form) and for bamboo *fan* (frame) is used. All of those are methods/law. The *History of the Han* [Dynasty 206 BC–220 AD] also uses the expression making molds (以木曰橫。以金曰鎔。以土曰型。以竹曰笵。皆法也。漢書亦作橅). Duan Yucai’s approach is primarily phonological. See (Duan 1988).

[21] Its location is given as *Dongerchang jie* (Southeast between the Nixiang Gate and the Zhaohua Gate, facing south).

[22] The Ming imperial kitchen was inside the palace in three rooms of the Dongpei Hall 東配堂 (東配殿) and the cooking stove was in the southern room at the gable where fires regularly broke out as reported in eds. Zhongguo diyi lishi dang’an guan, Gugong bowuguan (2014, vol. 225, Daoguang 道光 25, month 5, day 23).

[23] “明年方向有礙, 擬先勘估, 如後年方向相宜, 即行修建。” 此具燙樣, 即大約爲當時樣式房奉旨而做呈上御覽的。

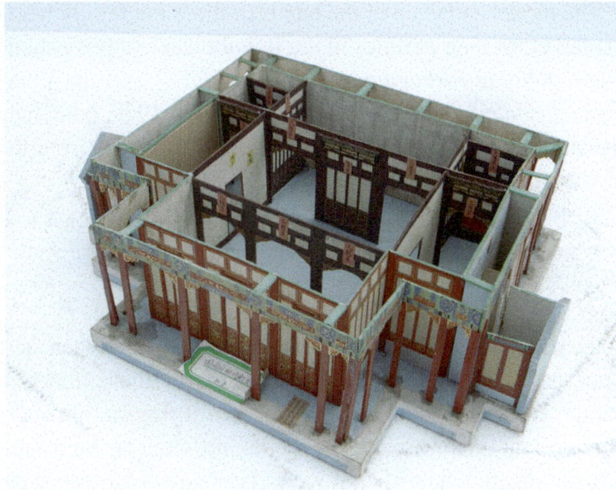

Fig. 2 The Yanxi model, made of cardboard, whose five parts come apart to provide a contextual view. Qing Dynasty; 48 × 40 × 24 cm; photographer: Tian, Mingjie; copyright: The Palace Museum

On the still extant model several yellow strips can be found which were loosely glued to the piece and could be easily removed. The strips gave different information. One usually identified the model, others added measurements. This could be the size of single rooms, usually the width and lengths, or the size of specific elements such as the peripheral column (*yanzhu* 檐柱) (height) and the height of the platform (*taiming* 台明). Other strips summarized information:

> The big hall's space covers five rooms. The outer room has a width of 1 *zhang*, 7 *chi*, 7 *cun*. The secondary rooms each have widths of 1 *zhang* 6 *chi* 5 *cun*. The two little rooms each have a width of 1 *zhang*, 2 *cun* and the distance from entrance to rear is 1 *zhang* 7 *cun*.
> 蓋大殿一座五間明間面寬一丈7尺7寸二次間各面寬一丈六尺五寸二稍間各面寬一丈二寸進深一丈七寸.

These measurements noted on replaceable slips—similar to modern post-it notes—rather than inscription, suggests re-use of models, or variable uses of the model ranging from display to negotiations over construction details. Stored over long periods, models could change meaning and purpose per se, and the yellow slips added adaptable information: such as color or material mix.

The yellow strips identify the model as part of a bureaucratic process with many aims. On a model of the Changchun palace this is even spelled out as the yellow strip attributes the model to a memorial (*zun zhi zhao Changchun gong shiyang* 遵旨照長春宮式樣).

Architectural models, once approved, took on a normative status. This was also the case when the court, having approved the reestablishment of the Yanxi halls, decided in the end not to put the plan into action and instead transformed the reserved space into a garden with a water pond. Similarly, the Crystal Palace (*Shuijing gong* 水晶宮), once planned with three levels of nine rooms each, was originally planned with pavilions at each of its four corners. It was designed with copper (*tong* 銅), ridge pools and walls of glass, open to four sides so that one could see the fish swimming in the pond. But the work was never completed. In 1931 the palace museum constructed archive storage on top of the unfinished palace.

In the case of the Qing period, the three dimensional model was a mandatory part of the administrative process as the emperor had to approve of it before construction plans were made, the procedure of work (*gongcheng zuofa* 工程做法) and accounting could take place and finally the construction itself could begin. The making was defined. The Yanxi model whose five parts provide a contextual view, is scaled to 39 × 28 × 8 cm following Song standards (Liu et al. 1984).

The model ended up in the hands of the Yang family who sold it when they realized that there was no market for their abilities in the Republican era. It was handed over to the National Beiping Library (*Guoli Beiping tushuguan* 國立北平圖書館) and the library of the Zhongfa University (*Zhongfa daxue tushuguan* 中法大學圖書館). After 1949 both handed all models to the Forbidden City Museum. 80 models are still in existence. Most are from the Tongzhi era (1861–1875) and concern the Summer Palace (*Yuanming yuan* 圓明園).

8 Conclusion

At the peak of his career between 1796 and 1799 Heshen, acting as the mouth and ear of the Qianlong Emperor, is said to have influenced primarily those mid-level bureaucratic positions with which the empire was ruled (MacMahon 2015, 45). It was on this operational level that models actually played an important role. Through the model, different social groups shared information and discussed how to make practical knowledge work. Heshen's theft also draws attention to models as tools of power beyond this era's well-known bureaucratic paper machine. In their attempt to affect Chinese culture also materially, the Qing court elites systematically instrumentalized material displays, samples and miniatures, to communicate aesthetic preferences and technical demands across regions and time. Legacies such as that left by the imperial architects of the Lei family further demonstrate the eclectic approaches to the knowledge circulation of actors standing at the intersections of the arts and crafts, state concerns on material production and construction works (Ruitenbeek 1996, 50; Schäfer 2010, 3–5).

Jade craftsmen were reluctant to consider moving to the capital, but some key figures were eventually persuaded to bring their expertise to the court. In this regard communication schemes within jade production differed from those of interior design or silk production, where sketches, models or written explanations often had the sole role of conveying ideas and designs. Jade carvers became close to and were valued by the emperor, often consulted by him on a daily basis, whereas the silk weavers, reelers and draw loom pattern makers who settled in the South remained an indistinguishable mass to the emperor. Conversely it also seems the case that the models used in jade carving became an important way for the emperor to participate in this trade. After a thorough investigation, the court dispatched models and defined on the raw material how to cut and shape the stone. These intricate instructions turned out to be a stop-gap solution because the best jade carvers, situated in Suzhou, were unwilling to work at the court in Beijing, but it also meant that a class of middle man could emerge that helped translate ideas from practice into words and texts. In this sense, the Qianlong Emperor's close involvement with jade carvers might well be read as a compensatory move, rather than one driven by an obsessive interest in jade techniques or a manipulation of arts and crafts. In the end a lack of local craft expertise was the main reason why Qianlong's court increasingly employed sketches, models and artifacts to organize production across the empire.

In both architecture and jade carving, representational methods included sketching and modeling. Bureaucratic handling promoted the unification of descriptive and prescriptive modes. This development may have been an important factor facilitating the aesthetic cross-fertilization over different crafts that marks this era. How it influenced technical development is still open to debate. Clearly, in both crafts the models elaborated procedures of production and design, while clerks explained material, size and color schemes on added yellow strips. Such methods suggest repeated use of in-between devices for production and reference. Workshops in the palace must have archived designs. In the fields of jade and architecture the workshops seem to have functioned more as design bureaus than as producing sites per se.

References

Akcetin, Elif. 2007. *Corruption at the frontier: The Gansu fraud scandal*. PhD thesis, The University of Washington.

Bray, Francesca. 1998. Machines for living: Domestic architecture and the engineering of the social order in Late Imperial China. *Cambridge Anthropology* 20 (1–2): 92–110.

Cammann, Schuyler V.R. 1982. *Miniature art from old China*. Freehold: Monmouth Graphics.

Chayet, Anne. 2004. Architectural wonderland: An empire of fictions. In *New Qing imperial history. The making of inner Asian empire at Qing Chengde*, ed. James A. Millward, Ruth W. Dunnell, Mark C. Elliott, and Philippe Foret. Abingdon: Routledge Curzon.

Chen Kaijun 陳愷俊. 2013. Yiwei jishu guanliao de ziwo xiuyang—dutaoguan Tang Ying de wenhua chuangzuo he jiyi xide 一位技術官僚的自我修養—督陶官唐英的文化創作和技藝習得 (The Self-cultivation of Technocrats: Tang Ying's (1682–1756) Cultural Production and Skill Acquisition). *The National Palace Museum Monthly of Chinese Art* 364: 92–102.

Chen Kaijun 陳愷俊. 2014. The rise of technocratic culture in High-Qing China: A Case Study of Bondservant (Booi) Tang Ying (1682–1756). PhD thesis, Columbia University.

Cheng Huanwen 程焕文. 1992. Jigu youwen, zhan gai xuelin 稽古右文, 沾溉学林 (Studying the ancient ways and enjoying literature – the benefit of a Scholarly Forest). *Tushuguan luntan* 4: 79–81.

Dai Zhen 戴震. 1982. *Mengzi ziyi shuzheng* 孟子字義疏證 (An evidential study of the meaning and terms of the Mengzi). Beijing: Zhonghua shuju.

Daston, Lorraine. 2006. Appearances all the way down: The glass flowers as scientific models. In *Kunstformen des Meeres. Zoologische Glasmodelle von Leopold und Rudolf Blaschka 1863–1890*, ed. Karlheinz Wiegmann and Meike Niepelt, 60–67. Tübingen: Stadtmuseum Tübingen.

Duan Yucai. 1988. *Shuowen jiezi zhu* 說文解字注 (Annotated explanations of characters and texts). Shanghai guji chubanshe.

Elman, Benjamin. 1984. *From philosophy to philology: Intellectual and social aspects of change in Late Imperial China*. Cambridge, MA: Harvard Univeristy Press.

Fallan, Kjetil. 2010. *Design history: Understanding theory and method*. Oxford: Berg Publishers.

Feng, Jiren. 2007. Bracketing likened to flowers, branches and foliage: Architectural metaphors and conceptualization in tenth to twelfth-century China as reflected in the Yingzao Fashi. *T'oung Pao* 93 (4–5): 369–432.

Furth, Charlotte. 2007. Introduction. In *Thinking with cases: Specialist knowledge in Chinese cultural history*, ed. Charlotte Furth, Judith Z. Zeitlin, and Ping-chen Hsiung, 1–27. Honolulu: University of Hawaii Press.

Guo Fuxiang 郭福祥. 2010. Gongting yu Suzhou: Qianlong gongting li de Suzhou yugong 宫廷與蘇州:乾隆宫廷里的蘇州玉工 (Royal Court and Suzhou: jade craftsmen from Suzhou at Qianlong's court). In *Gongting yu difang: Shiqi zhi shiba shiji de jishu jiaoliu* 宮廷與地方：十七至十八世紀的技術交流 (The court and the locality: 17th to 18th century technical knowledge circulation in China), ed, Gugong bowuyuan, and Bolin Mapu xuehui kexueshisuo, 169–220. Beijing: Zijincheng chubanshe.

Hay, Jonathan. 1999. Culture, ethnicity, and empire in the work of two eighteenth-century 'eccentric' artists. *RES: Anthropology and Aesthetics* 35: 201–223.

Hongli 弘歷 (Qing era). 1993. *Qing Gaozong Yuzhi Shiwen Quanji* 清高宗御制識文全集 (The complete collection of Imperial poems). Beijing: Zhongguo renmin daxue chubanshe.

Jing Qingsheng井慶升. 1985. *Qingshi da muzuo caozuo gongyi* 清式大木作操作工藝 (Qing dynastic carpentry arts and crafts), Part 1. Beijing: Wenwu chubanshe.

Jing Qingsheng井慶升. 1988. Xie Shan jianzhu damu qojia moxing zhizuo jishi (yi) 歇山建筑大木屋架模型制作纪实一 (Record of the Roof Truss Model of the Xieshan construction). *Gu Jianyuan lin jishu* 19 (2): 15–19.

Kahn, Harold. 1985. A matter of taste: The monumental and exotic in the Qianlong Reign. In *The elegant brush. Chinese painting under the Qianlong emperor, 1735–1795*, eds. Ju-hsi Chou and Claudia Brown. Phoenix: Phoenix Art Museum.

Kleutghen, Kristine. 2015. *Imperial illusions: Crossing pictorial boundaries in the Qing palaces*. Seattle: University of Washington Press.

Liu Dunzhen 劉敦楨 et al. 1984. *Zhongguo gudai jianzhu shi* 中國古代建築史 (History of architecture in China), vol. 6. Beijing: Zhongguo jianzhu gongye chubanshe.

Liu Jinzao 劉錦藻 (*jinshi* 1847) et al. 2000. *Qingchao Xu Wenxian tongkao* 清朝續文獻通考 (Qing dynastic continuation of comprehensive treatises based on literary and documentary sources). Hangzhou: Zhejiang guji chubanshe.

Long, Pamela O. 2001. *Openness, secrecy, authorship: Technical arts and the culture of knowledge from antiquity to the Renaissance*. Baltimore/London: Johns Hopkins University Press.

MacMahon, Daniel. 2015. *Rethinking the decline of China's Qing dynasty: Imperial activism and borderland management at the turn of the nineteenth century*. New York/Abingdon: Routledge.

Mannoni, Laurent. 2000. *The great art of light and shadow. Archaeology of the cinema*. Exeter: University of Exeter Press.

Needham, Joseph et al. 1965. *Science and civilization in China,* vol. 4, part 2. Cambridge: Cambridge University Press.
Peterson, Willard. 1975. Fang I-chih: Western learning and the *investigation of things*. In *The unfolding of Neo-confucianism*, ed. Wm T. Bary et al., 369–411. New York: Columbia University Press.
Popplow, Marcus. 2002. *Models of machines. A "missing link" between early modern engineering and mechanics*. Berlin: Max Planck Institute for the History of Science, Preprint 225.
Rheinberger, Hans-Jörg. 2010. *An epistemology of the concrete: Twentieth-century histories of life*. Durham: Duke University Press.
Ruitenbeek, Klaas. 1996. *Carpentry and building in Late Imperial China: A study of the fifteenth century carpentry's manual Lu Ban Jing*. Leiden: Brill.
Schäfer, Dagmar. 2010. Introduction. In *Gongting yu difang: shiqi zhi shiba shiji de jishu jiaoliu* 宫廷興地方: 十七至十八世纪的技术交流 (The court and the locality: 17th to 18th century technical knowledge circulation in China), eds, Gugong bowu yuan, Dagmar Schäfer, 3–5. Beijing: Zijincheng chubanshe.
Siu, Victoria M. 2013. *Gardens of a Chinese Emperor: Imperial creations of the Qianlong Era, 1736–1796*. Bethlehem: Lehigh University Press.
Spence, Jonathan. 1988. *Ts'ao Yin and the K'ang-hsi Emperor: Bondservant and master*. New Haven/London: Yale University Press.
Sutton, Donald S. 2007. Introduction: Ritual, cultural standardization, and orthopraxy in China. Reconsidering James L. Watson's ideas. *Modern China* 33 (11): 3–21.
Wang Qiheng 王其亨, Wu Cong 吴葱, Bai Chengjun 白成军. 2006. *Gu jianzhu cehui* 古建築測繪 (Ancient architecture). Beijing: Zhongguo jianzhu gongye chubanshe.
Watson, James. 1985. Standardizing the gods: The promotion of T'ien Hou along the South China Coast. In *Popular culture in Late Imperial China*, ed. David Johnson, 292–324. Santa Barbara: University of California Press.
Watson, James. 2007. Response: Orthopraxy revisited. *Modern China* 33 (1): 154–158.
Xianggang zhongwen daxue wenwuguan 香港中文大学文物馆, Zhongguo diyi lishi dang'an guan 中国第一历史档案馆, ed. 2005. *Qinggong neiwufu zaobanchu dang'an zonghui* 清宮內務府造辦處檔案總匯 (Complete compilation of the archival materials of the Qing Dynasty Imperial Workshops). Beijing: Renmin chubanshe.
Yang Boda 杨伯达. 2006. Qing Kangxi chao Yao shi ranyufa—"hupo tang" 清康熙姚氏染玉法—"琥珀烫" (The Yao family jade coloring technique in the Kangxi period of the Qing dynasty Amber red). In *Yang Boda Lun Yu: Ba Zhi Wen Xuan* 杨伯達論玉:八隻文選 (Yang Boda discusses jade: selection of eight anthologies), ed. Yang Boda. Beijing: Zijincheng chubanshe.
Yoon, Wook. 2012. Prosperity with the help of villains, 1776–1799: A review of the Heshen Clique and its era. *T'oung pao* 98 (4–5): 479–527.
Zhai Tunjian 翟屯建. 1991. Dai Zhen shuping 戴震述評 (A review of Dai Zhen). *Dongnan wenhua* 2: 276–279.
Zhang Guangwen 张廣文. 1986. "Qing chu chuqi zhizuo de si jian rengong ranse yuqi 清初制作的四件人工染色玉器 (Four pieces of colored jades of the Early Qing Era). *Gugong bowu guan yuankan* 2 (2): 74–76
Zhang Yibing 張一兵. 2005. *Mingtang zhidu yanjiu* 明堂制度研究 (Research on the Mingtang system). Beijing: Zhonghua shuju
Zhao Erxun 赵尔巽 (1844–1927). 2012. *Qing Shigao* 清史稿 (Draft history of the Qing). Tianjin: Tianjin guji chubanshe.
Zhongguo diyi lishi dang'an guan中国第一历史档案馆, Gugong bowuguan 故宫博物院馆, eds. 2014. *Qinggong neiwufu zouxiaodang* 清宮內務府奏銷檔 (Report archive of the Qing Palace Imperial Household office). Beijing: Gugong chubanshe.
Zhu Qingzheng 朱慶征. 2004. "Wanfang anhe" tangyang "萬方安和"燙樣 (The Ironed paper model "Wanfang anhe"). *Jinjin cheng* 3: 43–44.

Hybrid Experts

11

Ursula Klein

Abstract

This paper studies eighteenth-century "experts" and a particular sub-group of experts—"hybrid experts"—who were also recognized as men of science. "Experts" had characteristic epistemological features that distinguished them from the majority of early modern artisans and craftsmen, and they inhabited particular socio-economic niches: the big technological enterprises of the time, the luxury market in connection with global trade, and the medical and pharmaceutical professions. The hybrid scientific-technical experts personally bridged the worlds of science and of industry. Following their career and activities allows new insights into truly combined and interacting, technical and scientific knowledge and practices. The foundation of the new technological schools in eighteenth-century continental Europe and their amalgamation of "theory and practice" was to a large extent the achievement of hybrid experts.

"Practical knowledge" had many faces in early modern Europe, as is the case today in highly industrialized countries. Unlike today, however, the majority of early modern practitioners were peasants, artisans, craftsmen, and women occupied with household production, while engineers, inventors and other "experts" formed a small social group. This contribution is concerned with eighteenth-century "experts" and with a particular sub-group of experts who were also recognized as men of science.[1] I will first examine the epistemological characteristics of "experts"—as distinguished from ordinary artisans and craftsmen—as well as their distinct socio-economic niches in early modern Europe. Then I present two examples of experts from Prussia: first a "laboratory worker" (*Laborant*) and arcanist (*Arcanist*) of the Royal Prussian Porcelain Manufactory in Berlin; his training, experiments and inventions demonstrate the uncommon dynamics of experts' knowledge production. My second example is

[1]On early modern experts see also (Ash 2010; Klein 2012a; Klein and Spary 2010). For architects and engineers, see (Lefèvre 2004 and Chap. 9); for practical mathematicians, see (Bennett 1991; Johnston 1996).

U. Klein (✉)
Max Planck Institute for the History of Science, Wilmersdorf, Berlin
e-mail: klein@mpiwg-berlin.mpg.de

M. Valleriani (ed.), *The Structures of Practical Knowledge*, DOI 10.1007/978-3-319-45671-3_11

a Prussian mining official who was also recognized as a mineralogist. As he represents the figure "hybrid expert," we pay particular attention to his combination of technical work, technological research and learned natural inquiry.

1 Niches for Experts

In Renaissance and early modern Europe, the arts and crafts provided significant resources and stimuli for the learned world. But not all arts and crafts were relevant in this respect. As Pamela Long has recently observed, shoemaking, for example, had little impact on learned natural knowledge (Long 2011, 128). Inversely, traditional shoemakers could go on with their ways of knowing and making without carrying out experiments and taking notice of mathematics, mechanics, chemistry or any other science. Seen from a broader social and economic perspective, the interfaces between technical and scientific activities, and practical and learned knowledge in early modern Europe, were delimited to relatively small niches. Yet such niches were of crucial importance for the development of technical expertise as well as for the natural and technological sciences.

The term expert—in German *Sachkundiger* or *Sachverständiger*—was frequently used in eighteenth-century Europe to denote men (there were no women!) who possessed outstanding empirical knowledge about uncommon practical fields such as the construction of fortifications, the organization of deep mining, and porcelain manufacture. Compared to the large groups of craftsmen and artisans, the number of experts was small. Their individual career paths differed from each other, the military engineer and gunner excepted. As Wolfgang Lefèvre's study about "architects" demonstrates (Chap. 9), it was often the case that even their names were ambiguous. Generally speaking, the early modern experts did not fill a well-defined place in the classificatory system of practitioners. On the one hand, they were artisans—they possessed tools and also performed some handiwork—but, on the other hand, they were for the most part not members of guilds. Experts were not professionals in the sense of practitioners whose training and occupation would have been clearly defined by the rules of such guilds. Their knowledge neither stemmed exclusively from traditional apprenticeship training nor from the ordinary body of guild knowledge. On the contrary, from an epistemological point of view, experts had some outstanding features: they also borrowed knowledge from books; they were experimenters and inventors; they often kept experimental notebooks; and they also gained experience through travelling and through conversations with experienced merchants, men of science and all kinds of researchers and inventors. Experts did not carry out hard physical work, but all of them did some handiwork, for example, in the context of measuring, testing, experimenting and so on. And all of them mobilized propositional knowledge for organizing and carrying out their practical enterprises. Thus the social figure of the expert undermined the ancient ideological distinction between hand and mind.

The figure of expert was by no means a novelty of the eighteenth century. From the late Middle Ages through to the early eighteenth century, experts traveled from place to place, the courts belonging to the most important sites where they could make a living. In the mid-eighteenth century, however, experts entered the milieu of state bureaucracy. They became the consultants of ministers and leading state officials. In mercantilist countries such as France and Prussia, the emerging technical departments of state administration also hired experts for long-term public service. This new social status rendered the experts' life more settled and their research more sustained. Furthermore, in the eighteenth century, new technological schools and academies, such as the *École des Ponts et Chaussées* and the Mining Academy of Freiberg (*Bergakademie*), were established in order to ensure the training and sustained recruitment of experts.

There were three main socio-economic niches for experts in early modern Europe: the large technological enterprises of the time; the luxury market in urban centers connected with global trade; and the medical and pharmaceutical professions. The next section provides an overview of these niches.

1.1 Large Technological Enterprises

Since the late Middle Ages, there have been several large-scale technological enterprises in Europe that were marked by advanced machinery and an elaborate system of division of labor. The division of labor created a demand for experts concerned with the various tasks of planning and management, which also included practical activities such as experimentation, measurement, and quality tests of materials. Belonging to this early-modern "high tech" sector were activities such as: silver mining and smelting in the old mining regions of Saxony, the Harz Mountains and so on; complex construction projects such as the building of cathedrals, warships with heavy gunnery, fortifications, and siege engines; hydraulic engineering projects such as the drainage of marshland and the construction of canals, harbors, and dikes.

These industries were often financed and organized by the state, ranging earlier from princely courts and city councils all the way up to the eighteenth-century technical state departments. In some cases, they were also promoted by clerical commissioners, groups of private investors, or by cooperating clerical and state commissioners, as in the case of the construction of cathedrals. In Great Britain, the navy excepted, it was mostly inventive entrepreneurs who were at the forefront of high-tech enterprises.[2] The drainage of the Breemster Lake north of Amsterdam between 1608 and 1612, and the construction of Breemster polder, is another case in which wealthy urban merchants raised capital, sold stakes, and organized a board of overseers that managed and supervised work (Fleischer 2007). In most of the countries in continental Europe, however, it was the state in its various historical forms that organized big technological enterprises along with schools for the acquisition of related expertise.[3] Even in liberal, eighteenth-century Great Britain the state sometimes encouraged, protected, and regulated such endeavors (Ashworth 2007; Schaffer 2007; Mauskopf 2010).

The famous Renaissance engineers and artist engineers have long been a subject of detailed studies about experts and expertise. So let me exemplify here the role of experts in the context of early modern mining and metal production. Most of the early modern mining in continental Europe was organized by the state, in particular the mining of silver. A complex system of division of labor had been established in silver mining since the sixteenth century. In addition to the mining laborers (*Bergarbeiter*), such as hewers (*Häuer*) and haulers (*Haspelknechte)*, who received a relatively short apprenticeship training and did hard physical work, this system of division of labor also included better-trained craftsmen, such as smelters and carpenters, as well as "experts," who had received particularly long personal training. To the latter group belong the mining surveyors (*Markscheider*), the assayers (*Probierer*), the masters of smelting (*Hüttenmeister*), the *Kunstmeister* (later mining engineers), the mining masters (*Bergmeister*), and the mining councilors (*Bergräte*), all of whom were state officials in German states.

The mining surveyors (*Markscheider*) were skilled in measuring sites of mining above ground and underground as well as in drawing mining plans (*Grubenrisse*), and they implemented geometrical knowledge in these performances. The assayers in the mines and smelting works were specialists familiar with chemical methods used to test the quantity of silver in ores and during the smelting process. Masters of smelting (*Hüttenmeister*) possessed advanced metallurgical knowledge, often

[2]See, for example, (Ash 2007; Jones 2008; Miller 2009; Tomory 2012, 2013).

[3]See also (Ash 2010; Long 2011).

acquired through experimentation. From the seventeenth century the mining of silver in German states was embedded in a true technological system consisting of pumps, water wheels, mills, artificial pounds and canals, established to drain water from deep mines. The technical control and improvement of this technological system involved distinct technical experts—*Kunstmeister* (later mining engineers)—who were knowledgeable in hydraulics, machinery, and mechanics.

Whereas the former four types of mining experts belonged to the group of lower mining officials, the mining masters and mining councilors were high officials. They were fiscal accountants, managers as well as technical experts. The mining masters directed the local mining districts. In so doing, they regularly inspected mines, organized prospecting (searches for new ore deposits), and decided about technical improvements.[4] In the second half of the eighteenth century, many of them were experimenters and inventors who also possessed erudite mineralogical and geological knowledge. Mining councilors performed similar activities on a higher organizational level of mining administration, and they, too, had often received some formal education in the natural sciences.

Generally speaking, all of these mining experts contributed to the production process as a whole. In distinction to the various groups of mining laborers, they did not perform hard physical work, but rather highly skilled handiwork combined with measuring, drawing, chemical testing and so on, which presupposed specialized competence. These experts further wrote reports about their experiments and technical endeavors, and sometimes published technical treatises about mining. Carrying out their distinct tasks with expertise (*Sachkenntnisse, Sachverstand*) clearly presupposed physical skill as well as articulated empirical knowledge.

1.2 The Luxury Market

The expanding luxury marked and overseas trade provided another niche for early modern experts. To these belonged: the making of majolica and porcelain, alchemical gold and artificial gems, crystal glass and large mirrors, expensive tapestry and cloths, distilled liqueurs and exotic food; the public display of fireworks, and electrical experiments; the growing of tea, coffee, tobacco, exotic trees and flowers in gardens; collections of curious objects from overseas in cabinets and museums.[5] The early modern courts hired practical mathematicians, alchemists, fireworkers, and theater engineers for entertainment and the production of luxury goods, such as Kunckel's gold ruby glass (Chap. 6). As Viktoria Tkaczyk observes in her contribution to this volume (Chap. 4), theater engineers often possessed substantial geometrical knowledge, which they implemented in stage design. In addition to the courts, big port cities and marketplaces trading with goods from overseas were hothouses for experts. Artist engineers, alchemists, apothecaries, inventive goldsmiths, carpenters and so on, living in Venice, Florence, Nuremberg, Antwerp, or Paris were makers and providers of luxury goods for wealthy people. For example, in a recent study, Bruno Belhoste observed that the Parisian carpenter (*menuisier*) André-Jacob Roubo, patronized by the Duke of Chaulnes, possessed exceptional skills and further conversed with the learned elite and contributed to their publications.[6] In another recent study, Emma Spary showed that among the distillers in eighteenth-century Paris producing luxury

[4]Alexander von Humboldt, for example, was a Prussian mining master; see (Klein 2012c). Dowsers, discussed in Pamela Smith's contribution (Chap. 14), were occasionally, but by no means always, involved in prospecting. They did not belong to the corps of mining officials. As early as the sixteenth century, dowsing was a highly contested activity in the German silver mining regions. See (Agricola 1556, book 2).

[5]There is an enormous body of literature on this issue. To mention only a few: (Dauxois 1997; Dupré 2014; Dupré et al. 2014; Dupré and Lüthy 2011; Findlen 1994; Findlen and Smith 2002; Klein 2013; Smith 2004; Moran 1991; Müller-Wille 2004; Spary 2000; Werrett 2010; Weyer 1992).

[6]Roubo contributed to the *Descriptions des arts et métiers*; see (Belhoste 2012).

goods such as liqueurs, many were knowledgeable in chemistry and were the authors of books on distillation (Spary 2010). The studies by Harold Cook, Sven Dupré and Christoph Lüthy on the Renaissance and early modern Netherlands have illuminated networks of merchants, artisans and artists promoting trade with luxury goods but also with related knowledge (Cook 2007; Dupré and Lüthy 2011). In the early modern German lands, Nuremberg was a similar center and hometown of the goldsmith Wenzel Jamnitzer (1508–1585), studied by Pamela H. Smith, whose artistic work presupposed outstanding talent as well as some natural knowledge (Smith 2004; Chap. 14). To give an example from England, in Elizabethan London, colonial trade and navigation created a market for scientific instrument makers, who also studied and taught some mathematics and carried out astronomical and physical observations (Bennett 1987). The Royal Exchange was another center of economic life in London around 1600, which Deborah Harkness has described as a place full of expensive luxury items, exotic drugs, mathematical instruments and so on. As Harkness has also emphasized, reading, writing, and arithmetic were an advantage in such luxury markets (Harkness 2007).

Luxury production in early modern cities took place for the most part in small workshops and laboratories. For example, city apothecaries produced their precious chemical remedies and cosmetics in their relatively small, private pharmaceutical laboratories, and Johannes Kunckel (1630–1703) produced his gold ruby glass for the court of Brandenburg, in his laboratory on the Pfaueninsel (Chap. 6; Klein 2007b). But some larger production sites existed as well: for example, glass production on the Venetian island of Murano, and the production of jewelry, confectionery, remedies and other luxury goods in the Palazzo degli Uffizi in Florence (Beretta 2009; Kieffer 2014). Large production sites divided labor among many specialized makers and thus stimulated the development of outstanding expertise. It was such sites, in particular, that provided incentives for the authors of technological treatises such as Vannoccio Biringuccio's (1480–1539) *De la Pirotechnia* (1540) and Antonio Neri's (1576–1614) *L'arte vetraria* (1612), who mixed technical descriptions with reflections upon the causes of technical processes.

In the eighteenth century, large-scale manufactories for glass, porcelain, silk clothing, and tapestry proliferated in several European countries. In Paris, the *Gobelins* was a place for the inventive design and dyeing of tapestry, but also for chemistry, as studies by Bruno Belhoste, Agustí Nieto Galan, and Sarah Lowengard have shown (Belhoste 2011; Nieto-Galan 2001; Lowengard 2005). The eighteenth-century silk industry in Lyons, studied by Liliane Pérez, is another example of large-scale manufacture along with the collective management of innovation, which also involved members of the local academy of science (Pérez 2008; Spary 2012). The porcelain manufactories of Meissen, Berlin, and Sèvres, and the Royal Glass Manufactory of Saint Gobain, which produced crystal glass à la façon de Venice and large mirrors, were also sites of outstanding chemical expertise.[7] As I will show below, in the Royal Prussian Porcelain Manufactory (Königliche Porzellanmanufaktur or KPM), laboratory workers (*Laboranten*) and arcanists (*Arcanisten*) frequently experimented with external chemical experts, most notably with the famous Prussian chemist Martin Heinrich Klaproth (1743–1817). After Klaproth discovered uranium in 1789, he immediately made arrangements to carry out experiments with a laboratory worker at the *KPM* to study the use of the "calx of uranium" as a pigment for ornamenting porcelain (Klein 2014b).

1.3 The Medical and Pharmaceutical Professions

The early modern medical and pharmaceutical professions gave rise to experts who traded with medicines and offered medical therapy, supplementing the services of university-educated physicians.

[7] For more information, see (Lehman 2012; Beretta 2012; Beretta 2014; Klein 2013, 2014a).

Among the surgeons, barbers, midwives, distillers, apothecaries, and women trading with herbs there were many individuals who possessed exceptional medical competence and learned knowledge.[8] In addition to knowledge acquired during apprenticeship training, these individuals also took up elements of medical, botanical and chemical knowledge distributed through books.

In eighteenth-century France and Germany, an impressive number of apothecaries had careers that began in the apothecary's shop and ended in the scientific academy. In Prussia, apothecaries knowledgeable in botany and chemistry were hired by the state to become court apothecaries, army apothecaries, or members of medical boards that organized and surveyed the medical system. Their membership in medical boards, in particular, manifests a recognition of these men's distinct expertise, which differed from that of the physicians and medical professors who had long dominated public medical service.

Apothecaries were artisans trained for many years. In the eighteenth-century German states, the pharmaceutical apprenticeship normally took 5–6 years, followed by 6–8 years of service as a journeyman. Apothecaries living in cities traded with hundreds of different herbs, exotic *materia medica*, self-made remedies, and other materials such as cane sugar, confectionary, coffee, tea, tobacco, alcoholic beverages, cosmetics, lacquer, and chemicals. Competing with merchants, distillers, and alchemists, close knowledge about materials and their various uses, ways of production, storage, and unambiguous identification was a commercial necessity for them. In this context, botanical and chemical knowledge was particularly useful (Klein 2015).

In the eighteenth century it was further common that such apothecaries living in cities owned a laboratory, where they prepared chemical remedies and other goods. Establishing and running a laboratory presupposed knowledge about chemical apparatus and techniques, most of which had been newly introduced to the pharmaceutical business during the seventeenth century. In addition to chemical instruction in the shop, many apothecaries learned chemistry through reading chemical journals and books. Beginning in the mid eighteenth-century, pharmaceutical-chemical boarding schools offered additional chemical and other scientific education, as did the *Collegium medico-chirurgicum* in Berlin, founded in late 1723. Performing chemical operations in the laboratory, testing new ways of preparing medicines, and reading chemical books often provided incentives to further extend chemical studies beyond the immediately useful. As I have shown elsewhere, the eighteenth-century chemical-pharmaceutical periodicals such as Lorenz von Crell's (1744–1816) *Chemische Annalen* and Göttling's *Almanach oder Taschenbuch für Scheidekünstler und Apotheker* invited apothecaries to publish their knowledge and to communicate with chemists (Klein 2007a). Several of these publishing apothecaries developed specific expertise in preparing chemical remedies (such as ethers) that even transcended the know-how of chemists (Klein 2010a, b). Furthermore, among the wealthy apothecaries many were owners of libraries, botanical gardens, and cabinets of curiosity. Belonging to an urban well-to-do elite also meant frequent conversations with physicians and men of science.

The eighteenth-century apothecaries thus inhabited a highly innovative milieu that was continuously changing through global trade, the implementation of new chemical technology, the transfer of botanical knowledge, and demands by the state. Their material and cultural environment enabled them to shift their activities from commercial to scientific endeavors and vice versa. As a consequence, in eighteenth-century Germany, around half of all chemists were apothecaries, and this was largely the case in France as well (Hufbauer 1982; Klein 2007a; Tomic 2010). Prussia's most famous chemists of the second half of the eighteenth century—Andreas Sigismund Marggraf (1709–1782), Sigismund Friedrich Hermbstaedt (1760–1833) and Martin Heinrich Klaproth—were also apothecaries for most of their career (Klein 2015). These apothecary chemists were often state consultants, who further engaged in a number of additional useful projects organized by the state.

[8] See, for example, (Klein 2007a, b, 2008, 2010a; Leong and Rankin 2011; Park 1985; Pugliano 2012; Siraisi 1990; Tomic 2010).

2 Experts: The Laboratory Worker Friedrich Bergling

The manufacture of porcelain was a new technology in Europe, introduced in the early eighteenth century. In 1709, the German apothecary and alchemist Johann Friedrich Böttger (1682–1719) invented the first European hard-paste porcelain, with support from Ehrenfried Walther von Tschirnhaus (1651–1708), mathematician and director of the chemical laboratory of the Saxon Elector, Friedrich August I (1750–1827). By the end of the eighteenth century, the Holy Roman Empire hosted some 20 porcelain manufactories, among which the state manufactories in Meissen, Vienna and Berlin were the most famous. The Royal Prussian Porcelain Manufactory in Berlin was founded in 1763, at the end of the Seven Years' War. Unlike a traditional pottery, it was a large manufactory with some 400 workers (by 1780) and a high degree of division of labor. Notwithstanding the fact that the making of porcelain resembled the production of ceramics and glass, it also differed from these more traditional practices in many important aspects. Hence, there was no tradition of artisanal knowledge and skills in Europe, and no system of guilds, that would have engendered the kind of knowledge involved in this new practice.

All of the workers and officials of the Royal Prussian Porcelain Manufactory were educated and trained in the Manufactory. Around 1770, about 41 % of the Manufactory's "workers" (*ouvriers*) were highly qualified artisans and artists who had been apprenticed to a master of the Manufactory for generally 6–7 years (Fig. 1). To this group belonged the instrument makers, the artistic workers forming the porcelain objects, the painters decorating the porcelain, and the laboratory workers (*Laboranten*) who prepared the pigments for decorating. In addition to workers there was also a group of some thirty officials (by 1780), among them the two directors, who organized and supervised the different parts of labor.[9] To this group of officials also belonged the Manufactory's most important technical experts: the arcanists (*Arcanisten*) and, after the Manufactory's reorganization in 1786/1787, the laboratory workers (*Laboranten*) as well.

The arcanists knew the recipes for making the porcelain paste. They also performed quality control of the raw materials and carried out chemical analyses in this context. These men were knowledgeable in chemistry, and some of them were explicitly recognized as chemists. The laboratory workers knew the recipes for making pigments, fluxes, and oils to be used for painting on porcelain. They worked in the Manufactory's laboratory (Fig. 2), where enameling also took place. Both the laboratory workers and the arcanists were experimenters and inventors. The laboratory workers mainly sought new variations of colors, while the arcanists' tried to invent new varieties of paste. After the reorganization of the Manufactory in 1786/1787, they were obliged to keep notebooks and to report their experimental results in written form. Thus, both the laboratory workers and the arcanists exemplify the figure of an early modern expert, as outlined above.

In what follows I give a brief overview of the career of a laboratory worker at the Manufactory, named Friedrich Bergling (Died 1797), who was further trained to become an arcanist. The case provides insight into the practices as well as the chemical education and training of laboratory workers and arcanists after 1786/1787.[10]

In January 1788, the board of the Manufactory was looking for a new assistant in the laboratory to be further trained as a "laboratory worker" (*Laborant*). It decided to hire the apothecary Friedrich Bergling, whose abilities had previously been examined by the chemist Martin Heinrich Klaproth. Klaproth was Prussia's most famous chemist at the time, but he was also an apothecary who ran his shop until 1800, when he became the director of the laboratory of the Royal Prussian Academy of Sciences. In addition, Klaproth was a member of the highest medical board of Prussia, and he was the

[9] For more details, see (Siebeneicker 2002, 87–117).

[10] For further details, see (Klein 2013, 2014a, 2015).

Fig. 1 A man in the foreground of the *left* building is shown plunging porcelain earth, which is then dried on a wall depicted in the background; firing is shown on the *left* part. The second building on the *right* depicts the crushing of raw material and the forming of porcelain in the background. From (Milly 1774, plate 1)

most important consultant of the Minister Friedrich Anton von Heinitz (1725–1802), who was responsible for the Department of Mining and Smelting Works and for the Royal Prussian Porcelain Manufactory. As Klaproth was a member of an inspection committee of the Manufactory's laboratory, Minister von Heinitz had asked him to examine Bergling's "chemical knowledge" and send him an evaluating report.[11]

During the next 5 years, Klaproth remained Bergling's chemical mentor, and he further supervised his training in the Manufactory's laboratory. Bergling attended Klaproth's chemical course in the lecture series organized by the Department of Mining and Smelting Works, and he read chemical books on Klaproth's recommendation.[12] For Bergling's practical training Klaproth recommended keeping a notebook. He pointed out that in the notebook he should "not only report the work itself but also pay attention, in accordance with chemical principles, to the reason for procedures, their success, and the observed phenomena."[13] Bergling's written reports, presumably based on these notebooks, are preserved in the archive of the Manufactory. They provide unique insight into the chemical education and training of laboratory workers and arcanists after the Manufactory's reorganization in 1786/1787.

[11] KPM archive II 1, vol. 1, fol. 28; KPM archive V 6, fol. 2.

[12] This lecture series is also known as the Berlin *Bergakademie*. Based on archival material, I have shown elsewhere that this new institution was not a true *Bergakademie*, comparable to those in Freiberg and Schemnitz. A *Bergakademie* was founded in Berlin only in 1860; see (Klein 2010b, 2015).

[13] KPM archive V 6, fol. 3.

Fig. 2 On the *left* side of the illustration the grinding of pigments and painting is depicted; in the *middle* enameling in a muffle furnace is shown. From (Milly 1774, plate 8)

In spring 1789, after a year of intensive training, Bergling moved on to the position of a laboratory worker and official of the Manufactory. One of his first obligations was to write a report on the pigment laboratory. Thus he received the order to "apply the chemical knowledge and experience hitherto acquired to the perfection of the preparation of pigments, the invention of new pigments and fluxes and the most advantageous use of enamel fire." In addition, he had to examine the preparations of pigments in the laboratory, write down the recipes, and "clearly explain the reasons for the applied techniques."[14] A few weeks later, on 16 June 1789, Bergling completed his report. It included a list of items to be improved in the laboratory: the purity of the pigments and some additional materials, the cleanliness of the instruments and vessels, and the orderly storage of materials.[15] Clearly, his report showed that he had learned his lessons from Klaproth. Chemists were used to experimenting with a plethora of different kinds of substances, and they were experienced with the problems caused by impurity and disorder. With the exception of apothecaries, there were no artisans at the time who would have worked with a similarly broad range of materials and possessed comparable values for purity and cleanliness.

In April 1791 Bergling finished yet another long report for the Manufactory's board, concerned mainly with the preparation of pigments and fluxes.[16] Around this time, he became the most important experimenter on the Manufactory's inspection committee. In his report he presented a total of 40 recipes for different pigments, fluxes and oils, which relied on numerous experimental trials and repetitions

[14] Ibid., fol. 10.

[15] KPM archive XVII.12, fols. 15–6.

[16] Ibid, fols. 36–49.

of trials, in which he had varied proportions of ingredients or techniques. In several cases Bergling supplemented his recipes with explanations of the chemical processes taking place when preparing a pigment. His reference to phlogiston in some of his explanations even manifests his familiarity with chemical theory.

From 1791 Bergling and his mentor Klaproth also began to exchange ideas about inventions and improvements to existing pigments. Klaproth made comments on Bergling's suggestions, written on the margins of the latter's report, or in the committee's own report. But Bergling, too, commented on Klaproth's suggestions for improvement. Despite these exchanges the two men still had a student-teacher relationship. Bergling asked Klaproth for advice and help when problems occurred, and Klaproth continued to recommend books to him. One of their important co-inventions was "uranium yellow." In September 1789 Klaproth discovered uranium.[17] In the context of his discovery, he immediately examined the practical usefulness of the "uranium calx" (later uranium oxide) for coloring glass and porcelain. Bergling continued Klaproth's earlier experiments, describing their results in three reports in 1792.[18] Based on his trials, uranium calx yielded "a nice yellow color." Bergling's and Klaproth's "uranium yellow" (*Urangelb*) was soon included in the KPM's tables of porcelain colors.[19]

At the end of 1790, Bergling further started a series of entirely new experiments to make himself familiar with the preparation of the porcelain paste and the quality control of its ingredients, that is, with the work of an arcanist.[20] In this context, he also carried out quantitative chemical analyses of porcelain earths, feldspar, quartz and some other kinds of stones. Again, he reported on these experiments in written form a couple of weeks afterwards. On 6 and 27 January 1791, he presented his first reports on the wet chemical analysis of four samples of porcelain earth stemming from different deposits.[21] Three weeks later, he also performed wet quantitative analyses with feldspar, quartz and a few additional stones.[22] Thus he was learning by doing, but he also received continual instruction from one of the most advanced chemical analysts of the time: Klaproth. In June 1793, Bergling was promoted to the position of an arcanist. He was then the Manufactory's most important technical expert overseeing both the making of the paste and the pigments.[23]

It took some 5 years of intensive experimental training on the job along with chemical education by Klaproth for Bergling to become a person possessing the distinct knowledge and competences first of a laboratory worker and then of an arcanist. Bergling regularly performed experimental trials in the Manufactory's laboratory, wrote reports, made suggestions for technical improvements, and carried out invention work in collaboration with Klaproth.

If we compare Bergling and Klaproth, there are similarities and dissimilarities. The two men were experimenters employing the most advanced analytical techniques of eighteenth-century chemistry; they possessed chemical knowledge; they were experts concerning materials and techniques involved in the manufacture and decoration of porcelain; they shared the goal of improving techniques and inventing new, useful materials. And they even had the same social background: they had never

[17] Klaporth reported his discovery of uranium to the Royal Prussian Academy of Sciences on 24 September 1789; see Registres de l'Académie, in *ABBAW* I-IV-32, fol. 81. His related publications are: (Klaproth 1789a, b). See also (Klein 2014b).

[18] *KPM archive* XVII. 12, fols. 54–9.

[19] Ibid., fol. 58. Uranium yellow (*Urangelb*) is mentioned in a table of porcelain colors from 1838 by the Manufactory's director Georg Friedrich C. Frick; see (Köllmann and Jarchow 1987, 323 – color number 28 in the table).

[20] For further details concerning these experiments see (Klein 2014a).

[21] *KMP archive* XVII. 12, fols. 66–7.

[22] *KMP archive* XVII. 12, fol. 68.

[23] Ibid., fols. 76–9.

attended a university but had been apprenticed to a master apothecary. However, unlike Bergling, the apothecary Klaproth was also a famous teacher of chemistry, and he became a member of the Royal Prussian Academy of Sciences in 1788. In 1800, he sold his apothecary's shop and became director of the laboratory of the Royal Prussian Academy of Science. When the Berlin University was founded in 1810, he was nominated its first professor of chemistry. In addition to differences in social status, there were also epistemic differences between the two men, but these were differences only in degree. Klaproth was knowledgeable in all fields of the chemical discipline, and in his teaching he went far beyond the parts of chemistry that were useful for porcelain manufacture. He was recognized both as a man of science, or "natural researcher" (*Naturforscher*), and as an expert for a variety of techniques involving chemical processes. By contrast, Bergling was an expert whose chemical knowledge and skills were more narrowly tuned to porcelain manufacture. In the next part, I further illuminate the figure of the "hybrid expert" who, like Klaproth, combined the roles of experts and natural researchers.

3 Hybrid Experts: The Mining Official and Mineralogist Carl Abraham Gerhard

Mining and metal production was another socio-economical niche in eighteenth-century Europe where experts were thriving. In the second half of the eighteenth century, the Prussian state supported natural researchers who combined learned inquiry into nature with technical work and technological research. Members of the physical and mathematical classes of the Royal Prussian Academy of Sciences were regularly involved in state enterprises such as the manufacture of porcelain, mining and metal smelting, surveying and hydraulic engineering, the installation of lightening conductors, the cultivation of exotic plants, and the production of beet sugar.[24] These academicians were truly hybrid experts living in the worlds of both state-directed manufacture and scientific academies. Promoting the transfer of knowledge originating in these different worlds, they also combined technical and scientific knowledge in programs for the science of mining and other "useful sciences" (Klein 2016).[25] They were teachers of the useful sciences and contributed to the institutionalization of this novel type of science in mining academies, academies of civil architecture, academies of forestry and so on. In the following section I provide another example of the figure of the hybrid expert. We will follow the training and the various technical and scientific activities of the mineralogist and Prussian mining official Carl Abraham Gerhard (1738–1821). A close look at Gerhard's travels of inspection, in particular, will exemplify the way in which state-directed mining contributed to mineralogy and the fledgling science of geology.

Gerhard was the first academic mining official to make a lifelong career in the Prussian state administration. Immediately after the Department of Mining and Smelting Works was founded in 1768, he was nominated mining councilor and elected to the Prussian Academy of Sciences. As the Prussian Academy of Sciences frequently collaborated with the state administration, this deal presented no conflict. Gerhard's case is particularly instructive with respect to the figure of hybrid experts. He had been educated in medicine and then became a physician, a publisher on medical issues, and a

[24] For an overview see (Klein 2015). For surveying and civil architecture in Prussia see (Olesko 2009; Strecke 2000).

[25] As "experts" were not identical with "craftsmen" (see above), my argument differs from the scholar-craftsman thesis that Hall (wrongly) ascribed to Robert Merton; the thesis has also been ascribed to Zilsel. It should be noted, however, that in their studies Merton and Zilsel did not focus on ordinary craftsmen but rather on engineers and inventors. See (Hall 1963, 1959; Merton 1970; Zilsel 2000; Klein 2016).

private scientific lecturer in Berlin.[26] Beginning in 1771, that is, only shortly after he had entered Prussian service, he began to publish regularly on mineralogy and related geological and chemical issues. At the same time, he also acquired practical knowledge and skill in mining and metal production. Gerhard developed his technical expertise in mining and his scientific reputation in tandem. The hybrid nature of his inspections becomes manifested in his simultaneous reports to the mining administration and to the Royal Prussian Academy of Sciences.

As a mining councilor, Gerhard performed a variety of administrative paper-and-pencil tasks, ranging from accounting to writing reports and drafting letters and regulations for Minister von Heinitz. But he was by no means a mere administrator and fiscal official. He was regularly involved in various kinds of technical work, such as the inspections of mines and smelting works, the quality control of ores, and prospecting. From 1771 on, he was the leading Berlin official responsible for the ironworks in parts of Prussia (in the Kurmark, Neumark, and Pomerania), and he developed a particular expertise for work with cast iron, which was of great importance for the Prussian army. He also began to cooperate directly with the Prussian artillery. Furthermore, he carried out long series of scientific and technological experiments at the laboratory of the Department of Mining and Smelting Works, which he had established himself in 1774/1775 (Klein 2010b, 2016). To his most important technological experiments belonged the preparations of new varieties of glass, ceramics, and alloys and studies of their properties. Gerhard reported the results of these experiments not only to Minister Heinitz and the king, but also to the Royal Prussian Academy of Sciences.

As part of his regular obligations, Gerhard undertook long inspection trips to the Prussian mining regions every year, usually starting in April and ending in October. These trips provided numerous opportunities for natural studies, as we will see next.

3.1 Travels of Inspection Alongside "Geological" Inquiry

Travels of inspection to the Prussian mining provinces were an important part of Prussia's mercantilist policy. The Prussian mining districts were far away from the administrative center in Berlin, extending from the Mark in the west (the region around the Ruhr River and the city of Bochum, ca. 450 km from Berlin) to the district around Halle and Magdeburg in the south, to the northern parts of the Neumark in the province of Brandenburg, and to Silesia in the east, some 300 km from Berlin. After the Seven Years' War (1756–1763), Prussia had introduced the so-called "direction principle" (*Direktionsprinzip*) in mining, following the model of the advanced German mining countries, such as Saxony and Brunswick-Wolfenbüttel in the Harz region. This principle stipulated that state authorities were responsible for all significant mining equipment and organization, including technical improvement.

In the autumn of 1768 Gerhard undertook his first travel of inspection, which led him to his homeland Silesia. Silesia became a formal part of Prussia after the Seven Years' War, and in 1768 a new mining law for Silesia was under preparation. Hence, the Department of Mining and Smelting Works in Berlin wished to obtain an overview of the state of mining and metal production in this old mining area. On his first inspection trip, Gerhard was accompanied by the experienced Privy Finance Councilor (*Geheimer Finanzrat*) Heinrich Wilhelm Reichardt. During their inspection, Gerhard and Reichardt visited the royal ironworks in Malapane and in Kreuzburg, which had been established in 1754/1755.

On 30 October 1768, Gerhard completed an official report on these two ironworks, which was openly critical. In both Malapane and Kreuzburg the iron ores were not sufficiently washed and ground, he observed. In Malapane, in particular, the iron ore was not roasted, and thus the sulfur was

[26] For more details on Gerhard's career see (Klein 2012b, 2015).

not removed, so that the iron produced there was of poor quality. As Gerhard pointed out further, the iron ores used in Malapane and Kreuzburg contained zinc, which would have been of great commercial advantage if it were separated from the slag. But since this was not the case, he explained, the precious zinc was lost. A third point of criticism was the enormous loss of iron during its refinement in the smelting works; a fourth one, that the bellows for the smelting works' furnaces were not strong enough, and so on. The report demonstrates the kind of technical issues a mining councilor had to evaluate during an inspection. Gerhard must have prepared himself for this task through the available literature on iron production. Knowledge stemming from technological texts was then supplemented by information from Reichardt and the Silesian mining officials as well as observations of his own and tests on the job.

Another purpose of this first inspection trip was to search for unexploited ore deposits. But as winter was approaching, the two commissioners had to travel back to Berlin without completing their inspection. In May 1769, Gerhard set off for a second inspection in Silesia. In preparation for this second trip, he bought instruments for surveying (*Markscheiden*), mining drills for prospecting, and a portable smelting furnace for chemical assaying. The inspection committee was now considerably enlarged through seven simple miners (*Bergleute*), four foremen, and an experienced mining master (*Oberbergmeister*), named Elster, who came from the mining town of Zellerfeld in the Harz region. In the 5 months that followed Gerhard and Elster would rarely be separated. They would prospect for useful minerals and visit mines and smelting works, almost always together. In this way, Gerhard enjoyed a kind of personal apprenticeship similar to the practical training that the students at the Freiberg Mining Academy received.

The main official goal of this second journey was to search for useful minerals by examining abandoned mines and prospecting. For this purpose, the committee proceeded as follows. Gerhard and Elster first sent a foreman and one or two miners to abandoned mines for a rough check. Several times, old men and miners living in the area showed them the promising places. As many mines were in bad condition, more than once the miners first had to drain an abandoned gallery, repair a shaft and clear it from boulders before they could enter. After these preparations, Gerhard and Elster visited the place and decided whether they wanted to examine it further. They took samples of ores, checked their observable properties and then submitted them to chemical assaying, using their portable smelting furnace. In this way they identified the mineral species contained in an ore sample and their relative quantities. In addition to examining abandoned mines, Gerhard and Elster also prospected for new deposits. Using the new mountain drills Gerhard had bought, they drilled boreholes, sometimes more than 10 m deep, to take samples of materials. They then performed assaying tests to separate and identify the useful minerals they contained. In this way, Gerhard and Elster found ores of copper, lead, silver, and tin as well as arsenic in the first couple of weeks. In the weeks to follow, they also found cobalt ore, coal, iron ores, alum, calamine, chrysoprase, vitriol, and precious stones.

On 10 November 1769, Gerhard and Reichardt reported the results of their inspection to the Minister. The written report presented a comprehensive list of all active Silesian mines and smelting works, including fiscal information. It also contained suggestions about ways to attract new investors, such as public advertisement and financial benefits for the investor, something which had long belonged to the inventory of mercantilist policy. And it made yet another interesting proposal: to establish a mining academy in Berlin. In 1765 the Freiberg Mining Academy had been founded, whose main purpose was the education and training of competent Saxon mining officials, but Prussia lacked a similar institution. In January 1770, Gerhard elaborated a plan for this institution.[27] As a result, a series of lectures on "useful science" for would-be officials responsible for mining, civil

[27] *GStA PK*, I. HA, Rep. 121 Ministerium für Handel und Gewerbe. Abt. Bergwerks-. Hütten und Salinenwesen, No. 7957, fols. 2–6. On the concept of "useful science" see (Klein 2010b, 2016).

architecture, and forestry was established, which was organized and financed by the Department of Mining and Smelting Works.

During his second trip to Silesia, Gerhard also improved as a natural researcher. He collected minerals and further tested them for their identification and classification. His portable assaying furnace allowed him to examine their chemical composition. In the years to come, analytical knowledge of chemical composition would become a characteristic feature of his approach to mineralogy. Furthermore, Gerhard studied local mountains and valleys, wells and rivers, types of rocks and ore deposits. He determined the direction of mountain ranges, measured the altitude of mountains as well as the atmospheric pressure and the air temperature. For his measurements he carried with him a barometer and thermometers. We have insight into the second, naturalistic side of his travel of inspection from reports to the members of the Prussian Academy of Sciences and from publications. Gerhard's reports about his inspection to both the Department of Mining and Smelting Works and the Royal Prussian Academy of Sciences highlight his interconnection of technical work and scientific inquiry.

Gerhard reported on his travel experiences to the academicians several times. In September 1769, while he was still staying in Silesia, he sent a letter about the "natural history of Silesia," which the mathematician Jean Castillon (1708–1791) read at a meeting on 28 September. After his return to Berlin, he personally reported on his inspection at a meeting on 17 November 1769, also demonstrating his collection of Silesian minerals. The secretary recorded, as usual in French: "M. Gerhard a présenté des minéraux qu'il a rapportés de son dernier voyage de Schlésie."[28] An oral *mémoir systématique sur les Minéraux* followed a week later, which discussed issues of mineral classification. [29] In a second oral memoir, presented on 15 February 1770, Gerhard added "physical and mineralogical" observations on the rocks and mountains of Silesia.[30] In a published academic memoir of 1771, entitled *Observations physiques et minéralogiques sur les montagnes de la Silésie*, he elaborated on the latter issue (Gerhard 1771). This memoir was his first attempt at a comprehensive geological description of a mountain region, supplemented by causal explanation. Its very title—*Observations physiques et minéralogiques*—manifests his ambitious goal to combine different approaches in the current studies of the Earth: the study of minerals, "physical geography," and "subterranean geography" or what Abraham Gottlob Werner (1749–1817) later designated "geognosy" (Fig. 3).[31] Gerhard's references to other naturalists and natural philosophers further show that he assimilated his personal observations to communal natural knowledge. At the same time, Gerhard also informed the academicians about the uses of Silesia's minerals and the state of Silesian mining.

Gerhard's inspection trips provide an example of the way in which the hybrid, scientific-technical experts combined technical work with natural inquiry. Inspections of mines and smelting works were part of the administrative and technical obligations of a mining councilor. But they provided opportunities for observations of the Earth, chemical analysis, and the collection of minerals. In the second half of the eighteenth century, the rapidly expanding field of "chemical mineralogy" was closely linked with state-directed mining. The chemical analysis of minerals, and the study of their components, was encouraged by state mining administrations, particularly in Sweden, Saxony, and Prussia, where chemists and chemical mineralogists such as Axel Frederic Cronstedt (1722–1765), Torbon

[28] See the "Registres" of the Academy of Sciences from 21 August 1766 to 17 August 1786, *ABBAW*, I-IV-32, fol. 67. In 1780 Gerhard's mineral collection would include some 4,000 specimens.

[29] Ibid., fol. 68 ("mémoir systématique sur les Minéraux").

[30] Ibid. fol. 70. ("M. Gerhard a lu des Remarques physiques & minéralogiques sur les montagnes de Schlésie.").

[31] On these fields of knowledge that contributed to the emerging science of geology, see (Rudwick 2005). Contrary to Rudwick's argument, Gerhard's example demonstrates that both geognosy and physical geography were pursued in the context of Prussian mining; on the latter see (Klein 2012c).

Fig. 3 Depiction of layers of rocks. From (Trebra 1785, plate 1). Courtesy of the Library of the Max Planck Institute for the History of Science, Berlin

Olaf Bergman (1735–1784), Johann Gottschalk Wallerius (1709–1785) as well as Johann Friedrich Henckel (1678–1744), Andreas Sigismund Marggraf (1709–1782), Johann Heinrich Pott (1692–1777), and Klaproth introduced new analytical techniques along with numerous new analytical reagents.

Mining was also an important milieu for the development of geognosy and geology.[32] We get a glimpse of this fact from Gerhard's scientific activities during his travels of inspection in Silesia. Underground observation in mines opened up a horizon of new things that were only rarely observed above ground. These were the large masses of rocks, the veins of ores, the layered structures of rocks, and the fossils (in the modern sense). All of these entities became major objects of inquiry

[32] For more information, see (Guntau 1984; Laudan 1987; Oldroyd 1996; Rudwick 2005).

in geognosy (later geology) and the theories of the Earth's history. In the process of the epistemic constitution of these objects of inquiry, the miners' perspective and language played an important role. It helped to focus the researchers' attention on objects that were absent in the older tradition of mineralogy. Miners' "mountains" (*Gebirge*), for example, meant large masses of rocks, from which they extracted ores. Transferred into the context of geognosy, these large-scale objects acquired new meaning through their links with concepts of the "structure" of the Earth's crust and its development in history.

4 Conclusion

In this essay, I first presented an overview of the socio-economic niches that gave rise to a distinct early modern practitioner: the "expert." The three niches outlined above—large-scale technological enterprises, the luxury market, and the medical and pharmaceutical professions—were sites for the evolution of extraordinary expertise that was not provided by the guild system. The early modern experts drew together knowledge, methods, and instruments from many different sites to cope with technical challenges. They were involved, directly or indirectly, in the production of things, and they measured, experimented and analyzed objects for predominantly practical purposes. In so doing, they also used written sources of natural and technical knowledge. Furthermore, they often reported their trials and experiments in written form to their superiors in the state administration. From this, the step towards more systematic natural and technological inquiry was possible under favorable social conditions. Thus the apothecary and consultant Martin Heinrich Klaproth also became a well-known chemist and mineralogist.

Processes of "sharing" and interconnecting practical and academic knowledge have recently been studied with respect to "trading zones"—such as princely courts, arsenals, and mines—where the learned and the skilled communicated with each other and exchanged, or "traded" with, "substantive knowledge or expertise" (Long 2011, 94–5). Likewise, Harold Cook, Sven Dupré and Christoph Lüthy have studied exchanges and interactions between early modern artisans, merchants, and learned men in various social networks (Cook 2007; Dupré and Lüthy 2011). Eric Ash has pointed out that the rise of the early modern expert was paralleled by the development of the state, which needed experts for trade and manufacture and various kinds of public service. He has further observed that "many of the natural philosophers and the experts who served the early modern state were really pursing the same goal: a superior understanding of the natural world based in part on empirical experience, which would in turn enhance their operative knowledge of nature and improve their ability to produce valuable goods and effects." Hence, there was "significant overlap," he has argued, between the philosophers' and the experts' knowledge and activities (Ash 2010, 22).

While I have studied "trading zones" and "overlaps" between the goals and interests of technical experts and natural researchers (*Naturforscher*)[33]—namely the cooperation between the laboratory worker Bergling and the chemist Klaproth, and between the mining master Elster and the mineralogist Gerhard—I have further highlighted a distinct persona who personally combined the roles of technical expert and natural researcher and brought them into fruitful interaction: the hybrid technical-scientific expert. The chemist, apothecary and chemical expert Klaproth was such a persona, as was the mineralogist and mining official Carl Abraham Gerhard. Likewise, Peter Damerow, Jürgen Renn and

[33] It should be noted that the term "natural philosopher" had a different meaning in the German context. It meant scholars such as Christian Wolf and Immanuel Kant, who were concerned to a considerable extent with metaphysical questions.

Matteo Valleriani have proposed that Galileo would best be characterized as a "scientist engineer."[34] In eighteenth-century continental Europe, most of these hybrid experts were consultants of the state, or state officials, and at the same time members of scientific academies, teachers, and authors of technological and scientific texts. Some of these men had completed apprenticeship training in the guild system and then paved their own ways into the academic world, while others had studied at universities, mostly at medical faculties, and acquired technical competence afterwards. With respect to eighteenth-century England, Peter Jones and David Miller have described a similar figure, the inventive "savant fabricant," exemplified by James Watt (1736–1819) and Matthew Boulton (1728–1809).

Following the career and activities of hybrid experts, I argue, allows new insights, not just into "shared" knowledge between people of different social and educational backgrounds, but in truly combined, and interacting, technical and scientific knowledge and practices.[35] The eighteenth-century hybrid experts transformed mines, smelting works, construction sites, pharmaceutical laboratories, porcelain manufactories and so on, into sites of systematic natural observation and experimentation. At meetings of academies and in their publications they reported what they had experienced at sites of practice, unfamiliar to most academicians. Drawing on methods and experience acquired in practical contexts, they highlighted the importance of experimentation, measuring, calculation, and analysis. In so doing, they promoted the exact and analyzing natural sciences. Inversely, in their technical work and technological research, these men also implemented instruments, materials and knowledge that had originated at academies and universities. Thus, the figure of the hybrid expert epitomizes the breakdown of ancient ideological boundaries between hand and mind.

Among the eighteenth-century scientific-technical experts many were teachers in the context of the new technological academies and schools. Comparable to the "scientist-engineer" Galileo and his teaching of military officers on fortification (Valleriani 2010), they selected elements of sciences and mathematics that they considered to be useful in practice, and they further combined these scientific building blocks with knowledge that originated in practice. In so doing, they elaborated a new type of science: the so-called useful sciences, an early form of the nineteenth-century technological science (Klein 2016). The foundation of mining academies and other technological schools in eighteenth-century continental Europe and their amalgamation of "theory and practice" was to a large extent the achievement of hybrid scientific-technical experts.

References

Manuscript Sources

Archive of the Berlin-Brandenburgische Akademie der Wissenschaften (*ABBAW*).
Archive of the Königlich Preussische Porzellanmanufaktur (*KPM archive*).
Geheimes Staatsarchiv Preussischer Kulturbesitz (*GStA PK*).

Printed Sources

Agricola, Georg. (1556) 1977. *De re metallica libri XII* (Basel); Zwölf Bücher Vom Berg- und Hüttenwesen. Trans. and ed. Carl Schiffner. München: dtv.

[34](Damerow and Renn 2010; Jones 2008; Miller 2009; Valleriani 2010). On Galileo see also (Lefèvre 2001). For French examples see (Klein 2012a; Lehman 2012).

[35]See Peter Jones' discussion of this problem in (Jones 2008, 16–7).

Ash, Eric H. 2007. Amending nature: Draining the English fens. In *The mindful hand: Inquiry and invention from the late Renaissance to early industrialization*, ed. Lissa Roberts, Simon Schaffer, and Peter Dear, 117–143. Amsterdam: Koninklijke Nederlandse Akademie van Wetenschappen.

Ash, Eric H. 2010. Introduction: Expertise and the early modern state. In *Expertise: Practical knowledge and the early modern state*, ed. Eric H. Ash. Osiris 25: 1–24.

Ashworth, William. 2007. The intersection of industry and the state in eighteenth-century Britain. In *The mindful hand: Inquiry and invention from the late Renaissance to early industrialization*, ed. Lissa Roberts, Simon Schaffer, and Peter Dear, 349–378. Amsterdam: Koninklijke Nederlandse Akademie van Wetenschappen.

Belhoste, Bruno. 2011. *Paris Savant, parcours et rencontres au temps des Lumières*. Paris: Armand Colin.

Belhoste, Bruno. 2012. A Parisian craftsman among the savants: The joiner André-Jacob Roubo (1739–1791) and his works. *Annals of Science* 69: 395–411.

Bennett, Jim. 1987. *The divided circle: A history of instruments for astronomy, navigation, and surveying*. Oxford: Phaidon Press.

Bennett, Jim. 1991. The challenge of practical mathematics. In *Science, culture and popular belief in Renaissance Europe*, ed. S. Pumfrey, P.L. Rossi, and M. Slawinski, 176–190. Manchester: Manchester University Press.

Beretta, Marco. 2009. *The alchemy of glass: Counterfeit, imitation, and transmutation in ancient glassmaking*. Sagamore Beach: Science History Publications.

Beretta, Marco. 2012. The rise and fall of the glassmaker Paul Bosc d'Antic (1753–1784). *Annals of Science* 69: 395–412.

Beretta, Marco. 2014. Unveiling glass's mysteries: Lavoisier, Loysel and the first chemical treatise on glass (1765–1799). In *Objects of chemical inquiry*, ed. Ursula Klein and Carsten Reinhardt, 1–20. Sagamore Beach: Science History Publications.

Cook, Harold J. 2007. *Matters of exchange: Commerce, medicine, and science in the Dutch golden age*. New Haven: Yale University Press.

Damerow, Peter, and Jürgen Renn. 2010. The transformation of ancient mechanics into a mechanistic worldview. In *Transformationen antiker Wissenschaften*, ed. Georg Töpfer and Hartmut Böhme, 239–263. Berlin/New York: De Gruyter.

Dauxois, Jacqueline. 1997. *Der Alchimist von Prag. Rudolf II. von Habsburg. Eine Biographie*. Düsseldorf: Artemis & Winkler.

Dupré, Seven. 2014. Introduction. In *Laboratories of art: Alchemie and art technology from antiquity to the 18th century*, ed. Sven Dupré, VII–XIX. Heidelberg/New York/Dordrecht/London: Springer.

Dupré, Sven, and Christoph Lüthy (eds.). 2011. *Silent messengers: The circulation of material objects of knowledge in the early modern low countries*. Münster: LIT Verlag.

Dupré, Sven, Dedo von Kerssenbrock-Krosigk, and Beat Wismer. 2014. *Kunst und Alchemie, das Geheimnis der Verwandlung*. Munich: Hirmer.

Findlen, Paula. 1994. *Possessing nature: Museum and collecting in early modern Italy*. Berkeley: UC Press.

Findlen, Paula, and Pamela H. Smith (eds.). 2002. *Merchants and marvels: Commerce, science, and art in early modern Europe*. London/New York: Routledge.

Fleischer, Alette. 2007. The Beemster Polder: Conservative invention and Holland's great pleasure garden. In *The mindful hand: Inquiry and invention from the late Renaissance to early industrialization*, ed. Lissa Roberts, Simon Schaffer, and Peter Dear, 145–166. Amsterdam: Koninklijke Nederlandse Akademie van Wetenschappen.

Gerhard, Carl Abraham. 1771. Observations physiques et minéralogiques sur les montagnes de la Silésie. *Nouveaux Mémoires de L'Académie Royale des Sciences et Belles Lettres* 1771: 100–111.

Guntau, Martin. 1984. *Die Genesis der Geologie als Wissenschaft: Studie zu den kognitiven Prozessen und gesellschaftlichen Bedingungen bei der Herausbildung der Geologie als naturwissenschaftliche Disziplin an der Wende vom 18. zum 19. Jahrhundert*. Berlin: Akademie Verlag.

Hall, Rupert A. 1959. The scholar and the craftsman in the Sientific Revolution. In *Critical problems in the history of science*, ed. Marshall Clagett, 3–23. Madison: University of Wisconsin Press.

Hall, Rupert A. 1963. Merton revisited or science and society in the seventeenth century. *History of Science* 2: 1–16.

Harkness, Deborah E. 2007. *The jewel house: Elizabethan London and the Scientific Revolution*. New Haven: Yale University Press.

Hufbauer, Karl. 1982. *The formation of the German chemical community (1720–1795)*. Berkeley: UC Press.

Johnston, Stephen. 1996. The identity of the mathematical practitioner in 16th-century England. In *Der „mathematicus:" Zur Entwicklung und Bedeutung einer neuen Berufsgruppe in der Zeit Gerhard Mercators*, ed. Irmgard Hantsche, 93–120. Bochum: Universitätsverlag. Dr. N. Brockmeyer, 1996.

Jones, Peter M. 2008. *Industrial enlightenment, science, technology and culture in Birmingham and the West Midlands, 1760–1820*. Manchester: Manchester University Press.

Kieffer, Fanny. 2014. The laboratories of art and alchemie at the Uffizi Gallery in Renaissance Florence: Some material aspects. In *Laboratories of art: Alchemie and art technology from antiquity to the 18th century*, ed. Dupré Sven, 105–127. Heidelberg/New York/Dordrecht/London: Springer.

Klaproth, Martin Heinrich. 1789a. Mémoire chimique et minéralogique sur l'urane. *Mémoires de L'Académie Royale des Sciences et Belles Lettres* 1786 (87): 160–174.

Klaproth, Martin Heinrich. 1789b. Chemische Untersuchung des Uranits, einer neuentdeckten metallischen Substanz. *Annalen der Chemie* 1789 (Teil 2): 387–403.

Klein, Ursula. 2007a. Apothecary-chemists in eighteenth-century Germany. In *New narratives in eighteenth century chemistry*, ed. Lawrence M. Principe, 97–137. Dordrecht: Springer.

Klein, Ursula. 2007b. Apothecary's shops, laboratories and chemical manufacture in eighteenth-century Germany. In *The mindful hand. Inquiry and invention from the late Renaissance to early industrialisation*, ed. Lissa Roberts, Simon Schaffer, and Peter Dear, 246–276. Amsterdam: Koninklijke Nederlandse Akademie van Wetenschappen.

Klein, Ursula. 2008. The laboratory challenge: Some revisions of the standard picture of early modern experimentation. *Isis* 99 (4): 769–782.

Klein, Ursula. 2010a. Blending technical innovation and learned natural knowledge: The making of ethers. In *Materials and expertise in early modern Europe: Between market and laboratory*, ed. Ursula Klein and Emma C. Spary, 125–157. Chicago/London: The University of Chicago Press.

Klein, Ursula. 2010b. Ein Bergrat, zwei Minister und sechs Lehrende, Versuche zur Gründung einer Bergakademie in Berlin um 1770. *NTM* 18 (4): 437–468.

Klein, Ursula. 2012a. Introduction: Artisanal-scientific experts in eighteenth-century France and Germany. *Annals of Science* 69 (3): 303–306.

Klein, Ursula. 2012b. Savant officials in the Prussian mining administration. *Annals of Science* 69 (3): 349–374.

Klein, Ursula. 2012c. The Prussian mining official Alexander von Humboldt. *Annals of Science* 69 (2): 27–68.

Klein, Ursula. 2013. Chemical experts at the Royal Prussian Porcelain manufactory. *Ambix* 60: 99–121.

Klein, Ursula. 2014a. Depersonalizing the Arcanum. *Technology and Culture* 55: 591–621.

Klein, Ursula. 2014b. Klaproth's discovery of uranium. In *Objects of chemical inquiry*, ed. Ursula Klein and Carsten Reinhardt, 21–46. Sagamore Beach: Science History Publications.

Klein, Ursula. 2015. *Humboldts Preußen, Wissenschaft und Technik im Aufbruch*. Darmstadt: Wissenschaftliche Buchgesellschaft.

Klein, Ursula. 2016. *Nützliches Wissen: die Erfindung der Technikwissenschaften*. Göttingen: Wallstein.

Klein, Ursula, and E.C. Spary. 2010. Introduction: Why materials? In *Materials and expertise in early modern Europe: Between market and laboratory*, ed. Ursula Klein and E.C. Spary, 1–23. Chicago: The University of Chicago Press.

Köllmann, Erich, and Margarete Jarchow. 1987. *Berliner Porzellan*, Textband, München: Klinkhardt & Biermann.

Laudan, Rachel. 1987. *From mineralogy to geology. The foundations of a science, 1650–1830*. Chicago/London: The University of Chicago Press.

Lefèvre, Wolfgang. 2001. Galileo engineer: Art and modern science. In *Galileo in context*, ed. Jürgen Renn, 11–27. Cambridge: Cambridge University Press.

Lefèvre, Wolfgang. 2004. Introduction. In *Picturing machines 1400–1700*, ed. Wolfgang Lefèvre, 1–10. Cambridge, MA: MIT Press.

Lehman, Christine. 2012. Pierre-Joseph Macquer: An eighteenth-century artisanal-scientific expert. *Annals of Science* 69 (3): 307–333.

Leong, Elaine, and Alisha Rankin (eds.). 2011. *Secrets and knowledge in medicine and science, 1500–1800*. Farnham/Burlington: Ashgate.

Long, Pamela O. 2011. *Artisan/practitioners and the rise of the new sciences, 1400–1600*. Corvallis: Oregon State University Press.

Lowengard, Sarah. 2005. *The creation of color in eighteenth-century Europe*. New York: Columbia University Press.

Mauskopf, Seymour H. 2010. The crisis of English gunpowder in the eighteenth century. In *Materials and expertise in early modern Europe: Between market and laboratory*, ed. Ursula Klein and E.C. Spary, 288–320. Chicago: The University of Chicago Press.

Merton, Robert K. 1970. *Science, technology and society in seventeenth-century England*. New York: Harper Torchbooks (First published in *Osiris*, 1938).

Miller, David. 2009. *James Watt, chemist: Understanding the origin of the steam age*. London: Pickering & Chatto.

Milly, Comte de and Nicolas-Christiern de Thy. 1774. *Die Kunst das ächte Porcellan zu verfertigen*. Trans. and annotated Daniel Gottfried Scherber. Königsberg/Leipzig: Johann Jakob Kanter. Reprint: Leipzig: Zentralantiquariat der Deutschen Demokratischen Republik, 1977.

Moran, Bruce T. 1991. *Chemical pharmacy enters the university: Johannes Hartmann and the didactic care of chymiatria in early seventeenth century*. Madison: American Institute of the History of Pharmacy.

Müller-Wille, Staffan. 2004. Nature as a marketplace: The political economy of Linnean botany. In *Oeconomies in the age of Newton*, ed. Neil De Marchi and Margaret Schabas, 155–173. Durham: Duke University Press.

Nieto-Galan, Agustí. 2001. *Colouring textiles: A history of natural dyestuffs in industrial Europe*. Dordrecht: Kluwer.

Oldroyd, David. 1996. *Thinking about the Earth: A history of ideas in geology*. London: The Athlone Press.

Olesko, Kathryn M. 2009. Geopolitics and Prussian technical education in the late eighteenth century. *Actes d'Història de la Ciència de la Tèchnica, Nova Època* 2 (2): 11–44.

Park, Katharine. 1985. *Doctors and medicine in early Renaissance Florence*. Princeton: Princeton University Press.

Pérez, Liliane. 2008. Inventing a world of guilds: Silk fabrics in eighteenth-century Lyon. In *Guilds, innovation, and the European economy, 1400–1800*, ed. S.R. Epstein and Maarten RoyPrak, 232–263. Cambridge: Cambridge University Press.

Pugliano, Valentina. 2012. Specimen lists: Artisanal writing or natural historical paperwork? *Isis* 103: 716–726.

Rudwick, Martin. 2005. *Bursting the limits of time. The reconstruction of geohistory in the age of revolution*. Chicago/London: The University of Chicago Press.

Schaffer, Simon. 2007. The Charter'd Thames: Naval architecture and experimental spaces in Georgian Britain. In *Mindful hand: Inquiry and invention from the late Renaissance to early industrialization*, ed. Lissa Roberts, Simon Schaffer, and Peter Dear, 279–305. Amsterdam: Koninklijke Nederlandse Akademie van Wetenschappen.

Siebeneiker, Arnulf. 2002. *Offizianten und Ouvriers, Sozialgeschichte der Königlichen Porzellan-Manufaktur und der Königlichen Gesundheitsgeschirr-Manufaktur in Berlin 1763–1880*. Berlin: De Gruyter.

Siraisi, Nancy. 1990. *Medieval and early Renaissance medicine: An introduction to knowledge and practice*. Chicago/London: The University of Chicago Press.

Smith, Pamela H. 2004. *The body of the artisan: Art and experience in the Scientific Revolution*. Chicago/London: The University of Chicago Press.

Spary, Emma C. 2000. *Utopia's garden: French natural history from Old Regime to Revolution*. Chicago/London: The University of Chicago Press.

Spary, Emma C. 2010. Liqueurs and the luxury marketplace in Paris. In *Materials and expertise in early modern Europe: Between market and laboratory*, ed. Ursula Klein and Emma C. Spary, 225–255. Chicago/London: The University of Chicago Press.

Spary, Emma C. 2012. *Eating the enlightenment: Food and the sciences in Paris, 1670–1760*. Chicago/London: The University of Chicago Press.

Strecke, Reinhart. 2000. *Anfänge und Innovation der preußischen Bauverwaltung, von David Gilly zu Karl Friedrich Schinkel*. Cologne: Böhlau.

Tomic, Sacha. 2010. *Aux origines de la chimie organique: Méthodes et pratiques des pharmacien et des chimistes (1785–1835)*. Rennes: Presses Universitaire de Rennes.

Tomory, Leslie. 2012. *Progressive enlightenment: The origins of gaslight industry, 1780–1820*. Cambridge: MIT Press.

Tomory, Leslie. 2013. Fostering a new industry in the Industrial Revolution: Boulton & Watt and gaslight 1800–1812. *British Journal for the History of Science* 46 (2): 199–229.

Trebra, Friedrich Wilhelm Heinrich von. 1785. *Erfahrungen vom Innern der Gebirge*. Dessau/Leipzig: Verlagskasse für Gelehrte und Künstler.

Valleriani, Matteo. 2010. *Galileo engineer*. Dordrecht: Springer.

Werrett, Simon. 2010. *Fireworks: Pyrotechnic arts and sciences in European history*. Chicago/London: The University of Chicago Press.

Weyer, Jost. 1992. *Graf Wolfgang II. von Hohenlohe und die Alchemie. Alchemistische Studien im Schloß Weikersheim 1587–1610*. Sigmaringen: Thorbecke.

Zilsel, Edgar. 2000. *The social origins of modern science*. Forwarded by Joseph Needham, with an Introduction by Diederick Raven and Wolfgang Krohn, ed. Diederick Raven, Wolfgang Krohn and Roberts S. Cohen. Dordrecht: Kluwer.

Utilitas astronomiae in the Renaissance: The Rhetoric and Epistemology of Astronomy

12

Pietro D. Omodeo

Abstract

This essay will consider the views of several prominent Renaissance scholars regarding the practical embedment of astronomy. Certain rhetorical commonplaces are frequently returned to in their writings, in works related to the *utilitas astronomiae* (astronomy's utility). Such texts can be grouped within the Renaissance genre of the *encomia scientiarum et artium*, and include rhetorical pieces from the Averroist philosopher Girolamo Fracastoro (1478–1553) in his *Homocentrica* (1538), the Wittenberg ephemerist Erasmus Reinhold (1511–1553) in the dedicatory letter of his edition of Ptolemy's *Almagest* (1549), the renowned Jesuit mathematician Christophorus Clavius (1537–1612) in the preface to his commented edition of Sacrobosco's *De sphaera* (1570), and the Scottish mathematician Duncan Liddel (1561–1613) in a eulogy on mathematics delivered at Helmstedt in 1591. According to these sources, the practical realms benefiting from astronomy ranged from navigation and agriculture, to medicine, pedagogy, theology, and even civil affairs, where the discipline was relied upon, for instance, in the computation of calendars to schedule religious festivities. In his *Scholae mathematicae* (1569), the Paris *lecteur royale* Petrus Ramus (1515–1572) turned the treatment of the utility of mathematics into a reflection on the practical origins of science in general. Particular attention is devoted to Girolamo Cardano (1501–1586) as a scholar committed to an empirical and practical conception of knowledge. According to him, astronomy had a range of practical uses, whether in navigation or medicine, or even in the prediction of future events by means of astrology. His *Encomium astrologiae* (1543) is compared with his other eulogies concerning mathematical sciences and the arts, and with small tracts in which he developed a sort of epistemology of practical knowledge.

P.D. Omodeo (✉)
Max Planck Institute for the History of Science, Berlin, Germany
e-mail: pdomodeo@mpiwg-berlin.mpg.de

M. Valleriani (ed.), *The Structures of Practical Knowledge*, DOI 10.1007/978-3-319-45671-3_12

> *If any mathematician should be indignant at our suggesting that his "pure science" has any earthly import, we shall merely ask him to inform us why mathematics is studied by the merchants' sons in the commercial high schools, the would-be agronomists in the agricultural schools, the would-be engineers in the engineering schools....*
>
> Nikolai Bukharin (Bukharin 2011, 223)

> *Mathematicas artes popularis utilitatis caussa repertas esse.... Arithmetica tractandae mercaturae, Geometriam dimetiendae terrae gratia.*
>
> Pierre de la Ramée (Ramée 1569, 44)

It is no easy task to assess the practical application of astronomy in the Renaissance. Unlike other disciplines emerging directly from the practical concerns and technological advances of the time (e.g. mechanics, ballistics, theater engineering, and mineralogy),[1] astronomy rested on a well-established and very ancient tradition. It had a well-defined disciplinary status among the mathematical disciplines, and had been the subject of epistemological reflection for centuries, if not millennia. Its origins are lost in antiquity and historians, today just as in the Renaissance, could only speculate about its inception on the basis of philosophical assumptions on the *nature* of science rather than on historical evidence.

These preliminary remarks should not discourage us however from facing the challenge of investigating *practical astronomy*. We can in fact devise several approaches to shed light on various facets of this issue. One such approach would be to consider the implementation and transformation of the astronomical tradition as dependent on socio-economic factors and technological advances. A case in point is the development of astronomy in response to the needs of navigation and the colonial ambitions of the Europeans (Chap. 16). Another manner of addressing practical astronomy would be to focus on astronomical *practices*, such as the computation of ephemerides, the testing and correction of astronomical tables by means of observation, their astrological interpretation, or the employment and perfection of mathematical instruments up to the introduction of the telescope (Chap. 13).

I have chosen a different and complementary approach. I will let Renaissance scholars speak for themselves by taking into account their reflections on the usefulness of astronomy. Renaissance astronomers often extolled the practical dimension of their discipline, especially in the prefaces to their works and in dedicatory letters. The context in which the theme of astronomy's utility emerges is mostly in rhetorical praise of that science. Typically, the celebration of the practical dimension of astronomy was accompanied by an encomium of its noble origins, as well as by considerations on the sublimity of its subject of enquiry and about the certitude of its demonstrations. As a result, it is not always possible to disentangle these different dimensions of Renaissance rhetorical assessments of astronomy, namely the praise of its origins, nobility, and usefulness.[2]

A twofold difficulty presents itself to us when encountering the rhetoric of astronomy in the Renaissance. Firstly, the modern reader must identify the commonplaces that recur in eulogies of astronomy. Secondly, they must discern the author's original viewpoint, embedded within the canon of eloquent rhetoric in use. For this reason, it is expedient to initially identify the continuities visible in the texts and then, on this basis, to point out the variations in the rhetorical motifs. These variations will be understood as clear signs of the respective author's intentions and perspective. In fact, it is precisely the reliance on themes familiar to the literate elites that allowed Renaissance scholars to legitimate themselves and the views they brought forward.

---

[1]On the practical origins of mechanics, see (Renn 2001; Valleriani 2010, 2013; Büttner 2008). For theater engineering and mineralogy, see (Chaps. 4 and 14).

[2]On humanistic eulogies of astronomy and mathematics, see especially the studies on Regiomontanus from (Byrne 2006; Zinner 1990, 69–74; Malpangotto 2008, 42–3, 75–94). An overarching study on the historical perspectives presented in the rhetorical context under scrutiny is given by (Goulding 2010).

In the following, I will consider and compare several rhetorical eulogies of astronomy and mathematics, in order to highlight the common features of this genre and the differences emerging from different treatments of the *utilitas astronomiae*.[3] I will review influential examples taken from different origins and contexts: the first chapter of the *Homocentrica* (Venice 1538) by the Averroist philosopher and physician Girolamo Fracastoro (1483–1553); the dedicatory letter of a commented edition of the first book of Ptolemy's *Almagest* (Wittenberg 1549) by the Wittenberg ephemerist Erasmus Reinhold (1511–1553); the preface to a commented edition of Sacrobosco's *De sphaera* by the Jesuit mathematician Christophorus Clavius (1537–1612); and an oration on mathematics delivered by the Scottish mathematician Duncan Liddel (1561–1613) in 1591 at the Philippist University of Helmstedt. As an outlier, I will discuss the reduction of "dignity" to "utility" by the Calvinist *lecteur royale* Pierre de la Ramée (1515–1572) in his *Scholae mathematicae* (Basel 1569).

Finally, I will zoom in on a scholar who might be thought of as the embodiment of *practical knowledge* in the Renaissance, namely the physician and mathematician, astrologer and natural philosopher, engineer and erudite practitioner, Girolamo Cardano (1501–1576). This famous polymath often stressed his predilection for practice over pure theory. Cardano championed astronomy, not least for the variety of practical purposes for which it could be used. These ranged from navigation to medical applications, and the prediction of future events by means of astrology. Since Cardano composed a eulogy of astronomy, as well as eulogies of mathematical sciences and medicine, a comparative reading of these rhetorical texts will help us assess *practical astronomy* according to his viewpoint.

1 Girolamo Fracastoro on Astronomy's Usefulness in the *Homocentrica*

In 1538, the Veronese physician Girolamo Fracastoro of the Averroist School of Padua published an astronomical work, *Homocentrica*, aimed at resurrecting the astronomy of concentric spheres developed by Callippus, Eudoxus, and Aristotle. As Miguel A. Granada and Dario Tessicini have suggested, in the short run this project of a homocentric substitution of Ptolemaic eccentrics and epicycles was theoretically concurrent with Copernicus's *De revolutionibus* (1543) (Granada and Tessicini 2005).[4] Both works were dedicated to the same pontiff, Paul III, but the dedicatory letters were very different in tone. Their opposing historical viewpoints contrast strikingly. Whereas Copernicus did all he could to present his new planetary ideas (terrestrial motion and heliocentrism) as the reappraisal of ancient doctrines, Fracastoro emphasized the novelty of his post-Scholastic astronomy. Instead of reverence toward the wisdom of the past, he stressed the progressive character of scientific advancement. Fracastoro's dedicatory letter begins precisely with an affirmation of the idea of 'scientific progress:'

> Highest and very blessed Pontiff, that what the mortals discovered in this life were not all discovered by one nor came to light all at once. Rather, some [were discovered] in a certain age and some in another one, according to the varying conditions of times and situations. (Fracastoro 1538, fol. A1r)[5]

[3]Historians of science have become aware of the intricacies, but also the relevance of early modern rhetoric of science. Among others, Mario Biagioli pointed out the interconnection between style and etiquette in the scientific production of early modern academies; Robert S. Westman dealt with the complex relation between proof and rhetoric in Nicolaus Copernicus's work; while Granada and Tessicini reconstructed both the rhetorical structure as well as the sociological background of dedicatory epistles to astronomical works such as Copernicus's *De revolutionibus* and Fracastoro's *Homocentrica*. See (Biagioli 1996; Westman 1990; Granada and Tessicini 2005).

[4]On Renaissance homocentrism, see (Di Bono 1990).

[5]"Quae in vita mortales invenere summe Pont. Beatissime, nec ab uno omnia inventa sunt; nec in lucem simul omnia prodiere, sed haec una aetate, illa alia, ut dies et rerum conditio tulit."

Fracastoro presented science as a collective enterprise, continually developing through *inventiones* (discoveries). The Ancients were in a privileged position, since for them *everything* had yet to be discovered. By contrast, the Moderns were in a rather more ambivalent situation. On the one hand, much knowledge acquired in ancient times had been irremediably lost. On the other, the Moderns made many new discoveries, especially in the fields of geography and astronomy. Ancient knowledge about the *orbis terrarum* was rather limited compared with the flourishing of geography in the age of worldwide exploration. As for astronomy, Fracastoro adds that the Ancients had no knowledge of many celestial movements, nor of the existence of *orbes* accounting for them.

The first chapter of the *Homocentrica* is entitled *De dignitate utilitateque astronomiae* (On the Dignity and Utility of Astronomy). Fracastoro argues that this *nobilissima* (most noble) discipline itself should be regarded as possessing the same magnificence as its objects of study, *divina illa corpora* (divine celestial bodies) (Fracastoro 1538, 3). Hinting at the ongoing debates on the *certitudo mathematicarum*, he added that astronomy achieves the certainty of its conclusions thanks to the mathematics at its core (De Pace 1993). However, in Fracastoro's treatment, the emphasis is not on astronomy's 'dignity,' but rather on its utility. He deems this discipline to be *necessary* in many aspects of life and in the practice of many arts and sciences. Firstly, a knowledge of astronomy is essential when making prognostications in the fields of medicine, navigation, and agriculture. Secondly, this discipline forms the basis for measuring time and thus arranging civil affairs. Thirdly, astronomy is necessary in the study of geography (Chap. 16). Fourthly, Fracastoro also emphasizes astronomy's moral or anagogical relevance. From astronomy we learn balance and measure. In fact, we understand how insignificant we are in comparison to the firmament and the dimensions of the cosmos. Indeed, humankind derived the first notion of eternity from the contemplation of the heavens. To this list of the many ways astronomy may be utilised, and the potential benefits it may bring, Fracastoro adds some final remarks on the usefulness of his homocentric astronomy. These remarks are essentially theoretical. His combinations of concentric spheres, or *homocentrica*, account for celestial motions from a natural viewpoint, which is in agreement with basic assumptions presented in Scholastic philosophy. In this case, he observes that 'truth' is the most useful and desirable of all things. In this manner, at the end of this text, he virtually merges the *utilitas* and the *dignitas astronomiae*.

In summary, Fracastoro embeds his praise of astronomy in his celebration of scientific progress in general. In his eyes, astronomy parallels geography, because in these two scientific fields the Moderns supersede the Ancients. As to the *dignitas* of the discipline, his assessment is based on the double commonplace where both its object—heavenly bodies—and its mathematical method are instances of the sublime. As far as the *utilitas* is concerned, he refers to astrology by mentioning three arts that were traditionally considered tied to it: medicine, navigation, and agriculture. Furthermore, his reference to the reckoning of time and the measurements of geography points to areas in which astronomical expertise is necessary. Finally, Fracastoro underlines the ethical relevance of astronomy. Scholars are taught moderation and modesty, and the distinction between astronomy's practical usefulness, and the study of astronomy *per se*, becomes blurred.

2 Erasmus Reinhold's Melanchthonian Approach to the Usefulness of Astronomy

Erasmus Reinhold, the calculator of the first Copernican tables and professor of higher mathematics at the Lutheran University of Wittenberg, was a close collaborator of the *praeceptor Germaniae*, Philipp Melanchthon (1497–1570), whom he also advised on mathematical and astronomical issues.[6]

[6]See (Gingerich 1973; Omodeo and Tupikova 2013; Thüringer 1997).

He edited several textbooks on astronomy, which were reprinted many times. The most important among these were his edition of Georg von Peuerbach's (1423–1461) *Theoricae novae planetarum* (Wittenberg 1542) and a commented and illustrated edition of the first book of Ptolemy's *Almagest* (Wittenberg 1549). In the dedicatory letter of this latter work, he expanded on topoi connected to the usefulness and dignity (where the term *dignitas*, but also *bonitas*, are used) of astronomy.

This writing is marked by Melanchthonian humanism and pedagogical concerns. It begins with allegations of impiety directed against detractors of astronomy (Reinhold 1549, fol. A2v). Rude and vulgar opponents of the discipline are compared to the Giants in Greek mythology, who engaged themselves in a war "against the heavens." The study of the heavens boasts piety as the heavens reveal the divine. From such perspective, astronomy has a clear moral, pedagogical, and political importance. These ethical concerns lead Reinhold to conflate the usefulness of astronomy with its dignity. Is salvation not the most useful of all things? The knowledge of divine truths nurtures morality and piety.

In the dedicatory epistle of his edition of Ptolemy, Reinhold first points out the medical application of astronomy. He hints at the way astrology and astronomical computation were central in Galenic medicine (Reinhold 1549, fol. A2v). Secondly, he remarks that astronomy is fundamental for understanding ideas of chronology and geography. For their part, these disciplines are fundamental for the study of history, and most importantly for Church history: "no history can come to light without the series of times and lands' description" (Reinhold 1549, fol. A3r).[7] Reinhold refers also to the relevance of time reckoning for civil and religious purposes. As to rulers, their support for astronomy makes them immortal in the memory of future generations. For instance, Julius Caesar won immortal glory for his calendar reform, and Alfonso X of Castile (1221–1284) for his astronomical tables (Reinhold 1549, fol. A4r). Moreover, astronomy has pedagogical import. It instils in human souls a sense of gravitas, a sense of honesty and seriousness. Measure and proportion expressed by the mathematical order of the heavens (and in mathematics *tout court*) are models on which to base a balanced existence as well as civil justice:

> In fact, justice, which is the guardian of the good order in life, derives from those arts the sense of measure but [due to the avidity of present times] it no longer takes into account the right proportion of work, dignity and usefulness of the arts and of the expenditure as shown in the mechanical constructions, and establishes the right payment according to that [proportion] (Reinhold 1549, fol. A5r).[8]

As with Fracastoro, the moral dimension of mathematical studies evidently bridges the separation between utility and dignity. Reinhold also hints at astrology (*pars divinatrix doctrinae de coelestibus corporibus*) being an important field of application for the science of the heavens. Yet, he does not expand on this point in his introduction to Ptolemy's *Almagest*, arguing that this work does not deal directly with astrology. For enthusiastic eulogies of astrology, one should rather look at his and Melanchthon's prefaces to the *Theoricae novae planetarum* or Melanchthon's introduction to Ptolemy's *Quadripartitum* (Kusukawa 1995). In his edition of the *Almagest* book one, Reinhold limits himself to stressing the application of astronomical knowledge to meteorology, or astrological weather forecasting, and to the Galenic doctrine of the temperaments.

It should be added that, in this preface, Reinhold mixes up the benefits descending from the study of the stars with benefits descending from the stars themselves. This confusion rests on the idea that both astronomy as a science and the stars themselves ultimately descend from God's Providence and

[7]"Nulla historia lucem habet sine temporum serie et regionum descriptione."

[8]"Nam iusticia, quae cum custos sit boni ordinis in vita, sumit ab his artibus proportionum gradus, hic non auditur, quae proportionem laborum, dignitatis et utilitatis artium et sumptuum in fabricandis machinis ostendit, et iuxta eam iustas mercedes flagitat."

Wisdom. Nature's teleology is discernible in celestial motions, producing day and night, as well as seasonal changes, for the benefit of terrestrial creatures and human beings (Reinhold 1549, fol. A5v). The finitude of the cosmos is another sign of divine benevolence toward humankind, an aspect of existence through which God manifests himself:

> Hence originate many more useful consequences. It is necessary to admit that the dimensions of this world are not infinite and, since God revealed himself in this sphere to very illustrious witnesses, [it is clear that] he really looks after the human race, which is located at the center of this [worldly] machine. (Reinhold 1549, fol. A6r)[9]

Thus, celestial order bears witness to God's Providence and refutes the impious doctrines of the atomists (Reinhold 1549, fol. A6v). Knowledge of the Creation fosters piety.

The theological and philosophical background behind these views can be found in Melanchthon's work. In particular, in his textbook on physics, *Initia doctrinae physicae*, written with Paul Eber (1511–1569), it is asserted that the aim of all natural enquiry is the disinterested contemplation of God through his Creation:

> The whole universe is like a theater of human ingenuity which has to be admired according to God's will. Therefore, He gave to human minds the desire to investigate nature and the desire which accompanies cognition. For that reason, the intelligent ones have a bias toward the study of nature, even though no benefit [*utilitas*] will result for them. (Melanchthon and Eber 1550, fol. 17v)[10]

The world, as Melanchthon writes in a large section on astronomy in the first book of the *Initia*, is called *cosmos*, which in Greek means 'decoration,' because it should attract human admiration:

> Just as the Latins say '*mondum*' [world] from decoration, the Greeks say κόσμον, because the beauty of the celestial and terrestrial bodies is admirable, and the human nature, for which such a great and so greatly adorned residence was created, and the distribution and the collocation of all single parts were made according to the supreme art. (Melanchthon and Eber 1550, fol. 37v)[11]

God's perfect creation, as Melanchthon argues at length, is finite in space and time, spherical and unique.

Reinhold makes many points in his discussion that mirror Fracastoro's text on the *utilitas astronomiae*. Both mention medicine, geography, and time reckoning as important applications of astronomical knowledge, although they differ in the order of presentation and emphasis. Ethics and pedagogy are central for Reinhold and less so for Fracastoro. Moreover, whereas the latter's reference to the *divinity* of the discipline is generic, Reinhold's theological concerns are more specific and better defined. He stresses that astronomy and astrology are sciences that permit a glimpse into God's Providence and Wisdom. Their teaching boasts piety and therefore has a central place in the Melanchthonian *curriculum studiorum*.

---

[9]"Deinde multae accedunt utilitates aliae. Necesse est sciri non esse infinitam molem hunc mundum, cumque in hoc orbe se illustribus testimoniis Deus patefecerit, genus humanum quod huic machinae inclusum est, vere ei cura esse."

[10]"Tota natura rerum velut theatrum est humani ingenii, quod Deus vult aspici. Ideo indidit hominum mentibus cupiditatem considerandarum rerum, et voluptatem, quae agnitionem comitatur. Haec causae invitant sana ingenia ad considerationem naturae, etiamsi utilitas nulla sequeretur."

[11]"Ut Latini Mundum ab ornatu, ita Graeci κόσμον appellant, propterea quod Miranda est pulchritude coelestium et terrestrium corporum, et humanae naturae, propter quam hoc tantum et tam ornatum domicilium conditum est, et distributio ac collocatio singularium partium summa arte facta est."

3 Christophorus Clavius "on the Utility of Astronomy" in His Edition of Sacrobosco

The introduction to Sacrobosco's *De sphaera* by the Jesuit mathematician Christophorus Clavius includes a discussion *de utilitate astronomiae* (Clavius 1581, 7–10). This is in fact the title of the fourth and final section of the preface, in which the author firstly outlines the history of the discipline, secondly, presents its partitions, and thirdly, extols its dignity (*de praestantia astronomiae*). Clavius's style is dry, but clear in its schematic form, which is typical of the Scholastic approach in vogue among the Jesuits.[12]

In the historical section *De inventoribus astronomiae*, Clavius presents a concept of scientific development which is the opposite of that embraced by Fracastoro. Whereas the latter emphasized innovation and the superiority of the Moderns in many fields, Clavius regards intellectual history as a cumulative process, in which glosses are added to sciences rooted in ancient wisdom. The original initiator of astronomy was of course Adam, the first man (Clavius 1581, 3). Clavius's account could not reach any further back in history. Astronomy, like all other disciplines, flourished among antediluvian people. After the deluge, Abraham transmitted astronomical knowledge to the Egyptians, whose knowledge passed on to the Chaldeans and the Greeks. This is a simplified overview of a commonplace history that has its beginnings among ancient civilizations, explaining the transmission of wisdom and knowledge from one culture to another.

As far as practical knowledge is concerned, Clavius defines practical astronomy *strictu sensu* as astrology. In the second section of his introduction to Sacrobosco, *De partibus astronomiae*, Clavius makes a distinction between theoretical and practical astronomy. The theoretical exposition is a treatment of the *machina mundi per se* (Clavius 1581, 3), and is concerned with the "number, magnitude, and motions of all celestial bodies." Practical astronomy, on the other hand, is concerned with astrology (*astronomia iudiciaria seu prognostica* or *astronomia divinatrix*) examining the heavens and their effects upon the sublunary sphere. Clavius defines astrology as the "application" of astronomical knowledge "to the purposes of human life" (*Practica vero astronomia … omnia ista ad usum vitae humanae accomodat*). Even though Clavius does not entirely reject astrology, he is cautious. He underscores the point that many superstitious beliefs corrupted practical astronomy, which therefore brought it into opposition with the Catholic Church. For this reason, he refrains from expanding on the subject:

> Many dared to greatly augment this part of astronomy with temerity and amplified this predictive part to the extent that it became completely superstitious and detestable. The Church rightly holds it in suspicion and Saint Augustine condemned it in admirable manner in his books on the Christian doctrine. For this reason, I believe that I do not have to discuss it at all.[13] (Clavius 1581, 5)

The section on the dignity of astronomy is entitled *De praestantia astronomiae* (On the Excellency of Astronomy). In this part of his work, Clavius resorts to the same commonplaces used by Fracastoro,

[12]On Clavius's astronomy, see (Lattis 1994) and (Pantin 1996). On the pedagogical impact of his work, see (Romano 1999). The "Prolegomena" to his renowned commentary of Euclid's elements followed the same standard pattern. It began with a systematic section, ("Disciplinarum mathematicarum divisio"), followed by a brief historical overview ("Inventores mathematicarum disciplinarum"), a discussion of the dignity of mathematical sciences ("Nobilitas atque praestantia sicentiarum mathematicarum"), and of their usefulness ("Utilitates variae mathematicarum disciplinarum") (Clavius 1591, 4r–63).

[13]"Verum quoniam huic astronomiae parti multi multa temerarie, ac perperam ausi sunt adiicere adeoque hanc partem prognosticam amplificare voluerunt, ut sit tam res omnino superstitiosa, exosaque, et merito ab Ecclesia suspecta habeatur, mirumque in modum a B. Augustino damnata in libris de Doctrina Christiana, propterea nihil omnino de ea nobis dicendum existimo."

but gives more detail. He writes how, according to Aristotle, the nobility of a science lay either in its subject (*ex praestantia subjecti, de quo agit*) or in its method of demonstration (*ex certitudine demonstrationum*) (Clavius 1581, 6). Astronomy is noble in both respects. One of Clavius's main arguments for the nobility of the celestial bodies is derived from a strict adherence to the Aristotelian conception of the superlunary sphere as pure and incorruptible. This is opposed to the sublunary world, which is the place of change, alteration, corruption, and death. In regard to the *modus demonstrandi*, Clavius reiterates Ptolemy's statement at the beginning of the *Almagest*, where mathematics is considered superior to the other two speculative sciences, at least in its methodology:

> According to Ptolemy at the beginning of the *Almagest*, astronomy is simply the first among the sciences. He says in fact that natural philosophy and metaphysics, if we consider their manner of demonstrating, should be called surmises rather than sciences, due to the multiplicity and discrepancy of theories [*opinionum*].[14] (Clavius 1581, 7)

The two theoretical sciences that are in fact opposed to mathematical knowledge in the *Almagest* are physics and theology. Clavius's translation of the term τὸ θεολογικόν as *metaphysica* is obviously intended to soften Ptolemy's provocation (Ptolemy 1984, 35).[15]

In the section *De utilitate astronomiae*, Clavius firstly underlines the point that astronomy precedes all other disciplines. Like Melanchthon, he regards theology as the ultimate goal of astronomical study. Unlike Melanchthon, he does not overtly link Divine Providence and astrological causality. Rather, he limits himself to generic considerations concerning celestial order as a sign of the excellence of the Creator of the heavens: "*Coeli enarrant gloriam Dei*" (Clavius 1581, 7). The list of the disciplines and arts benefiting from astronomy includes Aristotelian metaphysics, Galenic medicine, poetry, and navigation. Astronomy was also considered useful for the Church; Clavius refers here to his work on the emendation of the calendar for Gregory XIII (1502–1585) (Clavius 1581, 8). Astronomy is, moreover, essential for geography, civil administration, agriculture, and war.

As an example of the usefulness of astronomical knowledge, Clavius recounts the Spanish conquest of Jamaica. The "barbarians" living on that island almost defeated the Christian army. The Spaniards had no more food, and were facing their ruin when they conceived of a stratagem that would potentially save them. Knowing that a lunar eclipse should soon occur, they scared the native people by threatening to obscure the Moon if they did not surrender. When the eclipse did then occur, the *barbari* threw themselves at the feet of the *conquistadores*. "*Taceo multa alia exempla similia*" (I will silently pass over many other similar examples) Clavius concludes laconically (Clavius 1581, 8).

In the last part of his preface to Sacrobosco, Clavius reminds the reader that the Ancients held astronomy in high esteem. The Greek Thales of Miletus pursued astronomy for its own sake. When the Milesians criticized Thales for his 'useless' specialist knowledge, he was able to prove its value when he demonstrated how money could be made from being able to predict future events:

> He apprehended by means of astrology that that year would be extremely fertile, therefore he bought all olive trees in the countryside of Miletus before they began flourishing and he became very rich. On that occasion he wanted to show the Milesians that a sensible and wise man could make money, if he only wanted. (Clavius 1581, 9)[16]

---

[14] "Voluit Ptolemaeus ad initium Almagesti, Astronomiam simpliciter inter reliquas scientias esse primam. Ait enim philosophiam naturalem et metaphysicam, si modum demonstrandi illarum spectemus, appellandas potius esse conjecturas, quam scientias, propter multitudinem, et discrepantiam opinionum."

[15] "Aristotle divides theoretical philosophy, too, very fittingly, into three primary categories, physics, mathematics and theology. For everything that exists is composed of matter, form and motion…. The first two divisions of theoretical philosophy should rather be called guesswork than knowledge, theology because of its completely invisible and ungraspable nature, physics because of the unstable and unclear nature of matter" (Ptolemy 1984, 35).

[16] "Edoctus [Thales] miram illius anni fertilitatem ab Astrologia, omnes in agro Milesio oleas, antequam florere coepissent, coemisse dicitur, ditissimus evasisse; Qua in re ostendere Milesiis volebat, prudentem virum, et sapientem, pecuniam, si velit, facere posse."

Clavius also resorts to the commonplace where kings and imperators give patronage to astronomy, with reference to Caesar and Alfonso X, exactly as Reinhold does (Clavius 1581, 9). Clavius observes that the contemplation of the stars also affords great pleasure. "Plato—he writes—did not fear asserting that the eyes have been conceded to us only for the sake of astronomy" (Clavius 1581, 10).[17] He adds a quotation from Ovid's *Fasti* (I 297–8): "Ah happy souls, who first took thought to know | these things and scale the heavenly mansions!" (Ovid 1989, 23).[18] We will soon encounter the same quotation in Liddel's encomium of mathematics. Evidently, even the choice of classical passages followed conventions during the Renaissance.

In Clavius's praise of astronomy's dignity and utility, we can recognise many commonplaces already seen at play in the texts by Fracastoro and Reinhold. Notably, Clavius's attitude is closer to Reinhold's, regarding pedagogical and theological matters. Among the applications of astronomy, astrology is belittled since its practice is suspected of impiety. This is an exact reversal of Melanchthon's and Reinhold's assessments, in their positioning of astrology as the science of Divine Providence. Moreover, Clavius makes explicit reference to the Aristotelian background of his ideas. His conclusions on the essential dignity of the heavens follow Aristotelian physics. It shall also be observed that Clavius revives commonplaces by using examples taken from his own day. For instance, he shows the usefulness of time reckoning for the Church by referring to his own calendar reform. Furthermore, Spanish colonialism offers him examples of how astronomical knowledge can be used to impress naive peoples by predicting rare and impressive phenomena such as the aforementioned solar eclipse.

4 Duncan Liddel's *Encomium of Mathematics*

In 1591, on the occasion of his appointment as a professor of mathematics at the Philippist university of Helmstedt, an institution that followed Melanchthon's cultural guidelines (Omodeo 2016), Duncan Liddel gave a reading of his eulogy of mathematics. The patron of the university, Heinrich Julius, Duke of Brunswick had asked Liddel to deliver a presentation on the excellence of an academic discipline (*de alicuius doctrinae praestantia aliquid publice dicerem*). Mathematics was Liddel's chosen topic (Liddel 1591, fol. 36r). In his oration, he laid out traditional arguments in praise of mathematics, avoiding any pretence, declaring himself uninterested in surprising his audience with novelties (Liddel 1591, fol. 37v). This approach was fitting, for in contrast to Fracastoro, his humanistic audience at Helmstedt may not have been particularly receptive to superficial flourishes. The context of the occasion too may have forced Liddel to pay more attention to the elegance of his presentation than he would have in a book aimed at scholars or students.

Although his oration celebrates the dignity of mathematics as he saw it, a talk about the excellence of the discipline cannot forgo pseudo-historical considerations about its noble origins nor a reference to utility:

> I believe that I will accomplish the task [I took upon myself] if I demonstrate that among the human sciences this [mathematics] (about which I decided to talk) not only excels in perfection and pleasantness but also it is the most divine [science] and it is extremely useful and even necessary to all aspects of life and to the other sciences. (Liddel 1591, fol. 37v)[19]

[17] "Plato solum Astronomiae causa oculos nobis esse concessos, asserere non est veritus."

[18] "Felices animae, quibus haec cognoscere primum, | Inque domos superas scandere, cura fuit."

[19] "Quod me facturum esse existimo si humanarum disciplinarum hanc [mathematicam] de qua dicendi partes mihi imposui non modo perfectissimam iucundissimamque sed et divinissimam maximeque ad omnes vitae partes caeterasque scientias utilem et necessariam esse demonstravero."

In Liddel's vocabulary "dignitas," "praestantia," and "excellentia" are synonyms. He calls mathematics "sublimis," as well as "perfectissima" and "pulcherrima." Among the mathematical disciplines, astronomy occupies a prominent place, also at a symbolic level. In fact, Liddel employs metaphors linked with ascension and with astronomy to stress the excellence of mathematics in general. For instance, he counts mathematics among the *disciplinae sublimiores* (Liddel 1591, fol. 36r); he says that vulgar people are not able to raise their souls high enough (or '*attollere animos*') to give the discipline due consideration (Liddel 1591, fol. 36r), and that *mathesis* is akin to the Sun. Mathematics illuminates other sciences just as the Sun gives light to the planets:

> While the darkness is dispersed by the Sun's brilliance, the other stars are obscured. Similarly, compared to the certainty of the knowledge that mathematics offers to the human intelligence, all other sciences look as if they were a fog of errors and imbued of darkness. (Liddel 1591, fol. 39*v*)[20]

The historical explanation Liddel provides in his eulogy is based on commonplaces, too. As to the ancestral origins of mathematics in general, Liddel's account is similar to ideas offered by Clavius on the origins of astronomy. In antediluvian times, as one can gather from Liddel's speech, humankind benefited from a kind of pristine wisdom. This was however only partially transmitted to posterity. Mathematical-astronomical knowledge nevertheless passed from the Hebrews (*ad Abrahamum posterosque eius*) to the Chaldeans, and from the Chaldeans to the Egyptians and the Greeks (Liddel 1591, fol. 38v). This topos, referred to as *translatio sapientiae* or *translatio astronomiae*, is repeated in various iterations in many sources. It is ultimately derived from the *Antiquitates Judaicae* by Josephus Flavius.

Like Fracastoro and Clavius, Liddel affirms that mathematics derives part of its excellence from the certainty of its method. He mentions here Pythagoras, who even considered mathematics to be the basis for all sciences (Liddel 1591, fol. 39r). Liddel also picks up on the aforementioned Ptolemaic judgement, on the superiority of mathematical knowledge over that of other speculative sciences. Liddel's attention to the *certitudo* of mathematics as a distinguished and ennobling trait reveals a debt to Plato. Mathematics, as we read, is the basis of all other knowledge, since one cannot doubt the certainty of its demonstrations (Liddel 1591, fol. 39v). Mathematical truth, according to Liddel, fills the human mind with the most certain conclusions.

Unlike Clavius, Liddel overtly refers to theology in order to distance himself from Ptolemy. He makes the point that, when speaking of the superiority of mathematics, he is not reducing the status of theology, since the latter is based not on reason but rather on faith. On the centrality of faith and the rejection of rational theology, a Protestant like Liddel clearly has a different opinion than a Jesuit:

> In fact, since the other human sciences (in this respect, I always exclude the sacrosanct theology, which relates to faith and not to reason) have a rough and less exquisite methodology [*forma docendi*]. Therefore, they rest on probable rather than on true reasons. (Liddel 1591, fol. 39v)[21]

Thinkers as foundational as Aristotle, Galen, and especially Plato in this case, very much stress the wide-reaching relevance of mathematics (Liddel 1591, fol. 40v). Liddel shares this attitude—for him, mathematics serves as a basis for many other disciplines, especially jurisprudence and medicine. Even logic derives its first evident principles and the idea of necessity from mathematics (Liddel 1591, fol. 41v). Liddel too makes reference to Archimedes, whose work bears witness to the universal utility of

[20]"Ut enim dum Solis fulgore tenebrae discutiuntur reliqua sydera obscurantur; ita ad eam quam ingenio humano Mathesis certitudinem cognitionis affert reliquae scientiae comparata caeca errorum caligine densisque tenebris tecta conspiciuntur."

[21]"Nam quemadmodum caeterae scientiae humanae (sacrosanctam Theologiam quae fidei est non rationis hic semper excipio) formam docendi rudem minusque exquisitam habent: ita probabilibus magis quam veris nituntur rationibus."

mathematics in the arts. As an example of his skilfulness as a "practical mathematician," Liddel mentions Archimedes's construction of a wonderful planetary model: "that admirable sphere, in which the art of one single man produced with one turn the different motions of the seven planets just as happens in the world thanks to divine command" (Liddel 1591, fols. 41v–42r).[22]

In order to emphasize the spiritual importance of mathematics, Liddel refers to the same passage from Ovid's *Fasti* that is highlighted elsewhere by Clavius: "*Felices animae quibus haec cognoscere*...." (Liddel 1591, fol. 42v). As to the importance of astronomy (*syderum cognitio*), Liddel reminds his audience of the aforementioned episode of Thales, the wise Greek who forecast a fruitful olive harvest, and used this knowledge not only for financial gain but to demonstrate to his fellow countrymen the utility of astronomy.

Astronomy also brought immortal glory to its patrons. In so many cases the stars were named after rulers from previous eras: "Therefore, it looks appropriate that certain celestial bodies were named after some very ancient kings (e.g. Heracles, Orion, and Perseus)" (Liddel 1591, fol. 46v).[23] Besides Caesar and Alfonso X, Liddel also mentions Alexander the Great, who gave his name to the newly founded city of Alexandria, one of the most important centers of mathematical studies in Antiquity. Liddel adds the example of recent emperors, such as Frederick I, Maximilian I, and Charles V, of King Frederick II of Denmark, who supported astronomer Tycho Brahe (1546–1601), and Landgrave Wilhelm IV of Hessen-Kassel. Last but not least, he extols the merits of Duke Julius of Brunswick (1528–1589), the founder of the University of Helmstedt and the father of Liddel's own patron, Heinrich Julius (1564–1613) (Liddel 1591, fol. 47v–48r). As a conclusion to his encomium of mathematics, he praises Germany as the country where a series of modern mathematical astronomers flourished—Regiomontanus, Copernicus, Peuerbach, Johannes Stöffler (1452–1531), and Johannes Werner (1468–1522)—and expresses the wish that Helmstedt students will be inspired by their example. This celebration of German scholarship could have been influenced by the *Scholae mathematicae* from Pierre de la Ramée—known also as Ramus—where the French Calvinist, also for confessional reasons, gave particular praise to the land of Melanchthon and his school. Although Liddel did not agree with Ramus's intention to reform logic and philosophy, he owned a copy of the *Scholae mathematicae*, the historical section of which has been densely annotated, confirming Liddel's interest nevertheless.

In summary, Liddel's encomium of mathematics largely relies on commonplaces on the excellence and the utility of astronomy, as well as the discipline's history. These commonplaces interchangeably refer to mathematics in general or to astronomy in particular. This is proven by the close similarity between Liddel's oration and the praise of astronomy by Fracastoro, Reinhold, and Clavius. Yet differences should be noted as well. The most evident of those is circumstantial. Liddel's text is conceived for an oration before an academic audience familiar with humanist thought at a Melanchthonian university. The occasion and location of his talk did not require him to surprise his listeners with novelties. Hence he did not strive for originality but rather for eloquence.

In spite of the relevance that astrology had among Melanchthon's followers, Liddel did not particularly expand on the astrological application of astronomy. He limited himself to medicine, or more specifically, indirectly, to Galenic medicine. In his account, the construction of astronomical instruments—actually of a *sphaera*—is an example of mathematical applications from Antiquity. Pedagogy is also relevant in Liddel's eulogy. To traditional pseudo-historical accounts about the *translatio sapientiae* and glory deriving from the patronage of astronomy, he added the mention of his own patrons.

[22]"... sphaera illa admiranda in qua unius hominis artificio non secus atque in mundo imperio divino septem errantium diversos motus una conversio regebat."

[23]"Hanc ob causam antiquissimorum quorundam regum nomina certis syderibus (ut Hercules, Orion, Perseus) indita fuisse consentaneum videtur."

Moreover, he used the occasion of his speech to indicate that the school of philosophy he most appreciated was the Platonic-Pythagorean. Platonism could be understood as a reaction to Scholasticism, which was particularly in vogue among Catholics, especially in theology. Moreover, Liddel made his confessional beliefs clear. In fact, he distinguished the certainty in mathematics from a faith in a revealed truth upon which theology relies. Whereas mathematics offers rational and demonstrative certainty, in the case of theology—of *reformed* theology—only faith holds value in the final analysis. Such distinction could also serve as a useful separation between the competences of theology, on the one hand, and those of mathematical and natural enquiry on the other.

5 Petrus Ramus's Practical Conception of Mathematics

Petrus Ramus, or Pierre de la Ramée was a French Calvinist philosopher and the leading professor at the *Collège Royal* in Paris from the 1550s until his death.[24] He proposed a daring reform of the liberal arts aiming at logical simplification. As far as astronomy was concerned, he prompted astronomers to develop a science *sine hypothesibus* (without hypotheses), that is to say, a practical science dealing with the computation of celestial motions without the aid of natural philosophy. He regarded such an approach as a restoration of Chaldean and Babylonian astronomy. On the other hand, he held that the geometrical devices of Ptolemy as well as the material spheres of Aristotle were fictions that corrupted the original simplicity of the discipline. By contrast, he advocated a practical, open, and non-esoteric science aiming solely at precise calculations. These views emerge clearly from the first three books of his *Scholae mathematicae*, which he conceived as an exhortation to the study of mathematics (*exhortatio ad mathematicas artes*) (Ramus 1569, fol. a1v) and in which he deals with the history of mathematics, its utility, and a reform of its methodology. This broad introduction is followed by several chapters on arithmetic and on geometry.

Ramus's introduction to mathematics does not explicitly praise its dignity, properly speaking. However, he claims that the dignity of the discipline is evinced by its history, as one reads at the beginning of book two: "Our first book of the mathematical schools dealt with the first discoverers and authors, whence the dignity and excellency of the art can be comprehended" (Ramus 1569, 41).[25] Hence, Ramus reduces the theme of the *dignitas* to history, or rather, to pseudo-history. His account of the origins and development of mathematics is in fact rather similar to those we have so far considered.

Ramus picks up the theme of the *translatio sapientiae*, the transfer of wisdom from the Hebrews to the Egyptians and from there to later civilizations (Ramus 1569, 3–4). The story of Thales is mentioned, too (Ramus 1569, 5–6), as well as Caesar's support for Sosigenes for the accomplishing a reform of the calendar (Ramus 1569, 36). Two distinguishing elements emerge. First, Ramus, as *lector royal* of the *Academia Parisiensis*, treats the *translatio sapientiae* to France in certain detail (Ramus 1569, 13–4). According to him, pristine Gallican wisdom was imported from the British Isles twice. The first transfer occurred before the Roman invasion deprived the Gauls of their freedom and destroyed their accumulated knowledge (Ramus 1569, 14). The parallel between Caesar's occupation of France and Roman Catholic 'oppression' in present times is clear, given Ramus's Calvinist beliefs. A second transfer of knowledge from Great Britain to France occurred when Charles the Great called Alcuin to Gallia. This pseudo-historical account is directly followed by a eulogy of English monarchs,

[24] On the cultural relevance of Petrus Ramus, see (Feingold 2001; Hotson 2007). On his historical outlook see (Goulding 2010).

[25] "Primus scholarum mathematicarum liber nobis adhuc fuit de mathematicae primis inventoribus et authoribus, unde artis dignitas praestantiaque intelligeretur."

and of Elizabeth in particular. Ramus covertly invites the French Crown to follow the Anglican and the German examples.

A second mark of originality in Ramus's history of mathematics is the stress put on the practical dimension of the discipline. This is mirrored by the exaltation of Architas, seen as an ancient mechanical inventor, of Daedalus, and of Archimedes.[26] The practical dimension of mathematics excludes astrology, though, which Ramus, following Johannes Calvin (1509–1564), regarded as mere superstition: "According to Cicero, Eudoxus criticised the horoscopes of the Chaldeans. I believe that here one can already appreciate Eudoxus's intelligence as one who railed against those vanities" (Ramus 1569, 18).[27] Ramus's proto-illuminist commitment emerges not only from his criticism of horoscopes, but also from his criticism of numerologies and their application to theology. If it were true, as Proclus claims, that the utility of mathematics lies in disclosing certain theological arcana, then mathematics really would serve no purpose (Ramus 1569, 43–4).

> In fact the subject of Plato's *Timaeus* is theology. Aristotle in *De coelo* III 8 dismissed and destroyed that theologically aimed geometry with energy. Proclus once philosophized even more earnestly, as he argued that the mathematical arts were discovered to the benefit of the people. Arithmetic [was invented] for the sake of trade and geometry for the sake of land surveying. (Ramus 1569, 44)[28]

The roots of knowledge, according to Ramus, are always practical. However, systematization is as important as practically acquired and transmitted knowledge. Aristotle's secret is revealed: Logic enabled him to integrate and organize the knowledge derived from many different sources.

> Aristotle was never a fisherman, a bird catcher, hunter, ploughman, lawyer, nor governor. How could he then describe the art of such people? Those ignorant of the power of solid logic in explaining and dealing with all of the arts look stupefied and regard Aristotle, who wrote on so many arts, as divine and as some kind of natural miracle. Actually, these natural miracles are miracles of logic. All this force was logical skilfulness: [the ability] to distinguish the knowledge accumulated by practitioners or in the practitioners' commentaries into genres or species of genres; to treat and explain general issues in a general manner and particular issues in a particular manner according to their place and order. Certainly Aristotle mastered that art to make an entrance to all arts, by far the greatest and the most excellent. (Ramus 1569, 42)[29]

The second book of the *Scholae mathematicae* deals with the issue of *mathematum utilitates*.[30] The "utility" is discussed along two lines: from the viewpoint of "contemplation" (*ad contemplandum*) and that of "action" (*ad agendum*) (Ramus 1569, 44). The former aspect concerns the epistemological and methodological significance of mathematics as an *organon* (*instrumentum quoddam*) for other disciplines, similar to logic (Ramus 1569, 45–6).

---

[26] On Architas, Daedalus, and Archimedes see (Ramus 1569, 15–6, 26).

[27] "Disseruit Eudoxus (ait idem Tullius) contra Chaldaeorum praedictiones de natali die. In quo ingenium Eudoxi imprimis suspiciendum arbitror, qui in istas vanitates invectus sit."

[28] "Haec enim Platonis in Timaeo theologia est. Verum sacer ille Geometriae finis ab Aristotele 8. cap. 3 lib. De Caelo vehementer exagitatus et labefactatus est, longeque gravius Proclus ipse est antea philosophatus, cum mathematicas artes popularis utilitatis caussa repertas esse docuit. Arithmetica tractandae mercaturae, Geometriam dimetiendae terrae gratia...."

[29] "Aristoteles piscator, auceps, venator, apiaries, arator, in foro orator, rerumpub. gubernator nunquam fuit. Quomodo igitur hominum istorum artes describere potuit? Hic homines ignari quantum solida logica in explicatione et tractatione atrium omnium possit, attoniti obstupescunt, et Aristotelem tot atrium scriptorem divinitus, ac nescio quo naturae miraculo facto arbitrantur. Verum ista naturae miracula, logica miracula fuerunt: totaque vis ista, facultatis logicae fuit: materiae sibi ab opificibus vel opificum commentariis accumulatae, genera generumque species secernere: generalia generaliter, specialia specialiter suis locis atque ordinibus distinguere atque explicare. Artem igitur hanc Aristoteles habuit, faciendae artis atrium omnium longe maximam praestantissimamque."

[30] The expression mathematum utilitates occurs in (Ramus 1569, 26).

> I am not dealing with mathematics as a kind of ringing bell of a clock waking us up in the morning. [In that case] it would be of no use. By contrast, I affirm that it acts upon the eyes of the mind and converts them like the clearest light and, once they are open, it illustrates all aspects of life and intellect. (Ramus 1569, 45)[31]

The list of disciplines which rely on mathematics includes music, optics, physics, astronomy, medicine, politics, and ethics (Ramus 1569, 46–51). Even theology needs mathematics, since "God created everything with number, measure, and weight, i.e., according to the rules of arithmetic, geometry, and the science of weights" (Ramus 1569, 51).[32] As for physics, the main source Ramus quotes in support of his claim is Aristotle as the author of the *Mechanicae* (Ramus 1569, 47). In that work, equilibrium is derived from the properties of the circle. More generally, it shows that geometry is fundamental in accounting for natural and unnatural motion and rest. Since physics essentially deals with the motion and rest of natural beings, the relevance of geometry for that discipline is patently clear.

As to astronomy, Ramus regards it as a discipline that needs a thorough reform. In his opinion, the use of "hypotheses" corrupted its original purity (Ramus 1569, 49). The main problem in the use of hypotheses is the claim to account for phenomena through causes that cannot be empirically tested. This defect is common to the homocentric school of Callippus, Eudoxus, and Aristotle, as well as to the Pythagorean school of those using epicycles and eccentrics. Copernicus is also included in the list of vain astronomers looking for causes (Ramus 1569, 50). By contrast, Ramus supports a computational astronomy like that used by the Babylonians, the Egyptians, and the Greeks before Eudoxus. According to Ramus, this computational approach is also common among contemporary ephemerists. An epistemological reform of astronomy should take this tendency into account and base astronomy only on arithmetic and trigonometry. These disciplines, in turn, should be reduced to their logical foundations. In this manner, all discussion about celestial causes would be banished from astronomy.

In the second part of his discourse on the utility of mathematics, Ramus tackles the aspect of the subject's application, dealing first with arithmetic and then with geometry. The first discipline is fundamental for merchants, as well as for goldsmiths, public administrators, and military strategists (Ramus 1569, 54–6). As far as geometry is concerned, it serves several arts including geodesy, mechanics, and metallurgy (Ramus 1569, 57, 58, 62). Ramus gives particular mention to three "German inventions," *tres singulars artes mathematicae Germanicae*: ballistics, typography, and navigation (Ramus 1569, 65). Note that these are the Baconian technological innovations accounting for the advance of science and humanity. The same trilogy is to be found in Cardano's autobiography, *De vita propria liber*, as I will soon stress. The latter art is connected directly with the colonial expeditions to the antipodes by Christophorus Columbus (1451–1506) and Amerigo Vespucci (1454–1512). As to the claim these technologies originated in Germany, this is in line with the praise of German mathematicians that immediately follows, and has an evident confessional meaning, since it is conceived as an encomium of Melanchthon's Wittenberg and the scientific developments he encouraged (Ramus 1569, 66).

In summary, Ramus's *exhortatio ad mathematicas artes* resorts to the rhetoric genre of the *encomium mathematicae* and reiterates many traditional commonplaces. However, he introduces fundamental changes reflecting his own philosophical and confessional convictions. First of all, Ramus renounced the theme of *dignitas*, reducing it to ideas of the nobility of origins. Since the origins and the aims of science are both practical, *dignitas* is ultimately reduced to *utilitas*. Ramus discusses

[31] "Neque vero iam disputo de mathematis tanquam horologiis sonitu et tintinabulo mane nos e somno excitari, nihil praeterea juvari: sed affirmo tanquam clarissimo lumine oculos mentis agi convertique, atque apertos, ad omia vivendum et intelligendum illustrari."

[32] "Deus creavit omnia in numero, mensura, pondere, id est, ad regulam arithmeticam, geometricam, isorropicam."

utility from two angles: one epistemological, the other applicative. Mathematics and the logic underlying it are indispensable tools in all fields of knowledge. Moreover, arithmetic and geometry are relevant in countless arts, including civil administration. Ramus's assessment of astronomy is embedded in his practice-oriented philosophy of science. Since he dismisses astrology as superstition, time-reckoning and orientation in space remain the most important fields of application for astronomical knowledge.

6 Girolamo Cardano's Practically Oriented Science

Remarks pointing to the peculiarity of practice and the superiority of practical experience over bookish and abstract knowledge are scattered throughout Girolamo Cardano's work. He learned to appreciate practical knowledge very early on, through his father's acquaintances. As one reads in his autobiography:

> The only very intimate friend of my father's, in spite of a decidedly different calling, was Galeazzo de' Rossi, this last being his family name. He ... was a smith who became acquainted with my father through a certain community of habits and similar intellectual pursuits. It was he who had discovered the screw [or *cochlea*] of Archimedes before the works of Archimedes had been published. He invented a method of so tempering swords that they could be bent like lead, and yet would cut iron almost as if it were wood. What is more, he wrought steel breastplates which resisted the shots of the firearms which the regimental soldiers carried; one plate alone sufficed to withstand a blow of five times ordinary force, and suffered scarcely a dent. I myself was frequently an onlooker at this experiment as a young man. (Cardano 1962, 7–8)

Cardano often stressed the limits of theory disconnected from practice. Abstract theory, as one reads in the following medical aphorism, for instance, lacks exact quantification:

> In matters of practice [*in negotiis*], in contrast to matters of theory [*in artibus generis*] a vague acquaintance with the subject is not sufficient as it is necessary to know the quantities. One can do good by offering rhubarb to a patient sick with tertian fever even if one does not know the exact quantity, but it is better to keep silent or not to visit a patient if one has not become acquainted with the dose to be employed. (Cardano 1962, 267 transl. revised)[33]

But what was Cardano's attitude toward the usefulness of mathematics, and of astronomy in particular? I will start my treatment with his *Encomium astrologiae* and then move on to his other eulogies of sciences and arts.

6.1 Cardano's *Encomium astrologiae*

Anthony Grafton has qualified Cardano's praise of astronomy as a "flowery text" (Grafton 1999, 128). It was an erudite text, juxtaposed to Cardano's *Libelli duo. Unus, de supplemento Almanach. Alter, de restitutione temporum* (*Two Books: One, a Supplement to the Almanach and the Other, the Restauration of Times*, Nuremberg, 1543). Cardano's *Encomium astrologiae* was rather a rhetorical appendix to this collection of seminal astrological works (Cardano 1543, fols. dd3*v*–dd4*r*)

Cardano begins his celebration of *astrologia* with reference to religion. He claims that the 'science of the stars' is *praestantissima* (very excellent),[34] since it reveals the divine. Moses established

[33] See (Cardano 1966, vol. 1, 49a): "In negotiis non ut in artibus generis cognitio quidquam iuvat, sed quantitatem scire oportet: Possumus enim prodesse rhabarbarum exhibendo tertiana laboranti, etiamsi nesciamus quantitatem; at melius est tacere, aut non invisere clientem, si modum in actione adhibere nesciveris."

[34] I purposely use Robert Westman's expression "Science of the Stars" to indicate the Renaissance discipline bringing together mathematical astronomy and astrological prediction (Westman 2011, pt. 1, ch. 1).

religious festivities on the basis of the lunar cycle and the equinoxes. Moreover, he devised the week, the *septenarium*, by analogy with the number and order of the wandering stars. The great Jewish legislator supported the observation of the heavens, as their regularities bear witness to the way all nature respects God's laws.

After the Hebrews came the Greeks, the Egyptians, and the Phoenicians, who were "astrophiles," contemplating the heavens with more curiosity, "but not more piously" (Cardano 1966, vol. 5, 727–8).[35] Cardano surmises that kings who supported astronomy were deified after their death. This was the case with such mythical figures as Hermes Trismegistus among the Egyptians and Berosus among the Chaldeans. The same considerations relate to the pagan identification of the wandering stars with gods:

> Those who discovered and augmented [the science of the stars] were regarded as kings when they lived and, once they died, they were revered as gods so that, with all effort, their work and dignity was rewarded with due honors.... And they [the Ancients] venerated the stock of the heroes for their virtue and the nature of the planets that was akin to them. It is likely that Apollo, Jupiter, Mars, Mercury, Venus and the Moon were kings who discovered the path of those stars and their influences [*vim*]. (Cardano 1966, vol. 5 1663, 728)[36]

Here Cardano clearly picks up on the theme of the glory attained by the rulers who supported science.

He applies his demythologizing reading of mythology, moreover, to a wide range of figures and stories taken from what may be called the pagan classics. He interprets the musician Orpheus, whose lyre became a constellation, as the discoverer of the harmony of the universe as well as of the parallel between the seven wandering stars and the seven chords of his instrument. The myth of Apollo's son, Phaeton, is interpreted as the transposition of the story of an ancient investigator of the solar cycle. The figure of Endymion, Selene's beloved, refers to an ancient investigator of the Moon; whereas the giant Atlas, who carried the heavens on his shoulders, was in reality an observer of the fixed stars. The myth of Daedalus and Icarus should be traced back to the fortunes and misfortunes of two astronomers as well. As to Tiresias, his change of gender indicates the determination of the genre of the celestial bodies by a scholar of his name. The Argonauts seeking the Golden Fleece were actually astronomers looking for the exact moment of the equinox, when the Sun entered the zodiacal sign of Aries. Even the Homeric poem could be seen as hiding astronomical truths. According to Cardano's exegesis, the divine protection accorded to the heroes of the *Iliad* should be understood in the context of astrological influences.

From mythology Cardano moves on to ancient history. Astronomy, one reads, has always featured in legislative procedures. The Greeks and the Romans, from Lycurgus to Augustus, prescribed festivities and prohibited certain activities on particular days for astronomical reasons. This referral to astronomy even in drafting legislation illustrates the extent to which the discipline featured in politics and political decision-making in Renaissance culture (Azzolini 2013).

In Cardano's classicist encomium of astronomy, the excellence of the discipline is connected with divine revelation and a history based on mythology. The conclusion points to astronomy's usefulness, which is indeed astrological. As one reads, astronomy anticipates the future, and since prescience is a gift of God, this discipline is equated with a form of grace. Historians, one reads, are responsible for

[35](Cardano 1966, vol. 5, 728): "Curiosius vero Graeci, atque Aegyptij, tum Phoenices, non tamen maiore veneratione." For Cardano's *Encomium astrologiae* see (Cardano 1966, vol. 5, 727–8).

[36]"Nempe eos, qui illam [astrorum disciplinam] invenerant auxerantve, viventes quidem pro regibus habuerant, mortuos coluerunt pro Diis: ut sic omni studio et illorum operam et dignitatem meritis prosequerentur honoribus.... At Heronum genium ex virtute, et Errorum natura, quae illi similis erat venerabantur: nam Apollinem et Iovem Saturnumque, Martem ac Mercurium, tum Venerem et Lunam, reges fuisse credendum est, qui cursus harum stellarum vimque earum invenerint."

the past, while kings are responsible for the present. The future is exclusively for gods. Hence, astrology was the ambrosia of the pagan divinities.

> Astrology is by necessity the most pleasant and divine of all disciplines. Would anybody doubt that it is also the most useful? It teaches [us] how to go along with future events if they are propitious and to avoid them or to bear them with equanimity, if they are averse. In fact, he who foresees [bad events] long in advance will bear them more easily even if he cannot diminish them. (Cardano 1966, 728)[37]

6.2 The Origins, Nobility and Usefulness of Geometry, Music and Arithmetic

Besides his encomium of astrology, Cardano wrote and read to audiences a wide range of eulogies on the arts and mathematical sciences. In 1535, he delivered a speech in praise of geometry to the Academia Platina of Milan. In this *Encomium geometriae* he presented geometry as the highest science or as a sort of *prisca scientia*.[38] He contended that geometrical thought, based on quantity and proportion, is the basis for all arts and disciplines. According to Cardano, the approach *more geometrico* is the essence of rationality. Additionally, he regarded geometry as the ontological a priori of God's Creation.

This discipline, one learns, had very practical origins, as its name suggests. Geometry was invented in Egypt for the needs of land measurement after the flooding of the River Nile (Cardano 1966, vol. 4, 442b).[39] Geometry, according to Cardano, is the most ancient of the mathematical disciplines, since the practical realm from which it emerged historically preceded the practical realms from which astronomy and arithmetic arose, i.e., navigation and trade. Remarkably, such considerations about the practical roots of mathematics predate those offered by Ramus and directly connect Cardano to the tradition of a philosophy of science leading from Francis Bacon to the *Encyclopédie*.

It should be added that, in Cardano's eyes, geometry is a practical discipline also insofar as it includes statics, mechanics, and architecture as subordinate fields of study. In fact, the list of arts and sciences depending on geometrical rationality (committed to the study of quantities and proportions) is long. Geometry, as has been stated, is fundamental for arithmetic, music, astronomy, and optics. It is the *conditio sine qua non* of architecture (*non aedificare sine illa licet*) and of the plastic arts, painting, and sculpture. Geometry is necessary for the construction and understanding of clocks and machines (*horologiorum, machinarumque structura*). It is further presupposed by natural magic, by the science of weights, by aesthetics (*pulchritudo … tota geometrica ratione constet*), and by countless other fields of human activity and knowledge (Cardano 1966, vol. 4, 441b). Among the ancients, this science reached its acme thanks to Euclid and Archimedes; among the Moderns, Cardano praises Leonardo Pisano (1170–1240) and Luca Pacioli (1445–1517). In his eyes, Johannes Regiomontanus contributed to trigonometry more as a restorer than as an original thinker, but could be considered "*pene divinus*" (almost divine) in his work as an astronomer. As to Nicholas of Cusa's (1401–1464) theologizing mathematics, his speculations were subtle but his conclusions in error (Cardano 1966, vol. 4, 443b–444a).

[37] "Quamobrem necesse est Astrologiam omnia alia disciplina et iucundiorem et diviniorem esse. At vero etiam utiliorem fore quis dubitat? Cum et futura si prospera sint sequi doceat, si adversa vel vitare vel aequius ferre: nam quae diu quisque ante praeviderit, laevius feret, etiam si minuere nequeat."

[38] For Cardano's *Encomium geometriae recitatum anno 1535 in Academia Platina Mediolana* see (Cardano 1966, vol. 4, 440–5).

[39] "Ab his igitur dum Nilus inundatione cuncta conturbaret, decrescentibus aquis inventa Geometria, ut agri suis dominis restituerentur."

God himself is a geometrician: "*Deum ipsum maxime esse Geometram, Geometriamque intendere*" (God himself is a geometer in the highest degree and understands geometry) (Cardano 1966, vol. 4, 444a). The science of quantity and proportion is, in fact, necessary for justice and for natural order. Cardano's cosmogony anticipates Johannes Kepler's (1571–1630) *Mysterium cosmographicum* by arguing that the number of the planets and the proportions of their distances can be derived by geometrical reasoning, namely by inscribing and circumscribing geometrical solids in spherical orbs. In other words, geometry applied to astronomy can reveal the archetypes informing the structure of the cosmos.

> It is well known that God the Creator [*maximum Opificem*] preserved geometrical proportion in the structure of the world: Nothing should be considered more attentively in that construction of his, or better still only that [proportion should be considered]. Perhaps somebody would ask whether the proportion of the seven planets [*ratio erraticarum*] can be grasped once that [cosmological] proportion is unveiled. Yes, indeed! (Cardano 1966, vol. 4, 445b)[40]

However, Cardano's system of the geometrical world remains geocentric. Following Plato's *Timaeus*, Cardano also surmises that geometrical figures account for the structure and the properties of the elements. Earth, for instance, has a cubic structure that accounts for its essential immobility (Cardano 1966, vol. 4, 445a). At the level of the microcosm, human beings are able to partake of divine wisdom and creativity by means of geometry. As Cardano writes, "Everybody agrees that mortals can be outstanding, if they are able to fully discuss either geometrical ratios or geometrical machines" (Cardano 1966, vol. 4, 445b).[41]

Cardano also wrote his *Laus musicae et eius praestantia* (Eulogy of Music and Its Excellence) as the first chapter of a *Liber de musica* (Book on Music) included in his *Opuscula miscellanea*. This eulogy followed the standard analysis: antiquity, utility, excellence (Cardano 1966, vol. 10, 105).[42] Furthermore, Cardano's *Artis arithmeticae tractatus de integris* (Arithmetic Treatise on Integers) began with an introductory chapter on the nobility and utility of the discipline, in this case arithmetic (titled "De nobilitate et utilitate huius artis …"). For him, the connection of arithmetic with practice is self-evident. He claims that trade is the field of origin, growth, and the application of the science of numbers (or *numerandi scientia*). Cardano points out that those who only work *counting numbers* can most easily and quickly grow rich. He illustrates the connection between financial transactions and capital accumulation as follows:

> We have come to the point that orators, poets, jurisprudents, physicians, farmers, architects, army leaders, and actually all those who are somehow reputed for their works or for some other reason cannot accomplish anything without that [science of numbers or arithmetic]. In fact, many become wealthy only thanks to that activity [of counting numbers], among them bankers, money exchangers and traders. None grows rich more rapidly and more easily than they! Physicians and orators work their whole life to spare a quarter sesterce (and they do not often succeed), whereas those people can amass hundreds of sesterces in very few years. (Cardano 1966, vol. 10, 117)[43]

---

[40]"… constat summam Geometriae rationem maximum Opificem in mundi constitutione conservasse: nihilque magis illa in eius constructione, imo et solam illam spectari debere. Sed forsitan quis quaerat, num ex hac ratione aperta septem erraticarum ratio habeatur? Certe sic."

[41]"Omnes … putant mortales satis se ornatos esse, cum de Geometrici seu rationis seu machinis abunde disceptare noverunt."

[42]Regrettably, in the quoted edition chapter one has been shortened and only the incipit can be read (Cardano 1966, vol. 10, 105): "Musicam neque ab antiquitate nec ab inventorum gloria, quae duo ferme sint cunctis artium laudandarum proemiis communia, sed ab ipsa utilitate excellentia subtilitateque laudanda ac caeteris aliis tum artibus tum disciplinis praeferenda existimo. Non quod vel vetustate, vel inventorum dignitate quiquam sit inferior, nempe…." At this point the chapter is interrupted.

[43]"Itaque factum est ut Oratores, Poetae, Iurisconsulti, Medici, Agricolae, Architecti, Exercituum duces ac demum omnes quotquot sunt operibus suis aut aliqua ratione illustres nihil ab hac efficere possint, cum plures etiam hac sola industria locupletes evadant atque inter hos mensarii, collibytae, mercatores, quibus nemo vel promptius vel facilius ditatur. Laborant Medici Rhetoresque tota vita, ut quater sestertium (nec id frequentius) lucrifaciant, at hi quam paucis annis centies sestertium congregant."

6.3 Cardano's *Encomium medicinae*

I have so far considered Cardano's rhetorical appraisals of mathematical disciplines. But what does he say about the nobility and usefulness of medicine? Does this art differ, in his account, from mathematics? With these questions in mind, I turn now to his *Medicinae encomium* (Basle 1559).

Among the sciences, geometry, as we saw, has a certain chronological, ontological, and epistemological primacy. Among the arts, Cardano esteems medicine's ancient origins, which even reflect the divine, with medicine's invention attributed to Apollo and Aesculapius. Moreover, the divinity of this art, veiled in the Greek myths, is also testified to by Christ's healing practices:

> I will omit the tales about Apollo and Asclepius. Christ, whom everybody would without doubt see as God, did not accomplish anything more greatly and more frequently than the exercise of medicine (although in his own manner, that is, divinely) and healed sick people of all kinds. (Cardano 1966, vol. 6, 2)[44]

Cardano divides medicine's *praestantia* into three categories: *utilitas*, *pulchritudo*, and *nobilitas*. Such excellence, according to Cardano, stems from the discipline's practical utility, alongside its aesthetical and axiological elements: "It seems to me—he declared—that the usefulness [of this art] is twofold, as both humanity and its practitioners do benefit from it" (Cardano 1966, vol. 6, 1–7).[45] Indeed, medicine is an art that secures health and longevity. Hippocrates, for instance, lived to be almost 100 years old, and Avenzoar reached the age of 135 (Cardano 1966, vol. 6, 4). Neither does Cardano neglect the economic advantages of medical praxis: "Do you seek richness? It [medicine] will bring it to you so largely and so quickly that, if I told you some examples, you would think I am inventing stories" (Cardano 1966, vol. 6, 2).[46] The most astounding case of a family rising from the practice of medicine to a position of influence is the story of a Tuscan ruling dynasty, whose name betrays its origins—the "Medici," or 'physicians' (Cardano 1966, vol. 6, 3).[47]

Medical practice, as one learns, is supported by an intricate nexus of knowledge drawn from different disciplines. Knowledge of astrology, mentioned directly after theology, is considered particularly valuable:

> The physician inquires whether illness and recovery are given by God, as Hippocrates did when he directed his mind upwards and became a theologian. [The physician also inquires] what is the soul, whether it is mortal or immortal, what is the power of the celestial bodies, whether the illness or the recovery depend on them [i.e., on celestial influences] in the same manner and, if both depend on them [i.e., from celestial influences], [he inquires] what makes one of the two [illness or recovery] more likely. Therefore, [the physician] deals with a great deal of astrology by considering the motions of the stars, their risings and settings, as well as by reconstructing the time of the genitures [of his patients]. (Cardano 1966, vol. 6, 6)[48]

---

[44] "Omitto fabulosa de Apolline … et Aesculapio. Christus, quem nemo addubitat Deum fuisse, nihil maius egit, nihil frequentius quam quod medicinam, sed suo more (divino scilicet) exerceret, aegros omnis generis sanando."

[45] "Utilitas duplex mihi genus esse videtur: tum ut humano generi, tum ipsi qui eam exercent, maximum afferat emolumentum."

[46] "Divitias quaeris? Haec dat et tantas et tam repente, ut si exempla referre velim, fabulas narrare videar."

[47] "Quin etiam medicorum familiam, quae nunc in tota Haetruria regnant, a medico quopiam ortum habuisse verisimillimum est. Quamobrem hac arte etiam ad regna paratur aditus: quo fit, ut etiam nobilissima sit: neque reges ob id eam exercere puduit."

[48] "Quaerit medicus, an morbi, an sanationes a Deo immittantur, quemadmodum facit Hippocrat[es] sursumque mentem dirigens Theologus evadit. Idem quid sit anima, an mortalis an immortalis, idem quid astra possint, et eodem modo an ab illis morbi sanationesque pendeant; ab iisdem enim utrumque fieri, a quo alterum eorum fiat verisimile est. Ita motus siderum, et exortus considerans, atque occasus, tempora quoque geniturarum reditus non mediocrem Astrologiae partem tractat."

Medicine is the field of practical knowledge where important facets of *practical astronomy* and even *practical theology* emerge and combine. Among other disciplines deemed to be important for medicine, Cardano mentions meteorology as knowledge about climatic changes; architecture, since the construction of an accomplished building should take into account also the health of the inhabitants; and the art of food preparation. Moreover, he recommends that an expert physician should have extensive knowledge of plants, fruits, stones, metals, fish, birds, reptiles, different species of quadrupeds, about their generation and the regions where they live. The vast knowledge of a cultured physician would make a philosopher appear quite ignorant in comparison: "Hence, when a physician and a philosopher discuss on these matters, the philosopher really looks like a child compared to the former" (Cardano 1966, vol. 6, 7).[49]

6.4 Cardano on the Art of Discovery: An Epistemology for Practical Knowledge

Cardano's practice-oriented interpretation of the sciences and the arts ultimately rests on epistemological premises that deserve our consideration. According to him, mathematical reasoning would be void without sensible experience and the experience gained through exercise and practice. Cardano goes so far as to stress that the validity of arithmetic is not only rational but also sensible: "This art [of arithmetic] is indeed the most certain of all according not only to reason [*ratione*] but also to the senses. This is why it is the clearest among the mathematics" (Cardano 1966, vol. 10, 117a).[50] In an essay on technical inventiveness, *De inventione*, Cardano stresses the superiority of practice over theory, placing *inventio* before *sapientia* (wisdom):

> Invention itself is higher than wisdom. In fact, human wisdom is rare owing to life's brevity and to the many impediments occurring to men even in their happiest times. By contrast, invention satisfies infinite cases at once. For example, he who demonstrates the reason why the angles of a triangle are equal to two rectangles, [demonstrated this] for infinite triangles, no matter how different they might be. The same applies to [demonstrations linked to] universals [*species*].... Although the power of invention is infinite, it is only finitely actualized. Therefore this is the only thing almost capable of connecting finitude with infinity. It makes us resemble gods. (Cardano 1966, vol. 10, 90)[51]

Invention is the capacity to devise the means to achieve an aim (*inventio medii*). In astrology, the aim is action; in geometry it is action or production (*actio opusve*); in music it is rather contemplation (*contemplatio*). The different means and methods may vary. In music and medicine, there are three elements essential to achieving one's goal—demonstration, exercise, and using the correct instruments. Regarding music, Cardano explains that an accomplished musician must possess a certain understanding of harmony (*intelligit unde Concordia*). In a second step, musicians must be well exercised in their discipline. Thirdly, they must know how to play their instrument. These are all condi-

[49] "Ut hac parte medicus philosopho conferatur, videar sane philosophus parvulus, viro comparato."

[50] "Est vero hac ars [arithmetica] omnium certissima non tantum ratione, sed et sensu: unde et inter ipsas mathematicas ob hanc causam praeclarissima est."

[51] "Inventio ipsa sapientia praestantior est: quoniam sapientia humana res est perexigua, tum ob vitae brevitatem, tum ob tot impedimenta, quae hominibus occurrunt etiam in temporibus felicissimis: at inventio infinitis simul satisfacit velut qui de triangulo demonstravit, quoniam tres angulos habet duobus rectis aequales, de infinitis et licet numero tantum differentibus: etiam iuxta species idem continget.... Itaque inventio potestate quidem infinita actu vero finita: Igitur haec pene sola infinitum cum finito coniungit. Nosque cum Diis continuat." As for Cardano's *De inventione*, see (Cardano 1966, vol. 10, 90–9).

tions *sine qua non* for musical perfection. Similarly, an experienced physician is able to distinguish health from disease, the lack of health and how to restore it. Cardano also considers more complex fields of human activity, where more intricate methods are required to achieve a goal. Such complex areas especially concern activities that presuppose human interactions, for instance matters of trade *circa negocia.*

The wide range of applications of human ingenuity, according to Cardano, is apparent in the variety of inventions listed in his autobiography, *De vita propria*, chapter 44, "Quae in diversis disciplinis digna adinveni" (Notable Discoveries of Mine in Various Disciplines). The *inventiones* he ascribed to himself range from the reorganization of dialectics and its application to mathematics, his innovations in arithmetic, especially in algebra, the elimination of fire from the elements, the attribution to God of the qualification of *immensus*, the superiority of the East over the West, the manner of transferring the observation of nature to the arts and the production of works (*docui deducere contemplationem rerum naturalium ad artem et opus*), the moral doctrine that all living beings share the same condition, the study of many diseases and their treatment, as well as *experimenta* (experiments) on rare diseases such as epilepsy, insanity, and blindness. Cardano was proud of the title he received from the illustrious man of letters Andrea Alciato (1492–1550) who used to call him "*virum inventionum,*" that is, 'the man of inventions' (Cardano 1966, vol. 1, 39b–40a).

In the same tract, Cardano offers a brief description of his sensible epistemology as a process of abstraction from the empirical to the universal:

> [Knowledge] is … an understanding threefold in nature. First, there is knowledge gained by my senses through the observing of innumerable things.... Secondly, there is an understanding of higher things obtained through the examination of their beginnings and pursued by conforming to certain principles. This aspect of knowledge is called proof because it is derived from wider application of the subject under consideration, or to place it in a clearer light or to give a general application from the particular....The third stage of my knowledge is that of things intangible and immaterial, and by this I have come wholly as a result of the ministrations of my … spirit.... The use of amplification and lucidity of understanding I have acquired partly from practice and partly at the inspiration of my spirit, for I devoted myself persistently to perfecting that intellectual flash of insight for more than forty years before I mastered it. (Cardano 1962, 245–7)

Hence, Cardano's practical approach to the sciences was connected with a theory of knowledge in which the senses, observation, and experience are the beginnings of an inductive and generalizing process.

Let me conclude this overview of Cardano with a famous passage from chapter 41 of his autobiography on "Ostenta naturalia visa, rara tamen, de vita propria" (Natural Prodigies Observed, Rare Though, in My Life), extolling the superiority of the Moderns over the Ancients on account of their geographical exploration and their mastery of the three Baconian technologies, namely the compass, gunpowder, and typesetting:

> Among the extraordinary, though quite natural circumstances of my life, the first and most unusual is that I was born in the century in which the whole world became known; whereas the ancients were familiar with but little more than a third part of it.
>
> … We explore America … Brazil, a great part of which was before unknown, Terra del Fuego, Patagonia, Peru … Toward the East under the Antarctic we find the Antiscians … and some Northern people not yet known, as well as Japan … all discoveries sure to give rise to great and calamitous events in order that a just distribution of them may be maintained.
>
> The conviction grows that, as a result of these discoveries, the fine arts will be neglected and but lightly esteemed, and certainties will be exchanged for uncertainties. These things may be true sometime or other, but meanwhile we shall rejoice as in a flower-filled meadow. For what is more amazing than pyrotechnics? Or than the fiery bolts man has invented so much more destructive than the lightning of the gods?

> Nor of thee, O Great Compass, will I be silent, for thou dost guide us over boundless seas, through gloomy nights, through the wild storms seafarers dread, and through the pathless wilderness.
>
> The fourth marvel is the invention of the typographic art, a work of man's hands, and the discovery of his wit—a rival, forsooth, of the wonders wrought by divine intelligence. (Cardano 1962, 189–90)[52]

In this passage, Cardano brings together epistemological and axiological considerations on practical knowledge. Practice is not only his key to uncover the origins of arts and sciences. It is also the measure of progress and the criterion to establish the preeminence of the Moderns over the Ancients.

7 Concluding Remarks

Historians of early modern science have often neglected the theme *de utilitate astronomiae* in favour of its rhetorical counterpart, namely the *dignitas astronomiae*. The celebration of the latter served as a way to theologically and philosophically legitimate the science of the stars. Therefore, the motif of *dignitas* has attracted the attention of scholars dealing with intellectual history and with the cultural framework of scientific ideas. Almost paradoxically, passages exalting the former aspect, the practical advantages of astronomy or its utilisation, look less attractive to historians interested in the material roots of science. Views from the Renaissance on astronomy's usefulness were often conventional and repetitive. Apparently, they cannot be taken at face value as genuine reflections on the applications of astronomical knowledge during that period. However, one might ask why these doubts should affect the topos of the *utilitas* alone and not concern the *dignitas astronomiae* as well? As I have argued, the praise of astronomy's dignity also relied on commonplaces in the wider discourse, and did not necessarily present original philosophical views or clear theological convictions. Nevertheless, historians of ideas have not dismissed rhetoric as being devoid of any valuable content. In fact, interpretative difficulties shall not discourage us from reading these more or less conventional sources on the dignity and utility of astronomy as indications of early modern scientific culture and its stylistic conventions. Rhetoric was not self-referential, as a rhetorical discourse must also have some plausibility in the minds of those pronouncing it or listening to it. Hence, commonplaces had to be supported by some direct or indirect knowledge of actual practices. New knowledge and discoveries, for instance after European encounters with so far unknown lands and cultures, transformed traditional perspectives or reinforced established views. Furthermore, different scholars placed different emphases in their treatments of the *utilitas astronomiae* depending on their different tastes and interests.

Even though a profound gap could separate declarations of intent from actual reality, I believe that statements about the practical embedment of astronomy in Renaissance works *of* astronomy and *on* astronomy offer an indication of the direction an investigation into practical astronomy should follow. Hence, I regard an assessment of these views as the first step toward further considerations and reflections on the relationship between astronomy and its applications. However, sources suggest exercising

[52]The entire quotation here follows from the original (Cardano 1966, vol. 1, pp. 34b–35a).: "Inter naturalia ostenta, primum illud quod et rarissimum est, natum me esse hoc saeculo, in quo totus orbis innotuit, cum antiquis paulo plus Triente cognitum fuisse: Hinc lustramus Americam (nunc patrem propriam appello) Brasiliam, maximam incognitae antea portionem, terram ignitam, Patagones, Peru, Charcas, Parana, Acutia, Caribanam, Picoram, Hisapaniam novam, Quitum, Quiniram partem Occidentaliorem, Galliam novam, et magis meridiem versus Floridam, Corte realem, Estotitilant, Marata: Praeter has ad Orientem sub Antarctico Antiscio squasi Scyniam, Binarchiam, Amazones, et quae est e regione insulae Daemoniorum (si hae fabulosae non sint) hinc certe ut aequa fiat distibutio magnas sequuturas calamitates verisimile est: crevit opinio, minuerunt et contemnentur bonae artes, et certa pro incertis commutabuntur. Sed haec alibi interim nos florente prato gaudebimus. Nam quid mirabilius Pyrotechnia, et fulgure mortalium, quod perniciosius multo est quam Caelestium. Nec te filebo Magnes, qui nos ducis per vastissima maria, et noctes tenebrosas, et tempestates horribiles in pereginas, et incognitas regiones. Addatur quartum Typographica inventio: manibus hominum facta, ingenio illorum inventa, aemula divinis miraculis, iam quid deest amplius ni Coelum occupemus."

caution with the term 'practical,' since the perceived usefulness of a discipline such as astronomy rested not only on its application in material life and the solutions provided for technical problems, but also on the moral, pedagogical, and philosophical consequences of the study of the heavens. In many cases, the 'utility' and the 'dignity' of astronomy merged, especially in the eyes of educators and theologians.

In summarising the results of this comparison between different Renaissance sources on the usefulness of mathematics, the *utilitas astronomiae* clearly emerges as an idea to which scholars and thinkers consistently return. Together with the issue of 'usefulness,' the *dignitas* of the discipline is celebrated, and pseudo-accounts about its origins and historical provenance were of clear interest. The weight given to the different parts of this assemblage varied depending on the philosophical and theological convictions of the respective author. Scholars could derive arguments in support of certain topoi from different philosophical currents. For instance, whereas Clavius based his claims for the nobility of the heavens on Aristotle, Liddel was inspired instead by Platonic-Pythagorean conceptions. Pseudo-historical narratives about astronomy's origins and progression relied on certain topoi as much as the motifs of 'dignity' and 'utility.' Cardano, for one, proposed a hermeneutic of mythology to unveil the origins of astronomy. The conception of history itself could vary: Fracastoro presented the history of science as a history of discoveries and progress, whereas Clavius regarded it as a tradition that developed through accumulation and integration. For Cardano, the superiority of the Moderns over the Ancients is evident. Moreover, the same *auctoritas* could be judged differently. This is particularly the case with Aristotle: while Fracastoro looked especially at *De coelo*, *Physica*, and *Metaphysica*, Ramus above all appreciated Aristotle's logic and mechanics. Similarly, differences mark the assessments about the applications of astronomy and their relative import. For Melanchthon and Reinhold, astrology was a relevant field of astronomical application; for Cardano, it was possibly the most important. Melanchthonians even regarded it as a discipline that approached Divine Providence. For Clavius, as a Catholic, astrology had to be downplayed and for Ramus, as a Calvinist, it had to be rejected *in toto*. Moreover, different perspectives affected the evaluations of the relationship between the topos of the *utilitas* and that of the *dignitas*. For the followers of Melanchthon and the Jesuits, the most important field of application of astronomical knowledge was theology. Accordingly, the discipline's highest utility tended to coincide with its dignity. In fact, astronomy was judged to have a value *per se* or for speculation. By contrast, Ramus reduced the *dignitas* to the *utilitas*, either in the form of epistemological utility or to utility in a more practical sense. As a consequence of his negation of astrology and theologizing *mathesis*, astronomy's usefulness was reduced to time reckoning and spatial orientation. In the case of Cardano, the importance of the practical dimension of astronomy lay in its potential to foresee the future and the guidance that may be thus obtained, as well as in its support for several practical disciplines beginning with medicine. His conception ultimately rested on an epistemological view in which observation, experience, and practice together formed a solid ground upon which all theory should be erected. In a sense, Cardano's conception is revealing of the growing awareness during the Renaissance about the centrality of practical knowledge and the distance between the Moderns and the Ancients, a gap that was ever increasing through scientific and technological progress.

References

Azzolini, Monica. 2013. *The duke and the stars: Astrology and politics in Renaissance Milan*. Cambridge, MA: Harvard University Press.

Biagioli, Mario. 1996. Etiquette, interdependence and sociability in seventeenth-century science. *Critical Inquiry* 22 (2): 193–238.

Bukharin, Nikolai. 2011. *Historical materialism: A system of sociology*. New York: Routledge.

Büttner, Jochen. 2008. Big wheel keep on turning. *Galilaeana* 5: 33–62.
Byrne, James Steven. 2006. A humanist history of mathematics? Regiomontanus's Padua Oration in context. *Journal of the History of Ideas* 67 (1): 41–61.
Cardano, Girolamo. 1543. *Libelli duo. Unus, de supplemento Almanach. Alter, de restitutione temporum*. Norimbergae: Apud Ioannem Petreium.
Cardano, Girolamo. 1962 [1581]. *De vita propria liber*. Trans. Jean Stoner. New York: Dover Publications.
Cardano, Girolamo. 1966 [1663]. *Opera omnia*. Lugduni: Sumptibus Ioannis Antonii Huguetan & Marci Antonii Ravaud. Reprint, Stuttgart-Bad Cannstatt: Fromann-Holzboog.
Clavius, Christophorus. 1581. *In Sphaeram Iohannis de Sacro Bosco Commentarius*. Romae: ex officinae Dominici Basae.
Clavius, Christophorus. 1591. *Euclidis elementorum libri XV*. Coloniae: Expensis Ioh. Baptistae Ciotti.
De Pace, Anna. 1993. *Le matematiche e il mondo: Ricerche su un dibattito in Italia nella seconda metà del Cinquecento*. Milan: Franco Angeli.
Di Bono, Mario. 1990. *Le sfere omocentriche di Giovan Battista Amico nell'astronomia del Cinquecento*. Genova: Consiglio Nazionale delle Ricerche. Centro di Studio sulla storia della tecnica.
Feingold, Mordechai (ed.). 2001. *The influence of Petrus Ramus: Studies in sixteenth and seventeenth century philosophy and sciences*. Basel: Schwabe.
Fracastoro, Girolamo. 1538. *Homocentrica sive de stellis*. Venetiis.
Gingerich, Owen. 1973. The role of Erasmus Reinhold and the Prutenic tables in the dissemination of the Copernican theory. *Studia Copernicana* 6: 43–62.
Goulding, Robert. 2010. *Defending Hypatia. Ramus, Saville, and the Renaissance rediscovery of mathematical history*. Dordrecht: Springer.
Grafton, Anthony. 1999. *Cardano's Cosmos. The worlds and works of a Renaissance astrologer*. Cambridge, MA: Harvard University Press.
Granada, Miguel Ángel, and Dario Tessicini. 2005. Copernicus and Fracastoro: The dedicatory letters to Pope Paul III, the history of astronomy, and the quest for patronage. *Studies in History and Philosophy of Science* 36 (3): 431–476.
Hotson, Howard. 2007. *Commonplace learning: Ramism and its German ramifications 1543–1630*. Oxford: Oxford University Press.
Kusukawa, Sachiko. 1995. *The transformation of natural philosophy: The case of Philip Melanchthon*. Cambridge: Cambridge University Press.
Lattis, James M. 1994. *Between Copernicus and Galileo. Christoph Clavius and the collapse of ptolemaic cosmology*. Chicago: University of Chicago Press.
Liddel, Duncan. 1591. [*Oratio de praestantia mathematicarum*], Niedersächsisches Landesarchiv/Staatsarchiv di Wolefenbüttel, *coll.* 37 Alt 379. Transcribed in *Duncan Liddel (1561–1613): Networks of Polymathy and the Northern European Renaissance*, ed. Pietro Daniel Omodeo, 227–236. Leiden: Brill.
Malpangotto, Michela. 2008. *Regiomontano e il rinnovamento del sapere matematico e astronomico nel Quattrocento*. Bari: Caucci.
Melanchthon, Philipp, and Paul Eber. 1550. *Initia doctrinae physicae*. Witebergae: Per Iohannem Lufft.
Omodeo, Pietro Daniel (ed.). 2016. *Duncan Liddel (1561–1613): Networks of polymathy and the Northern European Renaissance*. Leiden: Brill.
Omodeo, Pietro Daniel, and Irina Tupikova. 2013. The post-copernican reception of ptolemy: Erasmus Reinhold's commented edition of the *Almagest*, book one (Wittenberg, 1549). *Journal for the History of Astronomy* 44: 235–256.
Ovid. 1989. *Fasti*. Cambridge, MA: Harvard University Press.
Pantin, Isabelle. 1996. Is Clavius worth reappraising? The impact of a Jesuit mathematical teacher on the Eve of the astronomical revolution. *Studies in History and Philosophy of Science Part A* 27(4): 593–598.
Ptolemy. 1984. *Almagest*, ed. Gerald J. Toomer. London: Duckworth.
Ramus, Petrus (Ramée, Pierre de la). 1569. *Scholarum mathematicarum libri XXXI*. Basileae: Per Eusebium Episcopum et Nicolai Fratris haeredes.
Reinhold, Erasmus. 1549. *Ptolemaei mathematicae constructionis liber primus*. Wittebergae: Ex officinal Iohannis Lufft.
Renn, Jürgen (ed.). 2001. *Galileo in context*. Cambridge: Cambridge University Press.
Romano, Antonella. 1999. *La contre-réforme mathématique: Constitution et diffusion d'une culure mathématique jésuite à la Renaissance*. Rome: École Française de Rome.
Thüringer, Walter. 1997. Paul Eber (1511–1569): Melanchthons Physik und seine Stellung zu Copernicus. In *Melanchthon in seinen Schülern*, ed. Heinz Scheible, 285–321. Wiesbaden: Harrassowitz.
Valleriani, Matteo. 2010. *Galileo engineer*. Dordrecht: Springer.

Valleriani, Matteo. 2013. *Metallurgy, ballistics and epistemic instruments: The Nova scientia of Nicolò Tartaglia*. Berlin: Edition Open Access.

Westman, Robert S. 1990. Proof, poetics and patronage: Copernicus's preface to De revolutionibus. In *Reappraisals of the scientific revolution*, ed. David C. Lindberg and Robert S. Westman, 167–206. Cambridge: Cambridge University Press.

Westman, Robert S. 2011. *The Copernican question: Prognostication, skepticism, and celestial order*. Berkeley: University of California Press.

Zinner, Ernst. 1990. *Regiomontanus: His life and work*. Amsterdam: North-Holland.

Incunable Almanacs and *Practica* as Practical Knowledge Produced in Trading Zones

13

Richard L. Kremer

Abstract

This paper traces the development of practical knowledge in early modern astronomy and astrology by considering two literary genres apparently born in the fifteenth-century print shop, the annual broadside almanac and the annual multi-leaved practica. Taken together nearly 1000 editions printed before 1500 are extant, written in Latin or several vernacular languages. These texts combine astronomical, computistic, astrological and medical material. Examination of the term "practica" as an actors' category, of the social locations of more than one hundred authors named on the imprints, and of the content of these texts suggests that this pragmatic literature arose in sites Pamela O. Long has called trading zones. Exemplified by the early modern arsenal, mine, princely court, or print shop, these zones enabled artisans and scholars to exchange expertise and experience. The resulting almanacs and practica offered readers timely information about the church calendar, astronomical phenomena, and medical and astrological prospects for the coming year.

Manchester's Rylands Library holds an illuminated manuscript that is dated 1478 in its colophon and usually entitled simply *Astrologia*.[1] A letter of presentation introduces the text, explaining that one Christianus Prolianus was offering the work to the children of Antonello Petrucci, secretary to

For their many stimulating questions and comments I am indebted to the participants of the working group on "The Structures of Practical Knowledge," organized by Matteo Valleriani at the Max Planck Institute für Wissenschaftsgeschichte, where earlier versions of this paper were presented. I also benefited from conversations after presenting versions to faculty at Sciences, Philosophie, Histoire (SPHERE), Université Paris Diderot (Paris 7) and the Medieval Seminar, Dartmouth College. For their generous help with many queries I thank my colleagues Monica Azzolini, Ann Blair, Oliver Düntze, Bruce Duncan, Falk Eisermann, Margaret R. Graver, Jonathan Green, John Hodgson, Jason Moschella, Alexandra Ohlenschläger, Gianna Pomata, A. Kevin Reinhart, Michael Shank, Roberta Stewart, Alexandre Tur, Steven van den Broecke, and Jarosław Włodarczyk.

[1] Manchester, John Rylands Library, Latin MS 53.

R.L. Kremer (✉)
Dartmouth College, Hanover, NH, USA
e-mail: Richard.L.Kremer@dartmouth.edu

M. Valleriani (ed.), *The Structures of Practical Knowledge*, DOI 10.1007/978-3-319-45671-3_13

a b

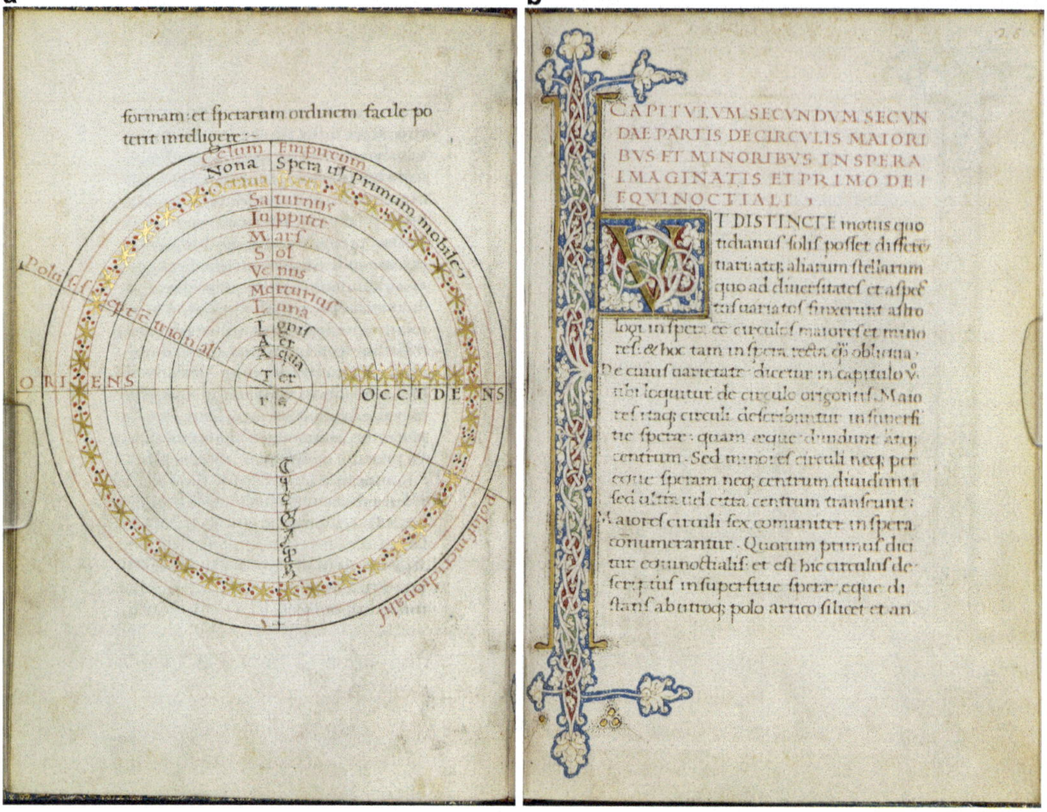

Fig. 1 The Aristotelian cosmos, details rendered in gold leaf and interlace. Christianus Prolianus *Astrologia*, manuscript copy for the Neapolitan court, 1478. Manchester, John Rylands Library, Latin MS 53, fols. 27v–28r. Copyright of the University of Manchester

the King of Naples, Ferrante I (1423–1495). Ostensibly, the children had found the standard university introductory text, Sacrobosco's *Sphaera*, too difficult to follow. So Prolianus had reduced to simple rules knowledge of the structure of the cosmos, the geometry of the essential celestial circles and spheres, the theorica of planetary motions, and the sizes and distances of the celestial bodies from the Earth.[2] The resulting "compendium of astrology," as Prolianus denoted his work, offered nothing new; nonetheless, he hoped that Petrucci's children would find the small book both useful and entertaining.

The Rylands codex, the only known manuscript of this text, is on vellum, written in a lavish humanist script. It is handsomely illuminated. Large swatches of gold leaf fill the scale drawings of the planets (Fig. 1). In the nineteenth century, the pre-Raphaelite artist, poet and socialist, William Morris, owned the manuscript and is thought to have been deeply influenced by its filigree designs. At the bottom of the opening folio, the illuminated border displays two putti holding the arms of the King of Naples. Given its opulence, we might guess that the Rylands manuscript was a presentation copy, intended not merely

[2]Prolianus extracted the cosmic dimensions explicitly from the thirteenth-century *Theorica planetarum* of Campanus of Novara (Prolianus 1478, fols. 32v–35v; Benjamin and Toomer 1971, 356–63).

for a secretary's children but for the Aragonese court.[3] Perhaps Prolianus, otherwise completely unknown except for a short text on an astronomical quadrant preserved in a single manuscript, hoped to secure a position as mathematician or astrologer at the court?[4] A rather plain and poorly printed edition of the *Astrologia* was also issued by a Naples printer, with its colophon dated August 1477, today extant in 18 copies according to the Gesamtkatalog der Wiegendrucke (Prolianus 1477). Perhaps the printed copies were meant to publicize Prolianus's gift, another tactic in his quest for court patronage?[5]

This quotidian story of gift-giving and courtly knowledge, reflecting common cultural practices in fifteenth-century Europe, is abruptly confounded, however, if we turn beyond the colophons of the Rylands manuscript and the incunable imprint. There we find an additional 15 folios filled with computistic tables, from which one can extract the dates of Easter, and astronomical tables giving the times of each new and full moon (syzygies) and the times and magnitudes of eclipses for every year from 1478 to 1515, calculated "from sunset for the meridian of Naples" (Fig. 2). Brief canons describe how to use the latter tables but do not indicate how they were computed. Suddenly we are dealing not with simplified Sacrobosco but with the most complex computations in all of Ptolemaic mathematical astronomy (Chabás and Goldstein 1997; Kremer 2003a). Surely Petrucci's children could not have understood, let alone used, these tables. Why are these two types of material combined into a single treatise? Indeed, I know of no other fifteenth-century text, in manuscript or print, that combines such diverse materials.

The only other incunable imprints offering syzygy and eclipse times over a such long range of years were the ephemerides (for 1475–1506) and calendars (for 1475–1531) prepared and printed in Nuremberg by Johannes Regiomontanus (1436–1476), the leading mathematician and astronomer of the fifteenth century. Printers both north and south of the Alps would reissue these tables in many editions well into the sixteenth century. In 1477–1478, the Ulm astronomer Jacob Pflaum printed several editions, under his name, of syzygy and eclipse times for 1477–1552, essentially copying Regiomontanus's data until 1531 but shifting his meridian from Nuremberg to Ulm. And in 1499, Pflaum and the Tübingen-based astronomer, Johannes Stoeffler (1452–1531), printed in Ulm a continuation of Regiomontanus's ephemerides, covering the years 1499–1531. All of these ephemerides and calendars, requiring literally tens of thousands of computations, were calculated from the Alfonsine Tables, the comprehensive set of astronomical tables developed in Castile in the 1270s, reformulated in Paris in the 1320s, and by the fifteenth century the most widely used set of tables across Europe.[6] Calculating ephemerides would have been laborious, mind-numbing, and fraught with pressure to police errors.

[3]The fifteenth-century library of the Court of Aragon in Naples has been well studied, with the Rylands manuscript considered as a presentation copy made for the court. The collection included very few astrological, astronomical or mathematical texts. Petrucci himself also had a large personal library of manuscripts and printed books, in Greek, Latin and Italian, a typical humanist set of classical literary, historical and philosophical texts with only one known astrological text, Guido Bonatti's *Introductio ad judicia stellarum* (now Paris, BnF lat. 7326). See (Marinis 1947–52, ii: 136; 210; Mazzatinti 1897, 48–52; Ruggiero 2009).

[4]Paris, BnF lat. 10263, fols. 137r–139r (Thorndike 1951; Poulle 1963, 51; Monaco 1999–2000; Aguilar and Marrero 2005).

[5]For a recent study suggesting that the manuscript was copied from the printed edition, see (Blair 2014). I think it more likely that both derived from a now lost *Urquelle* since the manuscript includes eclipse diagrams not in the incunabulum. For the King Ferrante I's patronage of the royal library, music, humanist translation, printing and other literary projects in Naples, see (Bentley 1987, 62–80).

[6]Only two fifteenth-century authors published multi-year tables of syzygy times not computed from the Alfonsine Tables. The Barcelona astronomer, Bernat de Granollachs, computed a *Lunaria* with times for the years 1485–1550 that was printed in at least 60 editions (mostly Rome and Spain but five in Naples) from 1485 into the sixteenth century. And the Hebrew astronomer Abraham Zacut in Lisbon prepared an *Almanach perpetuum*, printed in several incunable editions in the late 1490s, that provided syzygy times for the years 1478–1508. Both of these works are based on the fourteenth-century, non-Alfonsine tables of Jacob ben David Bonjorn, a Jewish scholar supported by the King of Catalonia and Aragon.

Fig. 2 Prolianus's times, listed to days, hours and minutes, of all the new moons for the years 1478–1485. Manchester, John Rylands Library, Latin MS 53, fol. 60r. Copyright of the University of Manchester

Prolianus's syzygy times, however, are not copied from these fifteenth-century ephemerides. My analysis suggests that he (or his source) independently computed the times from the Alfonsine Tables for a longitude of about 103 min of time east of Toledo.[7] He then converted those times from astronomical hours, counted from noon, to Italian hours, counted from sunset. Prolianus's is the only multi-year table of syzygy times specified in Italian hours that I have found in fifteenth or sixteenth-century sources. And the computational "noise" in his times, although roughly three times greater than what I find in Regiomontanus's syzygy computations (the gold standard for fifteenth-century computation), reveals a careful, competent mathematical astronomer at work.

Although the text of Prolianus's *Astrologia* presents elementary, descriptive material designed (at least rhetorically) for juvenile readers, its astronomical tables are unique for the fifteenth century and provide long-term, quantitative data that must have required considerable experience and computational skill to generate. These rows and columns of numbers would carry little meaning for a textbook reader. But for a court astrologer charged with preparing annual astrological predictions, editions of which by the 1470s were pouring from Europe's printing presses, Prolianus's tables of syzygy and eclipse times would have been invaluable.

As far as I know, Prolianus after 1478 disappears from documented memory. He apparently did not author annual astrological prognostications for the Naples court. Perhaps he died of the plague, in 1478 particularly acute in Naples (Biraben 1975–76, i: 396). Perhaps his attempt to secure patronage failed.[8] But the curious bifurcated content of his *Astrologia* raises questions that will be central for this essay. Why are these two types of knowledge combined into a single book? What types of readers would have wanted each type? And most importantly, can the categories of "theoretical" and "practical" help us understand Prolianus's text? Might his book have arisen in a trading zone where learned and artisanal practitioners of the astral sciences came together in the 1470s to exchange their expertise?

In her 2011 Horning Lectures, Pamela O. Long argued that a new set of empirical approaches to knowledge of the natural world arose in late medieval and early modern Europe, in physical settings she describes as "trading zones" where skilled artisans and unskilled learned men (often humanists) came together for communication and exchange. Many of these zones, Long noted, overlapped with the interests of rulers, powerful nobles, or states. Sharpening and making more specific a historiography of "scholar" and "craftsman" that she traced back to Marxist scholars like Edgar Zilsel (1891–1944), the philologist Leonardo Olschki (1885–1961) and the sociologist Robert Merton (1910–2003), Long asked how trading zones such as princely courts, printing shops, instrument makers shops, arsenals, mines and metal working sites, and even cities with extensive public construction (e.g., Rome) became sites for innovation not only in the making of things but also in the codification of knowledge in written texts. According to Long, the printed technical treatise, a genre that flourished in the sixteenth century, was not a "how-to" manual for building a ship, raising an obelisk, or pumping water from a mine; neither was it a textbook for training apprentices or craftsmen. Rather, these texts were written for powerful elites, investors, or state bureaucrats, offering a new type of what Long called "empirical knowledge" (Long 2011).

In this paper, I want to ask whether two new printed genres of the late fifteenth century, the almanac and the astrological practica, might offer another example of practical knowledge produced in a

[7] The table of geographical locations in the *editio princeps* of the Alfonsine Tables (1483), a table found in many earlier manuscript copies of this set, specifies the meridian of Naples as 104 min of time east of Toledo, close to Prolianus's value. On the other hand, this table gives the latitude of Naples as 40;20°; yet the Prolianus's table of day lengths is computed for a latitude of 42°, assuming a value of the obliquity (23;30°) widely used around 1500.

[8] By 1487, Petrucci and his two sons, Francesco and Gianantonio, would be dead, executed by Ferrante for having led a revolt of the barons against the King in 1485. See (Bentley 1987, 30–3).

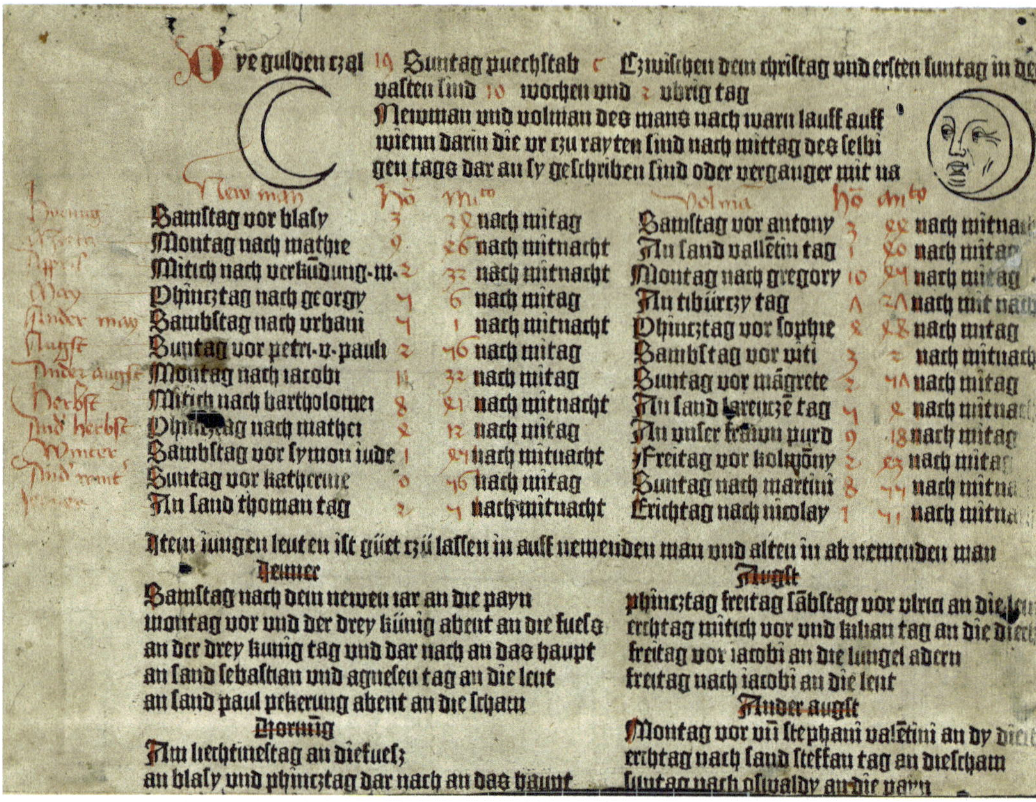

Fig. 3 The earliest extant printed broadside almanac, preserved in a unique fragment. *Almanac for the year 1462* (Vienna 1461), VE15, A-145, Princeton, Scheide Library, SC4.3.13, fragment. Courtesy of the Scheide Library, Princeton University

trading zone of interest to the state as well as to a wide swatch of European readers. Both genres flourished only with the advent of printing; both coalesced fairly quickly into a standard, periodic content, with each edition valid for a single calendar year. The almanac offered readers, in a broadside format, computistic information on the moveable feasts of the Christian calendar for the year, astronomical information on the dates and times of the new and full moons for the year (syzygies), and medical/astrological information on the most favorable days for bloodletting (Fig. 3). Practica, usually single-quire pamphlets of 8–10 leaves, listed not only syzygies but also times of eclipses and of the Sun's entry into the four cardinal points (seasons) of the year (Fig. 4). From horoscopes cast for these times, this genre offered a broad range of general astrological predictions for weather, agriculture, war and peace, plague and disease, kingdoms and cities, and various social groups (*Stände*) such as knights, soldiers, kings, bishops, clerics, merchants, women, learned and commoners, etc. Historians of early printing refer to these texts as pragmatic literature. Like papal bulls, city ordinances, or book seller advertisements, these imprints were date sensitive; consumed over the course of a year, they generally were discarded when no longer topical. Hence, they exist today in very few copies; many are uniquely preserved, often as fragments in book bindings.

This paper will have two parts. First I will explore the terms used by authors to denote these new genres, especially the practica with its suggestive link, at least etymologically, to "practical knowledge." Although genres of "practica" appear by the twelfth century in fields like medicine or geometry, astronomy/astrology would not get their "practica" until after 1450, a late emergence that might tell us something about the status of the "practical" in those fields. Second, I will consider the kinds

Fig. 4 One of the earliest printed, multi-leaved, annual prognostications to bear the title "Practica," authored by the first professor of mathematics and astronomy at the University of Ingolstadt Johannes Engel, *Die dewtsch practick … auf das Jar lxxxviii* (Nuremberg 1488), GW, 01903. Munich, Bayerische Staatsbibliothek, 4 Inc.c.a. 506 m, fol. 1r

of knowledge required for the drafting of astronomical almanacs and practica. Prolianus's astrological compendium, I shall suggest, would have provided some of the material needed to produce almanacs and practica. Although toolboxes of astronomical and astrological materials can be found in literally hundreds of 13th–16th century manuscripts now preserved in European libraries, the profusion of printed almanacs and practica after 1460 required a particular set of materials that not only the little-known Prolianus but also the well-known mathematician, Johannes Regiomontanus, would provide. Were bookish tools, however, sufficient to write the almanacs and practica? Or might experience generated in trading zones of medical practice, astrological consultancy, astronomical computation or printing shops, also have been required? What got traded, we shall ask, in the making of these texts during the early years of European printing?

1 Actors' Categories, or Theorica and Practica

Although Long and other scholars have warned against employing overly coarse dichotomous categories (craftsmen/scholar, unskilled/learned, art/nature, or practice/theory), attention to actors' categories shows that the theorica/practica distinction enjoyed a long history in medieval Latin before astronomical and astrological texts with these names started being printed in the 1470s.

In a 1961 study of the use of the Greek term *θεωρικος* in Latin, Olaf Pedersen found no evidence that the terms theorica and practica had belonged to the classical Latin idiom (Pedersen 1961).[9] Writers like Cicero and Boethius used the terms to explain the Greek philosophically inflected distinction between the speculative and the active. Latin church fathers and hagiographical writers described the *theorica vita* as the monastic, spiritual life, separated from the cares of the world. The Greek distinction of *theorica/practica* found new life in twelfth-century pedagogical contexts as Greek and Arabic texts increasingly became available in Latin.[10] As Danielle Jacquart has shown, two of the earliest commentaries used in lectures at the school of Salerno (late twelfth century) divide medicine into two parts, theorica and practica. Combining Aristotelian and Arabic materials, these authors presented the division as fundamental to the classification of knowledge. "The theoretical part [of the *Pantegni*, a translation of a large Arabic medical encyclopaedia] is the perfect knowledge of things that can be apprehended by the intellect alone, subject to the memory of things to be performed. Practice consists in demonstrating theory, subject to the sensory field and to manual operation, according to a previous understanding of theory." Practica is "the science that conserves health and treats disease by regimen, medication and surgery" (Jacquart 1988, 414).[11]

The practica section of the *Pantegni* includes literally hundreds of short chapters describing the sensible properties of individual medicaments (color, odor, texture, etc.) and the visible symptoms of diseases. Likewise, Bartholomew of Salerno wrote a stand-alone practica, probably mid-twelfth century, which divides its subject matter into two parts, conserving health and curing disease. Each malady and its therapy are briefly described in simple, empirical terms. For example, Bartholomew described the symptoms of several types of epilepsy, the foods and medicines that should be taken by people suffering these symptoms, and often provided "recipes" (his term) listing quantitative proportions for the various ingredients (1/2 of this, 1/6 of that, etc.) (Bartholomew of Salerno 1852–59, iv: 355–6). Many of the short paragraphs are written in the imperative voice: do this, do that. Bartholomew's practica ends with a chapter on how to impede conception.

[9] See (Du Cange 1883–87, vi: 445–6) for scattered examples in medieval Latin of the term practica (e.g., as realization, regular procedures, causes of action, treason or conspiracy, familiarity), all unrelated to theorica.

[10] We cannot here review the many Greek usages of the terms θεωρια and πραξις.

[11] See also (Jacquart 1992).

By 1400, medical professors at the Italian universities were being appointed as teachers of practica or theorica, which generated demand for textbooks and commentaries mirroring the dichotomy. By the 1570s, the lectures on *Medicina practica* of the famous professor at the University of Padua, Girolamo Mercuriale (1530–1606), would fill five massive volumes when printed (Antry 1989; Siraisi 2008). However, as Nancy Siraisi has noted, at the Italian universities both *theorica* and *practica* "were academic disciplines transmitted by the study of texts … [both] could be taught by commentary, and could involve much rationalization, systematization, and reliance upon authorities" (Siraisi 1987, 297). University-trained medici studied both. The medical practica, despite their recipe-driven structure, may not have arisen within a trading zone where two distinct social classes of knowledge-makers or knowledge-users came together.

The twelfth century also saw the introduction of the theorica/practica dichotomy in geometry, with those terms appearing in a text by Hugh of St. Victor (ca. 1096–1141) (Baron 1955). In his prologue, Hugh emphasized that his *Practica geometriae* was not new; rather it collected "older, scattered materials" that historians have traced back to Roman surveyors (*agrimensores*) or even to Heron's Greek manual for fixing boundaries, measuring land and making maps. The tradition began with technical concerns in the Greco-Roman world and appears to have developed without relying on the works of Euclid, Archimedes or Apollonius. According to Hugh, geometry is:

> … either theoretical (speculative) or practical (active). Theoretical geometry uses sheer intellectual reflection to study spaces and intervals of rational dimensions. But practical geometry uses instruments, and gets its results by working proportionally from one figure to another. There are three kinds of practical geometry: altimetry, planimetry and cosimetry (Homann 1991, 33–4).[12]

Explaining in detail the astrolabe, "the surveyor's most important instrument," Hugh described how this tool can be used to measure heights and planar areas (Homann 1991, 40). Most of his examples are discrete, as are the instructions to the "surveyor" who manipulates the rod, rule and alidade on the astrolabe:

> Suppose the height on a hill, the surveyor in a valley. Let him turn the astrolabe to the hill and unknown height. With medicline [alidade] set on the median line, let him note carefully through both apertures the spot opposite on the hillside corresponding to the median line. The surveyor should then turn the medicline up the hillside to the top of the object, and take as perpendicular the whole length above the horizon. The ratio of this composite length to the length of the intervening space is that of the medicline marker reading on the lower square side to the whole side…. (Homann 1991, 50–1)

Hugh's third section moves from the everyday to the cosmic and seems addressed to scholastics rather than to artisanal surveyors. Cosimetry, or "measurement of the cosmos, starts at its center, evaluates diameters and circumferences of the celestial sphere and its inner orbits with the help of proportion theory, and then probes the significance of these discoveries. It delimits all cosmic objects, and sets a specific ratio and value for each distance" (Homann 1991, 57). Hugh reported the results of Eratosthenes and Macrobius who measured the Earth's circumference with the "gnomen method" (which assumes the Earth is spherical and the solar rays hitting the Earth are everywhere parallel). He then described how the "Egyptians" computed the Earth-Sun distance by assuming the Earth is flat and the solar rays diverge from a point source. Hugh presented several other methods for measuring these distances that are, according to his modern editor, inconsistent and "flawed" by Euclidean standards (Homann 1991, 27–30, 57–64). Was Hugh merely compiling methods of his predecessors, without seeking to compare their results? The cosimetry of his *Practica geometriae* reads quite differently from the chatty instructions for student surveyors provided in the sections on altimetry and planimetry. Hugh's practica, his editor concludes, "provided practitioners and schoolmen with

[12] Dominicus Gundissalinus (1151–1190), too, had split geometry into the theoretical (for teaching) and the practica (for acting) and had divided the latter into altimetry, planimetry and cosmimetry. See (Pedersen 1961, 156–7).

vocabulary, problems, a sense of structure, an agrimensorial legacy, and an inchoate spirit of inquiry" (Homann 1991, 30).[13] Long would surely call Hugh's text the result of a trading zone.

Our pursuit of the genres of theorica/practica takes us next to alchemy and the thirteenth century. As William Newman has shown, the alchemical debate of that century produced the first Latin apologists for an operative alchemy in which practitioners not only transmute metals but also prepare dyes, medicines and other products by manipulating a wide variety of substances. One of the leading defenders of alchemy in this debate was Paul of Taranto, who around 1200 composed a lengthy *Theorica et practica*. Although Newman argues that Paul wrote for a scholastic audience, the content of the text reveals a trading zone between the Aristotle-inflected classrooms of the newly founded Italian universities and the hearths of working alchemists (Newman 1986; Perifano 1990–91).

Little is known about Paul beyond the colophon of the *Theorica et practica* that identifies him as a "lecturer of the Franciscan friary in Assisi," a claim Newman could not confirm in archival sources. The text, neither printed nor edited before Newman's 1986 dissertation, reveals an author well acquainted with Aristotelian physics and divisions of knowledge. Paul began by distinguishing "matter of intellect" from "instruments of art." The former requires theorica. It deals with the "speculation of alchemical truth," with knowledge of metals as natural things. It investigates their essences, quiddities, and intrinsic principles, their elementary forms, the causes and principles of their transmutations. The latter pertains to "operations" performed by the human artisan. Art (Paul mentions sculptors, painters, horticulturists, physicians, and alchemists) moves and directs nature. Painters and sculptors deal with accidental, extrinsic form; physicians, horticulturists and alchemists deal with substantial, intrinsic form. In both cases, it is nature itself that does the making, in response to the artisan's directing (Newman 1986, iii: 1–17 (Latin), iv/2: 1–12 (English)).

Paul's practica didactically describes ten operations used in the art of alchemy (Newman 1986, iv/2: 93–114),[14] discussing in considerable detail how hot to make the fire, how to cover and uncover vessels, how long to continue given processes. It then provides recipes for preparing dozens of materials required for the alchemist's art. In each case, the material is briefly characterized empirically (color, texture, odor, etc.) and then the operational procedures are listed. For example, in Newman's translation:

> How sal alkali should be made. Let the herb from which glass is made be taken, which looks almost like rosemary, except that it has longer leaves; and with a hole dug in the earth such as glassmakers make, let it be burned up. Cinders having been made from it, which are called clavellated cinders, let them be boiled with the prepared urine of boys according to this art, in the same quantity as the cinders, first imbibed in that <urine> until it be thickened. This will all occur in <one> vessel; let it <then> rest, for it will be good, useful sal alkali.

Not finding this recipe in earlier alchemical texts, Newman suggested that it "probably reflects the author's own experience (Newman 1986, iv/2: 124 (English))."

Paul's *Theorica et practica* derived from a trading zone where scholastic readings of Aristotle were combined with operational recipes drawn either from earlier recipes or directly from experience. One might ask whether the recipes could be followed without any knowledge of the theorica. Yet Paul discouraged such an attempt. The "unskilled" (inexpertus) cannot practice alchemy without recognizing the hidden as well as the manifest natures of things. The "artisan" (artifex), despite his "long experience and subtle investigation of nature alone," cannot arrive at the "true effect," especially regarding gold, silver and other minerals, without notice of "proper principles and causes." Only the "true artificer" (in veris artificibus), employing knowledge of both theorica and practica, will be able to "perfect true gold and silver from the principles of nature" (Newman 1986, iii: 15–6, iv/2: 10–1 (English)).

---

[13] For the further development of the genre of practical geometry into the sixteenth century, see (Victor 1979; Tura 2008).

[14] Separation, sublimation, fixation, calcination, solution, distillation, reduction, ceration, imbibition and assation.

Was Paul's a practical manual, thumbed by the alchemist at the hearth? Or was it a scholastic book, discussed by Franciscan monks while reading Aristotle in their library? It may have been both; in any case, Newman located only four manuscripts (all fifteenth-century or later) that contain Paul's text, only three of which contain both the theorica and practica.

According to Robert Halleux's typology of alchemical genres, practica, which he defines as similar to what we've seen in Paul's text, would become the largest part of the alchemical literature. Yet none of the classic texts Halleux lists, dating from the twelfth to the fifteenth centuries, include that term in their incipits or short titles. Perhaps the term practica was less self-consciously deployed in medieval alchemy than it was in medicine and geometry? (Halleux 1979, 79–83)

The term practica, albeit without its companion theorica, also was sometimes associated with a mathematical text on computation with Hindu-Arabic numerals, the *Algorismus* authored by al-Khwārizmī with many later adaptations by John of Seville, Sacrobosco, Jordanus de Nemore, Leonardo Fibonacci (ca. 1175–), John of Lignères, the anonymous *Algorismus ratisbonensis*, and others (Vogel 1954; Allard 1991; Busard 2000). Occasionally the *Algorismus* was accompanied by long lists of computational problems taken from daily commercial activities, such as buying and selling, converting currencies or other units of measure, determining interest, etc. I have not systematically examined all these texts. But a commentary on Sacrobosco's *Algorismus*, written by Peter Philomena of Dacia, c. 1300, preserved in a mid-fourteenth-century manuscript bears the explicit "… algorismus sive arismetrica practica." The mid-fifteenth century *Algorismus ratisbonensis* is divided into three sections. A manuscript written around 1450 presents in two sections the rules of computation. The scribe then inserted "Explicit textus algorismii ratisbonensis. Sequitur Tractatus tercius de practica," followed by more than 300 quotidian problems, written in both Latin and German (Curtze 1897, v; Vogel 1954, 27). For example:

> You have 730 horses to feed and you give 1 measure of feed to 40 horses. I now ask how many measures will you need to feed 730 horses. Or if 40 horses require 1 measure, how many will be required for 730 horses? Answer: 18 and ¼ measures (Vogel 1954, 68).

Although our term apparently does not again appear in the fifteenth-century manuscripts of the *Algorismus ratisbonensis*, this scribe considered the quantitative tasks of daily life as practical. The earliest incunabula edition of an algorismus, printed in Venice in 1478, includes some daily problems, describing itself on the opening folio: "Here begins a practica, very good and useful for anyone who wants to employ the arts of the merchant, commonly called the art of the abacus" (*Arte dell'abbaco* 1478, fol. 1r).[15]

In these four cases, practica deal with the quotidian, the specific, the operative, with medicines and diseases, with measuring and surveying, with manual manipulations at the hearth, with computational problems of quantity, measure, and value. Although medieval readers might have encountered these texts in classrooms, their contents undoubtedly were shaped by trading zones located beyond those classrooms, by spaces occupied by physicians and their clients, by surveyors, by alchemists, and by shopkeepers, bankers and farmers. Does the astrological practica fit this pattern?

The genre of the astrological practica would emerge considerably later than the four examples discussed above, although astronomical theorica had appeared already in the twelfth century. In the 1170s, Roger of Hereford composed a canon in 32 chapters on the use of astronomical tables that, in one manuscript, begins: Incipit theorica Rogeri Herefordensis (Haskins 1915, 66; Kren 1970–1980; Burnett 2004). In the thirteenth century, a more comprehensive summary of Ptolemaic astronomy (that does not mention Ptolemy's name) circulated under the title *Theorica planetarum*. Extant in

[15] "Incomminicia vna practica molto bona et vtile a ciaschaduno chi vuole vxare larte dela merchandantia, chiamata vulgarmente larte de labbacho."

more than 210 manuscripts and published in at least four incunable editions, this text, along with Sacrobosco's *Sphaera*, quickly became the standard introduction to astronomy at most European universities. Although variously attributed in the manuscripts to Alfonso of Castile (1221–1284), Gerard of Cremona, Robert Grosseteste, Michael Scott, John of Seville and Sacrobosco, its modern historian Olaf Pedersen refused to assign an author. The text in five chapters gives "theorica" for the Sun, Moon, head/tail of the dragon, superior and inferior planets. In each case, Ptolemaic models are presented with consistent Latin terms but with no numerical parameters. A sixth chapter describes planetary retrogradation and stationary points; a seventh jumbles together remarks on planetary latitudes, precession, geographical latitudes, and how to use observations to determine the Sun's eccentricity and apogee (this is the only mention of observations, made with an astrolabe that is attributed to al-Battānī). The final chapter very briefly describes planetary aspects but does not comment on their astrological significance. This popular textbook mentions nothing about astrological medicine, bloodletting, or horoscopes (Pedersen 1981, 1992; Crowther et al. 2015).

A longer, more complex *Theorica planetarum* would appear in the 1260s, authored by Campanus of Novara (ca. 1220–1296). Extent in 60 manuscripts and also intended for the university, Campanus's *Theorica* explicitly mentions Ptolemy and describes his planetary models, with numerical parameters, in their full technical splendor. The primary goal of the treatise is to provide instructions for constructing seven equatoria, or physical instruments composed of rotating disks of metal or wood and strings that replicate Ptolemy's models and enable planetary longitudes to be computed geometrically. As Campanus explained in his prologue, calculating planetary positions with astronomical tables is:

> … not only difficult for the untrained, but wearisome even for the trained and the wise…. Therefore, in order that all who, either through occupation with affairs or lack of training or weakness of understanding … may have the means to find out the exact positions of the planets … and that they may be able to see with their eyes by an instrument which is perceptible to the senses and which brings about a motion similar to the rotation of the heavenly bodies, I have striven to manufacture such a material instrument … and if I am not mistaken, anyone who sees it and learns the mode of its operation will both feast his gaze on its beauty and delight his intellect with its utility (Benjamin and Toomer 1971, 140–1).

Interestingly, in his previous paragraph, Campanus described two parts of astronomy, a "theoretical" part that determines the geometrical figures and relative sizes of the celestial circles via "infallible methods" and a "practical" part that uses numbers and arithmetic to "find the exact positions of all the planets for any given time" (an exact description of the two parts of Prolianus's 1487 treatise!) by means of tables or equatoria (Benjamin and Toomer 1971, 138–41). Yet although Campanus occasionally told readers where to place their ruler or compass when drawing out an equatorium, his general tone mirrors that of the *Almagest*. Campanus's *Theorica* is not a treatise on machines; it is not a construction manual for equatoria. Despite Campanus's brief remark about practical astronomy, he did not entitle his text *Practica planetarum*. Although there are Arabic predecessors, Campanus's was the earliest equatoria treatise known in the West; many subsequent proposals for more complex equatoria would appear, into the sixteenth century (Poulle 1980). Like Campanus's *Theorica*, they were intended for the university classroom not the instrument makers' shop.

In the 1450s the Viennese master Georg Peurbach (1423–1461) would author a *Theorica novae planetarum*, expanding the traditional *Theorica* by describing three-dimensional models for the planets, offering more detailed images of these models, and correcting several mistakes in the earlier version. This text, first printed by Regiomontanus around 1472, would provoke many printed editions and commentaries well into the sixteenth century. Like its predecessor, Peurbach's *Theorica* restricts its focus to Ptolemaic planetary models and ps.-Thābit's medieval models for the motion of the eighth sphere; astrological materials are not included (Pedersen 1978; Aiton 1987; Malpangotto 2012; Ragep 1993, 400–8). And Peurbach did not mention Campanus's distinction between practical and theoretical astronomy.

Hence, we come finally to the question of a *Practica planetarum*. Can we identify an astronomical or astrological genre that reflects the divide we have seen in medieval texts of medicine, geometry, alchemy and algebra? Although such a title might well exist somewhere in the massive collections of astronomical manuscripts preserved in European libraries, scholars generally have concluded that the genre of what eventually would be called the astrological practica was born in the fourteenth century, would be mandated by Italian university statutes by 1405, and eventually would flourish in fifteenth-century print shops (Tur 2014; Malagola 1888, 264; Sorbelli 1938; Matthäus 1969, 1001–6; Green 2012b).[16] The term practica, applied to this genre, arose in German, not Latin, linguistic contexts.

The multi-leaved annual prediction, based on astronomical and astrological content, would initially be described in university statutes as an *iudicum* and *tacuinum*. The former referred to the prediction of events to occur during the coming year, the latter to its medical and agricultural prospects. The term *tacuinum* (table) may have derived from the *Tacuinum sanitatis*, translated in the twelfth century from the *Taqwīm as-Ṣiḥḥa*, an Arabic treatise on hygiene, exercise and foods compiled in the mid-eleventh century by Ibn Buṭlān, a Christian physician in Baghdad. In his preface, Ibn Buṭlān wrote that he would present the information in tabular columns. George Sarton found the Arabic term *Taqwīm* more generally employed in texts of law, history, geography and logic from the twelfth through the fourteenth centuries, referring to the synthetic presentation of content in tabular form; *dispositio per tabellas* is Sarton's rendering of the term in Latin (Bovey 2005; Hoeniger 2006; Thorndike and Sarton 1928).

For whatever reason, by the time the annual predictions began to appear in print, the term *tacuinum* had essentially disappeared. According to the *Gesamtkatalog der Wiegendrucke*, only one incunabulum bears this title, a four-leaved imprint by an otherwise unknown Master of arts and medicine (Pavia) Johannes Strimer. This four-leaved *Tacuinum pro anno domini 1485* does indeed organize its content in tabular format, listing dates and times of syzygies, compustistic material, and felicitous and dangerous dates in each month for a wide variety of activities ranging from marrying and building houses to starting wars and holding disputations at universities. Strimer provided no hints as to how he had assembled these lists of days but did mention the widely circulated text, *De pronosticationibus egritudinum secundum motuum lune* (also known as the *Astrologia Ypocratis*), that correlates diseases and therapies to the Moon's zodiacal sign. Presumably he also drew on the Roman tradition of lucky and unlucky days. Such prescriptive lists of good and bad days would not, however, gain traction in fifteenth-century print culture (Strimerius 1485; Sudhoff 1908; Thorndike 1960; Kibre 1977; Keil 1957; Demaitre 2003).[17] Rather, such content would be woven into the prose formats of the practica.

The earliest extant imprints containing material that resembles what would become the genre of the practica were printed in Venice for 1474. In their incipits or explicits these Latin texts describe themselves as "iudicum sive pronosticum astrologicum" (Guasconus 1474; 1475). As European printers began issuing more editions in Latin and Italian, the terms prognosticon and iudicum quickly became standard in the colophons or incipits. None of the incunable editions printed south of the Alps that I have been able to examine (either in the original, scans, or from detailed bibliographic descriptions of the GW) use the term "practica;" these mostly Latin (a few Italian) imprints invariably identify themselves as prognostica or iudica. Late fifteenth-century Italian authors did not use the Latin polarity of theorica/practica to situate their work.

The name practica, for the annual astrological prediction, emerged in German print shops of the 1470s. Interestingly, the term first appeared not as a noun but as a verb, in a somewhat different

[16]For a London astrologer's personal prognostication for 1452, see (Page 2001).

[17]Nothing further is known of Strimerius. See (Thorndike 1923–1958, iv: 455) who nicely describes the production of annual prognostica at the fifteenth-century Italian universities but only lists printed editions without examining either contents or university archival materials.

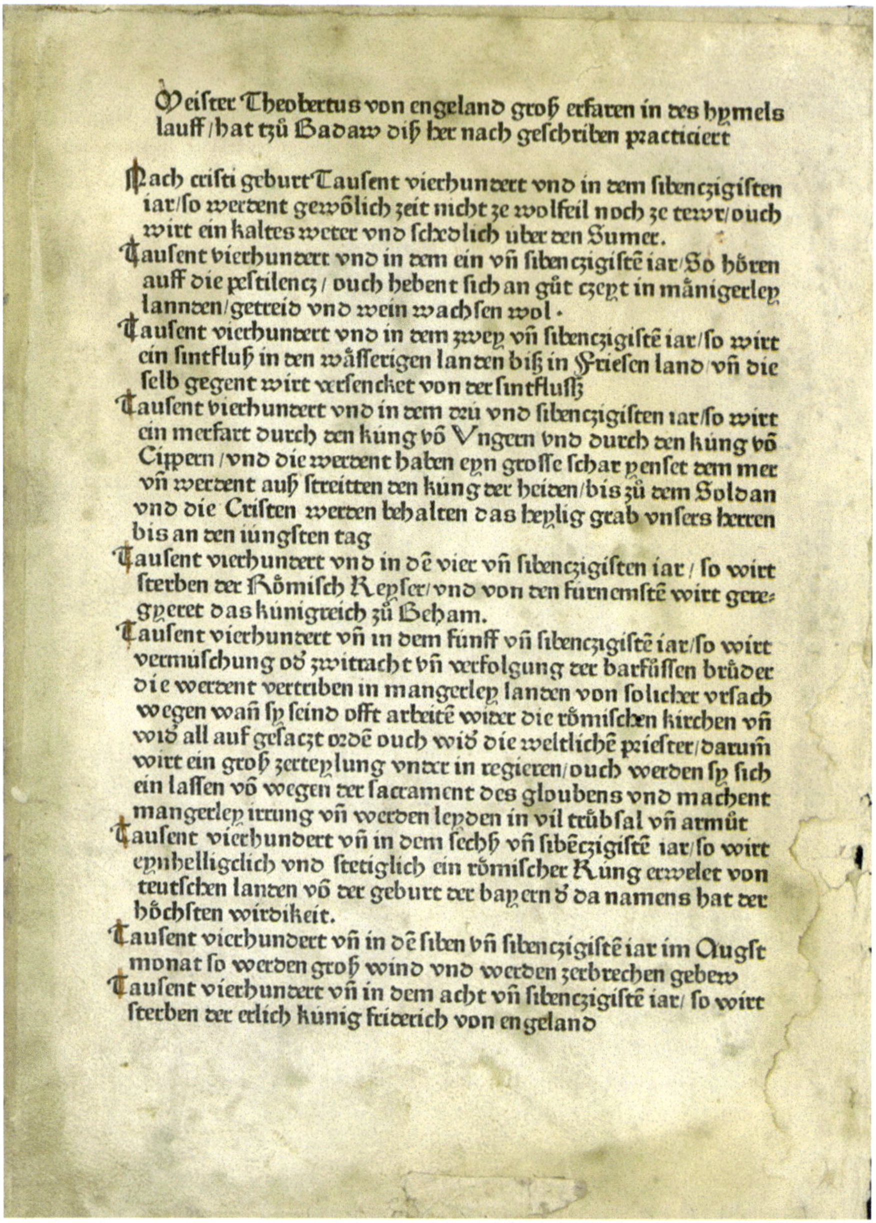

Meister Theobertus von engeland groß erfaren in des hymels
lauff / hat tzů Badaw diß her nach geschriben practiciert

Nach cristi geburt Tausent vierhundert vnd in dem sibentzigisten
iar / so werdent gewölich zeit nicht ze wolfeil noch ze tewr / ouch
wirt ein kaltes wetter vnd schedlich uber den Sumer.
Tausent vierhundert vnd in dem ein vñ sibentzigistē iar / So hören
auff die pestilentz / ouch hebent sich an gůt tzeyt in mänigerley
lannden / getreid vnd wein wachsen wol.
Tausent vierhundert vnd in dem zwey vñ sibentzigistē iar / so wirt
ein sintfluß in den wässerigen landen biß in Friesen land vñ die
selb gegent wirt versencket von der sintfluss
Tausent vierhundert vnd in dem drü vnd sibentzigisten iar / so wirt
ein merfart durch den küng vō Vngern vnd durch den küng vō
Cippern / vnd die werdent haben eyn grosse schar yenset dem mer
vñ werdent auß streitten den küng der heiden / bis zů dem Soldan
vnd die Cristen werden behalten das heylig grab vnsers herren
bis an den iüngsten tag
Tausent vierhundert vnd in dē vier vñ sibentzigisten iar / so wirt
sterben der Römisch Keyser / vnd von den furnemstē wirt gere=
gyeret das künigreich zů Beham.
Tausent vierhundert vñ in dem fünff vñ sibentzigistē iar / so wirt
vermüschung od̄ zwitracht vñ verfolgung der barfüssen brůder
die werdent vertriben in mangerley landen von solicher vrsach
wegen wañ sy seind offt arbeitē wider die römischen kirchen vñ
wid all auf gesatzt ordē ouch wid die weltlichē priester / darum̄
wirt ein groß zerteylung vnder in regieren / ouch werden sy sich
ein lassen vō wegen der sacrament des gloubens vnd machent
mangerley irrung vñ werden leyden in vil trůbsal vñ armůt
Tausent vierhundert vñ in dem sechß vñ sibētzigistē iar / so wirt
eynhelligclich vnd stetiglich ein römischer Küng erwelet von
teutschen landen vō der geburt der bayern d̄ da namens hat der
höchsten wirdikeit.
Tausent vierhundert vñ in dē siben vñ sibentzigistē iar im Augst
monat so werden groß wind vnd werden zerbrechen gebew
Tausent vierhundert vñ in dem acht vñ sibentzigistē iar / so wirt
sterben der edlich künig friderich von engeland

Fig. 5 The earliest known appearance of the German verb "practicieren" in print. Theobertus of England, *Prophesy for the years 1470–1478* (Augsburg, c. 1470), VE15, T-8. Württembergische Landesbibliothek Stuttgart, Signatur: Inc. fol.15475, fol. 1r

context of prophecy. Around 1470 Augsburg's first printer, Günther Zainer, issued a broadside bearing the title "Meister Theobertus von engeland groß erfaren in des hymels lauff, hat tzu Badaw diß her nach geschriben practiciert" (Fig. 5).[18]

[18] Since most of the broadside almanacs are anonymous and many, preserved only as fragments, lack incipits, I shall cite these imprints as designated in (Eisermann 2004), i.e., VE15, T-8, reproduced as Abb. 92. Zainer's earliest known imprint dates to 1468 (Geldner 1968–1970, i: 132–7).

Nine short paragraphs follow, prophesying events for years 1470 through 1478. Floods, plague, death of the Holy Roman Emperor, even a splitting of the Franciscan Order are mentioned, all without reference to astronomical or astrological phenomena. Prophecy, purporting to bring God's words to contemporary readers, had of course richly diverse medieval roots; with printing the genre flourished (Reeves 1969; Green 2012a). By 1470 it would appear as if prophets "practiced" their craft, at least in German.

Four presumably earlier manuscript versions of this text are known. The two Latin exemplars list the years as "lxx" to "lxxviiij" and "71" to "78." Karin Schneider has suggested that these texts refer to events for the years 1370–1379, as they accurately predict several events in that period.[19] Neither Latin manuscript employs the term "practica." The German manuscripts push the prophesies forward a century, to the years 1470–1478, as does Zainer's broadside. Munich, Clm 3586, fol. 260r–v provides a version of Theobertus's text that is very close to that in the broadside. A Latin phrase skeptically introduces the text, using our term as a noun: "Practica magistri theoberti anglici sequitur quam Padue calculat sed parum habens verum in se ut experimento probatum est." This *Sammelhandschrift* contains many theological, humanistic and monastic texts, variously dated to the 1470s and linked to the University of Erfurt. The other German manuscript, Innsbruck, Servitenkonvent I b 28, fols. 176v–77v, offers a slightly more elaborate text with a German incipit: Es ist zu wissen, daß ain mayster mit namen Teobart In astronomey auß engellandt der prodicirt [!] auß den waren castellacion der planeten von den lauffen der Jaren…." This manuscript dates to the 1460s; the verb "produciren" may well be a scribal error for "practiciren;" never have I seen the former in such a context. In any case, the German versions of Theobertus's prophecy, probably written in the 1460s, are referred to as "practica," as something that is calculated or derived from the true positions of the planets. Although none of the extant versions offer any astrologically-founded prophecies, the German scribes associated the term "practica" with prediction.

The earliest German imprints of the annual prediction, authored by Johannes Glogoviensis (1445–1507) and printed in Nuremberg for 1477 and 1478, are extant only in fragments that provide no evidence of their designations. Johannes's next German edition, printed in Leipzig for 1481, combined the Latin and German designations; its incipit reads: "… Judicien … durch magistrum Johannes von großen Glogaw gepracticzireth." Wenzel Faber (ca. 1455–1518), who would become the most prolific incunabula author of both almanacs and practica, described his earliest known imprint of the latter, for 1482 printed in Leipzig, as "eyn lobliche weissagung der wirckung der planeten von der vorburgenheyt [Verborgenheit] der vorbürgen gestirn nach erfarung der alden weiszen welhe disz Jaers." Faber described his 1484 German edition as "was der himmelouff … vorheisschet" (Glogoviensis 1477, 1478, 1481; Faber de Budweis 1482, 1484; Green and Duntze 2013).

An otherwise unknown author Konrad Rengart, who referred to himself as a former student at the university in Paris, wrote a Latin prognostication for 1483, printed in Augsburg by Christmann Heyny (closely associated with G. Zainer) that named itself at the top of the first folio "Practica Parisiensis." Several lines later on this same page we read: "Ego Conradus Rengart pronosticationibus illis patefacere." The German edition of this text, printed in Augsburg by Hermann Kästlin (also associated with Zainer), enthusiastically entitles itself "Practica des hochberumpten meisters in d[er] astronomia genant meister Cunrat Renngart zu pareiß" and concludes "Hie endet dise practica gepracticiert vnd gemacht zü pareiß." Copying his quantitative, astronomical data from Regiomontanus with the meridian adjusted to

[19] The deaths of "Urbanus" (Pope Urban V died in 1370) and an English King "Edwardus" (Edward III died in 1377). See Bayerische Staatsbibliothek, Clm 18770, fol. 189r; Cgm 754, fol. 201r; (Schneider 1984, 262–3). In a personal communication to the author, Jonathan Green wondered whether this prophesy might have been produced in 1378, at the death of Holy Roman Emperor Charles IV, by someone in Bavaria lobbying for a Wittelsbach successor. In the German versions authored presumably a century later, the English King Edward is replaced by "Bertholdus" (Bayerische Staatsbibliothek, Clm 3586 fol. 260v, a manuscript from Augsbrug!), by "Eberhart" (ULB Tirol, Innsbruck, Servitenkonvent I b 28, fol. 177v), or by "Friderich" (VE15, T-8). See (Rauner 2007, 378–96).

Paris, as per Regiomontanus's instructions, Rengart presumably intended his title to signal the place for which his predictions were applicable. Nothing in the content of Rengart's text, however, differentiates it from the Latin or German prognostica of the previous decade. Presumably Rengart or his Augsburg printers understood the terms "practica" and "prognosticum" as equivalent (Rengart 1483a, b). Judging from extant incunabula, we conclude that the German practica was born in Augsburg in 1483.

Several years later, Rengart's terminology appears in a 1485 German text by Leonhard Seybold, printed in Augsburg with the incipit: "Practica über die stat augspurg yedoch andere land vnd stet darmit eingeschlossen" (Seybold 1485).[20] The next year, a German "Practica von Ingelstat" was printed in Augsburg by Anton Sorg (ca. 1430–1493) and a German "Practica lipcensis" was printed in Nuremberg. In 1487, Faber for the first time entitled his German edition, printed in Augsburg by Sorg, "Practica von Leipßck;" his Latin editions for that year, printed in Leipzig, were entitled "Iudicium lipsense" (Götz 1486; Pollich von Mellerstadt 1486; Faber de Budweis 1487a, b).[21] Hence, only after more than 10 years and more than 140 editions would the term "Practica" emerge as the standard title for the German editions of the genre.[22] Augsburg printers seemed to have played a critical role in the emergence of this title. Interestingly, none of the incunable practica I have seen reflexively discuss the meanings of the titles they give themselves. And the term practica, as introduced in astronomy/astrology during the final quarter of the fifteenth century, does not seem to have been paired with the long-established theorica for those disciplines.

For the remainder of the incunabula period, use of the term "practica" generally tracks the language of the imprint. Of the 83 German editions extant for the years 1490–1501, 63 employ the term "practica" and only 5 refer to themselves as "prognostica." Of the 119 Latin editions for those years, 83 are named "iudicium" or "pronosticon" and only 9 are "practica" (all printed north of the Alps). After 1500 as Latin editions disappear, the noun "practica" would come to dominate the title pages and hence to give its name to the genre (Green 2012b, 4). I have found only two incunable editions that deploy a verb form of our term. An otherwise unknown author signed his colophon for 1491: "Gepractiziert vnd geendet durch maister Bartholomeo von aulon in artzny doctore." Johann Glogoviensis signed his edition for 1500: "… vnd den menschen czu nucz gepractiziert" (Scherenmüller 1491; Johannes Glogoviensis 1500).[23]

According to the dictionaries of the Grimm brothers and Matthias Lexer, the verb *practicieren* and the noun *Practik* entered Middle High German at the end of fifteenth century precisely in the prognosticatory, calendrical context we have described above. Citing du Cange (ref. 10 above), the Grimms suggested that the German usage was borrowed from the medieval Latin term "practica," referring to the exercise of any procedure (Grimm 1854–1960: 7, 2052). The lexicographers found that the German verb soon came to mean any regular activity or art that required sustained or special training, such as that performed by a calculator (*Rechenkünstler*), calendar-maker, lawyer, physician, surgeon, cleric, poet, notary or astrologer. Interestingly, in the conflict-riven, polemical print worlds of the early sixteenth century, meanings could also tip into the pejorative; one who practiced was sly, conspiratorial, not trustworthy, disreputable, a trickster. Although I have not exhaustively examined early usage of the term *practicieren* beyond prognostication, several examples illustrate this pejorative inflection, undoubtedly reflecting learned assumptions about social status and hierarchy.

---

[20] Since Seybold included no quantitative data, I cannot determine whether he prepared the practica for the meridian of Augsburg.

[21] For both editions in this year, Faber's syzygy times are computed very imprecisely compared to those in other years, for a meridian incorrectly located about 10 min of time west (rather than east) of Nuremberg. See (Kremer 2003b).

[22] For examples for the years 1488 and 1489, see GW 1903, M34698, M16475, M34700 (Low German), M50715 and M4084320.

[23] Use of the terms "almanac," "practica" and "prognostication" would continue shifting over the subsequent centuries. By 1600 in England, for example, multi-leaved annual pamphlets were often entitled "A New Almanacke and Prognostication." See (Kassell 2011).

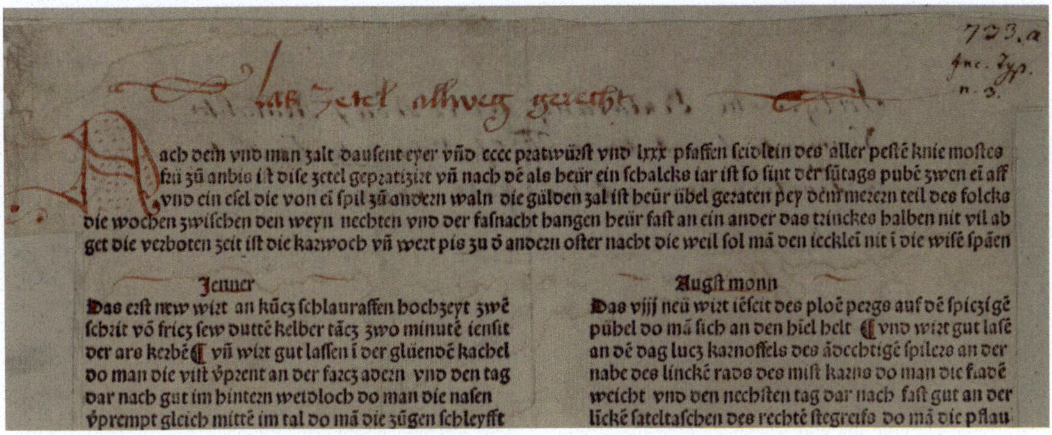

Fig. 6 A broadside almanac in standard format with parodied content. Hans Folz, *Satirical almanac for the year 1480* (Nuremberg, 1480), VE15, F-44. Munich, Bayerische Staatsbibliothek, Einbl.Kal. zl, fol. 1r

We begin with parodies of the annual predictions that deploy our term practica. Already in the early 1470s, just as printed broadside almanacs began to appear but before the earliest known printed practica, a parody of both forms appears, extant in a single manuscript. Consisting of 12 paragraphs for the new moons of an unspecified year, this anonymous text opens with what soon be a formulaic incipit: "Dieser Almenach ist gemacht vnd practizie…." But the text continues: "… dry milen vor vasnacht VIII tag neben der leymgruben. Die suntagsbüchstaben sind gauchs fliegen. Die gulden zal ist vngerechnot won sy ist klein … die wochen zwüschen dem Cristage vnd phaffenfasnacht sint VII brâdwürst vnd zü einer ietlichen 1 mas wins."[24] Each of the computistic elements found at the head of the broadside almanac—the Sunday letter, golden number, number of days between Christmas and Lent—appear in this manuscript, but the content of those elements is parodied (Werner 1888; Pfister 1990, 49). A similar broadside parody, printed in 1480 in Nuremberg by the well-known German *Meistersinger* and satirist Hans Folz (ca. 1437–1513) (Fig. 6) also uses the verb practizieren to describe itself:

> Nachdem und man zalt dausent eyer unnd cccc pratwürst und lxxx pfaffenseidlein des aller pesten kniemostes frü zum anbis ist dise zetel gepratizirt. und nachdem als heür ein schalcksiar ist, so sint der suntagspuben zwen: ein aff vnd ein esel, die von eim spil zum andern waln. die gülden zal ist heür übel geraten pey dem merern teil des folcks … VE15, F-44, transcribed in (Fischer 1961, 438). [25]

Folz's sheet continues with one paragraph per month, overflowing with irreverent nonsense words and hapax legomena. At the end it announces two eclipses for the year, one for nuns, the other for monks. Such parodies of the calendrical genres, Germanist Silvia Pfister has suggested, might be read less as attacks on the idea of astrological almanacs or practica and more as criticism of those who do not properly ply these genres; she labels the parodies "Katechese e negativo" (Pfister 1990, 312–409, quote at 314; Huey 2012, 16–7).

[24]"This almanac is made and practiced three miles before Shrove Tuesday [and] eight days next to the clay pit. The Sunday letters are flying fast, the golden number has not been calculated because it is small … the weeks between Christmas and Quinquagesima are seven bratwurst and one measure of wine for each." This manuscript was written by Gallus Kemli (1417–1481), a peripatetic Benedictine monk and avid book collector and copiest, especially of theological material. Cf. *Vefasserlexikon*, s.v. Kemli.

[25]Cf. *Verfasserlexikon*, s.v. Folz; (Pfister 1990, 221–4). "This sheet is made [practiced] for one thousand eggs and 400 sausages and 80 priest's measures of the best knee wine [pun on new moon?] for early breakfast, and since this is a fool's year [pun on leap year], the Sunday boys [pun on letters] are two: a fool and a donkey that makes pilgrimages from one game to another. The golden number this year looks bad for most people…."

Our term appears ever more pejoratively in the texts of later satirists, such as Sebastian Brant's (1457–1521) *Narrenschiff*, first printed in 1494. His chapter, "Von achtung des gstirns," chides those "fools" who seek to learn the future from the planets' courses, who do not acknowledge God's providential aims (Brant 1494, sig. l2r–l3r; 1944, 217–9).[26] The critique of astrology is conventional:

Eym kristen menschen nit zů stat	A Christian true should never heed
Das er mit heyden künst vmb gat	Base heathen arts of any creed:
Vnd merck vff der planeten louff,	One can't by scanning planets say
Ob dyser tag sy gůt zům kouff,	If this be our propitious day
Zů buwen, krieg, machung der ee,	For business, war, or marrying,
Zů früntschafft, vnd des glychen me	For friendship or for anything.
All vnser wort, werck, tůn vnd lon	Whate'ver we've done, where'ver we've trod,
Vfz gott, jun gott, alleyn sol gon[.]	Our conduct should rely on God.

Printers are also to blame, continued Brant, as are the "disgraceful" practica they distribute. And of course the true problem lay with the readers who want to be deceived:

Inn narrheyt ist all welt ertoubt	With folly all the world's insane,
Eym yeden narran man yetz gloubt,	It trusts the fool, the most inane.
Vil practick vnd wissagend kunst	In practica and arts of prophecy
Gatt yetz vast vfz der drucker gunst	In these the printers profit see,
Die drucken alles das man bringt	They'll print whatever one may bring
Was man von schanden sagt vnd singt	Yes, e'en the most disgraceful thing.
Das gott nůn als on straf do hyn	Such things go scot-free, are believed,
Die weltt die will betrogen syn.	The whole world wants to be deceived.

Our term also appears in non-astronomical contexts. An anonymous broadside, printed in Leipzig around 1498, bears the title: "Die beste Practica ich mein, trifft an alle menschen gemein" (The best practica that I have in mind affects everyone equally). Two facing woodcuts depict an evil and a good angel, each holding a mirror before well-clad on-lookers (Fig. 7). No image appears in the evil angel's mirror and a German verse below urges the reader to enjoy the present and not think on death or priests. A human skull appears in the good angel's mirror. Her verse warns that the devil's mirror kills the soul and that one should always look to God. In this moralizing sheet, the title might be the evil angel's voice and the "beste practica" his dishonest mirror.

Other well-known early sixteenth-century German authors used our term. For example, Martin Luther (1483–1546) in a 1520 public letter to Pope Leo X (1475–1521) castigated his theological adversary, Johann Eck (1486–1543), "eynen sonderlichen feynd Christi vnd der warheyt," who the previous year was widely regarded to have bested Luther's defender Andreas Karlstadt (1486–1541) in the Leipzig Disputation (Luther 1520b: sig. a4v). Luther's German text uses our term:

> Denn Eck mit seinen Lugen, Sendbrieffen vnd heymlichen bractiken die Sache alßo vorpitteret, vorwerret vnd zurschellet, das auff wilch seytt das Ortheyll gefallen were, eyn grösser Fewr, on Zweyffel sich erzundet hette. Denn er sucht Rum vnd nit die Warheyt.... (Luther 1520b: sig. B1v)[27]

[26] I have lightly revised the latter's translation.

[27] "Eck with his lies, open letters and secret practica had embittered, hindered and smashed the whole affair, so that regardless of the judgment [at the disputation], a great fire undoubtedly would have been ignited since Eck sought fame and not the truth...."

Practica … mein trifft an alle menschē gemein.

Angel' der bösz Engel:
Sich disen spiegel frölich ann.
du schões wyb du stoltzer mā.
Sich wie fün du bist geschaffē
Folg mynē rot vñ nit dē pfaffē
Zier dich mit cleid vñ ouch am
dz dir dy welt am̄ütig blib/lib
wo du fröud findest die behab
wān du alt würst so lasz dā ab
Hab fröud vñ lust jn diser zyt.
Bisz dz die welt dir vrlob gyt
Hab kein gedenckē an den dot
Bysz güter ding dz ist min rot
Noch ere vñ güt soldu streben.
Du würst noch lange jor lebē
Losz dir den dot nit bilden jn
Du kōst noch wo so es müsz/
syn.

Angel' der güt Engel
O mēsch betracht zů aller frist
Also wirdstu als disz ding ist.
Nitt ker dich an desz tüfels rot
Sin spiegel bringt der selen dot
So du wenst am bestē sin/ hin
Dan komt d' dot vñ zuck dich
wer disen spiegel schowet eben
Der midet sünd vñ fidt dz lebē
Da würdt er got allzit schowē
O mēsch des mastu dich frowē
Beschow den gerner flissiclich.
wz du do finst dē würstu glich
Veracht die weltlich vppikeit.
So würdt din sel zů gott bereit
Dā mag din hertz jn friidē stā
Vnd gibt dir got desz hymels
kron.

Fig. 7 The mirror as a moralizing "practica," revealing viewers' inner dispositions. *Practica des bösen und guten Engels* (Leipzig, c. 1498), VE15, P-242. Württembergische Landesbibliothek Stuttgart, Signatur: Inc. fol.13312 b, fol. 1r

Luther's Latin text, however, employs somewhat different terms: "… quandi Ecij mendaciis, simulationibus, technis, omnia ubique errant turbatissima, exulceratissima, confusissima…." (Luther 1520a, sig. a3v). Luther, who prepared his own translation, apparently understood *practica* as *techna*, inflecting both terms negatively as referring to cunning, trickery, and artifice rather than the exhibition of skill or craftsmanship.

The *Meistersinger* Hans Sachs (1494–1576), in a 1536 carnival play about Queen Esther, associated the term with Hamon, the personification of evil in this biblical story (Keller 1870–1908, i: 132):[28]

Zumb dritten bey dem Hamon secht	Thirdly, see Hamon
Aller listigen schmeichler art	Using all kinds of sly, flattering means
Sich streichen umb die herrschafft hart	To ingratiate himself with the rulers
Durch practic, lüg und hindterlist,	Through practice, lies and wile,
Der denn zu hoff unzalbar ist	Who then becomes indispensable at court
Und manchen mann auff dfleisch-banck geben,	And puts many a man on the torture rack,
Bringen umb ehr, gut, leib unnd leben,	Depriving them of honor, goodness, body and life,
Das doch zu-letzt an in außgeht,	But which finally comes to an end with him
Weyl nichts nit lang verborgen stet.	Because nothing can long remain concealed.

Like Luther, Sachs associated our term with unsavory characters and behaviors. The Grimms provided additional examples of such usage that we need not explore here. Our brief etymological review suggests that the German term *practica* emerged in the final third of the fifteenth century. The printed genre it denoted was not paired with the much older astronomical *theorica* but rather described a new, pragmatic, periodic literature that offered astrological, political, religious and social predictions for a single year. Parodies of the genre appeared almost immediately, testifying surely to the speed with which the texts had flooded the new print culture.[29]

The term used to describe our second genre, the almanac, has a longer history in Latin, even if its presumed Arabic roots remain unknown. Although the al- prefix seemingly points toward Arabic or Hebrew origins, Moritz Steinschneider could find no relevant roots in those languages and guessed that a thirteenth-century Latin author may have invented an Arab-sounding term as a synonym for the Greek ephemerides or a table of astronomical positions at regular intervals of time (days, months, years, etc.). George Sarton wondered whether the term al-manâkh might have Egyptian origins, meaning climate or calendar. In any case, by the twelfth century, the term appears in al-Andalusian tables listing for regular intervals of time the positions of the luminaries and the planets.[30] After the creation of the Alfonsine Tables in the 1270s, the term regularly can be found, in the Alfonsine corpus, in headings for tables of positions at regular intervals: "Almanak solis pro 4 annis scil. 1341–1344" (W. Reed) or "Almanach … annis 1336–1380 accomadatum" (John of Saxony) (Steinschneider 1888; Thorndike and Sarton 1928; Vida 1934; Millás 1947, 119; Benjamin and Toomer 1971, 374–5; Chabás and Goldstein 2000, 16–9; Chabás and Goldstein 2003, 282–9).

The anonymous authors of the earliest printed broadside almanacs described their sheets as tables: "Tafel der newen vnd vollen mond" (1470 in Augsburg, 1473 in Nuremberg) or "Tabula conjunctionum et oppositionum" (1473 in Nuremberg); and the syzygies times were listed in rows and columns. The term almanac first appears in a Latin sheet for 1474 printed in Erfurt: "Almanach continet in se omnes coniunctiones et oppositiones vera sol et lune cum diebus electis minucionum nec non farmacia," a title that nicely describes the medical and astronomical contents of the sheet. Our term does not appear again until 1481 in a German edition printed by Peter Drach (1455–1504) in Speyer (… disses

[28]I thank Bruce Duncan for help with this translation.

[29]I cannot here explore the conceptual and social links between the astrological practica and the "books of secrets" tradition that would emerge in the sixteenth century, featuring vernacular books of recipes, natural magic, and individual self-improvement for artisans in which the term "practical" frequently appears (Eamon 1994).

[30]According to (Benjamin and Toomer 1971, 375), the earliest known use of the term is found in a text attributed to Abraham Ibn 'Ezra: "Sed vere sunt tabule que singulis diebus docent coequare planetas vel a tempore determinato dant rationes componendi almanac, id est tabulas per quas semel factas per totum annum planetas coequatos habebis."

Almanach inhelt wan ein iglich liecht new oder vol wird darzu die vßwelten tag zu aderlassen oder ertzenneij zunemen) and as a title the next year in his German edition (Almanach anno domini VCCCCLXXXII nach dem waren lauff Sonen vnd mon). By the mid 1480s, the term regularly appears in the top lines of both German and Latin broadsides although "Tavel" also is used (e.g., 1486 and 1487 in Bamberg). Apparently, the term "almanac," referring to a list of astronomical phenomena at regular time intervals (syzygies in the broadsides) and a list of bloodletting rules, was borrowed from Alfonsine tabular manuscripts of the previous century (VE15, A-146, A-156, A-157, A-162, A-225, S-233, A-280, A-292, A-284).

Generally, incunable broadside almanacs do not include general predictions for the year. But I have found two examples of hybrids. A large, closely printed sheet for 1488 offers syzygy times, bloodletting rules, and extensive weather and other predictions for disease, *Stände*, etc. Its colophon names both author and printer: "Den Almanach hat gemacht meister Marx Schynagel/Dyse practick hat getruckt Erhart Ratdolt." For 1490, this author offered a similar edition with the colophon: "Die practica vnd almanach hat gemacht magister Marcus Schinagel" (VE15, S-9, S-10).

This review of titles and descriptions suggests that the authors of the genres of the practica and almanac, which began to flourish in the 1470s with the advent of printing, saw their work as embodying new types of knowledge, different from the astronomical *theorica*. Although the almanac had roots in the computational procedures of fourteenth-century Alfonsine astronomy, the practica was quite new, both in content and name. And the latter name emerged in a German rather than in the Latin context of the *theorica/practica* dichotomy.

2 Fifteenth-Century Almanacs and Practica and Their Trading Zones

Our questions in this section are simple to state but rather difficult to answer with the surviving source materials. What did the early almanac or practica makers need to know? Did they rely on theoretical or scholarly knowledge recorded in books and transmitted in university lectures? Did they employ empirical knowledge, encoded in written (or oral) recipes or collections of rules? Did they engage in "sharing practices" among discrete, different worlds of knowledge? Were special "how-to" manuals created from this sharing to ease the task of preparing these new printed genres? In sum, can we consider the incunabula almanacs and practica as examples of practical knowledge produced in Long's trading zones?[31]

Rarely do the extant incunable almanacs or practica explicitly discuss their sources or their epistemic foundations. And when such traces do appear, they can be vague. For example, a German practica for 1486 opens: "Practica von Ingelstat. Nach dem vnd spricht der sinnreichest maister vnd lerer Arestotiles das an ainem wesenlichen ding himel vnd die gancz natur hangen ist…." A meandering defense of celestial influence, under God's rule, follows. Before outlining the six chapters of the practica, the author identifies himself: "… so ich johannes götz von augspurg nach der weisen erfarung vnd regel beschriben was das gestirn den vndersten mit kummlichait verhaischet" (Götz 148, fol. 1r–v).[32] But readers are not told what comes from experience and what comes from rules, presumably textual sources. This is Götz's only known imprint (preserved as a fragment); we know only that he completed a baccalaureate in Ingolstadt and in 1482 matriculated at the university in Leipzig. In a practica for 1494, the most prolific incunable author of both almanacs and practica, Wenzel Faber of

[31] For a useful introduction to some of these issues, see (Grafton 1999).

[32] "Practica from Ingolstadt. Here speaks the ingenious master and teacher Aristotle about the essential thing on which heaven and all of nature depend … I Johannes Götz of Augsburg describe, according to wise experience and rules, what the stars conveniently promise the lower beings," See (Erler 1895–1902, ii: 279).

Budweis, similarly described his method: "Hec ergo brevissime in prognostico dicta ex philosophorum regulis atque experimentatis sciencie astrologice accepta sunt et collecta" (Faber de Budweis 1494, sig. B4v).[33] Faber, in 1494 a professor at the University of Leipzig, clearly suggests that both (text-based?) rules and experience shape the practica. In what trading zones did Götz, Faber and their fellow authors participate as they drafted their pragmatic texts?

To explore this question we turn first to the authors of the incunable almanacs and practica. The most striking difference between our two genres, beyond the printed format, arises in their authorship. Printed practica generally name their (single) authors whereas only about one in four of the extant almanac editions are signed. This difference, we might guess, undoubtedly points to different social spaces occupied by these genres. A preliminary prosopography of the named authors may reveal some contours of these spaces.[34]

Of the 466 known incunable practica editions, only 23 cannot be linked to a particular author. Of these, 22 are extant only as fragments, lacking the initial or final pages where authors' names generally appear; presumably complete copies of these editions would name authors. Hence, only one intact practica edition is known to be anonymous, a German *Pratik von Paris* for 1487, printed in Augsburg (*Practica von Paris, dt.* 1487). In the other extant editions, a total of 75 different authors are named. Over half of these men specified their academic titles in the practica; a third of them either named themselves as professors or are known to have taught at a university. Six universities north of the Alps and seven Italian universities are represented among the practica authors.[35] As noted above, the genre of the annual prognostication probably arose at the Italian universities, whose statutes from the early fifteenth century increasingly require the masters (usually of mathematics, astronomy or medicine) to prepare an annual prognostication for the students as well as local townspeople. On the other hand, practica by some of the most widely published university professors could find printers far beyond the author's location. For example, prognostica by Petrus Bonus Avogarius (1467–1506), longtime professor in Ferrara, were printed not only in Ferarra but also in Bologna, Venice, Milan, Florence, Perugia and Rome. Nonetheless, the university clearly was a leading site for the production of fifteenth-century practica, especially those prepared by the most prolific authors (see Table 1).

The second most significant site of practica production was the court. Fifteenth-century popes, holy Roman emperors, kings, counts, dukes and lesser nobility often patronized astrologers and/or personal physicians as advisors, healers and preparers of local calendrical material. At least 13 of the known practica authors enjoyed such relationships, serving (or at least claiming to serve) patrons ranging from Pope Alexander VI (Bonetus de Latis, d. ca. 1515), emperors Frederick III (Johannes Canter, 1424–1497; Johannes Lichtenberger, ca. 1440–1503, Johannes Basilius Augustinus, fl. 1489) and Maximilian I (Johann Stabius, 1460–1522; Josephus Grünpeck, 1473–1532) and kings Ferdinand of Aragon (Laurentius Bonincontrius, 1410–1502) and Alfonso of Aragon (Antonio Arquato) to the duchy of Urbino (Jacobus de Spira, Paul of Middelburg, 1445–1534) and Henry the Elder of Münsterberg, Count of Kladsko (Georg Leimbach). As has long been known, these astrologers often played convoluted, yet crucial political and medical roles in court culture.[36]

[33] "Therefore these things that have been most briefly declared in this prognostication are taken and collected from the rules of the philosophers and experience in the science of astrology."

[34] For this preliminary survey, I have not explored university or city archival records but simply have relied on standard biographical reference works, including (Thorndike 1923–58; *Verfasserlexikon*; Ghisalberti 1960–; Polska Akademia Nauk 1935–) and the hundreds of earlier biographical dictionaries indexed by de Gruyter 2016.

[35] Cracow (5 authors), Ingolstadt (2), Vienna (2), Leipzig (2), Tübingen, Heidelberg, Bologna (5), Padua (2), Cremona, Pavia, Ferrara, Florence, and Rome. One otherwise undocumented author, Guilemus Parronus, is described in a London-printed colophon as "in medicinis doctoris ac in astrologia professoris" but I have not been able to determine whether he was involved with any university (Parronus 1499, fol. 12v).

[36] The literature here is extensive; most recently see (Hayton 2015; Azzolini 2013; Grafton 1999).

Table 1 Most prolific authors of incunabula practica and almanacs

Practica	Career	Lat	Ver	Almanacs	Career	Lat	Ver
Faber, Wenzel	Prof Leipzig	27	30	Faber, Wenzel	Prof Leipzig	15	16
Virdung, Johannes	Prof Heidelberg	12	18	Angelus, Joh.	Prof Ingolstadt	4	7
Avogarius, Petrus	Prof Ferarra	13	16	Wonnecker, J.	City physician, Basel	0	10
Manfredis, H.	Prof Bologna	11	12	Hord, Jobst	Master ??	0	8
Paul of Middelburg	Prof Padua[a]	20	3	Virdung, Joh.	Prof Heidelberg	4	3
Scribanarius, M.	Prof Bologna	10	8	Eck, Paulus	Matriculated Leipzig	4	2
Polich, Martin	Prof Leipzig	8	9	Muncz, Joh.	Prof Vienna, canon	4	1
Dominicus, Maria	Prof Bologna	10	6	Mansfeldt, B.	City physician Munich	0	5
Joh. Glogoviensis	Prof Cracow	6	9	Schinnagel, M.	Master Cracow, clergy	1	3
Laet, Johannes	Physician Liége	8	5	Valentin	Monk OESA	1	3

Some of these authors pursued several "careers" and may not have engaged in those listed here during the entire span of their published editions. "Lat" and "Ver" (German, Italian, Low German, Dutch, Czech and French) refer to the languages of their extant editions

[a]Paul held the chair for astronomy in Padua from 1479 to 1481; he then entered the service of the dukes of Urbino, where he spent the remainder of his life. See (Heilen 2015)

At least ten practica authors served the church as canons, priors, monks or clergy. Another seven earned their living as physicians, city physicians, or personal physicians to particular patrons. One is known only as a printer's apprentice (Johannes Götz). Six are known only by their birthplace (e.g., Adam Eckstein from Schwäbisch Hall, Baptista Gemmatus from Cesena). And seven appear to have left no biographical traces in published records.[37] We might guess that these less well-documented figures would take us away from the university and court settings. Yet of the 29 editions authored by these 13 little-known authors, 19 are in Latin. Overall, more than half of the practica editions are in Latin (254); the vernacular editions include German (121), Italian (60), Low German (21), Dutch (7, Czech (1) and French (1). As measured by both social location of their authors and language, the incunabula practica seem learned in their orientation.

On the other hand, fully three-quarters of the known almanac editions are anonymous; in most cases the sheets provide neither printer's nor author's name. A total of 33 individual authors are named on the 128 signed incunable almanac editions. Thirty-two of these authors signed themselves with an academic title (e.g., master). Eight were city physicians, 12 university professors (or university related), 5 court astrologers or physicians, 4 clerics or monks, 1 printer, and 1 a "moler [Maler] vnd burger zu strossburck" (Hans Schrotbanck, probably a book illustrator and designer of woodcuts). Overall, two-thirds of the extant almanac editions are in the vernacular. A larger percentage (73 %) of the anonymous than the signed almanacs (58 %) are in the vernacular, which probably suggests fewer university-trained makers for the unsigned sheets. And the anonymous sheets show slightly less sophisticated astronomical computation than the signed sheets.[38] The incunabula almanacs, we conclude, are less Latinate and less learned than are the prognostications.

What might the content of the imprints tell us about the social and epistemic spaces of their making? Almanac-like materials appear soon after the beginning of typographic printing. Around 1457–58, the unknown "printer of the 36-line Bible"—an itinerant who had worked with Johannes Gutenberg (died 1469) in Mainz—issued three one-sided prints, a German cisiojanus (mnemonic verses in which

[37]Camillus Durantius, Benedictus Mysol, Agapitus Portius (Porcius, Porzio), Konrad Rengart, Rabbi Abba Salvadore and Carolus Susena.

[38]Twenty-three percent of the signed editions copy their syzygy times verbatim from Regiomontanus, without shifting his times to the local meridians of the places for which the almanacs were intended to be used; thirty-five percent of the unsigned editions copy these times verbatim. For issues of vernacularization, see (Crossgrove 1998; Demaitre 1998).

successive syllables point to the 365 feast days of the Christian calendar), a so-called "astrological planetary table," also German, a strange text whose historiography is too convoluted to discuss here, and a sheet of medical rules for the year 1457 (VE15, A-144, A-515, C-26). The latter contains three components that would define the genre of the broadside almanac: computistic information on the moveable feasts, syzygy times to hours, and bloodletting and medicine-taking rules for given days of each month, all specified for the year 1457 (i.e., an almanac becomes obsolete after 31 December). Presumably, this early printer sought to test the marketability of various calendrical broadsides. Only four incunable cisiojanus editions would appear; never again were "astrological tables" printed. Perhaps printers realized that the annual periodicity of the almanac would bring them more sales than a cisiojanus or sheets with extended shelf lives (Eisermann and Honemann 2000; Wehmer 1948)

The next extant printed almanac is for 1462, probably issued in Vienna since its meridian is specified as "auff Wienn" (see Fig. 3).[39] The sheet, preserved in a unique copy, divides the three types of content (computistic, astronomical, medical) into separate sections of the printed surface and defines what would become the standard format for the incunable almanac. It also presents two images of eclipses but the text does not refer to such phenomena and the images are not representational. Later incunable almanacs would regularly include realistic depictions of the eclipses predicted for the year (Fig. 8).[40] The sheet names neither its printer nor author, a practice that, as noted above, would mark most of the incunabula almanacs.

How might these almanac makers have assembled the content presented in their broadsides? Presumably, the computistic information was extracted from manuscript or printed sets of computistic or calendrical tables, such as John of Gmunden's (ca. 1380–1442) from ca. 1430, or Regiomontanus's printed calendars and ephemerides from the 1470s. Extensive instructions or canons usually accompany such tables, which often graphically display the cyclical information. Essentially, one inserts the given year into this graphical and tabular material and reads out the Sunday letter, the golden number, and the dates of the moveable feasts; we might describe this as a recipe-following procedure. Computus was a bookish activity that required only a rudimentary understanding of astronomical theory and the Christian calendar; the computist did need access, however, to the relevant tables.

Assembling the astronomical information for almanacs was also a bookish activity. The 1462 sheet specifies the day, hour and minute (more precision than the 1450s imprints) of the twenty-five syzygies for the year, according to true motions of the luminaries (Gmunden's calendar gives these times for mean motions). These data are presented in a tabular format that quickly became standard for the incunable almanacs, in which first the date and then the time is given for each syzygy, usually with the new moons in one column and the full moons in another (Fig. 3):

New man [manuscript in red ink]	New Moons
Samstag vor blasy 3:34 nach mittag	Saturday before St. Blasius, 3:34 p.m.
Montag nach mathie 9:46 nach mitnacht	Monday after St. Matthew, 9:46 a.m.
Mitich nach vekundung m[arie] 2;33 nach mitnacht	Wednesday after Annunciation, 2:33 a.m.
Phincztag nach georgy 5:06 nach mitag …	Tuesday after St. George, 5:06 p.m.

The numbers specifying each of the times are manuscript in the 1462 sheet, suggesting that the printer lacked access to the numerical data when setting the type (VE15, A-145).

[39] Assigning a name to its anonymous printer remains problematic and need not detain us here.

[40] The Alfonsine Tables predicted a partial lunar eclipse for 11 June 1462 that would have been visible in Vienna, but at its maximal extent, the horns of the crescent would point down, not to the left as depicted on the 1462 almanac. The Alfonsines predicted a partial solar eclipse for 20 November 1462. I have not computed to see whether it would have been predicted as full (as shown in the 1462 almanac) in Vienna. See (Mooney 1998, 155, 161).

Fig. 8 A broadside almanac with a woodcut depicting a nearly total lunar eclipse and its two ruling planets, Venus and Mercury. Wenzel Faber, *Almanac for the year 1500* (Leipzig 1500), VE15, F-30. Munich, Bayerische Staatsbibliothek, Einbl.Kal. 1500 am, fol. 1r

Computing syzygies (and eclipses) is perhaps the most difficult mathematical task in Ptolemaic astronomy. The intervals of time between successive mean syzygies are constant. But times of true syzygy can vary by up to 14 h from the times of mean syzygy, and over that duration the Moon's speed varies appreciably (at a precision of minutes of time or longitude). Syzygy computations thus require approximation and iteration (Chabás and Goldstein 1997; Kremer 2003a, b). The best minds of medieval Arabic, Hebrew and Latin astronomy sought mathematical tricks to reduce the complexity of these computations, but the task remained very labor-intensive.

What did the maker of the 1462 almanac know about these astronomical matters? Rather much, it seems, as we learn from an analysis of the syzygy times inserted by hand into that sheet. As a young student in Vienna, Regiomontanus in 1448 began computing for each year the times of syzygy (he also generally cast horoscopes for these times, charts used for weather prognostication). An autograph codex now in Vienna, ÖNB lat. 4988, contains his work for the years 1448 through the first 3 months of 1462. From the beginning, Regiomontanus showed himself to be a careful, even fastidious calculator, finding times of true syzygy computed with a double-entry table prepared by his teacher, Georg Peurbach, corrected to apparent solar time (sundial time) for the meridian of Vienna (80 time minutes East of Toledo). Regiomontanus would present similarly computed data in his calendars and massive ephemerides, printed c. 1474 and covering the years 1475–1506 and 1475–1532, respectively.

The 1462 almanac times match (to the minute!) Regiomontanus's syzygy times from ÖNB lat. 4988 (covering only January-March); the remaining 20 times for the year are computed but for a slightly different meridian (77 time minutes East) and a slightly different adjustment to apparent solar time. This leads me to conclude that the maker of the 1462 sheet was not Regiomontanus (as some scholars have suggested) but another Viennese mathematician who knew how Regiomontanus, who in September 1461 had left Vienna for Italy, worked. We need not call this astronomical knowledge "theoretical;" Campanus had called it "practical." Using astronomical tables to compute planetary positions (a procedure defined by Babylonian astronomers at least six centuries before Ptolemy) is recipe-following. But the syzygy- and eclipse-computing recipes contain so many branches, they require so many decisions along the way that depend on understanding the geometry (theory) of the models, that we might conclude that such computation would require, in addition to rule-following, some theoretical knowledge of the geometrical models of celestial motion. In any case, knowledge of times of syzygies is not empirical; by direct observation of the sky one cannot detect the time of syzygy to a precision better than perhaps half a day. Certainly one cannot observe such times to minutes of precision.

Hence, almanac-makers working before the printing of Regiomontanus's calendars would have needed access to the Alfonsine Tables or to a table, such as Peurbach's, for the correction of times from mean to true syzygy. My analysis of all the extant incunabula almanacs indicates that by 1477, their makers increasingly copied their astronomical information directly from Regiomontanus's handbooks, although many did shift the times to their local meridians (Kremer Forthcoming). With the aid of these handbooks, preparing the astronomical material for the almanacs would have been as straightforward and rule-following as was preparation of the computus material. In each case, the rules were well defined and explicit; in nearly every case, my recomputation of these data, following what I believe to be the relevant rules, generally matches (often verbatim!) the content found in the incunable sheets. The fifteenth-century makers and I appear to be working in the same algorithmic space, using the same fifteenth-century astronomical tables.

This is not the case, however, for the medical content of the almanacs. The greatest portion of the 1462 almanac is devoted to what I shall call bloodletting rules, i.e., to specifying dates for particular medical interventions. Offering only one general principle ("let the blood of young people during a waxing, of old people during a waning moon"), the sheet then becomes highly specific. For January (see Fig. 3) we read:

Samstag nach dem newen iar an die payn	Saturday after the New Year on the leg
Montag vor vnd der drey künig abent an die fuess	Monday before & evening of Epiphany on the foot
An der drey kunig tag vnd dar nach an das haupt	On Epiphany and thereafter on the head
An sand sebastian vnd agnesen tag an die lent	On St. Sebastian and Agnes on the loins
An sand paul pekerung abent an die scham	On the evening of St. Paul's Conversion on the genitals.

The almanac provides similar instructions for the remaining months of the year. How did our almanac maker formulate these rules? Medical texts on phlebotomy and astrology, of course, had circulated widely since the twelfth century. Some emphasized the zodiacal sign of the Moon, others the aspects (angles) of Mars and Saturn to the Moon, still others the passage of the seasons. A non-astrological tradition stressed the "Egyptian" or unlucky days, either 24 or 40 days of the year on which no medical interventions (bloodletting or medicine intake) were to be attempted.[41] Images of the zodiac man (Fig. 9), showing which zodiacal signs rule which parts of the human body, had circulated widely since the twelfth century.[42] From such (theoretical?) materials, authors (called

[41] Among others, these authors included Arnold de Villanova, Pietro d'Abano, Marus of Salerno, William the Englishman, Ramón Lull, Guy de Chauliac, and the ps.-Hippocratic *Astronomia ypocratis* (Bauer 1966; Gil-Sotres 1994; Chapman 1979; Keil 1957; Siraisi 2007; Wallis 1995).

[42] For a recent attempt to trace this image back to Babylonian sources, see (Wee 2015).

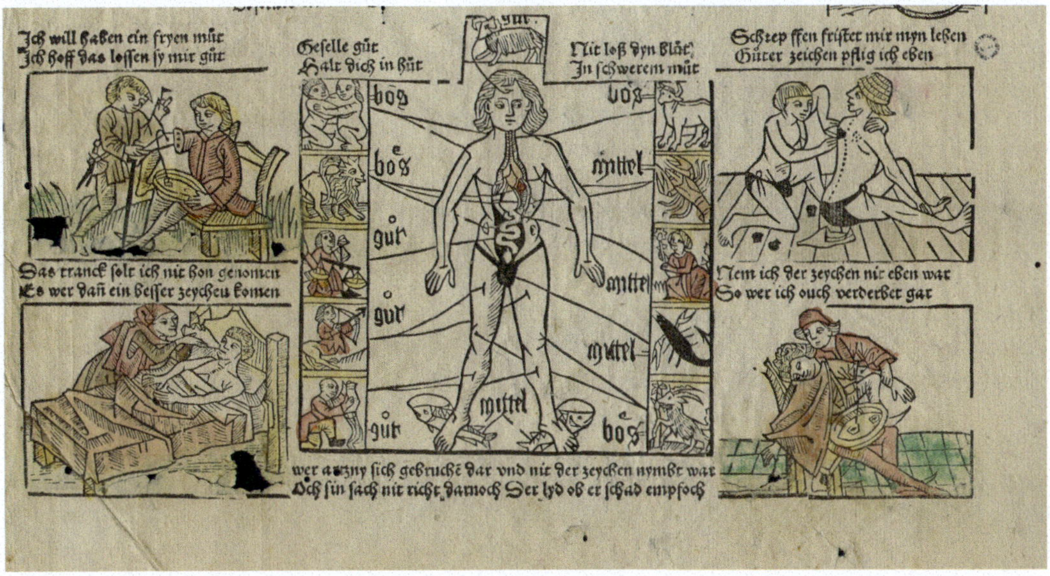

Fig. 9 The zodiac man and other instructions for bloodletting in Johannes Wonnecker, *Almanac for the year 1499* (Basel 1499), VE15, W-61. Courtesy of Basel, Universitätsbibliothek, AU V 13: 13/14 Folio, detail

"untheoretical" and "relatively unprofessionalized" by Faith Wallis) crafted monthly calendars of health advice. "In the month of March, one should drink sweet beverages, use pungent food and root vegetables, and chew one's food well. One should take sweat baths, undergo bloodletting, but not take a purgative, because a purgative produces chills" (Wallis 1995, 113; Riha 1985). For medicine grounded in an astrology of the Moon's motion or in unlucky days, dates and times mattered; the physician ignorant of astronomy, asserts the thirteenth-century *Astrologia Ypocratis*, is like a blind man stumbling about without a staff (Thorndike 1960, 106–7).

Yet since astrology offered multitudes of time-based medical recipes, disputes could arise among practitioners who emphasized different predictive features. Lynn Thorndike, for example, tells a delightful story of two Paris masters disagreeing over whether 2 January 1437 would or would not be a favorable day for bleeding. One had emphasized the Moon's zodiacal sign, the other its aspect to Venus. The university rector appointed two theologians to adjudicate, and they stressed a third astrological feature, the "terminus of Saturn" (Thorndike 1923–58, iv: 139–41). Hence, the incunabula almanac maker could not simply consult an authoritative recipe book to construct bloodletting rules in concert with the celestial motions of a given year. Rather, he had to combine quantitative astronomical data from the Alfonsine corpus with recipes selected from astrological medical texts to generate a year's set of bloodletting rules. As might be expected, I have found that in years for which several almanacs were issued, the editions often closely agree on the astronomical content yet diverge dramatically in astrological-medical content. Unlike its computistic and astronomical information, an almanac's medical materials cannot easily be reproduced by following an appropriate algorithm. Although Jonathan Green recently has called almanacs (and calendars) "factual," in contrast to the more "interpretative" practica, we must, I think, view the medical rules of the almanacs as derived from both recipe interpretation and experience (Green 2012b, 2).

How did these authors generate their bloodletting rules? Did almanac makers keep empirical records on bloodletting performed according to their annually revised rules? Might their own experience as surgeons have informed their construction of these rules? Did they "learn from experience," as our practica author Götz suggested in 1486? Might personal experience have shaped the choices of astrological features to emphasize in crafting the rules? To the best of my knowledge, no one has sought to unpack a given almanac maker's bloodletting rules, year by year, to see whether any significant changes appear. And as far as I know, we have no autobiographical records reporting how an incunable almanac maker drew on his own experiences.[43]

Another epistemic space for almanac making may have been national or local civic authorities. The French royal court in 1427 and again in 1467 issued decrees instructing the King's surgeon to prepare manuscript copies of the annual almanac to be sent to all master surgeons in France. In German regions, town councils or medical guilds sought to regulate the almanacs to be followed by their medical personal. In 1470, for example, the Basel Sternenzunft decreed that "each year a bloodletting almanac should be purchased from Strasbourg and attached to the wall of the guild hall so that every Master can copy it. If any Master bleeds at other times, he must as punishment provide all lords with 3 pounds of wax." (The earliest extant printed Strassburg almanac is for 1473.) In 1518, Basel's city council declared that no bloodletting almanac could be printed without prior approval by their university's medical faculty and the city physician. In 1498, the Constance city council ruled that members of the local barber guild (blood letters) could vote among themselves on which almanac to use but then must all use only that sheet. The Nuremberg city council's "Gemeine Ordnung der Wundärzt" required local astronomers and the city physicians to determine the permitted bloodletting days and to evaluate the various printed almanacs being sold. By 1527, so many competing almanacs were circulating that the council mandated the astronomer at the local Gymnasium, Johannes Schöner (1477–1547), and a physician to select one imprint for use within the city.[44] Unfortunately, no records have been found indicating how the regulatory bodies evaluated the merits of competing almanacs or intervened in their medical content. Even Monica Azzolini's wonderfully textured study of astrological activity at the ducal court in Milan, drawing on a rich set of preserved letters, does not expose cases of medical experience feeding back into prescriptive astrological rules for medicine.

The making of incunabula almanacs undoubtedly occurred in trading zones broader than bookish knowledge. Yet specifying the types of transactions within those zones is not simple, especially since we have so little evidence beyond the content on the printed sheets themselves.

We turn, finally, to the epistemic spaces in which the incunabula practica were generated. With content more varied than the almanacs and less standardized in their printing formats, the practica offered their authors (and printers) more flexibility in assembling material. I cannot here systematically survey the genre, even in the incunabula period. But many of these editions contain similar themes, divided into chapters, sections or other discrete units; I will limit my analysis of some of these themes.

Unlike the almanacs, many incunabula practica, especially those printed south of the Alps, open with a dedicatory letter addressed to local officials or a preface addressed to readers (separately formatted title pages do not start appearing until the 1490s). Many of these are signed and dated, which explains why we know so many more practica than almanac authors. These introductory materials usually defend the prognosticatory enterprise, offer either theological or humanist (citing Greek or Arabic authors) defenses for astrology, and emphasize the positive contribution of the practica to the common good. For example, one of the early prolific authors, Hieronymus Manfredis, a doctor of arts and medicine at the University of Bologna, dedicated his 1479 edition to the city of Bologna and the "use and profit" of its people. In 1497, Johannes Virdung, court astrologer for Philipp the Upright

[43] Systematic casebooks, richly documenting trading zones between bookish learning and encounters with clients, have been preserved for astrological physicians from the early fifteenth century. See (Page 2001; Traister 2001; Kassell 2005, 2014). None of these studies have asked whether the astrologer's interpretative practices changed over time.

[44] All these examples are discussed in (Hoffmann 1983).

(1498–1508), Elector Palatine of the Rhine, thanked his patron for "protection in all things" and promised to devote "the entirety of my art of the stars and all my activity [Handel]" to his faithful service (Manfredis 1479, sig. a1r; Virdung 1497, fol. 1v). Much of this prose seems boilerplate to me, as it did to Karl Schottenloher many years ago in his survey of dedications in sixteenth-century books (Schottenloher 1953). Preparing such introductions required knowledge of local politics, networks of patronage, and some of the standard medieval astrological texts. Writing these dedications would have taken authors into the public spaces of their cities, trading zones to be sure.

Practica usually grounded their prognostications for the coming year in the interpretation of horoscopes cast for the dates and times of the Sun's entry into the four cardinal points of the year (hinges of the seasons) and the year's eclipses. The chart for the vernal equinox, the "revolution of the year," was the most important component, for the astrologers determined the year's ruling planets from that configuration. As with the almanacs, the astronomical times were lifted directly from ephemerides or Regiomontanus's calendars, or more rarely were computed from the Alfonsine Tables. These data generally match, to within a few minutes, the times I compute using the same sources. Although practicing astrologers might claim that their reading of a horoscope was as "factual" as were its foundational astronomical data, historians have long noted that different authors' practica, prepared for the same year, can vary considerably in the ruling planets they select (Hellmann 1924, 5–6). I have found very few examples of practica authors publicly accusing their competitors of getting the planetary rulers "wrong."[45] Presumably fifteenth-century astrologers (and their readers) realized that interpreting a horoscope involved a mixture of personal experience and long-established rules for assigning meanings to individual planets, their aspects, and the astrological houses in which they were located at the given time.[46] Ptolemy himself had described this mixture in the opening book of the *Tetrabiblos*:

> If, then, a man knows accurately the movements of all the stars, the sun and moon … and if he has distinguished in general their natures as the result of previous continued study … and if he is capable of determining in view of all these data, both scientifically and by successful conjecture [φυσικῶς ἅμα καὶ εὐστόχως], the distinctive mark of quality resulting from the combination of all the factors, what is to prevent him from being able to tell on each given occasion the characteristics of the air from the relations of the phenomena at the time, … the general quality of his temperament from the ambient at the time of his birth, … and predict occasional events …? (Robbins 1940, 10–3)

The astrologer combines data, working in the manner of the natural philosopher and by well-aimed guessing. In his practica for 1485, the Augsburg astrologer Leonhard Seybold mentioned Ptolemy and asserted that to recognize the celestial influences required "long experience, gathered day and night without rest" (Seybold 1485, fol. 1r). In the same year, Wenzel Faber described his practica as written "according to the experience of the ancients and rules of the wise" (Faber de Budweis 1485, fol. 1v).[47] Merely reading books would not suffice.

Many, but certainly not all, incunable practica also list dates and times of the new and full moons in the context not of bloodletting (as in the almanacs) but of meteorology. By reading a horoscope cast for syzygy times an astrologer could predict weather patterns for the month. Often these predictions

[45] In a 1479 preface, the Bologna professor Hieronymus Manfredis did complain generically about the errors of "experts and astrologers" resulting from their carelessness and inadequate knowledge of the "principles of things." (Manfredis 1479, sig. a1r).

[46] (Green 2012b, 7) quotes from 1545 a rare complaint by a practica author about his overly critical readers: "Und dieweil ich weis/das viel seind die alle Practiken so sie bekommen können zu hauff keuffen/und gegen einander halten/ darinn sie denn offt finden/das einer regen setzet/der ander schnee/der dritte feucthe lufft etc." (And I know that many purchase all the practica they can find, in large piles, comparing them against each other and often finding that one predicts rain, another snow, the third moist air, etc.).

[47] "… habe uns … lange zeyt erfarung genommen tag vnd nacht mit unrü biß sy in sollicher ding erkentnuß kömmen seind." "… habe ich … nach der ersten erfaren vnd weisen regel beschriben, was das gestirne den vnderersten mit müglikeit verheischet."

were general: January would be cold, April wet, September would experience flooding, etc. But they could also be quite specific: February 6 would be windy, the 9th would see heavy snowfall, etc. I have not tried to compare various weather predictions for a given year to see whether different "interpretative styles" can be traced to given practica authors. I also have not read widely enough in the classical astrological-meteorological texts to understand the "rules" for such predictive work. In any case, untangling styles and rules from accumulated empirical experience undoubtedly would be difficult.

Many of the practica include separate chapters on a conventional set of topics, such as agricultural fertility; war and peace; and health, illness and plague (so-called mundane astrology). These predications might be grounded in particular astronomical phenomena, often the aspects of the malefic planets, Saturn and Mars, or their individual locations in the horoscope for the "revolution of the year." But more generally such chapters present only declarative statements with no hint of what might be "causing" the predicted phenomena. Such predictions could be local, restricted to the particular city for which the practica was made, especially when related to solar eclipses, visible in only narrow bands, or Europe-wide for lunar eclipses that are seen simultaneously across an entire hemisphere. Some of the predictions might be very discrete, surely pointing toward trading zones in which the practica authors were participating. Johannes Glogoviensis, in one of the earliest extant German editions, for 1477, predicted in his agricultural chapter that "nuts will be not very healthy to eat, lentils and honey will be plentiful enough, millet and beans less so with shortages."[48] An astrologer in Rome, Julianus de Blanchis, reported in 1482 on the prospects for wine, olive oil, honey, fruit, and trees (Blanchis 1482, fols. 3r–v). Wenzel Faber in Leipzig was even more specific about agricultural prospects for 1485. After describing general trends, he turned to particular crops:

> Wheat will be fertile and increase in monetary value, barley will thrive but its value will be middling. Grain [Korn] will be middling but its value will climb somewhat, the same with oats. Honey will increase but will be spoiled in many places because of the bees. Vineyards will grow well. Dairy products will be unhealthy and bring much illness, the waters [distilled liquor?] also. Peas and beans middling. Nuts will grow well and will not be excessively expensive. Crops of a cold nature, like lentils, pumpkin, chestnuts, apples, pears, etc. will flourish.[49]

On the other hand, Julianus's agricultural predictions for 1483 were bookish; Mars in an earthly sign means the destruction of forests by fire, "according to Albumasar's *Flores astrologiae* and Guido Bonatus's *De revolutionibus annorum mundi*, Chapter 89" (Blanchis 1483, fol. 8r). Do the lists of specific grains and foodstuffs reflect an astrologer's engagement with peasants or those engaged in agriculture?

Finally, the incunable practica also offer general predictions for particular social groups (generally north of the Alps) or particular places, kingdoms or cities (south of the Alps). Discussion of the *Stände* (estates and occupations) could become substantial, often filling over half of a practica's pages, as can be seen, for example, in Faber's many incunable editions. Faber appears to have organized his *Stände* around the groupings presented in the *Planetenkinder*, visual presentations of each planet and its related occupations that had circulated widely in fourteenth-century manuscripts. Deriving from astrological sources, these groupings also linked metals and colors to the planets (Faber de Budweis

[48] "… Nuß werden nitser gesunt zu essen. Linssen vnd honig wirdet vol genug steen. Hirsch vnd ponen mittelmessig mit abgang." Reproduced in (Green and Duntze 2013, 81).

[49] "Der weitz wirt diß iar fruchtbar sein vnd sein gelt wirt steigen. Gersten wirt wol wassen vnd ir gelt wirt mittelmessig sein. Das korn mittelmessig vnd sein gelt wirt etwaß steigen. Der haber auch also. Das hönig wirt zu nemen. Sunder an vil enden wirt es vorderben der pinen halben. Die weine werden wol wachsen. Milch speise werden vngesunt vnd vil krankheit bringen. Die wasser des gleiche. Erwise vnd bonn mittelmessig. Die nuß werden vol wachsen vnd nicht am gelde geacht. Die früecht kalter natur, alsz linzen körbis castanean Epfel birne vnd der gleich werden vol gedeien" (Faber de Budweis 1485, fol. 3v). For a Nuremberg merchant's claim, in the 1540s, to have an astrological scheme for predicting prices on the Antwerp spice market two weeks in advance, see (Ehrenberg 1896, ii: 15–7).

1485, fols. 7v–8v; 1490, fols. 4r–5r; 1497, sig. a4v–a6r). In his German editions, Faber issued predictions, first for Jews, Christians and Turks, then for:

- Pope, cardinals, legates, doctors of both laws, priests, kings, princes, nobility
- Knights, soldiers, physicians, goldsmiths, iron workers, all who love the color red [Mars]
- Masters of the liberal arts, students, astronomers, clever people, all who love the color blue [Jupiter; in the Latin editions, yellow = Sun]
- Women, virgins, singers, lovers, all who love the color green [Venus]
- Merchants
- Clerics, monks, elderly, manual laborers, all who love the color black [Saturn]
- Common people, peasants, all who love the color yellow [Sun, in the Latin editions, gray = Mercury]

Other practica place merchants into the liberal arts group, which is called grey. Clearly, links between the *Stände* and the *Planetenkinder* had not become standardized. Social historians might look for patterns in these groupings; historians of astrology could track them to various meanings assigned to planets in the medieval handbooks. For our purposes, these chapters undoubtedly reveal practica authors' knowledge both of the *Planetenkinder* tradition as well as of the social and economic hierarchies in their locales (Hauber 1916; Blume 2000).

Italian practica often focus general predictions on places rather than social classes. A 1489 practica by Laurentius Bonincontrius, printed in Rome, announces predictions for the kings of France, Spain, Portugal, Naples, Hungary, and Poland as well as localized reports for Milan, Venice, Florence, Calabria and Constantinople. A 1492 edition by Bernardinus de Luntis, also printed in Rome, connects the zodiacal signs with dozens of Italian cities and offers special predictions for the peoples of Venice, Florence, Bologna, Genoa and Perugia. A short practica of only four leaves, authored by Petrus Avogarius for 1498 and printed in Ferrara, makes predictions for the kings and dukes in Italy, France, Spain, and Hungary, as well as for the cities of Milan, Mantua, Urbino, Pisa, Ferrara, Tauro, Rimini, Venice and Florence (Bonincontrius 1489, fols. 5r–v; Luntis 1492, fols. 6v–10r; Avogarius 1498, fols. 2v–4r). Too few copies are extant to know whether the Italian editions actually circulated to all the places considered by their authors. But these examples illustrate, I think, the local and the particular that can be found in the practica. Yes, the mundane prognostications derived from eclipses, especially lunar eclipses that are simultaneously visible to an entire hemisphere on the Earth, were general and generic (floods, war and pestilence were favorites). But much of the practica content was local and must have been informed by the astrologers' involvements in trading zones as they plied their trade. The astrological handbooks of Guido Bonati and Māshā-allāh and the ephemerides of Regiomontanus do not include algorithms for generating such local predictions.

This sampling of the content of the incunable almanacs and practica has found tantalizing, but not definitive, evidence that their authors engaged in trading zones. First, our prosopography of the named authors finds that many described themselves as university-trained; a good number taught at universities. Yet a quarter of the practica authors cannot be further identified and fully three-quarters of the almanac editions are anonymous (or are preserved as fragments with no authors' names visible). The practica and almanacs were not written exclusively by scholars; and given many vernacular editions, their readers were not exclusively Latinate. Second, although self-reflexive remarks are rare, several authors described their task as requiring not only bookish knowledge of astronomy and astrology but also experience. Third, crafting the annual blood-letting rules for the almanacs probably required knowledge not only of texts on astrological medicine but also of surgical experience with patients. Fourth, moving from horoscopes, cast for the various critical dates of the year, to predictions for agriculture, war and peace, weather, *Stände* and particular regions or cities required, as even Ptolemy recognized, knowledge of astrology's bookish rules and a personal style of selecting and combining that only experience could shape. It is, therefore, not surprising that historians of the book refer to the incunable almanacs and practica as "pragmatic literature" (Palmer 1997, 40–91; Honemann et al. 2000).

3 Conclusion

As argued in the first section of this paper, what came to be called "practical" knowledge seems to have entered astronomy/astrology rather later than it did the worlds of medieval medicine, geometry, alchemy, and *algorismus*. In astronomy, the term arose in German rather than Latin and coincided with the spread of fifteenth-century printing. In the second section, we have suggested that the incunable almanacs and practica, preserved in nearly 1000 separate editions (out of a total of c. 29,000 incunable editions), might well have emerged in trading zones where bookish knowledge in astrology and astronomy interacted with the experiences of not only surgeons, physicians, astrologers, astronomers and town councils but also patients, peasants, and readers, both Latin and vernacular.

Unlike trading zones such as mines or arsenals that left many archival traces, the incunable print shop left few records regarding its production of pragmatic literature. To the best of my knowledge, we have no surviving records documenting how individual authors, working with printers, prepared their almanac or practica editions or combined material from printed, oral or experiential sources. The imprints themselves have survived in very few copies, often in fragments, and show minimal, if any, signs of early use. The trading zones in which the incunabule almanacs and practica arose are thus murkier than we might like.

This problem brings us back to Prolianus's 1477 astrological compendium with its mixture of general theory and specific computational syzygy and eclipse data. As noted in our introduction, Prolianus's presentation manuscript failed to secure patronage for its author; and his incunable volume attracted little if any attention. But two years earlier in Nuremberg, Regiomontanus had published a massive ephemerides and calendars, filled with just the quantitative, computational data that the almanac and practica makers would need. These works undoubtedly served as working handbooks, as tools, for everyone in the astronomical trading zones of the fifteenth century. Not surprisingly, Regiomontanus's ephemerides and calendars are the most reprinted astronomical/astrological texts, after the almanacs and practica themselves, of the incunabula period.

The texts of Prolianus and Regiomontanus clearly served the trading zones in which almanacs and practica were made. But their astronomy remained strictly Alfonsine, that is, traditional. Prolianus disappeared after 1477. Regiomontanus died in 1476, before he could embark on the "reform" of mathematical astronomy he earlier had announced, a reform he hoped to ground in new observations of the planetary motions. The next explicit attempt to reform astronomy, offered by the almanac and practica maker Johann Engel in 1510, also foundered on the early death of its author. One reform of astronomy, also launched around 1510, would of course gain traction, viz., that offered by Nicholas Copernicus. Although scholars disagree about the significance of the astrological context for Copernicus's work, practica makers after 1550 certainly played central roles in the early, empirical evaluation of Copernicus's new algorithms for calculating planetary positions (Kremer 2010; Dobrzycki and Kremer 1996; Westman 2011; Lerner 2012; Kremer 2006).

However, the content of the astrological and medical handbooks required to interpret horoscopes or to craft rules for bloodletting remained essentially unchanged during the fifteenth and sixteenth centuries. Not until 1610 would Kepler propose a major revision of Ptolemaic astrology by adding several new planetary aspects, based on his years of collecting weather observations, to its interpretative framework. Again, practica makers would take the lead in debating the appropriateness of the new aspects (Strauß and Strauß-Kloebe 1981; Bialas 2004; Boner 2011).

Thus, the trading zones in which European print shops, during the fifteenth and sixteenth centuries, issued literally thousands of editions of annual astrological, medical and calendrical texts did occasionally produce significant innovations in the practical knowledge exchanged there. But we must also emphasize that, over these two centuries, very little changed within the calendar-making enterprise. The social backgrounds of the calendar makers, the format and content of the imprints, and presumably the expectations of the reading public remained largely fixed, even into the seventeenth century.[50] Given this stable trading zone, the practical knowledge of medicine, astrology and astronomy, required to produce the annual almanacs and practica and in turn codified in those imprints, enjoyed a long run in European societies.

References

Manuscript Sources

Innsbruck. Universitäts- und Landesbibliothek Tirol, Servitenkonvent I b 28.
Manchester. John Rylands Library, Latin MS 53.
Munich. Bayerische Staatsbibliothek, Cgm 754; Clm 3586; Clm 18770.
Paris. BnF, lat. 7326; lat. 10263.
VE15, A-144; A-145; A-146; A-156; A-157; A-162; A-225; A-233; A-280; A-284; A-292; A-515; C-26; F-30; F-44; P-242; S-9; S-10; T-8; 2–61.
Vienna. Österreichische Nationalbibliothek, lat. 4988.

Printed Sources

Aguiar Aguilar, Maravillas, and José Antonio González Marrero. 2005. Un tratado latino sobre los usos del cuadrante de senos: edición de los cánones del tratado de Christianus de Prolianus conservado en el ms. lat. n. 10263 de la Biblioteca nacional de Francia. *Faventia* 27: 113–123.

Aiton, E.J. 1987. Peurbach's *Theoricae novae planetarum*: A translation with commentary. *Osiris* 3: 5–43.

Allard, André. 1991. The Arabic origins and development of Latin algorisms in the 12th century. *Arabic Sciences and Philosophy* 1: 233–283.

Antry, Theodore James (ed.). 1989. *A critical edition of the 'Practica medicinalis' of Thomas of Wrocław, prémontré bishop of Serepta (1297–c. 1378)*. Wrocław: Ossolineum.

Arte dell'abbaco. 1478. Treviso: Michele Manzolo.

Avogarius, Petrus Bonus. 1498. *Prognostikon auf das Jahr 1498, lat.* Rome: Petrus de Turre.

Azzolini, Monica. 2013. *The duke and the stars: Astrology and politics in Renaissance Milan*. Cambridge: Harvard University Press.

Baron, Roger. 1955. Sur l'introduction en Occident des termes 'geometria, theorica et practica.' *Revue d'histoire des sciences et de leurs applications* 8: 298–302.

Bartholomew of Salerno. 1852–1859. Practica. In *Collectio salernitana*, ed. Salvatore de Renzi, iv: 321–406. Naples: Tipografia del Filiatre-Sebezio.

Bauer, Josef. 1966. *Geschichte der Aderlässe*. Munich: Werner Fritsch.

Benjamin Jr., Francis S., and G.J. Toomer. 1971. *Campanus of Novara and medieval planetary theory: Theorica planetarum*. Madison: University of Wisconsin Press.

Bentley, Jerry H. 1987. *Politics and culture in Renaissance Naples*. Princeton: Princeton University Press.

Bialas, Volker. 2004. *Johannes Kepler*. Munich: Beck.

Biraben, Jean Noël. 1975–1976. *Les hommes et la peste in France et dans les pays européens et méditerranéens*. Paris: Mouton.

Blair, Ann. 2014. Reflections on technological continuities: Manuscripts copied from printed books. *Bulletin of the John Rylands Library* 96: 7–33.

de Blanchis, Julianus. 1482. *Prognostikon auf das Jahr 1482, lat.* Passau: Benedikt Mayr.

[50] (Herbst 2012) provides a useful introduction to the literature on these topics.

de Blanchis, Julianus. 1483. *Prognostikon auf das Jahr 1483, lat.* Rome: Eucharius Silber.
Blume, Dieter. 2000. *Regenten des Himmels: Astrologische BIlder in Mittelalter und Renaissance*. Berlin: Akademie Verlag.
Boner, P. 2011. Kepler's imprecise astrology. In *Kepler: La physique celeste, autour de l'Astronomia nova (1609)*, ed. Édouard Mehl, 115–132. Paris: Les Belles Lettres.
Bonincontrius, Laurentius. 1489. *Prognostikon auf das Jahr 1489, lat.* Rome: Stephan Plannck.
Bovey, Alexe. 2005. *Tacuinum sanitatis: An early renaissance guide to health*. London: Sam Fogg.
Brant, Sebastian. 1494. *Das Narrenschyff.* Basel: Johann Bergmann.
Brant, Sebastian. 1944. *The ship of fools, translated into rhyming couplets with an introduction and commentary by Edwin H. Zeydel*. New York: Columbia University Press.
Burnett, Charles. 2004. Hereford, Roger of. In *Oxford dictionary of national biography*. Oxford: Oxford University Press.
Busard, H.L.L. 2000. Über zwei Algorismus-Schriften aus dem 13. Jahrhundert. In *Sic itur ad astra: Studien zur Geschichte der Mathematik und Naturwissenschaften, Festschrift für Paul Kunitzsch zum 70. Geburtstag*, ed. Menso Folkerts and Richard Lorch, 91–137. Wiesbaden: Harrassowitz.
Chabás, José, and Bernard R. Goldstein. 1997. Computational astronomy: Five centuries of finding true syzygy. *Journal for the History of Astronomy* 28: 93–105.
Chabás, José, and Bernard R. Goldstein. 2000. *Astronomy in the Iberian peninsula: Abraham Zacut and the transition from manuscript to print*. Philadelphia: American Philosophical Society.
Chabás, José, and Bernard R. Goldstein. 2003. *The Alfonsine tables of Toledo*. Dordrecht: Kluwer.
Chapman, A.A. 1979. Astrological medicine. In *Health, medicine, and mortality in the sixteenth century*, ed. Charles Webster, 275–300. Cambridge: Cambridge University Press.
Crossgrove, William. 1998. The vernacularization of science, medicine and technology in late medieval Europe. *Early Science and Medicine* 3: 81–87.
Crowther, Kathleen M., et al. 2015. The book everybody read: Venacular translations of Sacrobosco's Sphere in the sixteenth century. *Journal for the History of Astronomy* 46: 4–28.
Curtze, Maximilian. 1897. *Petri Philomei de Dacia in Algorismum vulgarem Johannis de Sacrobosco commentarius*. Copenhagen: A. F. Høst & Fil.
Demaitre, Luke. 1998. Medical writing in transition: Between ars and vulgus. *Early Science and Medicine* 3: 88–102.
Demaitre, Luke. 2003. The art and science of prognostication in early university medicine. *Bulletin of the History of Medicine* 77: 765–788.
Dobrzycki, Jerzy, and Richard L. Kremer. 1996. Peurbach and Maragha astronomy? The ephemerides of Johannes Angelus and their implications. *Journal for the History of Astronomy* 27: 187–237.
Du Cange, C.D.F. 1883–1887. *Glossarium mediae et infimae latinitatis*, enlarged ed. Niort: L. Favre.
Eamon, William. 1994. *Science and the secrets of nature: Books of secrets in medieval and early modern culture*. Princeton: Princeton University Press.
Ehrenberg, Richard. 1896. *Das Zeitalter der Fugger: Geldkapital und Creditverkehr im 16. Jahrhundert*. Jena: G. Fischer.
Eisermann, Falk. 2004. *Verzeichnis der typographischen Einblattdrucke des 15. Jahrhunderts im Heiligen Römischen Reich Deutscher Nation*. Wiesbaden: Reichert Verlag.
Eisermann, Falk, and Volker Honemann. 2000. Die ersten typographischen Einblattdrucke. *Gutenberg-Jahrbuch* 75: 88–131.
Engel, Johannes. 1488. *Die dewtsch practick … auf das Jar lxxxviii*. Nuremberg: Marx Ayrer.
Erler, Georg. 1895–1902. *Die Matrikel der Universität Leipzig*. Leipzig: Giesecke & Devrient
Faber de Budweis, Wenceslaus. 1482. *Prognostikon für Leipzig auf das Jahr 1482, dt.* Leipzig: Markus Brandis.
Faber de Budweis, Wenceslaus. 1484. *Prognostikon für Leipzig auf das Jahr 1484, dt.* Leipzig: Markus Brandis.
Faber de Budweis, Wenceslaus. 1485. *Prognostikon für Leipzig auf das Jahr 1485, dt.* Nuremberg: Friedrich Creussner.
Faber de Budweis, Wenceslaus. 1487a. *Prognostikon für Leipzig auf das Jahr 1487, dt.* Augsburg: Anton Sorg.
Faber de Budweis, Wenceslaus. 1487b. *Prognostikon für Leipzig auf das Jahr 1487, lat.* Leipzig: Markus Brandis.
Faber de Budweis, Wenceslaus. 1490. *Prognostikon für Leipzig auf das Jahr 1490, lat.* Nuremberg: Friedrich Creussner.
Faber de Budweis, Wenceslaus. 1494. *Prognostikon für Leipzig auf das Jahr 1494, lat.* Leipzig: Martin Landsberg.
Faber de Budweis, Wenceslaus. 1497. *Prognostikon für Leipzig auf das Jahr 1497, dt.* Leipzig: Martin Landsberg.
Fischer, Hanns. 1961. *Hans Folz, Die Reimpaarsprüche*. Munich: Beck.
Geldner, Ferdinand. 1968–1970. *Die deutschen Inkunabeldrucker: Ein Handbuch der deutschen Buchdrucker des XV. Jahrhunderts nach Druckorten*. Stuttgart: Hiersemann.
Ghisalberti, Alberto M. 1960–. *Dizionario biografico degli Italiani*. Rome: Instituto della Enciclopedia Italiana.
Gil-Sotres, Pedro. 1994. Derivation and revulsion: The theory and practice of medieval phlebotomy. In *Practical medicine from Salerno to the Black Death*, ed. Luis García-Gallester et al., 110–155. Cambridge: Cambridge University Press.

Glogoviensis, Johannes. 1477. *Prognostikon auf das Jahr 1477, dt.* Nuremberg: Johann Sensenschmidt.

Glogoviensis, Johannes. 1478. *Prognostikon auf das Jahr 1478, dt.* Nuremberg: Johann Sensenschmidt.

Glogoviensis, Johannes. 1481. *Prognostikon auf das Jahr 1481, dt.* Leipzig: Markus Brandis.

Glogoviensis, Johannes. 1500. *Prognostikon für Krakau auf das Jahr 1500, dt.* Leipzig: Martin Landsberg.

Götz, Johannes. 1486. *Prognostikon für Ingolstadt auf das Jahr 1486, dt.* Augsburg: Anton Sorg.

Grafton, Anthony. 1999. *Cardano's cosmos: The worlds and works of a Renaissance astrologer*. Cambridge: Harvard University Press.

Green, Jonathan. 2012a. *Printing and prophecy: Prognostication and media change, 1450–1550*. Ann Arbor: University of Michigan Press.

Green, Jonathan. 2012b. Printing the future: The origin and development of the 'Practica.' *Archiv für Geschichte des Buchwesens* 67: 1–18.

Green, Jonathan, and Oliver Duntze. 2013. Johannes von Glogau and the earliest German practicas: On the dating and authorship of fragmentary prognostications. *Gutenberg Jahrbuch* 88: 68–85.

Grimm, Jacob, and Wilhelm Grimm. 1854–1960. *Deutsches Wörterbuch*. Leipzig: Hirzel

de Gruyter 2016. *World biographical information system*. http://db.saur.de/WBIS/welcome.jsf. Accessed 15 Dec 2015.

Guasconus, Franciscus. 1474. *Prognosticon anni 1474, lat.* Venice: Konrad von Paderborn.

Guasconus, Franciscus. 1475. *Prognosticon universale anni 1475, lat.* Venice: Christophorus Arnoldus.

GW. 2015. *Gesamtkatalog der Wiegendrucke*. http://www.gesamtkatalogderwiegendrucke.de. Accessed 15 Dec 2015.

Halleux, Robert. 1979. *Les textes alchimiques*. Turnhout: Brepols.

Haskins, Charles H. 1915. Reception of Arabic science in England. *English Historical Review* 30: 56–69.

Hauber, A. 1916. *Planetenkinderbilder und Sternbilder*. Strassburg: Heitz & Mündel.

Hayton, Darin. 2015. *The crown and the cosmos: Astrology and the politics of Maximilian I*. Pittsburgh: University of Pittsburgh Press.

Heilen, Stephan. 2015. Paul of Middelburg's prognosticum for the years 1484 to 1504. In *From Māshā'allāh to Kepler: The theory and practice of astrology in the Middle Ages and the Renaissance*, ed. Charles Burnett and Dorian Geiseler Greenbaum, 231–278. Ceredigion: Sophia Centre Press.

Hellmann, Gustav. 1924. *Versuch einer Geschichte der Wettervorhersage im XVI. Jahrhundert*, Abhandlungen der pruessischen Akademie der Wissenschaften, Physik.-math Klasse 1, 1–54.

Herbst, Klaus-Dieter (ed.). 2012. *Astronomie-Literatur-Volksaufklärung: Der Schreibkalender der frühen Neuzeit mit seinen Text- und Bildbeigaben*. Jena: HKD.

Hoeniger, Cathleen. 2006. The illuminated Tacuinum santitas manuscrits from Northern Italy ca. 1380–1400: Sources, patrons, and the creation of a new pictorial genre. In *Visualizing medieval medicine and natural history, 1200–1500*, ed. Jean A. Givens, Karen M. Reeds, and Alain Touwaide, 51–81. Aldershot: Ashgate.

Hoffmann, Leonhard. 1983. Almanache des 15. und 16. Jahrhunderts und ihre Käufer, *Beiträge zur Inkunabelkunde, 3.* Folge 8, 130–143.

Homann, Frederick A. 1991. *Practical geometry [Practica geometriae] attributed to Hugh of St. Victor, translated from the Latin*. Milwaukee: Marquette University Press.

Honemann, Volker, et al. (eds.). 2000. *Einblattdrucke des 15. und frühen 16. Jahrhunderts: Probleme, Perspektiven, Fallstudien*. Tübingen: Max Niemeyer Verlag.

Huey, Caroline. 2012. *Hanz Folz and print culture in late medieval Germany: The creation of popular discourse*. Farnham: Ashgate.

Jacquart, Danielle. 1988. Aristotelian thought in Salerno. In *History of twelfth-century Western philosophy*, ed. Peter Dronke, 407–428. Cambridge: Cambridge University Press.

Jacquart, Danielle. 1992. 'Theorica' et 'practica' dans l'enseignement de la médecine à Salerne au XIIe siècle. In *Vocabulaire des écoles et des méthodes d'enseignement au moyen âge: Actes du colloque, Rome, 21–22 octobre 1989*, ed. Olga Weijers, 102–110. Turnhout: Brepols.

Kassell, Lauren. 2005. *Medicine and magic in Elizabethan London: Simon Forman, astrologer, alchemist, and physician*. Oxford: Clarendon.

Kassell, Lauren. 2011. Almanacs and prognostications. In *The Oxford history of popular print culture. Vol. 1: Cheap print in Britain and Ireland to 1600*, ed. Joad Raymond, 431–442. Oxford: Oxford University Press.

Kassell, Lauren. 2014. Casebooks in early modern England: Astrology, medicine and written records. *Bulletin of the History of Medicine* 88: 595–625.

Keil, Gundolf. 1957. Die verworfenen Tage. *Sudhoffs Archiv* 41: 27–58.

von Keller, Adelbert, ed. 1870–1908. *Hans Sachs*. Tübingen: H. Laupp.

Kibre, Pearl. 1977. Astronomia or Astrologia Ypocratis. In *Science and history: Essays in honor of Edward Rosen*, ed. Erna Hilfstein, Paweł Czartoryski, and Frank D. Grande, 133–156. Cracow: Ossolineum.

Kremer, Richard L. 2003a. Thoughts on John of Saxony's method for finding true syzygy. *Historia Mathematica* 30: 263–277.

Kremer, Richard L. 2003b. Wenzel Faber's table for finding true syzygy. *Centaurus* 45: 305–329.

Kremer, Richard L. 2006. Copernicus among the astrologers: A preliminary study. In *Astronomy as a model for the sciences in early modern times*, ed. Menso Folkerts and Andreas Kühne, 225–252. Augsburg: Rauner.
Kremer, Richard L. 2010. War Bernard Walther, Nürnberger astronomischer Beobachter des 15. Jahrhunderts, auch ein Theoretiker? In *Astronomie in Nürnberg*, ed. Gudrun Wolfschmidt, 156–183. Hamburg: Tredition.
Kremer, Richard L. Forthcoming. *The practice of Alfonsine astronomy in the fifteenth century: A survey of incunable calendars and almanacs.*
Kren, Claudia. 1970–1980. Roger of Hereford. In *Dictionary of scientific biography*. New York: Sribners.
Lerner, M.P. 2012. Planetary order in the long sixteenth century. *Journal for the History of Astronomy* 43: 233–238.
Long, Pamela O. 2011. *Artisan/practitioners and the rise of the new sciences, 1400–1600*. Corvallis: Oregan State University Press.
de Luntis, Bernardinus. 1492. *Prognostikon auf das Jahr 1492, lat.* Rome: Stephan Plannck.
Luther, Martin. 1520a. *Epistola lutheriana ad Leonem decimum summum pontificiem*. Wittenberg: Rhau-Grunenberg.
Luther, Martin. 1520b. *Eyn sendbrieff an den Bapst Leo den czehenden*. Wittenberg: Rhau-Grunenberg.
Malagola, Carlo. 1888. *Statuti delle Università e dei Collegi dello Studio Bolognese*. Bologna: Nicola Zanichelli.
Malpangotto, Michela. 2012. Les premiers manuscrits des Theoricae novae planetarum de Georg Peurbach: Présentation, description, évolution d'un ourvage. *Revue d'histoire des sciences et de leurs applications* 65: 339–380.
de Manfredis, Hieronymus. 1479. *Prognostikon auf das Jahr 1479, lat.* Verona: Giovanni und Alberto Alvise.
Marinis, Tammaro de. 1947–1952. *La biblioteca napoletana dei re d'Aragona* U. Milan: Hoepli.
Matthäus, Klaus. 1969. *Zur Geschichte des Nürnberger Kalenderwesens: Die Entwicklung der in Nürnberg gedruckten Jahreskalender in Buchform*, Archiv für Geschichte des Buchwesens 9:cols, 965–1396.
Mazzatinti, Giuseppe. 1897. *La biblioteca dei re d'Aragona in Napoli*. Rocca S. Casciano: Licinio Capelli Editore.
Millás, J.M. *El libro de los fundamentos de las Tablas astronómicas de R. Abraham ibn 'Ezra: Edición critica, con introducción y nota*. Madrid: Instituto Arias Montano.
Monaco, Giuseppe Giovanni. 1999–2000. Christianus Prolianus balbanensis e l'Astrologiae compendium. *Bollettino storico della Basilicata* 15–6: 113–44
Mooney, Linne R. 1998. *The Kalendarium of John Somer*. Athens: University of Georgia Press.
Newman, William R. 1986. *The summa perfectionis and late Medieval alchemy: A study of chemical traditions, techniques, and theories in thirteenth century Italy*. Dissertation, Harvard University.
Page, Sophie. 2001. Richard Trewythian and the uses of astrology in late medieval England. *Journal of the Warburg and Courtauld Institutes* 64: 193–228.
Palmer, Nigel F. 1997. The high and later Middle Ages (1100–1450). In *The Cambridge history of German literature*, ed. Helen Watanabe-O'Kelly, 40–91. Cambridge: Cambridge University Press.
Parronus, Guilemus. 1499. *Prognostikon auf das Jahre 1499, lat.* London: Richard Pynson.
Pedersen, Olaf. 1961. Theorica: A study in language and civilization. *Classica et mediaevalia* 22: 151–166.
Pedersen, Olaf. 1978. The decline and fall of the *Theorica planetarum*: Renaissance astronomy and the art of printing. In *Science and history: Studies in honor of Edward Rosen*, ed. Erna Hilfstein, Paweł Czartoryski, and Frank D. Grande, 157–185. Wroclaw: Ossolineum.
Pedersen, Olaf. 1981. The origins of the 'Theorica planetarum.' *Journal for the History of Astronomy* 12: 113–123.
Pedersen, Olaf. 1992. The Theorica planetarum and its progeny. In *Filosofia, scienza, e astrologia nel trecento Europeo*, ed. Graziella Federici Vescovini and Franceso Barocelli, 157–186. Padua: Il Poligrafo.
Perifano, Alfredo. 1990–1991. Theorica et practica dans un manuscrit alchimique de Sisto de Boni Sexti da Norcia, alchimiste à la cour de Côme 1er de Médicis. *Chrysopoeia* 4: 81–146.
Pfister, Silvia. 1990. *Parodien astrologisch-prophetischen Schrifttums 1470–1590: Textform—Entstehung—Vermittlung—Funktion*. Baden-Baden: Verlag Valentin Koerner.
Pollich von Mellerstadt, Martin. 1486. *Prognostikon für Leipzig auf das Jahr 1486, dt.* Nuremberg: Friedrich Creussner.
Polska Akademia Nauk, Instytut Historii. 1935–. *Polski slownik biograficzny*. Cracow: Sklad glówny w ksieg, Begethnera i Wolffa.
Poulle, Emmanuel. 1963. *La bibliothèque scientifique d'un imprimeur humaniste du XVe siècle: Catalogue des manuscrits d'Arnaud de Bruxelles à la Bibliothèque National de Paris*. Geneva: Librarie Droz.
Poulle, Emmanuel. 1980. *Les instruments de la théorie des planètes selon Ptolémée: Équatoires et horlogerie planétaire du XIIIe au XVIe siècle*. Geneva: Libraire Droz.
Practica von Paris, dt. 1487. Augsburg: Johann Schönsperger.
Prolianus, Christianus. 1477. *Astrologia*. Naples: Heinrich Alding.
Ragep, F. Jamil. 1993. *Naṣīr al-Dīn al-Ṭūsī's Memoir on astronomy*. New York: Springer-Verlag.
Rauner, Erwin. 2007. *Katalog der lateinischen Handschriften der Bayerischen Staatsbibliothek München, Die Handschriften aus Augsburger Bibliotheken, Bd. 1: Stadtbibliothek, Clm 3501–3661*. Wiesbaden: Harrassowitz Verlag.
Reeves, Marjorie. 1969. *The influence of prophesy in the later Middle Ages: A study in Joachimism*. Oxford: Clarendon.
Rengart, Konrad. 1483a. *Prognostikon für Paris auf das Jahr 1483, dt.* Augsburg: Hermann Kästlin.

Rengart, Konrad. 1483b. *Prognostikon für Paris auf das Jahr 1483, lat.* Augsburg: Christmann Heyny.
Riha, Otrun. 1985. *Meister Alexanders Monatsregeln: Untersuchung zu einem spätmittelalterlichen Regimen duodecim mensium mit kritischer Textausgabe.* Würzburg: Böhler.
Robbins, F.E. (ed.). 1940. *Ptolemy, Tetrabiblos.* Cambridge: Harvard University Press.
Ruggiero, Raffaele. 2009. 'Homines talem scribendi qualem vivendi formulam tenent:' La biblioteca di Antonello Petrucci, 'secretario' ribelle. In *Biblioteche nel regno fra Tre e Cinquecento: atti del convegno di studi, Bari, 6–7 febbrario 2008*, ed. Claudia Corfiati, 171–192. Lecce: Pensa multimedia.
Scherenmüller, Bartholomaeus. 1491. *Prognostikon auf das Jahr 1491, dt.* Reutlingen: Michael Greyff.
Schneider, Karin. 1984. *Die deutschen Handschriften der Bayerischen Staatsbibliothek München, Cgm 691–867.* Wiesbaden: Otto Harrassowitz.
Schottenloher, Karl. 1953. *Die Widmungsvorrede im Buch des 16.* Münster: Jahrhunderts.
Seybold, Leonhard. 1485. *Prognostikon für Augsburg auf das Jahr 1485, dt.* Augsburg: Johann Blaubirer.
Siraisi, Nancy. 1987. Changing concepts of the organization of medical knowledge in the Italian universities: Fourteenth to sixteenth centuries. In *La diffusione delle scienze islamiche nel Medio Evo europeo*, ed. Biancamaria Scarcia Amoretti, 291–321. Rome: Accademia nazionale dei Lincei.
Siraisi, Nancy. 2007. *History, medicine, and the traditions of Renaissance learning.* Ann Arbor: University of Michigan Press.
Siraisi, Nancy. 2008. Medicina practica: Girolamo Mercuriale as teacher and textbook author. In *Scholarly knowledge: Textbooks in early modern Europe*, ed. Emidio Campi, Simone de Angelis, Anja-Silvia Goeing, and Anthony Grafton, 287–305. Geneva: Librairie Droz.
Sorbelli, Albano. 1938. Il 'Tacuinus' dell' Università de Bologna e le sue prime edizioni. *Gutenberg-Jahrbuch* 109–14.
Steinschneider, Moritz. 1888. Über das Wort Almanach. *Biblioteca mathematica, n.s.*, 2: 13–6.
Strauß, H.A., and S. Strauß-Kloebe. 1981. *Die Astrologie des Johannes Kepler: Eine Auswahl aus seinen Schriften [1926]*, 2nd ed. Felbach: Bonz.
Strimerius, Johannes Baptista. 1485. *Tacuinus pro anno domini 1485.* Parma: Deiphobus de Oliveriis.
Sudhoff, Karl. 1908. Medizinische Monatsregeln für Aderlass, Schröpfen, Baden, Arzneigebrauch und Auswahl der Speisen und Getränke aus einer Pariser Handschrift des 14. Jahrhunderts. *Archiv für Geschichte der Medizin* 2: 136–139.
Thorndike, Lynn. 1951. Sexagenarium. *Isis* 42: 130–133.
Thorndike, Lynn. 1960. Three Latin translations of the pseudo-Hippocratic tract on astrological medicine. *Janus* 49: 104–129.
Thorndike, Lynn. 1923–1958. *History of magic and experimental science.* New York: Macmillan.
Thorndike, Lynn, and George Sarton. 1928. Tacuinum, taqwīm. *Isis* 10: 489–493.
Traister, Barbara Howard. 2001. *The notorious astrological physician of London: Works and days of Simon Forman.* Chicago: University of Chicago Press.
Tur, Alexandre. 2014. *À l'entrée du Soleil en Bélier: Les prédictions astrologiques annuelles latines dans l'Europe du XVe siècle (1405–1484).* Thèse diplome d'archiviste-paléographe, École national des Chartes.
Tura, Adolfo. 2008. *Fra Giocondo & les textes français de géométrie pratique.* Geneva: Droz.
VE15. See Eisermann 2004.
Verfasserlexikon. See *Die deutsche Literatur des Mittelalters, Verfasserkexikon.* 1977–2008. 2nd Rev. ed. Berlin: de Grutyer.
Victor, Stephen K. (ed.). 1979. *Practical geometry in the high Middle Ages: Artis cuiuslibet consummatio and the Pratike de geometrie.* Phildelphia: American Philosophical Society.
Vida, Giorgio Levi della. 1934. Appunti e questi di storia litteraria araba 5: Almanacco. *Revista degli studi orientali* 14: 265–270.
Virdung, Johann. 1497. *Prognostikon auf das Jahr 1497, dt.* Nuremberg: Peter Wagner.
Vogel, Kurt (ed.). 1954. *Die Practica des Algorismus Ratisbonensis: Ein Rechenbuch des Benediktinerklosters St. Emmeram aus der Mitte des 15. Jahrhunderts nach den Handschriften der Münchner Staatsbibliothek und der Stiftsbibliothek St. Florian.* Munich: Beck.
Wallis, Faith. 1995. Medicine in medieval calendar manuscripts. In *Manuscript sources of medieval medicine: A book of essays*, ed. Margaret R. Schleissner, 105–43. New York: Garland.
Wee, John Z. 2015. Discovery of the zodiac man in cuneiform. *Journal of Cuneiform Studies* 67: 217–233.
Wehmer, Carl. 1948. *Mainzer Probedrucke in der Type des sogenannten astronomischen Kalenders für 1448: Ein Beitrag zur Gutenbergforschung.* Munich: Leibniz Verlag.
Werner, Jakob. 1888. Kalenderhumor. *Alemannia* 16: 181–183.
Westman, R.S. 2011. *The Copernican question: Prognostication, skepticism, and celestial order.* Berkeley: University of California Press.

14 The Codification of Vernacular Theories of Metallic Generation in Sixteenth-Century European Mining and Metalworking

Pamela H. Smith

Abstract

This essay explores a number of structuring frameworks for the practical knowledge of mining and metalworking, including the place and economic structures of material production in mining, the entanglement of the human body with matter in the landscape, and the constraining and affording properties of the materials themselves. Larger belief and knowledge systems, such as understandings about the relationship of earth and the heavens, as well as analogies to the human body and its processes, also structured practices and practical knowledge. Objects, too, could structure and reinforce practical knowledge systems.

The essay then turns to case studies that demonstrate the structures of practical knowledge formed by a vernacular knowledge system emerging from the complex of activities and practices that made up mining. This knowledge system was shaped by the methods of prospecting for ores, as well as by strong interest in a theory of the genesis of metals on the part of civic, territorial, and economic stakeholders. The essay ends by examining the codification of these vernacular "theories" in the books of scholars, where they found their way into the growing discourse of experimental philosophy (an increasingly powerful structuring framework over the course of the early modern period), and from there into modern science.

What are the structures of practical knowledge? There are a variety of ways in which this question might be answered. We might think first in the terms of practical knowledge itself being structured by inductive processes in which a set of experiences is generalized to form a more widely applicable rule. The productive knowledge of a craft workshop, often regarded as proceeding according to "rules of

P.H. Smith (✉)
Department of History, Columbia University, New York, NY, USA
e-mail: ps2270@columbia.edu

M. Valleriani (ed.), *The Structures of Practical Knowledge*, DOI 10.1007/978-3-319-45671-3_14

thumb" handed down from master to apprentice, may form an instance of this sort of structuring (at least in part; this will be discussed further below). Working procedures in a craft workshop can also be structured by song, by symbolic rituals (the reciting of a pater noster, for example, to measure time), or by practices of sociability. The meanings of these songs, practices, rituals, and other such structuring devices are amplified and deepened by the larger frameworks of knowing within which they are embedded. This larger framework, or knowledge system, may be explicated in a propositional form, or as a doctrine, or it may simply be dwelt or lived within as a lifeworld. In any case, the meanings that practices possess within a knowledge system are understood differently by different communities in a social or cultural grouping.

At another level, the structures of practical knowledge are institutional: practical knowledge is framed within and structured by institutions, such as guilds and academies, or princely courts and territories, and by legal codes. Practical knowledge is also structured by places of work and the ways in which work processes are organized and administered, such as in the large Brussels workshop of the painter Rogier van der Weyden (1400–1464) in the 1440s (Roy et al. 1997), or in the arsenals of Venice and Ragusa (Dubrovnik). Urban space can structure practical knowledge, determining the modes and manner of exchange among different workshops and realms of knowledge. The concentration of practitioners of different crafts in close proximity, such as at the courts of the rulers of Padua or the Grand ducal court of the Medici, brought about the intersection of different techniques and expertises, and, in some cases, caused the transfer of goals and techniques from one group of expert practitioners into the materials and practices of another area (Hermens 2013; Kieffer 2014).

Practical knowledge is also structured by texts, both by the texts in which it is codified and disseminated (when it is in fact capable of being disseminated by texts), and by pre-existing written codifications. Such an example of the latter case is offered by medieval and early modern Europe, where both within the universities and in humanist circles, practical and experiential knowledge was assimilated to and integrated into texts written centuries and even millennia previously. The integration of practical knowledge in this way was both of a practical type, being assimilated, for example, to recipe-like instructions codified sometimes long before, and of a theoretical type, or, rather, the practical knowledge was assimilated and integrated into theoretical frameworks (or "knowledge systems"), such as the system of elements and qualities of Aristotelian physics, or the Hippocratic-Galenic-Avicennian medical framework and its system of the four humors. Thus a recipe could be a reflection of practices, but it could also be underpinned by a theoretical or philosophical structuring framework. Such systemic bodies of knowledge were not always explicitly articulated by all practitioners or possessors of the knowledge, rather, these larger systems could, as noted before, simply be the basis of practices of life and work. Material objects also structured practical knowledge, as they formed models for imitation, emulation, and competitive rivalry. Practice might even be structured by matter, as the properties of the materials could determine, develop, and challenge the skills and techniques of the craftsperson in particular ways, and, as such, could cause the emergence of certain belief and knowledge systems (Lehmann 2012, 2009).

Perhaps before laying out some of the structuring frameworks for practical knowledge in the foregoing pages, I should have asked the question "what is practical knowledge?" And, "how do practice and practical knowledge differ from each other?" Practices and techniques are one component of practical knowledge. Practical knowledge itself seems to lie in the ability to cause external and real world effects, usually in the production of a tangible thing. It usually means engaging with the materials that humans find around them in their natural or human-cultivated or-influenced environment. To move from individual "practices" to "practical knowledge," there must be an ability to produce with an expectation of success, that is, with a fair degree of reliability. Thus, the proof of practical knowledge is, in the end, a tangible object or production that is the expected and sought-for outcome of a series of techniques and practices. This series of techniques and practices itself—in order to be rendered reliable—is the result of much trial and experimentation with materials (especially if the materials are not standardized), and it constitutes, among other things, an acculturation of the practitioner to the behavior of the materials (by which practical *knowledge* is learned or attained), as well as the development of "skill." In order to produce practical knowledge reliably over the long durée (i.e., at least more than the span of one lifetime, the duration of which we might consider purely "personal knowledge"), this series of techniques must be transmittable. A person who is literate in reading and writing, has been schooled mainly from texts and written codifications of knowledge, and understands knowledge to reside in the written word might believe that this knowledge must be codified and transmitted by a written or image-based account of procedures and practices. In contrast, a person who is not literate in words, writing, and text would understand knowledge as inhabiting or residing in an expert practitioner, such as a master craftsperson (and sometimes as that master channeling divine or revealed knowledge), or in the particular organization of the workshop's practices and procedures. It is clear that the very long durée existence and transmission of practical knowledge is a matter of historical fact.

If we turn to consider our position as historians of this practical knowledge, we must acknowledge that, for historians, practical knowledge is elusive. This is so because, while document-based historical sources can show us some of the frameworks of this knowledge, they cannot give us much insight into the knowledge itself. We have the products produced by that knowledge, we have images of that work, we have lists of materials and processes in the form of technical recipes, and in some cases we possess descriptions of process. From the 1400s, we have increasingly frequent statements about the power and capabilities of that knowledge, and we have guild records, judicial processes and legal codes, but the knowledge itself—that is, the ability to produce something tangible with reliable certainty—has disappeared. Even if we come to understand and recognize the structuring frameworks of practical knowledge, this will not actually allow us to learn, or even to recognize that knowledge.

This especially came home to me when I became interested in the actual techniques used and the skills possessed by artisans and in the knowledge systems that underpinned their practices. I marveled at the results of their skill that now survive in museums, and I tried to glean some sense of the nature of those skills and the content of their knowledge system by studying the objects they produced. This did allow a glimpse of knowledge systems, but was no help at all in capturing, learning, or understanding skill. I turned then to their writings, which were produced more frequently from the late fourteenth and early fifteenth centuries. The knowledge of metalworking and mining was among the first practical knowledge to be codified in writing, for example, *Von Stahel und Eisen* in the fifteenth century; *Distillierbücher* in the late fifteenth century; *Bergbücher* beginning around 1500; the *Probierbüchlein*, or assayer's books of the sixteenth century; *Das Feuerwerkbuch*, published in 1529 after circulating for 100 years in manuscript; mine manager Vannoccio Biringuccio's (1480–1537) *De la pirotechnia* (1540); and then compilations of technical recipes containing extensive information about metalworking, such as the pseudonymous Alessio Piemontese (probably Girolamo Ruscelli) (ca. 1518–1566), *De secreti* (1555), republished numerous times. But these books do not give access to skill, despite the fact that they often contain recipes and some instructional sections. In what follows, I employ some of these sources to consider the various structuring frameworks of practical knowledge, using the historically specific example of sixteenth-century Central European silver mining.

When historians of science consider mining in early modern Europe, they often think first of the *texts* of Greek teacher, St. Joachimsthal city physician, and wealthy mining shareholder, Georgius Agricola (1493–1555) and his immense *De re metallica*, or *Twelve Books on Metals*. Agricola apparently never saw his magnificent *De re metallica*, filled with the remarkable woodcut illustrations by Blasius Weffring of St. Joachimsthal, in print, for it was finally published by the learned Basel printer Froben only in 1556, the year after Agricola's death, delayed by his work as an envoy for the Duke of Saxony and possibly also by the Duke's suspicions that the book would give away the secrets of his territory's immense wealth gained in its booming silver mines. In this book, and in his earlier 1546 *De natura fossilium* (*About the Nature of Things Dug Up*), Agricola provided a new view of how metals and minerals form in the earth. Rather than beginning with the *books* of Agricola, I will use the emergence of a written corpus of practical mining knowledge in early modern Europe to gain access to the various structuring frameworks for practical knowledge, including vernacular conceptions about the generation of metals and minerals as these theories emerged from the physical experience of miners laboring in the earth, and, only at the end of this essay will I turn to the books of Agricola.

This essay offers no more than an admittedly schematic treatment of each of the structuring frameworks, which aims to give a sense of the many structures that framed the practical knowledge of mining, among which I will consider place, materials, the engagement of the human body with materials, the vernacular knowledge systems that emerge out of this engagement with the subterranean world, the systems of belief into which subterranean phenomena are placed, and, finally, the codification of these systems of belief and knowledge in texts.[1] These texts then carry the practical knowledge into the world of the present, where it forms part of a narrative of the history of science, shaping the present-day frameworks for scholars such as ourselves, as well as the general public about their own culture and knowledge systems in relation to those of the past, and, in the process, forms the identity by which we understand ourselves as different from the past.

1 Structures of Place: Mining in Central Europe

From 1460 until the 1550s (when New World silver brought down the value of precious metals), the princes and merchants of central Europe enjoyed a tremendous surge in prosperity fueled by innovations in mining. Silver production increased five-fold in this period. Although silver had been mined in central Europe since ancient times, a turning point came in the 1450s when the technique of separating silver from copper ores by means of lead was introduced into central Europe (apparently from China where it had been discovered in the eleventh century). This made profitable the greater capital investment in the large and complex machines needed for digging deep shafts to follow rich veins and the new methods of drainage and ventilation that made possible these shafts. Simultaneously, the rediscovery and spread of the blast furnace meant more effective cast iron tools for digging and breaking up the ore could be produced at a lower cost. In the wake of large-scale capital investment, independent partnerships of miners and smelters gave way to managers operating mines on behalf of absentee shareholders, including nobles, large merchants, even monasteries and universities (Nef 1987).

This very rapid expansion of the mining industry in Central Europe, and the phenomenon of boom towns that it produced, in which people of varied social groups, educational level, expertise, and background came together, caused an outpouring of cultural production around mining—images, manuscripts, printed books, and the recording in writing and print of mining songs, mining sermons, as well as many unprecedented objects, such as *Handsteine* (more on these below), that represented the materials and practices of mining and metalworking. Many of these books sought to articulate and codify the *practical knowledge* of mining and metalworking in writing. In this way, *place* formed one structuring framework for the production of practical knowledge.

As depictions of mining, such as that contained in the *Kutna Hora Gradual* indicate (Fig. 1), the subterranean nature of mining made more intriguing the practices of miners and the causes of the development of ores and metals because they were hidden from view. But a knowledge of the genesis of ores and metals was no doubt doubly intriguing because it promised tremendous—and sometimes sudden—wealth.

[1]This analysis draws evidence from previous articles, including (Smith 2015, 2014a, b).

Fig. 1 This gradual is one of several musical manuscripts produced at the end of the fifteenth and beginning of the sixteenth centuries, probably by scribes and illuminators working in Prague who had close ties to the mining town of Kuttenberg (now Kutná Hora, Czech Republic), where the manuscripts became part of local church possessions. This opening page of the gradual depicts the entire technology of mining, including miners working deep underground. This underground imagery was drawn in part from an antiphonal completed in 1471. Silver mine. Kutná Hora gradual, anonymous, ca. 1490, Hungarian National Museum, Budapest. Public Domain. Wikipedia Commons. © Photo by Miraceti

The problem of the genesis and location of metals and ores was made more acute by the fact that the formation of mineral deposits is extremely variable. The particularity of geological phenomena led ancient authors to lament the impossibility of a scientific study of stones, earths, and minerals because the unruly variety of these materials made them incapable of being marshaled into systematic generalized knowledge.

Practical knowledge was structured by this variability, and miners had to come to know the environment in which they worked and the signs of veins and "mother lodes," a kind of local knowledge that required living within and observing the landscape. Just as a hunter tracked the signs of his quarry through a forest, so the miner prospected his way over the surface of the land in order to gain clues about what lay below. Prospecting allowed miners to decide where in the landscape to begin to dig and tunnel. It consisted in recognizing signs based on prior experience of fruitful ore strikes, and it involved the use of all five senses—looking, touching, tasting, smelling, and even hearing—to sample and sound out the earth that lay on the surface or in exploratory diggings (Dym 2006, 169–74.). Looking carefully and observing the loose rocks as well as outcroppings obviously formed a primary tool of prospecting. Miners made a close study of rocks and the ground, looking carefully at the *Geschüb*, or ["alluvial fragments"]—a flood of rocks washed out of a vein which they followed into the earth. In addition, they looked for the effects of the underground vapors, such as damps lying low over the ground, discolored leaves and plants, stunted growth, and twisted roots. The founder and mine manager, Vannoccio Biringuccio, visualized minerals growing under the earth like the branches of a tree, striving continually skyward and drawing up the adjacent material, until they "arrive at the summit of the mountain and emerge with clear sign, sending forth, in place of leaves and blossoms, blue or green fumosities" (Biringuccio 1943, 13).

Tasting was central to prospecting—either by placing the earth directly on the tongue or in a vessel full of rain water, which was then sipped. Odor could also be useful in identifying the presence of *fossiles*, for certain minerals have an odor when struck with iron, for example, as Georgius Agricola noted, Hildesheim marble gave off the odor of burning horn. Minerals could also be differentiated by touch. As Agricola reported, some veins do not feel hot to the miners, but when the ore is brought to surface and a person buries his hands in it, it burns (Agricola 1955b, 8). Even hearing was a tool in the prospecting of ores; for example, good iron was indicated by presence of a red, soft, fat earth that did not make any crackling noise when squeezed between the teeth (Biringuccio 1943, 67).

This sensory data was then transformed into a generalisable system; we think of this generalisable system as being codified in books like those of Agricola, but this knowledge could also be codified in a vernacular system, such as the dowsing stick. As the historian, Warren Dym has shown, the dowsing rod seems to have been employed since ancient times to locate water and minerals, and was seen in early modern European mining as component of the knowledge and expertise of the miner. A man (never a woman) who attained this expertise was known as a *Bergverständiger*—he could speak the lingo, the *Bergsprache*, and possessed "mountain science:" *Bergwissenschaft.* Dowsing involved mind and body, and dowsers were asked to focus their "impressions and thoughts." Although the efficacy of this practice was controverted by some in the early modern period, it continued to be taught at the Bergakademie in Freiberg into the eighteenth century, according to Warren Dym (Dym 2010). How might we people of the present think about this practice of the past that we no longer believe to possess efficacy (even when well-diggers continue to employ it today)? One way of understanding it is to think about the skill that miners acquired by long acquaintance with their environment: Dowsers generally were experienced miners who no doubt possessed a reservoir of knowledge and observational skills that allowed them to attend to and integrate many particulars of the landscape—plant growth, rock outcroppings, gravel wash and silt in streams, the taste and smell of water running through the land, direction of mineral veins, and so on. Perhaps the dowsing stick helped focus their attention into a state of subsidiary awareness of all these particular observations which made them able to discern more likely locations of ore, in the way that Michael Polanyi in *Personal Knowledge* envisages skilled practice occurring. He sees skill as "a structural change achieved by repeated mental effort" (Polanyi 1998, 62). Polanyi draws a distinction between subsidiary awareness and focal awareness, with repeated practice allowing the skilled practitioner to move from a focus on particular components of the skill to an increasing unconscious of the particular actions, which results in attainment of the ability to hold in subsidiary awareness the particulars while performing a series of integrated movements and procedures to bring about a whole skilled performance or result (Polanyi 1998, 64).

In addition, the dowsing rod, as a recognizable symbol of social and epistemic authority, enabled the dowser to declare his knowledge more authoritatively because, in its visible dipping movement, the rod provided an external demonstration of his sensory skills and his local knowledge—what amounted ultimately to personal knowledge—to investors who wanted to know (or wanted to believe they knew) with some reliability where to sink their large capital investments.

In these ways, the structuring framework of *place* shaped the practical knowledge of mining in Central Europe.

2 The Structures of Materials

All those who labor on natural materials engage bodily with matter. This entanglement of the human body with matter is especially clear in the labor of mining and metalworking, in which the bodies of miners and smelters are penetrated by the very matter in which they labor. Their bodily striving brings the materials to the earth's surface, then draws out the metals from the matrix of their ores, and, in this process, the spirits and humors of the metals and minerals mix with those of the metalworkers, and they affect and shape each other in distinctive ways. Grounding his account on German physician Daniel Sennert's work, Joannes Jonston wrote in his 1657 *A History of the Wonderful Things of Nature*: "Truly the Flux of Veins hath somthing proportionable to vegetable nature; and the relation of a Physitian of Friberg, that, in the Lungs of such as use to dig in Mines, their bodies being opened when they are dead, you shall find the same Mettals grown hard, wherein they labored being alive (*Sennertus lib de consensu et diss. Chymicorum et Galenicorum)* seems to intimate as much" (Jonstonus 1657, 92).[2] A vivid emblematic evocation of the co-penetration of human body and the material of the earth can be seen in an early sixteenth-century altarpiece relief commemorating a mine accident that shows the limbs and a head of dead miners protruding from the mass of rock and earth of the collapsed shaft (Fig. 2).

This representation shows the endpoint of the struggle with matter that shaped the bodies of miners. Like other craftspeople, their specific bodily labor formed their musculature and posture, and the matter they worked marked their bodies by distinctive diseases, which is why Paracelsus (1493–1541) and others produced practical knowledge and published books on the diseases of miners.

This physical engagement with the materials of nature marked the bodies of the miners and metalworkers: the bodies of metalworkers and the very matter upon which they labored interpenetrated each other: bad breath could prevent the adhesion of metal gilding, and, conversely, metal fumes were known to shorten the lives of metalworkers. Records of their practice give evidence that they ate bread and butter before starting work because bread was viewed as a perfectly tempered food that filled the stomach and prevented the subtle metal vapours from being drawn into the body. At the same time, the hot, wet qualities of butter counteracted the miasmic exhalations of the minerals and metals rising up from the ground or billowing out from the smelting furnace. In eating butter, these artisans carried out their activities in a framework of the Greco-Roman-Arabic health worldview based on an

[2] My thanks to Daniel Margócsy for this reference.

Fig. 2 This fragment of an altar portrays a mass being said for the victims of a mine accident, which is depicted outside the frame of the church, where the heads and limbs of trapped miners can be seen under the debris of the collapsed shaft. Hans Leinberger, ca. 1510–1520, Mass for the victims of a mine accident. Limewood, Bode Museum, Staatliche Museen zu Berlin

Aristotelian framework of the four elements and qualities; that is, all things were composed of mixtures of the four elements—earth, water, air, and fire—and exhibited varying characteristics according to their qualities—wet, cold, dry, and hot. In this scheme, butter was hot; metal vapours were cold.[3] This was also the framework for Biringuccio's explanation for what occurred when bronze was poured into a cold mould: the cold, moist mould overcame the nature of bronze and turned it back to its primordial earthy-watery state (Biringuccio 1943, 250).

The human body and natural materials shared many properties. To take one example from metalworking: the idea of *temper* was crucial in the mental world of early modern Europeans. 'Temper' meant to balance by mixing, and a person's temperament was determined by a balance of the four qualities—hot, wet, dry, cold—and their instantiation in the humours: black bile, phlegm, blood, and yellow bile. Each individual's unique combination of the four qualities and humours could be tempered by the non-naturals of diet, exercise, purging, and so on. This process of tempering was crucial to human health. Metals too partook in this system and their balance of qualities could be rectified by tempering, as steel was tempered (sometimes using the urine of pre-pubescent boys), and copper was

[3]For a fuller statement of and references to the foregoing discussion, see (Smith 2014a).

given temper and tone by tin in making the alloy, bronze. Minerals and humans alike received their temper from the movements of the heavens, for the sun was a source of growth for all living things—gold grew better along riverbanks warmed by the sun and in south-facing veins, and, in the common understanding of health and identity in early modern Europe, the celestial spheres and bodies determined the temperament of human beings at their conception and birth, just as they did for metals (Smith 2014a).[4]

3 Structures Formed by Systems of Belief and Knowledge

The most widely disseminated and expansive of frameworks for practical knowledge are probably the structures of belief and knowledge systems. Such frameworks can be extremely pervasive. One such structure is what I have called elsewhere the "first universal law of nature," that is, that the heavens exercise influence over the earth. The motions of the heavenly bodies that correlate to the waning and waxing of the seasons and the agricultural cycles, the setting and rising of the sun, the ebb and flow of the tides in relation to the phases of the moon all gave common sense evidence of such a relationship (Chap. 16). This was both a lived reality, experienced and subscribed to by individuals, and a theorized relationship codified in religious rituals, megalithic monuments marking the movements of the heavenly bodies, and eventually in writing (Smith 2011). This most general structuring framework is found in metalworking as well, both in written accounts and in practices. For example, on establishing a mine, the shareholders and miners first had the site baptized by a priest "in the name of God and a fortunate outcome."[5] When loading a kiln, potters offered prayers "to God with the whole heart, ever thanking Him for all that He gives us." But they also kept their eye on "the state of the moon" for if the firing took place under a waning moon, the fire would not burn as brightly (Piccolpasso 1980, 109). This cosmic relationship is manifested also in the practice of miners praying in the chapel before going down the shafts—a practice that continued into the twentieth century (Schreiber 1962, 79–80) (Fig. 3).

Such practices all express a mental world in which the divine heavenly influence—as a 1505 treatise puts it "heaven in its course, emanations, and influence" (*Bergbüchlein*, 1505, aiir)—is felt intimately on earth. The course of the planets through the zodiac, the waxing and waning of the moon and the seasons determined all growing things, including metal ores. Like the authors of other metalworking treatises, Biringuccio viewed the characteristics of each ore as determined by the particular influence rained down upon it from its planet. Gold was linked to the sun, silver to the moon, and so on. The hierarchy of the metals mimicked that of the planets: The noblest metal, gold, sat at the apex, its power having been given it by the influence of the sun, and it was the purest and least corruptible of the metals (Biringuccio 1943, 27). A goldsmith or assayer around 1600 discussed the shaping of stones by the climate and influences of the heavenly bodies, writing:

> The pretious stones are eyther Gems, or Marble, Gems [are] grown for the most parte in the east Countries, because the temperature of the Clymatt is there fitt to generate them.... Sometyme they grow amontest the Rockes, Sometyme in the Mynes: ... and the influence of the heavens by continewance of tyme makes it a most pure stone; They are bred of other Stones, as pearles are of Sea oysters, ... wee may eselie coniecture that the caelestiall bodies are prodigall in transfusing & bestowing their vertues upon them (H.G. c. 1604, fol. 59r).

[4]See also (Smith 2011).

[5]The initiation of mining activities with an invocation to God is evident in *Das Schwazer Bergbuch*, a compilation of mining ordinances, mining information and images from ca. 1490–1556. See (Bartels et al. 2006) and (Biringuccio 1943, 17).

Fig. 3 Miners, identifiable by their distinctive hoods, emerge from a chapel before entering two shafts, pictured in the lower half of the illustration. From *Speculum metallorum* (1575), fol. 117r, *Ain Schicht* (the beginning of the shift). Stadtarchiv Calw. From (Kirnbauer 1961)

A related structuring framework, just as pervasive, was the view that the materials of the earth, and the earth itself were capable of growth, even if not fully alive or ensouled. Perhaps the most eloquent expression of this view occurs in Paracelsus' discussion of the diseases that afflict miners:

> The seeds of the metals and the minerals have been sowed in the earth. They have their fall and their harvest in order to sprout sooner or later according to the arrangement of the godly order. In the same manner, roses then cherries, pears, nuts, grapes etc. so long until the year has passed and it was a year. Although one comes later than the other, thus here too, now gold blooms in a region, now silver, here iron, there lead etc., that is past, that is present and the other future, that one's spring is past, that one's May, the other's hay month etc. for those who were there at the beginning of the world, they have reached to gold and silver of the spring with the violets, their successors have taken the gold and silver with the clover and the crowfoot etc.... (von Hohenheim 1941, 108)

Fig. 4 Native silver formations, found especially in Saxon silver mines, were prized by collectors and *Handstein* makers for their branching shapes, giving striking visual evidence to the conception of metals growing and ripening over time in the mines. Native Silver. Freiberg District, Erzgebirge. Saxony, Germany. Houston Museum of Natural Science. Public Domain. Wikipedia Commons. © Photo by sulla55

The pastor of St. Joachimsthal Johann Mathesius (1504–1565) suggested that mines and mining towns "had their time." He reported it a matter of simple experience that shafts grew smaller over time, as the minerals and earths grew within them. When miners struck bismuth they believed they "came too early" before the ore was ripe. With this human disturbance, they had ruined their chances to find the precious ore (Dym 2006, 180–1). Biringuccio speculated that antimony might be "a material that is about to reach metallic perfection, but is hindered from doing so by being mined too soon" (Biringuccio 1943, 91).

Practices and practical knowledge were also structured by analogies with the configuration and processes of the human body. Formations brought out of the *veins* in the body of the earth, such as the pure strands of native silver were self-evident proof for the growth and ripening of metals in the earth (Fig. 4).

Artisanal manuals and recipe collections mixed procedures for grafting and growing plants, fermenting liquids, and healing humans and animals with instructions for producing objects from ingredients we would now call inorganic because they were viewed as operating according to the same principles (Chap. 3). Blood and metal were related as well. We know from the work of Michael Cole and Francesca Bewer that blood and bronze were often brought into symbolic and material relationships (Bewer 2001; Cole 1999, 2002, 2003). Blood also appears among the ingredients and operations of numerous metalworking recipes: the blood of goats, oxen, foxes, rams, of a menstruating woman, and the red resin of the dragon tree.[6] The color red appears to have been seen as possessing many of the attributes of blood, and it too is mentioned frequently in both medicinal and metalworking recipes.[7]

4 Structures of Practical Knowledge Expressed in and Reinforced by Objects

Objects too could provide structures for practical knowledge, for example, the *Handstein* (Fig. 5). *Handsteine* began to be sought by noble patrons at the beginning of the sixteenth century but appear to have been preceded by vernacular arrangements of mineral specimens made by miners.[8] They were works of art and nature that gathered together interesting mineral or native metal formations, often including dendritic silver that appeared to have "grown." Such formations were seen as nature's art, "gifts of God,"[9] which often did not need to be worked further than the state in which the powers emanating from the heavens had caused them to be formed. In some cases the specimens were reshaped to make clearer the parallel between the divine order, in which God was creator of ores, and miners and smelters the creators of metals. In one *Handstein*, the blood of the crucified Christ rains down over the mining district, redeeming the world, just as the miners and smelters redeem the metals from the dross of the ore. The cosmic order and its signs can also be found in another piece of evidence: A *Bergkreye* (mining song) printed in 1530 that used the analogy of a Handstein to exemplify the connection between heaven and earth.

> Then I turn my heart and eyes alone to God
> Who shows me Handsteine/native white silver ore
> From whence my help will come/from God the lord
> Who to make us pious, created heaven and earth.[10]

This understanding is also illustrated in a 1575 manuscript, the *Speculum metallorum*, which shows, in direct analogy to the *Handstein*, the blood of Christ working through the influence of the seven

[6]On different sorts of blood, see for example the recipes in the collection written down in the Monastery of Tegernsee in the fourteenth and fifteenth centuries: (Bartl et al. 2005) and especially the discussion by (Krekel and Lautenschlager 2005, 675).

[7]I treat this in (Smith 2010).

[8]Johannes Mathesius, a pastor in St. Joachimsthal in the 1560s, mentioned that miners had given him all kinds of minerals and rock specimens: (Mathesius 1562, Vorrede, n.p.) See also, (Schreiber 1962) who gives examples of remarkable mineral specimens being used as votive offerings and in church ornament (p. 621) in the sixteenth through eighteenth centuries, as well as a miraculous mineral specimen which exhibited an image of Mary with the baby Jesus, found in 1669 (621–2). Very little work has been done on these extraordinary works of art and nature. For an introduction to Handsteine, see (Huber 1995) and further, (Haug 2012).

[9]In an inventory, a stone is described as illustrating "die gab gottes" (Slotta et al. 1990, 566) quoting from 1596 inventory of Ferdinand II's (1578–1637) Kunstkammer in Schloß Ambras. Interestingly, Heather Lechtman writes of an analogous practice among Andean miners as well: "Miners selected, safeguarded and venerated certain exceptional ore specimens they extracted from individual mines. These metallic minerals were called *mamas*. *Mamas* were distinguished by some special quality of suite of qualities: richness of the metal content of the ore, extraordinary size, vivid color, unusual shape, rare beauty. *Mamas* were *huacas,* sacred stones; each mine had its own *huaca*, regardless of the kind of metallic ore it represented: gold, silver, copper, mercury." (Lechtman 1999, 228).

[10]"Denn wendet mein hertz und augen/erhaben alleine zu Gott/

Der lies mich handsteine schauen/gedigen ertz güldigrodt/

Von wannen mire hülff wird komen/die ist von Gott dem Herrn/

Der uns allen zu fromen/gemacht hat himel und erdn." (printed in Zwickau bei Meierpeck in 1530), quoted in (Slotta, et al. 1990, 563). The information on *Handsteine* comes from (Slotta, et al. 1990, 562–3).

Fig. 5 Handstein with mine scene and Crucifixion, Bohemian, mid-1500s, silver, gilding, wood, mineral specimens, including proustite, argentite, marcasite, lautite, malachite, quartz, fluorite (27.1 cm high). The *handstein* portrays huts covering mine shafts, and, similar to the next illustration from the *Speculum metallorum*, it shows the Blood of Christ raining down, redeeming the world, just as the miners and smelters work in the bowels of the earth to redeem the metals from the dross of the ore. PA 4167 Böhmisch, Mitte 16. Jh., Handstein mit Bergwerk und Kruzifix. KK 4167, KHM-Museumsverband

planets on the earth-water-air globe of the earth to form the seven metals at its center. The seven metals are represented by horizontal and vertical bands of different colors (Fig. 6).

A remarkable example of the expression and reinforcement of belief systems in an object can be found in the depiction by the master goldsmith of Nuremberg, Wenzel Jamnitzer, of the moment at which Daphne was turned into a laurel tree in order to escape the unwanted attentions of Apollo. In his work, Jamnitzer (1508–1585) philosophizes about metamorphosis and transformation, as he notes the flow of silver and gold from the metal ores at the base, through the blood-red coral arms, from which tiny green leaves cast from life begin to sprout. The sculpture embodies multiple levels of meaning: the tiny fragments of ore at the base of the sculpture allude to the transformation of raw ores by human art into precious metals, many of the mineral specimens in the base also refer to metamorphosis, the coral of the arms is seen as a "juice of the earth" that causes stones and metals to grow in the earth, and coral is the result of metamorphosis likened in the sixteenth century to blood flowing in the veins.[11] The sprouting leaves, cast from life, represent the new life in which all these processes of transformation culminate (Figs. 7, 8 and 9).

[11] See (Bimbenet-Privat and Kugel 2007) and (Grasskamp 2013).

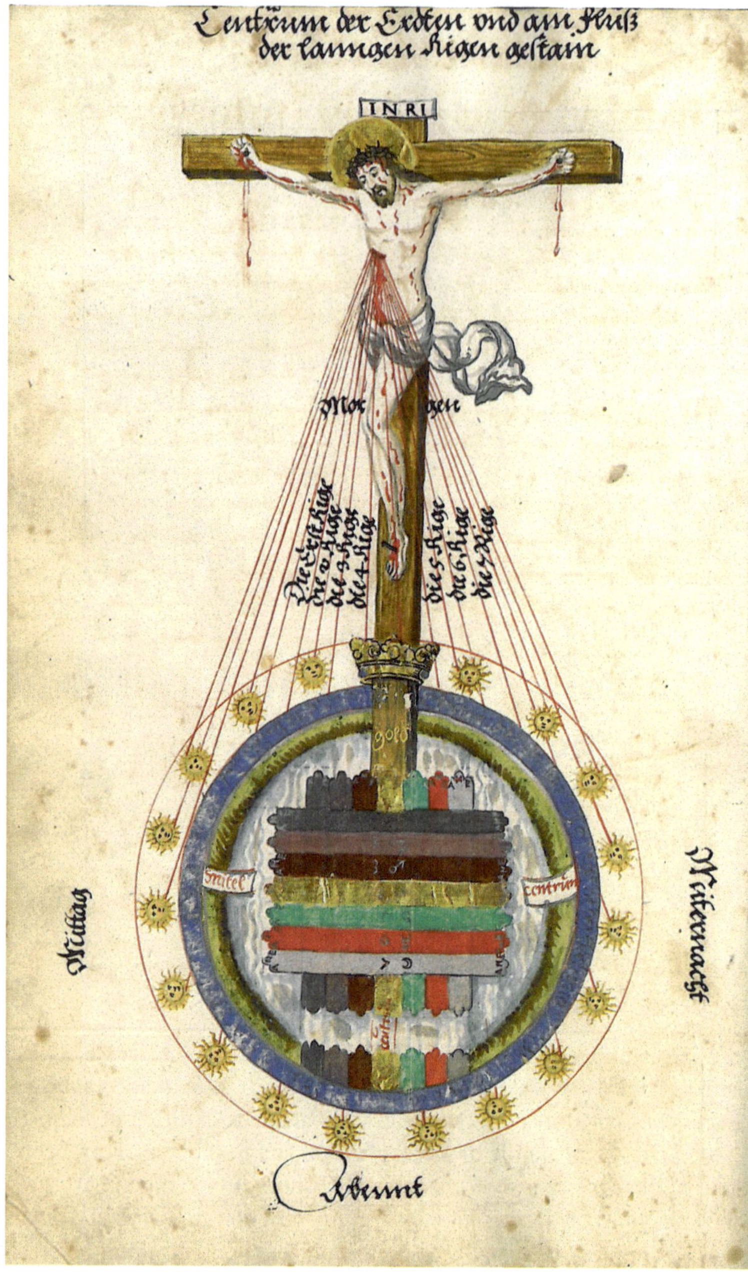

Fig. 6 The blood of Jesus on the Cross forming minerals in the earth. From *Speculum metallorum* (1575), fol. 56v Stadtarchiv Calw (Kirnbauer 1961)

Such objects not only expressed and codified a system of belief and knowledge,[12] but they also testified to the material, physical, and philosophical engagement of craft with the generative and transformative powers of nature.

[12]For a discussion of the synoptic form of knowledge contained in an object, see (Haug 2014).

Fig. 7 Gilded silver, coral and semi-precious stones. Wenzel Jamnitzer (1508–1585), *Daphne* (around 1570–1575). H.665 m.E. CL 207750. Courtesy of the Musée national de la Renaissance, Château d'Ecouen, France

Fig. 8 Detail of base with "generative" stones. Wenzel Jamnitzer, *Daphne* (around 1570–1575). H.665 m.E. CL 207750. Detail. Courtesy of the Musée national de la Renaissance, Château d'Ecouen, France. © Photo by Pamela H. Smith

Fig. 9 Detail of leaves (cast from life). Wenzel Jamnitzer, *Daphne* (around 1570–1575). H.665 m.E. CL 207750. Detail. Courtesy of the Musée national de la Renaissance, Château d'Ecouen, France. © Photo by Pamela H. Smith

5 The Structuring Frame of Polarity

Many recipes in early modern metalworking books attempt to "fix" volatile materials such as mercury to make them endure the fire and become malleable or workable, and conversely to make solid materials plastic and capable of being molded. Recipes for softening hard substances, such as liquefying stone or horn in order to be able to cast them, and the related process of making brittle materials malleable, as happens in the transformation of iron into steel, are to be found in large numbers in such recipe collections. This emphasis on changes of state indicates both the nature of many of these recipes—they aimed at varying the qualities of metals by putting them through phase processes, sometimes by slow, constant, and long-term heating, such as that of an egg sitting under a broody hen or by the heat that is given out by the slow fermentation of manure, while others produced changes of state, such as dissolution in acid, or transformation through heating, or tempering as in the production of steel. The polar oppositions of Aristotelian qualities—hot, dry, wet, and cold—were a fundamental structuring framework for all knowledge—practical and otherwise—in early modern Europe. An examination of craft writings shows that we can add to these qualities the polarities hard-soft, brittle-malleable, and lean-fat as well. Many metalworking recipes aimed at products that tempered, if not always reversed, the antipathetic qualities of lean and fat, hard and soft, wet and dry, and so on (Krekel and Lautenschlager 2005, 673–8).[13] Gunners believed that gunpowder exploded when contraries were mixed (Werrett 2009, 27 and 33). For Biringuccio, who disavowed what he called the alchemical idea of sulfur and mercury as the underlying principles of metals, believing instead with Aristotle that metals were composed of water, and who discussed metals in terms of their qualities (which precisely corresponded, incidentally, to the qualities of sulfur and mercury) viewed metals, earths and minerals as either lean (brittle or dry) or fat (unctuous and more liquid or metallic) (Biringuccio 1943, 46.). Like other metal workers, he believed all metals could be rectified of their "corruptions" through tempering, that is, by the addition of other substances, by treatment with another metal or with fluxes, or tempering by heating or cooling. Tin could alter the "tractable sweetness" of other metals, corrupting

[13] An excellent example of these kind of recipes that attempt to bring about opposite states can be found in (*Künstbüchlin gerechten gründlichen gebrauches aller kunstbaren Werckleüt* 1535).

them, which in some cases was desirable, such as when tin was added to copper in bell metal to make it more sonorous, "just as if it puts the spirit there and vitalizes the substances" (Biringuccio 1943, 61). For Biringuccio, making harder was a corruption of the metal, for hardness or brittleness reduced the essential metallic quality of metals, that is, their malleability. When a metal was lean, it caused porosity in casting, and the metal needed to be tempered with an unctuous substance. Conversely, a mold made of a lean substance would receive the unctuousness of metal particularly well (Biringuccio 1943, 324).

In *The Nature and Art of Workmanship*, David Pye regarded this effort at tempering qualities as part of an aesthetic system (which, of course, also forms another structure for practical knowledge) and as expressing a demonstration of artisanal skill:

> Much of the pleasure these things give us comes from the very fact of 'soft' properties being expressed in a hard material to which they are quite foreign. In a similar way, perhaps the most constant and delightful aesthetic phenomenon throughout the history of sculpture has been this very expression in hard stone of the properties of soft materials like flesh, hair, and drapery. The stone remains recognizably stone yet the hair is recognizable of hair and the cloth as cloth. (Pye 1968, 47)

According to Pye, this effort constitutes the essence of skill, or "workmanship," as the craftsperson strives to produce objects that display diversity, durability, and equivocality (by which he means a metal showing both its properties of liquid and solid, for example, in a sculpture). We might add that the craftsperson also strives to imitate and even compete with the artifice of nature in such an effort, and such imitation and rivalry possessed a long and important genealogy in European craft and art, embodied above all by the enduring use in literature and works of art of Ovid's *Metamorphoses* to allude to both material and symbolic transformation. But such competition with nature can also be regarded simply as a fundamental part of exploring, testing, trying, and coming to understand the properties and behavior of natural materials. These polarities in metalworking recipes, then, set forth a framework for working and thinking through the properties of materials and the processes and products they made possible. And, they articulated a quotidian mode of work by which an artisan struggled with his materials to effect a transformation in their properties and characteristics towards their opposites.

6 The Structuring Framework of Vernacular Knowledge Systems and Their Codification in the Texts of Scholars

Let us return, finally, to the framework for practical knowledge that we find in the texts of scholars working in the mining boom in the towns of St. Annaberg, Joachimsthal, Chemnitz, Zwickau, Freiberg, and in neighboring Bohemia, where silver-bearing copper ores were mined on an unprecedented scale. Two scholar-physicians living and working in these boom towns, *Stadtphysicus* and eventual *Bürgermeister,* Ulrich Rülein von Calw (ca. 1465–1523) in his single *Bergbuchlein* (published anonymously ca. 1500), and Georgius Agricola (1493–1555) in his numerous works on mining and metals published between 1530 and 1556, both practiced as physicians and sought to codify knowledge about mining. As silver production increased five-fold in this period, and towns grew from small hamlets to tens of thousands of inhabitants in a matter of years, von Calw and Agricola were confronted in their medical duties as city physicians and in the material conditions of their lives in these boom towns, with a world far different from that which their training in ancient languages and classical medical authors had prepared them. As city physicians both these men dealt with the illnesses of miners, and as municipal officials and mine shareholders they were naturally interested in the discovery of rich ores as the basis of both civic and personal wealth. The discovery of ores necessitated an understanding of where metals "grew" in the earth. In theorizing about the origin and genesis of metals, these two physicians could draw upon various structuring frameworks from the texts of authorities: Aristotle's very brief discussion in *Meteorologica* that rocks and metals were formed by the exhalation of hot vapors which then congealed in the cold of the earth; Theophrastus on earths and stones; Pliny's very brief mentions of mining; medieval gem books; Albertus Magnus's (1206–1280) *De Mineralibus*; Avicenna and other medical authorities, as well as alchemical authors who posited the "principles" of sulfur and mercury as the underlying substratum of all metals. Apart from Albertus, none of these works, however, treated mining or had very much to say about prospecting and the genesis of metals in the earth. Instead, both von Calw's single *Bergbuchlein* and Agricola's books had to abandon the texts of *auctoritates* for the experience of mining, and their books are marked by their exchange with miners. The *Bergbuchlein* is written as a dialogue between Daniel, a *Bergverständiger* and his apprentice (Pieper 1955; Sisco and Smith 1949). Agricola modeled his first work on mining, *Bermannus, sive de re metallica dialogus* (1530), directly on the *Bergbuchlein*, writing it as a dialogue between a "learned miner, *Bermannus*" and two traditionally trained physicians, one expert in Latin and one in Arabic and Greek. The miner teaches about mining terms and concepts which gives rise to conversation among the scholars about the relationship of these new terms to ancient texts, practices, and medicines. In this work, Agricola displays philological scholarship of the highest standard, but he also repeatedly employs the observations of miners to give credence to his ideas.

For example, in his volume on subterranean life, *De animantibus subterraneis liber* of 1549, Agricola discussed the sprites and spirits that miners believed to be the cause of underground "weathers" or *Witterungen*, often referred to as *Dünste, Broden, Dämpfe,* and *exhalationes*. These phenomena were believed to be given off as the minerals grew in the earth, and stirred up into a wind by the air that moved through the earth, an idea that was derived in part from Aristotle's work.[14] Agricola classified the spirits into evil and simply mischievous, recounting an example from the Rosenkranz mine in which an evil spirit killed twelve miners through its miasmic breath. The other less destructive *Geister*, which were known by the Germans as *Kobolden,* imitated humans, giggled "as if from joy," and amused themselves with harmless deceptions. Some called them *Bergmännlein,* because they were dwarf-like, and, while they looked like old men and were dressed like miners, they occupied themselves with hurrying around in the mines, appearing to do work, but actually doing nothing. Although they occasionally annoyed the workers with throwing gravel, they hardly ever did any harm, unless ridiculed by the miners. In this, Agricola writes, they were similar to the spirits who appeared seldom to humans but every day did a part of housework and took care of the animals. The *Berggeister* worked mainly in mines that had proven themselves or about which investors were hopeful, thus the miners were not afraid of them, but take it as a good sign and "are of good spirit and work all the harder" (Agricola 1961, 199–200).[15]

Besides ideas about mining spirits, Agricola also appears to have assimilated into his theory about the genesis of metals under the earth a vernacular knowledge system, that is, a structure or paradigm of practical knowledge current among miners and metalworkers that was expressed in the terms of "lean" and "fat." Miners spoke of *Guhr* or *Steinmark* (Dym 2006, 179–80), which was translated by generalizing theorists such as Agricola as the "marrow of stones" (Agricola 1955a, 158). This was analogous to the fatty substance of bone marrow, and was seen to be responsible for the generation of mineral ores. The paradigm of lean and fat seems to have formed a common currency in mines, smelting workshops, and foundries. In *De la pirotechnia*, Biringuccio explained the strength and durability of crucibles by reference to fat and lean, writing that use of lean clay that "seems to be arid" will make the crucible more able to resist the fire. He also employs fat and lean extensively in relation to alloying

[14]For an overview, see (Dym 2006, 170–2).

[15]Paracelsus, too, wrote that underground "weather" could be caused by the spirits and sprites of the subterranean world. Paracelsus noted that in mines all over Germany "*Bergmännlein*" made their homes (Webster 1982).

metals and often advises throwing in some lard or other animal fat in alloying (Biringuccio 1943, 296). He notes that in casting bronze the channels through which the metals flow into the mold should be greased with pig fat or tallow in order to make the metals more fluid (Biringuccio 1943, 257). In casting bronze cannons, he explains the need to have a quantity of bronze at the mouth of the mould so that it presses down and makes the other parts "fat." He advises adding tin to the furnace or putting it in the channels down which the bronze flows so that the feeding head becomes "fat with tin" where it will press down and give "fatness to the muzzle," thus producing a cast that is "close, dense, and safe, and pleasing to the sight" (Biringuccio 1943, 260). Without this, the metal would remain "lean and brittle." Again, in casting guns, tin must be added right at the end of the pour to add "unctuousness," so that the metal mass is solid and without small holes, which would result from a lean mixture (Biringuccio 1943, 297). Today, we would term this process "inverse segregation," in which the addition of tin compensates for the fact that in the last parts of the alloy to solidify, tin content is lower and thus the metal does not flow and often is more brittle at that point (Biringuccio 1943, 260n.). For Biringuccio, animal fats, tallow, tin, fluxes such as borax and crushed glass, all kinds of gum, ram's horn scrapings, and the smoke of pine resin and turpentine were all unctuous and fatty, and thus the cause both of the metals running more fluidly and the ability to produce a denser cast (Biringuccio 1943, 332–3.).

Agricola employs fat and lean extensively. In *De natura fossilium*, he classified earths and minerals by their fat or lean qualities, with many grades of dryness and unctuousness (Agricola 1956, 124). In order to test the leanness of an earth, it should be softened with water, thoroughly worked with the hands, then a small ball formed and thrown on ground. If it cracks, it is "meager." If not, it is unctuous. Some earths are so unctuous they melt like butter on the tongue (Agricola 1955b, 22). As well as drawing from Pliny and Theophrastus, Agricola appears to have studied the practices of farmers, detailing the soils cultivated by farmers and those used by potters, fullers, painters, carpenters and other artisans. Farmers, he wrote, select unctuous sandy soil for growing sesame; indeed, he noted that farmers classify soils first according to fertility, sterility or intermediate qualities, then according to denseness and porosity, and finally according to taste (Agricola 1955b, 23). Agricola details twenty-one species of earths of interest to farmers.

The polarity of fat and lean informed Agricola's theory of the genesis of minerals in the earth as well. In *De ortu et causis suberraneorum libri V* (Froben, 1544), Agricola asks what causes the heat that miners encounter as they dig down into the earth? He notes that there are many theories about this put forward by learned authors, and they include heating by the sun; wind penetrating into the earth and heating up by its constant motion; water becoming hot by falling on the rocks, but miners, Agricola writes, can testify every working day that none of these happen (Agricola 1956, 95). Rather, Agricola says that by carefully observing the veins, one can conclude that water flows along the veins and deposits its generative component into receptacles (the *matrix*) in which the veins then grow and burst out onto the surface in outcrops. Agricola saw veins as generative and formative receptacles (like wombs or matrices) in which the ores form. Miners knew that every vein had a floor and a ceiling, it had clefts (at which cross-shaped intersections the vein could be especially rich), and outcroppings where it burst onto the surface. As the water ran through the earth and settled into these vessels, it took on certain characteristics. For example, if it ran through soft rock, such as *Kalkstein*, it produced fat earths such as chalk, clay, marl, and when it ran through sandstone, lean earths were created. Once deposited in the veins, it was cooked by a fire kept burning by a fat substance, such as bitumen or sulfur (Agricola 1956, 98). Sometimes this fire burned so hot that it drove out the stone marrow from the earth, such as when sulfur protruded from the earth in volcanic regions (Agricola 1956, 140).[16] For

[16]In this theory (Agricola 1956, 133), Agricola was arguing against Albertus, *De mineralibus*, who wrote that the growth of different minerals is to be attributed to the power of the stars (Agricola 1956, 180), and against Paracelsus'

Agricola, fat earths resulted from slow fermentation in the vessels and matrices of the earth. In this idea of the generation of fat earths, Agricola might have been following the "secrets" of metalworkers, as recorded by Benvenuto Cellini (1500–71), who made a particularly unctuous clay (resembling a medicinal "salve") for small molds by mixing clay and wool with water until it resembled bread dough, beating it with an iron rod, and fermenting it for some months in the summer. This fermentation was viewed as giving rise to the life-giving fat (Cellini 2005, 169). This paradigm of fat and lean (and the generative nature of putrefaction) thus appears to have arisen not only in observation by and of farmers, but also from the practices of foundrymen in which it played a central part in the making of molds and crucibles, and in the alloying of metals. Fat earths were also central in mine prospecting, as miners sought out *Steinmark* or *Steinomarga*, "stone marrow," and its extrusion in ore-bearing veins. In all these instances, the paradigm of fat and lean emerges out of a practice that works, producing something of value, and thereby informing vernacular theory, which is then codified in the writings of the scholar Agricola.

In the seventeenth and eighteenth centuries, chemical philosophers such as Johan Rudolf Glauber and Herman Boerhaave identified stone marrow, which they called by the miners' term, *Guhr*, with the generative substance of metals,[17] and it remained an object of investigation into the nineteenth century. For seventeenth-century (al)chemical theorizers, "Fatty earth" (*terra pinguis*) would also come to be seen as a foundational and transformative element in many material processes, informing the concept of "phlogiston" and then "oxygen". Even Isaac Newton (1642–1726) was interested in "Gur" (Alfonso-Goldfarb and Ferraz 2013).[18]

7 Conclusions

This essay began by reviewing briefly how practical knowledge itself has been understood to be structured in logical terms, i.e., by generalizing efforts that might sometimes be called "theory". The rest of this chapter has explored a number of structuring frameworks for the practical knowledge of mining and metalworking, including the place and economic structures of material production, the entanglement of the human body with matter in the landscape, and the constraining and affording properties of the materials themselves. Larger belief and knowledge systems, such as understandings about the relationship of earth and the heavens, as well as analogies to the human body and its processes, also structured practices and practical knowledge. Objects could structure and reinforce practical knowledge systems, as we saw in the example of the *Handstein*. The essay then considered the structuring of practical knowledge via vernacular knowledge systems embodied, for example, in the paradigm of lean and fat, and other polarities. All these structures of practical knowledge existed as part of the complex of mining itself, in which multiple actors regarded prospecting and a theory about the genesis of metals as crucially important for civic, territorial, and economic ends.

Finally the essay examined the codification of these vernacular "theories" in the books of scholars, where they found their way into the growing discourse of experimental philosophy (an increasingly powerful structuring framework over the course of the early modern period), and from there into modern science, itself an unprecedently powerful structure for practical knowledge.

statement in *On the Miners' Sickness*, that "The seeds of the metals and the minerals have been sowed in the earth" (Agricola 1956, 108).

[17] On Glauber's equation of '*Gur*' with sulfurous spirit of vitriol (sulfuric acid), see (Roos, 2007, 41, 46). On Boerhaave's use of *Guhr*, see (Powers 2007).

[18] For Isaac Newton on 'Gur,' see (Newman 2007, 91).

References

Agricola, Georgius. 1955a. *Bermannus, sive de re metallica*. Basel: Froben, 1530. Trans. Helmut Wilsdorf. Berlin: VEB Deutscher Verlag der Wissenschaften.

Agricola, Georgius. 1955b. *De natura fossilium*. Ed. and Trans. Mark Chance Bandy and Jean A. Bandy. New York: Geological Society of America.

Agricola, Georgius. 1956. *De ortu et causis subterraneorum libri V.* Froben, 1544; republ. 1555. *Ausgewählte Werke*. Trans. Georg Fraustadt and Hans Prescher. Berlin: VEB Deutscher Verlag der Wissenschaften.

Agricola, Georgius. 1961. *De animantibus subterraneis liber*. Basel: Froben, 1549, revised 1556. *Ausgewählte Werke*. Trans. Georg Fraustadt and Rolf Hertel. Berlin: VEB Deutscher Verlag der Wissenschaften.

Alfonso-Goldfarb, Ana Maria, and Marcia H.M. Ferraz. 2013. Gur, Ghur, Guhr or Bur? The quest for a metalliferous prime matter in early modern times. *The British Journal for the History of Science* 46: 23–37.

Anon. 1535. *Kunstbüchlin gerechten gründlichen gebrauches aller kunstbaren Werckleüt*. Augsburg: Heinrich Steyner.

Anon. [Ulrich Rülein von Calw]. 1505. *Ain wolgeordnetß unnd nuetzlichs Büchlin wie man bergwerck süchen und erfinden sol [Das Bergbüchlein]*. Augsburg: Erhart Ratdoldt. Gallica, Bibliothèque nationale de France: https://patrimoine.mines-paristech.fr/document/Bergbuchlein Accessed 31 May 2016.

Bartels, Christoph, Andreas Bingener, and Rainer Slotta. 2006. *"1556 Perkwerch etc." Das Schwazer Bergbuch*, vol. 1–3. Bochum: Deutsches Bergbau-Museum Bochum.

Bartl, Anna, Christoph Krekel, Manfred Lautenschalger, and Doris Oltrogge. 2005. *Der "Liber illuministarum" aus Kloster Tegernsee*. Stuttgart: Franz Steiner Verlag.

Bewer, Francesca G. 2001. The sculpture of Adriaen de Vries: A technical study. In *Small bronzes in the Renaissance*, ed. Debra Pincus, 159–193. Washington, DC: Center for Advanced Study in the Visual Arts.

Bimbenet-Privat, Michèle, and Alexis Kugel. 2007. La Daphné d'argent et de corail par Wenzel Jamnitzer au musée national de la Renaissance. *La Revue du Louvre et des musées de France* 4: 62–74.

Biringuccio, Vannoccio. 1943. *The Pirotechnia*. Trans. Cyril Stanley Smith and Martha Teach Gnudi. New York: Basic Books.

Cellini, Benvenuto. 2005. *Traktate über die Goldschmiedekunst und die Bildhauerei (I Trattati dell'oreficeria e della scultura di Benvenuto Cellini)*. Trans. Ruth and Max Fröhlich, als Werkstattbuch kommentiert und heraugegeben von Erhard Brepohl. Cologne/Weimar/Vienna: Böhlau Verlag

Cole, Michael W. 1999. Cellini's blood. *The Art Bulletin* 81 (2): 215–235.

Cole, Michael W. 2002. *Cellini and the principles of sculpture*. Cambridge: Cambridge University Press.

Cole, Michael W. 2003. The Medici *Mercury* and the breath of bronze. In *Large bronzes in the Renaissance*, ed. Peta Motture, 129–153. Washington, DC/New Haven: National Gallery of Art/Yale University Press.

Dym, Warren Alexander. 2006. Mineral fumes and mining spirits: Popular culture in the *Sarepta* of Johann Mathesius (1504–1565). *Renaissance and Reformation Review* 8: 161–185.

Dym, Warren Alexander. 2010. *Divining science: Treasure hunting and earth science in early modern Germany*. Leiden: Brill.

Grasskamp, Anna. 2013. Metamorphose in Rot. Die Inzenierung von Korallen-fragmenten in Kunstkammern des 16. und 17. Jahrhunderts. In *Metamorphosen,* a special issue of *Tierstudien*, eds. Jessica Ullrich and Antonia Ulrich, 4: 13–24.

H. G. ca. 1604. The Goldsmith's Storehouse. V.a. 179. Folger Shakespeare Library. http://luna.folger.edu/luna/servlet/detail/FOLGERCM1~6~6~828866~154160:The-goldsmiths-storehouse--manuscri. Accessed 31 May 2016.

Haug, Henrike. 2012. 'Wünderbarliche Gewechse:' Bergbau und Goldschmeidekunst in 16. Jahrhundert. *Kritische Berichte* 3: 49–63.

Haug, Henrike. 2014. Artificial interventions in the natural form of things: Shared metallogenetical concepts of Goldsmiths and Alchemists. In *Laboratories of art: Alchemy and art technology from antiquity to the eighteenth century*, ed. Sven Dupré, 79–103. New York: Springer.

Hermens, Erma. 2013. The *botteghe degli artisti:* Artistic enterprise at the della Rovere and Medici courts in the late sixteenth century. In *The Renaissance workshop*, ed. David Saunders, Marika Spring, and Andrew Meek, 105–113. London: Archetype.

Huber, Peter. 1995. Die schöneste Stuffe: Handsteine aus fünf Jahrhunderten. *ExtraLapis* 8: 58–67.

Jonstonus, Joannes. 1657. *A history of the wonderful things of nature*. London: John Streater.

Kieffer, Fanny. 2014. The laboratories of art and Alchemy at the Uffizi gallery in renaissance Florence: Some material aspects. In *Laboratories of art: Alchemy and art technology from antiquity to the 18th century*, ed. Sven Dupré, 105–127. New York: Springer.

Kirnbauer, Franz. 1961. *Speculum metallorum 1575*. Wien: Montan-Verlag.

Krekel, Christoph, and Manfred Lautenschlager. 2005. Bearbeitung von Glas, Edelstein, Bein und Horn. In *Der "Liber illuministarum" aus Kloster Tegernsee*, ed. Anna Bartl, Christoph Krekel, Manfred Lautenschalger, and Doris Oltrogge, 673–678. Stuttgart: Franz Steiner Verlag.

Lechtman, Heather. 1999. Afterword. In *The social dynamics of technology: Practice, politics, and world views*, ed. Marcia-Anne Dobres and Christopher R. Hoffman, 223–232. Washington/London: Smithsonian Institution Press.

Lehmann, Ann-Sophie. 2009. Wedging, throwing, dipping and dragging – How motions, tools and materials make art. In *Folded stones*, ed. Barbara Baert and Trees de Mits, 41–60. Ghent: Institute for Practice-Based Research in the Arts.

Lehmann, Ann-Sophie. 2012. How materials make meaning. *Netherlands Yearbook for History of Art* 62 (1): 6–26.

Mathesius, Johannes. 1562. *Sarepta oder Bergpostill*. Nuremberg: Johann von Berg and Ulrich Neuber.

Nef, Jon U. 1987. Mining and metallurgy in medieval civilisation. In *The Cambridge economic history of Europe. Vol. 2: Trade and industry in the middle ages*, 2nd ed, ed. M.M. Postan and Edward Miller, 691–761. Cambridge: Cambridge University Press.

Newman, William R. 2007. Newton's theory of metallic generation in the previously neglected text 'Humores minerales continuo decidunt.' In *Chymists and chymistry. Studies in the history of alchemy and early modern chemistry*, ed. Lawrence M. Principe, 89–99. Sagamore Beach: Watson Publishing International.

Piccolpasso, Cipriano. 1980. *The three books of the potter's art (ca. 1558)*, 2 vols. Trans. Ronald Lightbown and Alan Caiger-Smith. London: Scholar Press.

Pieper, Wilhelm. 1955. *Ulrich Rülein von Calw und sein Bergbüchlein, Freiberger Forschungshefte*. Berlin – Ost: Akademie Verlag.

Polanyi, Michael. 1998. *Personal knowledge: Towards a post-critical philosophy* (1962) reprinted. London: Routledge.

Powers, John. 2007. Scrutinizing the Alchemists: Herman Boerhaave and the testing of chymistry. In *Chymists and chymistry: Studies in the history of alchemy and early modern chemistry*, ed. Lawrence M. Principe, 227–238. Sagamore Beach: Science History Publications.

Pye, David. 1968. *The nature and art of workmanship*. Cambridge: Cambridge University Press.

Roos, Anna Marie. 2007. *The salt of the earth: Natural philosophy, medicine, and chymistry in England, 1650–1750*. Leiden: Brill.

Roy, Ashok, Lorne Campbell, and Susan Foister. 1997. The materials and technique of five paintings by Rogier van der Weyden and his workshop. *National Gallery Technical Bulletin* 18: 68–86.

Schreiber, Georg. 1962. *Der Bergbau in Geschichte, Ethos und Sakralkultur, Wissenschaftliche Abhundlungen der Arbeitsgemeinschaft für Forschung des Landes Nordrhein-Westfalen, 21*. Cologne/Opladen: Westdeutscher Verlag.

Sisco, Anneliese Grünhaldt, and Cyril Stanley Smith. 1949. *Bergwerk- und Probierbüchlein; a Translation from the German of the Bergbüchlein, a sixteenth-century book on mining geology, by Anneliese Grünhaldt Sisco; and of the Probierbüchlein, a sixteenth-century work on assaying*. New York: American Institute of Mining and Metallurgical Engineers.

Slotta, Rainer, Christoph Bartel, Mitarbeit von Heinz Pollmann, and Martin Lochert. 1990. *Meisterwerke bergbaulicher Kunst vom 13. bis 19. Jahrhundert. Katalog zur Ausstellung des Deutschen Bergbau-Museums Bochums und des Kreises Unna auf Schloß Cappenberg vom 6.9.–4.11.1990*. Bochum: Selbstverlag des Deutschen Bergbau-Museums.

Smith, Pamela H. 2010. Vermilion, mercury, blood, and lizards: Matter and meaning in metalworking. In *Materials and expertise in early modern Europe: Between market and laboratory*, ed. Ursula Klein and Emma Spary, 29–49. Chicago: University of Chicago Press.

Smith, Pamela H. 2011. Science. In *A concise companion to history*, ed. Ulinka Rublack, 268–297. Oxford: Oxford University Press.

Smith, Pamela H. 2014a. Making as knowing: Craft as natural philosophy. In *Ways of making and knowing: The material culture of empirical knowledge*, ed. Pamela H. Smith, Amy R.W. Meyers, and Harold J. Cook, 17–47. New York/Ann Arbor: Bard Graduate Center/University of Michigan Press.

Smith, Pamela H. 2014b. Between nature and art: Casting from life in sixteenth-century Europe. In *Making and growing: Anthropological studies of organisms and artefacts*, ed. Elizabeth Hallam and Tim Ingold, 45–63. Farnham: Ashgate Publishing.

Smith, Pamela H. 2015. The matter of ideas in the working of metals in early modern Europe. In *The matter of art: Materials, practices, cultural logics, c. 1250–1750*, ed. Christy Anderson, Anne Dunlop, and Pamela H. Smith, 42–67. Manchester: Manchester University Press.

Von Hohenheim, Theophrastus (Parcelsus). 1941. *On the miners' sickness and other miners' diseases*. Trans. George Rosen. In *Four treatises of Theophrastus von Hohenheim called Paracelsus*, ed. Henry E. Sigerist. Baltimore: The Johns Hopkins University Press.

Webster, Charles. 1982. Paracelsus and Demons: Science as a synthesis of popular belief. In *Scienze, credenze occulte, livelli di cultura*, 3–20. Florence: Leo S. Olschki.

Werrett, Simon. 2009. *Fireworks: Pyrotechnic arts and science in European history*. Chicago: University of Chicago Press.

Preserving the Cutting Edge: Traveling Woodblocks, Material Networks, and Visualizing Plants in Early Modern Europe

15

Bruce T. Moran

Abstract

This study examines changes in the epistemic status of recycled botanical woodblocks brought about by shifts of social, economic, and intellectual purpose. Unmoored from their original intent, some woodblocks, once created, began to travel, finding their way into sites remote from their places of origin, and influencing processes of visualizing the world by fitting, with renewed purpose, into the space between the objects of nature and the imagination. As traveling objects of representation and as embodiments of plant depictions, woodblocks figure into the material culture of shared practices in the early modern era, transmitting specific images for diverse purposes while communicating the artistic and artisanal practices embedded in them as artifacts. They constitute in this way a relatively stable material source of visual imagery; physically fixed but culturally mutable objects that help revise and reshape the perception of the natural world.

An additional dimension of the study employs features of network theory to explore the "small world" of artists, draftsmen, cutters, printers and publishers that made and then reused woodblocks, focusing especially upon the conditions of woodblock use and reuse in two traditions. The first relates to the creation of woodblocks depicting plants commissioned by Leonhard Thurneisser (1531–1596) which were later reused by Thomas Panckow (1622–1665) and Batholomaeus Zorn (1639–1717). The second examines the creation of the woodblocks used by Pietro Andrea Mattioli (1501–1577) in sixteenth century editions of his *Commentaries on Dioscorides* and which continued to be reused into the eighteenth century. I argue that the reuse and repurposing of woodblocks is an obvious economic feature of shared practices among the processes of plant illustration

B.T. Moran (✉)
Department of History, University of Nevada, Reno, NV, USA
e-mail: moran@unr.edu

M. Valleriani (ed.), *The Structures of Practical Knowledge*, DOI 10.1007/978-3-319-45671-3_15

in the early modern era. In making woodblocks, and in the practices of printing images from them, knowledge that related to scribal, print, and xylographic traditions intersected. Social roles also overlapped as artists, draftsmen, cutters, and printers joined in associations linking their labors and materials to create compelling and reuseable "working objects" as a way to envision specific parts of nature.

Sometime in the early 1650s a Brandenburg court physician named Thomas Panckow (1622–1665) came into the possession of what he described as "a whole pile" of artfully cut woodblocks that naturally portrayed the images of plants and herbs (Panckow 1656, Günstiger Leser).[1] The blocks were left over from a project begun 80 years earlier by the Berlin court alchemist and physician Leonhard Thurneisser (1531–1596). Aided by a number of skilled assistants, Thurneisser produced, in the 1570–1580s, a steady flow of curiously illustrated books that followed in most part from the medical cosmology of Paracelsus (1493–1541) and focused upon the revelation and use of natural subtleties, the spiritual powers embedded in parts of nature that accounted for their vitality and medicinal efficacy (Morys 1982, 25–42). In regard to the investigation of plants, one book stands out prominently, his *Historia sive descriptio plantarum omnium*, published in both Latin and German editions in 1578 with privileges from the Holy Roman Emperor, the King of Poland, and the Duke of Tuscany. In its preface Thurneisser announced that 1921 images of foreign and domestic plants had already been cut and awaited printing (Thurneisser 1578a, Vorrede). As was typical of Thurneisser volumes, the book aimed to entertain as well as to inform. In recounting ancient, Paracelsian, and magical uses of plants and in connecting individual plants to the *tria prima*, sulfur, salt, and mercury, Thurneisser and his associates fashioned a paper theater, one that made the familiar strange and contained wonders, remedies, and stories involving the use of plants sometimes for purposes too scandalous to recount in full. For added effect and amusement, the book included curious images of animals representing various types of distillation (*per descensum, per ascensum, per latus*) and extraction. It also fascinated *via* blog-like entries that specified the treatment of age and gender specific diseases and advised about the management of problems peculiar to the male and female reproductive systems (Fig. 1).

The cost of production must have been considerable—Thurneisser employed well known artists including Hermann tom Ring (1521–1597) and Jost Amman (1539–1591)—and while ten volumes were planned, only the first saw the light of day, despite an additional appeal for funds from the king of Denmark (Schumacher 1758–1759, 298–300). But while the project vanished, the blocks did not. Panckow found 1363 of them, as he says, "all mixed up, just as though they had been left where they had grown in a field." He decided to identify the images as best he could and to publish them. The blocks, however, now served a different purpose. Panckow turned the images made from them away from curiosities related to wonder and Paracelsian alchemy and repositioned them as images related to utility and pragmatic domestic use. He wanted a book that required no theoretical commitment at all and one that was useful to apothecaries, surgeons, and students of medicine on the one hand, but also to gardeners and to every *Hausvater* on the other. He called his book a *Herbarium portatile* (Fig. 2).

Galenic qualities and their degrees were of no concern. Nor would there be any mention of what kind of signature a plant might have, or of the planet or sign under which it stood. One could read the books of Paracelsus, Oswald Croll (1493–1541), August Etzler (1563–1609), or Guilelmus Fabricius (1560–1634) for that sort of thing. His original idea was to publish just images to help medical

[1]On Panckow, see (Vitae Thomae Pancovii 1725, 1–3).

Fig. 1 *Angelica Maiore (Levisticum)* and animal image of distillation. From (Thurneisser 1578a, 28). Courtesy of the Economic Botany Library of Oakes Ames, Harvard University. © Photo by Bruce T. Moran

students identify useful plants, as Matthias de Lobel (1538–1616) had done in his much admired picture book of 1587 (sic) (Lobel 1581). To that end, he instructed, his book should be bound together with others, like the *Phytologia* of Johann Schröder (Schröder 1644, Liber IV) and the *Quadripartitum* of Simon Paulli (Paulli 1639), in order for students to discover their medicinal virtues (Panckow 1656, Günstiger Leser).[2] But a book containing only pictures left the images unanchored, and, encouraged by others at court, Panckow decided finally to add to the volume references to the names, properties, and preparations of the most useful medicinal herbs. Whether his idea or a decision of the publisher, the book also included the *Kräuter-Taffel* of Theophilus Kentmann (Helm 1966), the son of the better known Johannes Kentmann (1518–1574) (Helm 1971; Kentmann 2004), designating the times and places in which the *Kräutersuchende und liebhabende Leser* might search out specific plants.

Panckow's reuse of the Thurneisser woodblocks is a notable example of changes in the epistemic status of a certain kind of material object brought about by a shift of social, economic, and intellectual purpose (Dupré and Lüthy 2011). Unmoored from their original intent, such objects nevertheless remained in the world, sometimes residing undetected in private places, at other times traded and let out for hire. In both instances, however, some objects, in this case woodblocks, began to

[2] In another venture, however, Panckow retained the astrological connection of some of the Thurneisser woodblocks, including plant images, recycling them within the context of almanachs oriented toward domestic use; e.g. *Alter/Neuer und Römischer Historien-Haus-Kräuter und Schreib-Calender* (Chap. 13).

Fig. 2 Collecting local plants near Berlin. From (Panckow 1656, frontespiece). Courtesy of the Library of Arnold Arboretum, Harvard University

travel, finding their way by means of exchange or rediscovery into sites remote from their places of origin, and influencing the process of visualizing the world by fitting, with renewed purpose, into the space between the objects of nature and the imagination (Bender and Marrinan 2010). As traveling objects of representation and as embodiments of plant depictions, woodblocks figure into the material culture of shared practices in the early modern era,[3] transmitting specific images for diverse purposes while communicating the artistic and artisanal practices embedded in them as artifacts

[3]For an examination of collaborative efforts involving both artifacts and practices leading to the codification of technique in regard to chiascuro woodcuts (Chap. 7).

(Morgan 2011). They constitute in this way a relatively enduring material milieu that works aesthetically not to replicate natural objects but to envision them. Their effect could not be anticipated, yet some survived as a more or less stable source of visual imagery in the unsteady traffic of competition and criticism that shaped the representation of natural subjects. As repurposed and re-contextualized within different networks of knowledge, traveling woodblocks, largely unchanged since their fabrication, play a physically fixed but culturally mutable role among the processes and practices that help revise and reshape the perception of the natural world. They thus expanded the process of codification into a material genre while extending the practical activities embedded in their construction into unforeseen networks of exchange (Chap. 16).

1 Networks, Objects, Texts, and Images

Despite recent criticisms concerning the practical difficulties of narrative stemming from the complexity of networks (Shenk 2014), some features of network theory help bring to light social and technical aspects of the circulation of knowledge based in the production and reuse of woodblocks. Especially what has been called the "small world" model is an apt heuristic for the pattern of relationships that produced the material underpinning of plant images in early modern Europe. The model suggests that even a few ties between "actors" or "nodes" are sufficient for connectivity over an entire system of interrelated interests and practices (Watts and Strogatz 1998; Malkin 2011). The small world that made and then reused woodblocks as they traveled and created images of plants was made up of artists, draftsmen, cutters, printers and publishers. Some highly connected locations became hubs of exchange, places in which woodblocks, once created, were collected, bought and sold, copied, rented, and reclaimed.

Duncan Watts has noted the way in which well connected nodes are more likely to attract new links, the so-called "Matthew effect," and has emphasized as well contexts of affiliation as a means of enacting social ties (Watts 2003, 108–15). Although it is usually the activities of people that are thought of as establishing interactive structures, non-human elements as well figure into the contexts and affiliations of social life. As used and reused, shared and repurposed, material objects, during the course of their lifetime, become material members of intersecting groups, contexts, and networks. Just as technology, viewed as the rules and procedures for making things, mixes into the structures of social connectivity (Castells 2000), so too do the things that technology makes, especially as those things enter into the matrix of commercial production and consumption. The thing itself thus gains a social life with a flexible social standing as it is used, reused or exchanged within intersecting contexts. It is in this sense that a woodblock can have a "career," possessing at the end of it a kind of "biography" defined and redefined by its travels and connections to the nodes of commercial life (Kopytoff 1986).

The careers of woodblocks could, of course, be cut short, or, more poignantly, cut in half. Some practices undermined their individuality and status. In the process of seeking mercantile advantage within the aggressive small business world of print publishers, printers retooled and remarketed blocks in imaginative commercial ventures. The popularity of Sacrobosco's *Sphaera* moved printers to combine, re-cycle, and copy woodblocks to create frequent editions (Hamel 2014; Chap. 16). Printers also altered blocks by sometimes sawing them in half and combining one image with another, and sometimes by drilling out sections and inserting plugs that, for medical illustration, could then be cut to represent a specific part of the body (Shotwell 2011). Both techniques found their way into Hieronymus Brunschwig's (1450–1512) *Cirugia* (1497) printed at Strasburg by Johann Grüninger (1455–1531). There twenty-nine of the forty-nine full page plates in the volume were composed by combining blocks (sometimes of different sizes) to construct dissimilar scenes

(Herrlinger 1970; Hind 1963, vol. 1, 339–44). And, of course, woodblocks could be copied, imitated, and reissued for the purpose of making new, usually less expensive, books that mimicked other editions and connected publishers to new readers.

Practices of this sort sometimes ended up in court. The publication of Otto Brunfels's (ca. 1488–1536) *Herbarium vivae icones* by the Strassburg printer Johann Schott (1477–1548) included illustrations by Hans Weiditz (1495–1537) that depicted individual herbs as examples of specific species. In 1552, the Frankfurt printer, Christian Egenolff (1502–1555) (Grotefend 1881; Jäcker 2002), published a volume whose illustrations looked remarkably similar, *Herbarum, aborum, fruticum, frumentorum ac leguminum … imagines*,[4] and Schott cried foul. Egenolff, Schott contended, had stolen his images. The matter ended up in the chambers of the *Reichskammergericht*, and the arguments of the Frankfurt publisher tell us something important about the changing relationships between natural objects and those who represented them in the early modern era.

Egenolff objected to the court's authority in that he had not been accused of a violent crime. He admitted having modeled his book after another that had been composed and illustrated earlier, but he objected to Schott's assertion that he had copied the Brunfels volume word for word. Even if there were overlaps, he argued, it was for the public good, especially given the urgency and frequency of epidemics. His main defense, however, was that the images in his own publication were not knock-offs of Schott's images, but representations of the plants themselves which, because of the characteristic shape of the plant, could not be rendered any other way. "And even if the herbs resemble one another somewhat, Your Honor should consider that one can never paint or represent rosemary, asphodel, or another herb in another form or shape than what it itself is."[5] The legal dispute between Schott and Egenolff was, in its narrowest sense, a market-based controversy and a matter of plagiarism. However the quarrel also signaled a changing relationship between objects and stylistic agency. The objects in this case were both natural subjects like plants and manufactured objects like images. Making the latter involved a network of agents (artists, draftsmen, cutters, publishers) and, whatever its legal merits and self-serving rationale, Egenolff's argument was intellectually compelling and socially reorienting. There could be no patent or exclusive right in depicting a specific natural thing (i.e. to making a "counterfeit" image) based solely upon stylistic claims of representation.[6] There really was only one style from this point of view, the form of the object itself.

Certain printers not only commissioned woodblocks but saved them, anchoring networks of woodblock reuse. Wendel Rihel in Strasburg supplied Conrad Gesner (1516–1565) with blocks from Rihel's publication of the herbal of Jerome Bock (1498–1554). The second Czech edition of Mattioli's herbal recycled blocks used by Joachim Camerarius (1500–1574) that had come to Camerarius following Gesner's death (*zum Theil schon geschnitten/zum Theil allein gerissen*) (Camerarius 1590, Vorred an den günstigen Leser). The Czech editors rented the blocks from Camerarius's printer and since they needed to return them by a certain prescribed date, admitted running, not just rushing, to get the volume into print. When the publishing house that had printed Camerarius's herbal gave up business, blocks attributed to both Mattioli and Camerarius were purchased by the Ulm bookseller, Johann Görlin (1635–1663), who hired the German physician and alchemist, Johann Joachim Becher (1635–1682) to provide new descriptions of the plants they represented, now almost 75 years after Camerarius's edition (Becher 1663, dedication: Dem hochwürdigsten Fürsten und Herrn/Herrn Johann Philipsen/Dess. Heil. Stuels zu Mayntz Ertz-Bischoven). An extreme case of woodblock recycling concerns the famous English *Herbal* of John Gerard (1545–1612), a largely plagiarized translation of Rembertus Dodonaeus (1517–1585) published in 1597. The images were printed from blocks

[4] *Herbarum, Arborum, Fruticum;* also, *Herbarium imagines vivae.*

[5] Quoted in (Koerner 1993, 215–7).

[6] On the appropriation of nature by the artist in the northern European Renaissance see (Koerner 1993).

borrowed by the book's publisher, John Norten, from Nikolaus Basse, the publisher, in 1588, of Jacob Dietrich's (Tabernaemontanus's) (ca. 1525–1590) *Kreuterbuch* at Frankfurt am Main (Theodorus 1588–1591).[7] The blocks were then returned for printing subsequent editions of Tabernaemontanus's herbal. The result of the borrowing, however, was a mangled affair, and Gerard's seventeenth century editor, Thomas Johnson (ca. 1604–1644), noted about the images that "my author, who (as it seems) having no great judgment of them, frequently put one for another … and oft times by this means so confused all, that none could possibly have set them right, unlesse they knew this occasion of these errors" (Gerard 1633, To the Reader). Johnson tried to repair the injury reusing, as he tells the reader, the blocks of Dodoneaus, Lobel, and Clusius, most likely borrowed from Christoph Plantin (1520–1589) in Antwerp (Hatton 1909, 2–7).

Sachiko Kusukawa has recently discussed with particular insight the emergence of visual arguments in early modern scientific texts and has emphasized ways in which images, texts, and objects interrelated to make claims about natural knowledge (Kusukawa 2012). In reading texts through images arguments about nature and the body clearly emerged, and yet, the relationship between text and image in early modern herbals was not built in free space, but came into being as a result of other relationships, social as well as material in nature, constructed upon associations between objects, artists, authors, and commerce. And these relationships were fluid and shifting, as can sometimes be seen in the printed works that resulted from them.

Images appeared, were renamed, and re-described in books that assigned meanings fashioned by ancient authority and humanist assumptions about the nature of textual truth (Ogilvie 2006; Kusukawa 1997; Kemp 1990; Reeds 1976). Sometimes the text came first and images had to catch up to be added later. The printed sheets of the *Stirpium adversarium nova* of Pierre Pena (ca. 1535–1605) and Matthias Lobel, published by Thomas Purfoot in London in 1570 and then sold to Plantin for a reprinting in 1576 before seeing another edition back in London in 1605 (Pena and Lobel 1570; 1576; 1605), required inserting images, such as the images of *Cicerorobeum* and *Nicotiana Gallorum* (i.e. the Tabacco plant), in spaces kept vacant or, in the cases of *Melampyrum* and *Thymalus*, adding images where no space in the text had been provided for them. In the latter case, images were printed on little paper slips and hinge-pasted into narrow margins. As the sheets of the text traveled back and forth between London and Antwerp, the slips and added images tagged along to be added post production. Images also changed according to editorial purpose and the need for correction. Dodoens's *Stirpium historiae pemptades sex*, published by Plantin in 1583 (Dodoens 1583), required an iconographical errata sheet. In the case of *Leontopodium* the end user was to cut out and paste one image over another and was to make changes also to the accompanying text, renaming the plant and identifying Mattioli instead of Lobel as its descriptive source.

On other occasions images drove the text. Otto Brunfels complained at the beginning of the German *Contrafayt Kreuterbuch* that he only knew what was to be included in the book, and thus what should be arranged and described, when the artists and cutters had produced the blocks (Brunfels 1532, *Vorred*). What was most disturbing for Brunfels, however, was not the role of images themselves in fashioning and organizing his text. Of greater concern was a matter of relationship and influence. The association of the publisher, Johannes Schott, his artist, Hans Weiditz, a student of Albrecht Dürer, and the unknown cutters of woodblocks had created a "small world" bundle of agency that dissolved the focus upon Brunfels himself as the book's creative force.

Publishers and printers not only created the associations from which woodblocks and their images emerged but also manufactured projects that continued to shape visual knowledge through their

[7] The publication of the *Kreuterbuch* was financially supported by the Elector of the Palatinate who "would pay for all the expenses incurred by the sketching of the plants, the cutting of the blocks, and the wages for printing."

material reuse.[8] Within specific sites woodblocks passed from project to project, visually reinforcing descriptions of various kinds. The Antwerp publishing house of Christoph Plantin was one such site, and has long been considered a major hub in a network that manufactured, collected, rented, traded, and reused woodblocks. Plantin supported both artists and cutters. Among the artists, Pieter van der Borcht was the most active in Plantin's behalf. Among the cutters, the most prominent were Gerard van Kampen, Cornelius Muller, and Arnold Nicolai. Plantin paid around 11 sous for each block (five sous for the artist, six for the cutter) and the results supplied a shared material source for visual references underscoring descriptions of plants in books by Lobel, Dodoneaus, and Clusius. Among the cutters, van Kampen was especially energetic, cutting many, and possibly all, of the 220 blocks for Plantin's 1574 edition of Dodoneaus's book of purgative and medicinal plants (Dodoens 1574). Once produced, however, the same woodblocks served other commercial agendas. Plantin reused around thirty of the blocks for his edition of Clusius's book of Spanish plants two years later (Clusius 1576; López Terrada 2011, esp. 57–68), and then recycled many yet again in another, much larger, edition of Dodonaeus's *Stirpium* (1583) (Meerbeck 1841; Louis 1950; Nissen 1966, vol. 2, 49–50).

Plantin's blocks traveled with what Mary Morgan, in reference to other artifacts, has called "integrity" (Morgan 2011, 13–8), hardly being altered at all as they passed from one publication project to another. Borrowed woodblocks sometimes returned home, but sometimes did not, having been consigned to others and then permanently relocated. Some remained at work within different networks of artisans, scholars and entrepreneurs over lengthy periods. As repurposed objects of inscription they represented a unique sort of visual genre, itinerant material objects of visual knowledge that remained more or less untouched and unaltered as they moved from one place and context to another to perform various sorts of representative labor.

2 Clear-Cut Shapes and Adaptable Images

Panckow's herbal appeared in two editions, in 1654 and 1656, and, in 1673, a physician who had studied in Wittenberg and Altdorf before setting up a 50-year practice in Berlin named Bartholomaeus Zorn (1639–1717), expanded the Panckow volume, reusing the Thurneisser blocks yet again. He also added extra images and much more text, primarily references to the medicinal uses of plants found in earlier pharmaceutical works. Of the hundreds of woodblocks that Thurneisser commissioned and received for his grand project involving plants, only thirty six were used by him. These produced images printed in the first and only volume of his *Historia* and characterized plants that Thurneisser believed possessed specific celestial powers and were mostly of the umbelliferous sort (which Thurneisser associated astrologically with the Sun and Mars). The images themselves, and the images in the hundreds of blocks that Thurneisser never used, were, however, not original but were copied and cut from images that had already appeared in other books of plants, specifically in the herbals of Jerome Bock (Tragus) (1546), Mattioli (1554 and 1563), and Lobel (1576) (Baumann 1998, 85–9). These were the images printed by Panckow in his *Herbarium*. Zorn recycled them and added others, some apparently copied from Dodonaeus (1583), some drawn from plants sent to him by his former teacher at Altdorf, Maurice Hoffmann, and some cut from sketches based on his own observations. To distinguish these from the images initially produced by Thurneisser, used by Panckow, and reused in his own book, he designated them with two little stars (Baumann 1998, 86; Zorn 1673, Gewogener Leser) (Fig. 3).

In some instances repositioning the images meant giving them new names. What Thurneisser called *Imperatoria* (*Meisterwurtz* or *Peucedanum ostruthium*), for instance, becomes *Levisticum* in

[8] Even Johann Schott found ways to reuse images, sharing anatomical illustrations and exploiting images commissioned for one purpose to serve another. See (Carlino 1999, 88–93). Also (Shotwell 2011).

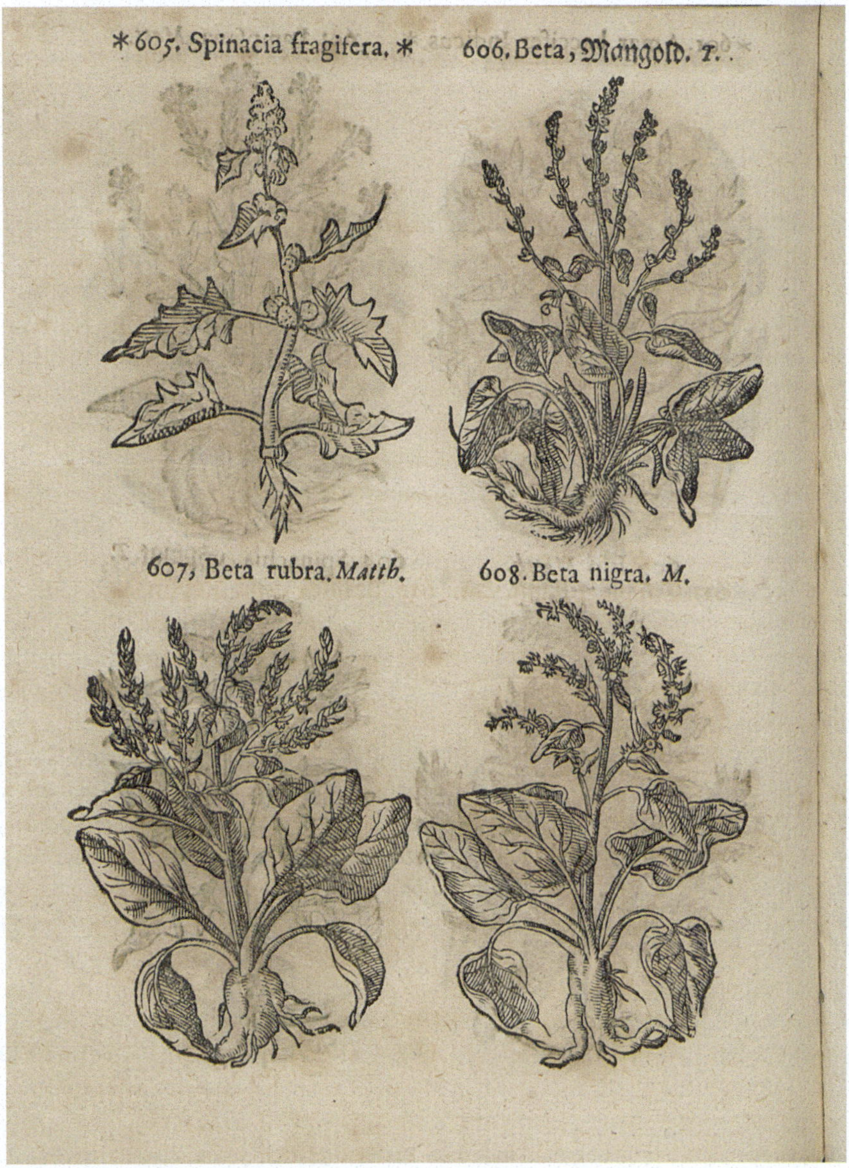

Fig. 3 An image designated with two stars (number 605) added by Zorn to the images used by Panckow. From (Zorn 1673, 605–8). Courtesy of the Staatsbibliothek Berlin, Preußischer Kulturbesitz

the Panckow text. In other instances, repurposing blocks and images within a local domestic network became a means of naming plants whose images could not be found, and thus identified, in contemporary herbals. Images in this sense did not so much communicate visual knowledge as gather it. Panckow made that relationship serve an interactive purpose, requesting any reader who recognized an unlabeled image from his or her own prior acquaintance with the plant, to "write down the name and send it to me" (Panckow 1656, Günstiger Leser). That sort of reverse reading of images (in which the reader rather than the text supplied relevant experiential knowledge) worked by means of confirming what the viewer already knew. Images became a way to verify that the plant whose properties and medicinal uses were being publicly discussed was one with which the reader was already familiar within a private domain. Images were acted upon and reinterpreted according to individual experience, and woodblocks, and the images they produced, were particularly suitable for such exchanges of observation and meaning.

What is it that is peculiar to woodblocks themselves that allowed them to function as specific yet adaptable sources of visual knowledge in regard to plants, and that, as a result, made them especially suitable for traveling and repurposing? Part of the answer, I think, can be found in the way that woodblocks are made. As opposed to drawing and engraving, the woodblock cutter exposed the line of the image by never touching it. Woodblock cutting is, after all, a process of carving and displacement (Ayres 1993; Griffiths 1996; Lambert 2001), and the woodcut, as David Rosand has observed, "involves the liberation of the line from the wood, a consequence of which may be a certain transformation of the line itself." Defined externally, that line becomes "a more fully independent and deliberately bounded mark" (Rosand 2002, 145–81, quoted 158; Treviranus 1949). What the cutter of plant images created, then, was a relief image that was, so to speak, clear cut, and that, as a result, made statements that were explicit without being literal. Such images have been discussed in terms of simplified abstractions that gave rise to a new sort of natural realism (Ogilvie 2006, 192–206) in the late Renaissance and within which could be found early stirrings of scientific objectivity by way of truth-to-nature visual statements (Daston and Gallison 2010, chapter 2). Yet, the very nature of woodblock imagery, clear-cut shapes that did not necessarily call attention to particularized details, made for a visual logic adaptable to various contexts and forms of thinking and reasoning. More schematic than realistic, the woodcut image afforded correlation between object and representation and broadened the experience of knowing by inviting the viewer to fill in the gaps between clear-cut lines and the thing itself (Bender and Marrinan 2010).[9]

Relief prints were, of course, more than contours, and the techniques of some cutters created compelling impressions of shading, plasticity and depth. Resemblance was important, but there were other techniques that did a better job in conveying similitude. By the end of the sixteenth century copperplate engravings were replacing woodblocks as a more precise way to communicate the details of natural subjects. For a long time, the technology of nature printing, while capable of only a few reproductions, had also been widely known due especially to the descriptions of Girolamo Cardano (1501–1576) and the "Book of Secrets" of Alexis of Piedmont.[10] Those who made woodblocks certainly knew of these techniques. Theophilus Kentmann, whose *Kräuter-Taffel* Panckow added to his *Herbarium*, was a well-known nature printer and assembled a collection of nature prints at Torgau in 1583, employing techniques most likely learned by way of his father's close friendship with Conrad Gesner (1516–1565). Whatever the tricks of reproductive technique, however, woodcuts provided the most efficient means of conveying general information by way of vivid, yet schematic, replication. They also filled a niche of increasing demand in the sixteenth century, namely a desire for individualized experience through private reading and evaluation. The medium of the woodcut, Hans Körner has argued, aided, early on, private religious devotion by freeing images from fixed liturgical requirements and ecclesiastical points-of-view (Körner 1979; Field 2005, 30–1). Images of plants might also be freed of required meaning, especially as they appeared sometimes without any adjoining text at all, published as *compendia*, extracts, or, as in the case of Lobel's *Stirpium icones*, as straight-forward picture books intended as guides to finding and naming plants. On these occasions woodcut images left it to the reader, or consumer, to provide relevant discussion and debate, often within a local context among family and friends.[11]

[9] On "affordances," see (Gibson 1979; Pfaffenberger 1992; Chemero 2003).

[10] On the early modern history of nature printing see (Cave 2009, 21–42).

[11] See the critical discussion by (Schmidt 2005, 41–3).

While the drawn line, Paul Kristeller observed, produced a monologue "in which only the artist himself intends to gain clarity concerning the problems that occupy him" (Kristeller 1922, Vorwort), the woodcut, in its repeated uses and replications, became the source for dialogue, both publicly and privately, and the material means by which the expressive nature of the image inspired discernment and criticism. Through replication, Walter Benjamin observed, the control of an image passed to the user who, by means of a process of what he called making, or bringing, the image up to date (*Aktualisierung*) separated the figure from whatever original meaning had been attached to it and adjusted its reality according to current cultural and political practices (Benjamin, 1955). The technology of woodblock reproduction, he argued, was particularly relevant in this regard as it "struck at the root of the quality of authenticity," becoming the instrument of "differentiating and gradating" claims to what was original and visually genuine (Benjamin 2002, 271, note 4). Benjamin's focus was upon the cultural and technological processes by which the notion of authenticity itself emerged in art. The legitimacy of woodblock images did not rest upon claims to artistic authenticity. Nor did it depend upon visual declarations of equivalence to natural objects, even though the images of some plants were advertised as "living," as in Brunfels's *Herbarium vivae icones*. Depicting a plant at one moment in its life cycle required statements of resemblance rather than precise similitude (Allmon 2007; Kusukawa 2006). In gaining that resemblance, however, the material statements made by artists, draftsmen, and cutters inspired new readings of the book of nature, even when removed by time and distance from the original site and purpose of their production (Dewey 1934, 86). Traveling woodblocks, like procedures, recipes (Chap. 3), and other artifacts, reified a formal point of view (Chap. 16), but there was also something different in the way they were used. An original and continuing material matrix now made images useful in creating a new angle of perception. The new images were, however, never orphaned from the material practices that originally gave them birth. Their maternal template remained, and, although cast into functional realms far from home and detached from their original purpose, the images made from repurposed woodblocks were, in terms of their underlying and enduring materiality, never alone in the world, never *Mutterseelenallein*.

3 The Mattioli Connection: Personality, Partnerships, and the Constraint of Agency

Panckow mentioned only one cutter for all the Thurneisser blocks, a mysterious Peter Holzmeyer. Despite repeated references in major art historical *lexica*,[12] however, Holzmeyer's role in the Thurneisser's network is unclear, and the name itself, it has been claimed, perhaps emerged as a conflation of names vaguely remembered by the Brandenburg court officials who brought the half-forgotten "pile" of Thurneisser blocks to Panckow's attention. Thurneisser's projects were corporate in scope, involving an assembled company of assistants and artisans, and the cutters he used for his various projects were spread out over a commercial complex linked to several cities—Prague, Kuestrin, Frankfurt an der Oder, Goerlitz, Leipzig, Halle, and Nuremberg among them. A few of their names are mentioned in his correspondence, among them Peter Hille and Wolfgang Stürmer, the latter known to have accepted a commission for 600 images for the Thurneisser herbal of which over 300 were completed in 1574 (Moehsen 1783, rept. 1976, 214–5; 220–30). Stürmer combined the roles of *Formschneider* (cutter) and publisher, organizing others for specific projects and business schemes that required the participation of a group of associates who knew how to exploit ink, paper, and wood. Although small in scale his workshop gained attention from on high. One of those to take note was

[12]Cf (Neues allgemeines Künstler-Lexicon, vol. 9, 232–233; Le Peintre-Graveur, vol. 4, 230–1; Allgemeines Lexikon der bildenden Künstler von der Antike bis zur Gegenwart, vol. 17, 422).

the Emperor, Maximilian II, who commissioned Stürmer to render, cut, and print images of coins as part of a merchant handbook that included evaluations of their rate of monetary exchange (Stürmer 1572). To Thurneisser, Stürmer noted that, for the project, he had made use of fourteen cutters, "good ones and bad ones," each receiving one half Taler for every completed block (Moehsen 1783, rept. 1976, 107, note g).

Thurneisser's network was bigger, involving literally hundreds of participants—secretaries, translators, copyists, proofreaders, type founders, type setters, engravers, cutters, collectors, glass painters and laborants—many living together with their families at or near Thurneisser's residence-workshop at the *Graues Kloster* in Berlin (Boerlin 1976). In fact, the "Grey Cloister" constituted a network terminal in which shared peripheral practices linked to design, publishing, and printing were brought largely under one roof. There Thurneisser shaped and coordinated unique projects related to the diagnosis of illness by means of the distillation of urine, the preparation of medicines and medical ingredients, and the collection for sale of curious and rare *naturalia* and *artificialia*. One of those to connect, as a cutter, to this nexus of design and practice, albeit for a short time, was Wolfgang Meyerpeck (the Younger) (1505–1578), an artist living at the time in Leipzig who had been trained in printing, print making, and drawing while working in the print shops of his father, of the same name, in Zwickau and Freiburg.[13] Meyerpeck was multi-talented and two letters recommending him to Thurneisser, one from Zwickau and the other from Leipzig, emphasized abilities in painting and drawing as well as etching and engraving. The association was brief,[14] and may have had led to little or nothing in the way of actually produced woodblocks. Nevertheless, from a network point of view the connection to Meyerpeck, if indeed the same person, may have linked Thurneisser to an agent of one of the biggest and most important projects of plant illustration of the sixteenth century involving the preparation of images for both the small and large format editions of Pierandrea Mattioli's *Commentaries on Dioscorides*.

Printed versions of Dioscorides's herbal already existed in Latin and Greek at the end of the fifteenth century, and an Italian and a new Latin translation by Jean Ruel (1474–1537) both appeared in 1516. Mattioli (1501–1577) compared Ruel's translation with other editions and with Greek manuscripts recently brought back from Constantinople, translated the text into Italian, added his own observations, and published his *Commentaries on Dioscorides* in 1544.[15] Other editions quickly followed, each time with added information sent to Mattioli from contacts within a web of correspondents that took shape, on one account, as part of a "republic of letters" (Findlen 1994; 1999). The flow of information within the network, however, was, for the most part, centered upon Mattioli. It thus resembled in some respects an exacting personal empire more than a cooperative system of exchange (Palmer 1985). Whatever the manner of give and take, the resulting "cascade" of descriptions and natural objects generated new contributors and correspondents by means of the momentum achieved through previous contacts.[16] The Commentaries expanded as a result, and, in 1554, images, printed from over 500 woodcuts, were added to the text, now in Latin, published in Venice by the internationally aligned Vincenzo Valgrisi.

Writing to Ulisse Aldrovandi (1522–1605) the year before, Mattioli noted that "his painter" had been sick in bed for a month and a half, and that the illness had prevented sending an image of a sesame plant to Bologna (Mattioli 1906, 16).[17] The painter was probably Giorgio Liberale

[13]On the identities of Wolfgang Meyerpeck, father and son, see (Claus 1986, esp. 10–5).

[14]Thurneisser complained of Meyerpeck's lethargy, perhaps the result of illness. See (Moehsen 1783, rept. 1976, 105).

[15]For discussion of the publication history of the Commentaries see (Nutton 2001; Ferri 1997; Toresella 1993; Ullmann and Degen 2009).

[16]For a lucid description of the dynamics of "information cascades" and other aspects of self-organization within networks see (Malkin 2011, chapter 1).

[17]See also (Tongiorgi Tomasi 1997, 369).

(ca. 1527–1579/1580) from Udine, a town near Gorizia where Mattioli, since 1542, served as chief physician. Liberale had been a pupil of Martino da Udine, also called Pellegrino da San Daniele (1467–1547), and lived in Gorizia, and later in Prague, while working for the Archduke Ferdinand II (1529–1595) to paint the fish and other marine animals of the Adriatic. The task required 17 years to complete and, when realized, gained a place in the archduke's *Kunstkammer* at the castle Ambras (Cronberg 1971; Irblich 1996; Auer 1995). When Meyerpeck joined the illustration team is uncertain, but he was most likely onboard by 1560 when a debate erupted, primarily between Mattioli and Conrad Gesner, that concerned the plant *aconitum pardalianches* (claimed to be a variety of wolfsbane) and for the illustration of which he was named as the artist (Delisle 2004; Tongiorgi Tomasi 1997).[18] By that time Mattioli had been appointed imperial physician and had been living in Prague since early 1555 as an imperial courtier. The controversy was thus entangled in the defense of courtly status and reputation, and became correspondingly nasty. It also emerged at the time of a major collaborative undertaking joining Mattioli and his artists with the imperial family, the Bohemian nobility, and the city of Prague in the preparation of a plant book whose woodblock constructions were designed to convey images of uniquely vivid representational power. Those images, painted and drawn primarily by Meyerpeck and Liberale, appeared first in an herbal emphasizing Bohemian plants, written in Czech and published in 1562.

Plans for the herbal had already begun the year before Mattioli arrived at the imperial court and took shape thereafter with the financial support of the Emperor, Ferdinand I (1503–1564), the Bohemian estates, and the city of Prague. Mattioli's network sprang into action. Late in 1558, he noted in a letter to Aldrovandi that "there are enough of those who continually send me plants; [and] everyone, having heard of the work that I do, offers me help." In addition, he now had "two painters, most skillful draftsmen, at home [*a casa*] who do nothing other than work continually, and I have five carvers [*intagliatori*] in Vienna who [also] work for me" (Mattioli 1906, 41).[19] But this was a project far beyond the dimensions of even Mattioli's widespread web of correspondents and contacts. The network that produced the volume was a structure built of imperial, noble, and urban components as well. Cutters, for instance, were spread out over several cities beyond Vienna and the archduke urged, through personal letters, the city fathers of Augsburg, Nuremberg, and Strasburg to find qualified *Formschneidern*, to prepare fifty woodblocks of plant sketches for the project (Bohatcová 1985). The text in Czech was prepared by Tadeáš Hájek von Hájek (after 1525–1600), a nobleman, astronomer, mathematician and physician with important connections to academic and courtly circles, whose noble coat of arms accompanied twenty seven others at the front of the volume. That the structure of a network can influence the success or failure of an innovation has long been observed (Watts 2003, 244–50). Networks not only of creators but also of consumers underpinned the success of the Mattioli

[18] According to Dioscorides, the plant had a tail like a scorpion, glistened like alabaster, stupified a scorpion that touched it and was poisonous to wild beasts. Nevertheless, it could be used as an analgesic in medicines for the eyes. In the 1560 edition of the Commentaries Mattioli added an appendix (555–6) in which he responded to the criticisms of Gesner, and named Giorgio Liberale as the painter who drew the plant. The same reference, within an appendix, appears in the Latin edition of the same year. However, in further editions, as the story of finding and drawing the plant moved from an appendix into the text of the Commentaries, Meyerpeck is given the credit, although the story is missing (along with the plant) from the Czech edition of 1562. While not sending Gesner the actual plant (it was too much of a journey, he explained, to return from Bohemia to the valley where the plant was found) Mattioli emphasized the living witnesses (including the mountains themselves and the person who dug out the plant) as testimony to its existence. "And that no one should think that I go about with baseless fables, I have exhibited the herb to the personal inspection of many learned men … [who] … have considered all the signs and marks [of the plant] against the image and description [of it], and they unanimously acknowledge that it is the pure truth. Also the master is still at hand, whose name is Wolfgang Meyerpeck from Freiburg, who drew and painted the image from the living plant as it here stands" (Mattioli 1563, 469r–469v). Further reference to Aconitum in relation to a discussion of poisons may be found in (Mattioli 1561, 131; 301–2).

[19] See also, (Palmer 1985, 153).

volumes, and both generated, by means of re-editions and the influence of patrons and correspondents, a "percolating cluster" of client judgment and interest. Mattioli's volumes, especially those with pictures, became fashionable.

Mattioli's text, and its comprehensive nature, including descriptions of pharmaceutical preparations by means of distillation (Milanesi 2002), represented a vast knowledge of plants, animals, and personal experience, as well as a mastery of humanist commentary. However, the importance and popularity of the book, and of the ever expanding editions to follow, ought not to be attributed solely to Mattioli's intellectual display. Its celebrity resulted also from the actions of a community of artists, cutters, printers, and patrons. That point was underscored in a unique way by the books printer, Georg Melantrich (1511–1580). Under his printer's device at the end of the volume, picturing a knight dressed in armor and holding a torch, Melantrich noted in a short Latin poem the means, as he saw it, by which ancient writings were brought out of darkness. Here, at least, that agency is not attributed to humanist scholarship solely, but to "the skillful minds of men … who produce books by the typographic art" (Mattioli 1562, Insignia Typographi, Georgii Melantrichi ab Auentino).[20] Indeed, Melantrich's own talents of material production were not lost on Mattioli, who acknowledged his printer's labors in forging [*cudendo*] the Czech edition and in casting [*conflandis*] its most elegant type (Mattioli 1562).

Nevertheless, even acknowledging the shared practices that figured into the production of the volume and its successors, it was Mattioli who stood out as the prodigious presence whose expertise informed a monumental volume. The "iconification" of Mattioli was altogether planned and materially reinforced as signs of artisanal practice, cut originally as monograms into the woodblocks of the Prague volume, were gouged away thereafter. The Czech edition contained almost a hundred images signed "GS," appearing usually with a little knife cut either horizontally or vertically into the block (Fig. 4).

Whoever "GS" may have been,[21] almost every reference to "GS" is erased in the reuse of the images in the following, German edition, the *New Kreüterbuch*, published just a year later, also in Prague. Business practices of the book's two publishers, Georg Melantrich in Prague and Vincenzo Valgrisi in Venice, supplanted artisanal technique, cutting ties quickly with the volume's Czech origins and repurposing the book and its images within a print network of international proportions.[22] The two had already collaborated in producing Mattioli's book of letters, *Epistolarum medicinalium libri quinque* in 1561 and Valgrisi and Melantrich now joined forces once again for further editions of Mattioli's herbal, beginning with the publication of a German translation.

The German *Kreüterbuch* was an expanded version of the Czech herbal, and its publication had to be reconsidered from the point of view of a different patronage cohort and a public culturally sensitive to the mysteries linking the macro and microcosm. To recognize the new patronage milieu Mattioli rewrote his acknowledgements, and, in his *Epistola nuncupatoria*, now drew attention to a subsidy network that included "emperors, kings, electors of the Holy Roman Empire, archdukes, cardinals, bishops, dukes, and princes." He noted in particular, apart from the beneficence of the Imperial family, the financial assistance of August, the duke of Saxony (1526–1586), Frederick, count Palatinate of the Rhine, Albert, duke of Bavaria (1528–1579), William, duke of Cleves (1516–1592), Cristoforo Madruzzo, the cardinal-prince of Trent (1512–1578), and Johannes Jacobus, the prince-bishop of Salzbug (Mattioli 1563, Epistola nuncupatoria).

[20] See also (Savage 1921).

[21] A possible identity may be Georg von Sichem, whose monogram looks very similar. See (Die Monnogrammisten, ed. Nagler 1863, vol. 3, 112–6). Reasons for doubt, however, have been suggested by (Bohatcová 1985, 176).

[22] Removing obvious marks probably also advanced the credibility of the images as visual statements of nature. In such cases, emphasizing a view from nowhere, without stylistic or artistic references, may have been a convenient means of making images more commercially acceptable everywhere.

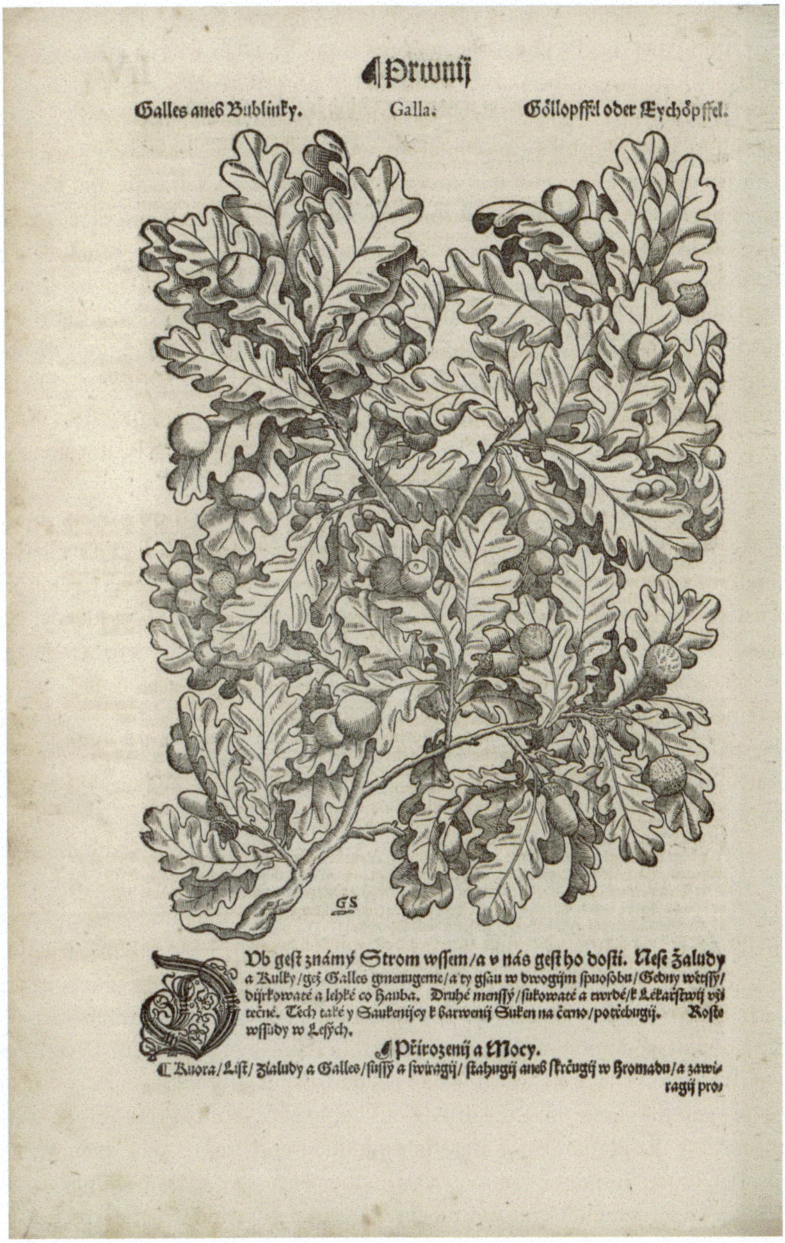

Fig. 4 "GS" plus knife. From (Mattioli 1652, 55v). Courtesy of the Library of Arnold Arboretum, Harvard University

Although written originally by Mattioli in Latin, the Mattioli-Melantrich-Valgrisi *Kreüterbuch* materialized as a triumph of the German language. Its translator, Georg Handsch noted that Latin and "Welsch" (i.e. Italian) editions of Mattioli's *Commentaria in Dioscoridem* had been in such demand in southern Europe and other lands, including parts of Asia, that over 32,000 copies had been made since the book first appeared almost 20 years earlier. A German edition with a domestic-vernacular market that had so far been kept at a distance from all those copies made obvious economic sense, and Handsch emphasized the prominent German potentates who had assisted in making sure that the images of plants drawn and cut for the book would come to good use in the German lands. He also emphasized the creation of the human being as a microcosm such that "the whole great world, that is the heavens and the earth, is observed in the human being in its likeness and qualities … [and] … everything that dwells under the heavens is put together from a mixture of the four elements together with the influence of the stars" (Mattioli 1563, Zum Leser).

The Czech *Herbář* had appeared with 589 illustrations. The *New Kreüterbuch* appeared with over 200 more, some of them stuck in at the end of the book since they were received too late to be given their proper place in a text that was zooming into print. Liberale and Meyerpeck continued to create new images, but, with the exception of a few marks remaining in places where their extraction would have disturbed the line or shading of depicted plants (even "GS" remains in this way), the cutters who did the work of producing the blocks mostly filled the ranks of "invisible technicians" (Shapin 1989). By the time of the next large image edition of the Commentaries, a Latin volume published at Venice in 1565, the number of images had grown to over 900 representations of plants and nearly a hundred images of animals.[23] And here, for the first time, Mattioli finally publicly acknowledged both Giorgio Liberale, "a most experienced man in the art of painting, and after him Wolfgang Meyerpeck … who spare no labor and diligence while they trace out assiduously the images of plants and animals" (Mattioli 1565, Epistola Nuncupatoria).[24]

Meyerpeck is often described solely as a cutter, yet, Mattioli's every reference to him is as a draftsman. This, of course, does not mean that drawing is all he did, and we have to be careful not to let a single model of the division of labor or collaboration define every relationship. Meyerpeck was an artist, draftsman, and was trained in cutting as well, and there is evidence, at least in one view, that for some of the blocks prepared by Liberale and Meyerpeck, the artist (painter) and the draftsman were one and the same (Quinby 1958, vol. 1, 97–8). Roles separated and combined by necessity within networks of production. In the working practices of artists, draftsmen, and cutters, versatility in training and integration of design, including skills of copying, were desirable traits.[25] In this regard, a draftsman might become a cutter, and an artist and a draftsman might be one and the same, depending upon the project.

The extent to which scribal practices such as drawing crossed paths and were shared with practices of execution like cutting is as much debated by art historians as by social historians and historians of science. Tilman Falk has cast the debate as a dispute between "pragmatists" and "idealists" (Falk 2004, esp. 211–4),[26] and has described its dimensions in terms of different models of collaboration. One view argues for professional collaboration and sees woodcuts emerging as a result of a partnership between artists and cutters. Another emphasizes collaboration of a different sort in the form of an association of shared practices. In this view, the "idealist" version, the person who designs and the person who cuts woodblocks needed, especially in the early years of woodcut production, to be one and the same since the artist could not successfully draw for the woodcut without experience concerning the properties of materials and their constraints. The agency involved in accommodating practice to the limitations of materials in the process of making things, in this case woodcut images, becomes the essential cultural means of originality and artistic creativity.[27] "Pragmatists," on the other hand, argued from a social perspective, emphasizing collaboration by means of the division of

[23] An Italian edition followed 1568, and in the end the blocks themselves settled into the hands of the Venetian publisher, Vincenzo Valgrisi, and his heirs, serving further editions and remaining in Venice until the mid 18th c. On Valgrisi religious context see (Baudrier 1921; Grendler 1977; Martin 1993).

[24] An Italian edition was finally published in 1568, also by Valgrisi, and other re-editions appeared until well after Mattioli's death in 1577.

[25] For working practices in the 18th c., including the copying and modification of previous illustrations, see (Nickelsen 2006).

[26] See also (Field 2005, 24–5).

[27] For the larger sense in which material agency and practice, by means of "the dialectic of resistance and accommodation" may be viewed as aspects of scientific culture, see (Pickering 1995).

labor and insisting that the experience of the artist and the experience of the cutter existed in different social and cognitive domains. The latter view, it seems fair to say, reflects also the sense of cultural compartmentalization that has separated knowing how and knowing that, and has made incommensurate, according to some (Sennett 2008), the agencies of design and material production. Innovations are however, from a network point of view, inevitably the result of shared knowledge, and innovations in woodcuts and image production are no exception. Cutters, after all, did not live in isolation, but were surrounded by design—in glass and panel painting, in the illumination of manuscripts, in woodcutting and sculpting. Design and the techniques of execution might accommodate each other and, in doing so, expand the range of innovation and representation in a reciprocal way as pen and knife, scribal and material cultures, intersected (Rosand 2002, 148–58).

4 Duhamel, Traveling Blocks, and the Utility of Wood

Images of the Mattioli Commentaries, large and small, continued to be copied and added to various books of plants well into the 18th c. The *Erbario del Mattioli* published by Niccolò Pezzana at Venice in 1744 is one example, Friedrich Zwinger's (1707–1776) *Theatrum botanicum*, published at Basel in the same year (a re-edition of the Mattioli-based *Kreüterbuch* of his father, Theodor) is another. In one case, however, some of the original woodblocks themselves endured yet another printing, this time in the *Traité des Arbres et Arbustes* (1755) of Henri Louis Duhamel du Monceau (1700–1782) (Duhamel du Monceau 1755).

Woodcuts were ideally suited to preparing large editions. This is undoubtedly one of the reasons that led Mattioli and his patrons to choose them. They resisted wear better than copperplates and yet, even given their relative endurance, it is surprising that many of the large figure woodblocks used in the Mattioli editions still survive. A large number came to light in 1956, and have since entered the public marketplace (Quaritch (Firm), c. 1988; Bidwell 2003, 9–10). These and others that remain from the Mattioli editions reveal a great deal about the materials and technologies involved and shared in their production. Cutters used knives and gouges of various sizes, usually of their own making and always kept exceedingly sharp on one side. When applied by the cutter, tools and technique combined to produce a material matrix of deep veining, texturing, hatching, and cross hatching (Fig. 5) that yielded such impressions of physical depth in the printed image that Mattioli felt it unnecessary to color them.

Carved reliefs are deeply cut from pear wood with the grain running, as is usual, along the length of the block. In examining surviving blocks, others have noticed instances in which cutters themselves made decisions about which line to follow, accommodating a knot or other obstruction by wrapping roots around it for instance. Because materials were not uniform, the blocks are of various thickness, and signs of the printer's adjustment can be seen on the back of some where metallic slugs or pieces of type were used to support the block or to raise it to the height of the typeface. Wood, of course, is unforgiving to the cutter, and although there are ways to fix mistakes, even a master's hand could not control everything. One woodblock (one of three at the Houghton Library, Harvard University) produced the image of *Fragaria* (i.e. strawberry), for the earliest, Czech, edition of Mattioli's herbal and for the Valgrisi editions in German, Latin, and Italian to follow. Closely examining the block reveals line breaks that become more noticeable in printed images over time (Figs. 6 and 7).[28]

Nevertheless, despite inevitable deteriorization, the graphic quality produced in wood by the partnership of design and craftsmanship in the Mattioli blocks was certainly one of the reasons why Duhamel turned to them once again to do the work of illustrating a project of his own in the mid eighteenth century.

[28] On line breaks and the deterioration of woodblocks and other printing materials see (Hedges 2006).

Fig. 5 Woodblock detail for *Beta Nigra* (*black beet*), TypR-87(1). Courtesy of the Houghton Library, Harvard University. © Photo by Bruce T. Moran

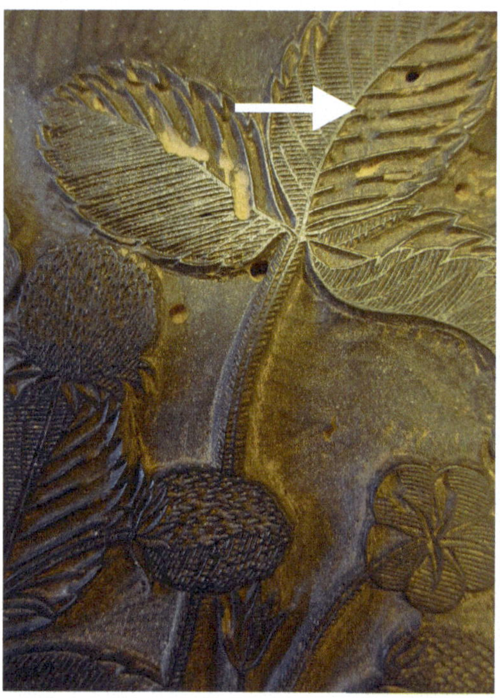

Fig. 6 Woodblock detail for *Fragaria* (strawberry) with *line breaks*, TypR-87(2). Courtesy of the Houghton Library, Harvard University. © Photo by Bruce T. Moran

Fig. 7 Printed image of *Fragaria* with *breaks*. From (Mattioli 1653, 437r). Courtesy of the Library of Arnold Arboretum, Harvard University. © Photo by Bruce T. Moran

Duhamel's interests ranged widely, and Duhamel himself stands out as one of the most prominent French agronomists, naturalists, and technical encyclopedists in the age of Buffon and Lavoisier. Part experimentalist, part engineer, Duhamel tied together in his writings an emphasis upon observation, problem solving, and general education. When the saffron harvest of France was threatened by a disease called *la mort*, he traced its origins to an independent living thing, a fungus, that thrived on its host, was capable of multiplication, and caused its host to become infectious. To prevent its spread he suggested excavating the soil around infected plants (Duhamel du Monceau 1981).[29] The soil was the means by which the fungus was transmitted, he argued, and the nature and health of soil became a primary focus, leading finally to recommendations for the introduction of a new system of agronomy fashioned on the views of the English agronomist Jethro Tull (1674–1741) (Bourde 1952, chapters 3 and 5). The success of the saffron studies earned Duhamel public renown, and largely on the basis of those studies he secured as well, in 1730, appointment as associate botanist of the *Académie Royale des Sciences*.

Experimental studies, especially observations concerning the mechanisms of plant physics, continued thereafter at the Duhamel estates at Denainvilliers, Vrigny, and Pithiviers. The trials became part of a family collaboration that involved his brother, Alexandre, and a nephew, Auguste Denis Fougeroux de Bondaroy (1732–1789). The practical interests of Fougeroux and Duhamel complemented one another and the two comprised a creative and productive pair with Fougeroux constructing his own web of correspondents and later inheriting his uncle's notes. These he enriched with his own observations on, among other matters, poisons, steam engines, chemistry, metallurgy, taxidermy, and paleontology.[30] Among projects in agronomy and plant chemistry at the family estates, Duhamel and Fougeroux took up experiments in sylviculture that aimed to respond to questions put to the Academy in 1732 by one of the king's ministers, the Comte de Maurepas (1701–1781). Among other duties, Maurepas managed the royal navy and was especially concerned with the propagation and strengths of different kinds of wood, the practices of cordage, and the relation between types of wood and limitations imposed upon the construction and the long-term usefulness of ships. Duhamel's investigations concerning silviculture, agricultural reform, and plant pathology had already led to a number of *mémoires* and these led now to his appointment as *Inspecteur Général de la Marine*.

The estates of Duhamel and Fougeroux were centers of botanical knowledge and practical information related to commerce, industry, and curiosity (Allard 1970; Dupont de Dinechin 1999; Corvol 2001). Both men were sensitive to building networks of exchange and communication and were privy to the flood of information flowing through the *Académie* from correspondents within and outside Europe. Seeds, drawings, dried specimens, and native nomenclatures passed through channels connected to the stations of colonial emissaries, and Duhamel gave advice, published later, to those "who, for their own utility, or for that of others," sought "to transport plants or seeds, or other natural curiosities, from one place to another at a great distance" (Duhamel du Monceau 1753, 1–2; first edition, 1752; Parsons and Murphy 2012). The advice was really nothing new, and appeared as a single edition as well as in three installments of the practical minded periodical, the *Journal économique ou memoires* (1752), before an expanded version reflecting additional remarks of naturalists, gardeners, and other correspondents appeared in 1753. Future correspondents were encouraged to construct catalogs of their plants, organizing information according to specific columns that included advice about the use of plants in medicine and the arts. They were also advised to take advantage of the *Journal* as a

[29] Emphasizing its utility to health, industry and the arts, Duhamel notes that the pistil of saffron "is considered to be a precious thing. It belongs to the spices used in cooking; it serves the painters of miniatures; it provides the dyers with a very fine colour; the physicians apply it quite usefully against various diseases …" (Duhamel and Louis 1981, 9).

[30] A large manuscript collection primarily related to trees, shrubs, and plants of various species, together with correspondence and essays on a variety of subjects composed by Fougeroux, may be consulted at the American Philosophical Society, Philadelphia; Mss.B.D87,580.D881. See (Chinard 1957; Ewan 1959).

convenient way to interact within the grid of practical botanical exchange and communication (Duhamel du Monceau 1753, Avertissement, i).

While matters of theory prompted attention and inspired observational projects, Duhamel did not write specifically for theoretical botanists (*Méthodistes*). His interest was largely in what was useful, and his experiments in regard to applied botany and botanical physics, and his efforts toward the reform of agriculture, served communities far beyond those in France. Seven substantial studies related to practical topics appeared between 1754 and 1768, and many of these, as well as other writings, appeared also in English, German, Spanish, Italian, and Dutch. Theory mattered less, but this does not mean that it mattered not at all, and some theoretical discussions helped also to shape his interactions and writings. In this regard, Duhamel returned frequently to a synthetic view of living things, emphasizing the unity of the living world and pointing out similarities between the physical mechanisms of animals and plants. He noted resemblances in the growth of wood and the growth of bones in the body, and argued for the existence of valves in plant structures that raised water against gravity and that operated in a similar fashion to the way valves in the veins worked to circulate blood in the body (Hartman 2001; Regouby 2014).

Between 1752 and 1777 Duhamel contributed essays also to Diderot's *Encyclopédie* and the *Dictionnaire des Arts et Metiers* that ranged in content over metallurgy, the printing of playing cards, the processes of making candlesticks, soap, and locks, as well as the methods of refining sugar and fishing. For maritime and building purposes Duhamel studied the growth and strength of wood and published the results of his findings in his *Traité des Arbres et Arbustes*. Forestry and the cultivation of trees were important to both marine and agricultural interests, and required rethinking the use of land and measures for the protection and expansion of forests (Schütz 2001).[31] This was an issue of significance to the state, and in studying the propagation, growth, and strength of trees Duhamel joined the company of others, like Louis de Froidour (ca. 1625–1685) (Froidour 1683), John Evelyn (1620–1706) (Evelyn 1664), Samuel Hartlib (1600–1662) (Hartlib 1651), and Hans Carl von Carlowitz (1645–1714) (Carlowitz 1713), who emphasized the planting of trees for the sake of national economic security. Evelyn considered attention to the "culture of forest-trees" to be among the great *disiderata* of his time, and encouraged "noble persons and gentlemen … at their first coming to their estates, and as soon as they get children," to "seriously think of this work of propagation also" (Evelyn 1664, To the Reader).

The practical value of forests for industry and the arts lay at the heart of Duhamel's *Traité*. There the object was to make known the characteristics of trees and shrubs to landowners, gardeners, water and forest officials, architects, builders, and the many others who used wood in building, and who had neither time nor inclination to study botany. For them, Duhamel announced, he had used every means possible to render the subject intelligible. Just as faces or shapes speak directly to the eye, he had chosen to represent the parts of trees and shrubs by means of exact illustrations and through the technique of *intaglio* engraving (*dans des vignettes gravées en taille-douce*), "since it would be otherwise very tedious and difficult to convey such a rendering by means of description [alone]" (Duhamel du Monceau 1753, vol. 1, preface, ix–x).

Brian Ogilvie (2006) and Kärin Nickelsen (2006) have discussed the shifting relations between botanical illustration and textual narrative within networks of botanical communication in the eighteenth century. In the case of the *Traité*, where botanical texts of the period were becoming less committed to imagery, Duhamel added images not solely to illustrate his treatise but to educate a reading gentry, as well as to enlighten communities of engineers and manufacturers. Equally important as potential consumers were also those who could afford to build libraries, e.g. physicians, officers of finance,

[31] An eight-volume treatise, *Traité complet des Bois et des Fôrets*, appeared between 1755 and 1767 in which Duhamel detailed the planting, growth, maintenance, and transportation of trees (Licoppe 1996).

representatives of the judiciary, and members of church and local agricultural societies. Among those with libraries he sought to influence especially gentlemen cultivators; and in this way reflected views expressed earlier by John Evelyn who advised his gentleman readers to "not easily commit themselves to the dictates of their ignorant hinds [i.e. farm workers] and servants, who are, generally speaking, more fit to learn than to instruct." The culture of trees, Evelyn continued, required "a deeper search than they are capable of … [for] … the business of planting is an art or science" (Evelyn 1664, To the Reader).

Joining art of the visual sort with the science of plants, and using both to instruct gentleman cultivators and bibliophiles, was one reason that the Mattioli blocks made a re-appearance in Duhamel's *Traité*. Duhamel recognized practical talent when he saw it. Perhaps because of his attraction to the practical and artistic uses of wood, but also because of the strong, clear-cut, images that they had produced, when he discovered that the Mattioli woodblocks were still extant in Venice, he bought them, and then, after identification, relabeling, and a bit of touching up, reused some of them.[32] 154 of the woodblocks related primarily to trees were made to do new work having been, at this time or earlier, cleaned up, more generally reworked, or slightly altered (a straightened line here, a missing leaf there) (Figs. 8 and 9).

New blocks were also added by Duhamel's publisher. Nevertheless, repurposing the talents of Liberale, Meyerpeck and the unnamed cutters of the Mattioli woodblocks served Duhamel's designs relating to the utility of craft and trade and emphasized how a common material, and a well-worn object, could be made distinctive and serve, by repurposing, the needs and knowledge of an industrious

Fig. 8 Detail of plant image with leaf. From (Mattioli 1565, 958). Courtesy of the Library of Arnold Arboretum, Harvard University. © Photo by Bruce T. Moran

Fig. 9 Detail of plant image with missing leaf. From (Duhamel and Louis 1755, vol. 1, Plate 69). Courtesy of the Library of the Gray Herbarium, Harvard University. © Photo by Bruce T. Moran

[32] J'ai eu le bonheur de recouvrer Presque toutes les planches de la belle édition latine du Matthiole de Valgrise: les Imprimeurs de mon Ouvrage ont fait graver avec soin celles qui y manquoient … (Duhamel du Monceau 1753, vol. 1, preface, p. x.).

public. In this regard, not only *what* was visualized but *how* it was visualized became relevant as Duhamel conveyed both a sense of the material and manner of production along with the representation itself.

5 Conclusion

The reuse and repurposing of woodblocks is an obvious economic feature of shared practices among the processes of plant illustration in the early modern era. In making woodblocks, and in the practices of making images from them, knowledge that related to scribal, print, and xylographic traditions intersected as technologies connected to design (painting and drawing) encountered those of execution (cutting and printmaking) (Needham 2009). This is the sense in which visual knowledge communicated through printed images constitutes, as Bruno Latour observed, "the fine edge and final stage" of a process of "mobilized practical techniques and resources" (Latour 1990, 40). Social roles also overlapped as artists, draftsmen, cutters, and printers joined in associations linking their labors and materials in replicable functions that resulted in compelling representations of the natural world (Castells 2000; Golinski 1990). In some instances the careers of woodblocks did not end with their initial use, however. Some began to travel, carrying the material font of images originally embedded in them along different cultural trajectories and, while remaining stable in substance, changing in epistemic status and trans*formed* through reuse (Valeriani 2011, 43). Their performance was not predictable. Purpose and function transmuted the visual knowledge that they bore. More succinctly, the representational knowledge carried by the thing changed, while the thing did not.[33] This was an itinerant "working object" forever falling, within different contexts, between reality and the imagination and prompting further involvement in the process of visual thinking.[34]

Among the structures of practice in the early modern era there is, then, a transformative material context consisting of a more or less enduring woodblock milieu that figures nevertheless into the shifting critiques and processes of picturing nature. The careers of woodblocks, like other "made things," sometimes jumped cultural parameters, bringing the agency of makers to bear upon new narratives in support of unforeseen visual agendas. While direct signs of agency may have worn away or had been intentionally erased while the object traveled, what the object expressed was still due to the "rendering practices" and know how of skilled artistic labor (Lynch 1985). That labor and the objects it created influenced more than the world that initially gave them meaning. Some possessed a further career, and although separated from their origins and repurposed over considerable time and distance, enabled nevertheless the further study of objects in nature, created visual displays for relationships that served assertions of both theory and utility, and reinforced strategies in shifting claims to natural knowledge.

References

Manuscript Sources

American Philosophical Society, Philadelphia; Mss.B.D87,580.D881.

[33] On claims to agency embedded knowledge within material things see (Baird 2004).

[34] See (Daston and Galison 1992; Daston and Gallison 2010, 19–35). Also (Bender and Marrinan 2010).

Printed Sources

Allard, Michel. 1970. *Henri-Louis Duhamel du Monceau et le Ministrère de la Marine*. Ottawa: Les Editions Leméac Inc.

Allgemeines Lexikon der bildenden Künstler von der Antike bis zur Gegenwart. 1907–1950, ed. Ulrich Thieme and Felix Becker, 37 vols. Leipzig: W. Engelmann.

Allmon, Warren D. 2007. The evolution of accuracy in natural history illustration: Reversal of printed illustrations of snails and crabs in pre-linnaean works suggests indifference to morphological detail. *Archives of Natural History* 34: 174–191.

Auer, Alfred. 1995. Erzherzog Ferdinand II (1529–1595) Renaissance Fürst und Herr über Rüstkammers, Kunstkammer und Bibliothek aus Schloß Ambras. In *Natur und Kunst: Handschriften und Alben aus der Ambraser Sammlung Erzherzog Ferdinands II*, ed. Alfred Auer and Eva Irblich, 13–78. Vienna: Kunsthistorisches Museum.

Ayres, Julia. 1993. *Printmaking techniques*. New York: Watson-Guptill Publications.

Baird, Davis. 2004. *Thing knowledge: A philosophy of scientific instruments*. Berkeley: University of California Press.

Baudrier, Henri-Louis 1921. Jean Vaugris. In *Bibliographie Lyonnaise. Recherches sur les imprimeurs, libraires, et fondeurs de lettres de Lyon au XVIe siècle*. 10: 457–464.

Baumann, Susanne. 1998. *Pflanzenabbildungen in alten Kräuterbüchern. Die Umbelliferen in der Herbarien- und Kräuterbuchliteratur der frühen Neuzeit*, Heidelberger Schriften zur Pharmazie- und Naturwissenschaftsgeschichte, 15. Stuttgart: Wissenschaftliche Verlagsgesellschaft.

Becher, Johann Joachim. 1663. *Parnassus medicinalis illustratus. Oder: ein Neues/und Dergestalt/Vormahln nie Gesehenes Their-Kräuter und Berg-Buch....* Ulm: In Verlegung Johann Görlins.

Bender, John, and Michael Marrinan. 2010. *The culture of diagram*. Stanford: Stanford University Press.

Benjamin, Walter. 1955. Das Kunstwerk im Zeitalter seiner technischen Reproduzierbarkeit. In *Schriften*, ed. Theodor Adorno, et al. Frankfurt a.M.: Suhrkamp Verlag. English edition: Arendt, Hannah. 1968. The work of art in the age of mechanical reproduction. In *Illuminations* (trans: Zohn, Harry). 217–252. New York: Harcourt, Brace, and World.

Benjamin, Walter. 2002. The work of art in the age of its technological reproducibility (third version). In *Walter Benjamin, selected writings, vol. 4, 1938–1940*, ed. Eiland Howard and Michael W. Jennings, 251–283. Cambridge, MA: Harvard University Press.

Bidwell, John. 2003. *Mattioli's herbal: A short account of its illustrations, with a print from the original woodblock*. New York: Pierpont Morgan Library.

Boerlin, Paul H. 1976. *Leonhard Thurneisser als Auftraggeber: Kunst im Dienste der Selbstdarstellung zwischen Humanismus und Barock*. Basel: Birkhäuser Verlag.

Bohatcová, Mirjam. 1985. Prager Drucke der Werke Pierandrea Mattiolis aus dem Jahren 1558–1602. *Gutenberg-Jahrbuch* 60: 167–185.

Bourde, André J. 1952. *The influence of England on the French Agronomes*. Cambridge: Cambridge University Press.

Brunfels, Otto. 1532. *Contrafayt Kreuterbuch....* Strassburg: Hans Schotten.

Camerarius, Joachim. 1590. *Kreutterbuch deß hochgelehrten und weitberühmten Hern D. Petri Andreae Matthioli....* Frankfurt am Main: Johan Feyrabendt in Verlegung Peter Fuchers und Heinrich Dacken Erben.

Carlino, Andrea. 1999. *Paper bodies: A catalogue of anatomical fugitive sheets, 1538–1687*. Trans. Noga Arikha. London: Wellcome Institute for the History of Medicine.

Carlowitz, Hans Carl von. 1713. *Sylvicultura oeconomica, oder hauswirtschaftliche Nachricht und naturmäßige Anweisung zur wilden Baum-Zucht....* Leipzig: verlegts Johann Friedrich Braun.

Castells, Manuel. 2000. Materials for an exploratory theory of the network society. *The British Journal of Sociology* 51: 5–25.

Cave, Roderick. 2009. *Impressions of nature: A history of nature printing*. New York: Mark Batty Publisher.

Chemero, Anthony. 2003. An outline of a theory of affordances. *Ecological Psychology* 15 (2): 181–195.

Chinard, Gilbert. 1957. Recently acquired botanical documents. *American Philosophical Society Proceedings* 101: 508–522.

Claus, Helmut. 1986. *Die Zwickauer Drucke des 16. Jahrhunderts, Teil 2 Wolfgang Meyerpeck (1530–1551)*. Gotha: Forschungsbibliothek Gotha.

Clusius, Carolus. 1576. *Rariorum aliquot Stirpium per Hispanias Observatarum Historia....* Antwerp: ex officina C. Plantini.

Corvol, Andrée 2001. *Duhamel du Monceau 1700–2000: un européan du siècle des lumières*. Orléans: Académie d'Orléans.

Cronberg, Guglielmo Coronini. 1971. Giorgio Liberale e i suoi fratelli. In *Studi di storia dell'arte in onore di Antonio Morassi*, ed. Arte Veneta, 85–96. Venice: Alfieri.

Daston, Lorraine, and Peter Galison. 1992. The image of objectivity. *Representations* 40: 81–128.

Daston, Lorraine, and Peter Gallison. 2010. *Objectivity*. New York: Zone Books.

Delisle, Candice. 2004. The letter: Private text or public place? The Mattioli-Gesner controversy about the Aconitum. *Gesnerus* 61: 161–176.
Dewey, John. 1934. The expressive object. In *Art as experience*, reprint 2005, 85–109. New York: Perigree.
Die Monogrammisten und diejenigen bekannten und unbekannten Künstler aller Schulen. 1863, ed. G.K. Nagler, 5 vols. Munich: G. Franz.
Dodoens, Rembert. 1574. *Purgantium aliarumque eo facientium tum et radicum, convolvulorum et deleteriarum herbarum historiae libri IIII*. Antwerp: Chr. Plantinus.
Dodoens, Rembert. 1583. *Stirpium historiae Pemptades sex, sive Libri XXX*. Antwerp: ex officina Christophori Plantini.
Duhamel du, Monceau, and Henri Louis. 1753. *Avis pour le Transport par Mer des Arbres, des Plantes Vivaces, des Semences et de Diverses Autres Curiosités d'Histoire Naturelle Seconde Édition, Considérablement Augmentée*. Paris: de L'Imprimerie Royale.
Duhamel du, Monceau, and Henri Louis. 1755. *Traité des arbres et arbustes qui se cultivent en France en pleine terre*. Paris: H.L. Guerin et F.L. Delatour.
Duhamel du, Monceau, and Henri Louis. 1981. In *Mr. Duhamel's 1728 treatise on the violet root rot of saffron crocus: Physical explanation of a disease that perishes several plants in the Gastinois, and saffron in particular*, ed. J.C. Zadoks. Wageningen: H. Veenman and Zonen B.V.
Dupont de Dinechin, Bruno. 1999. *Duhamel du Monceau: un Savant Exemplaire au Siècle des Lumières*. Luxembourg: CME.
Dupré, Sven, and Christoph Lüthy. 2011. Introduction: Silent messengers. The world of goods and the circulation of knowledge in the early modern Netherlands. In *Silent messengers: The circulation of material objects of knowledge in the early modern low countries*, ed. Sven Dupré and Christoph Lüthy, 1–12. Berlin-Münster: Lit Verlag.
Evelyn, John. 1664. *Sylva: Or a discourse of forest trees....* London: printed by J. Martyn and J. Allestry.
Ewan, Joseph. 1959. Fougeroux de bondaroy (1732–1789) and his projected revision of Duhamel du Monceau's Traité (1755) on trees and shrubs. An analytical guide to persons, gardens, and works mentioned in the manuscripts. *American Philosophical Society Proceedings* 103: 807–818.
Falk, Tilman. 2004. Formschneider. In *Reallexikon zur deutschen Kunstgeschichte*, vol. 10, ed. Zentralinstitut für Kunstgeshcihcte, 190–224. Munich: Beck.
Ferri, Sara. 1997. Il 'Dioscoride', I 'Discorsi', I 'Commentarii:' gli Amici e i Nemici. In *Pietro Andrea Mattioli. Siena 1501 – Trento 1578: la vita le opere*, ed. Sara Ferri, 15–48. Perugia: Quattroemme.
Field, Richard S. 2005. Early woodcuts: The known and the unknown. In *Origins of European printmaking: Fifteenth century woodcuts and their public*, ed. Peter Parshall et al. New Haven: Yale University Press.
Findlen, Paula. 1994. *Possessing nature: Museums, collecting, and scientific culture in early modern Italy*. Berkeley: University of California Press.
Findlen, Paula. 1999. The formation of a scientific community: Natural history in sixteenth-century Italy. In *Natural particulars: Nature and the disciplines in Renaissance Europe*, ed. Anthony Grafton and Nancy Siraisi, 369–400. Cambridge, MA: MIT Press.
Froidour, Louis de. 1683. *Ordonnance des eaux et forêts concernant les fonctions et devoirs des gardes et Traité pour server d'instruction aux gardes des eaux et forêts de la grandemaîtrise de Toulouse*. Toulouse: impr. De J. Pech.
Gerard, John. 1633. *The Herball, or, Generall Historie of Plants ... very much Enlarged and Amended by Thomas Johnson, Citizen and Apothecarye of London*. London: printed by Adam Islip, Joice Norton, and Richard Whitakers.
Gibson, James J. 1979. *The ecological approach to visual perception*. Boston: Houghton Mifflin.
Golinski, Jan. 1990. The theory of practice and the practice of theory: Sociological approaches in the history of science. *Isis* 81: 305–492.
Grendler, Paul F. 1977. *The Roman Inquisition and the Venetian press, 1540–1605*. Princeton: Princeton University Press.
Griffiths, Antony. 1996. *Prints and printmaking*. Berkeley: University of California Press.
Grotefend, Hermann. 1881. *Christian Egenolff der erste ständige Buchdrucker zu Frankfurt a. M. und seine Vorläufer*. Frankfurt a. M: K. Th. Völcker's Verlag.
Hamel, Jürgen. 2014. *Studien zur "Sphaera" des Johannes de Sacrobosco. Acta Historica Astronomiae 51*. Leipzig: Akademische Verlagsanstalt.
Hartlib, Samuel. 1651. *Samuel Hartlib his Legacie: Or an enlargement of the discourse of husbandry used in Brabant and Flanders....* London: printed by H. Hills for Richard Wodenothe.
Hartman, Claude. 2001. Duhamel du Monceau et les Sciences du Vegital: "Apports et Impacts.". In *Duhamel du Monceau 1700–2000: un Européen du Siècle des Lumières*, ed. Andrée Corvol, 55–76. Orléans: Académied'Orléans.
Hatton, Richard. 1909. *The craftsman's plant book....* London: Chapman and Hall.
Hedges, Blair. 2006. A method for dating early books and prints using image analysis. *Proceedings: Mathematical, Physical, and Engineering Sciences* 462 (2076): 3555–3573.
Heller, Joseph. 1823. *Geschichte der Holzschneidekunst von den Ältesten bis auf die Neuesten Zeiten....* Bamberg: im Verlage bei Carl Friedrich Kunz.

Helm, Johannes. 1966. Theophilus Kentmann (1552–1610): sein Leben und Werk, vor allem seine bisher unbekannten botanischen Schriften. *Beiträge zur Biologie der Pflanzen* 42: 1–112.

Helm, Johannes. 1971. *Johannes Kentmann, 1518–1574: ein Sächsischer Arzt und Naturforscher*. Wiesbaden: Steiner.

Herbarum, Arborum, Fruticum, Frumentorum ac Leguminum ... Imagines: ad Vivum Recens Depictae. 1552. Frankfurt am Main.: Egenolphus.

Herbarium Imagines Vivae: der Kreuter Lebliche Contrafaytung. 1972. West Germany: Editions Medicina Rara.

Herrlinger, Robert. 1970. *History of medical illustration from Antiquity to a.d. 1600*. London: Pitman medical and scientific publishing co.

Hind, Arthur M. 1963. *An introduction to a history of woodcut with a detailed survey of work done in the fifteenth century*, vol. 2. New York: Dover.

Gelehrter Männer Briefe an die Könige in Dännemark, vom 1522–1587. 1758–59, ed. Andreas Schumacher, Kopenhagen/Leipzig: bey Friedrich Christian Pelt.

Irblich, Eva. 1996. Natur studien Erzherzog Ferdinand II: zur Kunstkammer auf Schloß Ambras bei Innsbruck. In *Thesaurus Austriacus: Europas Glanz im Spiegel der Buchkunst, Handschriften und Kunstalben von 800 bis 1600*, ed. Irblich Eva, 209–225. Vienna: Das Nationalbibliothek.

Jäcker, Carsten. 2002. Christian Egenolff: Leben und Wirken eines Frankfurter Meisters des frühen Buchdrucks aus Hadamar. In *Christian Egenolff, 1502–1555: ein Frankfurter Meister des frühen Buchdrucks aus Hadamar*, ed. Kulturvereinigung Hadamar, 25–46. Limburg: Glaukos.

Journal Économique ou Mémoires, Notes et Avis sur l'Agriculture, les Arts, le Commerce, et Tout ce qui peut Avoir Rapport à la Santé, Ainsi Qu'à la Conversation & à l'Augmentation des Biens des Familles. 1752. Paris: chez Antoine Boudet.

Kemp, Martin. 1990. Taking it on trust: Form and meaning in naturalistic representation. *Archives of Natural History* 17: 127–188.

Kentmann, Thomas. 2004. In *Das Kräuterbuch des Johannes Kentmann von 1563*, ed. Thomas Bürger. Munich: Prestel.

Koerner, Joseph Leo. 1993. *The moment of self-portraiture in German Renaissance art*. Chicago: University of Chicago Press.

Kopytoff, Igor. 1986. The cultural biography of things: Commoditization as process. In *The social life of things: Commodities in cultural perspective*, ed. Arjun Appadurai, 64–91. Cambridge: Cambridge University Press.

Körner, Hans. 1979. *Der früheste deutsche Einblattholzschnitt*. Mittenwald: Mäander Kunstverlag.

Kristeller, Paul. 1922. *Kupferstich und Holzschnitt in Vier Jahrhunderten*. Berlin: Verlag von Bruno Cassirer.

Kusukawa, Sachiko. 1997. Leonhart Fuchs on the importance of pictures. *Journal of the History of Ideas* 58: 403–427.

Kusukawa, Sachiko. 2006. The uses of pictures in the formation of learned knowledge: The case of Leonhrad Fuchs and Andreas Vesalius. In *Transmitting knowledge: Words, images, and instruments in early modern Europe*, ed. Sachiko Kusukawa and Ian Maclean, 73–96. Oxford: Oxford University Press.

Kusukawa, Sachiko. 2012. *Picturing the book of nature: Image, text, and argument in sixteenth-century human anatomy and medical botany*. Chicago: University of Chicago Press.

Lambert, Susan. 2001. *Prints: Art and techniques*. London: V and A Publications.

Latour, Bruno. 1990. Drawing things together. In *Representations in scientific practice*, ed. Lynch Michael and Woolgar Steve, 19–68. Cambridge, MA: MIT Press.

Le Peintre-Graveur. 1860–64, ed. J.D. Passavant, 6 vols. Leipzig: Rudolph Weigel.

Licoppe, Christian (ed.). 1996. *La formation de la pratique scientifique" le Siscours de l'expérience en France et en Angleterre (1630–1820)*. Paris: Ed. La Découverte.

Lobel, Matthias de. 1581. *Plantarum seu stirpium icones*. Antwerp: ex officina Christophori Plantini.

López Terrada, María Luz. 2011. Flora and the Hapsburg crown: Clusius, Spain, and American natural history. In *Silent messengers: The circulation of material objects of knowledge in the early modern low countries*, ed. Sven Dupré and Christoph Lüthy, 43–68. Berlin-Münster: Lit Verlag.

Louis, Armand. 1950. La Vie et l'Oeuvre Botanique de Rembert Dodoens (1517–1585). *Bulletin de la Societe Royale de Botanique de Belgique* 82: 271–292.

Lynch, Michael. 1985. Discipline and the material form of images: An analysis of scientific visibility. *Social Studies of Science* 15: 37–66.

Malkin, Irad. 2011. *A small greek world: Networks in the ancient mediterranean*. Oxford: Oxford University Press.

Martin, John. 1993. *Venice's hidden enemies: Italian heretics in a Renaissance city*. Berkeley: University of California Press.

Mattioli, P.A. 1560. *I Discorsi di M. Pietro Matthioli Senese....* Venice: appresso Vincenzo Valgrisi.

Mattioli, P.A. 1561. *Epistolarum sedicinalium libri quinque*. Prague: in officinal Georgij Melantrichij ab Auentino, ad instantium Vincentij Valgrisij.

Mattioli, P.A. 1562. *Herbář: ginak Bylinář....* Prague: Melantrycha z Awentynu.

Mattioli, P.A. 1563. *New Kreuterbuch*. Trans. Georg Handsch. Prague: G Melantrich and Venice: V. Valgri.

Mattioli, P.A. 1565. *Commentarii in sex libros Pedacci Dioscoridis....* Venice: ex officina Valgrisianna.
Mattioli, P.A. 1906. In *Lettere di P.A. Mattioli ad Ulisse Aldrovandi, Estratto dal Bulletino Senese di Storia Patria, anno XIII fasc. I–II*, ed. Raimondi Carlo. Siena: Tip. e Lit. Sordomuti di L. Lazzeri.
Meerbeck, P.J. van. 1841. *Recherches historiques et critiques sur la vie et les ouvrages de Rembert Dodoens (Dodonaeus)*, reprint 1980. Utrecht: Hes Publihers.
Milanesi, Claudio. 2002. *Arte & alchimia nei discorsi di P.A. Mattioli*. Siena: Terre de Sienne.
Moehsen, J.C.W. 1783. *Leben Leonhard Thurneissers zum Thurn*, reprint 1976. Munich: Werner Fritsch.
Morgan, Mary S. 2011. Travelling facts. In *How well do facts travel? The dissemination of reliable knowledge*, ed. Peter Howlett and Mary S. Morgan, 3–39. Cambridge: Cambridge University Press.
Morys, Peter. 1982. *Medizin und Pharmazie in der Kosmologie Leonhard Thurneissers zum Thurn (1531–1596). Abhandlungen zur Geschichte der Medizin und der Naturwisenschaften*, vol. 43. Husum: Matthiessen Verlag.
Needham, Paul. 2009. Prints in the early printing shops. In *The woodcut in fifteenth-century Europe*, ed. Parshall Peter, 39–91. Washington, DC: National Gallery of Art.
Neues allgemeines Künstler-Lexicon. 1835–52, ed. G.K. Nagler, 22 vols. Munich: Verlag von E.A. Fleischmann.
Nickelsen, Kärin. 2006. *Draughtsmen, botanists, and nature: The Construction of 18th c. Botanical illustrations*. Dordrecht: Springer.
Nissen, Claus. 1966. *Die Botanische Buchillustration: ihre Geschichte und Bibliographie*, vol. 3. Stuttgart: Hiersemann.
Nutton, Vivian. 2001. Mattioli and the art of the commentary. *Gli atti dell'Accademia delle Scienze di Siena detta de' Fisiocritici* 20 (suppl.): 133–45.
Ogilvie, Brian W. 2006. *The science of Describing: Natural history in Renaissance Europe*. Chicago: University of Chicago Press.
Palmer, Richard. 1985. Medical botany in Northern Italy in the Renaissance. *Journal of the Royal Society of Medicine* 78: 149–157.
Panckow, Thomas. 1656. *Herbarium portatile: oder, Behendes Kräuter- und Gewächs-Buch....* Leipzig: in Verlegung Christian Kirchners.
Panckow, Thomas. 1663. *Alter/Neuer und Römischer Historien-Haus-Kräuter und Schreib-Calender deß nach Christi Geburt 1663 Jahrs ..., Dritter Anhang. Von etlicher Kräuter/nützlicher Wirckunge....* Berlin: in Verlegung Rupertus Volckern.
Parsons, Christopher M., and Kathleen S. Murphy. 2012. Ecosystems under sail: Specimen transport in the eighteenth-century French and British Atlantics. *Early American Studies* 10: 503–539.
Paulli, Simon. 1639. *Quadripartitum de simplicium medicamentorum facultatibus: ad artis medicae pharmaceutics studioso*. Rostoch: ex Bibliopolio Hallervordiano.
Pena, Pierre, and Matthias de Lobel. 1570. *Stirpium adversaria nova....* London: T. Purfoetij.
Pena, Pierre, and Matthias de Lobel. 1576. *Stirpium adversaria nova*. Antwerp: apud C. Plantinum architypographum regium.
Pena, Pierre and Matthias de Lobel. 1605. *Dilucidae simplicium medicamentorum explicationes et stirpium adversaria*, appended to de Lobel, In *G. Rondelletii methodicam pharmaceuticam officinam animadversiones....* London: excudebat praelum Thomae Purfootij.
Pfaffenberger, Bryan. 1992. Technological dramas. *Science, Technology and Human Values* 17: 282–312.
Pickering, Andrew. 1995. *The mangle of practice: Time, agency, and science*. Chicago: University of Chicago Press.
Quaritch, Bernard (Firm). c. 1988. *The Mattioli woodblocks*. London: Bernard Quaritch Ltd.
Quinby, Jane. 1958. *Catalogue of botanical books in the collection of Rachel McMasters Miller Hunt*. Pittsburgh: The Hunt Botanical Library.
Reeds, Karen Meier. 1976. Renaissance humanism and botany. *Annals of Science* 33: 519–542.
Regouby, Lynnette. 2014. The sensitive plant: Refiguring the boundaries of botanical bodies in eighteenth century France. PhD dissertation, University of Wisconsin-Madison.
Rosand, David. 2002. *Drawing acts: Studies in graphic expression and representation*. Cambridge: Cambridge University Press.
Savage, S. 1921. A little-known Bohemian herbal. *The Library* s4–II(2): 117–131.
Schmidt, Peter. 2005. The multiple image: The beginnings of print making, between old theories and new approaches. In *Origins of European printmaking*, ed. Peter Parshall et al., 37–56. New Haven: Yale University Press.
Schröder, Johann. 1644. *Pharmacopoeia Medico-Chymica....* Ulm: sumptibus Johannis Gerlini.
Schütz, Jean Philippe. 2001. Le Role de Duhamel du Monceau dans la sylviculture Europeenne. In *Duhamel du Monceau 1700–2000: un Européen du siècle des Lumières*, ed. Andrée Corvol, 77–83. Orléans: Académie d'Orléans.
Sennett, Richard. 2008. *The craftsman*. New Haven: Yale University Press.
Shapin, Steven. 1989. The invisible technician. *American Scientist* 77 (6): 554–563.
Shenk, Joshua Wolf. 2014. *Powers of two: Finding the essence of innovation in creative Pairs*. Boston: Houghton Mifflin Harcourt.
Shotwell, Allen. 2011. New perspectives on Renaissance images: Surgery and anatomy. Unpublished Paper.

Stürmer, Wolff. 1572. *Niedersachssischer Valuation Druck, Allerhand Gober und Kleiner Müntzsorten … Vorzeichnus und Geprege der Groben und Kleinen Müntzsorten.* Leipzig: gedruckt und zu finden bei Wolff Stürmer, Formschneider.

Theodorus, Iacobus. 1588–91. *Neuw Kreuterbuch, mit Schönen …, Figuren unnd Conterfeyten….* Franckfurt am Mayn: Gedruckt durch Nicolaum Basseum.

Thurneisser, Leonhard. 1578a. *Historia und Beschreibung Influentischer/Elementischer und Natürlicher Wirckungen/ Aller fremden unnd Heimischen Ergewechsen/auch irer Subtiliteten….* Berlin: Michael Hentzsken.

Thurneisser, Leonhard. 1578b. *Historia sive descriptio plantarum omnium….* Berlin: Michael Hentzske.

Tongiorgi Tomasi, Lucia. 1997. Il problema delle immagini nei 'Commentarii.' In *Pietro Andrea Mattioli Siena 1501 – Trento 1578: la vita le opere*, ed. Ferri Sara, 369–376. Ponte San Giovanni: Quattroemme.

Toresella, Sergio. 1993. L'Erbario Inedito di P.A. Mattioli. In *Congresso Nazionale dell'Accademia Italiana di Storia della Farmacia*: Siena, *9–11 novembre 1990*. 287–305. N. P. N.P.

Treviranus, Ludwig. 1949. *Die Anwendung der Holzschnitten zur bildlichen Darstellung von Pflanzen.* Utrecht: De Haan.

Ullmann, Manfred, and Rainer Degen. 2009. *Untersuchungen zur Arabischen Überlieferung der Materia Medica des Dioskurides.* Wiesbaden: Harrassowitz.

Valeriani, Simona. 2011. Facts and building artifacts: What travels in material objects? In *How well do facts travel? The dissemination of reliable knowledge*, ed. Peter Howlett and Mary S. Morgan, 43–71. Cambridge: Cambridge University Press.

Vitae Thomae Pancovii D. et Aulae Friderici Wilhelmi Electoris Brandenb. Medici, Apospasma. 1725. In *Decad. IIdae actorum medicorum Berolinensium, in incrementum collectorum et digestorum in volumen V*, ed. Johann Daniel Gohl. Berlin: apud Godofredum Gedickium. Bibliopol.

Watts, Duncan J. 2003. *Six degrees: The science of a connected age.* New York: Norton.

Watts, Duncan J., and S.H. Strogatz. 1998. Collective dynamics of 'small world' networks. *Nature* 393: 440–442.

Zorn, Bartholomaeus. 1673. *Herbarium, oder Kräuter- und Gewächsbuch … mit Sonderlichen Experimentis und Observationibus der Bewehrtesten Scribenten….* Cölln an der Spreee: Druckts Georg Schultze.

The Tracts on the *Sphere*: Knowledge Restructured Over a Network

16

Matteo Valleriani

Abstract

The essay focuses on the early modern tradition of commentaries on the late Medieval work of Johannes de Sacrobosco, *Tractatus de sphaera*. Arguing that a new knowledge system founded on the geocentric worldview emerged during the thirteenth century, the work analyzes how and why such a knowledge system changed during the early modern period in correlation with the development of practical knowledge, which emerged in the framework of the phenomenon of journeys of exploration. It is shown that new subjects were added to the original texts and finally codified and printed in the early modern editions of the text, thus creating new structures of knowledge. By applying methods derived from social network analysis, the essay finally examines how a single new subject, namely a mathematical demonstration concerning the division of the Earth's surface into climate zones, entered the history of commentary treatises on the *Sphere* and how it spread over time and space by being adopted into further editions published in other places and by different printers. By means of this case study, this work concludes with the analysis of the process that leads from one abstract structure of knowledge to the next.

Nam absque perfecta Sphera Mundi cognitione, nec tota Mathematica ipsa, neque Naturalis, neque Divina Philosophia, neque etiam Medicina, nec Agricultura, neque Historia, neque Nauticam nec alia quaepiam Ars, sive Liberalis, sive Mechanica rite percipi, exercerique possunt.

Francesco Barozzi, *Cosmographia. In quatuor libros distributa*, Venetijs 1585

M. Valleriani (✉)
Max Planck Institute for the History of Science, Berlin, Germany
e-mail: valleriani@mpiwg-berlin.mpg.de

M. Valleriani (ed.), *The Structures of Practical Knowledge*, DOI 10.1007/978-3-319-45671-3_16

The present work focuses on the early modern phase of a commentary tradition of treatises that began with one tract that went under the name of *Tractatus de sphaera* by Johannes de Sacrobosco. This treatise is a short, qualitative tract on geocentric cosmology. It was written during the thirteenth century in the frame of the University of Paris but then continuously rewritten, commented on, expanded on and, after the diffusion of print technology, reprinted over 300 times by the end of the seventeenth century.

During the early modern period, Sacrobosco's group of treatises was just a subset of those dedicated to cosmology and its related subjects. The impressive total number of publications dedicated to these subjects during the Renaissance indicates that the results of this analysis represent the basis for understanding one of the most significant contributions to the emergence of a common scientific and, therefore, cultural identity in Europe during this era.

After having introduced the early modern version of Sacrobosco's treatise based on two examples, and by means of a comparison with the original treatise of the thirteenth century, the following sections investigate the fundamental steps in the transformation process such knowledge underwent and the reasons for this transformation process. The last two sections are dedicated to: first the analysis of the knowledge system pivoted around the geocentric worldview and its cosmology, considered as a network in order to explain how (and why) the knowledge on the *Sphere* was continuously restructured; second, an example of reconstruction of network dynamic concerned with the real knowledge network on the *Sphere* and, more precisely, with the spread of knowledge related to a renewed interest in the subdivision of the Earth's surface into climatic zones, as a consequence of the journeys of exploration. The work concludes with the interpretation of the achieved results in terms of structures of knowledge.

1 The Treatises on the *Sphere* and the Label "Sacrobosco"

The second edition of the treatise on the *Sphere* by Peter Ryff (1552–1629) in 1598 (Ryff and Lucii 1598)—a mathematician active in Basel—opens with a letter to the reader and two tables. The tables show the general disciplinary structure of mathematics in his time, the subjects of its various branches, and the position of the knowledge described in this specific treatise within that general structure. As specified in the cover of the same treatise, it had been written for use at the Studium of Basel, with which Peter Ryff was affiliated.

Like any other treatise entitled the *Sphere*, Ryff's treatise contains a qualitative description of the cosmos according to the geocentric view that early modern scholars attributed to Aristotle and Ptolemy.[1] Besides this, the treatise contains accurate descriptions of astronomical instruments, especially concerning the use of the astrolabe, and, finally, a substantial section dedicated to the technique of calculating distances between positions on the Earth's surface, such as the position of cities. This technique made use of the concepts of longitudes and latitudes as had become common since the reintroduction of metric geography during the fifteenth century.[2] If this treatise had to be catalogued according to a general disciplinary taxonomy, it would be classified as a book on cosmology, geography, and scientific instruments for both observation and the measuring of distances.

The treatise contains two tables. According to the first one (Fig. 1), the knowledge codified in this treatise belongs to practical mathematics, as opposed to abstract mathematics.

[1]The fundamental Medieval and early modern text on cosmology was Aristotle's *De caelo*. For more information, see (Grant 1994, 14).

[2]For the process that led to the reintroduction of metric geography during the early modern period, see (Valerio 2012). For the role played by metric geography in the frame of the activity of journeys of exploration, see (Randles 1990).

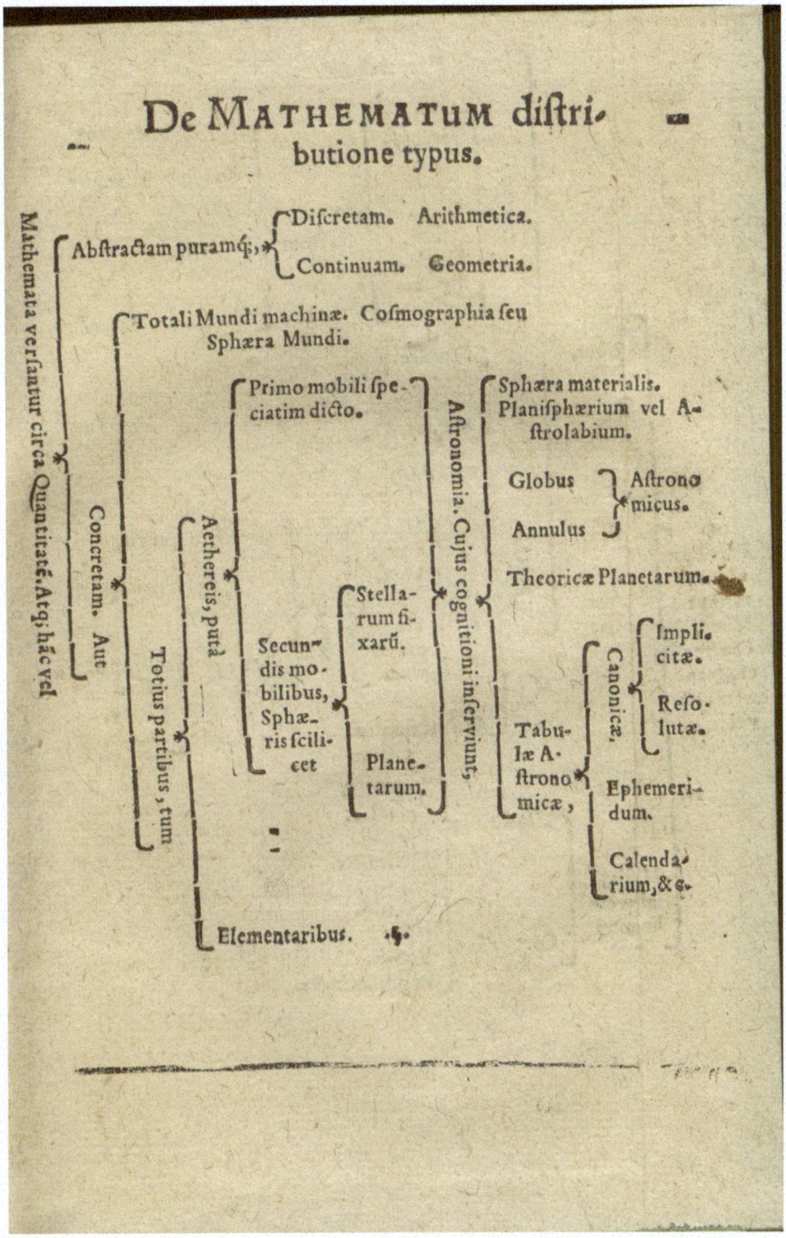

Fig. 1 Table of the subjects belonging to the mathematical discipline. From (Ryff and Lucii 1598). Courtesy of the Universitätsbibliothek Basel

Practical mathematics is in fact subdivided into the discipline that pertains to the entire cosmos or "world machine," and the discipline that focuses only on singular components of the cosmos, like stars and planets.

If the content of this particular treatise is taken into consideration, however, not all of the subjects mentioned in the tables are described in it. While the qualitative descriptions of the spheres—primum mobile, fixed stars and planets—are given at length, only a small portion of the subjects common in both practical mathematics and mathematical astronomy is discussed. This discussion corresponds to the section in the book that concerns the use of the astrolabe and, in general, the description of the armillary sphere. The remaining subjects pertinent to mathematical astronomy are the method for

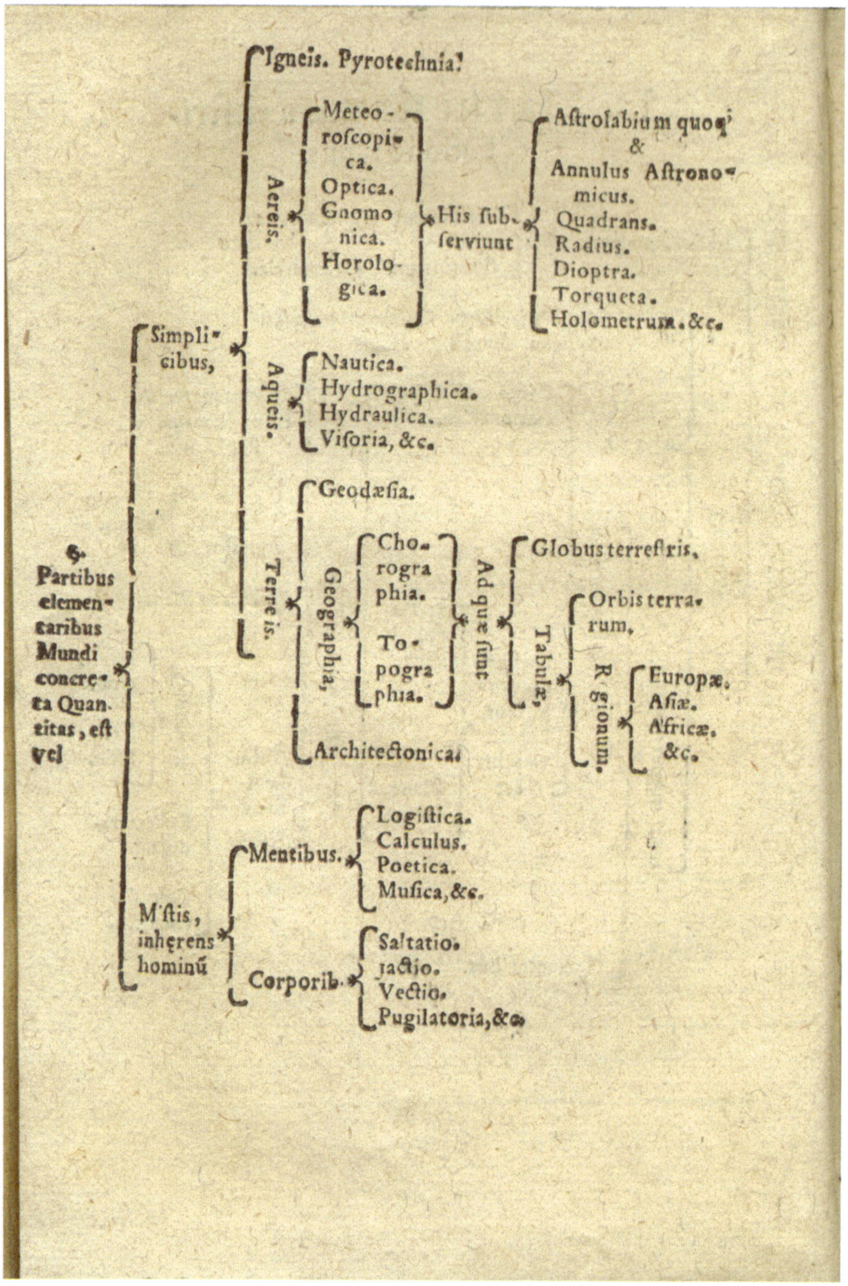

Fig. 2 Table of the subjects belonging to the branch of practical mathematics that deals with the specific disciplines and activities usually organized, on an abstract level, according to the four fundamental elements of which the cosmos is constituted. From (Ryff and Lucii 1598). Courtesy of the Universitätsbibliothek Basel

calculating the positions of the planets—*theorica planetarum*—and the methods used to calculate the values of a variety of tables, such as ephemerides, religious holidays and calendrical issues in general. These subjects are all mentioned in the table but not discussed in the treatise. In the frame of a scientific education, they were usually addressed after completion of a course on the *Sphere* and in the context of studies to train as an astronomer and/or astrologer.

The second table (Fig. 2) shows how a further series of subjects belonging to the branch of practical mathematics is structured. Most of the subjects are further grouped on an abstract level according to the four basic elements—Fire, Air, Water and Earth—here clearly used as a principle for organizing knowledge, with philosophical implications that are at most indirect and instrumental. The actual disciplines mentioned at this level are practical activities commonly performed during the early modern period, such as practical optics (measuring by sight), elements of geography (the calculation of distances between two points on the Earth's surface) and nautical astronomy. As mentioned, some of these subjects are also discussed in this specific treatise, but most of them are not.

The apparent discrepancy between the working program suggested by the tables at the beginning of the book and the number of subjects actually discussed in it was, however, probably not perceived at all at the end of the sixteenth century. The knowledge identified under the label *The Sphere* could indeed count on an extremely successful tradition based on the compilation of a very short treatise with the same title by Johannes de Sacrobosco, which was written in Paris in the thirteenth century.[3] The content of the tables therefore refers to the subjects that were treated within the frame of the entire tradition of these treatises.

1.1 The Tradition of the Treatise

The *Sphere* originally denotes a well-defined cluster of notions that qualitatively describe the cosmos according to a geocentric conception. Such a conception clearly goes all the way back to ancient cosmology and in particular to the Aristotelian worldview, according to which the cosmos is constituted of concentric spheres. Treatises on the *Sphere* were indeed circulating much earlier than the early modern period or the late Middle Ages, especially in the context of the Arabic tradition, as is evidenced for instance in the works of Thābit ibn Qurra and al-Bīrūnī and al-Fargānī. However, in the process of re-elaborating both ancient and Arabic scientific knowledge in the West during the thirteenth century, this knowledge tradition was reshaped. At that time new knowledge was usually produced in the form of commentaries on older works. Nevertheless, three different scholars engaged in producing new treatises on the *Sphere* during the thirteenth century, which were characterized by a new distribution of content-related aspects rather than by the introduction of new or different knowledge. These scholars were Johannes de Sacrobosco, Robert Grosseteste (1175–1253) and John Peckham. Their works are very similar to each other, though not identical. They had the same function, but presented differences concerning specific aspects and detailed content-related issues. As a matter of fact, these treatises were in concurrence with each other (Fig. 3).[4]

The original thirteenth-century short treatise of Sacrobosco's *Sphere* is divided into four chapters (Thorndike 1949). In Sacrobosco's words:

---

[3] According to sixteenth-century sources, Sacrobosco wrote his treatise in 1232. For a short biography of Sacrobosco among the historical sources, see (Sacrobosco et al. 1549). For an extensive biographical study, see (Pedersen 1985). A thesis states that Johannes de Sacrobosco originally came from Ulster, Northern Ireland. For more information, see (Drennan 1991).

[4] L. Thorndike, the historian who contributed most to the understanding of the tradition of treatises on the *Sphere*, counted the works of R. Grosseteste, J. Peckham and Campanus among Sacrobosco's rival texts. For more information, see (Thorndike 1949, 23). All of the subjects touched upon by Grosseteste are included in Sacrobosco's treatise but not vice versa. However, when Grosseteste describes the mechanism of the eclipses, he introduces the mathematical concepts of epicycles and equants, which are in turn absent from Sacrobosco's tract. This supports the thesis that Sacrobosco's tracts have a more qualitative character. For a German translation of Grosseteste's *De sphaera*, see (Baur 1912, 10–32).

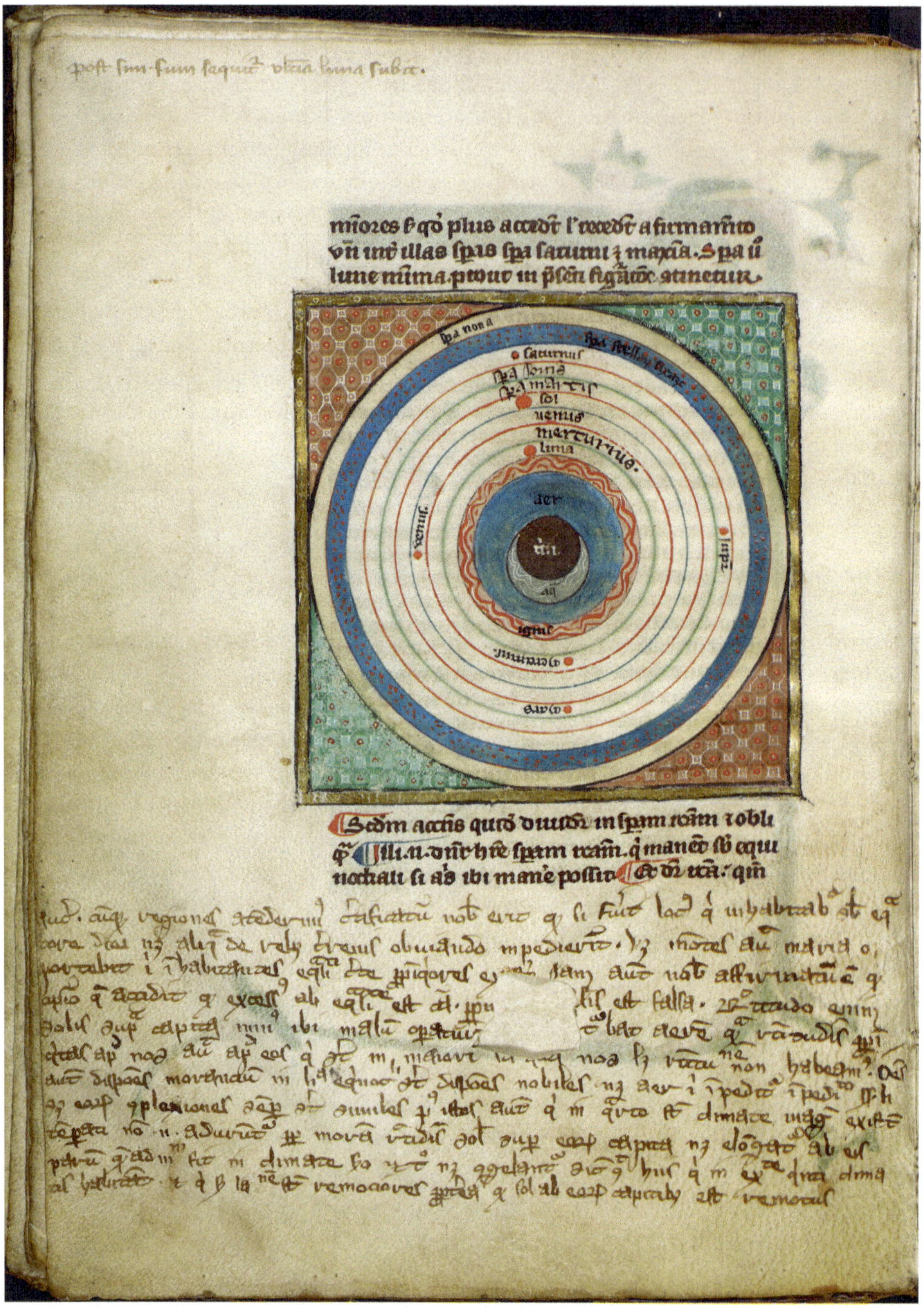

Fig. 3 Example of a folio of a thirteenth-century copy of Sacrobosco's *Sphere*. (Sacrobosco 1275–1300, f. 11v). Courtesy of Royal Library Copenhagen

The treatise on the sphere we divide into four chapters, telling, first, what a sphere is, what its center is, what the axis of a sphere is, what the pole of the world is, how many spheres there are, and what the shape of the world is. In the second we give information concerning the circles of which this material sphere is composed and that super celestial one, of which this is the image, is understood to be composed. In the third we talk about the rising and settings of the signs, and the diversity of days and nights which happens to those inhabiting diverse localities, and the division into climes. In the fourth chapter the matter concerns the circles and motions of the planets, and the causes of eclipses. (Thorndike 1949, 118)

The incipit of the book clearly leaves many options open concerning the textual details that could follow. For instance, by maintaining the same order of the exposition, content-related issues such as the number of spheres could easily be changed over time, so that the treatise indeed historically revealed itself capable of integrating new aspects of cosmological knowledge without changing its fundamental internal order through editing.[5] Moreover, there is a more fundamental aspect of the order given to the text by Sacrobosco that made it capable of spawning an editorial tradition four centuries long. The first part of the book is essentially related to cosmology. It outlines a simple scheme concerning the composition of the cosmos. This could be related to the knowledge required by all of the activities associated with the practice of traveling, where a rough understanding of the movements in the sky was required for a variety of reasons.[6] The second part of the book deals with the materiality of the sphere, but with two very different meanings. On one side, the treatise is a perfect example of deictic writing, that is, it was supposed to be read accompanied by a material instrument. In fact, the treatise can be considered as a description of an armillary sphere, to be positioned on the desk of the teacher. The second level, on which the treatise deals with materiality is concerned with the physics of the cosmos. In this respect the book was considered as a propaedeutic text to be read before works such as Aristotle's *De caelo*. Sacrobosco's text introduces here, for instance, the doctrine of elements. This is the foundation for the understanding of physics according to the meaning associated with this word in the frame of the Aristotelian doctrine, and thus also from the thirteenth century on. The third part of the book is the one that would be termed as astrology today. In fact, the issues explained in this section offered links between this knowledge and aspects related to fields such as medicine, agriculture, navigation and predictive astrology. This implies that this section, in turn, could later have offered links to the subjects of cosmography, metric geography, and cartography. The last part of the book, which is the one least developed in the work of Sacrobosco, represents a first step in what can be considered a study on mathematical astronomy. It introduces the fundamental qualitative notions that were eventually needed in order to learn, for example, how to calculate the positions of the planets or the occurrence of an eclipse.

Such an open design of the book is certainly the formal condition that allowed for the impressive historical phenomenon that the history of science associates with the treatise by Johannes of Sacrobosco. Although the *Sphere* compiled in the thirteenth century was not a commentary, it indeed immediately underwent a transformation process that took the form of a commentary, as was usually practiced in science in the period between the thirteenth and seventeenth centuries: the standard commentary and the *quaestio* (Grant 1994, 23). Initially, this commenting activity was necessary in order to render the text suitable for university classes (*lectio*). In fact, the *Sphere* became a mandatory course in all European universities up to the seventeenth century.

During the first 150 years after the appearance of the *Sphere*, moreover, the commentary works clearly show that of the three tracts on the *Sphere* compiled in the thirteenth century, the one produced by Johannes of Sacrobosco best met the intellectual and institutional requirements of that period. In fact, by the end of the fourteenth century new tracts on the *Sphere* were almost always associated, generally in their very title, with the original author: Johannes of Sacrobosco. From this point onward, a new scientific tradition was established, profoundly shaping the growing society of knowledge built around the growing number of universities, courts, and places where knowledge training was taking place, including artist workshops, *Abaco* schools, and religion-oriented seminars.

A most recent census of the tracts, which only took into consideration the prints compiled under the title the *Sphere* of Sacrobosco between the end of the fifteenth and the end of the seventeenth

[5]The first to recognize the open character in the design of Sacrobosco's treatise was (Johnson 1953). The innovative character of this work and others by Sacrobosco is also investigated by (Pedersen 1985).

[6]This subject is discussed in more depth in the next section.

century, counted over 300 editions.[7] Especially after the diffusion of the printing press, tracts on the *Sphere* eventually became one of the most common scientific tracts in many cities in Europe. The content of the tract, however, was transformed, and not only for didactic reasons. In the commentary works compiled after the end of the fourteenth century, the four chapters that describe the astronomical and cosmological systems represent only a portion of the whole work. The additions to the *Sphere* began ranging from pure supplementary commentaries, merely to explain an issue in other terms or to correct an aspect, to addenda on new subjects previously extraneous to the original work.

The *Sphere* became a mandatory course at universities in general, starting from the second half of the fourteenth century. Although the history of science has usually classified this text as a sort of introductory reading in preparation for the study of astronomy, and although in certain specific cases it was also preparation for reading fundamental texts such as Aristotle's *De caelo*, the texts used for teaching activities underwent a profound transformation process. Hundreds of printed texts and innumerable handwritten manuscripts testify to knowledge that began as a short treatise of around 30 folios on qualitative cosmology and ended up having an encyclopedic character, integrating a great variety of new subjects but maintaining the original design of the treatise in most cases.

The treatise by Peter Ryff examined above is just one, non-spectacular example picked out from the middle of this tradition and this process. The example of Ryff's text, moreover, is also peculiar because this text does not belong, strictly speaking, to the tradition of the treatises subsumed under the name of Sacrobosco, as the title of the work—*Elementa sphaerae mundi sive cosmographiae in usum Scholae Mathematicae Basilensiis*—does not contain his name.

Nevertheless, the treatise contains the original text and preserves the fundamental original design of the tract. Even most of the images which were used to illustrate it can be easily recognized as equivalent to the images used in most of the printed Sacrobosco treatises of his time[8] (Fig. 4).

This issue is remarkable because it recalls the existence of an innumerable amount of treatises compiled and published during the early modern period, which actually also contributed to the dissemination of the same knowledge. Whether they were entitled "sphaera mundi" or "cosmographia" or in a different manner, and whether or not they strictly followed the original design of the treatise of Sacrobosco, they also testify to an impressive interest in these subjects hitherto rather ignored due to the tendency to see these historical sources only within the constricted context of the history of astronomy. There is moreover a huge amount of handwritten material that was produced continually and intensively between the thirteenth and seventeenth century which should also be added to these treatises and to those that can be clearly ascribed to the Sacrobosco tradition. The same is also true for other traditions of *Sphere* treatises like the one authored by Pseudo-Proclus and those originating from earlier authors and in different cultures or epochs, such as the Arabic one and the one of classical

[7]Counting between 1472 and 1697 a total number of 376 treatises have been found. The majority of them contain the original text, either in Latin or in vernacular, and bear the name of Sacrobosco either in the title or on the cover. Other treatises are considered to be part of this group because they contain the text or passages of it, which are then commented upon, but do not bear Sacrobosco's name. In addition to these treatises, other works have been selected because they are works on geocentric cosmology and are designed exactly in the same way as the original text and their relation to it can be also recognized because of the usage of the same images. Finally, the *Quaestiones* on the *Sphere* of Sacrobosco are counted here as well, that is, those books used by the students for the specific purpose of preparing themselves for examinations during their studies. Fundamental for the harvesting of bibliographic information was the database published online by Roberto de Andrade Martins – http://www.ghtc.usp.br/server/Sacrobosco/Sacrobosco-ed.htm (Accessed April 2016) – and the bibliographic collection of Jürgen Hamel (Hamel 2014). In 1998, Trevor Lipscombe, the physical sciences editor for Princeton University Press, defined Sacrobosco's work as the "First Science Best Seller" (Lipscombe 1998).

[8]The most comprehensive research on the use and meaning of the illustrations in the tradition of treatises on the *Sphere* is (Crowther and Barker 2013).

Fig. 4 Illustration used in most of the treatises on the *Sphere* since it entered the era of printed technology, to explain the geometric properties of a sphere at the beginning of the original text. From (Ryff and Lucii 1598, 15). Courtesy of the Universitätsbibliothek Basel

antiquity which was translated, discussed and circulated in early modern Western culture during the period in which the tradition of the treatises of Sacrobosco was so strong. It is therefore necessary to recognize that approaching the historical knowledge tradition wrapped around the treatise tradition of the *Sphere* means approaching four centuries of history that involve such a large amount of historical sources that it can be considered one of the most important groups of sources that shaped the scientific identity of Europe.

1.2 The Label "Sacrobosco"

The present work focuses on the only subgroup of sources identified by means of (a) the name of Sacrobosco, (b) the presence of the original text, and (c) the preservation of the original design of the work. This subgroup of historical sources represents the largest subgroup of the entire amount of sources hitherto analyzed for this knowledge domain. It is therefore considered to be representative of all of them.

Before approaching the analysis of a sample of printed treatises of Sacrobosco's *Sphere*, however, it is necessary to understand how the original treatise was transformed by means of an example of a treatise published at the beginning of the seventeenth century, when the study of one "Sacrobosco treatise" was mandatory not only in almost all European universities for courses in mathematics at the faculties of liberal arts, but also as part of the education of noblemen and at the courts.

In 1604 the monk Francesco Pifferi (1548–1612) published a quite elegant edition of the treatise that stretches over 350 pages and is entitled *Sfera di Giovanni di Sacrobosco tradotta e dichiarata da Don Francesco Pifferi* (Sacrobosco and Pifferi 1604). Pifferi was appointed to the Tuscan court for the mathematical education of the Grand Duke's son Cosimo by his mother. After Galileo's astronomical discoveries he belonged to the army of scientists who were fighting in the frontline either *pro* or *contra* the new cosmology.[9] Within this social frame, he decided to publish a treatise on the *Sphere* of Sacrobosco in order to give an "entire overview of the mathematical sciences" (Sacrobosco and Pifferi 1604, Dedicatory letter). The treatise by Pifferi maintains the original four-chapter design of Sacrobosco's treatise, but the number of subjects discussed is impressively greater than in the original treatise and very closely resembles Ryff's tables of the subjects of the mathematical sciences. Moreover, Pifferi is aware of the fact that the treatise of the *Sphere* by Sacrobosco is the place where new subjects were to be added. In fact, before opening his disquisition, he offers a two-page table of

[9]Francesco Pifferi belonged to the small group of philosophers and astronomers, for instance, who were privileged to attend the meeting with Galileo Galilei in 1610 on Gianicolo in Rome, during which the new telescopic discoveries were first shown in Rome by the Tuscan scientist. For more information, see (Galluzzi 2014, 114).

the "new subjects," placing the actual complete table of contents at the end of the work. Some of the new subjects in Pifferi's treatise were actually already common in most of the treatises of his time, but others show that he was at the forefront of the long transformation process the treatise was undergoing. While it was not uncommon to find a chapter dedicated to plane geometry, it was less usual, for instance, to find an introduction to the geometry of three-dimensional objects, as the author expresses himself, which went well beyond the subject of the geometric sphere. The treatise contains a long section on practical optics which was discussed directly with Antonio Magini (1555–1617); an approach to the medical art, as this was needed to accompany the compilation of almanacs (Chap. 13); specific aspects of judiciary astrology, metrology; and even mechanics, that is the method for calculating the center of gravity of a given object. Other subjects listed here that were less new were connected to geography, cartography, meteorology, nautical astronomy and navigation techniques and, finally, information concerning the construction of solar clocks.

Between the table of the new subjects and the actual work, moreover, a few further texts were added, mostly due to the courtly character of this publication. But in one of them the author tells the reader which sources he used and, finally, that the book was intended as a re-elaboration in the Italian language of what he considered to currently be the best treatises on the *Sphere*, with a few new subjects added. It is interesting to note that the original treatise by Johannes de Sacrobosco is not listed among his sources, in spite of the title of the work. Apart from the already mentioned Antonio Magini, Pifferi quotes the treatises by Christophorus Clavius (1538–1612) (Sacrobosco and Clavius 1570),[10] Francesco Giuntini (1523–1590) (Sacrobosco et al. 1564),[11] Oronce Finé (1494–1555) (Fine 1524),[12] Proclus, Maurolicus,[13] and Francesco Barozzi (1537–1604) (Barozzi 1585).[14]

It thus becomes clear that Sacrobosco's name had the function of what modern marketing strategies would call a label. It helped to identify a certain framework, a certain domain of knowledge with a specific history, tradition and transformation process which, at that time, appeared as an improvement and enrichment process. Sacrobosco's treatises can therefore be considered the output of what was known as "knowledge on the *Sphere*" along its entire tradition, which includes a continuous process of codifying and integrating new subjects.[15]

2 The Taxonomy of the Transformation

To furnish an overview of the transformation process the treatise underwent, the investigation is narrowed down to the sixteenth century and, as this investigation requires the actual analysis of the content of the sources, the present results are based on a total of 35 treatises.[16] The analysis of the treatises

[10] Republished seventeen times by 1618.

[11] Republished in different collections twelve times up until 1660 and in Italian in 1582.

[12] Republished six times by 1542, both in collections and alone.

[13] Maurolicus compiled a long series of texts on cosmology and cosmography, two of which make direct reference to Sacrobosco. It is, however, impossible to determine exactly which of these texts was used by Pifferi. A list of Maurolicus's texts can be found on the website dedicated to "Progetto Maurolico:" http://www.dm.unipi.it/pages/maurolic/edizioni/cosmogra/intro.htm (last access: March 2016).

[14] Republished in 1598.

[15] In a paper published in 1990, Owen Gingerich, beginning his argument with the hypothesis that about one hundred different editions of Sacrobosco's treatises were produced during the early modern period—about one third of the real total—reaches the conclusion that there were about 200,000 of such material books circulating in that epoch. For more information, see (Gingerich 1990).

[16] The list of treatises analyzed is given in Appendix 1. To adjust the analysis, some treatises from the final years of the fifteenth century as well as of the early years of the seventeenth century have also been taken into account. Between 1500 and 1599, 263 treatises were printed.

is based on a taxonomic principle defined according to disciplinary areas. For instance, all treatises are considered to be treatises on qualitative geocentric cosmology, as this is what they contain by default. The further subjects defined while reading the treatises, that is, which the source analysis itself suggested, are: mathematical astronomy, calendrical calculations, the use of and in some cases instructions for constructing astronomical instruments, nautical astronomy and geography, cartography, meteorology, arithmetic, geometry, the use and construction of mathematical instruments to execute arithmetic calculations, judiciary astrology, literature, practical optics, and mechanics. In practice, all analyzed treatises were tagged following this taxonomy according to whether they contain subjects that belong to any of these disciplinary areas.[17]

2.1 The Taxonomy of Disciplinary Areas

The subject of mathematical astronomy is always associated to some aspects of the knowledge that was circulating in a codified form in Ptolemy's *Almagest*. Techniques for calendrical calculations, clearly related to the need to prepare calendars to regulate religious life, was a practice designated as *computus*. It required good mathematic skills that were often handled in separate treatises[18] (Chap. 13). Nevertheless, some of the later treatises tend to include this subject in the treatise itself. The use and the construction of astronomical instruments, such as the quadrant, the astrolabe and the Jacob's staff, represented a knowledge functional to almost all of the disciplinary areas that can be found in the later treatises. The subject of astronomical instruments was, however, closely related to the subjects of geography and nautical astronomy. As is well known, during the early modern period the discipline of geography changed radically. The medieval representations of geography, which aimed to illustrate cultural, political and religious knowledge, were abandoned during the early modern period. Based on the older tradition of Portolans and by means of the appropriation of ancient metric geography and especially Ptolemy's *Geography*, early modern geography quickly became a mathematically elaborated discipline with the purpose of creating a representation of the surface of the Earth that could be used practically, for instance, while traveling. For this reason, while in certain treatises it is possible to find sections genuinely dedicated only to techniques of nautical astronomy, as in the one published by Pedro Nuñes (1502–1578) in 1537 (Sacrobosco and Nuñez 1537), in most cases the subjects of nautical astronomy and geography largely overlap. Wherever either is addressed, it is possible to find at least an introduction to metric geography, sometimes one explaining the geometric structure of Ptolemy's *Geography*, or often a simplified rectangular grid more suitable for teaching purposes (Fig. 5).

The concepts of latitude and longitude are always introduced and explained, and generally it is also described how to calculate them. Obviously, longitude could only be calculated on the basis of well-prepared ephemerides and mostly only on the occasion of eclipses.[19] Certainly, there was no way to calculate longitude at sea. But plenty of expeditions and voyages also had the task of compiling

[17]The results of the taxonomic analysis of the 35 treatises considered in this section are given in the second part of Appendix 1. The selection of the 35 treatises was conducted with regard for the places of publication to comprise several geographic regions and the languages in which they are written, so that the sample would include at least a few examples of treatises written in vernacular. The taxonomy used here is dictated by the definition of disciplinary areas. While some of them clearly denote their actual historical meaning during the early modern period, others tend to overlap with each other or to be subareas of others. This issue will be taken into consideration below.

[18]Sacrobosco compiled a treatise for the teaching of the *computus ecclesiasticus*, too. For a later edition that contains the original text in Latin and only this, see (Sacrobosco 1547).

[19]For an extensive work dedicated to the history of the concept of longitude and of the methods used to calculate it, see (Andrews 1996).

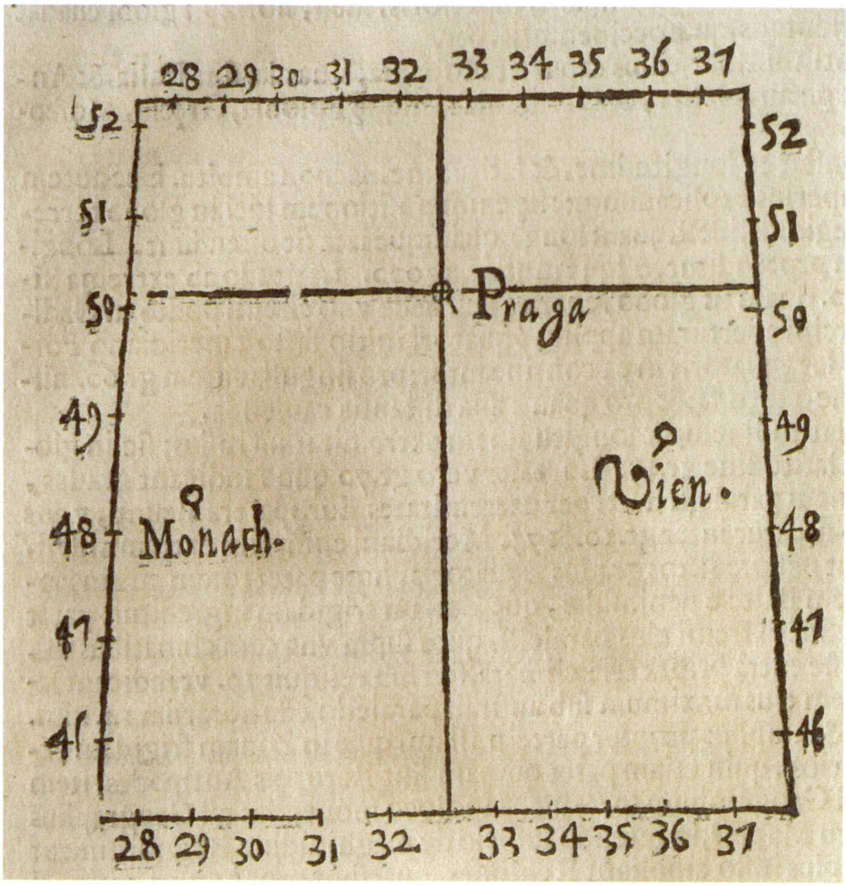

Fig. 5 Simple method to explain the geographic coordinate system. From (Biancani 1620, 204). Courtesy of the Library of the Max Planck Institute for the History of Science

precise maps of the territories visited by using the means at their disposal. For navigating, moreover, the calculation of the latitude as well as the altitude of some fixed stars at night, or of the sun during the day, were the information needed from the sky. Very frequently, even the general literature on the journeys at the time also shows long lists of data referring to these positions and the way in which they could be calculated.

There is therefore a sort of disciplinary continuum that spans from the subject of mathematical astronomy to navigation techniques, but for the sake of analysis it is admissible to break it down into two disciplinary groups, one associated with mathematical astronomy and mathematical algorithms to execute calendrical calculations, and the other related to geography, including cartography and nautical astronomy. Between the two groups there is the overlapping subject of the use and construction of astronomical instruments of observation.[20]

[20]Early modern mathematical instruments and the knowledge they embody often appear to be multi-layered epistemic instruments which assume different and partially overlapping meanings from one discipline to the next, or even from one domain of knowledge to the other, such as between practical and theoretical knowledge. For an example dealing with the quadrant for artillerists and its role in developing a theory from a practice during the first half of the sixteenth century, see (Valleriani 2013, Chap. 5).

The category "meteorology" is a distinct one. Even the original treatise of the thirteenth century contains a section on the division of the Earth's surface according to the climate zones, which was originally a taxonomy to distinguish between places where life was more comfortable from those where meteorological conditions made it more difficult or even impossible. This section was obviously greatly expanded as a consequence of the discovery of new territories and the reports that were reaching Europe. But meteorology was already a well-defined discipline during the early modern period thanks to the tradition of knowledge associated with Aristotle's homonymous work. In principle, therefore, all of the subjects discussed in the scholastic tradition of meteorology could have easily found their way into the tradition of the treatises on the *Sphere* as well. However, only one of these subjects was discussed at length and in many treatises, that is, the table of the winds, and, in some cases, how they were generated. While the Aristotelian doctrine represented the backbone from which this section was expanded in the treatises on the *Sphere*, the section concerning the winds, their directions, their seasonal behavior, their power and their geographic dislocations became a very central one and was obviously also expanded on the basis of travel reports. Occasionally, further subjects are included in this section, such as the generation of fog and its behavior. But in this case, too, the emphasis is on the relation between traveling at sea and the phenomenon of fog.[21]

A further disciplinary group comprises the subjects related to arithmetic, geometric and mathematical instruments. Arithmetic is mostly related to conversion operations, for instance, to convert currencies or rules of proportions, occasionally extending to the rules for solving square or cubic roots. If the rules for constructing instruments like the astrolabe are described, the corresponding section on mathematics often includes a more or less elaborated introduction to stereometry. If, moreover, the subject of judiciary astrology is included, then the section on arithmetic is usually more elaborate, as it will be shown below. Otherwise the arithmetic imparted via tradition of knowledge on the *Sphere* often corresponds to the arithmetic needed, for example, to use mathematical instruments like a geometric compass, which, in turn, partially overlaps with the mathematics taught at *Abaco* schools in the same period. The arithmetic found in these treatises, therefore, is generally the one that would have been learned by merchants and, later, military officers.[22]

The subject of geometry was already included in the original treatise, as Sacrobosco included the geometric description of a sphere. This section therefore was often expanded, sometimes quite extensively, in order to include an entire introduction to both two- and three-dimensional geometry. In this case as well, firm geometric knowledge was needed to efficiently approach the discussion concerning the metrics needed in geography and cartography, both two-dimensional (maps) and three-dimensional (globes). As is well known, mathematical and geometric procedures were introduced into many practical activities during the sixteenth century. Traditionally related to architecture, such methods spread especially because of the revolution of the art of war, which further implies a close connection to practical optics, that is, to the subject of measuring distances by sight. In this practice, a great number of instruments were conceived during the sixteenth century and in many cases their descriptions found their way into the treatises on the *Sphere* as well. There is therefore a further disciplinary group that can be generally defined by mathematics and that, again, seems to be intimately correlated with the instruments used in the frame of the practices mentioned above.

[21] The connection between meteorology and cosmology was further developed in different genres, such as the *Practica* (Chap. 13). The connection between meteorology and mathematics is discussed below in this paper.

[22] Arithmetic and geometry related to the use and functioning of geometric compasses have been investigated many times, for instance, also in reference to Galileo's military compass. For more information, see (Valleriani 2010, Chap. 2). For an overview of the mathematics for merchants taught at Abaco schools, see (Goldthwaite 1972) and, more generally, (2011).

The number of occurrences of subjects related to mechanics and literature is very low in the sample considered here and therefore can also be disregarded. It is sufficient to mention that the category "mechanics" applies mostly because of the presence of a chapter or a section concerning the rules for calculating the centers of gravity of more or less complex material bodies, while the category "literature" always means that the treatise contains a section that collects common quotations from classic works—such as those of Vergil, whose knowledge was needed in particular social contexts.[23]

Some final and different remarks are necessary for the subject of astrology. Clearly, the subjects of astrology, astronomy and cosmology were intimately related to each other in the period discussed here. In particular, every astronomical disquisition concerning the movement of celestial bodies implicitly carried a meaning about the relation between these movements and their consequences in the sublunar world. The very mechanism of the cosmos was conceived so that the external spheres acted on the internal ones, and therefore it was inadmissible that the events of the sublunar world would not be connected to the events of the superlunar one. This consideration, which Lynn Thorndike called the "universal law of astrology" (Chap. 14) before the era of Isaac Newton (1643–1727) (Thorndike 1955, 273), means that astrology, like cosmology, was always implicit in the treatises on the *Sphere*. Nevertheless, there is a difference between this consideration and the belief that the prediction or the analysis of astronomical events, such as the totality of positions of a certain number of celestial bodies, would allow for the prediction of events concerning human actions in the sublunar world. Including both judiciary astrology and astrological medicine, astrology was recognized and generally accepted only in reference to specific aspects and in specific circumstances, but not universally. The practice of calculating natal charts, for example, had been heavily and critically discussed in Western culture during the early modern period, as it seemed to violate man's free will.[24] The small sample of treatises considered here shows only that, when astrology is a subject discussed in the treatise or in a further treatise published together alongside it in a collection of commentaries labeled the *Sphere* of Sacrobosco, a relevant series of calculation techniques are described as well.[25]

3 The Motor of the Transformation

The chronology of the argument of the present paper begins with the occasion of the radical transformation of knowledge on the *Sphere* in the thirteenth century. As mentioned, this knowledge, and in particular the texts and their designs, were reshaped at this time so that they became more accessible.

The rapid and exceptional diffusion of the treatises compiled on the basis of this new design of the thirteenth-century *Sphere* can be read as the consequence of newly arising intellectual needs for a new knowledge system in astronomy and cosmology on a qualitative and descriptive basis. In turn, this need was most probably a consequence of the increased mobility of people during the thirteenth

[23] The original thirteenth-century treatise already contained several quotations from ancient works. They did not constitute, however, a specific and distinct chapter or section of the work. They were rather intended as mnemonic aids associated with specific aspects of the content.

[24] A clear manifestation of the relation between astrology and medicine is represented by the impressive number of almanacs produced in this same period (Chap. 13).

[25] For a lucid study about the Medieval and early modern discussion on the scientific character of astrology and its mathematics, see (Lerch 2015).

century in comparison to the previous period: when significant portions of society began to travel, their demand for knowledge concerning the world in which they were traveling increased.[26]

There were many new motives to travel which emerged between the thirteenth and fourteenth centuries:

- The foundation of universities on a larger scale, which, because of their "universal" character, soon developed into a strong network that allowed students and scholars to circulate freely from one university to the other (Rüegg 1992, vol. 1, 280–304).
- Building masters and many other kinds of artisans went through a social dynamic of specialization and became organized in mobile groups of skilled laborers moving from one large building site to the next (Binding 2014).
- Long-distance pilgrimage evolved as a slowly but constantly increasing activity, involving large numbers of people.
- Crusades and other large-scale military campaigns contributed decisively to a change in the societal mobility rate, especially during the thirteenth century when the number of crusades was the highest in history.
- Large-scale economic networks were established, also along the lines of the routes of the crusades, connected the Near East—mostly Palestine, Syria and Constantinople—to southern Europe—Genoa, Pisa, Venice—for the movement of spices and silk, and then from the south to the north of Europe between the Italian ports and Flanders, especially due to the demand for exchanging products from the East with northern European wool (Postan 1963–1989, vol. 3).

The very idea of "travel" was instantiated as it became the path for training, a way of showing off, and a sign of professionalism and knowledge.

Basic and qualitative cosmological knowledge, though it might sound odd today, was also absolutely necessary in earlier centuries to even understand the activity of traveling. The instruments used to calculate the time and one's position on the Earth's surface had been used since antiquity and during the later epochs on the basis of the geocentric spherical system described in the treatise. This aspect must have become extremely relevant when the era of exploration began.[27]

Amerigo Vespucci (1451–1512) describes one of the adventures he experienced during his trip from Cape Verde to Cabo de São Roque in Brazil in 1501 with these words:

> I mentioned that the distance between the newly found land and Cape Verde is almost 700 leagues, although I believed I had navigated for more than 800 leagues. This happened because of the terrible storm, of the unfortunate cases and of the ignorance of the helmsman, and all these things are the reasons why journeys become longer. We had reached a place where, had I not been an expert in cosmography, because of the ignorance of the helmsman, we would already have ended the course of our lives. In fact, no helmsman knew where we were,

[26] For an analysis of the transformation process of scientific knowledge as constantly requiring practical knowledge and activities in the background, see (Valleriani 2014b) and (2014a). While the argument explains the reasons for the increasing needs of cosmological knowledge in the thirteenth century, the success of the Sacrobosco treatise in particular might be due to other reasons. These are related to the necessity of learning cosmology in order to be able to compute calendars. During the thirteenth century, the Roman church centralized the right to define the liturgical calendar, which in turn had to be computed on a local level. As the treatise of Sacrobosco was used at the University of Paris and because of the central role of this institution in the catholic world, it might be inferred that the treatise spread outside the boundaries of the university in order to instruct computists in the religious institutions that ruled outsides the cities.

[27] For the role of cosmography not only in collecting traces of such development but also in promoting it, see the illuminating paper of Angelo Cattaneo (Cattaneo 2016).

Fig. 6 Amerigo Vespucci finding the Southern Cross. From (Straet 1580). Deutsches Museum, Münich

> even within a range of 50 leagues. And we were rambling and vagabonding without knowing where we were going until I took care of my health and that of my companions by means of the astrolabe and the quadrant, which are astrological instruments. (Ramusio 1978–1991, vol. 1, 672)

Amerigo Vespucci was a very skilled and notorious cosmographer, but the knowledge he displayed in 1501 actually corresponds to what was becoming common background knowledge, even for those not engaged in journeys of exploration. Navigating and, more generally, moving from one place to the other implies a constant need to know one's actual position. What Vespucci is referring to, is the procedure of position fixing, which was developed at the turn of the fifteenth century in order to navigate at long distances from the coast. Fundamentally, it presupposes the idea that the globe can be represented by a grid of coordinates and that these are the usual latitude and longitude (Fig. 5). In the Northern Hemisphere, the latitude could be easily determined by measuring the height of the North Star above the horizon. Although it had long been known, this procedure was actually rarely applied on ships before oceanic traveling began. Navigation was accomplished mostly by remaining close to the coasts, or at least by traveling distances that rarely required such calculations, as for instance, within the Mediterranean Sea. Constantly controlling the angular height of the polar star allowed the travelers to continually determine the north-south axis of their paths while navigating, for instance, from Lisbon to the American coast. The situation was different while traveling below the equator, that is, in the Southern Hemisphere. At the end of the fifteenth century, the stellar configurations visible

Fig. 7 Nautical astrolabe used to measure the altitude of the Sun. Dundee Museum and Art Gallery

from the Southern Hemisphere were indeed largely unknown to Europeans. Stars that could absolve the function of the polar star had not yet been determined. This task was then accomplished by the first explorers. The same Vespucci mapped the Alpha and Beta Centauri and the Crux constellations (Fig. 6), which were possibly known during antiquity, but then forgotten because of their disappearance due to the procession of the equinoxes. In absence of this possibility, the Portuguese navigators developed a method according to which the altitude of the Sun could be used as the basis for calculating latitude (Fig. 7), a method apparently derived from the domain of astrology (Randles 1998).

An institutional testimony to this exploratory function of the knowledge tradition of the *Sphere* is represented by the work executed at the *aula de esfera* in Lisbon toward the end of the sixteenth century. Starting with occasional courses in 1550, in 1590 the activity of the *aula de esfera* at the College of S. Antão in Lisbon became permanent until 1640. Organized by the Jesuits, the school did not follow the usual program of the Jesuit colleges, but rather overlapped with and integrated the program of the school for helmsmen, also located in Lisbon (Baldini 2000; Leitão 2008). In 1542, moreover, Alessandro Piccolomini (1508–1578) in his seminal *De la institutione*, one of the most widely disseminated ethic-pedagogic works of the early modern period, indicated that studies of the *Sphere* should begin between the ages of fourteen and eighteen—preferably on the basis of his own treatise first published in 1540 (Piccolomini 1553). This would have been preparatory to the study of astronomy, but especially useful in learning how to use the astrolabe,[28] so that both cosmography and geography could be learned as well (Piccolomini 1542, 59).

The nature of the knowledge connected with the *Sphere* was essentially practical. Obviously, the cosmological aspects of such knowledge were considered to have an ontological value during the thirteenth century. Later, however, knowledge concerning the *Sphere* became completely instrumental to practical needs. Even after Copernicus's doctrine had been widely endorsed, production of the *Sphere* continued undisturbed for a long time; only in a few exceptions did *Sphere* treatises include contemporary cosmological alternatives, because this aspect was actually secondary.[29] Astrolabes, for instance, were instruments conceived back in antiquity on the basis of a geocentric worldview. The advent of the Copernican doctrine did not change the habit or the reasons for using such instruments. Their functioning was based on a precise stereometric projection of the eighth sphere—the sphere of the fixed stars—onto the plate of the instrument. The use of the astrolabe therefore always presupposes the knowledge of the geocentric structure of the world (Danti 1569, 28).

The famous *Sphere* treatise by Pedro Nuñes, published in Lisbon in 1537, contains, beyond the usual original text of Sacrobosco, (1) a Portuguese translation of Georg von Peuerbach's (1423–1461) theoretical description of the motion of the Sun and the Moon (Regiomontanus and Peuerbach 1543),[30] instrumental to the use of the nautical astrolabe in the Southern Hemisphere, (2) a Portuguese translation of the first book of Ptolemy's *Geography*, which shows the early recognition of the relevance of the ancient techniques for metric geography, (3) a treatise to teach navigation on the ocean by means of charts and nautical astronomy (Sacrobosco and Nunes 1537).[31]

Around the mid-sixteenth century, the treatises on the *Sphere* were already starting to overlap with those on cosmography. Indeed, cosmography was the name of the discipline that described the relations between the Earth's surface and the skies. In practice, this was the original task of the thirteenth-century *Sphere*, which was fully accomplished in the frame of the travels of exploration. A clear example of this shift is the 1524 work of Petrus Apian, translated into Spanish in 1550 (Apian 1550) (Fig. 8).

The establishment of cosmography during the sixteenth century is in fact the result of the connection between metric geography and astronomy, so that for a time cosmography and the *Sphere* belonged to the same knowledge domain.

The more practical aspects of this knowledge domain are clearly related to proper navigation techniques, as is shown in the 1551 work of the Spanish cosmographer Martin Cortés, who published a book on the art of navigation as an addition to the *Sphere* (Cortés 1551). The practical knowledge and experience codified through the transformation process, involving the knowledge tradition of the *Sphere* during the first half of the sixteenth century was, therefore, specifically related to navigation techniques, nautical astronomy, geography and metric geography, cartography and aspects of meteorology such as the theory and description of the winds.

[28] Piccolomini suggests the reading of (Stöffler 1513). Piccolomini's indications are in agreement with E. Grant's more general considerations as regards the identity and the needs of the students who were supposed to use the treatise on the *Sphere* for their education. For more information, see (Grant 1994, 20).

[29] The argument concerned with the missing ontological character of the early modern treatises is developed in reference to Galileo's treatise on the *Sphere* in (Valleriani 2014c).

[30] Peuerbach's theory of planetary motion was first published in 1472 by Regiomontanus. Nuñes translated only the sections concerning the motions of the Sun and the Moon.

[31] For a modern edition of Nuñes's treatises, see (Nuñes and Leitão 2002).

Fig. 8 Petrus Apianus's World Map. From (Apian 1550). Courtesy of the Library of the Max Planck Institute for the History of Science

3.1 The Actors in the Transformation

Knowledge transformation is obviously also the result of a series of actions and, as such, a process related to real people: the authors and the readers of the treatises. While the former can be identified quite uniformly, a few, more elaborated remarks are needed to discuss the latter, who actually created the demand for such treatises.

The total number of authors involved in the production of the treatises on the *Sphere* amounts to almost 150. Although no systematic research has been concluded on this list of authors, it is quite evident that most of them were natural philosophers and mathematicians. Concerning the latter category, it is impossible to attach unique disciplinary professions to early modern scholars, so that it is actually misleading to define the authors simply as mathematicians. Most of them were also astronomers, or experts on mechanics, or even humanists.[32] What they had in common, however, was a clear

[32] Edward Grant identified Medieval and early modern cosmologists mostly with only natural philosophers (Grant 1994, 46). The addition of the category of mathematicians in this work is due to the particular focus on the book tradition of the *Sphere* of Sacrobosco and on the early modern period. As mentioned, however, the two categories largely overlap.

education and activity in the field of mathematics of the period, both practical and theoretical. A representative and well known example of an author of the treatise on the *Sphere* is Christophorus Clavius. Mathematician and astronomer at the Jesuit *Collegio Romano*, he published one of the most relevant treatises on the *Sphere*, which was republished 17 times over the course of 48 years! (Sacrobosco and Clavius 1570).

Identifying the readership is more challenging. First of all because it might have changed over time, and second because Sacrobosco's treatises were published in very different social contexts. Such variety is confirmed by the inventory of a sixteenth-century Spanish bookstore. It shows that treatises in vernacular were cheaper than treatises in Latin. The 1545 Spanish edition by Jerónimo de Chaves (1523–1574) (Sacrobosco and Chaves 1545), for instance, cost 51 *maravedì*, much less than the 471 needed to buy the fine collection published by Giunti in Venice in 1531 (Sacrobosco et al. 1531).[33] Certainly, students of the liberal arts faculties were using this book for their studies, mostly in the course for mathematics.[34] The same was true, of course, with respect to all the colleges, for instance those of the Jesuits, which were similar but not equivalent to universities. Galileo also produced a treatise on the *Sphere* for his public and private teaching activities. He was teaching young noblemen, who after their studies, were destined first for military careers, and who would later take over the family activities (Valleriani 2010, 71–113). The *Sphere* was therefore also part of a curriculum for military officers and, more generally, a text to learn practically useful knowledge for a variety of purposes. The most evident example is related to the above mentioned Portuguese *aula de esfera*, but generally speaking this knowledge was useful in many different disciplines and matters. Writers such as Gabriel Harvey (Wolfe 2009, 151), polymaths and alchemists such as John Dee (James 1921), physicians and astrologers like Achilles Pirmin Gasser (Niebler 1973, 19), and naturalists like Konrad Gessner (Leu 2008), also possessed editions of such treatises. The readership ultimately ranged from practically oriented professionals like military officers and admirals to philosophers, and included literates, collectors, merchants, functionaries, apothecaries, noblemen and noblewomen, and finally passionate readers of travel reports during the early modern periods. The demand for the treatise and especially for new subjects to be included into the treatises was therefore growing along with the size of the knowledge society.[35]

4 The Knowledge Network: Structure, Model and Dynamics

New subjects were codified in the treatises on a local basis. For instance, it seems clear that subjects related to navigation first appeared in the *Sphere* treatises on the Iberian Peninsula. As mentioned in the introduction, this work aims to investigate exactly those mechanisms that allowed new subjects to enter the tradition of such treatises on a much larger geographic scale. The analysis of a small number of historical sources, paradigmatically chosen in different regions and in different times, seems indeed to show that, once a subject had appeared in any *Sphere* treatise, it was often then adopted by all subsequent publications, irrespective of the place of publication and the origins and background of the actual author. However, such a transfer process often implied a transformation process nested within the mechanism used to appropriate a new subject into great portions of the tradition of treatises. These appropriation and transformation mechanisms over a large geographic area and a long time interval can best be described once the fundamental data available are known and have been analyzed—a

[33] For the inventory of the bookstore, see (Pettas 1995, 142, 70). A *maravedì* was at that time 1/34th of a Spanish *Real*.

[34] Concerning the teaching of mathematical sciences during the early modern period, see (Pantin 2006).

[35] For a beautiful analysis of the reading habits concerning mostly students and training elite and in reference to the treatises on the *Sphere* during the early modern period, see (Oosterhoff 2015).

purpose for which some aspects of network theory become necessary instruments.[36] In particular, the structure of the network, which offers an overview of the centers of productions of the treatises and their position in relation to each other, will be distinguished from the network dynamics, that is, a representation of the evolution of the knowledge network, which in turn depends on the network model.

The sample upon which this analysis is based comprises 376 printed treatises on the *Sphere*. The timespan taken into consideration is given by the publication date of the treatises themselves. They begin in the year 1472 and go on until 1697. As for the geographic distribution, they were printed in 45 different places distributed from the extreme western coast of Europe to the meridian that passes through Prague.[37]

4.1 Network Structure

While the timespan and geographical scale of the network are easy to define, the definition of the actual nodes of which the network is constituted is somewhat more demanding and decisive for understanding the structure of the knowledge network. In this specific case, the network needed is a network of knowledge and the nodes must therefore be identified with the contexts that produced that knowledge, that is, the treatises. The structure of the network therefore identifies the social and economic structure that produced such knowledge.

New knowledge arises in response to specific activities, for instance, related to local economy. Agriculture, medicine, navigation, and cartography are all disciplines that can be traced back to local roots, traditions and needs. As the material treatises in which the codified knowledge is printed are clearly the vehicles that traveled over the connections between nodes in the network, these network nodes are therefore the entities whose work produces new treatises, possibly one containing a new subject. In this knowledge network the node must therefore be an elaborated constellation between authors, institutions like universities, and economic activities like those executed by early modern printers. Such constellations are called here "micro-regions" (Chap. 15). The internal dynamic of a micro-region generated knowledge whose external representation took the form of a treatise. In other terms, a node in a knowledge network identifies the knowledge production dynamic of a "micro-region." The final product is the print and diffusion of a new treatise and the incurred transaction costs to maintain the network alive are clearly placed on the shoulders of the early modern printers. They are the key to the knowledge evolution dynamics of the early modern period (Chap. 13).

The difficulty of identifying a node arises also because this network has been arbitrarily identified in the past and, therefore, encounters the usual difficulties experienced in history writing. First of all, the sample can never be considered complete, as more treatises could be discovered, or simply catalogued so that they can be included in the analysis, and might influence the final results. Second, and more relevant, is the fact that large-scale networks are always constituted of subnetworks, in this case, of geographic areas within the European continent. For the analysis of the set of Sacrobosco's treatises a few criteria led to the definition of regions: the prevalent economic activities, the local language, and relevant geographic boundaries. In this way five regions were identified: (1) the Iberian

---

[36] A fascinating example concerning the use of network theory as developed in sociology for the purpose of history writing is (Malkin 2011). Malkin analyzes the emergence of the Greek world in the Archaic Era by means of network theory. An introduction to network theory, mainly devoted to its application in the framework of sociology, is (Watts 2003). For the relation between network theory and the concept of accumulation of knowledge, see (Renn et al. 2016). The following section is written mostly on the basis of these works. Preliminary results of this and the next section were also discussed with Claire Lemercier and Emmanuel Lazega, whom I would like to thank for their precious advice.

[37] The number of data collected concerning the treatise from the eastern regions of Europe is very low; however, this might be due to the difficulties in finding such data rather than to a real absence of publications.

Peninsula; (2) central Europe, that is, the region that runs parallel to the Alps until Frankfurt and that unifies both Germans and Francophones; (3) northern Europe, extending exactly over central Europe up to the coast; (4) eastern Europe, the region that begins to the east of Wittenberg in Germany; and (5) Italy, that is, south of the Alps.[38] The analysis was further enriched by the distinction between coastal and non-coastal areas in all regions where this distinction is meaningful. This is because of the assumed particular relevance of the coastal regions as a consequence of their prevalently economic activities and of the close connection between such activities and travel.

Because of the long time interval considered here—over 200 years—an additional distinction is necessary in order to evaluate more closely the dynamic of the network and its historical developments. Therefore, the period was first divided into four time intervals: the period before 1500, followed by two time intervals each of 50 years for the sixteenth century, and finally the seventeenth century.

4.2 Network Model

There are many possible network models. In his study on the archaic Greek world, for instance, Irad Malkin determines the network as a "many to many" network, that is, without a central hub (Malkin 2011). As will be shown, the knowledge network on the *Sphere* is characterized by a relatively small number of central hubs that change their relative dominance in the main network over time.

To determine which model the network of knowledge on the *Sphere* had, first the micro-regions, that is, the nodes, were grouped according to the places and the cities where they were situated. During the sixteenth century, most publications originated in northern Europe, followed by Italy and central Europe in the first half of the sixteenth century and then in the opposite order during the second half (Fig. 9).

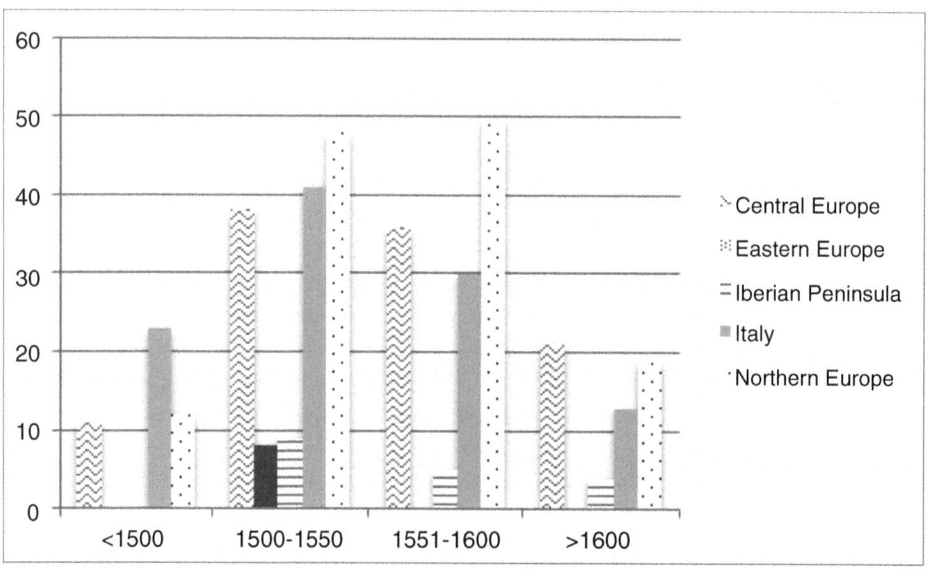

Fig. 9 Absolute values of the publication rates and their distribution over space and time

[38]For a precise subdivision of the regions and the list of places where Sacrobosco's treatises were published in each region, see Appendix 2.

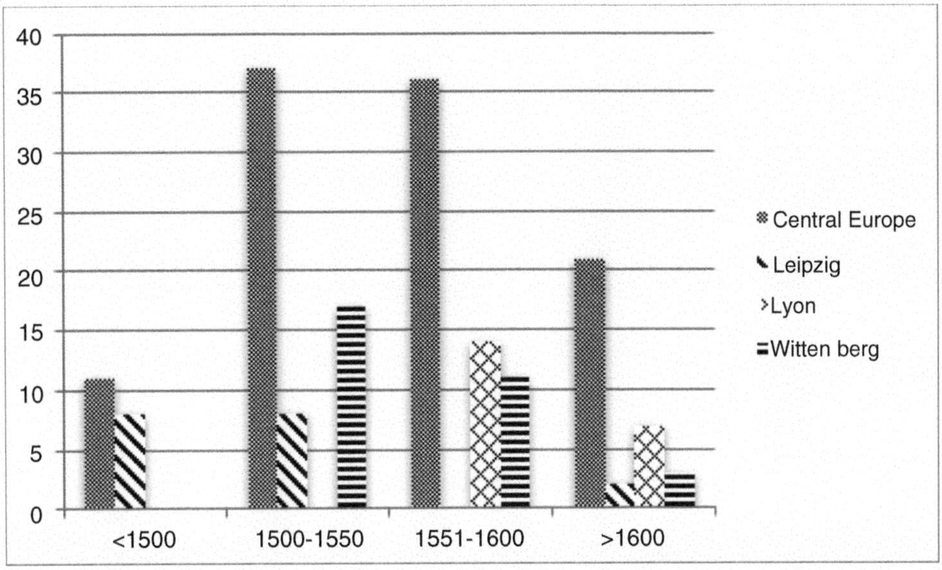

Fig. 10 Absolute values of the publication rates distributed over time for Central Europe and identification of the three major centers of publication

The relatively low number of publications from the Iberian Peninsula should, however, be compensated by the fact that most of the treatises on the *Sphere* compiled in this region during the sixteenth century were produced for institutions like the *aula de esfera* and remained in manuscript form.[39] Concerning the region of central Europe, it is possible to determine that the predominant center of publication was Wittenberg during the first half of the sixteenth century and Lyons during the second half (Fig. 10).

In southern Europe, the dominant center of publication is without a doubt Venice, though its relative predominance decreased in the second half of the sixteenth century (Fig. 11).

In northern Europe, the dominant place is clearly Paris, though its relative predominance also decreased during the second half of the sixteenth century in favor of the position of Antwerp, considered here as a coastal city (Fig. 12).

Paris, Venice, Antwerp and Wittenberg turn out to exhibit the numbers necessary to be designated as central hubs of the network. The relevance of the coastal cities as publication centers, finally, reached its peak during the first half of the sixteenth century (Fig. 13).[40]

Disregarding the temporal distinctions, seven of forty-five places of publications disseminated 72 % of all treatises, with 43 % dispersed in Paris and Venice alone. The network model that emerges, therefore, is a network whose nodes are micro-regions, organized at the level of continuously changing subnetworks, connected to each other mostly via the great hubs of Venice, Paris, Wittenberg, and Antwerp. Based on the example of the tradition of the treatises on the *Sphere*, it can first be concluded that the model, according to which the social and economic structure of the knowledge network was shaped, was quite heavily centralized.

[39] For a review of many of the texts on the *Sphere* used at the *aula de esfera*, see (Leitão 2008).

[40] I am profoundly indebted to Christoph Sander for his invaluable help while analyzing the data to determine the network model.

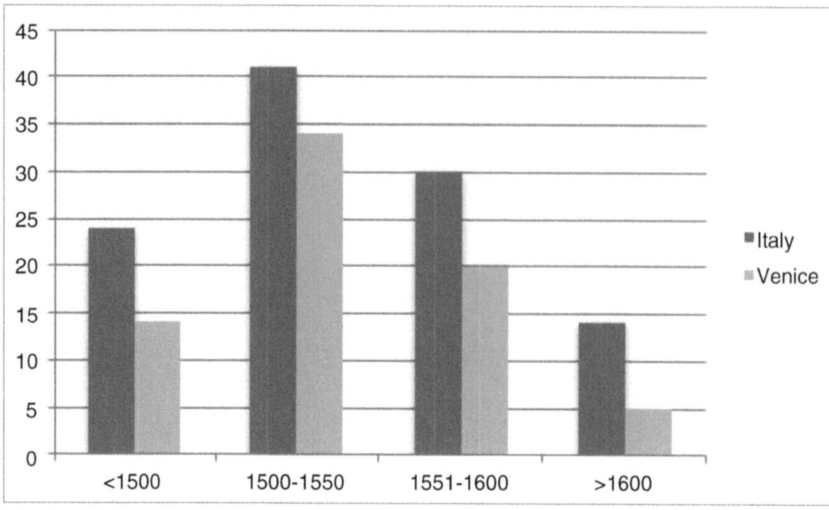

Fig. 11 Absolute values of the publication rates distributed over time for Italy and identification of one major center of publication

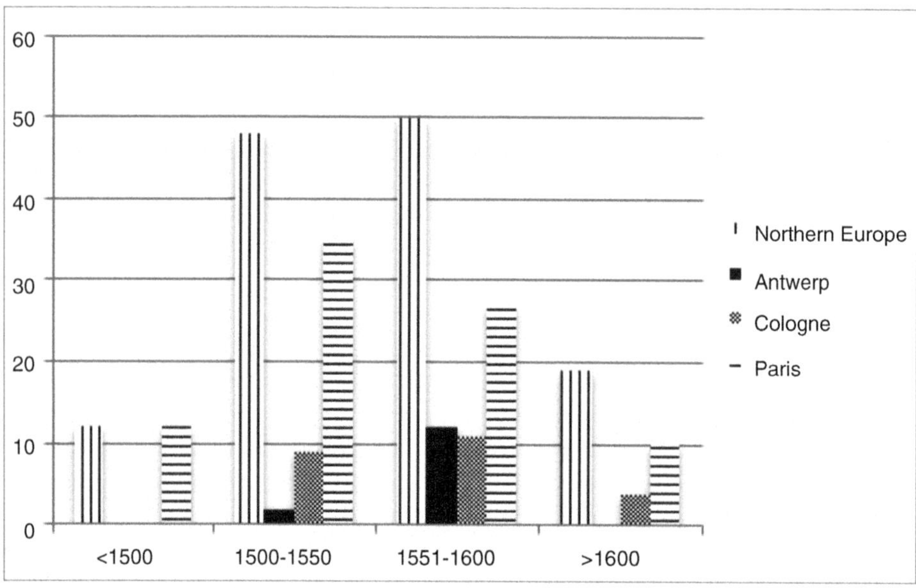

Fig. 12 Absolute values of the publication rates distributed over time for Northern Europe and identification of the three major centers of publication

4.3 Network Dynamics

To explain how it could happen that a new subject, perhaps codified in a treatise for the first time in Portugal during the first half of the sixteenth century, would become a new subject in many or all of the subsequent publications—regardless of the place, the provenience and the relevance of the new subject to the region where it appeared—it is now necessary to understand the dynamics of the network. In other words, the fact that the network had a specific model does not yet explain how it actually worked.

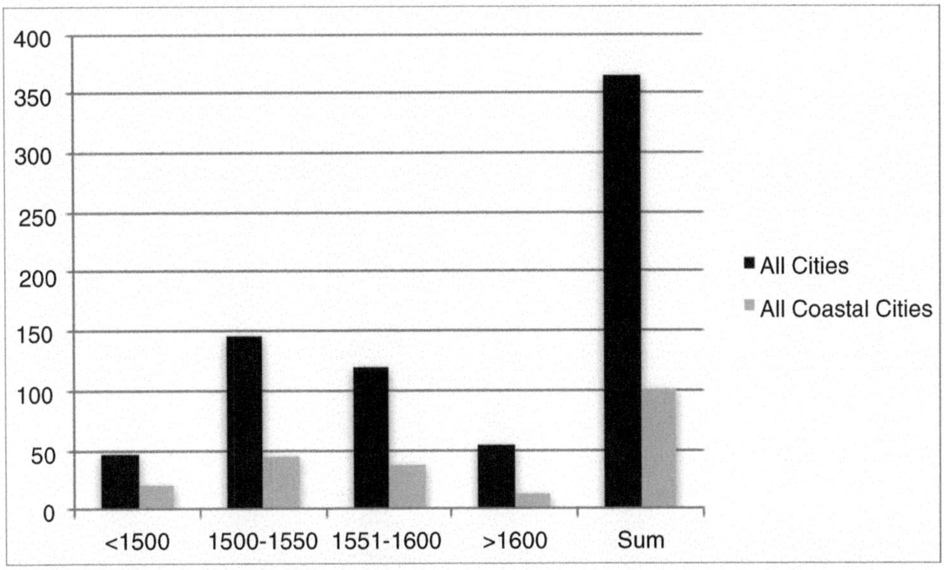

Fig. 13 Absolute values of the publication rates distributed over time and distinguished between publications realized in maritime centers and those realized in other locations in Europe

Network theory approaches this investigation in terms of connectivity between nodes. In particular, it developed a fascinating conceptual tool, denominated "small-world," to imply a phase of particular intensive connectivity between the nodes, that is, when the minimal number of steps are needed to connect each pair. In a small-world, regions of the network are connected so they can reach each other without necessarily being aware of this. This means that contacts between all regions of the network are possible, independently of the individual characteristics, ability, and knowledge of the nodes.

Any additional subject added to any treatise on the *Sphere* over time can be considered as an innovation. The network model that allows for the analysis of phenomena of the spread of innovations is the network of diffusion of innovation (Valente 1995). In the frame of an evolving network of diffusion, the small-world corresponds to a specific status of the network, more precisely, to the status of maximal connectivity. This means that when this status is reached, the network is able to redistribute knowledge in the fastest way. If the data can show the exact time interval during which the "*Sphere* knowledge network" remained in this state, it is possible to assume that this is the period during which the maximal transfer of knowledge from different regions over the network occurred.

A first indication, but not a demonstration, is given by the graph of distribution of the production of the treatises over time. Between 1472 and 1697, there are 149 years in which treatises were published, but these are not regularly distributed. In fact, 71 % of all treatises were printed during the sixteenth century. Looking at the exact distribution (Fig. 14), a first peak is reached around 1494 and the last around 1560.

From a mathematical perspective and basing the analysis on a random network, the curve of connectivity is usually about zero and moves up to the maximal connectivity over a very short period. In the case of the "*Sphere* knowledge network" this is clearly not the case. This aspect is to be explained mainly by considering that a historical and real knowledge network can only be analyzed by involving a series of factors and parameters in the argument that, for a time, are able to prevent connectivity from increasing. These factors mainly concern the dynamic internal to the regions or subnetworks, and, in particular, the requirement that a certain threshold must be reached before information exits the local area where it was created, to spread over the network via the highly connected great hubs.

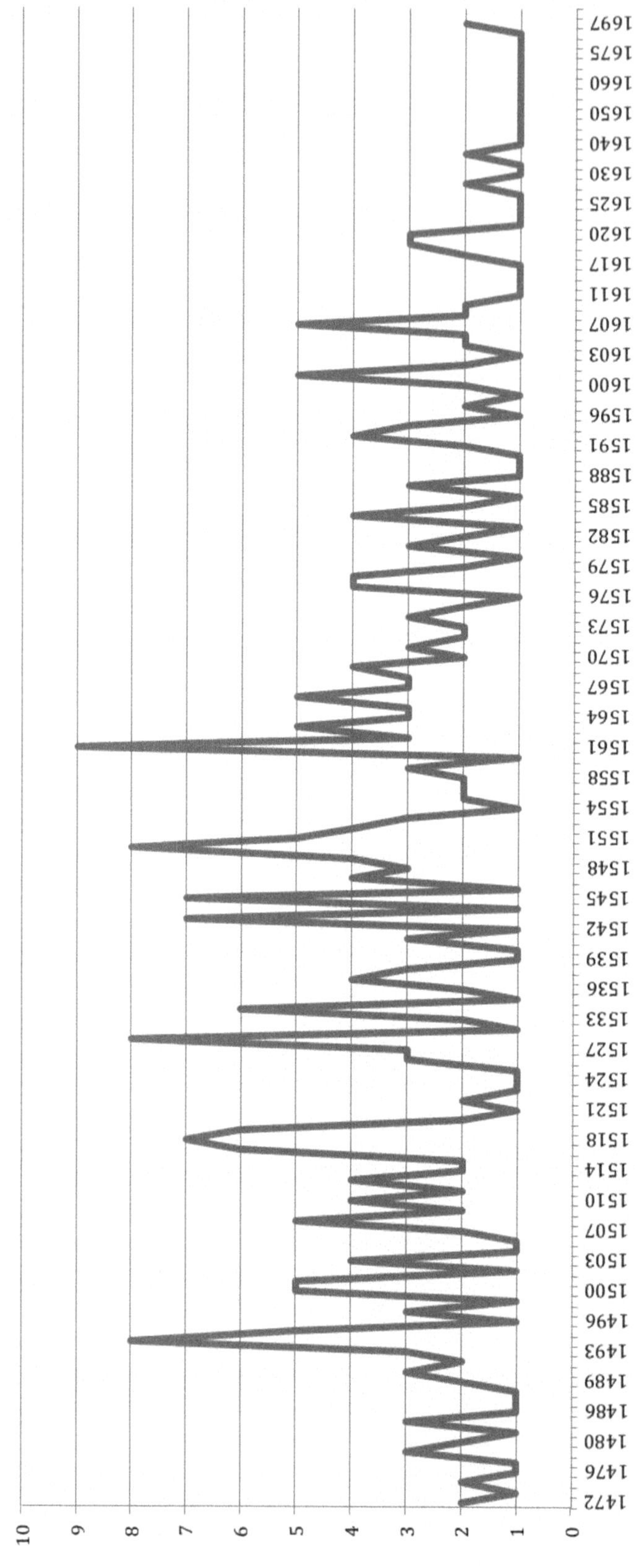

Fig. 14 Chart showing the distribution of all publications over time against a fine temporal grid

The peak in 1494 and the rapid decline after 1560 might therefore represent such an achievement and departure from the small-world status of the network.

If the 1494 peak corresponds to the surmounting of this threshold, it should be possible to see that not only many publications were realized in these few years, but they also appeared in many places and nodes of the main network. In fact, by considering the interval between 1493 and 1501, it turns out that the substantial number of 25 publications were realized in this period. This happened not only in Paris (7 publications) and in Venice (8 publications), but also in Bologna, Leipzig, Lisbon, and Cologne. If a later peak within the supposed period of the small-world state is considered, for instance between 1549 and 1553, the total amount of publications is 26, but their distribution is noticeably different. While Paris retains its predominant role (9 publications), many other nodes, places and micro-regions emerged in this period, thus testifying to the network's great dynamic. Further publications were indeed realized in Wittenberg, Venice, Salamanca, Florence, Antwerp, Basel, Frankfurt and Seville.[41] In accordance with this result, it also turns out that the treatise by Sacrobosco was published within the frame of collections of different works, such as those by Peuerbach and Regiomontanus, but almost only in publications realized in either Paris or Venice, thus increasing its dissemination effect.

In the state of maximal connectivity, it is also to be expected that all of those characteristics that could act as barriers would exert less of an impact. Barriers to knowledge transfer could be of a geographic, cultural, political or even linguistic nature. An interesting aspect to evaluate under this perspective concerns, therefore, the use of vernacular instead of Latin. As shown, the production of treatises after the sixteenth century decreases. During the entire seventeenth century, the publication of only 58 treatises is documented. The production of treatises in vernacular in this period, however, reaches its highest peak, that is, 29 % of the entire production during the same time interval. No other time interval shows such a high percentage. The same data for the first half of the sixteenth century is lower: only 14 %. However, if the absolute numbers for the production of the treatises in vernacular are considered, it turns out that as much as 40 % of the total number of publications in vernacular over the entire two centuries appeared in the first half of the sixteenth century alone. From a historical point of view, finally, it is well known that the use of vernacular in treatises of a scientific character increased during the early modern period, so that it is hardly surprising that 29 % of the production in vernacular took place during the seventeenth century. On the contrary, that as much as 40 % of the entire production in vernacular was realized during the first half of the sixteenth century is a clear indication that, during this period, the distribution of the knowledge on the *Sphere* reached regions, either geographic or social, that had not been touched before. The "*Sphere* knowledge tradition" was being shaped against a background of practical activities and practical knowledge during the small-world phase. It is in fact from this specific social milieu that the habit of using vernacular to write treatises concerning technology and science started in Western culture.[42] With regards to the regional distribution of the use of vernacular, it might suffice to show the large preponderance of its use on the Iberian Peninsula, where most of the production centers, that is, the nodes, are close to the coast and therefore likely to be closely related to the activity of journeys of exploration (Fig. 15).

As mentioned, according to the network model conceived here, a piece of information is created at the level of a micro-region. However, the emergence of new knowledge is only a *conditio sine qua non* in order for it to spread over the network. Its emergence does not guarantee that it will spread over the network. Moreover, networks like historical knowledge networks do not have a plan or goal concern-

[41] The bibliographic data of the treatises published during the two time intervals are listed in Appendix 3.

[42] Isabelle Pantin also identifies an increasing tendency to translate Latin treatises on the *Sphere* into vernacular languages. For more information, see (Pantin 2000). A recent paper analyzes four different editions of the *Sphere* of Sacrobosco written in vernacular. For more information, see (Crowther et al. 2015).

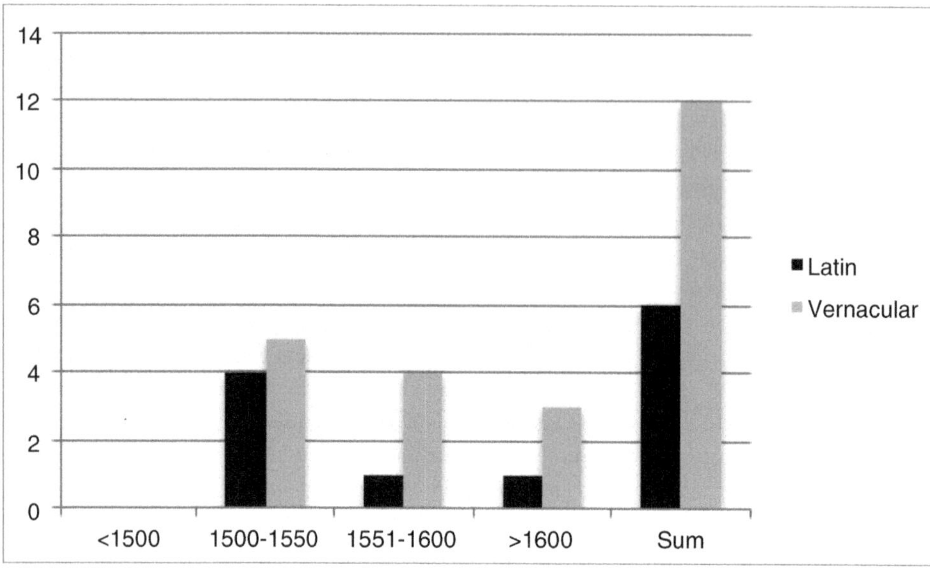

Fig. 15 Chart showing the distribution of languages used in the publications realized on the Iberian Peninsula

ing their evolution. Therefore, if the data are able to show that network-like knowledge transfers took place, the reasons for such transfers have to be found in the structure of the network itself.

Between the micro-regions and the great hubs there are other entities, such as politically or linguistically defined geographic areas. According to the network theory, these regions are subnetworks or affiliation networks, and it is to be expected that after maximal connectivity is reached within them, a piece of information spreads over the main network too. The way a piece of information spreads from an affiliation network is explained by referring to the mechanisms of weak ties (Granovetter 1973, 1978). In his seminal papers on the "strength of weak ties," Mark Granovetter was able to show that small-worlds are constituted by subnetworks connected to each other by means of a few "weak" connections, characterized by a long range and no necessity for constant activity. For this investigation, as the affiliation networks are defined in geographic terms, in order for a piece of information to be spread over the network, the region in which the information is created has to include at least one node which can connect at least haphazardly with at least one node of another region (long range). From a mathematical point of view, the maximal connectivity of the small-world in reference to the main network is reached when just a few weak ties are active. In this status, the local nodes which are able to activate weak ties become what are called great hubs, that is, hubs that are able to connect the entire network and to produce knowledge transfer on a large scale.

5 The Real Network of the *Sphere*: The Subject of Climate Zones

To examine how the real historical network of knowledge worked in the process of spreading knowledge of the subjects contained in the treatises, the primary focus lies on the process of diffusion of a specific mathematical demonstration concerned with the climate zones, first published by Pedro Nuñes in 1537 (Sacrobosco and Nunes 1537, 37–45).

The term 'climate zones' indicates the subdivision of Earth's surface into zones delimited by parallels to the equator and whose width depends on the number of sun hours per day. The concept of climate zone has ancient origins and entered Western culture through the appropriation of many different

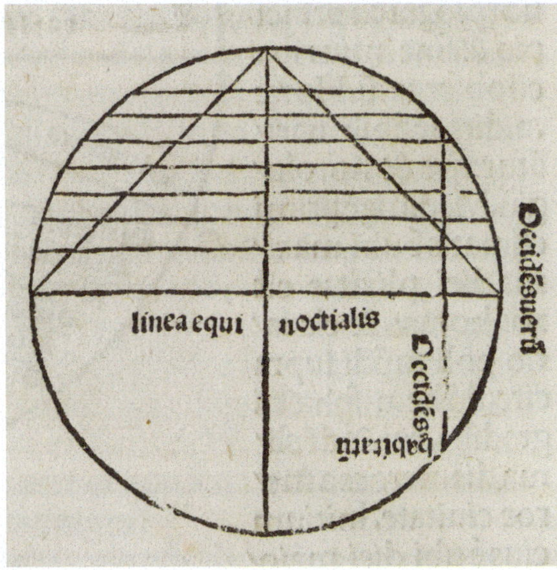

Fig. 16 Drawing representing the classical subdivision of the Northern Hemisphere into seven climate zones. The last one, close to the North Pole was not considered. This drawing was printed in nearly all the treatises on the *Sphere*. From (Sacrobosco et al. 1490, Section "De divisione climatum"). Courtesy of the Library of the Max Planck Institute for the History of Science

sources. Most relevant is Ptolemy's use of the concept in the *Almagest*, but classical works from the Roman time, such as those of Ovid and Pliny, also made use of this concept. Finally, late medieval European science appropriated the concept especially in the versions delivered by Arabic science. Sacrobosco appropriated the by then established view according to which the Northern Hemisphere was subdivided into seven climate zones, the largest one being the one closest to the equator and the narrowest one the one closer to the North pole (Fig. 16).

The knowledge concerned with the distribution of the climate zones was relevant for a variety of reasons during the late Middle Ages. Sacrobosco connected it especially with considerations of meteorological character and defined the inhabitable zones of the Earth based on them. He also associated specific characteristics of inhabitants as being dependant on the climate zone. Generally speaking there had been a stable association between medicine and the cosmological and geographic knowledge concerned with the climate zones ever since antiquity, expressed by Galen and the Medieval *regimen sanitatis*. This association certainly extended to astrological aspects, for instance concerning the definition of the temperament of individuals. This aspect became even more relevant during the early modern period and especially when methods of quantification were introduced into the medical practice at the beginning of the iatromechanical sciences, by the physician Sanctorius at the turn of the sixteenth century in Venice (Sanctorius 1614, 20–31, esp. aph. XLI; Temkin 1973, 85, 102).

Even more relevant, however, was the mentioned re-emergence of metric geography and cultural environment dominated by the new geographic discoveries and the journeys of exploration. While late Medieval texts defined the zones geographically, determining the parallels as those passing through specific cities, the early modern period could increasingly count on a well-developed geometry of the Earth's surface. The journeys of exploration, moreover, gave innumerable inputs into the re-discussion of the established subdivision of the climates while extending them to the new territories both toward West and East and in the Southern Hemisphere. This served to correct the original text concerning the inhabitability of the zones and to expand on ethnological topics.[43]

[43]For the intensity of the debate concerning the concept of climate zones during the age of the journeys of exploration, see (Cattaneo 2009).

During the sixteenth century, the values of the widths and the latitudes of the climate zones from the equator toward the poles were well known, but this knowledge was based only on observational experience and did not have any mathematical fundament. In other words, until 1537, no mathematical demonstration in the frame of cosmography, that is at the cross road between astronomy and metric geography, was available to show and predict the length of the artificial day (daytime) as depending on the latitude (or vice versa).

Such a demonstration was finally formulated by the Portuguese cosmographer Pedro Nuñes in 1537. As mentioned, his commentary on the *Sphere* of Sacrobosco was extremely innovative as he collected it together with a vernacular translation of the first book of Ptolemy's *Geography* and with a treatise on nautical astronomy. Seen from this perspective, the mathematical demonstration concerning the climate zones is one small innovation in the tradition of the treatises on the *Sphere,* if compared to the entire impressive work of Nuñes. As a matter of fact, however, and as surprising as it might appear nowadays, only the few pages on the mathematical demonstration entered the network of the treatises on the *Sphere* and spread all over Europe.[44]

In a network of diffusion of innovations, what counts most is not the first appearance of the innovation but its following adoptions. The second adoption, that is, the first time that Nuñes's demonstration was republished, was in 1556 (Sacrobosco et al. 1556). This work is a typical collective edition of treatises and sections all devoted to the *Sphere* of Sacrobosco. Apart from the text of Sacrobosco and the usual chapters on the *theorica* of the Moon and the Sun by Peuerbach and Regiomontanus (1436–1476), the treatise includes the commentary by Petrus Valerianus and the Latin translation of Nuñes' demonstration. The translator was the famous French humanist Elié Vinet (1509–1587). In this, as in the next 30 further editions containing Vinet's translation, such a section is announced in the very title of the book with the words "Et, Petri Nonii Salaciensis demonstrationem eorum, quae in extremo capite de Climatibus Sacroboscus scribit, de inaequali Climatum latitudine, eodem Vineto interprete." It is clearly a reference to the original tract from 1537, in which Nuñes added his demonstration in the form of an annotation at the end of the third chapter of the original work of Sacrobosco, that is, where the climate zones are discussed.

Elié Vinet mostly worked as a professor at the Collège de Guyenne in Bordeaux but he was in Coimbra, where Nuñes was teaching at the university, between 1547 and 1549 (Desgraves 1977, 158). Vinet never once published his translation in a book authored by himself only. This implies that he accomplished his work in accordance with the publisher of the 1556 edition of the treatise on the *Sphere.* The publisher was Guillaume Cavellat, one of the first, if not the very first early modern printer, who specialized in scientific publications.[45] In the introduction to a treatise on the construction on quadrants, published by Vinet in 1564, he mentions that the work on the quadrant was constituted of those statements that were founded on the most certain mathematical demonstrations "comme de mon grand ami Pero Nunes docteur Portugalois" (Vinet 1564, Section 5).[46]

[44]The first, who extensively worked on this demonstration and contextualized the general work of the Portuguese mathematician is Henrique Leitão. For more information, see (Leitão and Almeida 2012). I would like to express my gratitude to Henrique Leitão for the personal communication, from which this section of the work benefited greatly.

[45]It is evident from the bibliographic data that the publisher Cavellat in Paris played a significant role in promoting and supporting the knowledge network on the *Sphere,* thanks to his close institutional relations with the professors of the university. For more information concerning this publisher, see (Pantin 1998).

[46]On demand of the King of Portugal, Vinet went to Coimbra together with a delegation of professors of the same college in Bordeaux, where he was teaching, to help conceive and found a new college. The undertaking did not go as planned and some of Vinet's colleagues were even imprisoned for religious reasons. Vinet apparently maintained a good relationship with Nuñes. However, when he republished the treatise on the construction of the quadrants in 1583 together with another of his works, the paragraph where he mentioned Nuñes was substituted and the mention of the Portuguese mathematician disappeared (Vinet 1583, Section 6).

Eventually activated by the publisher, Vinet seems to be the weak tie that transported part of Nuñes's work to the great hub Paris and in particular to the node built around the publisher Cavellat. It is impossible to know how this adoption was conceived, who suggested it first and whose idea it was, unless further historical sources are found. The surprising choice to translate only the demonstration highlights, however, the relevance of the process of mathematization of sciences and, in this case, of cosmology and even more the increasing relevance of a mathematical, precise treatment of the climate subjects. It can only be conjectured, but the obvious need to know more about the life conditions of the newly discovered territories and to contour them geographically in a precise way could easily be one of the reasons for this choice. The list of Vinet's works shows that he only occasionally wrote about practical matters and only in this particular case about mathematical ones. When he did write about practical matters, he did so in French but when he worked on other subjects and published, he did so mostly in Latin. The choice of Vinet and Cavellat therefore also shows that mathematics justified a translation from vernacular to Latin, while all other great innovations introduced by Nuñes remained the subject of treatises in vernacular.

The distribution of the production of treatises that contain Vinet's translation of Nuñes's demonstration shows that it peaked during the 1560s.[47] The influence of such diffusion can be recognized by the increasingly spread change concerned with the number of climate zones. The original mathematical demonstration shows the latitudes for each climate zone as defined by an increase in daytime of 30 min. In the turn of a short period many authors and commentators began pondering the utility of subdividing the hemisphere into a greater number of zones as delimited by geometrically determined parallels, which would then have allowed the zones to be more precisely described. Soon a subdivision of 24 climate zones was suggested and generally accepted.[48] Many treatises, such as the one already mentioned by Pifferi (Sacrobosco and Pifferi 1604), adopted the new table of climate zones but did not furnish either the mathematical demonstration or the information that it had been accomplished by Nuñes (Fig. 17).

Other treatises, however, do mention the demonstration and its provenience while appropriating the new subdivision of climate zones. This is the case for instance for the influential treatise on the *Sphere* by Christophorus Clavius. Clavius had been a pupil of Nuñes and therefore he can be considered a second weak tie that brought this innovation into the main network of diffusion. The seventeen editions of Clavius's work show some peculiarities.[49] The first edition of Clavius's *Sphere* was published in Rome in 1570, that is, in a place that does not represent an important hub in the entire network, and fourteen years after the second adoption by Cavellat. The second edition of Clavius's *Sphere,* moreover, was published not less than eleven years later. Compared to the rhythm of publication of the treatises on the *Sphere*, or to Vinet's translation of Nuñes's demonstration, this appears to be a very long time interval. However, the seventeen re-editions of Clavius's work were then produced in the short turn of 37 years. The reason for this delay was most probably not a lack of interest in Clavius's work but an incredibly restrictive contract that he agreed upon with his first publisher,

[47] The list of the treatises containing Vinet's translation of Pedro Nuñes's demonstration is given in the first part of Appendix 4.

[48] The original description of Sacrobosco was not deleted or substituted. Usually the new subdivision was just added in form of commentary at the end of the third part of Sacrobosco's original tract, as Nuñes initially did. The Southern Hemisphere was subdivided into climate zones as well.

[49] The list of editions of Clavius's treatises is given in the second part of Appendix 4. There are two further treatises that list Clavius as an author. These are collections of treatises produced by the Jesuit publisher Colinus in Cologne, respectively in 1601 by the founder and in 1610 by his son: (Sacrobosco et al. 1601) and (Sacrobosco et al. 1610). In these texts, there are only short explanatory comments of Clavius on the original text of Sacrobosco. Both of these texts however also contain Vinet's translation of Nuñes's demonstration. For the relevance of Clavius's texts in the early modern scientific framework, see (Romano 1999).

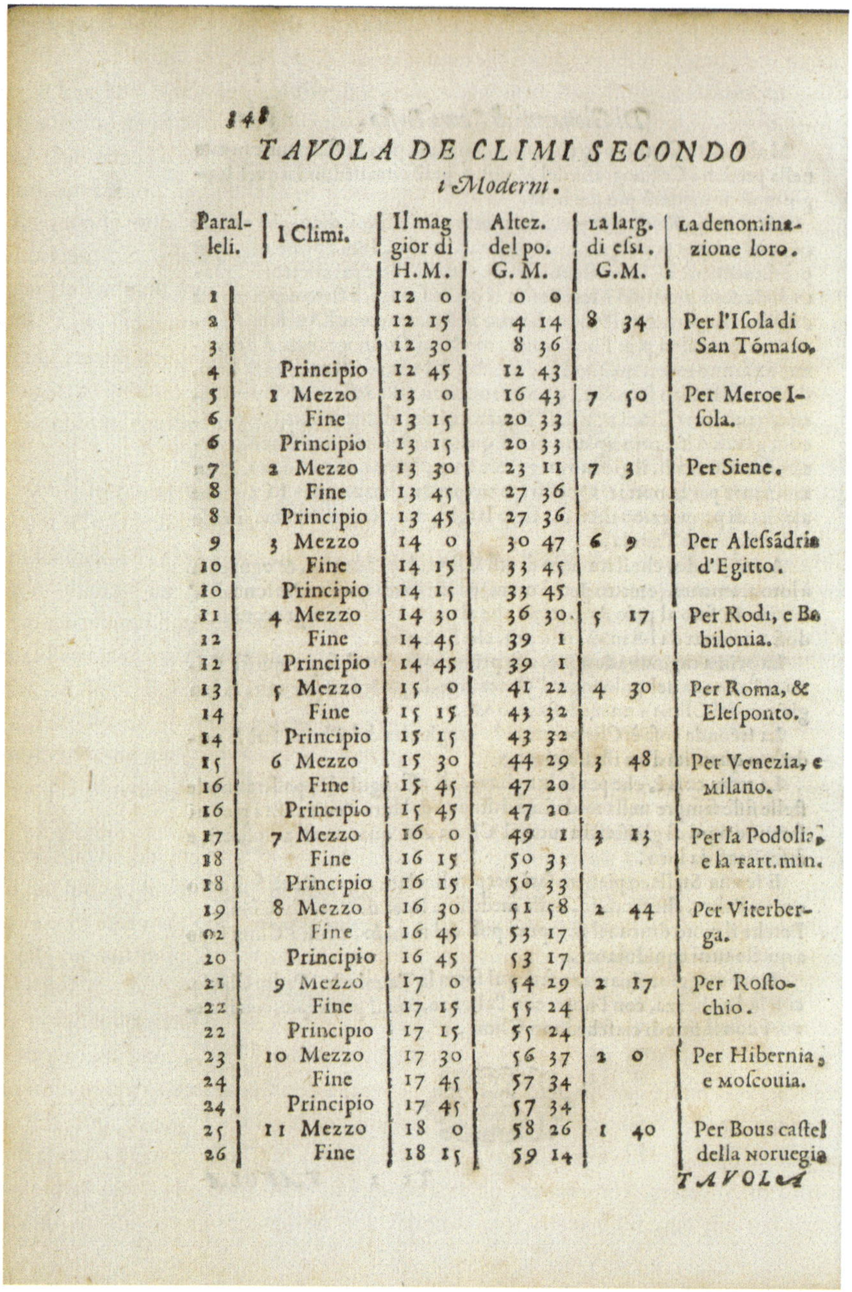

148

TAVOLA DE CLIMI SECONDO
i Moderni.

Paralleli.	I Climi.	Il maggior di H.M.	Altez. del po. G.M.	La larg. di essi. G.M.	La denominazione loro.
1		12 0	0 0		
2		12 15	4 14	8 34	Per l'Iſola di
3		12 30	8 36		San Tómaſo.
4	Principio	12 45	12 43		
5	1 Mezzo	13 0	16 43	7 50	Per Meroe I-
6	Fine	13 15	20 33		ſola.
6	Principio	13 15	20 33		
7	2 Mezzo	13 30	23 11	7 3	Per Siene.
8	Fine	13 45	27 36		
8	Principio	13 45	27 36		
9	3 Mezzo	14 0	30 47	6 9	Per Aleſsãdria
10	Fine	14 15	33 45		d'Egitto.
10	Principio	14 15	33 45		
11	4 Mezzo	14 30	36 30	5 17	Per Rodi, e Ba
12	Fine	14 45	39 1		bilonia.
12	Principio	14 45	39 1		
13	5 Mezzo	15 0	41 22	4 30	Per Roma, &
14	Fine	15 15	43 32		Eleſponto.
14	Principio	15 15	43 32		
15	6 Mezzo	15 30	44 29	3 48	Per Venezia, e
16	Fine	15 45	47 20		Milano.
16	Principio	15 45	47 20		
17	7 Mezzo	16 0	49 1	3 13	Per la Podolia,
18	Fine	16 15	50 31		e la Tart. min.
18	Principio	16 15	50 33		
19	8 Mezzo	16 30	51 58	2 44	Per Viterber-
02	Fine	16 45	53 17		ga.
20	Principio	16 45	53 17		
21	9 Mezzo	17 0	54 29	2 17	Per Roſto-
22	Fine	17 15	55 24		chio.
22	Principio	17 15	55 24		
23	10 Mezzo	17 30	56 37	2 0	Per Hibernia,
24	Fine	17 45	57 34		e Moſcouia.
24	Principio	17 45	57 34		
25	11 Mezzo	18 0	58 26	1 40	Per Bous caſtel
26	Fine	18 15	59 14		della Noruegia

TAVOLA

Fig. 17 Table of the 24 climate zones that Francesco Pifferi called "Modern Table of the Climate Zones." For each zone, the beginning, the middle parallel and the end is given. The entire table extends over two pages. From (Sacrobosco and Pifferi 1604, 348). Courtesy of the Library of the Max Planck Institute for the History of Science

Helianum in Rome, as it can be read in the imprint of the first edition. Helianum never republished any further treatise on the *Sphere* or any further new edition of the *Sphere* of Clavius. He probably was not connected to the network, at least what concerns publications of scientific nature, and the fact that Clavius conceded him all possible rights for ten years both for Italy and all other nations, prevented this book from spreading until those rights ceased. The situation changed radically ten years later and the subsequent editions must therefore be taken into consideration as well while analyzing the process of the diffusion of the innovation introduced by Nuñes.

To analyze such process of diffusion, the editions of the treatises that contain Nuñes's demonstration are listed according to their years of publication, printers, and the geographic location thereof (Fig. 18).

After the second adoption in 1556, the text was republished by the same Cavellat another four times up until 1562. Beginning in 1563, the knowledge began to spread over Europe, first to Venice then to Lyons and Antwerp. In the turn of 6 years, six further printers appropriated and redistributed Vinet's translation: the Venice and Lyons offices of the publisher Giunti, Colinus in Cologne, Birckmann and Richard in Antwerp and Scotus in Venice. In 1570 Clavius's work was published in Rome and eleven years later it further spread over some regions of Europe too. A total of 48 treatises, published between 1556 and 1620, contain the innovation.[50] The peaks of production occured in 1566 and 1594, when three different editions were printed respectively in each year. The number of printers involved amounted to twenty-one but, if the publications distributed after 1610, when the transmission process had clearly slowed down, are disregarded, the number of printers involved amount to only seventeen. In three cases—Cavellat and Cavellat D. in Paris, Giunti in Lyons and Venice, and Gabiano in Leiden and Lyons—the printers involved were closely related to each other. This reduces the number of printers to 14. These printers were localized in nine different cities. Excluding however those places, where only one or two editions were printed (S. Gervasio, Mainz and Leiden), there was a total number of only six cities that contributed most to the diffusion process.

In terms of network, this situation confirms the characteristics of the distribution of the production of all the treatises of Sacrobosco's *Sphere*, that is, that the network appears to be strongly centralized. As mentioned, the nodes of the network are the social micro-regions that pivoted around the printers and at each print-event. The edges of the network that connect the micro-regions can be of a different type—that is, they can be defined according to different attributes. First of all, edges can be taken into consideration that express the chronology of the process and therefore the history of the editions.

If the period between 1537 and 1564 is considered (Fig. 19), the innovation was appropriated and spread by only two printers and in two locations, though several times each. The zigzag line represents the weak tie. The other edges show that between the printers only one bi-directional relation was established. This is a dyad that goes in both directions and that expresses a certain instability. If for whatever reason, the connection is impeded, nodes remain excluded or isolated, in this case Giunti in Venice and half of the network established during the first 27 years.[51]

The peak of production during the 1560s changed the situation completely. If the time until 1571 is taken into consideration (Fig. 20), the network of diffusion constituted of the only chronological links shows the beginning of formation of triads. From the perspective of social network analysis, triads connect three nodes in such a way that if one link between two of them for some reason disappears, the connection between these two nodes is guaranteed by a third.

Formation of triads indicates that the knowledge is stably present and that the network might be moving toward a phase of small-world, for instance, by means of overlapping triads and the activation of further weak ties. The process of the accumulation of knowledge, as it is perceived while investigating the early modern period, is the result of the acceleration of the process of diffusion of innovations as codified in texts, images or artifacts. By 1571, triads began to take form, as always, spread by Cavellat in Paris, for instance, to Birckmann and Giunti, to Colinus and Giunti, and to Richard and

[50] At this stage, it cannot be totally excluded that other treatises of the tradition of commentaries on the *Sphere* might also contain this demonstration.

[51] I would like to kindly thank Claire Lemercier, as the following analysis, in its previous version, took advantage of her extremely relevant remarks and advice. See in particular (Lemercier 2012, 2015). See also (Lazega 2014).

	Lisbon	Paris	Lyon	Venice	S. Gervasio	Cologne	Mainz	Leiden	Antwerp	Rome
1537	Galharde									
1556		Cavellat								
1558		Cavellat								
1559		Cavellat								
1561		Cavellat								
1562		Cavellat								
1563				Giunti						
1564		Cavellat	Giunti							
1566						Colinus			Birckmann / Richard	
1567			Giunti							
1569		Cavellat		Scotus						
1570										Eliano
1571				Scotus						
1572		Cavellat								
1573									Bellerus	
1574				Scotus						
1577		Cavellat								
1581						Colinus				Zanettus
1584		Cavellat								
1585										Basa
1586				Scotus						
1591				Ciotti		Colinus				
1594			Gabiano			Colinus		Gabiano		
1596										Basa
1601				Ciotti		Colinus				
1602			Gabiano		Crespin					
1603				Ciotti						
1606			Gazeus							Zanettus
1607										Zanettus / Hugonis
1608		Cavellat D.			Crespin					
1610						Colinus				
1611							Hierat			
1617			Pillehotte							
1618			Rigaud							
1619		Quesnel								
1620				Scotus						

Fig. 18 Table showing the distribution of printers/cities/years of the publications of the treatises on the *Sphere* containing Pedro Nuñes's demonstration

Giunti. In the same year, the Venetian printer Scotus shows that it was becoming more central, as it produced editions alternating those of Cavellat and Giunti. In 1570, moreover, the second weak tie to Rome and the first edition of Clavius appeared. This offers the possibility to add another sort of edge. The edges considered until now are in fact not only an expression of the chronology of the editions.

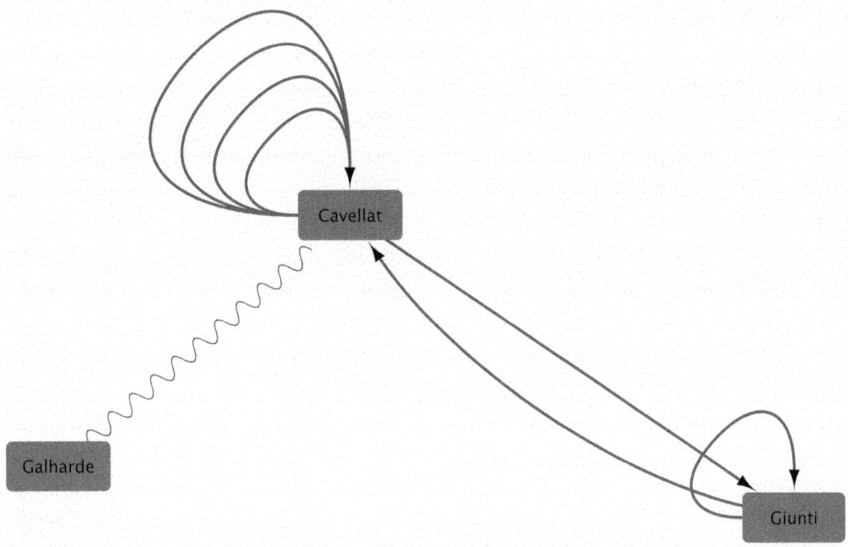

Fig. 19 Visualization of the network of diffusion of Nuñes's demonstration between 1537 and 1564. Formation of bi-directional dyads

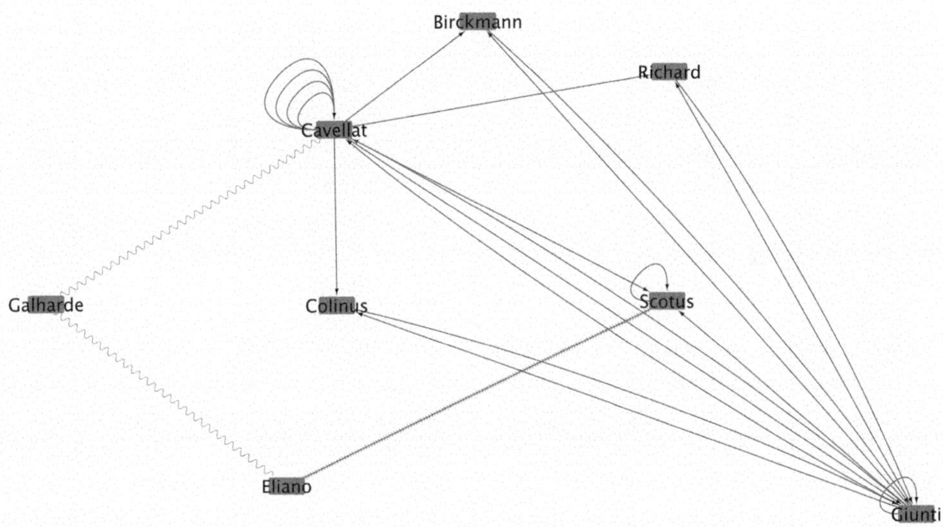

Fig. 20 Visualization of the network of diffusion of Nuñes's demonstration between 1537 and 1571. Formation of triads

Those treatises contained exactly the same text and therefore the edges are an expression of a certain sort of contact between the publishers. These contacts could be specific agreements or contracts, for instance, stipulated at book fairs, or the decision to republish the text might have been taken on the basis of marketing and concurrence reasons. The author himself might have acted as a mediator. The last editions of Vinet's translation published by Scotus in Venice, for instance, are clearly reprints of always the previous editions published by Cavellat in Paris, as it is reported in the covers of the books with the words "Ex postremo Impressione Lutetiae."[52] The same sort of edges connected then the

[52]Each time the historical sources furnish a new piece of information, this can be taken into consideration within the network by adding it as a specific attribute to the edges.

publishers that printed the work of Clavius, as this operation needed the acquisition of permissions to publish, sometimes from the author too. Finally, the edge that connects the printer Helianum, who published the first edition of Clavius's work, to the printer Scotus, represented by a continuous arrow, is of a different sort. This expresses pure temporality and it means that Scotus's publications of the translation of Vinet is the first action in the network after the publication of Clavius. Nevertheless, as both contain the same knowledge, though codified in a different text, the edge also indicates the evolution of the process of diffusion of this innovation on a more general term. In conclusion, besides the weak ties this network is constituted by three sorts of edges: those that connect the subsequent publications of Vinet's translation, those that connect the publications of Clavius's work in the same way, and those that express the pure temporality of the print-events that took place. While the latter is needed to analyze the evolution of the network from a general perspective, the first two are expressions of some kind of social contacts that must have taken place.

What is decisive for the understanding of the dynamics of real knowledge networks in history, is the decision concerning the basic time unit with which to measure the transformation processes over the network.[53] After the second adoption of Nuñes's demonstration, the further publications were certainly influenced by the fact that such a subject appeared in a collection published in the frame of a great and influential hub like Paris, and by a great and influential printer like Cavellat. Nevertheless, it is conceivable that the very first appearance of the innovation not only was a condition for its spread, but that it was also influential to the successive editions. On an abstract level, every previous edition could be linked to all subsequent editions.

The length of life of the material prints justifies the possibility of analyzing the entire process of diffusion of this innovation between 1537 and 1620 at once (Fig. 21).[54]

The two groups—editions of Vinet's translation and of Clavius's work—are visualized respectively on the right and on the left. In each of them several triads were shaped, though the spread of Clavius's work shows this characteristic to a lower degree. The edges that indicate the diffusion of the knowledge concerned with this innovation, but independent from the form in which it is codified, contribute to the formation of many further triads. It is at this stage, that the power of the network analysis can be best appreciated.

By shifting from the pure visualization of the network to the quantification of its parameters, it is possible to determine with precision further aspects through methods directly imported from the field of sociology. One such parameter is the betweenness centrality, visualized in the Figs. 21 and 22 by means of the dimension of the nodes. This reflects how central a node is in the process of diffusion of the innovation in the network.[55] The relevance of a printer in the network obviously depends on the number of editions it produced. But the characteristics and the influential role of its position as relative to one of the other nodes in the temporal evolution of the network does not depend only on this. It also depends on the circumstances that allowed the publisher to achieve a central position, for instance, between big hubs, like Cavellat and Scotus. This in fact seems to have been the case for the publisher

[53]The network analysis described in the following was previously discussed with Dirk Wintergrün, whom I would like to thank.

[54]The edges that connect any previous edition to all subsequent ones, on an abstract level, are not added to the network visualization here, as they would only render the network unnecessarily redundant.

[55]"A node's betweenness centrality is defined as the fraction of the number of geodesics passing through this node over the number of geodesics between all pairs of nodes in the network." From the glossary in the "Resources" of the site of "Archeological Networks:" https://archaeologicalnetworks.wordpress.com (last accessed April 2016). Networks of the diffusion of innovations are directed networks. A further relevant parameter is closeness centrality. This reflects the average closeness of one node to all other nodes in the network and therefore can be interpreted as particularly relevant in the process of diffusion. In this specific network, the analysis of the closeness centrality assigns the highest values to the nodes that have been assigned the highest values of betweenness centrality.

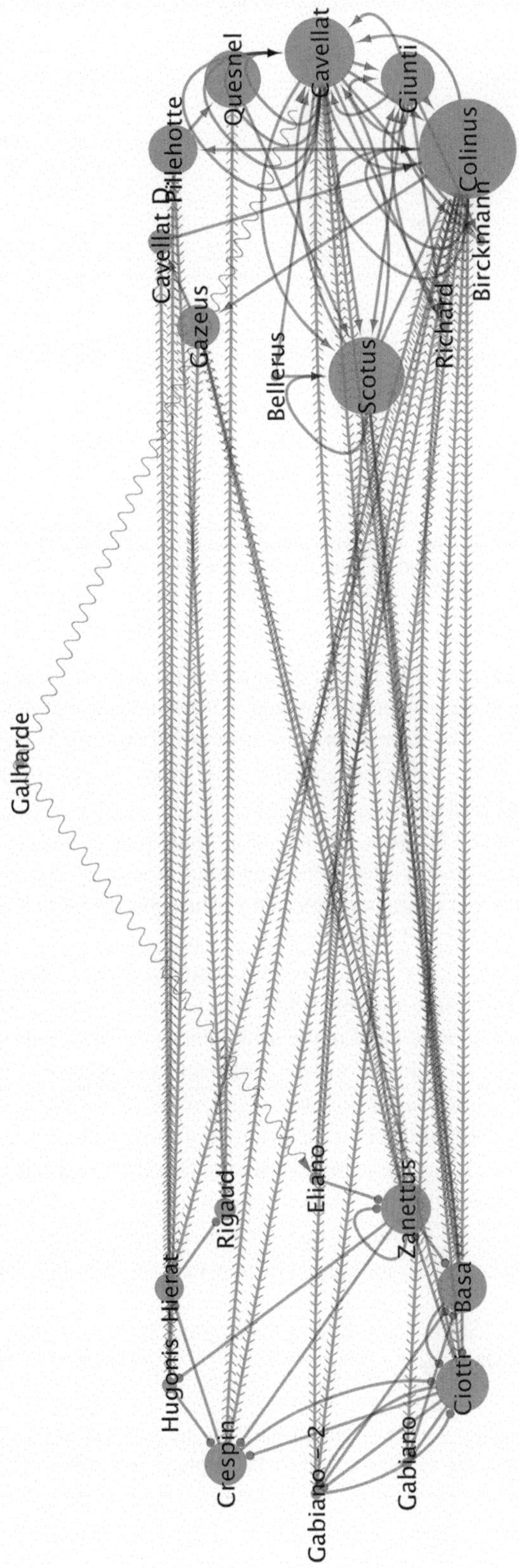

Fig. 21 Betweenness centrality of the early modern printers that published Vinet's translation of Nuñes's demonstration and/or Clavius's commentary on the *Sphere*. Visualization of the network of diffusion of Nuñes's demonstration between 1537 and 1620

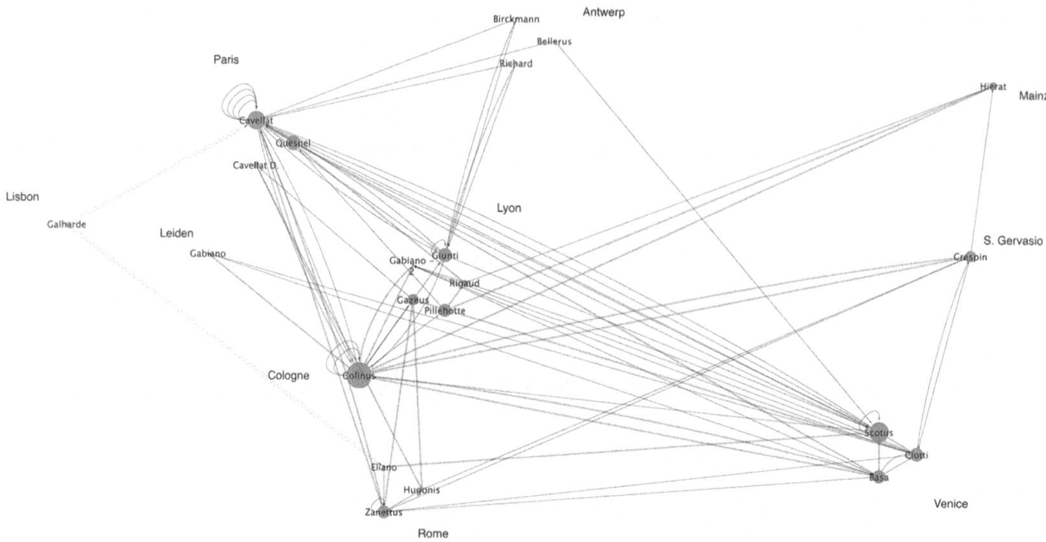

Fig. 22 Betweenness centrality of the early modern printers, which published Vinet's translation of Nuñes's demonstration and/or Clavius's commentary on the *Sphere*, grouped according to their geographic locations

Colinus in Cologne, which, in spite of only producing a minor number of editions, appears to be the conjunction point between several regions of the network. The heuristics that result from the application of the social network analysis therefore shows, in this case, that to understand the details of the process of diffusion of this knowledge, further investigation should move toward Cologne and the Colinus editions of the *Sphere* of Sacrobosco.

But history is not a network and a network is not a geographic map. Printers, authors, institutions and what have here been called micro-regions, were integrated in specific social and economic contexts that had both specific individual characteristics and further characteristics that they shared with other micro-regions of the same place and which were active in the same time. It is therefore historically correct to embed the network analysis into a more realistic historical chart that shows this relative closeness of specific groups of micro-regions (Fig. 22).

Micro-regions active in the same period and in the same place constitute sub-networks that can be further analyzed, both in reference to their internal dynamics and among each other against the background of the main network. It is however sufficient to realize the relevant singularity of the printer Colinus in Cologne, in spite of the dominance of the printers of Paris and Venice.

Such dominance is no surprise. It can be explained by the fact that the particular knowledge network analyzed here is in truth inserted in a historical continuum; it is well known that places like Venice and Paris had already become dominant hubs in knowledge networks earlier, and for other reasons. They were given, and as such also took over the new task of spreading this new knowledge.[56]

6 Structures of Knowledge

The case study of the knowledge network associated with the tradition of the treatises on the *Sphere* of Sacrobosco shows that knowledge emerges with a local character in order to respond to single issues, such as the need for a new instrument for navigation or a new mathematical algorithm to con-

[56] Within the frame of network theory, this is called the rich-get-richer effect.

vert values from one specific table into those of another.[57] Because of the need for the same knowledge within a local network, such knowledge is then codified, eventually institutionalized, and then spread, either freely or in a protected way. It was at this stage that the potential of the new design Sacrobosco had given to the treatise on the *Sphere* was exploited. Probably consistent with the original historical motivation for such an order of subjects, the treatise on the *Sphere* became not only the vehicle that disseminated knowledge, but also a nucleus attracting new codified knowledge that was related to, useful for or merely associated with the activity of traveling.

At this stage Nuñes restructured the knowledge concerned with the climate zones. Such knowledge existed already and had been spread through a myriad of manuscripts and prints on the *Sphere*. Nuñes structured it on a more abstract level by means of mathematics, linking it therefore with a branch of knowledge with which it had not been linked and integrated before. The same Nuñes acknowledged that this knowledge was already available but that merely a mathematical demonstration had been missing. As Henrique Leitão remarked, Nuñes's own motivation to proceed in this direction might be explained by his habit to furnish as precise as possible means for navigation, that is, to work at the boundary between technology and mathematics (Leitão and Almeida 2012). And in fact Nuñes published his commentary together with a section on nautical astronomy that was related to his activity as Chief Cosmographer for the Crown of Portugal, which in turn was related with the Portuguese well known activity in the journeys of exploration.

The *Sphere* of Sacrobosco was for Nuñes, and remained for the later commentaries, the natural place to mention and discuss this demonstration. As it opened with the description of the celestial spheres, the original treatise itself offered a hierarchical structure to which further branches could be added and thus subsumed under a unifying principle: cosmology. Because of the practical nature of this knowledge, there was no need to change this hierarchy or the content of the cosmological description even after the heliocentric system began collecting adepts. Corresponding to the incremental codification processes of practical knowledge during the early modern period, the treatises gradually assumed an almost encyclopedic character. But not all new subjects were or had to be included: there is no absolute necessity for a new subject to enter the network of knowledge exchange. The subjects that entered the structure dictated by the treatise were those which had a historical, institutional, and economic meaning in the context of the micro-regions in which they were adopted and produced. As soon as the micro-regions or the subnetworks stopped their activities or were no longer able to face the transition costs, the leadership of the network was taken over by other regions dominated by new hubs, potentially characterized by different needs for knowledge.

By means of personal contacts, the weak ties, Élie Vinet and Christophorus Clavius, Nuñes's knowledge migrated. But not all of it. Only a tiny portion. Maybe in agreement with Nuñes himself, the micro-region Cavellat-Vinet, strongly connected with several educational institutions in Paris and their professors, decided that only the mathematical demonstration was worth being appropriated and integrated into a new structure of knowledge on the *Sphere*, represented by a book constituted of several treatises and commentaries on the original texts and several additions. The subject of nautical astronomy was not touched upon in other sections of these treatises. It was just ignored. Largely, the other texts that were collected beginning in Paris in 1556, also contributed to shaping a new structure of knowledge characterized by an increased mathematization of the subject matter. The same sort of selection with the same sort of consequence was operated by Clavius, too. Clavius published his own encyclopedic commentary on the *Sphere* of Sacrobosco and, in his case too, the goal was to improve the mathematical understanding of the subject matter.

[57]The fundamental epistemic frame to understand the local-universal relation concerning processes of diffusion of knowledge is discussed in (Renn 2012, 27–52).

Nuñes's demonstration became part of a different area of knowledge and this in turn also changed because of this addition. From this point on, however, the new structure did not concern only erudite books for university use. The work of the printers continued and the renovated knowledge on the *Sphere* benefitted from a status of high connectivity achieved by the early modern publishers during the sixteenth century. The insertion of textually codified knowledge in the frame of the treatises on the *Sphere* led to the unification of knowledge domains into one more comprehensive semantic network of the *Sphere*. The example of the knowledge tradition on the *Sphere* indeed shows that once practical knowledge or knowledge derived from a practical activity is codified, it no longer belongs to practical knowledge alone. It immediately becomes an element that can be restructured and connected with other codified elements so as to create vast knowledge systems that, in turn, represent common scientific backgrounds over large spatio-temporal areas.

From the perspective of the knowledge network, this means that during the sixteenth century the social, institutional, and material networks, the last of which includes artifacts like astronomical instruments and the material books considered here, contributed to the emergence of a semantic network—a scientific culture and identity—that connected all of the subjects listed in the taxonomy mentioned in this work.

The fertility of the design of the treatise on the *Sphere*, that is, its capacity to expand and renovate the structure of the practical knowledge codified within it, certainly began its declining phase during the seventeenth century, possibly during the second half. The tradition of treatises had shown its ability to systematically include and integrate mathematically based knowledge such as mathematical astronomy, geometry, and metric geography. But the ongoing re-foundation of sciences on a mathematical basis, and the slow disappearance of the distinction between the super- and the sublunar world and of their distinct physics, progressively limited the creative function of this treatise in the restructuring of knowledge. By the end of the seventeenth century, a new astronomy and a new cosmology were unified and a new disciplinary taxonomy began to emerge.

When the tradition of the treatises on the *Sphere* came to an end, the early modern period was over too, and a new economic, political, and scientific order was emerging in Europe. But until that point the impressive phenomenon of the treatises on the *Sphere* and of their transformation had greatly contributed to the formation of a common scientific or perhaps techno-scientific identity that was shared all over the small-world of Europe during the sixteenth century.

Appendices

Appendix 1

Analyzed Sources from the Sixteenth Century and Their Taxonomy

Sacrobosco, Johannes de, Joannes Regiomontanus, and Georg von Peurbach. 1488. *Spaerae Mundi Compendium foeliciter inchoat. Nouiciis adolescentibus: ad astronomicam remp. capessendam aditum impetrantibus: pro breui recto et tramite a uulgari uestigio semoto: Iohanis de Sacro busto sphaericum opusculum una cum additionibus nonullis littera A sparsim ubi intersertae sint signatis Contra[que] et cremonensia in planetarum theoricas delyramenta Iohanis de monte regio disputationes tam acuratiss. atque utills. Nec non Georgii purbachii in erundem motus planetarum acuratiss. theoricae: dicatum opus: utili serie contextum fausto sidere inchoat.* Venice: Johannes Lucilius Santritter & Hieronymus de Sanctis.

Sacrobosco, Johannes de. 1489. *Iohannis de Sacro Busto Spericum opusculum vnacum vtilissimis figuris textu[m] declara[n]tibus.* Lipcz: Conradum Kachelouen.

Sacrobosco, Johannes de, Joannes Regiomontanus, and Georg von Peurbach. 1490. *Spaerae mundi compendium foeliciter inchoat. Nouiciis adolescentibus: ad astronomicam remp. capessendam aditum impetrantibus: pro breui rectoque tramite a uulgari uestigio semoto: Iohanis de Sacro busto sphaericum opusculum una cum additionib nonullis littera A sparsim ubi intersertae sint signatis Contraque cremonensia in planetarum theoricas delyramenta Iohanis de monte regio disputationes tam acuratiss. atque utills. Nec non Georgii purbachii in erundem mot planetarum acuratiss. theoricae: dicatum opus: utili serie contextum: fausto sidere inchoat.* Venetiis: [Bonetus Locatellus für] Octavianus Scotus.

Sacrobosco, Johannes de, Joannes Regiomontanus, and Georg von Peurbach. 1491. *Sphaerae mondi compedium feliciter inchoat. Nouiciis adoloscenbus: ad astronomicam remp. capessendam aditum impetrantibus: pro breui rectoq: tramite a uulgari uestigio semoto: Iohanis de Sacro busto sphaericum opusculum una cum additionibus nonullis lra A sparsim ubi intersertae sint signatis Contra q: cremonensia in planetarum theoricas delyramenta Iohanis de monteregio disputationes tam acuratiss. quem utills: Nec non Georgii purbachii: in erundem motus planetarum acuratiss. theoricae: dicatum opus: utili serie contextum: fausto sidere inchoat.* Venetiis: Per Magistrum Gullielmum de Tridino de Monteferrato.

Sacrobosco, Johannes de and Wenzel Faber von Budweis. 1500. *Opus sphericum Iohannis de Sacro Busto/figuris et p[er]utili co[m]mento illustratu[m].* Cologne: Henricum Quentell.

Sacrobosco, Johannes de, Pedro Sanchez Ciruelo, and Pierre d'Ailly. 1508. *Uberrimum sphere mundi commentum Intersertis etiam questionibus domini petri de aliaco Nuper magna cum diligentia castigatum.* Paris: Jean Petit.

Sacrobosco, Johannes de, Thebites b. Chorath, and Conrad Tockler. 1509. *Textus Spere materialis Joannis de Sacrobusto cum lectura Magistri Conradi Norici in florentissimo Lipsensi gymnasio nuper exarata. Verba Thebit acutissimi Astronomi de imagine totius mundi atque corporis sperici compositione: introducndis ipsis incrementum magnum prebentia: per eundem addita. Ordinatio spere materialis: erderem circulis huic operi inseriuens: per Magistrum Conradi Noricum nouitem addita et diligenter reuisa.* Leipzig: Martin Landsberg.

Sacrobosco, Johannes de, Jacques Lefèvre d'Etaples, Boethius, Bonet de Lattes, and Euclid. 1511. *Textus De Sphera Johannis de Sacroboseo, Cum additione (quantum necessarium est) adiecta: Aouo commentario nuper edito, Ad vtilitatem studentium Philosophice Parisien Academie illustratus, cum Compositione Anuli Astronomici Boni Latensis. Et Geometria Euclidis Megarensis.* Paris: Henri Estienne.

Sacrobosco, Johannes de, Pierre d'Ailly, and Pedro Ciruelo. 1515. *Habes lector Iohannis de sacro busto sphere textum vna cum additionibus non aspernandis Petri Ciruelli. D. (a vero tamen textu apparenter distinctis) cu[m] ipsiusmet sublimi [et] lucule[n]tissima expositio[n]e aliquot figuris nouiter adiu[n]ctis decorata; itersertis p(rae)terea q[uae] stio[n]ib[us] D[omi]ni Petri de Alliaco; omnia peruigili cura ad amussim castigata, et rursus coipressa.* Venundantur Parrhisius: A Iohan[n]e Paruo.

Sacrobosco, Johannes de, Jacques Lefèvre d'Etaples, Bonet de Lattes, Euclid, and Boethius. 1516. *Textus de Sphera Joannis de Sacrobosco. Cum additione (quantum necessarium est) adiecta: Novo commentario nuper edito. Ad vtilitatem studentium Philosophice Parisien. Academie illustratus cum Compositione Anuli Astronomiici Boneti Latensis. Et Geometria Euclidis Megarensis.* Paris: Henri Estienne.

Sacrobosco, Johannes de and Conradus de Megenberg. 1516. *Sphera materialis.* Nürnberg: Iobst Gutknecht.

Sacrobosco, Johannes de, Peter Apian, and Georg Apian. 1526. *Sphaera Iani de Sacrobusto astronomiae ac cosmographiae candidatis scitu apprime necessaria per Petrum Apianum accuratissima diligentia denuo recognita ac emendata.* Ingolstadii: in Apianis aedibus.

Sacrobosco, Johannes de and Antonio Brucioli. 1543. *Trattato della Sphera, nel quale si dimostrano, & insegnano i principii della astrologia raccolto da Giouanni di Sacrobusto, & altri Astronomi, & tradotto in lingua Italiana. Per Antonio Brvcioli. Et con nvove annotationi in piu luoghi dichiarato.* Venetia: impresso per Francesco Brucioli, & i Frategli.

Sacrobosco, Johannes de. 1547. *Sphæra Iohannis de Sacrobosco.* Lovanii: vaeneunt apud Iohannem Waen.

Sacrobosco, Johannes de. 1547. *Sphaera Iohannis de Sacrobosco.* Antverpiae: Apud Ioannem Ricardum in Sole aureo.

Sacrobosco, Johannes de and Hartmann Beyer. 1549. *Quaestiones in novae in libellum de sphaera Ioannis de Sacro Busto, in gratiam studiosae iuventutis collectae ab Hartmanno Beyer, recognitae & plaerisque in locis auctae.* Francoforti: ex officina P. Brubachij.

Sacrobosco, Johannes de, Philipp Melanchthon, al-Farghani, and Joannes Regiomontanus. 1549. *Sphaera Ioannis de Sacrobosco Typis auctior, quam antehac, atque ex diligenti manu scriptorum impressorumque codicum collatione castigator, Praemissa Philippus Melanchthon doctiß. praefatione, qua vtilitatem astrologicae scientiae et Christiano homini non negligendam scite probat.* Parisiis: apud Gulielmum Cauellat.

Sacrobosco, Johannes de and Mauro Fiorentino. 1550. *Annotationi sopra la lettione della spera del Sacro Bosco. Doue si dichiarano tutti e principii Mathematici et Naturali, che in quelle si prossan desiderare. Con alcune quistioni Notabili a detta Spera Necessarie et altri notandi et rari segreti, che in quella son afcosti. Con le infrascritte cose, cio, e, Vna nuoua & fedele (ad verbum) traduttione di detta Spera: Vna spera theologia dinis, & christiana: Vna spera platonica, con alcune eccitationi mathematiche, theologiche & diuine: Una nuova inventione, et astronomico instrumento per fabricare le dodici case celesti della figura astronomica, senza altri canoni, o, calculo, authore m. Mauro Fiorentino, Theosebo, Phonasco, & Philopanareto.* Firenze: Lorenzo Torrentino.

Sacrobosco, Johannes de. 1559. *Sphaera Ioannis de Sacrobosco.* Antuerpiae: apud Ioannem Richardum.

Catena, Petrus. 1561. *Sphaera Petri Catena.* Patavia: Apud Gratiosum Perchacinum.

Sacrobosco, Johannes de and Beyer Hartmann. 1563. *Quaestiones in libellum de sphaera Ioannis de Sacro Busto: in gratiam studiosae iuventutis olim in Academiae Witebergensi collectae/per Hartmannum Beyer.* Wittenberg.
Sacrobosco, Johannes de and Rodrigo Saenz de Santayana y Spinosa. 1567. *La sphera de Iuã de Sacro Bosco nueua y fielmente traduzida de Latin en Romance, por Rodrigo Saenz de Santayana v Spinosa. Con una Exposicion del mismo.* Valladolid: por Adrian Ghemart, a costa de Pedro de Corcuera.
Sacrobosco, Johannes de and Erasmus Oswald Schreckenfuchs. 1569. *Erasmi Osuualdi Schreckenfuchsii Commentaria, in Sphaeram Joannis de Sacrobusto accuratissima, qvibvs non solvmea qvae in autoris contextu sunt, sed alia etiam ad Sphaericam doctrinam necessaria, explicantur: Tabularum que constructio, ex suis principiis per demonstrationum seriem clare dilucide que docetur. His adiecti svnt eivsdem avtoris canones, quibus usus Tabularum, quae operi ex libro Directionum Ioannis Regiomontani, passim inseruntur, ad pulcherrimas inquisitiones Astronomicas, luculentisime continetur.* Basileae: ex Officina Henricpetrina.
Sacrobosco, Johannes de, Elie Vinet, Pedro Nuñes, and Francesco Giuntini. 1573. *Sphaera Ioannis de Sacro Bosco, emendata. In eandem Francisci Iunctini Florentini, et Elie Vineti Santons Scholia. Caetera pagina sequens indicabit.* Antverpiae: apud Ioannem Bellerum.
Sacrobosco, Johannes de and Francesco Giuntini. 1577. *Fr. Iunctini Florentini, sacrae theologiae doctoris, Commentaria in tertium et quartum capitulum Sphaerae Io. de Sacro Bosco. Ad nobilem virum D. Marcum Bonavoltam Florentinum. Ad nobilem virum D. Marcum Bonavoltam Florentinum.* Lvgdvni: apud Philippvm Tinghivm.
Sacrobosco, Johannes de and Alessandro Piccolomini. 1579. *La sfera del mondo di M. Alessandro Piccolomini. Di nuouo da lui ripolita, accresciuta, et fino à Sei Libri, di Quattro che erano ampliata, et quali per ogni parte rinouata, et riformata.* Vinegia: appresso Giovanni Varisco et Compagni.
Sacrobosco, Johannes de, Élie Vinet, Albertus Hero, Pierio Valeriano, and Pedro Nunes. 1581. *Sphæra Ioannis de Sacrobosco emendata. Elia Vineti Santonis scholia in eandem Sphaeram ab ipso authore restituta. Quibus nunc accessere scholia Heronis. Adiunximus huic libro compnedium in Sphaeram per Pierium Valerianum Bellunensem, et Petri Nonii Salacensis demosntrationem eorum, quae in extremo capite de climatibus Sacroboscus scribit de inaequali climatum latitudine, eodem Vineto interprete.* Coloniæ: apud Maternum Cholinum.
Sacrobosco, Johannes de and Francesco Giuntini. 1582. *La sfera del mondo, di M. Francesco Giuntini, dottore theologo: col testo di M. Giouanni Sacrobosco. Opera utile & necessaria à poeti, historiografi, naviganti, agricoltori, & ad ogni sorte di persone.* Lione: appresso Simforiano Beravd.
Sacrobosco, Johannes de. 1584. *Sphaera Ioannis de Sacrobosco, emendata. Eliae Vineti Santonis scholia in eandem Sphaeram, ab ipso auctore restituta. Adiunximus huic libro compendium in Sphaeram, per Pierium Valerianum Bellunensem. Et Petri Nonii Salaciensis demonstrationem eorum, quae in extremo capite de climatibum Sacroboscius scribit de inaequali climatum latitudine, eodem Vineto interprete. Post omnes omnium editiones, auctior er locupletior.* Paris: Hieronymus de Marnef Gulielmus Cavellat.
Barozzi, Francesco. 1585. *Cosmographia in qvatvor libros distribvta, svmmo ordine, miraqve facilitate, ac breuitate ad Magnam Prolemaei Mathematicam Constructionem, ad vniuersamque Astrologiam instituens. Francisco Barocio, Iacobi Filio, Patritio Veneto Avtore. Cum Praefatione eiusdem Autoris, in qua perfecta quidem Astrologiae Diusio, et enarratio Autorem Illustrium, et Voluminum ab eis conscriptorum in singulis Astrologiae partibus habetur: Ioannis de Sacrobosco vero 84 errores, et alii permulti suorum Expositorum, et Sectatorum ostenduntur, rationibusque; redarguuntur. Praecessunt etiam quaedam Communia Mathematica, necnon Arithmetica, et Geometrica principia, nonnullaeque: Propositiones, de quibus in toto opere saepe fit mentio: Ac demum locupletisimus Index eorum, quae ipsa Cosmographia coutinentur.* Venetijs: ex officina Gratiosi Perchacini.
Sacrobosco, Johannes de and André do Avelar. 1593. *Sphæræ vtrivsque; tabella, ad sphæræ huius mundi faciliorem enucleationem. Autore Andrea d'Avellar Olysipone, ac Philosophiae Magistro Conimbricensi Academia Mathematum professore.* Conimbricæ: apud Anton. Barrerium Typ. Reg.
Sacrobosco, Johannes de and Konrad Dasypodius. 1593. *Cunradi Dasypodii institutionum mathematicarum voluminis primi erotemata : logisticae, geometriae sphaerae, geographiae.* [Strassburg]: pro schola Argentinensi imprimebat Iosias Rihelius.
Ryff, Peter and Ludovicus Lucii. 1598. *Elementa sphaerae mundi sive cosmographiae in usum Scholae Mathematicae Basilensiis. Per Petrum Ryff, Mathmaticum, Art. ac Med. D. Logicis Legibus deducta. Denuo, partim limatiora, partim observationum utilium acceßione locupletiora nunc edita. Quibus deovo appensus est Tractatus De Structura et usu Planispherii seu Astrolabi. Cura M. Ludovici Lucii Basil.* Basiliae: Typis Johannis Schroeteri.
Sacrobosco, Johannes de and Gines Rocamora y Torrano. 1599. *Sphera del vniverso. Por don Gines Rocamora y Torrano, Regidor de la Ciudad de Murcia, y Procurador de Cortes por ella, y su Reyno. Dirigida a don lvis Faxardo, Marques de los Velez y de Molina, Adelantado mayor y Capitan general del Reyno de Murcia, y Marquesado de Villena.* Madrid: Iuan de Herrera.
Sacrobosco, Johannes de and Francesco Pifferi. 1604. *Sfera di Gio. Sacro Bosco tradotta, e dichiarata da don Francesco Pifferi sansa vino Monaco Camaldolense, e Matematico nello Studio de Siena. Misurato Intronato. Al Serensissimo don Cosimo Medici Gran Principe di Toscana. Con nuoue aggiunte di molte cose notabili, e varie demostrazioni vtili, e diletteuoli. Come nella seguente Tavola si vede.* Siena: Saluestro Marchetti.

Taxonomy

Cosmology only:
Leipzig 1489, Nuremberg 1516, Ingolstadt 1526, Venice 1543, Antwerp 1559.

Mathematical astronomy:
Venice 1488, Venice 1491, Leipzig 1509, Paris 1511, Paris 1516, Paris 1549, Florence 1550, Wittenberg 1563, Basel 1569, Antwerp 1573, Lyon 1577, Venice 1579, Lyon 1582, Venice 1585.

Calendric calculations:
Paris 1511, Florence 1550, Wittenberg 1563, Antwerp 1573, Lyon 1577, Venice 1579.

Use and construction of astronomical instruments:
Venice 1490, Paris 1549, Florence 1550, Padua 1561, Basel 1569, Antwerp 1573, Lyon 1577, Venice 1579, Cologne 1581, Lyon 1582, Basel 1598, Siena 1604.

Metric geography:
Cologne 1500, Paris 1508, Paris 1511, Paris 1515, Paris 1516, Louvain 1547, Antwerp 1547, Frankfurt 1549, Paris 1549, Florence 1550, Wittenberg 1563, Valladolid 1567, Antwerp 1573, Lyon 1577, Venice 1579, Cologne 1581, Lyon 1582, Venice 1585, Coimbra 1593, Strasbourg 1593, Basel 1598, Madrid 1599, Siena 1604.

Geometry (enlarged section):
Venice 1488, Venice 1490, Paris 1511, Paris 1516, Frankfurt 1549, Florence 1550, Venice 1579, Lyon 1582, Venice 1585, Coimbra 1593, Strasbourg 1593, Madrid 1599, Siena 1604.

Nautical astronomy:
Antwerp 1573, Venice 1585, Siena 1604.

Arithmetics, mathematical instruments, optics and cartography:
Paris 1511, Florence 1550, Antwerp 1573, Lyon 1582, Venice 1585.

Meteorology:
Paris 1508, Cologne 1581, Lyon 1582, Madrid 1599, Siena 1604.

Predictive astrology:
Valladolid 1567, Lyon 1577, Venice 1579, Coimbra 1593, Strasbourg, 1593, Madrid 1599, Siena 1604.

Appendix 2

Regions and Corresponding Lists of Places of Publication

Central Europe:
Augsburg, Avignon, Basel, Dijon, Frankfurt, Freiburg, Geneva, Ingolstadt, Leipzig, Lyons, Mainz, Nuremberg, Strasbourg, Tubingen, Wittenberg, Zurich.

Eastern Europe:
Frankfurt an der Oder, Krakow, Vienna.

Iberian peninsula:
Alcala de Henares, Coimbra, Lisbon, Madrid, Salamanca, Segovia, Seville, Valladolid.

Italy:
Bologna, Brescia, Ferrara, Florence, Milan, Padua, Perugia, Pesaro, Rome, San Gervasio Bresciano, Siena, Venice.

Northern Europe:
Antwerp, Cologne, Herborn, Leiden, Leuven, Paris.

Appendix 3

Treatises Published Between 1493 and 1501

Sacrobosco, Joannes de, Friar Alvaro da Torre, and Hieronymus Münzer. 1493. *Tractado da Spera do mundo tirada de latim em linguoagem portugues Com huma carta que huma grande doutor Alemam mandou a eliRey de Portugall dom Joam ho segundo.* Lisboa: Germam Galharde.

Sacrobosco, Johannes de. 1493. *Tractatus de sphera.* Paris: F. Baligault.

Sacrobosco, Johannes de, Georg Peuerbach, and Johannes Regiomontanus. 1493. *Tractatus de sphera. Novitiis adolescentibus ad astronomicam rempu. capessendam additum impetrantibus per brevi rectoque tramite a vulgari vestigio semoto. Iohannis de sacro busto sphericum opusculum.* Paris: Antoine Caillaut.

Sacrobosco, Johannes de, Joannes Regiomontanus, and Georg von Peurbach. 1493. *Nouitiis adole[s]centibus ad a[s] tronomicam rempu[blican]. cape[ss]e[n]dã additu[m] impet[r]ãtib[us]: p[er] breui rectoq[ue] tramite a v[u] lga[r]i ve[s]tigio [s]emoto. Ioh[annes]. de Sac[r]o bu[s]co [s]pericu[m] op[us]culu[m] Georgiiq[ue] Cu[r]bachii in mot[us] planeta accurati[ss].theo[r]ice. Necno[n] [con]t[r]a Cremone[n]fia i[n] eode[m] planeta theo[r]icas de lirama[n]ta Joha[n]nis de mo[n]te regio di[s]putatio[n]es ta[m] accu[r]ati[ss]. q[ue] vtili[ss].dicatu[m] op[us] vtili [s]erie [ue]textu[m] inchoat.* Parisiis: Georgius Mittelhus.

Sacrobosco, Johannes de. 1494. *Commentvm Electvm Et Praeciosvm Nvperrime Compositvm Svper Tractaetvm Sphaerae Mvndi.* Venice: Bonetus Locatellus or Bartholomaeus de Zanis.

Sacrobosco, Johannes de. 1494. *Opusculum Joha[n]nis de Sacro Busto spericum: cu[m] figuris optimis et nouis textu[m] in se. sine ambiguitate declarantibus.* Leipzig: Martin Landsberg.

Sacrobosco, Johannes de. 1494. *Uberrimum sphere mundi commentum.* Parisiis: G. Mittellius.

Sacrobosco, Johannes de and Wenzel Faber von Budweis. 1494. *Ante[que] ad textum tractatus Spere accedam[us].* Leipzig: Martin Landsberg.

Sacrobosco, Johannes de and Jacques Lefèvre d'Etaples. 1494. *Textus De sphera Johannis de Sacrobosco cum additione (quantum necessarium est) adiecta: Novo commentario nuper editio Ad utilitatem studentium philosophice Parisien. Academie illustratus.* Paris: [Johannes Higman, for] Wolfgang Hopyl.

Sacrobosco, Johannes de. 1495. *Sphaera mundi.* Venetiis.

Sacrobosco, Johannes de and Wenceslaus Fabri. 1495. *Opusculu[m] Johannis de sacro busto spericum cum notabili commento atq[ue] figuris textum declarantibus vtilissimis.* [Leipzig]: [Martin Landsberg].

Sacrobosco, Johannes de and Michael Scott. 1495. *Eximii at[que]: excellentissimi physicoru motuum cursus[que]: syderei indagatoris Michaelis scoti sua auctor sperae cum questionibus diligenter emendatis incipit expositio [con] fecta Illustrissimi Imatoris Dni. D. Fedrici precibus.* Bononiae: Iustinianum de Ruberia.

Sacrobosco, Johannes de, Georg Peuerbach, and Joannes Regiomontanus. 1496. *Sphericum opusculum contraque Cremonensia in planetarum theoricas delyramenta Joh. de Monteregio disputationes; necnon Peurbachii in eorundem modus planetarum theorice.* Venetiis.

Sacrobosco, Johannes de and Wenceslaus Fabri. 1497. *Opusculum Johannis de sacro busto spericum cum notabili commento atq[ue] figuris textum declarantibus vtilissimis.* Leipzig: Martin Landsberg.

Sacrobosco, Johannes de. 1498. *Opu[s]culum Johannis de [s]acro bu[s]to [s]pericum cum figuris optimis et nouis textum in se sine ambiguitate declarantibus.* Leipzig: Wolfgang Stöckel.

Sacrobosco, Johannes de, Pierre d'Ailly, and Pedro Sanchez Ciruelo. 1498. *Uberrimum sphere mundi commentum intersertis etiam questionibus domini Petri de aliaco.* Paris: Guy Marchant for Jean Petit.

Sacrobosco, Johannes de, Cecco d'Ascoli, Francesco Capuano, Jacques Lefèvre d'Etaples, Georg von Peurbach, Laurentius Donatus, and Carolus Bora. 1499. *Sphera mundi cu[m] tribus commentis.* Venetiis: per Simonem Papiensem dictum Biuilaquam (Simon de Bevillaqua).

Sacrobosco, Johannes de and Wenceslaus Fabri. 1499. *Opusculum Johannis de sacro busto sphericu cum notabili commento atq[ue] figuris textum declarantibus vtilissimis.* Leipzig: Wolfgang Stöckel.

Sacrobosco, Johannes de and Wenzel Faber von Budweis. 1500. *Opus sphericum Iohannis de Sacro Busto/figuris et p[er]utili co[m]mento illustratu[m].* Cologne: Henricum Quentell.

Sacrobosco, Johannes de, Jacques Lefèvre d'Etaples, Bonet de Lattes, Euclid, and Boethius. 1500. *Textus de Sphera Johannis de Sacro bosco Cum additione (quantum necessarium est) adiecta: Nouo commentario nuper edito Ad vtilitatem studentium Philosophice Parisien. Academie illustratus Cum Compositione Anuli Astronomici Boni Latensis. Et Geometria Euclidis Megarensis.* Paris: Wolfgang Hopyl.

Sacrobosco, Johannes de and Geogius de Ferrariis. 1500. *Figura sphere: cum glosis Georgii de Monteferrato Artium et medicine Doctoris.* Venetiis: Georgii de Mo[n]teferrato.

Sacrobosco, Johannes de. 1501. *Sphere: cu[m] glosis Georgii de Mo[n]teferrato Artiu[m] [&] medici[n]e Docto[r]is.* Venice: Giorgio de Monteferrato.

Sacrobosco, Johannes de and Wenceslaus Fabri. 1501. *Opus sphericum Johannis de sacro busto figuris et p[er] utili co[m]mento illustratu[m].* Köln: Henrich Quentell.

Sacrobosco, Johannes de, Joannes Regiomontanus, Georg von Peurbach, and Gherardo da Sabbioneta. 1501. *Sphaerae mvndi compendivm foeliciter inchoat. Nouitiis adolescentibus: ad astronomica[m] re[m]pu. capessenda[m] additu[m] impetrantibus: pro breui rectoq[ue] tramite a uulgari uestigio semoto: Ioãnis de sacro busto sphæricu[m] opusculum una cu[m] additionibus nonnullis littera A sparsim ubi intersertæ sint signatis. Contraq[ue] cremone[n] sia in planetarum theoricas delyramenta Ioannis de monteregio disputationes tam acuratis. q[uam] utilis: nec no[n] Georgii purbachii: in eorundem motus planetarum accuratis. theoricæ: dicatum opus: utili serie contextum: fausto sidere inchoat.* Venetiis: Io. Baptista Sessa.

Sacrobosco, Johannes de and J. Fabro Stapulense. 1501. *Textus de sphaera.* Venetiis.

Treatises Published Between 1549 and 1553

Sacrobosco, Johannes de. 1549. *Arielis Bicardi questiones in sphaeram Jo. de Sacro-Bosco.* Venetiis.

Sacrobosco, Johannes de and Hartmann Beyer. 1549. *Quaestiones in novae in libellum de sphaera Ioannis de Sacro Busto, in gratiam studiosae iuventutis collectae ab Hartmanno Beyer, recognitae & plaerisque in locis auctae.* Francoforti: ex officina P. Brubachij.

Sacrobosco, Johannes de, Philipp Melanchthon, al-Farghani, and Joannes Regiomontanus. 1549. *Sphaera Ioannis de Sacrobosco Typis auctior, quam antehac, atque ex diligenti manu scriptorum impressorumque codicum collatione castigator, Praemissa Philippus Melanchthon doctiß. praefatione, qua vtilitatem astrologicae scientiae et Christiano homini non negligendam scite probat.* Parisiis: apud Gulielmum Cauellat.

Sacrobosco, Johannes de, Philipp Melanchthon, and Erasmus Reinhold. 1549. *Ioannis de Sacrobusto libellus de sphæra: accessit eiusdem autoris computus ecclesiasticus, et alia quædam in studiosorum gratiam edita/cum præfatione Philippi Melanthonis.* Vitebergae: apud Vitum Creuzer.

Piccolomini, Alessandro and Jacques Goupyl. 1550. *La sphere du monde.* Paris: pour Guillaume Cavellat.

Sacrobosco, Johannes de. 1550. *Sphera Ioannis de Sacro Busto cum commentariis Petri a Spinosa.* Salmanticae: Ioannes Iunta.

Sacrobosco, Johannes de. 1550. *Sphaera Ioannis de Sacrobvsto: addita sunt quaedam ad explanationem eorum, quae in sphaera dicuntur facientia.* Venetiis: per Petrum de Nicolinis de Sabio, expensis Melchioris Sensæ.

Sacrobosco, Johannes de. 1550. *Quaestiones novae in libellum de sphaera Johannis de Sacro Bosco, in gratiam studiosae inventutis collectae ab Ariele Bicardo, et nunc denuo recognitae, adiectis tabulis, quae in priore editione desiderantur.* Wittebergae.

Sacrobosco, Johannes de and Mauro Fiorentino. 1550. *Annotationi sopra la lettione della spera del Sacro Bosco. Doue si dichiarano tutti e principii Mathematici et Naturali, che in quelle si prossan desiderare. Con alcune quistioni Notabili a detta Spera Necessarie et altri notandi et rari segreti, che in quella son afcosti. Con le infrascritte cose, cio, e, Vna nuoua & fedele (ad verbum) traduttione di detta Spera: Vna spera theologia dinis, & christiana: Vna spera platonica, con alcune eccitationi mathematiche, theologiche & diuine: Una nuova inventione, et astronomico instrumento per fabricare le dodici case celesti della figura astronomica, senza altri canoni, o, calculo, authore m. Mauro Fiorentino, Theosebo, Phonasco, & Philopanareto.* Firenze: Lorenzo Torrentino.

Sacrobosco, Johannes de and Philipp Melanchthon. 1550. *Sphaera Joannis de Sacrobosco, typis auctior quam antehac, atque ex diligenti manu scriptorum impressorumque codicum collatione castigatior, Praemissa Philippi Melanchthonis doctiß. praefatione, qua vtilitatem astrologicae scientiae, et Christiano homini non negligendam scite probat.* Parisiis: apud T. Richardum.

Sacrobosco, Johannes de, Philipp Melanchthon, and Erasmus Reinhold. 1550. *Ioannis de Sacrobusto libellus de sphaera: accessit eiusdem autoris computus ecclesiasticus, et alia quaedam in studiosorum gratiam edita. Cum Præfatione Philippi Melanthonis.* Vuitebergæ: apud Iohannem Cratonem.

Sacrobosco, Johannes de, Philipp Melanchthon, and Pieri Valeriano. 1550. *Sphaera Ioannis de Sacro Bosco, typis auctior, quam antehac, atq; ex diligeti manuscriptorum impressorumque codicum collatione castigator, cum annotationibus, quæ locis aliquot obscuris magnam lucem afferunt. Praemissa Philippi Melanchthonis doctiss. praefatione, qua utilitatem astrologicæ scientiæ, & Christiano homini non negligendam scitè probat. Adiunximus huic libro compendium in Sphæram, per Pierium Valerianum Bellunensem.* Parisiis: apud Gulielmum Cauellat.

Cortés, Martin. 1551. *Breve compendio de la sphera y de la arte de navegar, con nuevos instrumentos y reglas, exemplificado com muy subtiles demostraciones.* Sevilla: en casa de Anton Alvares.

Sacrobosco, Johannes de. 1551. *Sphaera Ioannis De Sacrobosco.* Antverpiæ: apud Ioannem Richardum.

Sacrobosco, Johannes de. 1551. *Sphaera mundi cum tribus commentariis Cicchi esculani, Francisci Capuani de Manfredonia, Jacobi Fabri stapulensis; Theoricae novae planetarum G. Purbachii, cum commento F. Capuani de Manfredonia.* Basiliae.

Sacrobosco, Johannes de and Hartmann Beyer. 1551. *Qvaestiones Novae In Libellvm De Sphaera Ioannis de Sacro Bosco, in gratiã studiosæ iuuentutis collectæ ab Ariele Bicardo [pseud.], & nunc denuo recognitæ, figuris mathematicis ac tabulis illustratæ, quæ in reliquis editionibus antehac desiderabantur.* Parisiis: apud Gulielmum Cauellat.

Sacrobosco, Johannes de, Pierio Valeriano, Philipp Melanchthon, Élie Vinet, Joannes Regiomontanus, and Proclus. 1551. *Sphaera Ioannis de Sacro Bosco, typis auctior quam antehac, atque ex diligenti manuscriptorum impressorumque codicum collatione castigatior, cum annotationibus, & scholijs doctissimi viri Eliae Vineti, quae locis aliquot abscuris magnam lucem afferunt. Praemissa Philippi Melanchthonis doctissimi praefatione, qua utilitatem astrologicae scientiae, & Christiano homini non negligendam scite probat. Adiunximus huic libro compendium in Sphaeram, per Pierium Valerianum Bellunensem.* Parisiis: apud Gulielmum Cauellat.

Piccolomini, Alessandro. 1552. *Editione Tertia. Della Sfera del Mondo di M. Alessandro Piccolomini, divisa in libri quattro, i quali non per via di traduttione, ne à qual si voglia particolare Scrittore obligati, ma parte da migliori raccogliendo, e parte di nuovo producendo, contengano in se tutto quel ch'intorno a tal materia si possa desiderare, ridotti à tanta agevolezza, et à cosi facil modo di dimostrare, che qual si voglia poco essercitato negli studij di Mathematica potra agevolißimamente, et con prestezza intenderne il tutto, Di nuovo ricorretta, et ampliata. De le Stelle Fisse. Libro uno con le sue figure, e con le sue Tavole, dove con maravigliosa agevolata potra ciascheduno conoscere qualunque stella delle XLVIII Imagini del Cielo Stellato, é le Favole loro integramente, et sapere in ogni tempo del'anno, à qual si voglia hora di notte, in che parte del cielo si trouvino, non solo le dette Imagini, ma qualunque stella di quelle.* Venetia: Nicolo de Bascarini.

Sacrobosco, Johannes de and Hartmann Beyer. 1552. *Quaestiones in libellum De sphaera Ioannis de Sacro Busto, in gratiam studiosae iuuentutis collectae ab Hartmanno Beyer & nunc denuo recognitae.* Francoforti: ex officina Petri Brubacchij.

Sacrobosco, Johannes de and Hartmann Beyer. 1552. *Quaestiones in libellum de sphaera Ioannis de Sacrobosco.* Parisiis: apud Gulielmum Cauellat.

Sacrobosco, Johannes de, Philipp Melanchthon, Elié Vinet, Pieri Valeriano, Johann Conrad Felix Josef Wallier, and Guillaume Cavellat. 1552. *Sphaera Ioannis de Sacro Bosco, typis auctior quam antehac, atque; ex diligenti manuscriptorum impressorumque codicum collatione castigatior, cum annotationibus, & scholiis doctissimi viri Eliae Vineti, quae locis aliquot obscuris magnam lucem afferunt. Praemissa Philippi Melanchthonis doctissimi praefatione, qua utilitatem astrologicae scientiae & Christiano homini non negligendam scite probat. Adiunximus huic libro compendium in Sphaeram, per Pierium Valerianum Bellunensem.* Parisiis: apud Gulielmum Cauellat.

Sacrobosco, Johannes de and Antoine Mizauld. 1552. *Antonii Mizaldi Monlvciani, de mundi sphaera, seu Cosmographia, Libri tres: figuris & demonstrationibus illustrati. Ad illustratissimam principem Margaretam Valesiam, Biturigum Ducem.* Paris: apud Guilielmum Cavellat.

Piccolomini, Alessandro. 1553. *Editione Tertia. Della Sfera del Mondo di M. Alessandro Piccolomini, divisa in libri quattro, i quali non per via di traduttione, ne à qual si voglia particolare Scrittore obligati, ma parte da migliori raccogliendo, e parte di nuovo producendo, contengano in se tutto quel ch'intorno a tal materia si possa desiderare, ridotti à tanta agevolezza, et à cosi facil modo di dimostrare, che qual si voglia poco essercitato negli studij di Mathematica potra agevolißimamente, et con prestezza intenderne il tutto, Di nuovo ricorretta, et ampliata. De le Stelle Fisse. Libro uno con le sue figure, e con le sue Tavole, dove con maravigliosa agevolata potra ciascheduno conoscere qualunque stella delle XLVIII Imagini del Cielo Stellato, é le Favole loro integramente, et sapere in ogni tempo del'anno, à qual si voglia hora di notte, in che parte del cielo si trouvino, non solo le dette Imagini, ma qualunque stella di quelle.* Venetia: Appresso Bartholomeo Cesano.

Sacrobosco, Johannes de. 1553. *Sphaera cum commentariis E. O. Schreckenfuchsii.* Basileae.

Sacrobosco, Johannes de and Philipp Melanchthon. 1553. *Libellus de sphaera.* Wittenberg: J. Crato.

Sacrobosco, Johannes de, Philipp Melanchthon, and Erasmus Rheinhold. 1553. *Ioannis de Sacro Bvsto Libellvs de Sphaera; Accessit Eivsdem Avtoris Computus Ecclesiasticus Et alia quædam in studiosorum gratiam edita. Cum Præfatione Phi. Melanthonis.* Witebergae: per Ioannem Cratonem.

Appendix 4

Treatises Containing Vinet's Translation of Pedro Nuñes's Demonstration on the Climate Zones

Sacrobosco, Joannes de and Pedro Nuñez. 1537. *Tratado da sphera com a Theorica do Sol & da Lua e ho primeiro livro da Geographia de Claudio Ptolomeo Alexandrino. Tirados novamente de Latim em lingoagem pello Doutor Pero Nunez Cosmographo del Rey Don Joano. Et acrescentatos de muitas annotaçiones i figuras per que mays facilmente se podem entender. Tratado que ho doutor per nunez fez em defensam de carta de marear. Item dous tratados quo mesmo Doutor fez sobra a carta de marear. Em os quaes se decrerano todas as principaes du vidas da navegação. Con as pavoa do movimento do sol: i sua declinação. Eo regimento de altura.* Lixboa: Galharde.

Sacrobosco, Johannes de, Èlie Vinet, Petrus Valerianus, Petrus Nuñes, and Joannes Regiomontanus. 1556. *Sphaera Ioannis de Sacro Bosco emendata. Eliae Vineti Santonis scholia in eandem Sphaeram, ab ipso auctore restituta. Adiunximus huic libro compendium in Sphaeram, per Pierium Valerianum Bellunensem, Et, Petri Nonii Salaciensis demonstrationem eorum, quae in extremo capite de Climatibus Sacroboscius scribit, de inaequali Climatum latitudine, eodem Vineto interprete.* Lutetiae: Gulielmum Cauellat.

Sacrobosco, Johannes de, et al. 1558. *Sphaera Ioannis de Sacro Bosco emendata. Eliae Vineti Santonis scholia in eandem Sphaeram, ab ipso authore restituta. Adiunximus huic libro compendium in Sphaeram, per Pierium Valerianum Bellunensem, Et, Petri Nonii Salaciensis demonstrationem eorum, quae in extremo capite de Climatibus Sacroboscius scribit de inaequali Climatum latitudine, eodem Vineto interprete.* Lvtetiae: apud Gulielmum Cauellat.

Sacrobosco, Johannes de, et al. 1559. *Sphaera Ioannis de Sacro Bosco emendata. Eliae Vineti Santonis scholia in eandem sphaeram, ab ipso authore restituta. Adiunximus huic libro compendium in Sphaeram, per Pierium Valerianum Bellunensem, Et Petri Nonij Salaciensis demonstrationem eorum, queen extremo capite de climatibus Sacroboscus scribit de inaequali climatum latitudine: eodem Vineto interprete.* Lvtetiae: apud Gulielmum Cauellat.

Sacrobosco, Johannes de, Élie Vinet, Pedro Nunes, Pierio Valeriano, and Johannes Regiomontanus. 1561. *Sphaera Ioannis de Sacro Bosco emendata. Eliae Vineti Santonis scholia in eandem sphaeram, ab ipso authore restituta. Adiunximus huic libro compendium in Sphaeram, per Pierium Valerianum Bellunensem, Et Petri Nonij Salaciensis demonstrationem eorum, queen extremo capite de climatibus Sacroboscus scribit de inaequali climatum latitudine: eodem Vineto interprete.* Lutetiae: apud G. Cauellat, sub pingui gallina, ex aduerso Collegij Cameracensis 1561.

Sacrobosco, Johannes de, Élie Vinet, Pierio Valeriano, and Pedro Nunes. 1562. *Sphaera Ioannis de Sacro Bosco emendata. Eliae Vineti Santonis scholia in eandem sphaeram, ab ipso authore restituta. Adiunximus huic libro compendium in Sphaeram, per Pierium Valerianum Bellunensem, Et Petri Nonij Salaciensis demonstrationem eorum, queen extremo capite de climatibus Sacroboscus scribit de inaequali climatum latitudine: eodem Vineto interprete.* Lvtetiae: apud Gulielmum Cauellat, sub pingui Gallina, ex adverso collegij Cameracensis.

Sacrobosco, Johannes de, Pedro Nuñes, and Elie Vinet. 1563. *Annotatio in extrema verba capitis de Climatibus Joannis de Sacrobosco.* Venetia: Francisco Juntas.

Sacrobosco, Joannes de, Pedro Nuñes, Elie Vinet, Pieri Valeriano, and Francesco Giuntini. 1564. *Sphaera Ioannis de Sacro Bosco emendata. Cum additionibus in margine, et Indice rerum et locorum memorabilium, et familiarissimis scholijs, nunc recent compertis, et collectis à Francesco Iunctino Florentino sacra Theologia Doctore. Interserta sunt Eliae Vineti Santonis egregia Scholia in tandem Sphaeram. Adiunximus huic libro compendium in Sphaeram, per Pierium Valerianum Bellunensem. Et Petri Nonij Salaciensis demonstrationem eorum, quae extremo capite de climatibus Sacroboscus scribit de inaequali climatum latitudine: eodem Vineto interprete.* Lvgdvni: apud hæredes Iacobi Iunctæ, excudebat Symphorianus Barbier.

Sacrobosco, Joannes de, Élie Vinet, Piero Valeriano, Pedro Nuñes, and Francesco Giuntini. 1564. *Sphaera Ioannis de Sacro Bosco, emendata. Eliae Vineti Santonis scholia in eandem sphaeram, ab ipso autore resituta. Audiunximus huic libro compendium in sphaeram per Pierium Valerianum Bellunensem, et Petri Nonii Salaciensis demonstrationem, eorum, quae in extremo capite de climatibus Sacroboscius scribit de inaequali climatum latitudine, eodem Vineto interprete.* Lvtetiae: apud Gulielmum Cauellat.

Sacrobosco, Johannes de, Francesco Giuntini, Élie Vinet, and Pedro Nunes. 1566. *Sphæra Ioannis De Sacro Bosco Emendata. Cum additionibus in margine, & Indice rerum & locorum memorabilium, & familiarissimis scholiis, nunc recenter compertis, & collectis à Francisco Iunctino Florentino sacrae Theologiae Doctore.; Interserta etiam sunt Eliae Vineti Santonis egregia Scholia in eandem Sphaeram. Adiunximus eiusdem auctoris Computum Ecclesiasticum. Et Petri Nonii Salaciensis demonstrationem eorum, quae in extremo capite de Climatibus Sacroboscius scribit de inaequali climatum latitudine, eodem Vineto interprete.* Antvverpiæ: apud Ioannem Richardum.

Sacrobosco, Johannes de, Francesco Giuntini, Élie Vinet, Pedro Nunes, and Philipp Melanchthon. 1566. *Sphaera Ioannis de Sacro Bosco emendata. Cum additionibus in margine, et Indice rerum et locorum memorabilium, et familiarissimis scholijs, nunc recent compertis, et collectis à Francesco Iunctino Florentino sacra Theologia Doctore. Interserta sunt Eliae Vineti Santonis egregia Scholia in tandem Sphaeram. Adiunximus eiusdem auctoris Compendium Ecclesiasticum. Et Petri Nonij Salaciensis demonstrationem eorum, quae extremo capite de climatibus Sacroboscus scribit de inaequali climatum latitudine: eodem Vineto interprete.* Antvverpiae: apud hæredes Arnoldi Birckmanni.

Sacrobosco, Johannes de, Pieri Valeriano, Pedro Nuñes, Elie Vinet, Joannes Regiomontanus, and Proclus. 1566. *Sphaera Ioannis de Sacro Bosco emendata. Eliae Vineti Santonis scholia in eandem Sphaeram, ab ipso authore restituta. Adiunximus huic libro compendium in sphaeram per Pierium Valerianum Bellunensem, Et Petri Nonii Salaciensis demonstrationem eorum, quae in extremo capite de climatibus Sacroboscius scribit de inaequali climatum latitudine: eodem Vineto interprete.* Coloniae: apud Maternum Cholinum.

Sacrobosco, Johannes de, Élie Vinet, Joannes Regiomontanus, Pierio Valeriano, Pedro Nunes, Francesco Giuntini, and Proclus. 1567. *Sphaera Ioannis de Sacro Bosco emendata. Cum additionibus in margine, & Indice rerum et locorum memorabilium, & familiarissimis scholijs, nunc recenter compertis, et collectis [DELETION: à Francesco Iunctino]*

Florentino sacra Theologia Doctore. Interserta sunt Eliae Vineti Santonis egregia Scholia in eandem Sphaeram. Adiunximus huic libro compendium in Sphaeram, per Pierium Valerianum Bellunensem. Et Petri Nonij Salaciensis demonstrationem eorum, quae in extremo capite de climatibus Sacroboscius scribit de inaequali climatum latitudine: eodem Vineto interprete. Lugduni: apud haeredes Iacobi Iunctae, excudebat Symphorianus Barbier.

Sacrobosco, Johannes de, Pedro Nuñes, Elie Vinet, and Pieri Valeriano. 1569. *Sphaera Ioannis de Sacro Bosco emendata. Eliae Vineti Santonis scholia in eandem Sphaeram, ab ipso authors restitute. Adiunximus huic libro compendium in Sphaeram per Pierium Valerianum Bellunensem, Et, Petri Nonii Salaciensis demosntrationem eorum, quae in extremo capite de climatibus Sacroboscius scribit de inaequali Climatum latitudinem, eodem Vineto interprete. Ex postrema impressione Lutetiae.* Venetijs: apud Hieronymum Scotum.

Sacrobosco, Johannes de, Élie Vinet, Pierio Valeriano, Pedro Nunes, Joannes Regiomontanus, al-Farghani, and Proclus. 1569. *Sphæra Ioannis De Sacro Bosco, emendata.* Parisiis: apud Hieronymum de Marnef, & Gulielmum Cauellat.

Sacrobosco, Johannes de, Élie Vinet, Pedro Nunes, and Pierio Valeriano. 1571. *Sphaera Ioannis de Sacro Bosco emendata. Eliae Vineti Santonis scholia in eandem Sphaeram, ab ipso authors restitute. Adiunximus huic libro compendium in Sphaeram per Pierium Valerianum Bellunensem, Et, Petri Nonii Salaciensis demosntrationem eorum, quae in extremo capite de climatibus Sacroboscius scribit de inaequali Climatum latitudinem, eodem Vineto interprete. Ex postrema impressione Lutetiae.* Venetiis: apud heredem Hieronymi Scoti.

Sacrobosco, Johannes de, et al. 1572. *Sphaera Ioannis de Sacro Bosco, emendata. Eliae Vineti Santonis scholia in eandem Sphaeram, ab ipso auctore restituta. Adiunximus huic libro compendium in Sphaeram per Pierium Valerianum Bellunensem, Et Petri Nonii Salaciensis demonstrationem eorum, quae in extremo capite de climatibus Sacroboscius scribit de inaequali climatum latitudine, eodem Vineto interprete.* Parisiis: apud Hieronymum de Marnef, & Gulielmum Cauellat, sub Pelicano.

Sacrobosco, Johannes de, Elie Vinet, Pedro Nuñes, and Francesco Giuntini. 1573. *Sphaera Ioannis de Sacro Bosco, emendata. In eandem Francisci Iunctini Florentini, et Elie Vineti Santons Scholia. Caetera pagina sequens indicabit.* Antverpiae: apud Ioannem Bellerum.

Sacrobosco, Johannes de, Élie Vinet, Pedro Nunes, Joannes Regiomontanus, al-Farghani, Proclus, and Pierio Valeriano. 1574. *Sphaera Ioannis de Sacro Bosco emendata. Eliae Vineti Santonis scholia in eandem sphaeram, ab ipso authore restituta. Adiunximus huic libro compendium in sphaeram, per Pierium Valerianum Bellunensem. Et Petri Nonij Salaciensis demonstrationem eorum, queen extremo capite de climatibus Sacroboscus scribit de inaequali climatum latitudine: eodem Vineto interprete. Ex postremo Impressione Lutetiae.* Venetiis: apud haeredem Hieronymi Scoti.

Sacrobosco, Johannes de, Élie Vinet, Pierio Valeriano, and Pedro Nunes. 1577. *Sphæra Ioannis de Sacrobosco emendata. Elia Vineti Santonis scholia in eandem Sphaeram ab ipso authore restituta. Adiunximus huic libro compnedium in Sphaeram per Pierium Valerianum Bellunensem, et Petri Nonii Salacensis demosntrationem eorum, quae in extremo capite de climatibus Sacroboscus scribit de inaequali climatum latitudine, eodem Vineto interprete.* Parisiis: apud Hieronymum de Marnef, & viduam Gulielmi Cauellat.

Sacrobosco, Johannes de, Élie Vinet, Albertus Hero, Pierio Valeriano, and Pedro Nunes. 1581. *Sphæra Ioannis de Sacrobosco emendata. Elia Vineti Santonis scholia in eandem Sphaeram ab ipso authore restituta. Quibus nunc accessere scholia Heronis. Adiunximus huic libro compnedium in Sphaeram per Pierium Valerianum Bellunensem, et Petri Nonii Salacensis demosntrationem eorum, quae in extremo capite de climatibus Sacroboscus scribit de inaequali climatum latitudine, eodem Vineto interprete.* Coloniæ: apud Maternum Cholinum.

Sacrobosco, Johannes de. 1584. *Sphaera Ioannis de Sacrobosco, emendata. Eliae Vineti Santonis scholia in eandem Sphaeram, ab ipso auctore restituta. Adiunximus huic libro compendium in Sphaeram, per Pierium Valerianum Bellunensem. Et Petri Nonii Salaciensis demonstrationem eorum, quae in extremo capite de climatibum Sacroboscius scribit de inaequali climatum latitudine, eodem Vineto interprete. Post omnes omnium editiones, auctior er locupletior.* Paris: Hieronymus de Marnef Gulielmus Cavellat.

Sacrobosco, Johannes de, Élie Vinet, Pierio Valeriano, and Pedro Nunes. 1586. *Sphaera Ioannis de Sacro Bosco emendata. Eliae Vineti Santonis scholia in eandem sphaeram, ab ipso authore restituta. Adiunximus huic libro compendium in sphaeram, per Pierium Valerianum Bellunensem: et Petri Nonij Salaciensis demonstrationem eorum, queen extremo capite de climatibus Sacroboscus scribit de inaequali climatum latitudine: eodem Vineto interprete. Ex postremo Impressione Lutetiae.* Venetiis: apud Haeredem Hieronymi Scoti.

Sacrobosco, Johannes de, Élie Vinet, Albertus Hero, Pierio Valeriano, and Pedro Nunes. 1591. *Sphæra Ioannis De Sacrobosco Emendata: Eliae Vineti Santonis scholia in eandem sphaeram, ab ipso authore restituta: Quibus nunc accessere scholia Heronis: Adiunximus huic libro compendium in Sphaeram per Pierium Valerianum Bellunensem et Petri Nonii Salaciensis demonstrationem eorum, quae in extremo capite de climatibus Sacroboscius scribit de inaequali climatum latitudine, eodem Vineto interprete.* Coloniæ: apud Gosiunum Cholinum.

Sacrobosco, Johannes de, Élie Vinet, Albertus Hero, Pedro Nunes, and Pierio Valeriano. 1594. *Sphæra Ioannis De Sacrobosco Emendata: Eliæ Vineti Santonis scholia in eandem sphæram, ab ipso authore restituta. Quibus nunc accessere scholia Heronis. Adiunximus huic libro compendium in Sphæram per Pierium Valerianum Bellunensem, et Petri Nnii Salaciensis demonstrationem eorum, quæ in extremo capite de climatum latitudine, eodem Vineto interprete.* Coloniæ: apud Gosuinum Cholinum.

Sacrobosco, Johannes de, Christoph Clavius, Albertus Hero, Pedro Nuñes, Petrus Valerianus, and Élie Vinet. 1601. *Sphæra Ioannis de Sacrobosco emendata, aucta et illustrata. Elia Vineti Santoris scholia in eandem Sphaeram ab ipso auctore restituta. Quibus praetear Alberti Heronis scholia nunc accessere R. P. Christoph. Clavij breves commentarij. Adiunximus huic libro compnedium in Sphaeram per Pierium Valerianum Bellunensem: et Petri Nonii Salacensis demosntrationem eorum, quae in extremo capite de climatibus Sacroboscus scribit de inaequali climatum latitudine, eodem Vineto interprete.* Coloniae: apud Gosuinum Cholinum.

Sacrobosco, Johannes de, Élie Vinet, Pedro Nunes, and Pierio Valeriano. 1606. *Sphaera Ioannis de Sacrobosco emendata: Eliae Vineti Santonis scholia in eandem spheram ... et Petri Nonii Salaciensis ... de inaequali climatum latitudine.* Lvgdvni: apvd Hvgonem Gazævm.

Sacrobosco, Johannes de, Elie Vinet, Jacobus Martinus, Pedro Nunes, and Pierio Valeriano. 1608. *Sphaera Ioannis de Sacro Bosco emendata. Eliae Vineti Santonis scholia in eandem Sphaeram, ab ipso authore restituta. Annotationibus Jacobi Martini Pedemontani acta. Petri Nonij Salaciensis annotatio in caput de climatibus, eodem Vineto interprete. Compendium in Sphaeram, per Pierium Valerianum Bellunensem. Post omnes omnium editiones, auctior & locupletior.* Parisiis: in officina H. de Marnef, apud Dionysiam Cavellat.

Sacrobosco, Johannes de, Élie Vinet, Francesco Giuntini, Albertus Hero, Christoph Clavius, Pedro Nuñes, and Petrus Valerianus. 1610. *Sphaera Iohannis a Sacrobosco emendata, avcta et illustrata. Elia Vineti Santonis, et Francisci Iunctini Florentini scholia in eandem Sphaeram restituta. Quibus praeter Alberti Heronis scholia nunc accessere R. P. Christoph. Clavij breves commentarij. Adiunximus huic libro compendium in Sphaeram, per Pierium Valerianum Bellunensem: et Petri Nonij Salaciensis demonstrationem eorum, queen extremo capite de climatibus Sacroboscus scribit de inaequali climatum latitudine: eodem Vineto interprete.* Coloniae Agrippinae: apud Petrum Chrolinum.

Sacrobosco, Johannes de, Jacob Martini, Pierio Valeriano, Élie Vinet, and Pedro Nunes. 1617. *Sphaera Ioannis de Sacro Bosco emendata. Eliae Vineti Santonis scholia in eandem Sphaeram, ab ipso authore restituta. Annotationibus Jacobi Martini Pedemontani acta. Petri Nonij Salaciensis annotatio in caput de climatibus, eodem Vineto interprete. Compendium in Sphaeram, per Pierium Valerianum Bellunensem. Post omnes omnium editiones, auctior & locupletior.* Lugduni: sumptibus Antonii Pillehotte.

Sacrobosco, Johannes de, Elie Vinet, Jakob Martini, Petrus Nonius, and Pierio Valeriano. 1619. *Sphæra Ioannis de Sacrobosco emendata. Eliæ Vineti Santonis Scholia in eandem Sphæram, ab ipso auctore restituta, et annotationibus Iacobi Martini Pedemontani aucta. Petri Nonij annotatio in caput de climatibus, eodem Vineto interprete. Compendium in Sphæram per Pierium Valerianum Belunensem.* Parisiis: Iacobus Quesnel.

Sacrobosco, Johannes de, Élie Vinet, Pierio Valeriano, Pedro Nunes, and Alfraganus. 1620. *Sphaera Ioannis de Sacro Bosco emendata. Eliae Vineti Santonis scholia in eandem sphaeram, ab ipso authore restituta. Adiunximus huic libro compendium in sphaeram, per Pierium Valerianum Bellunensem: et Petri Nonij Salaciensis demonstrationem eorum, quae in extremo capite de climatibus Sacroboscus scribit de inaequali climatum latitudine: eodem Vineto interprete. Ex postremo impressione Lutetiae.* Venetiis: apud hæredem Hieronymi Scoti.

Editions of Clavius's Commentary on the *Sphere of Sacrobosco*

Sacrobosco, Joannes de and Christophorus Clavius. 1570. *Christophorii Clavii Bambergensis, ex Societate Iesv, in Sphaeram Ioannis de Sacro Bosco, commentarivs.* Romæ: apud Victorium Helianum.

Sacrobosco, Johannes de and Christoph Clavius. 1581. *Christophori Clavii Bambergensis ex Societate Iesu In Sphaeram Ioannis de Sacro Bosco commentarius.* Romæ: apud Franciscum Zanettum ex officina Dominici Basæ.

Sacrobosco, Johannes de and Christoph Clavius. 1585. *Christophori Clauii Bambergensis ex Societate Iesu In sphaeram Ioannis de Sacro Bosco commentarius.* Romae: ex officina Dominici Basae.

Sacrobosco, Johannes de and Christoph Clavius. 1591. *Christophori Clauii ... In Sphaeram Ioannis de Sacro Bosco commentarius.* Venetiis: apud Ioan. Baptistam Ciottum Senensem.

Sacrobosco, Johannes de and Christoph Clavius. 1593. Christophori Clauii Bambergensis ex Societate Iesu, in Sphaeram Ioannis de Sacro Bosco. Commentarius. Nunc quarto ab ipso Auctore recognitus, & plerisque in locis locupletatus. Lugduni: sumptibus Fratrum de Gabiano/excudebat, Guichardus Iullieron.

Sacrobosco, Joannes de and Christophorus Clavius. 1594. *In Sphaeram Ioannis de Sacro Bosco. Commentarius.* Leyden: Gabianus, Iullerion.

Sacrobosco, Johannes de and Christoph Clavius. 1594. *Christophori Clavii Bambergensis Ex Societate Iesv, In Sphæram Ioannis De Sacro Bosco Commentarivs.* Lvgdvni: svmptibvs Fratrvm De Gabiano.

Sacrobosco, Johannes de and Christoph Clavius. 1596. *Christophori Clavii ... In Sphæram Ioannis De Sacro Bosco Commentarivs.* Venetiis: apud Bernardum Basam.

Sacrobosco, Johannes de and Christoph Clavius. 1601. *Christophori Clavii Bambergensis ex Societate Iesv, In Sphæram Ioannis de Sacro Bosco, Commentarius.* Venetiis: sub signo solis, apud Io. Baptistam Ciottum.
Sacrobosco, Johannes de and Christoph Clavius. 1602. *Christophori Clavii Bambergensis ex Societate Iesv, In Sphæram Ioannis de Sacro Bosco, Commentarius.* Lvgduni: svmptibvs Ioannis De Gabiano.
Sacrobosco, Johannes de and Christophorus Clavius. 1602. *Christophori Clavii, ... In Sphaeram Joannis de Sacro Bosco commentarius.* S. Gervasii: apud Samuelem Crispinum.
Sacrobosco, Johannes de and Christoph Clavius. 1603. *Christophori Clavii Bambergensis ex Societate Iesv, In Sphaeram Ioannis de Sacro Bosco Commentarivs.* Venetiis: Apud Io. Baptistam Ciottum, Sub signo Aurorae.
Sacrobosco, Johannes de and Christoph Clavius. 1606. *Christophori Clauii ... in Sphaeram Ioannis de Sacro Bosco commentarius.* Romae: sumptibus Jo. Pauli Gellii ... apud Aloisium Zannettum.
Sacrobosco, Johannes de and Christoph Clavius. 1607. *Christophori Clavii Bambergensis ex Societate Iesv, In Sphæram Ioannis de Sacro Bosco.* Lvgdvni: ex officina Q. Hvg. a Porta, svmpt. Io. de Gabiano/Ex Typographia Iacobi du Creux dict Molliard.
Sacrobosco, Johannes de and Christoph Clavius. 1607. *Christophori Clavii Bambergensis ex Societate Iesv, In Sphaeram Ioannis de Sacro Bosco commentarivs.* Romae: svmptibvs Io. Pauli Gellii. Apud Aloisium Zannettum.
Sacrobosco, Johannes de. 1608. *Christophori Clauii ... In Sphaeram Ioannis de Sacro Bosco commentarius.* S. Gervasii: apud S. Crispinum.
Sacrobosco, Johannes de and Christoph Clavius. 1611. *Christophori Clauii Bambergensis e Societate Iesv Opera Mathematica.* Moguntiae: sumptibus Antonij Hierat, excudebat Reinhard Eltz.
Sacrobosco, Johannes de and Christophorus Clavius. 1618. *Christophori Clavii Bamberg. ex Societate Iesv, In spheram Ioannis de Sacro Bosco commentarivs.* Lvgdvni: sumptibus P. Rigavd.

References

Andrews, William J.H. 1996. *The quest for longitude.* Cambridge, MA: Harvard University Press.
Apian, Petrus. 1550. *Cosmographia.* Antwerp: Gregorio Bontio.
Baldini, Ugo. 2000. L'insegnamento della matematica nel Collegio di S. Antão a Lisbona, 1590–1640. In *A companhia de Jesus e a missionção no oriente*, 275–310. Lisboa: Brotéria – Revista de cultura.
Barozzi, Francesco. 1585. *Cosmographia in qvatvor libros distribvta, svmmo ordine, miraqve facilitate, ac breuitate ad Magnam Prolemaei Mathematicam Constructionem, ad vniuersamque Astrologiam instituens. Francisco Barocio, Iacobi Filio, Patritio Veneto Avtore. Cum Praefatione eiusdem Autoris, in qua perfecta quidem Astrologiae Diusio, et enarratio Autorem Illustrium, et Voluminum ab eis conscriptorum in singulis Astrologiae partibus habetur: Ioannis de Sacrobosco vero 84 errores, et alii permulti suorum Expositorum, et Sectatorum ostenduntur, rationibusque; redarguuntur. Praecessunt etiam quaedam Communia Mathematica, necnon Arithmetica, et Geometrica principia, nonnullaeque: Propositiones, de quibus in toto opere saepe fit mentio: Ac demum locupletisimus Index eorum, quae ipsa Cosmographia coutinentur.* Venetijs: ex officina Gratiosi Perchacini.
Baur, Ludwig (ed.). 1912. *Die philosophische Werke des Robert Grosseteste, Bischofs von Lincoln.* Münster: Aschendorfsche Verlagsbuchhandlung.
Biancani, Giuseppe. 1620. *Sphaera mundi, seu Cosmographia demonstrativa ac facili methodo tradita.* Bononiae: typis Sebastiani Bonomii.
Binding, Günther. 2014. Bauwissen im Früh- und Hochmittelalter. In *Wissensgeschichte der Architektur*, vol. 3, ed. Jürgen Renn, Wilhelm Ostheus, and Hermann Schlimme, 9–94. Berlin: Edition Open Access.
Cattaneo, Angelo. 2009. Refléxion dur les climats et les zones face à l'expansion des XVe et XVIe siècles. *Le Monde des Cartes: Revue du Comité français de cartographie* 199: 7–17.
Cattaneo, Angelo. 2016. European medieval and Renaissance cosmography: A story of multiple voices. *Asian Review of World Histories* 4 (1): 35–81.
Cortés, Martin. 1551. *Breve compendio de la sphera y de la arte de navegar, con nuevos instrumentos y reglas, exemplificado com muy subtiles demostraciones.* Sevilla: en casa de Anton Alvares.
Crowther, Kathleen M., and Peter Barker. 2013. Training the intelligent eye: Understanding illustrations in early modern astronomy texts. *Isis* 104 (3): 429–470.
Crowther, Kathleen M., Ashley Nicole McCray, Leila McNeill, Amy Rodgers, and Blair Stein. 2015. The book everybody read: Vernacular translations of Sacrobosco's sphere in the sixteenth century. *Journal for the History of Astronomy* 46 (1): 4–28.
Danti, Ignazio. 1569. *Trattato dell'vso et della fabbrica dell'astrolabio: con l'aggivnta del planisferio del roias.* Fiorenza: Giunti.
Desgraves, Luis. 1977. *Elie Vinet. Humaniste de Bordeaux (1509–1587). Vie, Bibliographie, Correspondance, Bibliothèque.* Genève: Librairie Droz.

Drennan, Anthony S. 1991. Sacrobosco in Ulster. *Irish Astronomical Journal* 20 (2): 78–83.

Fine, Oronce. 1524. *Mundialis Sphere opusculum nuper emendatum una cum figuris cumque marginariis annotaciunculis recenter adjectis*. Paris: impressum Regnault Chaudière.

Galluzzi, Paolo. 2014. *«Libertà di filosofare in naturalibus». I mondi paralleli di Cesi e Galileo*. Roma: Scienze e Lettere. Editore commerciale.

Gingerich, Owen. 1990. Five centuries of astronomical textbooks and their role in teaching. In *The teaching of astronomy*, ed. J.M. Pasachoff and J.R. Percy, 189–211. Cambridge: Cambridge University Press.

Goldthwaite, Richard A. 1972. Schools and teachers of commercial arithmetic in Renaissance Florence. *Journal of European Economic History* 1: 418–433.

Goldthwaite, Richard A. 2011. *The economy of Renaissance Florence*. Baltimore: The Johns Hopkins University Press.

Granovetter, Mark S. 1973. The strength of weak ties. *American Journal of Sociology* 78 (6): 1360–1380.

Granovetter, Mark S. 1978. The threshold models of collective behavior. *American Journal of Sociology* 83 (6): 1420–1443.

Grant, Edward. 1994. *Planets, stars and orbs: The medieval cosmos 1200–1687*. Cambridge: Cambridge University Press.

Hamel, Jürgen. 2014. *Studien zur "Sphaera" des Johannes de Sacrobosco*, Acta Historica Astronomiae. Leipzig: AVA Akademische Verlagsanstalt.

James, Montague Rhodes. 1921. *Lists of manuscripts formerly owned by Dr. John Dee; with preface and identifications*. London: Printed at the Oxford University Press for the Bibliographical Society.

Johnson, Francis R. 1953. Astronomical textbooks in the sixteenth century. In *Science, medicine, and history: Essays on the evolution of scientific thought and medical practice written in honour of Charles Singer*, vol. 1, ed. Edgar Ashworth Underwood, 285–302. Oxford: Oxford University Press.

Lazega, Emmanuel. 2014. *Réseaux sociaux et structures relationelles*, 3rd ed. Paris: Presses Universitiaures de France.

Leitão, Henrique (ed.). 2008. *Sphaera mundi: A ciência na aula de esfera. Manuscriptos científicos do Colégio de Santo Antão nas colecções da BNP*. Lisboa: Biblioteca Nacional de Portugal.

Leitão, Henrique, and Bruno Almeida. 2012. A anotação de Pedro Nunes acerca da "largura" dos climas. In *António Estácio dos Reis. Marinheiro por vocação e historiador com devoção*, ed. Jorge Semedo de Matos, 123–142. Lisboa: Comissão Cultural de Marinha.

Lemercier, Claire. 2012. Formale Methoden der Netzwerkanalyse in den Geschichtswissenschaften: Warum und Wie? In *Historische Netzwerkanalyse*, ed. Albert Müller and Wolfgang Neurath, 16–41. Innsbruck: Studien Verlag.

Lemercier, Claire. 2015. Taking time seriously. How do we deal with change in historical networks? In *Knoten und Kanten III. Soziale Netwerkanalyse in Geschichts- und Politikforschung*, ed. Markus Gamper, Linda Rechke, and Marten Düring, 183–211. Bielefeld: transcript Verlag.

Lerch, Andreas. 2015. *Scientia Astrologiae. Der Diskurs über die Wissenschaftlichkeit der Astrologie und die lateinische Lehrbücher. 1470–1610*. Leipzig: AVA. Akademische Verlagsanstalt.

Leu, Urs. 2008. *Conrad Gessner's private library online: The revealing hand-written notes of an early modern polymath* [online text], Brill http://www.brill.com/publications/online-resources/conrad-gessner-s-private-library-online. Accessed 25 June 2015.

Lipscombe, Trevor. 1998. The first science best seller? What a thirteenth-century manuscript can teach science publishers today. *Journal of Scholarly Publishing* 29 (3): 167–173.

Malkin, Irad. 2011. *A small Greek world. Networks in the ancient Mediterranean*. Oxford: Oxford University Press.

Niebler, Klaus. 1973. *Bücher aus der Bibliothek des Augsburger Humanisten Achilles Pirmin Gasser (1501–1577) in der Stadtbibliothek Mainz: eine Untersuchung von Restbeständen der Bibliotheca Palatina*. Köln: Bibliothekar-Lehrinstitut des Landes Nordrhein-Westfalen.

Nuñes, Pedro, and Henrique Leitão (ed.). 2002. *Obras*, 6 vols. Lisboa: Fundação Calouste Gulbenkian.

Oosterhoff, Richard J. 2015. A book, a pen, and the «Sphere:» Reading Sacrobosco in the Renaissance. *History of Universities* 28 (2): 1–54.

Pantin, Isabelle. 1998. Les problèmes de l'édition des livres scientifiques: l'exemple de Guillaume Cavellat. In *Le livre dans l'Europe de la Renaissance: Actes du XXVIIIe Colloque international d'etudes humanistes de Tours*, ed. Bibliothèque Nationale, 240–252. Paris: Promodis, Editions du Cercle de la Librairie.

Pantin, Isabelle. 2000. Alessandro Piccolomini en France, le problème de la langue scientifique et l'évolution du genre du traité de la sphère. In *La réception des écrits italiens en France à la Renaissance: ouvrages philosophiques, scientifiques et techniques: actes de la journée d'étude de Paris, Institut culturel italien, 8 novembre 1997*, vol. 24, ed. Alfredo Perifano, 9–28. Paris: Universite Paris III, Sorbonne Nouvelle, Centre Censier.

Pantin, Isabelle. 2006. Teaching mathematics and astronomy in France: The Collége Royal (1550–1650). *Science & Education* 15: 189–207.

Pedersen, Olaf. 1985. In quest of Sacrobosco. *Journal for the History of Astronomy* 16 (3): 175–220.

Pettas, William. 1995. A sixteenth-century Spanish bookstore: The inventory of Juan de Junta. *Transactions of the American Philosophical Society* 85 (1): 1–247.

Piccolomini, Alessandro. 1542. *De la institutione di tutta la vita de l'homo*. Venetia: Scotus.
Piccolomini, Alessandro. 1553. *Editione Tertia. Della Sfera del Mondo di M. Alessandro Piccolomini, divisa in libri quattro, i quali non per via di traduttione, ne à qual si voglia particolare Scrittore obligati, ma parte da migliori raccogliendo, e parte di nuovo producendo, contengano in se tutto quel ch'intorno a tal materia si possa desiderare, ridotti à tanta agevolezza, et à cosi facil modo di dimostrare, che qual si voglia poco essercitato negli studij di Mathematica potra agevolißimamente, et con prestezza intenderne il tutto, Di nuovo ricorretta, et ampliata. De le Stelle Fisse. Libro uno con le sue figure, e con le sue Tavole, dove con maravigliosa agevolata potra ciascheduno conoscere qualunque stella delle XLVIII Imagini del Cielo Stellato, é le Favole loro integramente, et sapere in ogni tempo del'anno, à qual si voglia hora di notte, in che parte del cielo si trouvino, non solo le dette Imagini, ma qualunque stella di quelle*. Venetia: Appresso Bartholomeo Cesano.
Postan, Michael M. (ed.). 1963–1989. *Cambridge economic history of Europe*. 10 vols. Cambridge: Cambridge University Press.
Ramusio, Giovanni Baptista. 1978–1991. *Navigazioni e viaggi*, 6 vols. Torino: Einaudi.
Randles, W.G.L. 1990. The evaluation of Columbus' India project by Portuguese and Spanish Cosmographers in the light of the geographical science of the period. *Imago Mundi* 42: 50–64.
Randles, W.G.L. 1998. The emergence of nautical astronomy in Portugal in the XVth century. *Journal of Navigation* 51 (1): 46–57.
Regiomontanus, Johannes and Georg Peuerbach. 1543. *Ioannis de Monte Regio et Georgii Purbachii eiptome, in Cl. Ptolemaei magnam compositionem, continens propositiones & annotationes, quibus totum Almagestum, quod sua difficultate etiam doctiorem ingenioque praesentatiore lectorem deterrere consueverat, dilucuda & brevi doctrina ita declaratur & exponitur.* Basilelae: Henrichum Petrum.
Renn, Jürgen (ed.). 2012. *The globalization of knowledge in history*. Berlin: Edition Open Access.
Renn, Jürgen, Dirk Wintergrün, Roberto Lalli, Manfred Laubichler, and Matteo Valleriani. 2016. Netzwerke als Wissensspeicher. In *Die Zukunft der Wissensspeicher. Forschen, Sammeln und Vermitteln im 21. Jahrhundert*, ed. J. Mittelstraß and U. Rüdiger, 35–80. Universitätsverlag Konstanz.
Romano, Antonella. 1999. *La Contre-Réforme mathématique. Constitution et diffusion d'une culture mathématique jésuite*. Rome: École française de Rome.
Rüegg, Walter (ed.). 1992. *A history of the University in Europe*, 4 vols. Cambridge: Cambridge University Press.
Ryff, Peter, and Ludovicus Lucii. 1598. *Elementa sphaerae mundi sive cosmographiae in usum Scholae Mathematicae Basilensiis. Per Petrum Ryff, Mathmaticum, Art. ac Med. D. Logicis Legibus deducta. Denuo, partim limatiora, partim observationum utilium acceßione locupletiora nunc edita. Quibus deovo appensus est Tractatus De Structura et usu Planispherii seu Astrolabi. Cura M. Ludovici Lucii Basil.* Basiliae: Typis Johannis Schroeteri, Anno Salvatoris.
Sacobosco, Johannes de. 1275–1300. *De sphaera*. Königlichen Bibliothek Kopenhagen, MS NKS 275a.
Sacroboscο, Johannes de. 1547. *Ioannis de Sacrobusto Libellus, de anni ratione: seu, ut vocatur vulgo, Computus Ecclestiaticus*. Antverpiae: Apud Martinum Nutium Typographum iuratum, et à C as. M. admissum.
Sacrobosco, Joannes de, and Christophorus Clavius. 1570. *Christophorii Clavii Bambergensis, ex Societate Iesv, in Sphaeram Ioannis de Sacro Bosco, commentarivs*. Romæ: apud Victorium Helianum.
Sacrobosco, Johannes de, and Jerónimo de Chaves. 1545. *Tractado de la sphera que compuso el doctor Ioannes de Sacrobusto con muchas adiciones, Agora nuevamente traduzido de Latin en lengua Castellana por el Bachiller Hieronymo de Chaves: el qual annidio muchas figuras tablas, y claras demonstrationes: iunctamente con unos breves Scholios, necessarios á mayor illucidation, ornato y perfection di dicho tractado*. Trans. Jerónimo de Chaves. Seuilla: Iuan de Leon.
Sacrobosco, Johannes de, and Pedro Nunes. 1537. *Tratado da sphera com a Theorica do sol & da lua*. Lixboa: G. Galharde.
Sacrobosco, Joannes de, and Pedro Nuñez. 1537. *Tratado da sphera com a Theorica do Sol & da Lua e ho primeiro livro da Geographia de Claudio Ptolomeo Alexandrino. Tirados novamente de Latim em lingoagem pello Doutor Pero Nunez Cosmographo del Rey Don Joano. Et acrescentatos de muitas annotaçiones i figuras per que mays facilmente se podem entender. Tratado que ho doutor per nunez fez em defensam de carta de marear. Item dous tratados quo mesmo Doutor fez sobra a carta de marear. Em os quaes se decrerano todas as principaes du vidas da navegação. Con as pavoa do movimento do sol: i sua declinação. Eo regimento de altura.* Lixboa: Galharde.
Sacrobosco, Johannes de, and Francesco Pifferi. 1604. *Sfera di Gio. Sacro Bosco tradotta, e dichiarata da don Francesco Pifferi sansa vino Monaco Camaldolense, e Matematico nello Studio de Siena. Misurato Intronato. Al Serensissimo don Cosimo Medici Gran Principe di Toscana. Con nuoue aggiunte di molte cose notabili, e varie demostrazioni vtili, e diletteuoli. Come nella seguente Tavola si vede.* Siena: Saluestro Marchetti.
Sacrobosco, Johannes de, Joannes Regiomontanus, and Georg von Peurbach. 1490. *Spaerae mundi compendium foeliciter inchoat. Nouiciis adolescentibus: ad astronomicam remp. capessendam aditum impetrantibus: pro breui rectoque tramite a uulgari uestigio semoto: Iohanis de Sacro busto sphaericum opusculum una cum additionib nonullis littera A sparsim ubi intersertae sint signatis Contraque cremonensia in planetarum theoricas delyramenta Iohanis de monte regio disputationes tam acuratiss. atque utills. Nec non Georgii purbachii in erundem mot planetarum*

acuratiss. theoricae: dicatum opus: utili serie contextum: fausto sidere inchoat. Venetiis: [Bonetus Locatellus für] Octavianus Scotus.

Sacrobosco, Johannes de, et al. 1531. *Spherae tractatus Ioannis de Sacro Busto Anglici uiri clariss.* Venetiis: in aedibus Luce Antonii Juntae Florentini.

Sacrobosco, Johannes de, Philipp Melanchthon, al-Farghani, and Joannes Regiomontanus. 1549. *Sphaera Ioannis de Sacrobosco Typis auctior, quam antehac, atque ex diligenti manu scriptorum impressorumque codicum collatione castigator, Praemissa Philippus Melanchthon doctiß. praefatione, qua vtilitatem astrologicae scientiae et Christiano homini non negligendam scite probat.* Parisiis: apud Gulielmum Cauellat.

Sacrobosco, Johannes de, Èlie Vinet, Petrus Valerianus, Petrus Nuñes, and Joannes Regiomontanus. 1556. *Sphaera Ioannis de Sacro Bosco emendata. Eliae Vineti Santonis scholia in eandem Sphaeram, ab ipso auctore restituta. Adiunximus huic libro compendium in Sphaeram, per Pierium Valerianum Bellunensem, Et, Petri Nonii Salaciensis demonstrationem eorum, quae in extremo capite de Climatibus Sacroboscius scribit, de inaequali Climatum latitudine, eodem Vineto interprete.* Lutetiae: Gulielmum Cauellat.

Sacrobosco, Joannes de, Élie Vinet, Piero Valeriano, Pedro Nuñes, and Francesco Giuntini. 1564. *Sphaera Ioannis de Sacro Bosco, emendata. Eliae Vineti Santonis scholia in eandem sphaeram, ab ipso autore resituta. Audiunximus huic libro compendium in sphaeram per Pierium Valerianum Bellunensem, et Petri Nonii Salaciensis demonstrationem, eorum, quae in extremo capite de climatibus Sacroboscius scribit de inaequali climatum latitudine, eodem Vineto interprete.* Lvtetiae: apud Gulielmum Cauellat.

Sacrobosco, Johannes de, Christoph Clavius, Albertus Hero, Pedro Nuñes, Petrus Valerianus, and Élie Vinet. 1601. *Sphæra Ioannis de Sacrobosco emendata, aucta et illustrata. Elia Vineti Santoris scholia in eandem Sphaeram ab ipso auctore restituta. Quibus praetear Alberti Heronis scholia nunc accessere R. P. Christoph. Clavij breves commentarij. Adiunximus huic libro compnedium in Sphaeram per Pierium Valerianum Bellunensem: et Petri Nonii Salacensis demosntrationem eorum, quae in extremo capite de climatibus Sacroboscus scribit de inaequali climatum latitudine, eodem Vineto interprete.* Coloniae: apud Gosuinum Cholinum.

Sacrobosco, Johannes de, Élie Vinet, Francesco Giuntini, Albertus Hero, Christoph Clavius, Pedro Nuñes, and Petrus Valerianus. 1610. *Sphaera Iohannis a Sacrobosco emendata, avcta et illustrata. Elia Vineti Santonis, et Francisci Iunctini Florentini scholia in eandem Sphaeram restituta. Quibus praeter Alberti Heronis scholia nunc accessere R. P. Christoph. Clavij breves commentarij. Adiunximus huic libro compendium in Sphaeram, per Pierium Valerianum Bellunensem: et Petri Nonij Salaciensis demonstrationem eorum, queen extremo capite de climatibus Sacroboscus scribit de inaequali climatum latitudine: eodem Vineto interprete.* Coloniae Agrippinae: apud Petrum Chrolinum.

Sanctorius, Sanctorius. 1614. *Ars Sanctorii Sanctorii Iustinopolitani de statica medicina, aphorismorum sectionibus septem comprehensa.* Venetiis: apud Nicolaum Polum.

Stöffler, Johannes. 1513. *Elucidatio fabricae ususque astrolabii.* Oppenheym: in aedibus Iacobi Cobelii.

Straet, Jan van der. 1580. *Nova reperta.* Amsterdam.

Temkin, Owsei. 1973. *Galenism: Rise and decline of a medical philosophy.* Ithaca and London: Cornell University Press.

Thorndike, Lynn. 1949. *The sphere of Sacrobosco and its commentators.* Chicago: The University of Chicago Press.

Thorndike, Lynn. 1955. The true place of astrology in the history of science. *Isis* 46 (3): 273–278.

Valente, T.W. 1995. *Network models of the diffusion of innovations.* Cresskill: Hampton Press.

Valerio, Vladimiro. 2012. La «Geografia» di Tolomeo e la nascita della moderna rappresentazione dello spazio. In *Scienza antica in età moderna*, ed. Vanna Maraglino, 215–232. Bari: Cacucci Editore.

Valleriani, Matteo. 2010. *Galileo engineer*, Boston studies in the philosophy of science. Dordrecht: Springer.

Valleriani, Matteo. 2013. *Metallurgy, ballistics and epistemic instruments. The «Nova scientia» of Nicolò Tartaglia. A new edition.* Berlin: Edition Open Access.

Valleriani, Matteo. 2014a. Ancient pneumatics transformed during the early modern period. *Nuncius* 29 (1): 127–174.

Valleriani, Matteo. 2014b. Appropriation and tranformation of Ancient Science. *Nuncius* 29 (1): 1–8.

Valleriani, Matteo. 2014c. La nuova pratica del «Trattato della Sfera» di Galileo Galilei. In *Tintenfass und Teleskop*, ed. Andrea Albrecht, Giovanna Cordibella, and Volker Remmert, 321–336. Berlin: DeGruyter.

Vinet, Èlie. 1564. *La Manière de fere les solaires, que communement on appelle quadrans.* Poitiers: par Enguilbert de Marnef.

Vinet, Èlie. 1583. *L'arpanterie d'Elie Vinet, et la maniere de fere les solaires, que communemant on apele quadrans.* Bourdeaus: Par S. Millanges, Imprimeur ordinaire du Roi.

Watts, Duncan J. 2003. *Six degrees: The science of a connected age.* New York: Norton.

Wolfe, Jessica. 2009. *Humanism, machinery, and Renaissance literature.* Cambridge, MA: Cambridge University Press.

List of Contributors

Jochen Büttner

Jochen Büttner studied physics and philosophy at Universität Konstanz and the Freie Universität Berlin where he received his physics diploma. Since 1998, he has worked as a historian of science at the Max Planck Institute for the History of Science, Berlin, where he completed his Ph.D. thesis on the emergence of Galileo's new science. Jochen Büttner has studied the role of so-called challenging objects as mediators between practical and theoretical knowledge in early modern mechanics. Within the framework of the Excellence Cluster 264 Topoi, he is currently directing the junior research group "Between Knowledge and Innovation: The Unequal-Armed Balance."

Sven Dupré

Sven Dupré is professor of history of art, science, and technology at Utrecht University and the University of Amsterdam. He is the director of the project "Technique in the Arts: Concepts, Practices, Expertise, 1500–1950," supported by a European Research Council (ERC) Consolidator Grant. Previously, he was professor of history of knowledge at the Freie Universität Berlin and director of the research group "Art and Knowledge in Premodern Europe" at the Max Planck Institute for the History of Science, Berlin. In Spring 2015, he was Robert H. Smith scholar in residence for Renaissance sculpture in context at the Victoria and Albert Museum in London. His recent publications include *Embattled Territory: The Circulation of Knowledge in the Spanish Netherlands* (Academia Press, 2015), *Laboratories of Art: Alchemy and Art Technology from Antiquity to the 18th Century* (Springer, 2014), *Art and Alchemy: The Mystery of Transformation* (Hirmer, 2014), and *Translating Knowledge in the Early Modern Low Countries* (LIT, 2012).

Ursula Klein

Ursula Klein is senior research scholar at the Max Planck Institute for the History of Science in Berlin and professor of philosophy of science at Universität Konstanz. She is author of *Verbindung und Affinität* (Birkhäuser, 1994); *Experiments, Models, Paper Tools: Cultures of Organic Chemistry in the Nineteenth Century* (Stanford University Press, 2003); *Humboldts Preußen: Wissenschaft und Technik im Aufbruch* (WBG, 2015); and *Nützliches Wissen: Die Erfindung der Technikwissenschaften* (Wallstein, 2016) and coauthor of *Materials in Eighteenth-Century Science, A Historical Ontology* (MIT Press, 2007). Her current research project is concerned with hybrid experts and the "useful sciences" in the industrialization of Prussia.

M. Valleriani (ed.), *The Structures of Practical Knowledge*, DOI 10.1007/978-3-319-45671-3

Richard L. Kremer

Richard L. Kremer teaches history of science at Dartmouth College (USA) and curates that institution's collection of historic scientific instruments. His research deals with the "material culture" of Medieval and early modern European astronomy, its instruments, tables and computational tools, and products (calendars, almanacs, horoscopes, prognostications). Currently, Kremer is completing a book on astronomical ephemera printed in the fifteenth century and another on calendar-makers' responses to Copernican, Tychonic, and Keplerian astronomy in the sixteenth and seventeenth centuries.

Wolfgang Lefèvre

Wolfgang Lefèvre was senior research scholar at the Max Planck Institute for the History of Science until his retirement in 2006. His publications relevant to the volume's topic are *Naturtheorie und Produktionsweise* (Luchterhand, 1978), *Picturing Machines* (ed.) (MIT Press, 2004), and *Database Machine Drawings*, together with Marcus Popplow. (2008 ff.: http://dmd.mpiwg-berlin.mpg.de/home).

Elaine Leong

Elaine Leong leads the "Reading and Writing Nature in Early Modern Europe" research group at the Max Planck Institute for the History of Science, Berlin. She has published widely on early modern vernacular medicine and science and is the coeditor of *Secrets and Knowledge: Medicine, Science and Commerce 1500–1800* (Ashgate Publishing, 2011).

Pamela O. Long

Pamela O. Long is an independent historian who has published widely in premodern and early modern history of science and technology and cultural history. Her books include *Openness, Secrecy, Authorship: Technical Arts and the Culture of Knowledge from Antiquity to the Renaissance* (Johns Hopkins University Press, 2001) and *Artisan/Practitioners and the Rise of the New Sciences, 1400–1600* (Oregon State University Press, 2011). She is a John D. and Catherine T. MacArthur fellow (2015–2020).

Elizabeth M. Merrill

Elizabeth M. Merrill is a postdoctoral fellow at the Max Planck Institute for the History of Science, Berlin. She studied at Columbia University and at the University of Virginia, where she earned her Ph.D. Her research focuses on the role of the early modern architect, architectural design and building processes, and the structures of architectural collaboration and information exchange. Currently she is completing a book on the Sienese architect Francesco di Giorgio and the development of the early modern profession of architecture.

Bruce T. Moran

Bruce Moran is professor of history at the University of Nevada, Reno, where he teaches courses in the history of science and early medicine. He is the author of *Distilling Knowledge: Alchemy, Chemistry, and the Scientific Revolution* (Harvard University Press, 2005) and *Andreas Libavius and the Transformation of Alchemy: Separating Chemical Cultures with Polemical Fire* (Science History Publications International, 2007). Most recently, he has coedited *Bridging Traditions: Alchemy, Chemistry, and Paracelsian Practices in the Early Modern Era* (Truman State University Press, 2015). In progress are *Things Made and Things in the Making: Sentiment and Practice in Early Modern Alchemy and Paracelsus: Recreating Body and Soul.*

Pietro D. Omodeo

Pietro Daniel Omodeo is research scholar at the Max Planck Institute for the History of Science, Berlin and member of the Collaborative Research Center *Episteme in Motion* at the Freie Universität Berlin. His research focuses on science, philosophy, and literature in the early modern period, as well as on historical epistemology. He authored *Copernicus in the Cultural Debates of the Renaissance: Reception, Legacy, Transformation* (Brill, 2014) and coedited the volume *Duncan Liddel (1561–1613): Networks of Polymathy and the Northern European Renaissance* (Brill, 2016).

Eileen Reeves

Eileen Reeves is professor of comparative literature and an associate member of the program in the history of science at Princeton University. She took her Ph.D. in comparative literature from Stanford. She works at the intersection of early modern literary studies, the history of art, and the history of science. Much of her research has focused on the figure of Galileo Galilei and his relationship to astronomy, religion, optics, art, and a range of literary forms, including the scientific treatise and dialogue, poetry, dialect literature, journalism, and drama.

Her books include *Painting the Heavens: Art and Astronomy in the Age of Galileo* (Princeton University Press, 1997), *Galileo's Glassworks: The Telescope and the Mirror* (Harvard 2008), *On Sunspots* (with Albert Van Helden, Harvard University Press, 2010), and *Evening News: Optics, Astronomy, and Journalism in Early Modern Europe* (University of Pennsylvania Press, 2014).

She is currently working on a book that deals with the evolution of printed images in the context of early modern astronomy. Other recent projects include essays on the relationships between optical and musical instruments, between astrology and literature, and between the new sciences and the visual arts.

Dagmar Schäfer

Dagmar Schäfer is director of Dept. 3, managing director of the Max Planck Institute for the History of Science in Berlin and professor h.c. of the history of technology at the Technische Universität Berlin. Her main interest is the history and sociology of technology of China, focusing on the paradigms configuring the discourse on technological development, past and present. She has published widely on materiality, the processes and structures that lead to varying knowledge systems, and the changing role of artifacts—texts, objects, and spaces—in the creation, diffusion, and use of scientific and technological knowledge. Her monograph *The Crafting of the 10,000 Things* (University of Chicago Press, 2011) won the Pfizer Award (History of Science Society) in 2012 and the Joseph Levenson Prize (pre-1900) (Association for Asian Studies) in 2013. Her current research focus is the historical dynamics of concept formation, situations, and experiences of action through which actors have explored, handled, and explained their physical, social, and individual worlds.

Pamela H. Smith

Pamela H. Smith is Seth Low professor of history and director of the Center for Science and Society at Columbia University where she teaches courses in early modern European history and history of science. Her books include *The Business of Alchemy: Science and Culture in the Holy Roman Empire* (Princeton University Press, 1994; Pfizer Prize); *Merchants and Marvels: Commerce, Science, and Art in Early Modern Europe* (ed. with Paula Findlen, Routledge, 2002); *The Body of the Artisan: Art and Experience in the Scientific Revolution* (University of Chicago Press, 2004; Leo Gershoy Prize); *Making Knowledge in Early Modern Europe: Practices, Objects, and Texts, 1400–1800* (ed. with Benjamin Schmidt, University of Chicago Press, 2008); *Ways of Making and Knowing: The Material Culture of Empirical Knowledge* (ed. with Amy R. W. Meyers and Harold Cook, University of Michigan Press, 2015); and *The Matter of Art: Materials, Practices, Cultural Logics c. 1250–1750* (ed. with Christy Anderson and Anne Dunlop, Manchester University Press, 2015). She has published numerous articles on early modern European artisanal knowledge and culture, and, in current research, she is directing a large collaborative research and teaching initiative, *The Making and Knowing Project*, to reconstruct the vernacular knowledge of early modern craftspeople from a variety of disciplinary perspectives, including hands on work in a laboratory.

Viktoria Tkaczyk
Viktoria Tkaczyk leads the research group "Epistemes of Modern Acoustics" at the Max Planck Institute for the History of Science in Berlin and is professor at the Institut für Kulturwissenschaft, Humboldt-Universität Berlin. Her first book, *Himmels-Falten. Zur Theatralität des Fliegens in der Frühen Neuzeit* (Fink, 2011), was awarded the Ernst Reuter Prize in 2008 and the Book Award of the Amsterdam School of Cultural Analysis in 2012. From 2008 to 2010, she worked as a research fellow on the project "Theatrum Scientiarum. Performativity of Knowledge as Agent of Cultural Change" (Freie Universität Berlin), and in 2011, she carried out research as a Feodor Lynen fellow at the Atelier de Recherche sur l'Intermédialité et les Arts du Spectacle (CNRS) in Paris. Between 2011 and 2014, she was assistant professor of arts and new media at the University of Amsterdam and a Dilthey fellow at the Max Planck Institute for the History of Science. Viktoria Tkaczyk is a member of the Junge Akademie at the Berlin-Brandenburg Academy of Sciences and Humanities. Recent publications include *Listening in Circles: Spoken Drama and the Architects of Sound, 1750–1830* (Annals of Science 71/3, 2014), *The Shot is Fired Unheard: Sigmund Exner and the Physiology of Reverberation* (Grey Room 60, 2015), and *The Making of Acoustics around 1800, or How to Do Science with Words* (in *Performing Knowledge, 1750–1850*, ed. M. H. Dupree & S. B. Franzel, De Gruyter, 2015).

Matteo Valleriani
Matteo Valleriani is senior scholar at Dept. 1 of the Max Planck Institute for the History of Science. In his research, he investigates the process of the emergence of scientific knowledge in relation to its practical, social, and institutional dimensions. In this frame, and in reference to the early modern period, his major projects have been dedicated to Galileo's science (*Galileo Engineer*, Springer, 2010) and to Nicolò Tartaglia's ballistics (*Metallurgy, Ballistics and Epistemic Instruments: The "Nova Scientia" of Nicolò Tartaglia. A New Edition*, Edition Open Access, 2013). Affiliated to the Excellence Cluster Topoi, he investigated the emergence of hydromechanics in classical antiquity (*Hydrostatics and Pneumatics in Antiquity*, in G. L. Irby (ed.), *A Companion to Science, Technology, and Medicine in Ancient Greece and Rome*, Wiley-Blackwell, 2016). Matteo Valleriani worked on the processes of the appropriation of ancient science during the early modern period. He edited an issue of *Nuncius* (*Appropriation and Transformation of Ancient Science*, *Nuncius*, 29, 2014), which contains his final contribution to the study of the appropriation of ancient pneumatics through early modern garden technology (*Ancient Pneumatics Transformed During the Early Modern Period, Nuncius*, 29, 2014, 127–174: also: http://pratolino.mpiwg-berlin.mpg.de). His current major project is concerned with the evolution of the scientific knowledge system in Europe from the thirteenth to the seventeenth centuries that revolved around cosmological knowledge. In the frame of digital humanities, Matteo Valleriani implements methods of social network analysis in history writing.

Name Index

M. Valleriani (ed.), *The Structures of Practical Knowledge*, DOI 10.1007/978-3-319-45671-3

Q

R

S

Place Index

M. Valleriani (ed.), *The Structures of Practical Knowledge*, DOI 10.1007/978-3-319-45671-3

The manufacturer's authorised representative in the EU is Springer Nature Customer Service Centre GmbH, Europaplatz 3, 69115 Heidelberg, Germany. If you have any concerns regarding our products, please contact ProductSafety@springernature.com

Printed and bound by CPI Group (UK) Ltd, Croydon, CR0 4YY
05/10/2026
02228133-0014